R.M.T. Dahlgren    H.T. Clifford    P.F. Yeo

# The Families of the Monocotyledons

Structure, Evolution, and Taxonomy

In Cooperation with
R.B. Faden · N. Jacobsen · K. Jakobsen
S.R. Jensen · B.J. Nielsen · F.N. Rasmussen

With 225 Figures

Springer-Verlag
Berlin Heidelberg NewYork Tokyo 1985

Prof. Dr. ROLF M.T. DAHLGREN
Botanical Museum, University of Copenhagen
Gothersgade 130
DK-1123 Copenhagen

Prof. Dr. H. TREVOR CLIFFORD
Department of Botany
University of Queensland
St. Lucia, Brisbane
Australia 4067

Dr. PETER F. YEO
University Botanic Garden
Cambridge CB2 1JF
Great Britain

213525

ISBN 3-540-13655-X Springer-Verlag Berlin Heidelberg New York Tokyo
ISBN 0-387-13655-X Springer-Verlag New York Heidelberg Berlin Tokyo

Library of Congress Cataloging in Publication Data. Dahlgren, Rolf, 1932–    . The families of the monocotyledons. Bibliography: p.    . Includes index. 1. Monocotyledons–Classification. 2. Monocotyledons–Morphology. 3. Monocotyledons–Evolution. 4. Botany–Classification. 5. Botany–Morphology. 6. Plants–Evolution.
I. Clifford, H.T. (Harold Trevor) II. Yeo, Peter, 1929–    . III. Title. QK495.A14D33    1984    584    84-14089

Typesetting, printing and bookbinding: Universitätsdruckerei H. Stürtz AG, Würzburg
2131/3130-543210

# Preface

In this book we present an integrated account of the monocotyledons. The classification is supported by an extensive character analysis and by evolutionary models constructed on the basis of this analysis. An assessment of the character states as either primitive or derived, in the monocotyledons as a whole and in their constituent groups, is presented. These parts of the book have their forerunners in *The Monocotyledons: a Comparative Study* (DAHLGREN and CLIFFORD, 1982) and in *Monocotyledon Evolution: Characters and Phylogenetic Estimation* (DAHLGREN and RASMUSSEN, 1983).

Thus the presentation gives great weight to evolutionary considerations. We have aimed at doing away with old heterogeneous families, and arranging the new, smaller and more homogeneous ones according to their presumed relationships. Most taxonomists may regard us as "splitters". We believe, however, that a concept of Amaryllidaceae, for example, which includes Amaryllidaceae sensu stricto, Ixioliriaceae, Agavaceae pro parte, and, perhaps, Alstroemeriaceae, is of no help to the botanist seeking to recognize natural and comprehensible groups, and that it will prevent him from perceiving the evolutionary pathways which have led to the families and their genera. Again, it is certainly of no advantage to unite Liliaceae, as circumscribed here, with Alliaceae, Hyacinthaceae, Asphodelaceae, Hemerocallidaceae, Convallariaceae, Trilliaceae and other elements, if at the same time the Liliaceae are kept distinct from, for example, Alstroemeriaceae, Iridaceae or Philesiaceae.

Our classification has been based on an extensive body of evidence and the decisions taken are explained as fully as possible. It has been our intention to avoid the constraints of mere convention and to adopt an unbiased approach. Yet, in our concepts we have been greatly influenced by previous treatises, such as those by HUBER (1969) and HAMANN (1961), which were likewise based on extensive comparative studies.

The classification in most cases is sufficiently practical, we believe, for use in the herbarium and in the field, but convergent evolution has led to cases where it may be difficult to refer a genus to a particular family and where its position in that family is still uncertain. For some families supplementary studies are needed before we have a full understanding of their circumscriptions. All such cases are clearly indicated.

Because our orders and families are generally rather narrowly circumscribed the families are numerous, especially in Asparagales. Keys to the families of each order have therefore been provided. Further guidance can be obtained from the evolutionary models ("cladograms") supplied.

One of the authors (R.D.) is involved with the preparation of the monocotyledon volume in the series *The Families and Genera of Flowering Plants* (Editor-in-chief: K. KUBITZKI, Cambridge University Press). We are keen to stress that the present work has been able to benefit little from the larger project, the information for which has been kept separate. In fact, only few family treatments were available when the manuscript of the present book was completed. However, great and often indispensible help has been given to us by some colleagues (see below), and the texts of some of the families are written wholly or mainly by specialists.

*Acknowledgements*. The following have contributed substantial parts of the text: Dr. SØREN ROSENDAL JENSEN and BENT JUHL NIELSEN, the Technical University, Copenhagen: the chapter on chemical characters; Dr. ROBERT B. FADEN, Smithsonian Institution, Washington D.C.: the texts on Commelinaceae and Mayacaceae; the late Dr. KNUD JAKOBSEN, University of Copenhagen: the texts on Arecaceae and Pandanaceae; Professor NIELS JACOBSEN, The Royal Veterinary and Agricultural University, Copenhagen: the texts on Araceae and Lemnaceae; and Dr. FINN N. RASMUSSEN, University of Copenhagen: the texts on Apostasiaceae, Cypripediaceae and Orchidaceae.

Various specialists have carefully revised the texts for certain families, Dr. L. ANDERSSON, Gothenburg, for Marantaceae, Musaceae, Heliconiaceae and Strelitziaceae; Dr. C.D.K. COOK, Zürich, for families of the Alismatiflorae, Dr. J. DRANSFIELD, London, for Arecaceae, Dr. P. GOLDBLATT, St. Louis, for Iridaceae, Dr. P. LINDER, Cape Town, for Restionaceae, and Drs. D. and U. MÜLLER-DOBLIES, Berlin, for Amaryllidaceae and Typhales. Dr. D.F. CUTLER, Kew, has given us valuable information for Asphodelaceae (incl. Aloaceae) and Mr. G. KEIGHERY, Perth, for Dasypogonaceae and the heterogeneous group Johnsonieae in Anthericaceae. Dr. GERTRUD DAHLGREN has critically read the proofs.

Mr. BENT JOHNSEN, Copenhagen, made the original drawings for Figs. 46, 54, 59, 63, 64, 65, 66, 67A–I, 74, 77, 78, 81, 86, 88, 101, 108, 113, 114, 126 and 212. Many other illustrations, such as most of those for the orchids and the schematic and diagrammatic ones for Cyperaceae and Poaceae, have been redrawn by him. The illustration of *Alexgeorgia* (Fig. 213) has been placed at our disposal by Dr. J.P. JESSOP, Adelaide.

Most illustrations, however, have been taken from other sources. Many are from S. ROSS-CRAIG, *Drawings of British Plants,* for which paid copyright has kindly been granted by Bell & Hyman. M. CORREA has kindly permitted us to use many illustrations from *Flora Patagonica.* Further, Professor A. TAKHTAJAN has allowed us to use several illustrations from *Plant Life* Vol. 6; Dr. F.N. HEPPER several from *Flora of West Tropical Africa,* ed. 2, Vol. 3 (1); Dr. A. CRONQUIST several from *Intermountain Flora* Vol. 6; University of Washington Press several from *Vascular Plants of the Pacific Northwest* Vol. 1; Plenum Publishing Corporation some illustrations from *Evolutionary Biology* Vol. 16; Dr. A. EL-GADI some illustrations from *Flora of Libya;* and Dr. W. BURGER a plate from *Evolutionary Theory* Vol. 5. Illustrations come from many other works, all cited in the legends of the respective figures. We acknowledge gratefully our indebtedness to the persons, institutions and publishers responsible for all these publications.

The facilities available to us at the Botanical Museum, University of Copenhagen, where most of the work on the present book was carried out, were indispensible. The secretarial work has been carried out by Mrs. KIRSTEN HARDER and Mrs. LENE FUGMANN, and technical work by the Staff of the Museum. The generosity of the staff and colleagues in helping with this book is kindly acknowledged. We acknowledge similar help, though on a much smaller scale, from the University Botanic Garden, Cambridge, and the Botany Department of University of Queensland, St. Lucia, and their office staffs.

The support given by the Carlsberg Foundation, Copenhagen, for preparing the *Families and Genera of Vascular Plants* has also benefited the work for this book in the preparation of illustrations and in data retrieval.

The authors wish to acknowledge their deep gratitude and appreciation to all these individuals and institutions, and also to all others who have contributed valuable information.

ROLF M.T. DAHLGREN
H. TREVOR CLIFFORD
PETER F. YEO

# Contents

*Les monopétales régulieres constituent moins une famille qu'une
grande nation dans laquelle on compte plusieurs familles bien distinctes ;
en sorte que pour les comprendre toutes sous une indication commune,
il faut employer des caractères si généraux et si vagues que c'est paraître
dire quelque chose en ne disant en effet presque rien du tout. Il vaut
mieux se renfermer dans des bornes plus étroites, mais qu'on puisse
assigner avec plus de précision.*

J.-J. ROUSSEAU
Lettres sur la botanique. Lettre IV
19e Juin 1772.


The regular monopetals constitute less a family than a great nation in
which one may recognize several quite distinct families; so that in
order to describe them under a common heading, it is necessary to employ
characters so general and so vague that the heading when it appears to say
something is saying in effect almost nothing at all. It would be better
to restrict oneself within narrower boundaries that can be delimited with
greater precision.

# Introduction

# Morphological Concepts

In this book we present an integrated account of the monocotyledons. We have attempted to decide whether the character states of the monocotyledons as a whole, and those of their constituent groups, are primitive or derived. On the basis of these considerations, and with the use of some justifiable general assumptions, we present some evolutionary models in accordance with DAHLGREN and RASMUSSEN (1983). In doing this we use the elementary terms currently employed by the school of cladistics. Many of the character states and their distributions are presented in greater detail by DAHLGREN and CLIFFORD (1982).

The major part of the book is taken up by our classification of the monocotyledons, which is synthetic in the sense that it uses data of many different kinds and evolutionary in the sense that the evolutionary model is given as much weight as possible. The classification is sufficiently practical, we believe, for use in the herbarium and in the field, though for the latter assistance from artificial diagnostic keys will be required as well. Our orders and families are generally rather narrow and the families, consequently, are numerous, especially in Asparagales. Keys to the families of each order have therefore been provided.

We believe that a concept of Amaryllidaceae, for example, which includes Amaryllidaceae sensu stricto, Ixioliriaceae, Agavaceae, pro parte, and, perhaps, Alstroemeriaceae, is of no help to the botanist seeking to recognise natural and comprehensible groups, and that it will prevent him from understanding the evolutionary pathways within and around the family. Again, it is certainly of no advantage to unite Liliaceae, as circumscribed here, with Asphodelaceae, Hypoxidaceae, Tecophilaeaceae, and Trilliaceae, if at the same time the Liliaceae are kept distinct from Orchidaceae, Alstroemeriaceae or Iridaceae.

It has been our intention to avoid the constraints of convention and to adopt an unbiassed approach, using as wide a range of data as we could within the limitations of time (self-imposed), resources and competence.

The following short explanations of terms refer to the monocotyledons only and are not meant to be general definitions. They apply primarily to the concepts used in the chapters on character states and their distributions and in the taxonomic section of the book.

## Underground Parts

In monocotyledons the first root formed on the embryo, the radicle, is ephemeral and sometimes hardly distinguishable. The root system arises from the basal nodes of erect shoots or from any node in prostrate shoots. Roots produced by aerial shoots may be green and assimilatory (as are the roots of epiphytic orchids and aroids) or may form massive props, as in *Pandanus* and some palms.

Some of the cells of the root epidermis send out *root hairs*. These root-hair epidermal cells may resemble other epidermal cells of the root, but in some groups they are conspicuously shorter and are called *root-hair short cells* (Fig. 27).

The roots in some groups of monocotyledons are fusiform or tuber-like; these contain nutrients and function as *storage roots*.

The underground stem, when well-developed, may be a rhizome, corm or tuber. It is often elongate, and either horizontal or vertical, and then forms a *rhizome;* plants with a rhizome are described as *rhizomatous*. A short, compact underground stem filled with nutrients is a *corm,* which is synonymous with *rhizomatous tuber*. The corm axis generally extends over several internodes. It may be enclosed by a *tunic* of dry leaves or leaf bases which sometimes form a characteristically sculptured fibrous envelope; such tubers are called *tunicated corms* (Fig. 117B), and are particularly common in Iridaceae. Part of the underground stem may become inflated to form a globose storage organ and is then called a *tuber*. Frequently, as in many Dioscoreales and Arales, the tuber seems to be chiefly made up of the axis immediately below the cotyledon, the *hypocotyl,* and is then termed a *hypocotylar tuber*.

A *bulb* is defined as a short, often plate-like stem bearing a number of thick, fleshy leaves or leaf bases, which store water and nutrients; these are *bulb scales*. The bulb scales may vary from one to many, as in the Liliaceae s.str.

(*Pseudobulbs,* as found in many orchids, are dilated parts of the aerial stem that store water and nutrients. They are corms rather than bulbs.)

*Velamen* is a water-storing tissue in the outer layers of some roots. It has a parchment-like appearance and consists of one to several layers of non-living cells with thickened and lamellate cell walls. Velamen is particularly common in epiphytic orchids and aroids, but is also found in the roots of many other monocotyledons, in particular in the Liliiflorae (see BARTHLOTT 1976a).

## The Aerial Stem

Branching is *monopodial* when the shoot grows apically, the main shoot generally exceeding the lateral ones in length. It is *sympodial* when axillary branches successively take over the growth, as when the main shoot develops into an inflorescence or tendril, or dies off. A *sympodium* is a sequence of such lateral shoots which successively overtake their predecessors (Fig. 52 A).

In most monocotyledons the aerial stem is herbaceous, i.e. soft, usually green, and withering within a limited time; in other groups it becomes strengthened with lignin fibres and may be provided with bark. In some groups it becomes conspicuously thick and long-lived, as in palms and pandans, neither of which, however, has secondary thickening. In other groups the stems are lignified but slender, and these plants are genuine shrubs; they are common among the berry-fruited Asparagales.

In only a few families does *secondary thickening* occur, and this is of a kind different from that found in dicotyledons, as explained on p. 45. Most of these plants have a thick woody trunk.

The stems in a great many groups are hairy (see below) but thorns and spines are rare, being found in *Smilacaceae* and *Petermanniaceae*. The stems are climbing (*scandent*) in both of these families as well as in Dioscoreaceae and some other groups.

## Anatomical Concepts (Leaf and Stem)

The vascular tissue is mainly in the form of primary vascular strands that consist of *xylem* (with tracheids and/or vessels, see below), and *phloem,* which consists mainly of sieve tubes and their companion cells.

Although the vascular strands in monocotyledons initially arise in a single ring, secondary bundles soon develop so that the strands appear "*scattered*". No cambium is formed between the phloem and xylem where it is present in most dicotyledons. Such an organization of vascular strands is called an *atactostele,* and the condition *atactostely.* It is contrasted with the condition of *eustely,* found in dicotyledons. There, a single ring of primary vascular strands is formed. Then a cambium develops within and between strands, producing xylem on the inside and phloem on the outside, so that a cylinder of secondary tissue is produced, an *eustele.* Where secondary tissue occurs in monocotyledons a meristematic tissue produces new sets of isolated vascular strands outside those first produced.

The *xylem* consists of *tracheids* or tracheids and vessels. The former are elongate living cells, the cavities of which are without direct contact with one another. *Vessels* begin as living cells which are produced in continuous rows; they die, and acquire direct connections by means of *perforations* in the end walls, the *perforation plates.* The end walls in narrow vessels are oblique and have a row of transverse, narrow, slit-like perforations, separated by "bars", so-called *scalariform perforation;* but wider vessels, when present, have a less oblique, or even transverse, perforation plate with only a few perforations or one *simple,* circular *perforation.* Roots, rhizomes, aerial stems and leaves frequently differ in the type of xylem which they contain.

The sieve tubes of the phloem contain small *plastids* (leucoplasts). These may store starch and/or protein. The occurrence of protein [as compact bodies (*crystalloids*) or as an annular structure of thin filaments] and starch characterizes different groups of angiosperms. All monocotyledons have a number of triangular (*cuneate*) protein crystalloids in their plastids; a few also have starch and some have filaments. According to BEHNKE (1981), who has studied these structures, the monocotyledonous sieve tube plastids are of the PIIc Type (P = protein, II = type 2 according to BEHNKE, l. c., c = cuneate), with PIIcs and PIIcf representing

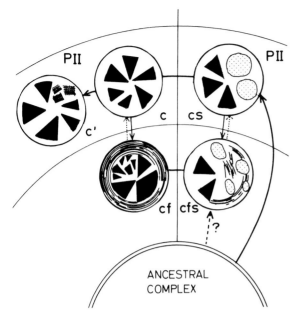

**Fig. 1.** Sieve tube plastids in monocotyledons and their presumed evolution, according to BEHNKE (1981a). The forms with cuneate protein crystalloids are called *PIIc* (=cuneate). Protein filaments (*f*) are found in certain monocotyledons (*PIIcf*) and in some there are starch grains (*s*) as well as the protein crystalloids (*PIIcs*). Sometimes other protein crystalloids (*c'*) are found beside the normal cuneate ones.

forms with starch and filamentous protein respectively (Fig. 1).

*Laticifers* are tubes or rows of elongate cells, containing a fluid of a somewhat milky appearance, in the monocotyledons only rarely coloured (as in *Dilatris* of Haemodoraceae).

*Crystals* of calcium oxalate occur in monocotyledons rather frequently, most commonly as *raphides*, i.e. bundles of thin crystalline needles (Fig. 28), which are contained in cells filled with mucilage. These cells may be almost isodiametric, but sometimes are elongated ("*sacs*") or form narrow tubes ("*raphide vessels*"). In some groups oxalate occurs in the form of thicker solitary needles, so-called *styloids* (*pseudoraphides*), which occur in *suberized (cork) cells*. These are known in Pontederiaceae and Philydraceae, where raphides are rare, and in Nolinaceae, Phormiaceae and Agavaceae, where raphides are lacking.

*Silica* is deposited in several major groups of monocotyledons, either as numerous small granules (*silica sand*) or as larger bodies of various sizes and shapes. These shapes are described on p. 63. The silica bodies are often deposited in special short epidermal cells, *silica short cells*.

*Stomata* (sing. *stoma*) are epidermal structures composed of two *guard cells* embracing a pore through which gas exchange takes place. The stomata may be surrounded by normal epidermal cells and are then *anomocytic,* but frequently they are surrounded by two or more cells differing in size and shape from other epidermal cells; these are called *subsidiary cells*. A stoma and its subsidiary cells make up the *stomatal complex*. When there are two subsidiary cells, one alongside each guard cell, as in many monocotyledons, the stoma is *paracytic;* when there are four or six subsidiary cells surrounding the guard cells the stoma is *tetracytic* or *hexacytic* respectively. As will be explained in the chapter on Distribution of Character Conditions the *ontogeny,* i.e. the individual development, of the stoma and the origin of the subsidiary cells is not always reflected in the appearance of the mature stomatal complex; a special terminology covering this situation is given in that chapter.

*Trichomes* are processes arising from the epidermis (hairs etc.). The hairs may be unicellular, i.e. comprise extensions of epidermal cells, but more often consist of a single row of cells. Sometimes, as in Bromeliaceae (Fig. 154I), the trichomes have a short row of cells at the base and are branched above to form a stellate (star-shaped) or peltate (shield-shaped) head. Some peltate hairs are of importance for water uptake. Multicellular hairs with a broad multicellular base occur on the leaf margins of *Luzula* (Juncaceae). *Microhairs* are small, generally bicellular thin-walled hairs found mainly in grasses (Fig. 193J–O). *Glandular hairs* are hairs where one or more cells, generally at the end of the hair, are enlarged and secretory.

*Intravaginal squamules,* here interpreted as trichomes, are non-vascularized multicellular processes situated in the leaf axils. They generally secrete mucilage, which is thought to protect the axils from micro-organisms (TOMLINSON 1982).

*Epicuticular wax* is wax secreted on the outer surface of the epidermis. It is frequently sculptured in different ways, typical of the main groups of monocotyledons (BARTHLOTT and FRÖHLICH 1983). See also the chapter on Distribution of Character Conditions.

## Leaves

The arrangement of the leaves on the stem is called *phyllotaxy*. When three or more leaves are placed at the same level, they are said to be *whorled (verticillate)*, when in pairs they are *opposite* and when solitary *alternate*. Where successive pairs of oppo-

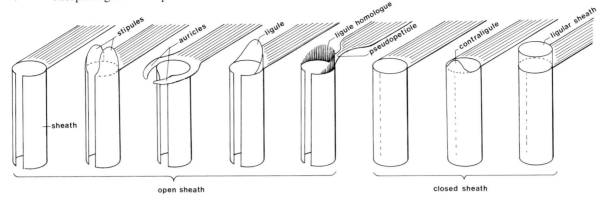

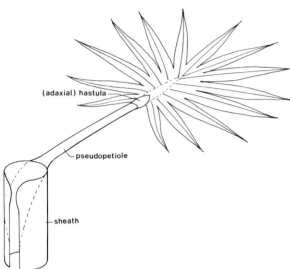

**Fig. 2.** Terms used for the bases of (chiefly grass) leaves (*above*) and for a palm leaf (*below*). (Orig. B. JOHNSEN)

Potamogetonaceae. The homology of stipule-like structures is not in every case established, for example, hyaline extensions and stipular lobes (as in *Joinvillea* and *Juncus* species) which are always lateral on the leaf near its base. A *ligule* is a hyaline extension of the leaf sheath on the adaxial side of the leaf; it is typical of most grasses but is also known in Pontederiaceae. A *contraligule* is an extension of the leaf sheath on the side opposite the lamina. It occurs in the palms, grasses (Fig. 2) and sedges. The *hastula* found in palms is at the distal end of the pseudopetiole, but somewhat resembles the ligule of the grasses. (See Fig. 2.)

A true petiole is present only in certain groups of monocotyledons. In other groups the leaves are linear and there is no differentiation at all between petiole and lamina. Some regard such leaves as *phyllodial,* consisting of a flattened petiole only, but it may be more correct to interpret them as the result of a particular mode of development of the leaf meristem. However, even these leaves are frequently differentiated secondarily into a narrower proximal part, a *pseudo-petiole,* and a broader distal part. In certain bamboos and other grasses and in palms, the pseudo-petiole may be long and clearly delimited.

When the leaf is differentiated into a distinct petiole and lamina, as in some Dioscoreales and Arales, the lamina may be lobed or even *compound.* *Pseudo-compound* leaves are leaves which when initiated are simple, but which in the course of development split along the nerves to become "compound", as in many palms and Cyclanthales. In Musaceae and some other families, simple leaves may become torn so as to appear compound. The leaf margins in monocotyledons are typically entire, but in some, mostly succulent species they

site leaves are set at an angle of 90° to their predecessors the leaves are called *decussate.* The points of attachment of successive solitary (alternate) leaves form a spiral, the angle between successive leaves often being constant. When the angle is 180°, i.e. two leaves to a revolution, they are *distichous* (standing in two rows), and when it is 120°, i.e. three to a revolution, *tristichous* (standing in three rows).

The leaves of some monocotyledons, e.g. *Dioscorea,* are differentiated into a short *leaf base,* a distinct *petiole* and a *blade (lamina).* However, the leaf base more often encloses the stem completely and forms a *sheath* of variable length which can be *open,* the two margins of the sheath being free, or *closed,* when the margins are fused with each other. Regardless of whether it is sheathing or not, the leaf base may be extended into *lateral lobes* or *stipules.* The base of a linear leaf, or leaf blade in the case of a leaf with a sheath, may be prolonged on either side of the stem as a lobe, the *auricle (ear).* The stipules may be displaced into the leaf axil to form a *stipular sheath,* as in many

are *serrate,* or beset with *spines* (*Aloë, Agave,* many genera of Bromeliaceae).

When distichous, the leaves, or their basal parts only, are sometimes strongly compressed from the sides and their adaxial surface (i.e. that facing the stem) merely forms a groove basally which acts as a sheath and is obliterated above. Such leaves are described as *ensiform* or *unifacial,* in contrast to the horizontally flat and *bifacial* normal leaves. They occur in several families, e.g. Iridaceae, Orchidaceae and Haemodoraceae.

In a few taxa the leaves are twisted through 180° at the base so that the morphologically abaxial side is adaxial; we refer to this condition as "*reversed* leaf blades" (it is sometimes called *resupinate*).

The terminology of the shape of the leaves follows general botanical practice.

*Ptyxis* is a term that refers to the folding or rolling of individual leaves in the bud stage. Types of ptyxis in monocotyledons are the *conduplicate, conduplicate-plicate, plicate, involute, supervolute,* and *explicative.* (Ptyxis is sometimes included in *vernation,* a term which covers the arrangement of parts in a bud with respect to each other).

*Venation* in monocotyledons is generally *acrodromous,* i.e. with longitudinal parallel or arched veins, converging at the apex, whereas the veins in many of the primitive dicotyledons are *camptodromous-brochidodromous,* with a midvein and, arising from it, lateral veins each of which is arched to meet the one arising next above it (see Hickey and Wolfe 1975). In leaves with a broad blade a minor, reticulate, vein system is intercalated between the main veins.

*Cataphylls* are scale-like leaves similar to the sheath of a foliage leaf. They may be present at the base of the shoot and as the first leaves (scales) of lateral branches, especially those of the inflorescence.

The shoot may continue into an inflorescence, or the inflorescence may be distinctly separated from the vegetative part, especially where there is a *leaf rosette,* and the inflorescences are borne each on a leafless peduncle. A *peduncle* arising from a basal rosette is known as a *scape* and the plant (or inflorescence) is described as *scapose,* a condition found in Hyacinthaceae and Amaryllidaceae, for example.

# The Inflorescence

*Inflorescences* are floriferous branch systems more or less distinctly delimited from the vegetative part of the plant. Inflorescences can be divided into *determinate* (closed), where the primary axis is terminated by a flower, and *indeterminate* (open), where the axis does not terminate in a flower.

Flowers of determinate inflorescences are normally actinomorphic, whereas indeterminate inflorescences can have either actinomorphic or zygomorphic (or even asymmetric) flowers (see below, Floral Symmetry).

Determinate inflorescences may be divided into panicles and cymes. In *panicles* there are lateral flowers, and branches terminated by flowers, at several levels below the terminal flower (as in *Tricyrtis*). Panicles may be many-flowered and complex or few-flowered. The panicle concept is also used for grasses, where the flowers are substituted by spikelets (which are indeterminate units). Panicles may be distally expanded and more or less flat-topped. Where the branches are very dense and the pedicels long this type may be *umbel-like.*

In *cymes,* lateral flowers are borne at only one level below the terminal flower, their stalks (pedicels) arising in the axils of prophylls (bracteoles) on the pedicel of the terminal flower. This process can be repeated indefinitely because every pedicel has its own prophyll(s). The number of flowers arising at the same level greatly affects the appearance of the inflorescence; in the cymes of monocotyledons there is generally only one, which makes the cyme a *monochasium.* As the prophyll subtending it is generally on that side of the pedicel which faces the preceding flower (the *adaxial* side) the cyme forms a zigzag and this is called a *rhipidium.* Where the prophyll and its axillary flower are not strictly adaxial the monochasium may resemble a *bostryx* (*helicoid cyme*). A bostryx in the strict sense results when each pedicel arises in the axil of a lateral prophyll (i.e. one set at right angles to the preceding one) and the direction of rotation is always the same. The umbel-like inflorescence of Alliaceae and Amaryllidaceae is probably derived, by condensation, from this type of cyme. Sometimes, each new flower emerges in the axil of an *abaxially* (opposite to adaxially) placed bracteole, resulting in a *drepanium,* a rare type of inflorescence (in some Juncaceae).

Indeterminate inflorescences may be divided into *thyrses,* where the pedicels of lateral flowers bear new flowers in the axils of their bracteoles (these lateral components are thus lateral cymes), and *racemose inflorescences* (*botrya*) where this is not so. The latter are classified according to the length and thickness of the main axis and the length of the pedicels: *racemes* have a long inflorescence axis and long pedicels, *umbels* have a short inflorescence axis and long pedicels, *spikes* have a long, slender inflorescence axis and very short pedicels or none, *spadices* (sing. *spadix*) have a long fleshy inflorescence axis and very short pedicels or none and *capitula* (*heads*) have a short inflorescence axis, and very short pedicels or none.

The inflorescences of many monocotyledons are extremely complicated, and painstaking analyses may be necessary to reveal their true nature. Thus the superficially simple "spike" or "spadix" of *Typha* and the "head" of *Sparganium* have proved to be complex, branched inflorescences (D. MÜLLER-DOBLIES 1968 and U. MÜLLER-DOBLIES 1969, respectively). The asymmetric flowers of Marantaceae are aggregated into complex thyrses (ANDERSSON 1976, 1981), and the dense "fascicles" of *Bobartia,* Iridaceae, into complex panicles (DAHLGREN, unpublished).

## The Flower

The flower is situated in the axil of a *subtending leaf* or *bract*. It consists of a *pedicel* (flower-stalk), which usually bears a *prophyll* (*bracteole*) on the adaxial side (i.e. the side opposite the subtending leaf and towards the parent axis). In the monocotyledons the prophyll is frequently two-ribbed, suggesting that it may have arisen by concrescence of two lateral prophylls (such as are present in most dicotyledons). Flower buds may also develop in the axil of the prophyll, as in cymose inflorescences. Rarely two or more prophylls are present on the pedicel.

The pedicel terminates in the *floral axis* (*receptacle*), on which the *tepals* (*perianth*), *stamens* (*androecium*) and *pistil(s)* (*gynoecium*) are inserted.

The flower is termed *complete* when perianth, stamens and pistil(s) are all present, *incomplete* if any of these parts are missing. It may be *bisexual* (*perfect, hermaphrodite*), having both stamens and pistil(s), or *unisexual* (*imperfect*), when it lacks either stamens or pistil(s). Species with unisexual flowers may be either *dioecious,* with *pistillate* (*female*) flowers and *staminate* (*male*) flowers on sep-

arate plants, or *monoecious,* with both kinds of flowers occurring on the same plant. *Sterile* (*neuter*) flowers lack both functional stamens and functional pistil(s); such flowers are often specialized for the attraction of pollinators.

In connection with the symmetry of the flower the concepts of the median and transverse planes are used. The *median plane* of a lateral flower is that falling through both the inflorescence axis and the pedicel and main axis of the flower. The *transverse plane* of a lateral flower is that cutting the floral axis at right angles to the median plane. Neither of these terms can be applied to a flower that is terminal on the inflorescence axis.

A flower is *actinomorphic* (*polysymmetric, radially symmetric, "regular"*) when three or more planes of symmetry (giving mirror images) can be placed through it. It is *bisymmetric* when two planes of symmetry can be placed through it, and *zygomorphic* (*monosymmetric*) when only one plane of symmetry can be placed through it. The zygomorphic flowers of monocotyledons always have a median plane of symmetry. The twisting of the pedicel or ovary through 180°, inverting the flower, is called *resupination;* it occurs in several groups and is prevalent in orchids. *Asymmetric* (*irregular*) flowers have no planes of symmetry at all (examples: Cannaceae, Marantaceae).

When describing a lateral flower, the lower side, which faces away from the inflorescence axis, is described as *abaxial,* and the upper, facing towards the inflorescence axis, as *adaxial.* These terms are also used to describe respectively the outer and inner sides of floral parts in relation to the floral axis (i.e. the centre of the flower).

*Floral diagrams* are constructed to illustrate and compare, in a uniform and schematic manner, the (transverse) plans of flowers. The floral components are placed in the diagram in such an order that the lowest and/or outermost parts are on the periphery and the uppermost/inner parts are in the centre (i.e. in the order bracteoles-tepals-stamens-pistils). It is usual to employ standard symbols for homologous parts. *Empirical floral diagrams* are those in which the components of the flower are shown in their position without any attempt at interpretation, whereas *theoretical floral diagrams* involve an interpretation, for instance they may indicate supposedly lost parts by crosses, and divisions and fusions by other suitable symbols.

# The Floral Axis

The floral axis (receptacle) in monocotyledons (see above, The Flower) is generally not strongly developed. Thus a *floral disc,* i.e. a disc-shaped or annular process developed from the receptacle, is extremely rare, whereas in dicotyledons this structure is common and often functions as a nectary. In dicotyledons the receptacle is also frequently urceolate, but urceolate structures of monocotyledon flowers are mainly formed by the fused tepals (and stamens). See below under perigyny.

# Numerical Conditions and Insertion of the Components of the Flower

A flower is described as *cyclic* when all the organs of the same type are in whorls (for *whorled,* see above, Leaves). This is the condition in perhaps all monocotyledons, whereas in many dicotyledons all or some of the floral parts are *spirally* set, the flowers being *acyclic* or *hemicyclic* respectively.

The number of whorls of floral parts (prophylls excluded) in the flower is indicated by the terms *pentacyclic* (with five whorls, which is a common and probably ancestral state in monocotyledons), *tetracyclic* (with four whorls; as when one whorl of stamens is lacking), *tricyclic* (with three whorls), etc.

A whorl of floral parts is classified according to its number of components (*merism, merous condition*). Thus nearly all monocotyledons are *trimerous* (parts in threes), but some are *dimerous* (parts in twos), or *tetramerous* (parts in fours).

According to the position of the tepals and stamens in relation to the ovary of the pistil (or the pistils), distinction is made between *hypogynous* flowers, where tepals and stamens arise from the floral axis "below the gynoecium", and *epigynous* flowers, where these parts arise above the ovary, "on the gynoecium", their basal parts being then fused with the pistil wall, taking part in the formation of the wall of the ovary and fruit. More rarely the flowers are *hemi-epigynous,* having the tepals and stamens inserted halfway up the ovary. *Perigynous* flowers, where there is a cup-shaped dilation of the receptacle, free from but surrounding the ovary or part of it, are common in some groups of dicotyledons but are strictly speaking not found in monocotyledons.

# The Perianth

The *perianth* (*perigone*) represents the floral envelope and consists of *floral* (*perigonal*) *leaves* (*tepals*). In most monocotyledons the perianth consists of two whorls of tepals which are either similar or dissimilar. Even when dissimilar they are often not readily divisible into outer green *sepals,* and inner contrastingly coloured *petals,* though this is so in some groups (taxa of Alismatales, Bromeliales, Commelinales, etc.). When green and sepal-like the tepals are described as *sepaloid,* and when of a colour other than green (white or bright colours) as *petaloid.* In dicotyledons the sepals are often collectively termed the *calyx* and the petals the *corolla,* and this is also possible in monocotyledons showing this kind of differentiation, but we shall not follow this usage here.

When the tepals in the two whorls differ conspicuously the perianth is termed *heterochlamydeous.*

Flowers in which the tepals are fused are called *syntepalous,* and the condition *syntepaly.* Tepals are often fused to form relatively narrow *floral tubes.* The tepals may taper basally into a stalk and are then described as *clawed* (*unguiculate*), a rare condition in monocotyledons.

Some or all of the tepals may be provided basally with a nectary known as a *perigonal nectary.* The nectarial area is sometimes recessed to form a *pouch* or *spur* (Fig. 108 H and N).

A median *labellum,* or lip petal, is present in some families. This is so in Orchidaceae, where it is the median, upper tepal of the inner whorl (though the flowers are usually resupinated), and in Zingiberaceae and Costaceae, where it consists of the two lower petaloid staminodes of the inner staminal whorl fused together. Though non-homologous, the labellum in each group has a similar function, namely as a landing place for pollen vectors.

When tepals are lacking the flowers are described as *atepalous* (*naked*) or, when the loss of the tepals is obvious, *apochlamydeous.*

The *paracorolla* (*corona*) is a structure derived from appendages of stamens or tepals as, for example, in Amaryllidaceae and Velloziaceae.

# The Androecium

*Androecium* is the collective term for the stamens and stamen homologues (i.e. also staminodes, see below). By definition, stamens are the floral structures that carry *microsporangia* in which the *microspores* are formed and subsequently develop into *pollen grains.*

*Diplostemonous* flowers, where two whorls of stamens are present (*diplostemony*), are here considered to be the ancestral state in the monocotyledons. Flowers with only one whorl of stamens are *haplostemonous,* the condition, *haplostemony,* being assumed to have resulted from loss of either the outer or the inner staminal whorl.

Stamens are numerous in several groups of monocotyledons, presumably by secondary multiplication ("dédoublement"), and the androecium is then described as *multistaminate* or *pleiomerous.*

The stamens consist of a generally slender stalk, the *filament,* and an *anther.* The anther consists of the *connective,* which is the continuation of the filament, and two *thecae,* each of which consists of two *microsporangia* (*locules, pollen sacs*). Rarely, by reduction, the anthers have only one theca and are called *monothecous,* or have thecae that consist of only one microsporangium which are called *unisporangiate.*

In some monocotyledons, e.g. many Dioscoreales, the stamens are flat and somewhat leaf-like, and in some members of this order as well as many others, the microsporangia are attached below the apex, which is then often described as a *connective tip* or *appendage.* In these kinds of stamens the anthers are not clearly set off from the filament and may be described as "*undifferentiated*" (SCHAEPPI 1931). Where the anther is attached at its base to the filament it is described as *basifixed* (*impeltate*). Basifixed anthers are called *sagittate* when the thecae are lobate and divergent from the connective at the base; otherwise they are *nonsagittate.* The anthers are described as *dorsifixed* (*peltate*) where the anther with its connective extends below its point of attachment, which is thus located somewhere along the mid-line of the anther. Peltate anthers can be divided into *epipeltate* when the part of the anther that is prolonged downwards beyond the attachment point of the filament faces *inwards,* and *hypopeltate* when this part faces *outwards* in relation to the centre of the flower. These two types have somewhat different distributions, the former being common in Liliales, the latter in Asparagales (HUBER 1969). (See Fig. 3.)

According to the position of the microsporangia in relation to the connective, the anthers can be divided into *introrse,* facing towards the centre of the flower, *extrorse,* facing away from the centre of the flower, and *latrorse,* facing laterally. Generally these are also the directions in which the microsporangia empty their pollen, which they do through longitudinal slits. In some groups, however, they dehisce by one, two or four apical pores and are then called *poricidal;* in these it is still possible to determine whether they are introrse or extrorse. When the anthers dehisce by longitudinal slits, the wall separating the two microsporangia of each theca has generally broken down beforehand, so that only a single line of splitting is required to open each theca.

In some monocotyledon flowers the filaments are laterally fused (connate) at the base to form a *staminal tube.* In male flowers of Araceae the stamens are often totally fused to form a *synandrium;* the fusion of the anthers is termed *synanthery.* Rarely, the filaments are fused with the style to form a *gynostemium (column).*

The terminology of the numerous specializations in the flowers of Orchidaceae appears under that family (pp. 255–259 and Figs. 119, 122, 124, 125, 126).

*Staminodes* are sterile homologues of stamens and may or may not have rudimentary anthers. Staminodes are considered to be derived (in the phylogenetic sense) from functional stamens, and stamens and staminodes together generally do not exceed six in any flower. When they are flat and brightly coloured, as in many Zingiberales, they are known as *petaloid staminodes* (Fig. 169 H).

# The Anther Wall and Tapetum

The layers of the microsporangium wall can be classified according to the behaviour of the two parietal cell layers which line the microsporangial epidermis and invest the microsporogenous tissue. There are four different types, named Basic, Dicotyledonous, Monocotyledonous and Reduced (DAVIS 1966).

The *Basic Type* is not known in monocotyledons and is rare in dicotyledons; it involves one periclinal division of each parietal layer to form four layers in all: an outer layer which becomes the endothecial layer with characteristic wall thickenings (see below), two middle layers and an inner layer which gives rise to the tapetum (see below).

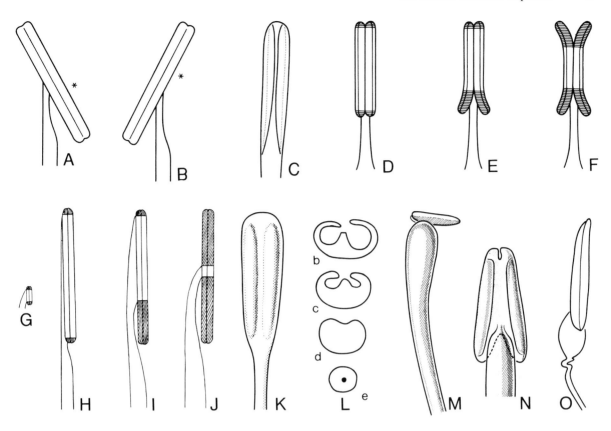

**Fig. 3.** Stamen characters. **A** and **B** comparison between hypopeltate (**A**) and epipeltate (**B**) anthers, the asterisk (\*) representing the morphological upper side. **C** "Undifferentiated" anther. **D** Basifixed, impeltate, non-sagittate anther. **E** Basifixed, impeltate, sagittate anther. **F** Basifixed, impeltate, x-shaped anther. **G–J** Sequence to explain derivation of adnate (**H**), semiadnate (**I**) and dorsifixed (**J**) anther types from an initial (**G**). **K** "Undifferentiated" anther, exemplified by *Sparganium*. **L** Series of transverse sections through an undifferentiated anther, exemplified by *Smilax*. **M–N** Stamens of *Yucca filamentosa*, **M** fully developed in lateral view, **N** in juvenile stage, front view. **O** Anther of *Dianella caerulea*, with filament dilated below anther. (All after SCHAEPPI 1939; from WEBERLING 1981)

In the *Dicotyledonous Type,* which is known among the monocotyledons only in *Tacca* (?), but is common in dicotyledons, the outer parietal layer divides as in the Basic Type, but the inner becomes the tapetal layer directly, so that there is only one middle layer. In the *Monocotyledonous Type,* which is by far the commonest in monocotyledons,

the *inner* parietal layer divides as in the Basic Type and the outer becomes the endothecium directly, so that again there is only one middle layer. In the *Reduced Type,* finally, neither parietal layer divides and there is no middle layer; the outer forms the endothecium directly and the inner the tapetum. This type is extremely rare in dicotyledons, and in monocotyledons is known only in *Najas* (Najadaceae) and in *Lemna* and *Wolffia* (Lemnaceae).

The *endothecium* is a layer in which the cells are provided with wall thickenings which play a part in the dehiscence of the microsporangial walls. The wall thickenings (DAHLGREN and CLIFFORD 1982) can be divided into two general types, the *Spiral* and the *Girdle* Types.

The *tapetal layer* (*tapetum*) is best developed when the microspores are in the tetrad stage. The tapetum then surrounds the spores and supplies them with nutrients. There are three types of tapetum (1) *secretory* or *glandular,* in which the tapetal cells remain in their initial position but lose their walls and finally degenerate in situ, (2) *amoeboid,* in which the tapetal cells persist for some time but eventually form a periplasmodium and invade the cavity occupied by the microspores, and (3) *periplasmodial,* where the cells dissolve at an early stage to form a periplasmodium. The last two types are often known collectively as the *amoeboid-periplasmodial* type (Fig. 32).

## The Pollen Grains and Their Dispersal

The interior of the microsporangium is occupied by the *archesporial cells,* which give rise to the *microspore mother cells;* these undergo meiosis, involving two successive divisions, to produce the *microspore tetrads.* This meiotic division is known as *microsporogenesis* and it proceeds in one or other of two modes, the Successive and the Simultaneous (Fig. 33). In the *Successive Type* a cell wall is formed after the first division, and another in each of the daughter cells after the second meiotic division; the cells of the tetrad usually lie in one plane and the arrangement is square, T-shaped or linear. In the *Simultaneous Type* the two perpendicular nuclear divisions proceed before wall formation starts, and wall formation occurs by the advance of constricting furrows from the periphery of the tetrad, which is tetrahedral. Both types are widespread in the monocotyledons, whereas in dicotyledons the Simultaneous Type is dominant. The Successive Type is sometimes known as the *Monocotyledonous Type.*

The terminology of pollen morphology (Fig. 4) here follows ERDTMAN (1952) as in DAHLGREN and CLIFFORD (1982). The pollen grains of monocotyledons generally have only one *aperture* (a thin part or hole in the *exine,* which is the outer layer of the pollen grain wall). The aperture can be located at the *distal* pole, i.e. the part of the pollen grain that is directed away from the centre of the pollen tetrad. When a distal aperture has the shape of a *slit* (*sulcus*) the pollen grain is called (*mono-*)*sulcate* and when it has the shape of a circular hole (*ulcus*) it is called *ulcerate.* These two types are the commonest in the monocotyledons. In other cases the exine is very thin or not even coherent, as in the Najadales of the Alismatiflorae, the Triuridiflorae, most Zingiberiflorae and most taxa of Orchidaceae of the Liliiflorae, in which case an aperture is naturally not visible, and the pollen grain is called *inaperturate,* although this is probably inaccurate in most cases and the grain is better described as omniaperturate (i.e. with the aperture covering the whole pollen grain) (THANIKAIMONI, 1978). Some aperture conditions that are rare or not definitely known in monocotyledons follow:

*trichotomosulcate,* where the sulcus is trilobate;
*spiraperturate,* where one or more furrow-like apertures are spirally arranged on the pollen surface;
*foraminate,* where a number of round apertures (*foramina*) are evenly distributed on the pollen surface;

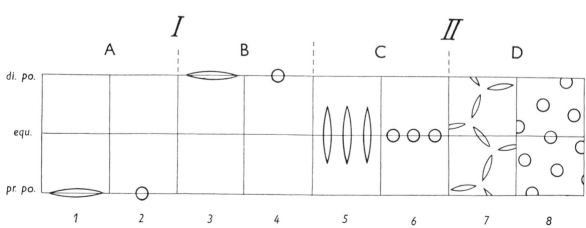

**Fig. 4.** Terminology of pollen grain apertures. Abbreviations: *di.po.* distal pole; *equ.* equator; *pr.po.* proximal pole; **I** apertures polar; **II** apertures non-polar; **IA** apertures in proximal pole; **IB** apertures in distal pole; **IIC** apertures equatorial; **IID** apertures global. The apertures, which are either slit-like or circular, are classified accordingly: (*1* laesura); (*2* hilum); *3* sulcus; *4* ulcus; (*5* colpi); (*6* pori); (*7* rugae); *8* foramina. Those placed in parentheses not found in monocotyledons. (After ERDTMAN 1952)

*sulculate,* where two (rarely three) slit-like apertures are present and lie in the equatorial plane of the pollen grain;

*zonisulculate* or meridionosulcate, where there is one continuous aperture all round the equator of the pollen grain, thought to have arisen through fusion of two sulculi, and

*porate,* where one or more round apertures are present in the equatorial plane of the pollen grain (this in its typical form is not found in monocotyledons but see, however, *Carludovica* in Cyclanthaceae).

*Colpate* pollen grains, where slit-like apertures (*colpi*) are present in the equatorial plane and directed at right angles to it, are probably not found in the monocotyledons.

The pollen is generally dispersed as separate grains but occasionally it is coherent in *tetrads* or *massulae.* In the mature pollen grain division of the microspore cell has taken place, so that either the pollen grain contains two cells, the *vegetative* and the *generative* cell, or, where the latter has divided further, as happens in certain groups, it contains three cells, the vegetative cell and two *sperm cells.*

The pollen may be dispersed by animals (*zoogamy*), wind (*anemogamy*) or water (*hydrogamy*). Dispersal by insects (*entomogamy*), birds (*ornithogamy*) and bats (*chiropterogamy*) are the most important types of zoogamy. Flowers show many specializations for these types of pollination. The list of specializations which characterize a particular method of pollination, some or all of which are expressed in all plants adapted to that method, is termed a *pollination syndrome,* for example wind pollination is characterized by the "*syndrome of anemogamy*". Some monocotyledons exhibit *hydrogamy* (water pollination) and in the more advanced marine exemplars of this method the pollen grains are almost completely devoid of exine and, although globose at first, they rapidly elongate to become filiform, in which condition they are more easily caught by the stigmas. This elongation may be equivalent to the development of a pollen tube in other groups.

## The Gynoecium

*Gynoecium* is a collective term for the *carpels* of a flower, which can be free and make up separate *pistils,* a condition called *apocarpy,* or fused into a single pistil, a condition known as *syncarpy.*

The number of carpels in the gynoecium is indicated by the terms *monocarpellary, bicarpellary, tricarpellary,* or *multicarpellary* (with one, two, three or many carpels, respectively), conditions which all occur in the monocotyledons.

The pistils, whether monocarpellary or consisting of two or more carpels, may comprise the following parts:

a *stipe,* which is the stalk of the pistil; it is generally very short but may in exceptional cases be long, as in *Ruppia* (Potamogetonaceae);

an *ovary,* the widened part enclosing one or more cavities (*locules*) containing the ovules and characterized as *unilocular, bilocular* or *trilocular* according to the number of locules (ovaries containing more than three locules are rare in monocotyledons); and

a *style,* the narrow upward prolongation of the ovary carrying the stigmatic papillae; the carpels of a pistil, when more than one, may each have a separate "style" which is then called a *stylodium,* or they may share a single common style (*style* in the strict sense) and this may be three-branched (*tribrachiate*), three-lobed (*trilobate*), or entire, when it may end in a head (being *capitate*) or a point (being *punctiform*); the style is generally terminal, i.e. situated on the top of the ovary, but in some groups it emerges from the side or even the base of the ovary (*gynobasic*), as in many Alismataceae and Triuridaceae; the stylodia may be bifurcate nearly to the base, as in some Najadales (Fig. 150 D) and some Iridaceae; when there are stylar lobes, these are generally opposite the midvein of the carpel;

*stigmatic papillae,* cells receptive to pollen tubes and situated on the distal parts of the stylodia, the style or its branches or lobes, being either restricted to their actual apex or, especially in wind-pollinated taxa, covering a considerable part of their surface.

Monocarpellary pistils in the monocotyledons always have a *unilocular* ovary, whereas the bi- and tricarpellary pistils generally have two and three separate locules respeetively, the locules being separated by partitions (*septa*). In several groups, however, the syncarpous pistils lack septa, or have incomplete septa, so that they are in effect unilocular.

The *placenta* is defined as that part of the ovary on which the ovules are borne. When the ovules are numerous the placenta is a richly vascularized tissue. The placenta is generally confined to a particular part of the locule (base, apex or central axis of the pistil).

When there is a single carpel placentation can be classified as:

*marginal,* with the ovules inserted along the carpel margin,

*apical,* with the ovules concentrated at the top of the locule,

*basal,* with the ovules concentrated at the bottom of the locule, and

*laminar (-dispersed),* with the ovules scattered over the whole or most of the inner wall of the carpel.

In syncarpous pistils the placentation is classified according to the position of the placentae in the ovary as follows:

*parietal,* with the ovules inserted on the wall of the ovary along the carpel margins, which is generally the case with unilocular ovaries (without septa),

*central* or *axile,* with the ovules on the central axis of the ovary,

*free-central,* with the ovules borne on a central column or at least centrally at the bottom of a unilocular pistil (e.g. *Luzula*).

By combining these two sets of concepts the placentation can be described adequately. Certain trends are observable in monocotyledons; thus, when the ovules are apical they tend to be orthotropous (see below) and when the ovules are extremely numerous the ovary is often unilocular and the placentation parietal (Orchidaceae, many Burmanniales, some Philydrales, Mayacaceae).

## The Ovule

An *ovule* consists of the following parts:

the *funicle,* the stalk by which the ovule is attached to the placenta; in curved anatropous ovules the funicle is fused with one side of the ovule and is visible as a rib or ridge, the *raphe,*

the *nucellus,* the central region of the ovule (often regarded as being equivalent to a megasporangium), and

one or two *integuments* which are two or a few cell layers thick and envelop the nucellus.

The *micropyle* is the more or less narrow, rarely completely obstructed, pore at the top of the integuments. The micropyle can be formed in both integuments or in either the inner or outer integument, if one or the other integument is delayed in development. It may have a "zigzag" form if the pore of the inner integument does not exactly coincide with that of the outer. Nearly all monocotyledons have two integuments, and no monocotyledon family constantly has only one. Ovules with two integuments are termed *bitegmic,* those with one integument are *unitegmic* and those in which the integuments have been lost (rare) are called *ategmic.* The *chalaza* (chalazal part) of the ovular body is generally that located at the end opposite to the micropyle.

In addition, tissues that develop into arils, strophioles, caruncles and analogous structures are present in scattered groups (see below under Seed). *Arils,* when present, may form a conspicuous envelope outside the outer integument as early as the ovular stage, e.g. in Asphodelaceae (Fig. 5S–T) and many Zingiberales.

According to the number of layers of cells surrounding the archesporial cell and embryo sac (see below) the ovule is termed *crassinucellate* (several or numerous layers) or *tenuinucellate* (one or few layers). Within the nucellar tissue one *primary archesporial cell* (rarely more) will differentiate. This may cut off one or more cells (*parietal cells*) at the micropylar end before the two meiotic divisions leading to the *megaspore tetrad* (cf. The Pollen Grains). The parietal cells, when present, generally divide further, forming *parietal tissue* between the embryo sac (see below) and the nucellar epidermis. In a number of groups a parietal cell is *not* cut off; this is often the case where the ovules are tenuinucellate. Sometimes the nucellar epidermis in the apical part of the nucellus divides by periclinical divisions to form what is known as a *nucellar cap.* If this is the case in ovules without parietal tissue some embryologists refer to such ovules as "*pseudocrassinucellate*" (here belong many grasses, most Cyclanthaceae except *Cyclanthus* and a few other groups).

According to their morphology ovules are divided into the following types:

*anatropous,* with a straight nucellar axis, and reflexed upon the elongated funicle, which is usually fused at the side to the ovular body (=raphe; see above); the micropyle is thus close to the funicle;

*hemianatropous (hemitropous),* with a straight nucellar axis at right angles to the funicle;

*campylotropous,* with a curvature of the nucellar axis, which brings the micropyle near the funicle and makes it basally directed; the embryo sac in this case is more or less straight, but if the nucellar axis is even more curved and the embryo sac is also curved the ovule is described as *amphitropous*; *orthotropous (atropous),* with a straight nucellar axis and the micropyle opposite the funicle.

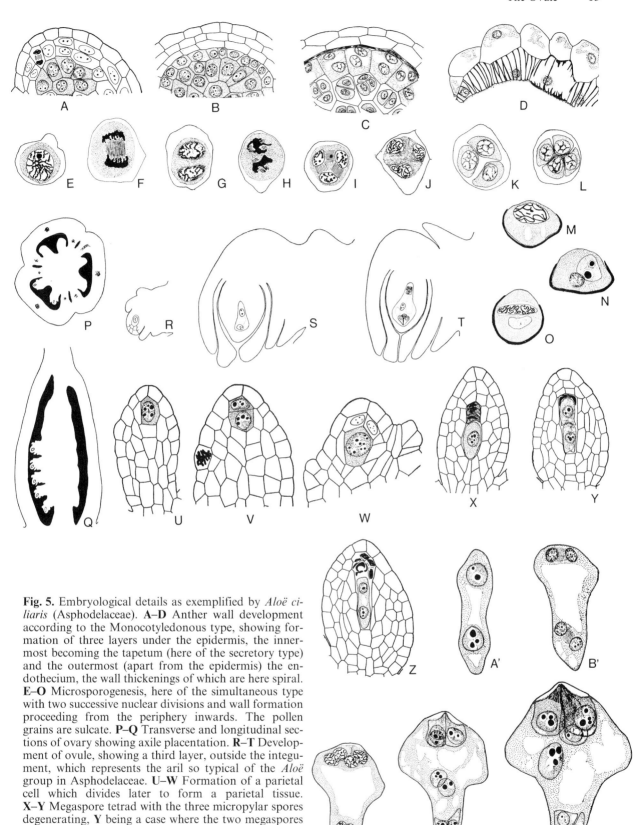

**Fig. 5.** Embryological details as exemplified by *Aloë ciliaris* (Asphodelaceae). **A–D** Anther wall development according to the Monocotyledonous type, showing formation of three layers under the epidermis, the innermost becoming the tapetum (here of the secretory type) and the outermost (apart from the epidermis) the endothecium, the wall thickenings of which are here spiral. **E–O** Microsporogenesis, here of the simultaneous type with two successive nuclear divisions and wall formation proceeding from the periphery inwards. The pollen grains are sulcate. **P–Q** Transverse and longitudinal sections of ovary showing axile placentation. **R–T** Development of ovule, showing a third layer, outside the integument, which represents the aril so typical of the *Aloë* group in Asphodelaceae. **U–W** Formation of a parietal cell which divides later to form a parietal tissue. **X–Y** Megaspore tetrad with the three micropylar spores degenerating, **Y** being a case where the two megaspores are still large. **Z–E′** Development of embryo sac according to the *Polygonum* Type, from the binucleate stage to the mature embryo sac. (GOVINDAPPA 1955)

## Embryo Sac Formation

The primary archesporial cell or the cell remaining after the parietal cells have been cut off (see above), functions as the *embryo sac mother cell;* this divides by the meiotic process (two nuclear divisions) to form four nuclei, the *megaspore nuclei.* Generally wall formation takes place after each division, producing four *megaspores,* only one of which takes part in the formation of the embryo sac (see below), which is then *monosporic.* Sometimes wall formation does not take place after the second division, so that two cells (*dyads*), each with two megaspore nuclei, are produced; one of these forms the embryo sac, which is then *bisporic.* Rarely, but especially in monocotyledons, no cell walls are formed in connection with meiosis and the cell containing all four megaspore nuclei develops into the embryo sac, which is *tetrasporic.*

The following main types of *embryo sac* occur in monocotyledons (see DAHLGREN and CLIFFORD 1982):

The *Polygonum* (*Normal*) *Type* (Fig. 5X–E'), a monosporic type in which the lowest (chalazal) megaspore undergoes three divisions to form an eight-nucleate embryo sac with one *egg cell* accompanied by two *synergids* at the micropylar end and three *antipodals* at the chalazal end. The remaining two nuclei, the *polar nuclei,* may fuse to form a central nucleus. This is the commonest and most widely distributed type in monocotyledons and also in dicotyledons.

The *Allium Type* (Fig. 50 I–P), a bisporic type, in which the lower chalazal dyad undergoes two divisions to form an eight-nucleate embryo sac similar in organization to the *Polygonum* Type. This is known in several groups of monocotyledons (e.g. Alismataceae, most Trilliaceae, and various genera of Alliaceae, Hyacinthaceae, Amaryllidaceae, etc.).

The *Endymion Type,* a bisporic type which differs from the *Allium* Type in that the upper (micropylar) dyad develops into the embryo sac. This is less common than the *Allium* Type and is found in a few Convallariaceae, Hyacinthaceae, etc.

The *Fritillaria Type* (Fig. 112 M–T), a tetrasporic eight-nucleate type, with two divisions following upon meiosis but accompanied in a characteristic way by nuclear fusions. The nuclei of the egg cell and synergids and the upper polar nucleus are haploid, and are all derived from the micropylar megaspore nucleus. The three antipodals and the lower polar nucleus are triploid instead of haploid, which is the result of a form of nuclear fusion of the three lower nuclei at the four-nucleate stage, followed by two successive divisions of this triploid nucleus. (*Haploid* means having one set of chromosomes and *triploid* means having three, the normal condition of the vegetative cells being *diploid,* with two sets.)

Further types, the *Clintonia, Adoxa* and *Drusa Types,* all tetrasporic, are rare and need not be defined here (see, however, DAHLGREN and CLIFFORD 1982).

## Endosperm Formation

The angiosperms are characterized by what is known as *double fertilization,* where the nucleus of one of the two *sperm cells* fuses with the nucleus of the egg cell (forming the *zygote*), and the nucleus of the other fuses with the central nucleus or the polar nuclei. The zygote (diploid) divides to form the *embryo;* the fusion product of the nucleus of the sperm cell and the central nucleus or polar nuclei (usually triploid) divides to form the *endosperm.*

Three main types of endosperm formation occur:

*cellular,* where the successive divisions of the endosperm nuclei are quickly followed by the formation of cell walls,

*helobial,* where the first division of the primary endosperm nucleus is followed by a wall division giving rise to a larger *micropylar* and a smaller *chalazal chamber* or division; in the micropylar chamber the succeeding nuclear divisions are followed by wall formation after a delay, whereas those in the chalazal chamber may or may not be followed by wall formation, and

*nuclear,* where the successive divisions of the endosperm nuclei are not accompanied by wall formation, which occurs considerably later.

The endosperm types are of importance in the study of monocotyledon phylogeny.

## Embryo Formation

Embryo formation (*embryogeny, embryogenesis*) is classified according to the planes of division of the zygote (see above, Endosperm Formation). The first division of this cell is generally transverse, and this is the case in all known monocotyledons.

This division gives rise to a *basal* (proximal) *cell* and a *terminal* (distal) *cell*. Further classification is based on whether the terminal cell divides transversely or longitudinally. Longitudinal division gives rise to the Onagrad and Asterad Types, while three main types arise from transverse division, the Caryophyllad, Solanad and Chenopodiad Types.

The Asterad and Onagrad Types are common in monocotyledons. They differ mainly in the role of the basal cell – in the *Asterad Type* the basal cell plays an important part in the formation of the embryo, whereas with the *Onagrad Type* its role in the formation of the embryo is insignificant.

The latter type is characteristic of most orchids and of Cyperales but is on the whole less common than the Asterad Type. A special modification of the Asterad Type in which the primary division is laid down obliquely is found in the grasses.

In the *Caryophyllad Type* (Fig. 135 L) the basal cell does not divide further but increases in size and forms a vesicular suspensor for the embryo which is formed wholly from the terminal cell and has a neck that varies in length. It is found in all (?) Alismatiflorae and most Ariflorae, but otherwise occurs only sporadically in monocotyledons.

The *Solanad* and *Chenopodiad Types* are only occasionally reported in monocotyledons and will therefore not be described here.

## The Fruit

The fruit, like the ovary, may be formed by one or several carpels. Where carpels are more than one, the whole set of monocarpellary fruits of the flower developed from an apocarpous gynoecium may be validly compared with the single fruit developed from a syncarpous gynoecium (the main difference being the degree of fusion of the carpels).

A distinction is usually made between an *aggregate fruit*, which is composed of a number of coherent individual fruits formed by separate pistils in the same flower (virtually absent in monocotyledons) and *multiple fruits*, which are a group of coherent fruits formed from adjacent flowers (as in *Ananas*, *Cyclanthus* and *Carludovica*). This distinction is lost when both of these are indiscriminately called syncarpia. The capsule-like fruit of *Gaimardia* in Centrolepidaceae (Fig. 215 M) is here considered

to be a double fruit of follicles (see below) belonging to adjacent monocarpellary flowers.

The fruit consists of the fruit wall or *pericarp* and the *seeds*. When the flower is epigynous, the wall of the fruit inevitably incorporates *accessory parts* derived from the perianth and androecium or from the receptacle, and its pericarp is therefore not fully homologous with that of a fruit developing in hypogynous flower.

Monocarpellary fruits in monocotyledons are classified as follows:

*follicle* – dry, usually dehiscing along the carpellary margin;

*achene* – small, hard, dry and indehiscent, with one or a few seeds;

*monocarpellary drupe* – pericarp fleshy on the outside and hard within; indehiscent and generally one-seeded;

*monocarpellary berry* – fleshy, indehiscent and with one or several seeds.

Syncarpous fruits (which in monocotyledons rarely consist of more than three carpels) are classified as follows:

*capsule* – dry and dehiscent, generally with few to numerous seeds. The capsules are *loculicidal* if dehiscent along the midribs of the carpels, *septicidal* when breaking open and dehiscing along the sutures of the septa, and *septifrage* when the outer walls are shed as valves leaving the septa and the (usually axile) placentae;

*pyxis* (*pyxidium, lid capsule*) – dry, circumscissile, i.e. splitting round the capsule so that the upper portion forms a lid;

*syncarpous nut* or *nutlet* – dry, more or less hard, indehiscent, and with one or a few seeds;

*caryopsis* – a syncarpous nutlet in which the pericarp and seed coat are fused (found in most grasses);

*schizocarp* – usually dry, splitting radially along the septa so that complete locules are detached (*mericarps*); these are generally nut-like. Schizocarps are very rare in monocotyledons but are known in Cyanastraceae and Juncaginaceae;

*syncarpous drupe* – like the monocarpellary drupe but consisting of two or more carpels;

*syncarpous berry* – like the monocarpellary berry, but consisting of two or more carpels.

The *utricle* should not be equated with the foregoing fruits; it consists of a nutlet which is tightly enclosed in a flask- or bladder-like envelope and is known in *Carex* and some other Cyperaceae. The term is also widely applied to the thin-walled fruits of some grasses, in particular *Eleusine* and its allies.

## The Seed

The *seed* consists of:

the *seed coat,* which is formed from the integuments,

*storage tissue,* which may consist of endosperm, perisperm, and/or chalazosperm (see below), or which may be lacking, and

the *embryo,* which represents the young plant which has become dormant.

The *seed coat* is here divided into two layers (following CORNER, 1976):

the *testa,* which is the outer layer, formed by the outer integument, and

the *tegmen,* which is the inner layer, formed by the inner integument.

These layers vary in relative thickness and in the shape of their cells from one group of monocotyledons to another. The epidermis of the testa may be black owing to the presence of a characteristic carbon-rich compound, *phytomelan,* which is known only in taxa of Asparagales (see Fig. 90 M– P). Rarely, the testa is fleshy, forming a *sarcotesta,* as in certain Araceae.

The storage tissue is generally made up chiefly of the endosperm, which is surrounded by a thin remnant of nucellar tissue. In most orchids endosperm is not formed, while in all Alismatiflorae and at least half of the Ariflorae it is absorbed during seed development. It is also weakly developed in the few groups which have a copious (i.e. rich) perisperm or chalazosperm. *Perisperm* is a storage tissue that consists of *nucellar tissue* which has undergone active cell division after fertilization of the ovule. It occurs only in families of Zingiberales and in the Hydatellales. *Chalazosperm* is a storage tissue that consists of proliferating tissue from the chalazal part of the ovule. It occurs together with perisperm in Costaceae, Cannaceae and Marantaceae (Zingiberales) and as the chief storage tissue in the family Cyanastraceae (Asparagales).

## The Embryo

The *embryo* consists of the following parts: radicle, hypocotyl, cotyledon and plumule. The *radicle* is the rudimentary root, which in monocotyledons ceases to grow at a relatively early stage of development. The *hypocotyl* is the region of the axis between the radicle and the cotyledon. The *cotyledon* is the first leaf and is often tubular; the exact relationship of the so-called cotyledon in monocotyledons to the two cotyledons in the dicotyledons has not been satisfactorily determined (see p. 44). The *plumule* is the stem bud and is found at the base of or inside the cotyledon and emerges from it through a lateral pore. It bears the first leaves after the cotyledon and in most monocotyledons has an initial distichous phyllotaxy.

In groups which lack endosperm the *hypocotyl* may be very thick and fleshy, serving as the storage reservoir. Embryos in this condition are termed *macropodous* and occur in many Alismatiflorae and Ariflorae.

The classification of the different types of embryo is outlined in the chapter Distribution of Character Conditions (section on Fruit and Seed Characters).

# Chemical Characters

S.R. JENSEN and B.J. NIELSEN

Plants are built up mainly from organic molecules and water. In the course of *primary metabolism* they synthesize large amounts of *sugars, amino acids, fats* and *lignans*. These are the basic units from which are built up the more complex substances of the cells such as cellulose, starch and enzymes (protein), end-products that constitute most of the organic material in roots, stems, leaves, flowers and fruits. Most of these basic substances are of no taxonomic interest since they are ubiquitous in higher plants.

A certain portion of the compounds formed by the primary metabolism is diverted into *secondary metabolites,* giving rise mainly to products involved in the basic vital processes of the cell by acting as regulators (hormones) of the primary metabolism. Some, however, may merely accumulate, particularly in certain tissues (e.g. the bark). Examples of such compounds are ethereal oils, bitter glycosides and alkaloids. Earlier botanists did not perceive any obvious role for these substances and therefore usually regarded them as mere waste products of the primary metabolism. However, it has been found that they often make the plants poisonous or disagreeable to feeding animals (*predators*) and in that way protect them. The presence of such compounds may thus be of positive selective value for the plants and may favour their survival. In this context it is usually assumed that the poisonous or repellent substances are not mere by-products and that, instead, their production imposes an energy cost on the plant which is "justified" by the protection gained. Various kinds of substances have been shown to be effective against micro-organisms either by their chemical action (as antibiotics) or by mechanically sealing wounds (gums). Others protect the plant from vertebrate and invertebrate herbivores.

There are also cases where herbivores have developed means of evading the harmful effects of certain compounds, sometimes even becoming dependent on them, so that they seek out the plants containing them. Thus the presence of secondary compounds is not unambiguously advantageous to plants, even though it may effectively reduce exploitation by the less specialized herbivores.

Chemistry has always been used in the classification of plants. The colour of flowers, as well as the taste and smell of fruits, seeds and other parts, are examples of chemical characters. Until recently the occurrence of specific compounds has been little known and even our present knowledge is based on studies of a very small proportion of existing species.

Chemists are generally most interested in economically important plants, and systematic surveys of substantial groups of related species have only recently been undertaken. Chemical methods have been improved tremendously in the last three decades, facilitating the isolation of compounds and the determination of their chemical structure. Thus a branch of biology that can be called *chemotaxonomy* or *biochemical systematics* has evolved, with the main purpose of elucidating relationships in plant groups, though some studies also consider the biological effects of secondary metabolites. The fields of botany, chemistry and pharmacology have benefited mutually from this development.

*The Definition of Chemical Characters.* For several reasons it is not possible to give a precise definition of a chemical character. If a certain compound is present in a plant this may be taken to constitute a character condition. But what one needs to know is whether this compound is present in a whole genus or family under study. Presence or absence in a single plant, on its own, may be taxonomically useless information. Reports of "absence" also have to be viewed in relation to the sensitivity of the method of detection used. Absence of a compound in some taxa within a group where the compound is otherwise present may represent a loss of the ability to produce it rather than an ancestral lack of it. *Only the verified presence of a compound constitutes a character condition that is of use for our purpose.*

The next problem is the significance of the presence of a compound (or group of biogenetically related compounds). Where, for example, two particular families between which there are other conspicuous similarities have the same compound in many (or at least several) of their members, we can certainly use this as another indication of relationship. Especially if a compound is of rare occurrence in the higher plants (or in the monocotyledons), chemists will be inclined to stress that its presence in, let us say, two families, is a strong indication of a close relationship. This chemical similarity, however, must be supported by other evidence (see our discussion of tricin in Poales, Cyperales and Arecales, pp. 90, 105). Naturally such a situation should make the taxonomist aware of the need to reconsider the previously accepted positions of the two families in the light of the total evidence. If one or both of the families has obscure relationships, or if a family seems to have equally strong affinities in different directions, the newly discovered chemical character may decide the issue. But this can only be done when all evidence is included in the consideration. In conclusion: *Chemical characters should only be used in conjunction with other characters.*

In constructing a phylogenetic classification, great care must be taken that the character states on which it is based are homologous, i.e. that they arose in a common ancestor. A chemical compound has a unique structure independent of its source, and its identity can be unequivocally established. As this is not so with most other characters used, some chemists are inclined to regard chemical characters as the most reliable ones. However, homology is not ascertained by the occurrence of the same compound, because its biosynthesis may follow different pathways in different plant groups. Moreover, even if the biosynthesis when known is proved identical in two groups of plants, the enzymes involved in this biosynthesis may not be homologous, i.e. the biosynthetic pathways may have arisen independently. Ultimately, these enzymes can be characterized, and if they are identical their homology is almost certain, since it reflects a particular gene sequence in the nucleus.

It is accordingly important to investigate the enzymes and the genetic material. This, however, is much more time-consuming and requires much more material, usually living, than the study of secondary metabolites, which requires less equipment and may even be done using small amounts of herbarium material. Nevertheless, investigations of enzymes and of RNA and DNA base sequences are now performed routinely, mainly on bacteria and some animals. Thus, for example, BOULTER and associates (e.g. BOULTER 1973a, b)

have studied the amino acid sequences in cytochrome c and plastocyanin.

So far, we have considered chemical characters mainly in the form of single compounds, but this can be extended to progenitors of a compound on its biosynthetic pathway, the only limit being that the progenitor is not ubiquitous or too widespread. One way of establishing whether compounds share the same biosynthetic pathway is to feed an isotopically marked precursor to a living plant, and subsequently isolate the resultant metabolite. A wealth of information is now available on chemical transformations likely to occur in biological systems (e.g. MANN 1980). This information is useful for postulating pathways, and in many instances the chemist, knowing the structure of a new compound, may predict its anabolic pathway.

Below, we present some important compounds, most of them secondary metabolites, grouped on the basis of biosynthetic criteria.

Retrieval of chemical information is facilitated through the serial publication Chemical Abstracts, in which both the Latin names of the plants and the names of the chemical compounds appear.

In several ways the data in the literature are not representative; thus north-temperate taxa are over-represented, as are those which are of use to man, and chemists tend not to report on plants in which the compounds sought are not found, negative results being generally noted only when of particular interest.

Some compounds can be recognized by their colour or smell, or in some other way; for example, they may be easily crystallized; their occurrence in nature tends to be better known than that of others, and some are mentioned below. It may appear that so little is known about most chemical substances or that our knowledge of them is so uncertain that they should be disregarded for taxonomic purposes. However, most chemical data are accurate, and with adequate background knowledge they can be highly useful as a complement to other characters.

### Glycosides (Carbohydrates)

As stated in the introduction, the carbohydrates belong mainly to the primary metabolism. Various compounds are often bound to one or more molecules of carbohydrates, however, and thus may appear in the form of glycosides. Most organic compounds are hydrophobic, i.e. they do not dissolve in water. This depends on the number of

dhurrin    glucose    hydrocyanic acid    hydroxy-benzaldehyde    oxalic acid

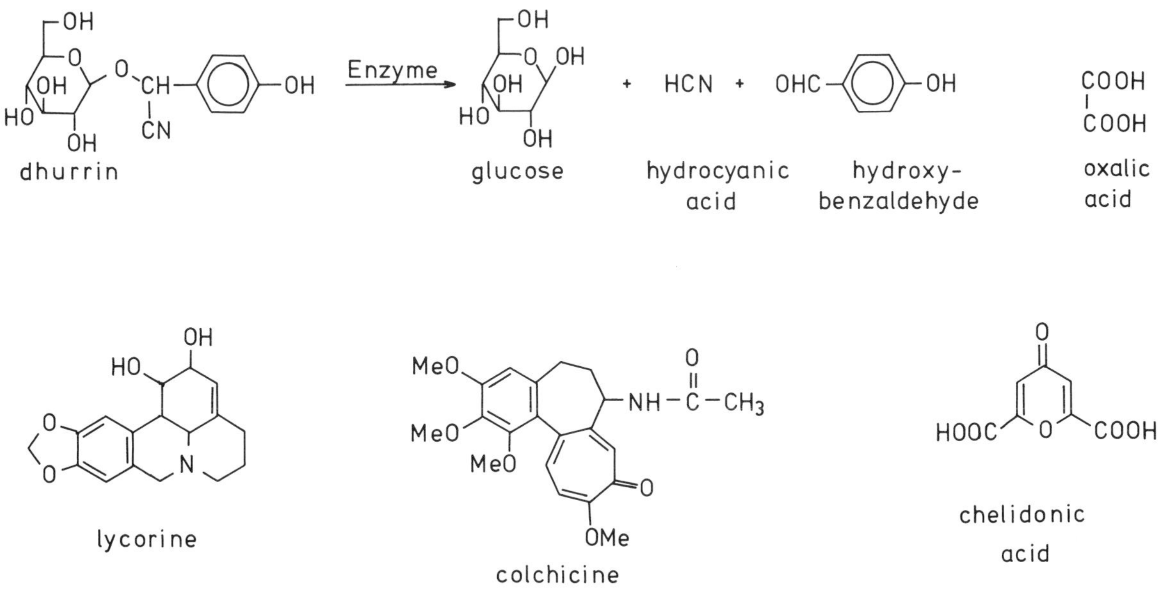

lycorine

colchicine

chelidonic acid

hydroxyl ($-$OH) groups attached to them. In glycosides, however, the hydrophilic sugar moiety is present and renders most compounds water-soluble. This enables these compounds to be transported in the sap of the plant. Another advantage of glycosidization is the masking of reactive centres in the compounds. Thus, by removal of the sugar moiety, cyanogenic glycosides become unstable and the noxious hydrocyanic acid is liberated.

*Cyanogenic Glycosides.* In the monocotyledons the cyanogenic glycosides comprise a rather uniform group of compounds and are derived from protein amino acids. Of the 30–40 different cyanogenic compounds known from higher plants, only 7 have been recorded from monocotyledons. These are derived from only three different amino acids, namely tyrosine, phenylalanine and leucine. Glucose is so far the only sugar found in these glycosides.

Hydrocyanic (prussic) acid is not liberated from the cyanogenic glycosides as long as the glycosidic linkage is intact. But on contact with the enzyme β-glycosidase this bond is broken and the poisonous hydrocyanic acid is formed. In plants this usually happens when tissues are broken or damaged. Enzymatic cleavage of dhurrin is shown in the scheme in Fig. 6.

The presence of cyanogenic glycosides in plants is easily detected in small amounts of plant material, either fresh or from a herbarium sheet. Therefore a large number of species has been examined and the coverage is considered fairly good. The taxonomic implications of these compounds are

**Fig. 6.** Examples of some compounds found in the monocotyledons. In the *upper line*, to the *left*, is shown the formation of hydrocyanic acid and a sugar from a cyanogenic glucoside, dhurrin. See further in the text.

discussed by SAUPE (1981). Nearly all cyanogenic compounds found in the monocotyledons are derived from the amino acid tyrosine, whereas in the dicotyledons other pathways are more common. As shown by HEGNAUER (1973, 1977) the presence of the tyrosine pathway indicates phylogenetic relationships with the Magnoliiflorae of the dicotyledons.

*Tuliposides* A and B are other examples of compounds bound as glycosides and dissolved in the cell sap. Breaking of plant tissues brings the glycosides into contact with an enzyme contained elsewhere in the cells, whereby the aglycone is liberated. This is known to cause allergic reactions in man. The tuliposides have so far been recorded in the related families Liliaceae and Alstroemeriaceae (Liliales) in the monocotyledons, but have been shown to be absent from genera in many other families in the Liliiflorae (SLOB et al. 1975).

## Organic and Inorganic Acids

Some organic acids that belong mainly to the primary metabolism are accumulated in a number of plants, and some of these appear to have an interesting distribution in the monocotyledons. The calcium salt of oxalic acid (Fig. 6) is almost insoluble in water. In many groups of monocotyledons it occurs in certain cells as bundles of fine needle-like crystals, called *raphides*. There are also other types of oxalate crystals. The inorganic silicic acid, which is also insoluble, is accumulated in a similar way in the form of silica bodies, which like the oxalate raphides may have a somewhat protective function for the plant against excessive grazing. The two kinds of crystals have an almost complementary distribution in the monocotyledons. On the ordinal level only the Bromeliales and Arecales consistently contain both kinds of crystals. Another organic acid with a taxonomically interesting distribution is *chelidonic acid* (Fig. 6). This acid is easy to detect by a qualitative test and many species have therefore been investigated. Members of Asparagales, Liliales and Haemodorales often contain this acid, whereas in other groups of monocotyledons it occurs only sporadically.

### Alkaloids

These compounds are generally biosynthesized from the essential amino acids, although exceptions are found. Alkaloids are more common in the dicotyledons than in the monocotyledons. The only alkaloids that seem to have any systematic interest in monocotyledons are those characteristic of the Amaryllidaceae (e.g. lycorine, Fig. 6) and those of the colchicine group (Fig. 6), found in Colchicaceae. Both groups are formed by combination of tyrosine and phenylalanine, but in apparently unrelated ways. From a systematic point of view, both Amaryllidaceae and Colchicaceae would be expected to contain steroidal saponins, but these are absent from the two families and have apparently been lost as superfluous in the protection of these plants. All genera of the Amaryllidaceae investigated contain alkaloids, while in the Colchicaceae, only about half of the genera contain them (WILDMAN 1968; WILDMAN and PURSEY 1968).

### Terpenoids

Compounds formed by the mevalonic acid pathway of biosynthesis are classified as terpenoids. They can often be recognized by the number of carbon atoms, which are multiples of 5, and by the regular positions of one-carbon substituents. Mevalonic acid contains six carbon atoms, but loss of one carbon atom gives rise to a $C_5$-unit with an iso-pentenyl skeleton, which by oligomerization forms *monoterpenes* ($C_{10}$), *sesquiterpenes* ($C_{15}$), *di-*terpenes ($C_{20}$), *triterpenes* ($C_{30}$), etc. Many of the terpenoids keep the original number and position of their carbon atoms and can thus be recognized as such, but others have rearranged carbon skeletons or have lost some carbon atoms, or both, and then are more difficult to identify as terpenoids. The differences in size of the molecules and in the functional groups give the compounds many different properties. Furthermore they may occur as glycosides, be esterified with organic acids, or be bound to other compounds of a different biosynthetic pathway.

*Hemiterpenes* are usually found as substituents in phenolic compounds, but these are not common in the monocotyledons and will not be further considered here.

*Monoterpenes* are the main constituents of many *essential oils*. They are more or less common in plants and can usually be smelled when they are abundant. In plants they may occur, in free form, in schizogenic oil ducts. Examples are the well-known products camphor (Fig. 7) and vegetable turpentine, the former consisting of a single crystalline compound, the latter being a liquid mixture of mainly α- and β-pinene (Fig. 7). The main constituent in lemon grass oil from *Cymbopogon* (Poaceae) is citral (Fig. 7).

*Sesquiterpenes* and *diterpenes,* being larger in molecular size than the monoterpenes, have higher boiling points and are more viscous. These properties make mixtures of the compounds more "resinous", and plant resins are often made up of this group of compounds. (Oxidized forms, for example the sesquiterpene lactones, are taxonomically important markers for certain dicotyledons, but conspicuously lacking in the monocotyledons).

*Triterpenes* are, from a taxonomic point of view, the most important terpenoids in the monocotyledons. They may be found in their free forms, called *sapogenins,* or as glycosides, called *saponins.* The saponins have the property of haemolyzing blood cells. In the monocotyledons true triterpenoid sapogenins (e.g. taraxerol, Fig. 7) appear to be rather scattered (but not as common as in the dicotyledons) and are not taxonomically interesting. *Steroidal saponins,* on the other hand, which have a degraded triterpenoid skeleton, show an interesting distribution: they are very common in Dioscoreales, Asparagales and Liliales and have only scattered occurrences elsewhere in the monocotyledons. The most common of them are those related to dioscin (Fig. 7). Other forms may be elaborated, namely *cardenolides* and *bufadienolides,* which occur in Convallariaceae and Hyacinthaceae (Aspar-

camphor          dioscin          taraxerol    α-pinene    citral

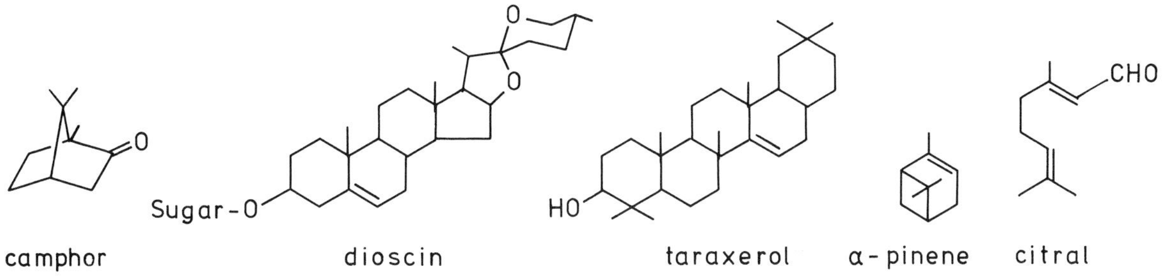

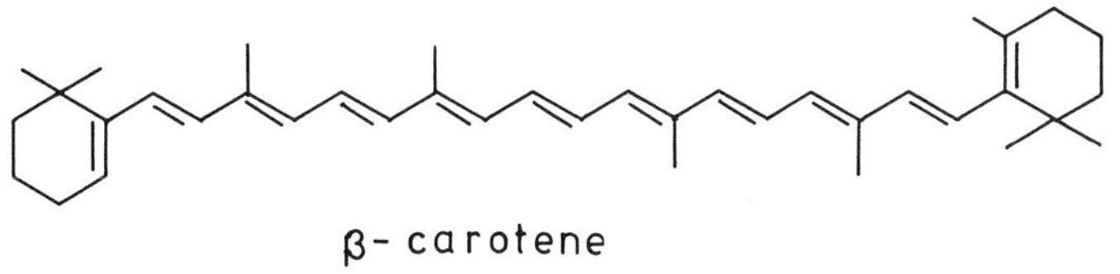

β-carotene

**Fig. 7.** Examples of terpenoids in monocotyledons. See further in the text.

agales). They occur at least in *Veratrum* of Melanthiaceae (Melanthiales). (Systematically, steroidal alkaloids also belong here). Diosgenin, the aglycone of dioscin, has for some time been used as raw material for the industrial production of sexual hormones. The compound is extracted from tubers of *Dioscorea* species; about 100,000 tons of tubers are used each year.

*Higher Terpenes.* Red and yellow flower pigments can either be anthocyanins (see below) or *carotenoids* (β-carotene, Fig. 7), which are $C_{40}$-terpenes, or mixtures of these. Milky sap or *latex* may consist of high polymeric terpenes emulgated in the sap; when dry it provides raw rubber which by treatment with sulphur (vulcanization) gives rubber.

*Flavonoids.* This is a large group of compounds biosynthesized partly through the shikimic acid pathway and partly through the acetate pathway.

They are thus of mixed origin. Flavonoids are ubiquitous in land plants and it appears that they play a role in protecting the cells against the UV fraction of sunlight. This is presumably their original importance, but they obviously serve various other purposes in plants being, for example, pigments in flowers, accounting for white, yellow, red and blue colours. Other, less conspicuous, flavonoids have been shown to be antibiotic, to serve as growth regulators, etc., and they may be found in all parts of the plants (HARBORNE et al. 1975).

To mediate so many different functions, the flavonoids must obviously show a large range of chemical diversity. This is mostly expressed by differences in the oxidation pattern on the same basic skeleton, but rearranged skeletons can also be found. Most of the flavonoids in plants are found as glycosides, often with glucose, but other sugars, such as galactose, rhamnose or arabinose, are also

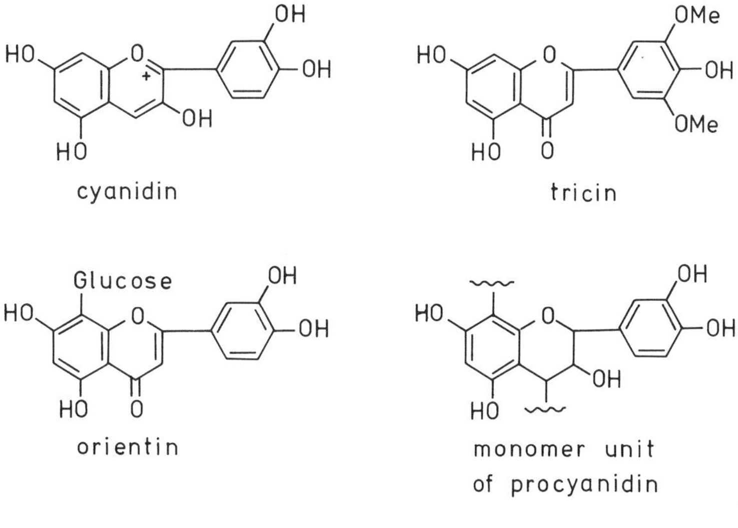

cyanidin          tricin

orientin          monomer unit of procyanidin

**Fig. 8.** Examples of flavonoids in the monocotyledons. See further in the text.

common. As the sugars apparently have limited systematic significance, only the parent flavonoids will be considered in the following.

Due to their ubiquity, structural diversity, stability and ease of detection the flavonoids show great promise for taxonomic use (GORNALL et al. 1979). A very large number of plants have been examined for their flavonoids by paper chromatography.

According to the oxidation pattern, the flavonoids are divided into a number of subgroups, and some of these will be commented here.

*Anthocyanins* are glycosides with an ionic structure (e.g. cyanidin, Fig. 8), making them red or blue according to the structure and allowing them to absorb light in the visible part of the spectrum. As their colours are of great ecological importance in pollination and subject to selective pressures they have become too variable to be of great taxonomic value.

*Flavones and Flavonols* are the commonest types of flavonoids, constituting white and certain yellow colours in flowers. Only a few individual compounds seem to be of taxonomic importance. In the monocotyledons the compound *tricin* (Fig. 8) has been found, through extensive surveys, to be frequent only in Poaceae, Cyperaceae and Arecaceae, and rare or absent in other monocotyledons.

*C-Glycoflavones* are compounds in which the sugar is bound to a carbon atom (e.g. orientin Fig. 8) instead of, as usual, to an oxygen substituent. In the monocotyledons glycoflavones are widespread and common, with the greatest concentration in Poaceae, Arecaceae, Araceae and Commelinaceae, a pattern reminiscent of that found for tricin.

*Flavonoid Sulphates,* in which the compounds are bound to sulphuric acid, are not very common in the angiosperms, except in certain families. In the monocotyledons, they are found in numerous taxa in Arecaceae, Juncaceae and Poaceae, and thus coincide in occurrence with the C-glycoflavones and tricin.

*Proanthocyanidins,* or condensed *tannins,* are polymeric compounds composed of flavonols which are linked by acid-labile bonds. They are not related to the anthocyanins in spite of their ability to produce red anthocyanidins by treatment with acid. The proanthocyanidins have astringent properties (and are common in unripe fruits) because they bind irreversibly with proteins, rendering these less palatable to many animals. This is also the property that is exploited in tanning hides. The proanthocyanidins are very widespread in the monocotyledons. A detailed chemical investigation (ELLIS et al. 1983) has shown that the composition of the polymers found in many monocotyledons is different from that found in the dicotyledons.

ELLIS et al. (1983) report that certain monocotyledons often contain proanthocyanidin polymers with 2, 3-*cis*-procyanidin units, which have a partly racemic mixture of 2 *R* (the normal configuration) and 2 *S* units. This enantiomerism occurs in Areciflorae (*Phoenix, Rhopalostylis, Cocos*), in Bromeliiflorae (Typhales: *Typha;* Bromeliales: *Ananas;* Haemodorales: *Anigozanthos*) and some Zingiberiflorae (Zingiberales: *Musa* and *Strelitzia;* but is lacking in taxa of Zingiberalean families with one functional anther), and one case of Liliiflorae, viz. in Dioscoreales (Smilacaceae: *Ripogonum*). The phylogenetic significance of this character is still uncertain but may prove to be great.

*Isoflavonoids* have a rearranged skeleton in relation to flavonoids and are rare in plants. In the monocotyledons, they have so far been found only in Iridaceae.

# Evolutionary Concepts, Classification

In taxonomy in general, and not least in monocotyledon taxonomy, there is often great difficulty in deciding whether morphological similarities depend on a shared ancestry or are the result of an evolutionary process by which ancestrally unrelated forms have acquired similar attributes.

This latter source of resemblance is not uncommon and is probably caused by strong selection acting in a constant direction, induced by climatic or special edaphic factors, or the influence of other biota. Alternatively, it may be due to chance factors, especially where there has been reduction, but with chance factors similarity is not likely to be present in more than a few, though sometimes conspicuous, features.

## General Concepts

Certain terms are in general use to describe phenomena in regard to evolutionary courses.

*Divergent evolution* (*divergence*) denotes a differentiation of structures, or taxa, that have originated from one common basic structure, or one single ancestral taxon, and have attained radically different appearances in the course of evolution along different lines, sometimes as a consequence of having been affected by different selective forces.

*Convergent evolution* (*convergence*) denotes the condition where more or less different ancestral structures (or different, generally distantly related taxa) have attained a similar appearance in the course of their evolution. Sometimes convergence occurs in non-homologous (homology, see below) structures so that superficially similar organs in different taxa, for example spines or tendrils, may prove to represent entirely or partially different parts of the plant. Convergence in homologous structures may result in a superficial similarity of plants, such as succulents, which are distantly related as judged on the total evidence available.

*Parallel evolution* (*parallelism*) implies that basically corresponding structures in two or more plant groups have evolved independently in a similar way so that the derived states show more resemblance to one another than to their ancestral states, e.g. loss of exine in the pollen grains of two different groups of plants, as in Arales and Alismatales.

Some additional concepts (Fig. 9), derived from the school of cladistics founded by HENNIG (1950),

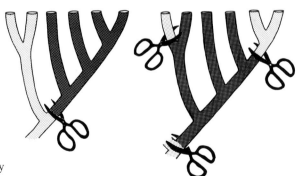

**Fig. 9.** The Cutting Rules. Concepts used in evolutionary biology and cladistics. Here are presented the terms monophyletic, paraphyletic and polyphyletic (*dark-shaded branchlets*). (Illustr. by H. RASMUSSEN; DAHLGREN and RASMUSSEN 1983)

Monophyletic

one and only one cut

One branch

Paraphyletic

one cut below the group and one or more cuts higher up

A piece of a branch

Polyphyletic

more than one cut below the group

More than one piece of a branch

are also necessary for understanding the chapter on Evolution Within the Monocotyledons.

A *monophyletic group* or *clade* is one that comprises all the derivatives of a single ancestral species, i.e. a group that comprises all the branchlets that are separated off with one cut in a cladogram (which is a sort of phylogenetic tree – see below under *character*). If the group requires more than one cut for its delimitation then it is *non-monophyletic*.

A *paraphyletic group* is one that can be removed from the cladogram with one cut, and is trimmed by one or more cuts further along the detached branch, i.e. it does not include all members of a monophyletic group. An alternative name for a paraphyletic group is *complementary group* because it will form a monophyletic group, i.e. a clade, if we add the pruned-off branchlet(s).

Groups that are compiled from different parts of the phylogenetic tree, i.e. those which cannot be obtained without making more than one cut, are called *polyphyletic*.

The clades that share an immediate ancestor are called *sister groups*.

A *character* is the sum of expressions (states) of a particular kind in a particular respect. Thus "red perianth" is a *character state* of the character "perianth colour". We shall, in this book, use the concept *character state* where the attribute is shared and we have reason to believe that it arose in a common ancestor. When it is believed to have arisen in independent lines "character state" will be replaced by the expression *character condition*. *Plesiomorphy* is a term proposed by Hennig (1950) for a state that is *primitive* for the clade under consideration, *apomorphy* being the *derived* state. When a derived character state is shared by two taxa with the same ancestor it is called a *synapomorphy*; a primitive character state shared by two taxa is called a *symplesiomorphy*. A derived character state that distinguishes a particular clade is said to be an *autapomorphy* – it is synapomorphic for the members of the clade. The term *cladogram,* used above, means a tree-like structure, the branching of which portrays hypothetically the evolutionary courses that have given rise to the extant taxa at its extremities. The construction of such a phylogenetic tree is based on the changes in character states, with the branching (bifurcations) representing the successive formation of new taxa. The preferred one of a number of possible cladograms is that which is the most parsimonious, i.e. which involves the fewest changes of character states (including convergences and reversals).

However, due consideration is given to the functional compatibility of states at each stage in the evolution. Cladograms with more information (e.g. fossil ancestors, chronology etc.) are usually referred to as phylogenetic trees.

The decision as to which character states are primitive (ancestral) and which derived in each clade, i.e. portion of a cladogram, usually rests mainly on a comparison with the groups most closely related to the group concerned (the study group), a procedure called *out-group comparison*. Outgroups may be sister-groups or complementary groups of the study group.

*Homology,* in the strict sense, is the term used for structures which have developed from corresponding primordia of a plant and which, in addition, arose in a common ancestor. However, the latter requirement is often not taken into consideration, and we may not always have done so in the present book either. Homologous structures are often not at all similar in appearance or function.

*Analogy,* on the other hand, is the term used in connection with structures that are similar in appearance and function, irrespective of whether they are homologous or not – in fact it is generally used when structures are *not* homologous. For example, in Orchidaceae the *labellum* (lip) is the medial inner tepal, whereas in Zingiberaceae the *labellum* represents two fused petaloid staminodes of the inner staminal whorl. The labellum is non-homologous in the two groups but as it may look alike and serve the same function in pollination it is indeed analogous. We may also observe that whereas *caruncles* or *strophioles* on seeds may look alike and have the same function, promoting seed dispersal, and are therefore *analogous,* they are often quite different in origin, for they can arise from the funicle, the raphe, or even the outer integument, thus being *non-homologous*.

The existence of similar non-homologous structures that have arisen independently, and of homologous structures that have become dissimilar by diversification, can make it extremely difficult to reconstruct sequences of evolutionary events and to determine phylogenetic relationships.

When classifying *taxa* (species, genera, tribes, subfamilies, families, orders, superorders) we must always keep in mind the purpose of the classification. One that serves *practical purposes* in grouping together taxa which share gross-morphological attributes may entirely fail to reflect the ancestry inferred from the whole body of evidence (including microstructural, micromorphological, cytological, physiological and phytochemical data). On the other hand a classification that does make use of these data may be highly confusing in a Flora. Classifications that are intentionally based on few practical characters are termed *artificial;* those where similarity in characters is used, without evolutionary interpretation, are termed *phenetic* or "natural" as used by 19th Century taxonomists. Those classifications that aim at taking both similarity and the inferred relationships into consideration are called *synthetic* (from synthesis, i.e. from putting together information from many sources) or *eclectic* (i.e. drawing freely on diverse sources). As the concept of evolution refers to both phenetic development (*anagenesis*) and phylogenetic branching (*cladogenesis*), this kind of classification is also called *evolutionary classification.*

*Cladistic classifications* are based solely on phylogenetic branching, and are therefore often referred to as *phylogenetic classifications.* The latter designation, however, has been used for all kinds of classifications that include phylogenetic aspects. *Cladistic classifications* are based on the Hennigian method of constructing an evolutionary tree (*cladogram*) from the distribution of character states and their classification into ancestral and advanced. All taxa in a cladistic classification must be clades.

Both the phylogenetic and cladistic methods of classification use extensive material from all possible sources but they differ in their operational procedure.

Because the available information is frequently inadequate or controversial, systematists often disagree as to which conditions are ancestral and which derived, and as to which characters are of most significance. They may also differ on the extent to which a classification should meet phylogenetic or phenetic requirements. These are the main reasons for the great differences existing between current classifications.

For a fuller treatment of concepts in monocotyledon evolution see DAHLGREN and RASMUSSEN 1983.

# Divergent Evolution

*Divergent* or *radiating evolution* is prevalent in all groups of living organisms. The evolution of a plant group is bound up with environmental constraints and opportunities affecting all stages of the life history.. The environment is constantly changing, and it does so partly because of evolution in the living world itself. By playing on innovations in the genetic constitution and the process of genetic recombination *natural selection* promotes adaptive change, diversification and specialization. However, the diversification of taxa is highly variable, apparently reflecting variation in the rate of evolutionary change, the length of time evolution has been going on in the group, and the success of its derivatives. Examples of taxonomically isolated, monotypic families are Butomaceae, Geosiridaceae and Petermanniaceae. Examples of isolated monotypic genera in larger families are *Pistia* in Araceae, *Nipa* in Arecaceae, *Cyclanthus* in Cyclanthaceae and *Anomochloa* in Poaceae. A number of families are unigeneric: Aponogetonaceae, Blandfordiaceae, Calochortaceae, Cannaceae, Cyanastraceae, Doryanthaceae, Eriospermaceae, Flagellariaceae, Hanguanaceae, Heliconiaceae, Herreriaceae, Ixioliriaceae, Joinvilleaceae, Mayacaceae, Najadaceae, Posidoniaceae, Scheuchzeriaceae, Sparganiaceae, Taccaceae, Thurniaceae and Typhaceae. There will always be different opinions as to whether some of these families should remain separate or be included in others, but the morphological distinctness and isolation of the included genus are indisputable.

By contrast some families are large and highly variable, showing an "explosive" speciation and seeming to have had their main diversification relatively late in the history of the angiosperms. Extreme examples are Orchidaceae and Poaceae. Given the narrow circumscription of families adopted in our classification, other particularly large families are Zingiberaceae, Commelinaceae, Araceae and Eriocaulaceae (tropics and subtropics), Iridaceae and Amaryllidaceae (largely South Africa), Bromeliaceae (largely South America), Cyperaceae and Juncaceae (worldwide), and Hyacinthaceae and Alliaceae (largely temperate regions). Such families are sometimes called *climax* families. The palms, which have radiated in tropical regions, were, however, already well differentiated at the end of the Cretaceous.

As regards the orders, the Asparagales have differentiated more than any other, and there are hardly any features in which the order is constant. The life-forms range from tiny herbaceous plants to huge branched trees; in addition, wood characters, stomatal types, inflorescences, floral size and embryological characters (microsporogenesis, endosperm formation, etc.) vary considerably. All this suggests that a pronounced differentiation took place among the ancestral forms of the order, perhaps at the beginning of the Tertiary.

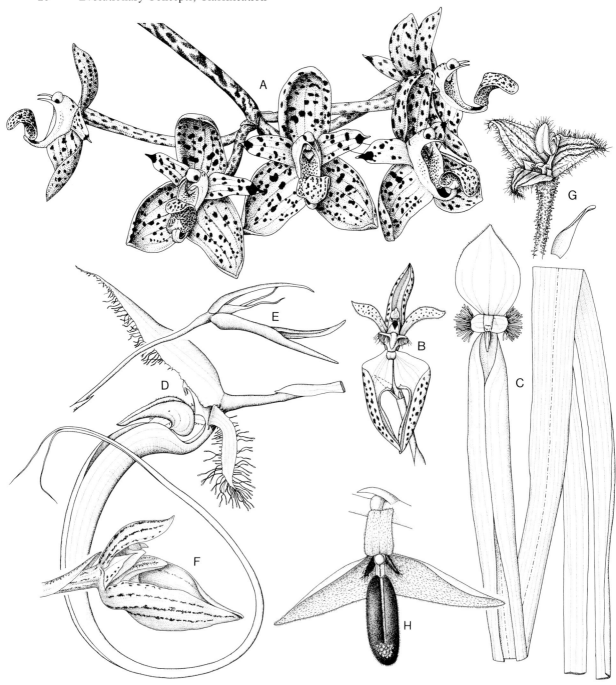

Marked differentiation in vegetative growth and size is encountered in genera such as *Potamogeton, Asparagus, Dracaena, Aloë, Tillandsia, Paepalanthus, Juncus, Carex* and *Pandanus*.

An example of complicated biological interaction is provided by *Ophrys* (Orchidaceae). The flowers of the different species resemble the females of their pollinating insects, which are non-social bees or wasps. The floral odours of the *Ophrys* species are very specific, having characteristic compounds in particular proportions. These have the same at-

**Fig. 10.** Diversification in a large genus of tropical orchids, *Bulbophyllum* in Southeastern Asia, selected examples from Thailand. **A** *B. violaceolabellum;* **B** *B. lasiochilum;* **C** *B. plumatum;* **D** *B. fascinator;* **E** *B. laxiflorum;* **F** *B. liliaceum;* **G** *B. lindleyanum;* **H** *B. angusteellipticum.* (**C** and **D** SEIDENFADEN 1973; **B, E, F** and **G** SEIDENFADEN 1979; **A** and **H** SEIDENFADEN 1981)

tractive effect as, and partly coincide with, certain chemical substances (alchohols and terpenes) emitted by the female insects and they act as long-

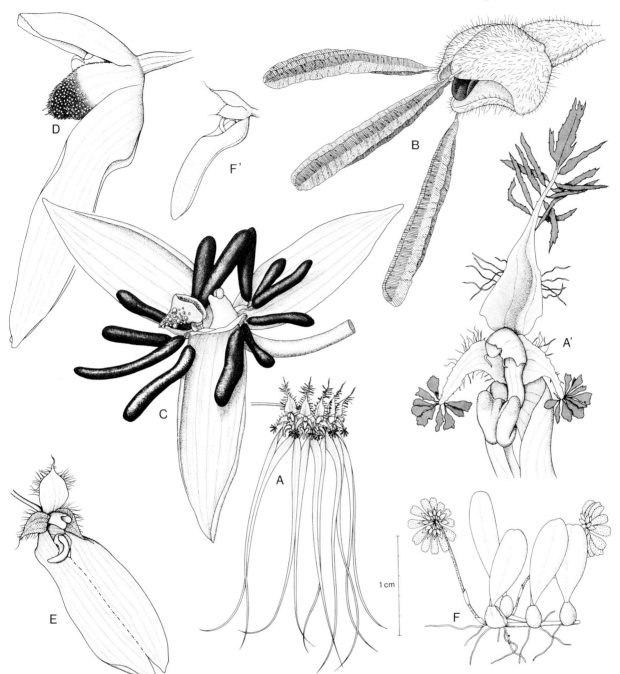

**Fig. 11.** Continuation of diversification in *Bulbophyllum*. **A** and **A'** *B. wendlandianum;* **B** *B. tripaleum;* **C** *B. adagense;* **D** *B. longidens;* **E** *B. lepidum;* **F** and **F'** *B. sikkimense.* (A–C Seidenfaden 1979; D–F Seidenfaden 1973)

distance attractants of the males. When closer, the males are attracted by the shape of the flower and its colour patterns and glossiness. The detailed construction of the flower with its distribution of pubescence etc., in combination with the chemical stimulus, evokes copulation behaviour in the male insect on the labellum of the flower. This behaviour makes it highly probable that the pollinia of the flower will adhere to some particular part, usually the head or back, of the insect, which then carries the pollinia to another flower and effects pollination. The pollination mechanism and pollinators are more or less specific for each species of *Ophrys*. The sterility barrier between species is weak, and hybrids are formed by the occasional occurrence of interspecific pollination. The hybrids, if fertile, may suffer strong negative selection, because they are unlikely to provide the right stimulus to attract potential pollinators.

Although *Ophrys* provides a particularly good illustration of differentiation related to pollination, much more pronounced differentiation of floral morphology, accompanied by prolific speciation, is found in many tropical orchid genera. For example, *Dendrobium* and the mainly fly-pollinated genera *Bulbophyllum* (Figs. 10, 11), *Masdevallia* and *Pleurothallis,* consist of over 1,000 species each.

*Bulbophyllum* Thouars is adopted in its broad sense by current specialists (e.g. Seidenfaden 1979) as a distinct taxonomic unit. It is characterized by the uninodal "pseudobulbs" that carry one or (more rarely) two leaves. The inflorescence rises from the base of the pseudobulb of the previous branch generation. The flowers conform to the basic orchidaceous type. The lateral tepals of the outer whorl are joined with the column foot into a distinct "mentum". The pollinia are invariably four in number, only very rarely with pollinium stalks (see the family treatment, p. 257). The conspicuous variation seen in Figs. 10 and 11 is mainly in the arrangement of the flowers on the inflorescences and in the size, shapes and proportions of the floral parts. This has no doubt profound influence on the pollination biology of the plants. Several attempts have been made to subdivide *Bulbophyllum* into smaller natural genera. Intermediate species, however, connect all conceivable subgroups. Long-established custom requires that genera are clearly separable, while subdivisions of genera (subgenera, sections) can be permitted to intergrade.

Other families besides the Orchidaceae show considerable differentiation in the floral region. Surprisingly, in Iridaceae, for example, there are septal nectaries in the subfamily Ixioideae, whereas other Iridaceae have perigonal nectaries. Within South African Ixioideae (Figs. 117–118) there is a continuous series between almost actinomorphic flowers and strongly zygomorphic ones, terminating in the large, bright red flowers of *Antholyza ringens,* a bird-pollinated species growing on the sandy flats of the western Cape. In Amaryllidaceae, a similar adaptive extreme is represented by the Mexican *Sprekelia formosissima,* which has a fleshy bright red, two-lipped flower.

Other kinds of differentiation connected with pollination can be observed in the superorder Alismatiflorae, in which the flowers are inconspicuous and pollination is by insects, wind or water. *Pota-*

*mogeton* (Fig. 146A) and *Triglochin* (Fig. 144A and E), for example, have inflorescences above water level and the pollen grains, globose and with a very thin exine, are dispersed by the wind. In some Hydrocharitaceae (Fig. 140A) pollination is by insects. A highly specialized member of this family is *Vallisneria spiralis* (Fig. 141G), which is a submerged aquatic with linear leaves and unisexual flowers, male and female on different plants. The male flowers are minute, numerous and aggregated together in an inflorescence enclosed by spathal bracts. Just before flowering, the spathes open, the male flowers become detached at the pedicel and float up to the surface of the water where the three tepals curve backwards. The small male flowers, with their two stamens exposed, drift about on the surface. The much larger female flowers are solitary on a spirally twisted pedicel which at anthesis untwists so that the epigynous flower just reaches the surface of the water with the base of the tepals, between which the recurved stylar branches extend. The male flowers, when sufficiently close to the female flower, slide down the sides of a depression in the surface film around the female flower and their anthers make contact with the stylar branches. After fertilization the pedicel of the female flower coils spirally again and the flower is drawn under the water where the winged capsule ripens. A similar, perhaps even more complex type of pollination occurs in *Enhalus* (Fig. 141A–F).

This kind of water-surface pollination contrasts with the true hydrogamy found in *Thalassia* (Fig. 142) and *Cymodocea* (Fig. 150), in which the pollen is dispersed as moniliform threads of spherical pollen grains or as already developed filiform pollen tubes; the female flowers are sessile and provided with long filiform styles.

Differentiation in floral biology in the family Araceae is based on the organization of the inflorescence (spathe and spadix) into a single functional unit. In many genera the female and male flowers are in zones at different levels on the spadix, sometimes with neuter flowers between them, and the spadix may end in a sterile appendix with special functions. The base of the spathe may form a tube around the spadix (Fig. 129), which acts as a trap to which insects are attracted by means of odours produced by the spadix. Various specializations on this theme are found in this family. The most conspicuous example of this type of inflorescence is the enormous inflorescence of *Amorphophallus titanum* from Sumatra, which smells like carrion and attracts beetles of the genus *Diamesus.*

# Reduction

Reduction is a process frequently postulated in botanical morphology and taxonomy. The notion covers reduced size, simplification, and/or loss of certain parts of the plant and its flowers. Reduction occurs particularly in connection with adaptation to certain modes of pollination (by wind or water), to particular habitats (for example the aquatic), and to saprophytism and parasitism, though the latter is extremely rare in monocotyledons. Reduction may affect only certain structures or complexes of structures of the plant, leaving the remainder well differentiated.

Reductions often occur in parts used as diagnostic characters for a group, with consequent difficulties in referring the plants to their nearest relatives.

It is obvious today that certain families and orders with reductions in floral structure have been incorrectly associated through artificial classification on the basis of a single striking peculiarity (see Typhales, below). Such difficulties have now begun to decrease as other less accessible characters (anatomy, pollen morphology, embryology, cytology, chemistry of secondary metabolites as well as proteins) are increasingly employed in phylogenetic evaluation. Nevertheless, there are still many unsolved problems which are the consequence of evolutionary reduction. For example, the relationships of the small family Hydatellaceae are not yet settled (see below).

Reductions that have resulted in the evolution of wind-pollinated plants have often been in the same direction leading to superficial similarity (see Convergence, below). For wind pollination neither a well-developed and petaloid perianth nor the secretion of nectar are necessary. The pollen grains are better dispersed when globular and smooth, and the type of aperture most compatible with this condition is circular; the pollen grains therefore tend to be ulcerate, as in most Poales, Cyperales and Pandanales. Long, pendulous stamen filaments and long stigma lobes facilitate, respectively, the dispersal and trapping of pollen grains. The chances that more than occasional pollen grains will become attached to the stigmas of each flower are still small; hence an ovary with several ovules is not economic, and decrease in ovary size and number of ovules are of selective advantage. Instead the flowers become numerous and more scattered. As a consequence of reduced ovule number there is less reason for the fruit to be dehiscent; usually, therefore, each fruit is separately dispersed and corresponds in function to a seed.

As we can thus see, several reductional changes may be of selective advantage for a wind-pollinated plant and will probably occur in the course of a long period of evolution. This is a case where the interdependence of character states is easily understood. The resultant character combination is not random but governed by selective advantage under particular conditions; such a character combination may be called a *syndrome* (STEBBINS 1974).

In the following pages we present some examples of monocotyledon groups with inflorescences and flowers that have become respectively contracted and reduced, thereby making difficult the recognition of relationships between taxa and the interpretation of the morphological structures concerned. Brief summaries are given first.

1. In Lemnaceae, a family that includes the smallest of all angiosperms, reduction affects the whole plant. The shoot is a flat plate, "frond", not differentiated into stem and leaf. The inflorescence consists of a monocarpellary pistil and one or two stamens, and there is nothing to show whether the number of flowers is one, two, or three.

2. In the Zosteraceae the flowers are distributed on a flattened axillary spike and have reached a state of extreme reduction, being naked and unisexual. The pollen is adapted to true (submerged) hydrogamy (unknown in dicotyledons) and lacks an exine.

3. Centrolepidaceae have extremely reduced flowers in small compact inflorescences. The reduction is so strong that there have been doubts as to whether the clusters of carpels and stamens are flowers in which the parts have been translocated to somewhat different levels, or are inflorescences of female flowers with one carpel each and male flowers with one stamen each, the former sometimes being fused pairwise to form a bilocular "capsule". In this family were formerly placed the two genera *Hydatella* and *Trithuria* which, in spite of their superficial similarity to the genera of Centrolepidaceae, show many differences from them and are now placed in a new family, Hydatellaceae. The similarities are perhaps adaptations to life in seasonally inundated habitats.

4. Reductions in the flowers comparable to those in ancestral Centrolepidaceae have also occurred in the Cyperaceae. In the tribe Mapanieae the flowers are regarded as naked and reduced either to single pistils or single stamens which are aggregated into spike-like pseudanthia. Still further reduction may have occurred in the subfamily Sclerioideae with their terminal pseudanthia reduced

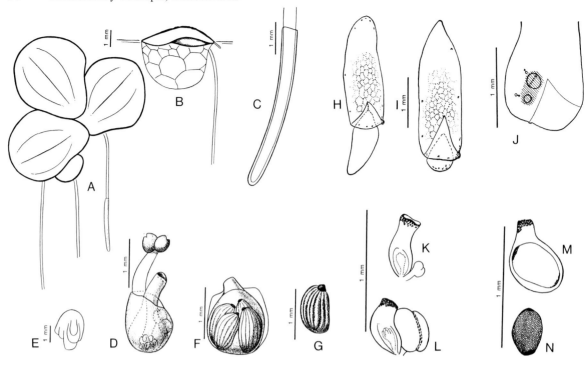

**Fig. 12.** Reduction. Among the smallest plants are the members of the genera *Lemna, Wolffiella* and *Wolffia.* Here are shown *Lemna gibba* (**A–G**) and *Wolffiella oblonga* (**H–N**). *Lemna* has one root per frond with a distinct root cap (**C**); the flowers, one female and two male, are enclosed in a membranous spathe (**D**). The anatropous ovules (**E**) are several per carpel and mature to longitudinally striate seeds (**F–G**). In *Wolfiella,* of which two sterile specimens are shown in the process of budding (**H–I**), the flowers are concealed in a pocket (**J**) and the spathe is lacking (**K**). There is only one (orthotropous) ovule per carpel (**K–L**), and the seed (**N**) has a minutely reticulate pattern. (All from CORREA 1969)

to single pistils. Such extreme reduction in some parts of the family has led to the ordinary flowers being regarded by some as pseudanthia.

5. In *Carex* (also Cyperaceae) the ovary is surrounded by a utricle in which is also enclosed a sterile axis, the rachilla. This suggests that this unit is a reduced female spikelet with a single naked flower subtended (largely enveloped, in this case) by a prophyll (the utricle).

6. In Typhaceae and Sparganiaceae, which are wind-pollinated, the superficial similarity to other wind-pollinated plants, Cyperales or Pandanales, may conceal the actual relationships. A list of mainly anatomical and cytological characters is given, the evidence of which points to a relationship with insect-pollinated groups.

7. A final example is the strong floral reduction accompanied by fusion of flowers that has resulted in the peculiar androecial and gynoecial rings in the inflorescence of the genus *Cyclanthus* of Cyclanthales.

## The Lemnaceae (Figs. 12, 132 and 133)

Lemnaceae is a small subcosmopolitan family consisting of six genera. All Lemnaceae are small free-floating aquatics (with the exception of one submerged species). They are not differentiated into stem and leaf but have a thallus or frond which can bud off new daughter fronds. Roots occur only in *Spirodela* (several roots) and in *Lemna* (one). The frond varies from elliptic or orbicular to elongate and may be thick, sometimes nearly globular. It lacks vessels although sometimes one or more veins are visible. Budding occurs in one or two pouches on the margins or in one pouch on the margin and one on the upper surface.

The inflorescence develops in one of the pouches. It consists of 1-2(-3) male flowers, each with one stamen, and one female flower: a unilocular pistil, perhaps consisting of a single carpel, with one or several ovules. The flowers are naked, borne on a short axis and sometimes surrounded by a rudimentary membranous spathe. The pollen grains are ulcerate. The fruit is a small utricle-like nutlet with one to four seeds (Fig. 12F and M), which are generally filled with a starchy endosperm (Fig. 133).

There is a clear trend within the family in that *Spirodela* is the most complete, and at least in some features most ancestral, genus, having (1) a larger vegetative body (2) raphides in the cells (3) several roots (4) a spathe (5) two to three male flowers per inflorescence (6) the Monocotyledonous Type of anther-wall formation (others have the Reduced Type) and (7) the *Polygonum* Type of embryo-sac formation (others have the *Allium* Type). *Lemna* agrees with *Spirodela* in having a spathe and two male flowers. The minute rootless, nerveless, spatheless fronds of the genera of the *Wolffia* group represent the climax of reduction, *Wolffia arrhiza* being the smallest of all angiosperms.

The affinities of the Lemnaceae have perhaps not been ultimately determined, but the general view is that they have their ancestors among extinct Araceae, the genus *Pistia* (Fig. 131) being the extant genus most closely resembling it. This genus consists of free-floating plants, with sessile obovate leaves and inflorescences reduced to one naked female flower and one male flower consisting of two fused stamens. The flowering shoot in *Pistia* is more complete than in the Lemnaceae, consisting of one basal cataphyll, one foliage leaf and the small inflorescence with its spathe. The fruit in *Pistia* is drupaceous.

The Lemnaceae have alternatively been associated with the Alismatiflorae (LAWALRÉE 1945), but their similarities with them are also shared by Araceae. The occurrence of raphides in *Spirodela*, the cellular endosperm formation, the ulcerate pollen grains, and the frequently endospermous seeds with starch indicate a decidedly araceous affinity.

In this case reduction has gone so far that very few character states are unaffected and micro-morphological and chemical characters such as serological profiles become crucial for indicating relationships.

## The Zosteraceae (Fig. 148)

The Zosteraceae are marine water-pollinated representatives of the Alismatiflorae. The family consists of three genera, *Phyllospadix*, *Heterozostera* and *Zostera*. The genera occur in temperate regions in both hemispheres. The plants are submerged and rhizomatous, growing rooted in sand, and are without vessels and stomata. The leaves are linear and distichous and are of two types: scale leaves and long, linear, foliage leaves with a rounded apex and an open or closed sheath. The inflorescence is a flattened spike or spadix situated in the axil of, and enclosed in the sheath of, a foliage leaf. One of the flat surfaces of the inflorescence axis bears flowers. These are unisexual, and the plants are either dioecious (*Phyllospadix*) or monoecious (*Heterozostera, Zostera*); in the latter case the flowers are in two rows and the male and female flowers alternate both longitudinally and transversely. The male flowers may be interpreted as consisting either of two bisporangiate anthers connected basally by a narrow flap of tissue or of one tetrasporangiate anther with a broad connective. They usually also have a lateral, flat appendage ("retinaculum"), which is marginal on the spadix. The female flower consists of a single ellipsoidal or crescent-shaped pistil, with a style divided nearly to the base into two lobes. The ovary is unilocular, but the evidence of the style permits it to be interpreted as monocarpellary or bicarpellary. It contains one apical, pendulous, orthotropous ovule. No parietal cell is formed, but the nucellar epidermis divides to form a nucellar cap. The pollen grains (devoid of exine), are dispersed in the water. They are globose when formed but elongate rapidly when liberated to become filamentous, and are then more easily caught by the stylar branches. Endosperm formation is nuclear, and the resultant fruit is small, drupaceous or nut-like. The seed is exendospermous, filled by the macropodous embryo having a sheathing cotyledon in a groove of the thick hypocotyl; the radicle is probably completely lost.

The above description indicates that, in their adaptation to aquatic life and aquatic pollination, the Zosteraceae are much reduced in the vegetative as well as reproductive parts. The consequent divergent interpretations of them show the difficulties in determining the homologies of their parts. We consider it most probable that the male flowers consist of a tetrasporangiate anther and the female of a single carpel. The assumption that the two stylar lobes are branches of one style finds its support in the fact that in several other Alismatiflorae the stylodia are brachiate (see *Cymodocea*, Fig. 150D). The "retinacula" have also been variously interpreted as bracts, tepals or outgrowths from the connective, the last alternative being supported by comparative studies with *Potamogeton*, *Posidonia*, etc.

In judging the relationships of a group as specialized as Zosteraceae, a number of intrinsic features not immediately connected with the habitat become increasingly important. These include the occurrence of intravaginal squamules, orthotropous ovules, the Caryophyllad Type of embryo formation and absence of endosperm in the seed, which are characteristic of the order or at least of that group of families associated with Zosteraceae. Even some of these characters, however, may represent adaptations to aquatic life.

On the basis of the collective evidence it seems that the family is most closely related to Posidoniaceae and Potamogetonaceae, being perhaps a sister-group of the former.

## The Centrolepidaceae and Hydatellaceae
(Figs. 215, 179)

The Centrolepidaceae (Fig. 215) comprise four genera of about 35 species centred in Australia, but also distributed over parts of South-East Asia, the Pacific Islands and the southern tip of South America. They chiefly grow in temporarily waterlogged sites.

They are small, tufted, mostly annual plants with a grass-like, or even moss-like, habit (Fig. 215J). They lack the silica bodies of the grasses but have paracytic stomata, and the stems contain vessels with scalariform or reticulate perforation plates. The stem, unlike that of many grasses, is solid. The leaves are tufted, with an open sheath and a lamina. The flowers are situated in a small spike or head and are subtended by larger bracts; further bracts may subtend the units of the inflorescence. Each flower is here considered to consist of a single stamen with a bisporangiate anther or a single carpel. Thus they are naked, unisexual and monomerous. They are clustered into bisexual pseudanthia, which could be mistaken for flowers. Sometimes two adjacent carpels are fused collaterally on the same level, and sometimes several carpels on different levels are fused together. The pollen grains are ulcerate and tricellular, as in Poaceae and Restionaceae. Each carpel contains an apical, pendulous, orthotropous ovule, which is tenuinucellate or nearly so, and in which no parietal cell is formed. Endosperm formation is nuclear. The carpels develop into follicles, which are fused into double or multiple fruits.

There is no doubt that these plants have undergone reduction from states present in ancestral Poales, and that monomery in androecium and gynoecium is secondary to trimery. The reduced size and subjection to temporary submersion are probable reasons for the loss of silica bodies and the failure of advanced types of vessels to be developed. Whereas the Centrolepidaceae are unique in the order in their monomery, they agree with most Restionaceae in their bisporangiate (monothecous) anthers and lack of silica. The unisexual flowers, the apical, orthotropous ovule in each locule and the ulcerate pollen grain with a somewhat irregular aperture are also attributes of Restionaceae.

We can conclude, then, that Centrolepidaceae are probably most closely allied to Restionaceae.

Until recently Centrolepidaceae was regarded as comprising two tribes, but as shown by HAMANN (1975, 1976) these are so different that they demand treatment as separate families. Accordingly, HAMANN raised the tribe Trithurieae to family rank with the name Hydatellaceae (Fig. 179), comprising the genera *Hydatella* (4) and *Trithuria* (3). These are restricted in their distribution to Australia, Tasmania and New Zealand, i.e. regions where Centrolepidaceae occur, and also grow in very much the same habit, namely temporarily inundated, silty soil.

HAMANN's investigations revealed that the two genera differed from other Centrolepidaceae in the following respects:

1. Vessels are confined to the roots (in Centrolepidaceae vessels occur in the stem).
2. The stomata are anomocytic (paracytic, of grass type, in Centrolepidaceae).
3. The bracts (hyaline) subtend only the collective inflorescence (in Centrolepidaceae there are generally also bracts within the inflorescence).
4. The anthers are tetrasporangiate (dithecous) and basifixed (in Centrolepidaceae bisporangiate and monothecous).
5. The pollen grains are (mono-)sulcate (ulcerate in Centrolepidaceae).
6. The pistil may be pseudo-monocarpellary, which is indicated by the three slits by which the "capsule" of *Trithuria* splits open and one or two rudimentary pistil bundles that may occur beside the main one (in Centrolepidaceae the pistil is monocarpellary).
7. The stigmatic papillae are in the form of a fascicle of two to ten uniseriate stigmatic hairs resting directly on top of the ovary (in Centrolepidaceae there is a long stylodium with a long, papilliferous stigmatic tip).
8. The ovules are pendulous-anatropous, with the micropyle directed upwards (in Centrolepidaceae it is pendulous-orthotropous with the micropyle directed downwards).
9. The embryo sac is proportionally much broader than in Centrolepidaceae.

10. The antipodals of the embryo sac are lacking or they degenerate early (in Centrolepidaceae they are distinct, and often have doubled nuclei).
11. Endosperm formation is cellular and the endosperm is few-celled (in Centrolepidaceae endosperm formation is ab initio nuclear and the endosperm becomes multicellular).
12. In the ripe seeds the endosperm is vestigial and does not function as a nutritive tissue (in Centrolepidaceae it fills up most of the seed and contains copious starch).
13. A perisperm is developed from the nucellar tissue, which makes up a nutritive tissue with copious starch (in Centrolepidaceae such a tissue is lacking).
14. The nucellar tip includes cells that have divided once or twice collaterally and which are not elongated (in Centrolepidaceae the nucellar cells do not divide transversely but becomes conspicuously elongated).
15. In the seed coat the inner part of the tegmen is retained and also, in particular, the outer part of the testa, in which the cells become large and thick-walled (in Centrolepidaceae the inner part of the tegmen makes up the main part of the seed coat).
16. There is an "operculum" or seed lid, which is formed by the inner part of the tegmen (in Centrolepidaceae there is no operculum).
17. The pericarp is indehiscent or dehiscent by three longitudinal slits (in Centrolepidaceae the monocarpellary fruits generally open by a dorsal longitudinal slit).

This information (HAMANN 1975) illustrates how misleading similarity in habit may be, and how plants which are superficially similar may be quite distinct in terms of their anatomy, embryology and other inconspicuous characters.

How should this evidence be interpreted? As we have concluded, the Centrolepidaceae are presumably related to Restionaceae and may even be derivatives of early Australian (East Gondwanaland) Restionaceae. But the Hydatellaceae have some attributes, notably the lack of vessels in the stem, the anomocytic stomata and the sulcate pollen grains, which make the family quite out of place in the Poales; anomocytic stomata are lacking throughout the entire superorder Commeliniflorae.

The characteristics of the Hydatellaceae may be partly explained as adaptations to submergence and thus largely ecologically conditioned. This could account for the lack of subsidiary cells and of vessels in the stem. The loss of antipodals may be consequence of a shortening of the developmental processes of the seed and, especially, the formation of a perisperm, which does not have to await fertilization as endosperm formation does.

However, this does not explain some features in which we consider Hydatellaceae to be more ancestral than the Centrolepidaceae (and, in the last two respects, than all the Poales), namely the te-

trasporangiate anthers, the sulcate pollen grains (now thought to be binucleate: HAMANN, personal communication), and the anatropous ovules. These all indicate a completely divergent ancestry, ruling out the possibility that the family is a sister-group of Centrolepidaceae.

Further alternatives are also possible. We may note that anomocytic stomata occur in a few Arales (e.g. *Arisaema*, Lemnaceae), where the stem lacks vessels, the inflorescence is subtended by a bract, the stamens are basifixed, the pollen grains may be sulcate, the pistils lack a style and are often monocarpellary, the ovules are often anatropous and the endosperm formation is cellular (the only other group of monocotyledons where this is so).

An affinity with the Arales, however, seems improbable because anomocytic stomata occur in so few Arales and because the Hydatellaceae have a totally different habit, a totally different (broad) type of embryo, non-Caryophyllad embryogeny and other discordant characters. It may therefore be relevant to seek the ancestors of Hydatellales among the Liliiflorae, Zingiberiflorae and early Commeliniflorae.

Against the Liliiflorae argue chiefly the broad embryo, the cellular endosperm formation, the occurrence of perisperm and possibly the fact that anemophilous plants are rare in the group.

The Zingiberiflorae are more likely, as they have perisperm and seed lid but if the whole syndrome of their characteristics is considered they show little sign of relationship.

If the lack of vessels and subsidiary cells, as well as the transfer of starch from the endosperm to a proliferous nucellus (perisperm), and the speeding up of wall formation in the early endosperm could all be explained as adaptations, then the Hydatellaceae might be considered to have evolved from early graminoids on the common Cyperalean-Poalean stem as in Fig. 40, thus remaining in the Commeliniflorae, but in a position remote from the Centrolepidaceae.

The Hydatellaceae, like the Cyclanthales and Pandanales, must still be considered unsettled as regards their closest relationships.

## The Reproductive Structures in Cyperaceae Tribus Mapanieae (Figs. 13, 185)

As noted in the family description of the Cyperaceae, the monotypic genus *Scirpodendron* (Fig. 13) in the tribe Mapanieae has a puzzling organi-

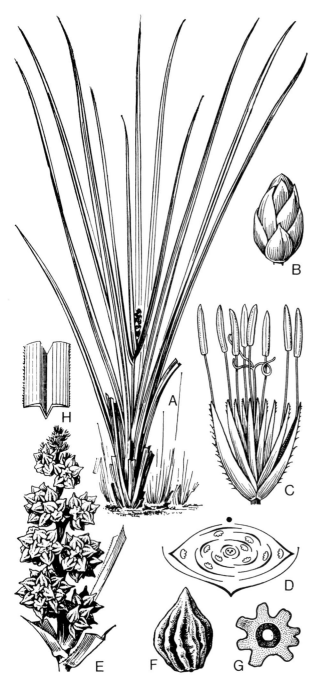

**Fig. 13.** *Scirpodendron ghaeri.* **A** plant, some leaves cut away. **B** Pseudanthium, bud stage. **C** Pseudanthium, floral stage. **D** Diagram of pseudanthium, interpreted as a terminal, naked, tricarpellary female flower and, below this, a number of spirally set bracts, each with a male flower consisting of one stamen. **E** Inflorescence, fruit stage. **F** Fruit. **G** Same, transverse section. **H** Leaf, transverse section. (TAKHTAJAN 1982)

*Euphorbia.* (The resemblance to *Euphorbia* even extends to the naked, terminal and tricarpellary female flower!) Current views favour the latter interpretation.

If the reproductive structures of *Scirpodendron* are pseudanthia rather than flowers, the same is likely to be true for related genera such as *Mapania, Lepironia, Hypolytrum* etc., where the pseudanthia are even more flower-like, with fewer bracts subtending the stamens (Fig. 185).

In the subfamily Sclerioideae the terminal pseudanthia are even further reduced than in the Mapanioideae and lack stamens in the axils of the scales.

The rather isolated position of the Mapanioideae and Sclerioideae within Cyperaceae is accentuated by their often non-conical silica bodies and the frequent occurrence of spherical, ulcerate, grass-like pollen grains in Mapanioideae.

Some morphologists have gone much further than this, considering virtually all so-called flowers in Cyperaceae to be pseudanthia. We do not find support for this view, but regard the Mapanioideae and Sclerioideae as isolated advanced groups within the family. However, other reductions certainly occur within the family, as shown below for the *Carex* utricle.

## The Inflorescence and Flower in *Carex*
(Figs. 184 and 189)

An interesting example of reductional processes has resulted in the conditions met with in *Carex*. In this genus the female spikelets of the ultimate order are greatly reduced: they are situated in the axils of "glumes" (spikelet bracts) that are often characteristic in size, colour and shape; their axis is very short and is generally completely concealed within the flask-like prophyll, the *utricle,* in the axil of which a single naked female flower is situated. This is an ovary consisting of three or two carpels, as indicated by the number of stylar branches (three or two) and the shape of the fruit (trigonous or biconvex). The fruit is a nutlet and

zation of the stamens and pistils. These are arranged in bisexual units consisting of a terminal tricarpellary pistil subtended by a great number of spirally inserted bracts, in the axil of each of which one stamen is inserted. It may be assumed either that the structure as a whole represents a flower in which the regularity of "tepals" and stamens has become severely disturbed or that it is a small inflorescence of reduced flowers reminiscent of the cyathium in the dicotyledonous genus

together with a sterile continuation of the axis (the *rachilla*) it is wholly enclosed in the utricle. The male flower is also naked and consists of three stamens in the axil of a subtending glume. The distribution of female spikelets (utricles) and male flowers is highly variable within *Carex* and in combination with other characters provides a basis for the subdivision of the genus.

It is naturally impossible to trace evolution backwards, but comparison with other genera makes possible a fairly good reconstruction. The rachilla of the utricle in *Carex* subgenus *Primocarex* may be more differentiated, suggesting that there has been a process of reduction from a better-developed ancestral rachilla carrying more flowers. In some species of the genus *Schoenoxiphium* (Africa, Asia) the rachilla extends far outside the utricle and bears a few scales with axillary male flowers, each consisting of three stamens. In other species of *Schoenoxiphium* the rachilla merely projects from the utricle (which it rarely does also in *Carex*, e.g. in *C. microglochin*, Fig. 184), but there is only a rudiment of the "male part" of the rachilla (HAINES and LYE, 1972). Similar conditions obtain in the genus *Kobresia*. A possibly more ancestral state is that in *Rhynchospora*, where the spikelets have bisexual flowers which are even provided with bristles homologous with the perianth.

Accordingly, the series *Rhynchospora – Kobresia/Schoenoxiphium – Carex*, although not forming a phylogenetic series, may reflect levels of reductional evolution in one part of the family Cyperaceae.

## The Typhaceae and Sparganiaceae
(Figs. 160, 161)

Typhaceae and Sparganiaceae are very closely related and could preferably be treated in one family, Typhaceae s.lat. (D. MÜLLER-DOBLIES 1970). They both consist of perennial, rhizomatous herbs with wind-pollinated, unisexual flowers in dense elongate spikes (*Typha*) or globose "heads" (*Sparganium*), thereby resembling the Pandanales, with which they have generally been, and are still occasionally, associated. Whereas the flowers have 3 + 3 tepals in *Sparganium*, the perianth of *Typha* is substituted by a variable number of fairly long trichomes which are generally broadest distally and not inserted at one level. The stamens are variable in number, 1–8 in *Sparganium*, (1–)3(–8) in *Typha*, in the latter genus being often fused basally. The carpels are generally reduced to one in

both genera. In *Sparganium* the carpel develops into a drupaceous fruit, in *Typha* into a longitudinally dehiscent fruit which may be best classified as a follicle.

On these gross morphological features alone the two families are extremely difficult to place. They have attained most features of the wind-pollination syndrome (see above, p. 29): small flowers, inconspicuous or reduced perianth, long filaments, ulcerate pollen grains, monocarpellary pistils with a long stigma, one ovule per carpel and, in *Sparganium*, an indehiscent fruit.

With this description we could, with about equal justification, place the families in Poales, Cyperales or Pandanales. Though few members of the Liliiflorae show this syndrome the tendencies even here are apparent in, for example, Dasypogonaceae in Asparagales.

In judging the relationships of Typhales, we shall here take into account such attributes as the occurrence of vessels with scalariform perforation plates in roots, rhizomes, aerial stem and leaves, and the occurrence of raphides, distichous phyllotaxy, paracytic stomata, amoeboid tapetum, successive microsporogenesis, bicellular pollen grains, anatropous and crassinucellate ovules, helobial endosperm formation (of a particular type with small chalazal chamber), starchy endosperm, *Strelitzia*-Type epicuticular wax, and organic acids in the cell walls giving ultraviolet fluorescence. These attributes, which are nearly all impossible to see unaided, together place Typhaceae and Sparganiaceae in the vicinity of Pontederiaceae, Haemodoraceae and Philydraceae in the superorder Bromeliiflorae. This position does not exclude the possibility that they also come near the ancestors of Cyperales or Pandanales, but it does suggest that they adopted wind pollination independently of the ancestors of Cyperales and Pandanales (assuming that wind pollination in these groups is derived).

## The Genus *Cyclanthus*, Cyclanthaceae
(Fig. 217)

A final example of thoroughgoing floral reduction is met with in *Cyclanthus*. This genus, with one species only, *C. bipartitus*, found in northern South America and Central America as far north as Guatemala, is rather aberrant in the family. It deviates from the other genera in the lack of tannin cells, the possession of true laticifers (WILDER and HARRIS 1982) and in details of root exodermis, cork, leaf epidermis, mesophyll (*Cy-*

*clanthus* lacks a palisade layer) and petiole (in *Cyclanthus* having air canals). The leaf in *Cyclanthus* appears to be bicostate (by almost total reduction of the central costa) but it is uni- or tricostate in the other genera. Thus there is ample evidence that *Cyclanthus* is a rather isolated genus, a fact which is also reflected in the embryology: for example, it deviates from all other Cyclanthaceae in lacking periclinal divisions in the nucellar epidermis.

What is interesting about *Cyclanthus* is the extreme degree of reduction and specialization of the flowers. These are generally situated in alternating rings (rarely in a spiral). The female flowers are fused laterally to form pistillate whorls, and the male flowers with their stamens are fused to form staminate whorls. The pistillate whorls are separated from the staminate ones by thin plates which may be homologous with a perianth. Inside these plates there are two larger, grooved staminal lamellae each bearing a row of more or less abortive anthers. The pistillate whorls have two alternating rows of carpels, which are so intimately fused that the ovaries form a coherent annular cavity. This ovarian cavity is almost completely lined with crowded placentae bearing numerous ovules (i.e.

it has laminar-dispersed placentation). When the inflorescence is in the fruiting stage, the fragmenting multiple fruits form dry, hollow rings (or spirals) filled with seeds which are probably dispersed by ants.

Obviously *Cyclanthus* forms the climax of an evolutionary branch lateral to that of the other Cyclanthales, and has undergone reduction and specialization independently. There are no reasons to doubt its relationships with other Cyclanthaceae, making up subfam. Carludovicoideae, but *Cyclanthus* is apparently much more isolated than is generally thought.

## Convergent Evolution

### General

Convergence is an evolutionary phenomenon of general occurrence. Any case where two taxa share a character condition as a result of evolution along separate pathways is classified as a convergence (Fig. 14). Certain conditions, such as epigyny, have arisen not twice but many times in the monocotyledons.

Convergence may be more or less obvious. It is more obvious where there are shared attributes which make plants or plant parts look very similar, while at the same time other attributes show fundamental differences. The best-known example is perhaps that of "cactoid" stem succulence in different groups of dicotyledons (Cactaceae, Euphorbiaceae and Asclepiadaceae tribus Stapelieae). A comparable but less well-known example from the

**Fig. 14.** Convergence in habit, exemplified by three different genera of Juncaceae and Cyperaceae from Patagonia, all presumably affected by selection under similar climatic and edaphic conditions. **A** *Patosia clandestina* (male plant) (Juncaceae). **B** *Carex sorianoi* (Cyperaceae: Caricoideae). **C** *Oreobolus obtusangulus* (Cyperaceae: Rhynchosporoideae). (All from CORREA 1969)

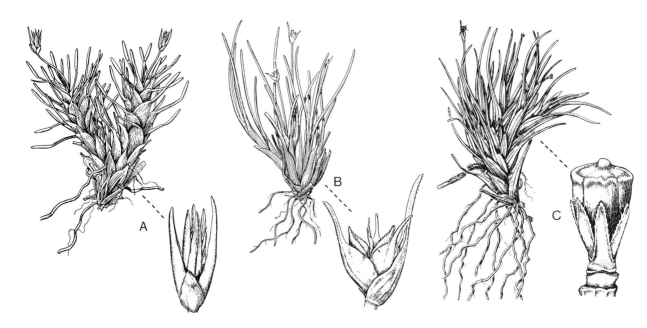

monocotyledons is that of the "rosette trees". This growth habit occurs in Cyperaceae (*Microdracoides* spp.), Bromeliaceae (*Puya* spp., Fig. 153I), Velloziaceae (*Vellozia* spp.), Asphodelaceae (*Aloë* spp.), Agavaceae (several genera), Nolinaceae (several genera), Xanthorrhoeaceae (*Xanthorrhoea* spp., Fig. 68A), Dracaenaceae (*Dracaena* spp., Fig. 60) and Dasypogonaceae (*Dasypogon* sp., *Kingia*, Fig. 68I). It is not certain that the ability to form a woody trunk was independently evolved in all of the last five families named but it probably did so in more than one, and it almost certainly did so in each of the first four, where in some cases it occurs only in one or a few species of a single genus.

Convergence is less obvious where several attributes are held in common, making an initial presumption of common ancestry stronger. The case of the Centrolepidaceae and Hydatellaceae described above is a good example. Another is the resemblance which is found between the elliptic floating leaves, long pedicels and the stipules of *Heteranthera* (of the Pontederiaceae in Bromeliiflorae) and the corresponding parts in *Hydrocharis, Aponogeton* and *Potamogeton* (see Figs. 159, 140, 136, and 146). The last three genera belong to three separate families of the Alismatiflorae, in each of which the resemblances in the leaves have probably evolved independently.

Similarly a "graminoid" (grass-like) appearance is met with in various groups of monocotyledons (e.g. *Alania,* Fig. 86), and the "shrublet" habit, with a branched woody base and very stiff, spinose leaves, is found in certain grasses (Poaceae), such as *Micrairia,* and in *Calectasia* (Calectasiaceae, Fig. 66), *Acanthocarpus* (Dasypogonaceae, Fig. 65F) and *Borya* (Anthericaceae: Johnsonieae, Fig. 85R). These shrublets are all inhabitants of Australia, suggesting that the vegetative similarity is the result of selective forces deriving from the climate, from herbivory or a combination of both.

Selection is also evidently responsible for convergence affecting the appearance of the flower. Thus lilac-coloured tepals in combination with bright yellow, poricidal anthers, which occur in the Australian genus *Calectasia* (Fig. 66), find their counterparts not only in the monocotyledons *Stypandra* and *Dianella* (Phormiaceae) but also in very distantly related plant groups among the dicotyledons, such as the insectivorous *Byblis* (Byblidaceae), and genera of Tremandraceae and Pittosporaceae, all probably pollinated by hoverflies (Syrphidae, see THORNE 1983).

Surprising cases of similarity have puzzled botanists from time to time. Thus one may be struck by the great similarity between the South American genus *Philesia* (Philesiaceae; see Fig. 53E) and members of Ericaceae, e.g. species of *Rhododendron*. These similarities cover the shrubby growth, the stiff, coriaceous, linear-elliptic leaves with recurved margins, the several bracteoles subtending the flower, the fact that the outer tepals are sepaloid in *Philesia* as in many dicotyledons, the funnelform-campanulate shape formed by the inner whorl of tepals (petals in Ericaceae), and their reddish, more or less variegated colour, etc. This is indeed an interesting case of convergence, and one where the combination of attributes has resulted from adaptive response to climate as well as from insect pollination.

Probable convergence in vegetative respects is met with in the achlorophyllous, generally very smallsized saprophytic plants of the Triuridaceae and Melanthiaceae-Petrosavieae (*Petrosavia* and *Protolirion,* Fig. 99J). In both groups, as in other achlorophyllous plants, the leaves are hyaline and bract-like, but the reduction of the vascular tissue and the anatomical construction (STANT, 1970; TOMLINSON, 1982) are also very similar in the two families. The fact that the carpels are nearly free in *Petrosavia* and *Protolirion* and wholly free in Triuridaceae (Fig. 134) is another circumstance that may have contributed to the impression that the two groups are closely related (they are treated together in the order Triuridales by CRONQUIST, 1968, 1981), but we are convinced from differences in the reproductive parts that the similarity is largely due to convergence as the result of the mycotrophy.

Cases of striking similarity may also appear in the construction of the plant. One such example is afforded by *Bowiea volubilis* (Fig. 91D) (Hyacinthaceae) in Southern Africa and *Thysanotus patersonii* (Fig. 85L) (Anthericaceae: Johnsonieae) in South-Western Australia. In both of these species the vegetative leaves soon wither, and the leafless inflorescence becomes profusely branched and twining, serving as assimilatory body and climbing device as well as floriferous organ. The flowers are quite dissimilar, however: those of *Bowiea* are greenish yellow, inconspicuous, those of *Thysanotus* violet, with fringed margins.

One of the more obvious cases of convergence is that between some members of Zingiberaceae (Fig. 167) and some Orchidaceae (Fig. 125), where the flower is provided with a *labellum* which is directed downwards, and where there is only one functional stamen. In some cases colour patterns and the habit of the plants further increase the similarities between the groups. A close study reveals, however, that the labellum in orchids consists of the median inner tepal, morphologically the upper one, which by resupination is usually directed downwards, whereas in the Zingiberaceae the labellum consists of two fused petaloid staminodes of the inner staminal whorl (no resupination here). In Orchidaceae, the functional stamen is the median one of the inner whorl, the resupination of the orchid flowers giving the orchid stamen a position corresponding to that in Zingiberaceae. Some botanists may feel inclined to regard the two families as phylogenetically related (see p. 95), but in doing so, the similarity of having a labellum and a single stamen must definitely be neglected, as these structures are only analogous.

A case of probable convergence (not wholly settled yet) is provided by the *Disporum* group versus Convallariaceae. BJÖRNSTAD (1970) found that *Disporum* (Fig. 108) and *Clintonia,* previously placed in Convallariaceae (Fig. 56) (Asparagales), have a different combination of embryological features (see DAHLGREN and CLIFFORD 1982, p. 307) and, besides, lack both raphides and septal nectaries, but have perigonal nectaries. In all these features they resemble *Tricyrtis* and other Liliales, although their habit, white flowers, and baccate fruit give a convallariaceous impression. It seems that they may be closest to the genus *Kreysigia* (with capsules) which, possibly incorrectly, has been placed in Colchicaceae. Whether all four of these genera and some more may be related and whether a family (Uvulariaceae) is justified for them, as here suggested, needs to be further investigated. The great similarity between Convallariaceae and some members of this group thus represents a case of convergence.

A further, similar, case is that of Hemerocallidaceae versus Liliaceae. The superficial resemblance between species of *Hemerocallis* (Fig. 81) and some yellow-flowered species of *Lilium* (Fig. 111 G) ("Day Lilies" and "True Lilies", respectively) is striking, but a close study reveals that *Hemerocallis* has rhizomes (*Lilium* bulbs), septal nectaries (*Lilium* perigonal nectaries), black, phytomelan-covered seeds (*Lilium* brown seeds), and that there are a number of other differences. The similarity in habit and flower shape are undoubtedly the result of convergence.

Poricidal anthers occur in some families of monocotyledons, e.g. a few genera of Commelinaceae, all Rapateaceae and Mayacaceae, some genera of Tecophilaeaceae and Cyanastraceae, and a few of Amaryllidaceae (*Galanthus* and *Leucojum*). There is little doubt that this condition has evolved independently in all or most of these families, thus representing convergence.

Sometimes there is a conspicuous combination of character conditions that may give misleading indications of relationships. One example is provided by the occurrence of flowers that have sepaloid outer tepals and petaloid inner tepals in combination with an amoeboid plasmodial tapetum, both conditions occurring in many Alismatales (Alismataceae, Fig. 135 I, Limnocharitaceae, and Hydrocharitaceae) and in Commelinales (Commelinaceae, Fig. 171 B–E). HUTCHINSON, in *The Families of Flowering Plants* (1934, 1959, 1973) gave great weight to the perianth character and placed these two complexes (together with some others which show a tendency to have a green or hyaline "calyx") in a separate "division", which he called Calyciferae (i.e. a group that bears a calyx). The tepaline and tapetal conditions have almost certainly evolved independently in Alismatales and Commelinales, because we find few other similarities between the two orders and therefore do not consider them closely related.

Quite often convergent evolution may raise difficulties of evolutionary interpretation. It has, for example, generally been thought that evolution goes in the direction of numerical reduction and therefore that the possession of numerous stamens is a condition more ancestral than possession of 3 + 3 or 3 stamens. Although the trend of numerical reduction may be valid in some groups such as the Magnoliiflorae it is not necessarily so in others, and probably not in monocotyledons. Thus, there is little reason to believe that the numerous stamens in single genera, such as *Ochlandra* in the grasses, or *Gethyllis* in Amaryllidaceae, represent an ancestral state. In the Alismatales the numerous stamens and carpels in many genera of Alismataceae (Fig. 139 B), Limnocharitaceae (Fig. 138 H) and Hydrocharitaceae were previously thought to be a state homologous to that found in most Nymphaeales, but this has been shown not to be so since the stamens are not spirally inserted in Alismatales as they are in Nymphaeales. SINGH and SATTLER (e.g. 1977) discovered that the primary carpel initials in Alismatales are three, and that further initials appear later. In a phylogenetic sense that is probably also true of the stamen initials, i.e. the multistaminal and multicarpellary condition in many Alismatales would be secondary.

It now seems most likely that multistaminal androecia in monocotyledons are always secondary and that they have appeared separately in a number of evolutionary lines, thus also representing convergence rather than relictual occurrences of an archaic state.

Some further cases in which there are several apparently important similarities shared by certain groups, and which are of great phylogenetic interest, will now be reviewed.

## The Spadiciflorae

The Typhales, Arales, Cyclanthales, Pandanales and Arecales have often been treated together in one major complex, Spadiciflorae, but it now seems highly probable that the Typhales are most closely related to the Pontederiales and Haemodorales (see p. 35, above) and that Arales have their closest relatives in the Alismatiflorae and Triuridiflorae (see p. 96). Some systematists have also begun to doubt seriously (THORNE 1983; DAHLGREN 1983a) that the remaining orders are closely related. The small densely aggregated flowers of Arales, Cyclanthales and some Arecales that are pollinated collectively with the help of highly efficient attraction devices (coloured spathe, strong odour and/or heat production) may have been selected as one of the most successful arrangements in a tropical forest, just as wind pollination is suited to more open habitats.

## Arales-Piperales

A striking case of convergence is afforded by the Arales (monocotyledons) and Piperales (dicotyledons). Both orders consist mainly of rain-forest herbs and vines with atactostely, petiolate and often cordate leaves with reticulate venation, often scattered stomata, which tend to be randomly oriented (KEITING, personal communication, for Arales), inflorescences in the form of spikes (*Piper, Pothos*) or more often a spadix, flowers which, though often strongly reduced, have a trimerous ground-plan, a sessile stigma (i.e. no style) and a baccate fruit (see DAHLGREN and CLIFFORD 1982, p. 336). These observations are by no means new, and have sometimes been regarded as providing significant evidence for phylogenetic relationship. LOTSY (1911) considered that the Spadiciflorae (see above) had evolved from Piperalean or Piperales-like ancestors, and similarly EMBERGER (1960) de-

rived the Piperales and "Spadiciflorae" from the same ancestors. Thus, both disregarded the traditional monocotyledon/dicotyledon boundary as a consequence of this similarity.

Further similarities, such as the frequently (mono-)-sulcate pollen grains and the cellular endosperm formation, can also be used as arguments in support of a close relationship between the two orders. In addition *Houttuynia,* in Saururaceae (Piperales), with its cordate leaves and its spike or spadix subtended by conspicuous, white bracts (often four), may strikingly resemble species of *Anthurium* in Araceae (Arales) where exceptionally there may be several spathal bracts.

Why is it then that most taxonomists do not follow LOTSY and EMBERGER, breaking down the monocotyledon/dicotyledon division at this point? Primarily, this is because of the strong tradition of using one versus two cotyledons as a principal distinction. This single character has been used to create a stable and orderly taxonomy. But it is also because it has subsequently been increasingly possible to associate the Piperales with the Magnoliiflorae and now (DAHLGREN and CLIFFORD 1981) with the Nymphaeiflorae. The sulcate pollen grains and trimerous flowers agree as well with the Magnoliiflorae as with the monocotyledons, and the general construction of the plant agrees with the former. This is supported also by chemical evidence. Thus the occurrence of cells with ethereal oils is a characteristic which the Piperales share with most Magnoliiflorae, and a few finds of benzylisoquinoline alkaloids in *Piper* are concordant with the common occurrence of these alkaloids in the Magnoliiflorae. The Piperales agree also with the Nymphaeales in vegetative characters and in details such as the construction of the perispermous seeds (see p. 53) and the S-type sieve tube plastids, the latter being wholly out of place in the monocotyledons.

On the basis of what is mentioned here and on pp. 53–54, we thus consider that the Piperales have their ancestry among dicotyledons with simple, trimerous flowers, sulcate pollen grains, eustely and cells with ethereal oils. The Arales on the other hand have their ancestors among monocotyledons with simple, trimerous flowers, sulcate pollen grains, atactostely and cells that lack ethereal oils. The two groups are derived from ancestral forms on opposite sides of the monocotyledon/dicotyledon border which may have been quite unlike each other. Adaptation through selection in a rain-forest habitat in both groups has influenced equally their vegetative construction, the condensation of

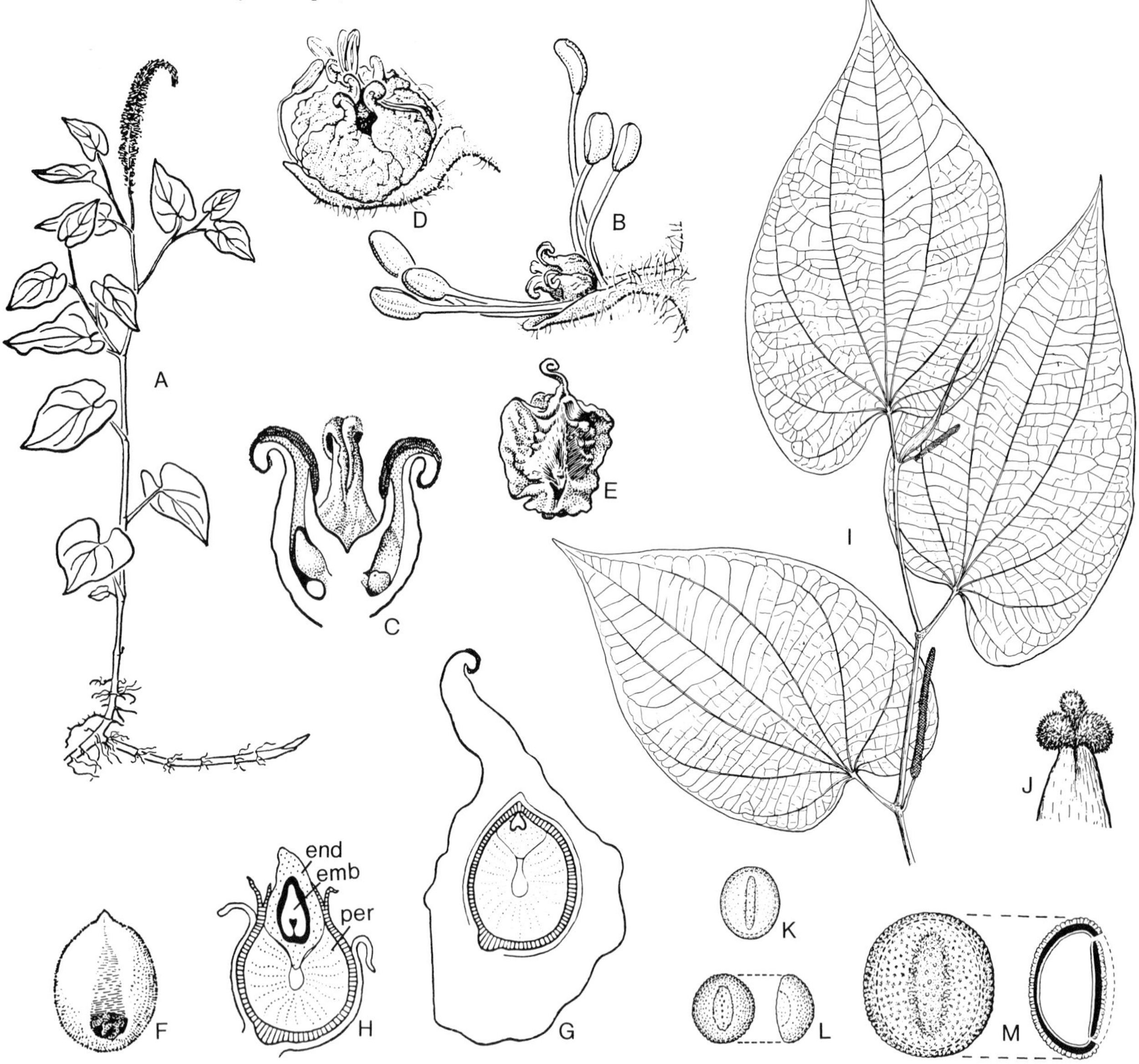

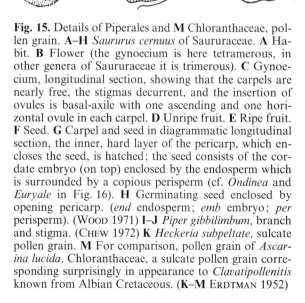

**Fig. 15.** Details of Piperales and **M** Chloranthaceae, pollen grain. **A–H** *Saururus cernuus* of Saururaceae. **A** Habit. **B** Flower (the gynoecium is here tetramerous, in other genera of Saururaceae it is trimerous). **C** Gynoecium, longitudinal section, showing that the carpels are nearly free, the stigmas decurrent, and the insertion of ovules is basal-axile with one ascending and one horizontal ovule in each carpel. **D** Unripe fruit. **E** Ripe fruit. **F** Seed. **G** Carpel and seed in diagrammatic longitudinal section, the inner, hard layer of the pericarp, which encloses the seed, is hatched; the seed consists of the cordate embryo (on top) enclosed by the endosperm which is surrounded by a copious perisperm (cf. *Ondinea* and *Euryale* in Fig. 16). **H** Germinating seed enclosed by opening pericarp. (*end* endosperm; *emb* embryo; *per* perisperm). (WOOD 1971) **I–J** *Piper gibbilimbum,* branch and stigma. (CHEW 1972) **K** *Heckeria subpeltate,* sulcate pollen grain. **M** For comparison, pollen grain of *Ascarina lucida,* Chloranthaceae, a sulcate pollen grain corresponding surprisingly in appearance to *Clavatipollenitis* known from Albian Cretaceous. (**K–M** ERDTMAN 1952)

**Fig. 16.** Nymphaeales, details. **A–N** Cabombaceae: *Cabomba.* **A** Branch. **B** Floral plan. **C** Flower. **D** Tepal. **E** Stamen. **F** Gynoecium (three free carpels). **G** Carpel opened to show laminar-dispersed placentation (only two ovules). **H** Pollen tetrad (after successive microsporogenesis). **I** Anther, tranverse section, for details see J. **J** Anther wall and microspores. **K** Anther wall, later stage with immature pollen grains. **L** Anther wall at stage of anthesis. **M** Ovule with embryo sac mother cell. **N** Nucellus after first meiotic division, the lower cell in stage of second division. (**A** and **C–G** HUTCHINSON 1926; **B** ZIMMERMANN 1965; **H–N** BATYGINA et al. 1982) **O–Z** and **A'–E'** Nymphaeaceae. **O** *Ondinea purpurea,* seed, longitudinal section (*op* operculum; *lf* plumular leaves; *end* endosperm; *cot* cotyledon; *per* perisperm; *ii* inner integument; *hyp* hypostase). (SCHNEIDER and FORD 1978.) **P–R** *Nymphaea alba.* **P** Habit. **Q** Floral plan. **R** Half-flower. (**P** and **R** LARSEN 1973 a; **Q** ZIMMERMANN 1965) **S–V** *Nymphaea stellata.* **S** Placentation. **T–V** Helobial endosperm formation. (**S** WEBERLING 1981; **T–V** BATYGINA et al. 1980) **X–Y** *Euryale ferox.*

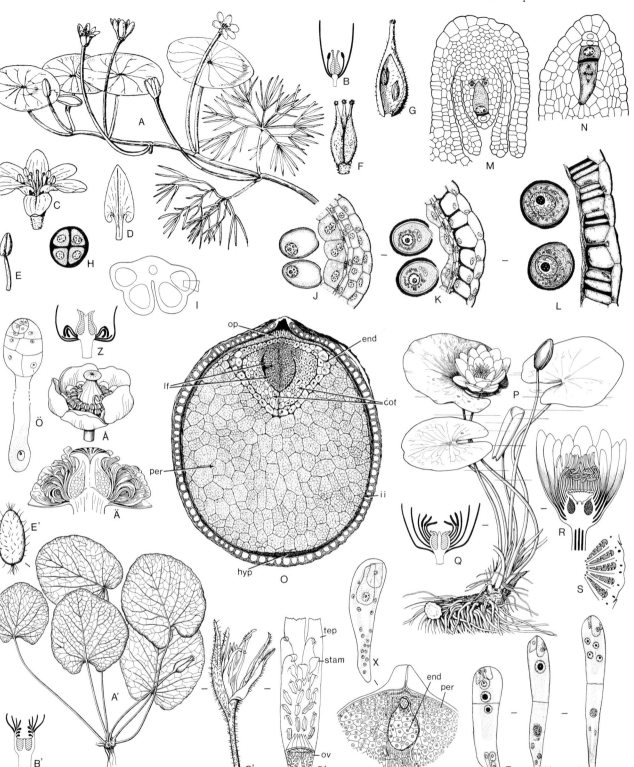

**X** Nuclear endosperm formation. **Y** Detail of unripe seed showing embryo, endosperm (*end*) and perisperm (*per*). (BATYGINA et al. 1980.) **Z–Ö** *Nuphar luteum.* **Z** Floral plan. **Å** Flower; **Ä** Half-flower; **Ö** Cellular endosperm formation, nucleus of micropylar chamber still undivided. (**Z** ZIMMERMANN 1965; **Å** LARSEN 1973; **Ä** ROSS CRAIG 1951; **Ö** BATYGINA et al. 1980) **A′–E′** *Barclaya motleyi.* **A′** Habit. **B′** Floral plan. **C′** Flower. **D′** Sagittal section through flower showing tepals (*tep*), staminodes (*stam*), stamens, stigmas and ovules (*ov*). **E′** Seed. (**A′** and **C′–E′** VAN ROYEN 1962; **B′** ZIMMERMANN 1965)

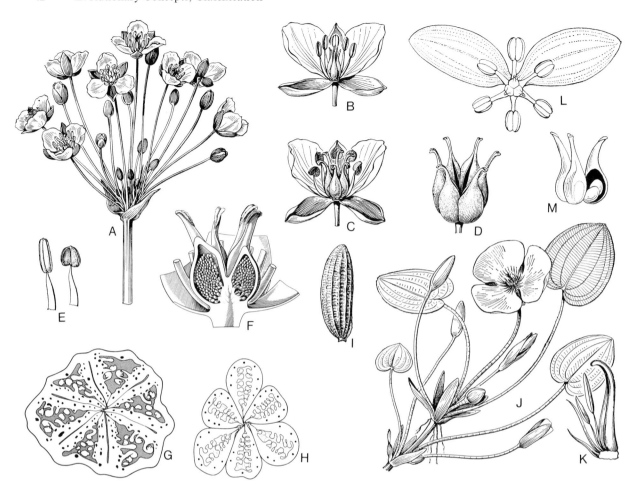

the flowers into a spadix, reduction in the individual flowers and the development of a fleshy fruit. If our conclusion is correct, the two orders supply one of the most outstanding examples of convergence found in the angiosperms.

## Alismatales-Nymphaeales

Hardly less striking than the case of the Arales-Piperales in terms of probable convergence is that of Cabombaceae in Nymphaeales (dicotyledons) versus Butomaceae-Hydrocharitaceae-Limnocharitaceae in Alismatales (monocotyledons). This case, too, has caused some serious doubts as to whether the monocotyledon/dicotyledon borderline has been correctly drawn; it has been elucidated by DAHLGREN and CLIFFORD (1981, 1982).

Of the two small genera of Cabombaceae, *Brasenia* shows the greater resemblance to Nymphaeaceae, approaching – at least superficially – *Barclaya* in the latter family. It has elliptic floating leaves and rather numerous stamens and carpels. *Cabomba*,

**Fig. 17.** Alismatalean taxa showing similarities to taxa of Nymphaeales, e.g. *Cabomba*, cf. Fig. 16. **A–I** *Butomus umbellatus* of Butomaceae. **A** Inflorescence. **B–C** flower during and after anthesis, one tepal of the inner whorl removed. **D** Multifollicle. **E** Stamens of different whorls. **F** Gynoecium, longitudinal section, to show laminar-dispersed placentation. **G–H** Gynoecium in transverse section, different stages and levels. (**A–E** and **I** TAKHTAJAN 1982; **F–H** WEBERLING 1981) **J–K** *Hydrocleys nymphoides* of Limnocharitaceae. **J** Branch with petiolate floating leaves. **K** Carpel, stamens and staminodes along a floral radius. (TAKHTAJAN 1982) **L–M** *Aponogeton ranunculiflorus* of Aponogetonaceae. **L** Flower. **M** Gynoecium, one wall partly removed to show basal-axile placentation. (GUILLARMOD and MARAIS 1972)

which has dissected, submerged leaves, has flowers with 3+3 tepals, 3+3 stamens and 3 carpels, which is the "basic" condition in monocotyledons. There are many other similarities between Cabombaceae and one or other of the members of the Butomaceae, Hydrocharitaceae and Alismataceae, including the aquatic mode of life, atactostely, the petiolate floating leaves (*Brasenia, Hydrocharis, Hydrocleys,* etc.), details of the root cap, the ephemeral radicle, the lack of vessels in stem and leaves, the trimerous flowers, the basifixed anthers, the sulcate pollen grains (e.g. *Butomus, Cabomba*), the free carpels, the laminar-dispersed placenta (Cabombaceae, Butomaceae, Limnocharitaceae), the helobial endosperm formation and the follicular fruit. In addition it can be noted that stipules occur in both Hydrocharitaceae (*Hydrocharis*) and Nymphaeaceae (*Nymphaea*).

It is not surprising that this evidence can appear very convincing when considering Alismatales and Nymphaeales as potentially closely related. It might be claimed that either group should be transferred to a place next to the other, either in the monocotyledons, as has been suggested by HAINES and LYE (1975), or in the dicotyledons. Both CRONQUIST (1968, 1981) and TAKHTAJAN (1969, 1980) use the evidence as a basis for considering the monocotyledons to be derived from dicotyledons via an evolutionary bottle-neck occupied by aquatic or semi-aquatic plants in the form of proto-Nymphaeales in transition to proto-Alismatales. During this stage the ancestral monocotyledons would have irreversibly lost their ability to form a ring-like cambium, giving rise to the secondary vascular tissue found in most dicotyledons.

We do not consider that this tempting hypothesis can stand up to a critical analysis. The presence of two cotyledons, the S-type sieve element plastids and the occurrence of ellagic acid and perispermous seeds in the Nymphaeales argue strongly against their position as a starting point for the monocotyledons, and none of these attributes occurs in the Alismatales. Instead, several pieces of evidence, for example the construction of the perispermous seeds (see p. 39), indicate that Nymphaeales are related to Piperales (see p. 53 and Fig. 21). In the same way there is strong evidence that the Alismatales and Zosterales are rather closely related to Arales and Triuridales (see p. 96).

Thus, if our reasoning is correct, the Nymphaeales/Alismatales case is another demonstration of multiple convergence. Only some of the similarities of the two groups (for example the sulcate pollen grains and trimerous flowers) are due to their shared ancestry; the others are the result of convergence. This has undoubtedly occurred as the response of both groups to selective forces in similar habitats. We may suppose that in the ancestors of each of Alismatiflorae and of Nymphaeales, "plasticity" in the form of mutations and virtually the same selection pressure could have been responsible for many of the similarities.

# Criteria for the Monocotyledons

We have accepted here the hypothesis that the monocotyledons are monophyletic on the basis of two character conditions: (1) they have a single cotyledon, or according to an alternative interpretation no cotyledon at all, whereas the dicotyledons have two (other numbers such as 0, 1, 3 or more are obviously secondarily derived); (2) their sieve tube plastids, where investigated, always accumulate protein in the form of triangular bodies (Fig. 1), a condition known in the dicotyledons in only two genera (*Saruma* and *Asarum*) in Aristolochiaceae.

Thus, with these two character conditions the monocotyledons can be unequivocally defined. Other criteria have exceptions, in that either they do not apply to all monocotyledons or they also apply to certain groups of dicotyledons, or both. As has been pointed out by some workers, e.g. HUBER (1977), the characters show a certain polarization in their expressions between those plants that we may regard as "extreme" monocotyledons, via monocotyledons with obvious dicotyledonous attributes, and dicotyledons with obvious monocotyledonous attributes, to dicotyledons which have relatively few attributes in common with the monocotyledons. On the basis of this, HUBER (1977) prefers to consider some of the monocotyledons and the "ranalean dicotyledons" as two wings of a single natural unit. Apart from what we presume to be a phylogenetic relationship between certain monocotyledonous groups and certain families in Magnoliiflorae, there are extensive morphological similarities between other pairs of groups of monocotyledons and dicotyledons, such as between Alismatales and Nymphaeales, and between Arales and Piperales, which we assume have evolved by convergence (see previous chapter).

Most or all monocotyledons differ from all or most of the dicotyledons in the following respects.
1. *The monocotyledons do not have two cotyledons, which is usually the case with the dicotyledons.* The leaf, which is considered to be the single cotyledon in monocotyledons, is frequently two-nerved, which supports the view that the cotyledon has originated by "intercalary concrescence" of the margins of the two cotyledons in the presumably dicotyledonous ancestors (STEBBINS 1974). The single cotyledon in monocotyledons is generally tubular and in most cases has a (pseudo-)terminal position on the embryo, and either the plumule or its first leaf generally breaks through the side of the cotyledonary tube. It is only in some taxa of Dioscoreales that the cotyledon is more or less lateral and the plumule approaches a terminal position.

Some alternative interpretations that have been presented are that one of two cotyledons has been suppressed, that the two cotyledon homologues have become separated from each other by an internode, and even that the monocotyledons do not have a cotyledon at all (JACQUES-FÉLIX 1982), cotyledons being exclusively characteristic of dicotyledons. We shall not analyze these hypotheses here. The evidence for the homology of the cotyledon of the monocotyledons with the two cotyledons or their petioles in the dicotyledons is that in many young monocotyledon embryos the single cotyledon is formed from a zone or group of cells corresponding to the common initial that gives rise to both cotyledons in the dicotyledons. This is discussed by JOHANSEN (1950) and STEBBINS (1974).

Reduction or loss of one of the two cotyledons is known in a number of dicotyledons. It is found in particular in plants with a hypocotyledonary tuber. In other cases, e.g. *Ficaria* of the Ranunculaceae, the two cotyledons are fused, so that a case of syncotyly, not monocotyly, is involved.

2. *The sieve tube plastids in all monocotyledons so far investigated contain a number of triangular protein bodies,* i.e. are of the PIIc type (2–3 and BEHNKE 1969, 1981; BEHNKE and BARTHLOTT 1983). Among the dicotyledons, sieve tube plastids of this type are known only in *Asarum* and *Saruma* (BEHNKE and BARTHLOTT 1983) in Aristolochiaceae. Most dicotyledons have sieve tubes with starch grains and no protein (S-type), but protein crystalloids of various shapes are known, some of which are characteristic of particular taxonomic groups (BEHNKE 1981b). The protein crystalloids are particularly variable in the Magnoliiflorae, where they are polygonal, square or triangular and

can be solitary or numerous. In addition, protein filaments are found in some families of this super-order. Protein may also be lacking, and starch may be present either together with protein or where protein is lacking. Protein bodies are also present in a variety of other dicotyledons. They occur frequently in Fabales and in Caryophyllales (where protein filaments are organized into an annulus). As pointed out above, the triangular protein bodies are so constant and uniform in monocotyledons and so rare in dicotyledons, that this constitutes a strong reason for considering the monocotyledons as monophyletic.

3. *The vascular system of all monocotyledons is an atactostele,* i.e. with scattered vascular bundles, where a cambium, which produces secondary vascular tissue, is never formed between the phloem and xylem elements. Thus the form of secondary growth typical of most dicotyledons is completely absent in monocotyledons. Where secondary thickening occurs it is produced by a cambium which is continuous with the primary thickening-meristem if the latter is discernible, but which functions in the part of the stem that has completed elongation. The cambium arises in the parenchyma outside the vascular bundles and produces vascular bundles and ground parenchyma towards the inside, and small amounts of parenchyma towards the outside (TOMLINSON and ZIMMERMANN 1969; TOMLINSON 1970; DAHLGREN and CLIFFORD 1982). This type of secondary thickening, which is rare in monocotyledons and lacking in dicotyledons, is known as "monocotyledonous thickening growth".

An atactostele with scattered vascular strands and no cambium, as in monocotyledons, is known in various dicotyledons, e.g. in all Nymphaeales, some Piperales and also in various Ranunculales, e.g. *Podophyllum.* The fact that the same condition is present in Nymphaeales has been one of the bases for the derivation of monocotyledons from dicotyledons via aquatic or semi-aquatic forms resembling extant Nymphaeales (see previous chapter; TAKHTAJAN, 1969; and CRONQUIST, 1981).

4. *The first root, the radicle, of the monocotyledons generally deteriorates at an early stage, its function being taken over by a system of adventitious roots.* In most dicotyledons the radicle continues its development and functions in the adult plant as a tap root, from which laterals are developed. The restricted life-span of the radicle in monocotyledons is connected with the lack of a cambium to supply the secondary vascular tissue of a tap root.

5. *The leaves of the monocotyledons are generally supplied with several or many veins that converge towards the leaf apex and the nodes have numerous leaf trace bundles.* The leaves vary considerably and there are exceptions to the above statement. Monocotyledon leaves are generally either well-differentiated petiolate leaves with a broad lamina or strap-like or ribbon-like leaves without this differentiation. Both types occur together in certain groups (Alismataceae, Potamogetonaceae, Pontederiaceae) and can sometimes be found at different levels on the same plant. The linear leaves can be regarded as less differentiated and largely phyllodial homologues of the well-differentiated leaves. This type of leaf is uncommon in the dicotyledons, but its equivalent is found in bud scales, basal scales, bracts, etc., collectively called cataphylls.

The linear leaves in monocotyledons develop in a different way from those of most dicotyledons. The leaf initial is proportionately broader and the intercalary meristem is developed closer to the leaf base. In other leaf initials intercalary growth is concentrated in the distal part, which has as a consequence a well-differentiated leaf lamina. Both kinds of leaves, and particularly the linear ones, are generally sheathing at the base. Linear or linear-lanceolate parallel-veined leaves sometimes have a constricted portion between the sheath and the broader part of the "lamina"; this is not to be regarded as homologous with an ordinary petiole and is called a pseudo-petiole (p. 4). In certain Poaceae-Bambusoideae, some Orchidaceae, Funkiaceae, Cyanastraceae, Convallariaceae, etc., this pseudo-petiole has considerable length. Leaves with a broad, often reticulate lamina, as in Dioscoreaceae and Smilacaceae, are truly petiolate and have a non-sheathing base.

The venation of monocotyledons cannot be readily distinguished from that in all dicotyledons, but in the latter the veins more rarely run longitudinally to meet towards the apex.

The stomata of the leaves of the monocotyledons are nearly always arranged with the long axis of the guard cells in the longitudinal direction of the leaves, whereas in dicotyledons they are generally scattered in various directions.

Stipules, which are common in dicotyledons, are rare in monocotyledons, but occur in certain Alismatales, Zosterales and Pontederiales. More or less hyaline lobes arising from the leaf bases occur in many groups.

6. *The flowers of the monocotyledons are with few exceptions trimerous,* whereas in the dicotyledons they are prevailingly tetramerous or pentamerous,

although other conditions occur. Exceptions to the trimerous condition in monocotyledons occur sporadically in families where trimery otherwise prevails, some exceptions being the tetramerous Cyclanthaceae, the tetramerous or dimerous Potamogetonaceae and dimerous Stemonaceae. Other numbers of floral parts are also met with in Trilliaceae (several genera) and Convallariaceae (*Maianthemum, Aspidistra*). *Pentastemona* (Stemonaceae, Fig. 48) is probably the only genus in all the monocotyledons with a clearly pentamerous perianth (VAN STEENIS 1982), i.e. excluding obvious cases with a hexamerous perianth with one member reduced or two fused. Reductions in the flower, however, often result in a reduced number of parts in one or more of the floral whorls. More than $3+3$ stamens and more than 3 carpels are also found in several groups, in particular in many Alismatales, but these conditions are apparently secondary wherever they occur in monocotyledons. (See Convergent Evolution in the previous chapter.)

Trimerous flowers also occur in some dicotyledons, in particular in many Magnoliiflorae, Nymphaeiflorae and Ranunculiflorae. Among these groups various Magnoliiflorae and Nymphaeiflorae show a number of similarities with monocotyledons, indicating a phylogenetic relationship and implying possession of the trimerous condition in their common ancestor. The taxa of Magnoliiflorae and Nymphaeiflorae that resemble monocotyledons in their numbers of floral parts include *Lactoris* (Lactoridaceae, Fig. 20) (P3 A3 + 3 G3); *Orophea* spp. (Annonaceae) (P3 + 3 + 3 A3 + 3 G3); *Saruma* (Aristolochiaceae, Fig. 19 J) (P3 + 3 A6 + 6 G3 + 3); and *Cabomba* (Cabombaceae, Fig. 16 C) (P3 + 3 A3 + 3 G3), the last of which agrees completely in the basic floral diagram with most monocotyledons.

7. *The pedicel of the monocotyledons, if provided with a prophyll (bracteole), generally bears it dorsally,* whereas in most dicotyledons, if prophylls are present, there are two placed transversely. The prophyll in monocotyledons frequently has two prominent veins or ribs and may be bipartite at the apex, suggesting that it corresponds to two transverse prophylls that have fused dorsally (by intercalary concrescence). A dorsal bicostate prophyll occurs among the dicotyledons in most Annonaceae and in certain other Magnoliiflorae.

8. *Anther wall formation in most monocotyledons is of the so-called Monocotyledonous Type* (see Morphological Concepts), a type which is uncommon in dicotyledons. Anther wall formation has

so far been studied in few monocotyledons and these have the Monocotyledonous Type except for *Tacca* in Taccaceae, Dioscoreales, where it is of the Dicotyledonous Type (may need to be re-studied), and *Najas* in Najadaceae, Zosterales, and *Lemna* and *Wolffia* in Lemnaceae, Arales, where it is of the Reduced Type. In dicotyledons the Monocotyledonous Type is reported in a number of taxa in Caryophyllales and Polygonales, and in some other orders.

9. *The pollen grains of most monocotyledons are either (mono-)sulcate or ulcerate with the single aperture at the distal pole,* whereas in the dicotyledons the apertures are generally situated in the equatorial zone or scattered over the pollen grain surface (zono- and pantoaperturate pollen grains, respectively).

Other types of pollen grains occur exceptionally in the monocotyledons, for example foraminate, sulculate, zoni-sulculate and spiraperturate, but these are generally derivable from the sulcate type and are of scattered occurrence.

Among the dicotyledons sulcate and ulcerate pollen grains are virtually absent, except in Magnoliiflorae and Nymphaeiflorae, where they are frequent and where zoni-aperturate, colpate types are rare. This indicates that the monocotyledons and the Magnoliiflorae and Nymphaeiflorae are phylogenetically related (see also trimerous flowers).

10. *Microsporogenesis in most monocotyledonous families is of the Successive Type, although the Simultaneous Type also has a wide distribution,* whereas in the dicotyledons microsporogenesis is generally simultaneous (see Morphological Concepts).

In the monocotyledons microsporogenesis is simultaneous in all Cyperales, most Arecales, many Liliales (e.g. Iridaceae and Orchidaceae), many Dioscoreales (e.g. Taccaceae and most Dioscoreaceae) and some families in Asparagales, but although these groups are numerous in terms of species they represent a minority in terms of families.

Among the dicotyledons successive microsporogenesis and transitional types of microsporogenesis are known in certain Magnoliiflorae and Nymphaeiflorae (taxa of Lauraceae, Myristicaceae, Annonaceae, Aristolochiaceae, Rafflesiaceae, Cabombaceae and Ceratophyllaceae), as well as in families in other orders where it has presumably evolved independently, e.g. in Podostemonaceae, Asclepiadaceae and Apocynaceae.

It is not unlikely that the Successive Type of microsporogenesis is ancestral in angiosperms; if so the

Simultaneous Type must have evolved independently in a number of lines of evolution in both monocotyledons and dicotyledons.

11. *Endosperm formation in the monocotyledons is either of the Helobial or the Nuclear Type,* the Cellular Type being restricted to Arales and Hydatellales. In the dicotyledons the endosperm is only rarely of the Helobial Type, while the Nuclear and Cellular Types are dominant (see Morphological Concepts).

12. *About half of the species of monocotyledons lack vessels in the stem, and even more lack them in the leaves,* whereas vessels are generally present in the roots. In dicotyledons vessels are generally present in the stem, although there are a number of exceptions, e.g. the Nymphaeales and certain woody families in the Magnoliiflorae. This character is of fairly low significance in characterizing the monocotyledons.

13. *In most insect-pollinated monocotyledons nectar is secreted from the walls of the carpels; in the syncarpous pistils these nectaries are situated in the septa of the ovary and are termed septal nectaries.* Exceptions are most members of Liliales, where septal nectaries are substituted by nectaries on the inner side of tepals.

Septal nectaries are very rare in dicotyledons, where nectar secretion generally takes place from floral discs.

Since a great number of monocotyledons are wind-pollinated, and certain insect-pollinated groups have perigonal nectaries or lack nectaries altogether, this character is of restricted significance in characterizing the monocotyledons.

14. *The monocotyledons sustain certain generalizations about their chemical qualities.* However, none of the chemical compounds involved affords a means of distinguishing absolutely between monocotyledons and dicotyledons.

Ellagic acid and ellagi-tannins are totally absent among the monocotyledons, while they occur frequently in certain major groups of dicotyledons. Thus ellagic acid is known in some members of Nymphaeaceae, a fact that, together with the type of sieve tube plastids, militates against the inclusion of the Nymphaeiflorae in the monocotyledons.

An isoenzyme of the type dehydroquinate hydro-lyase (DHQ-ase), activated by shikimic acid, is common in the Commeliniflorae and is found in certain other monocotyledons (BOUDET et al., 1977). It seems to be restricted to monocotyledons.

It is also important that benzylisoquinoline alkaloids are rare in the monocotyledons, while they occur in several of the families in Magnoliiflorae which exhibit the most numerous and conspicuous similarities to the monocotyledons, e.g. Annonaceae, Aristolochiaceae and Magnoliaceae. True tannins are rare in monocotyledons but common in certain groups of dicotyledons, while the "condensed" type of tannins occurs in both groups.

Iridoid compounds, betalains, glucosinolates and certain other groups of compounds are entirely lacking in monocotyledons, though occurring in some dicotyledons.

Saponins occur abundantly in certain groups of monocotyledons and are mainly *steroidal saponins.* Such saponins are relatively rare in the dicotyledons, although they occur in Asclepiadaceae, *Helleborus* (Ranunculaceae), *Trigonella* (Fabaceae), *Digitalis* and *Solanum* (Solanaceae), etc. In dicotyledons saponins are widely distributed and are mostly triterpene saponins, a type which is rare in monocotyledons.

It may also be mentioned that calcium oxalate, which is common in dicotyledons as well as monocotyledons, often occurs in the form of raphides in the latter group, which is more rarely the case in dicotyledons, although they are known in Dilleniaceae, Aizoaceae, Balsaminaceae, etc.

In actual practice combinations of the above-mentioned characteristics of monocotyledons have generally been sufficient to classify any unknown plant in the flowering stage as a monocotyledon (a few taxa, such as members of Stemonaceae, Lemnaceae, Hydatellaceae and Zosteraceae being exceptions). For example, the combination of herbaceous growth, linear leaves and trimerous flowers is generally sufficient to determine that a plant is a monocotyledon.

We shall, however, show that certain dicotyledonous families, such as Cabombaceae, Ceratophyllaceae, Saururaceae, Piperaceae, Lactoridaceae, Aristolochiaceae and Annonaceae possess a considerable number of the monocotyledon attributes enumerated above. Likewise, certain monocotyledons, such as Trichopodaceae, Dioscoreaceae, Taccaceae, Stemonaceae, Araceae and Alismataceae each resemble various different families that are assigned to the dicotyledons. In some cases similarities between two taxa on either side of the monocotyledon/dicotyledon border can be so strong that we are still in serious doubt as to the distinctness of the monocotyledons. We have also avoided classifying monocotyledons and dicotyledons as units of definite rank, such as class, in this presentation.

# Origin of the Monocotyledons

## General Remarks

The prevailing view of the origin of the monocotyledons is that they evolved from a primitive dicotyledonous group. The dicotyledons that have been considered as likely candidates in this context are almost without exception plants that we refer to one of the superorders Magnoliiflorae and Nymphaeiflorae (including Piperales) or, rarely, Ranunculiflorae.

The view that these superorders are primitive rests on what may be called the Ranalian or Strobilus Hypothesis (BURGER 1981). This is partly rooted in the theory of the origin of the angiosperms flower from a strobilus like that of cycads and Bennettitales with leaf-like reproductive structures (microsporangiophylls, megasporangiophylls) (ARBER and PARKIN 1907). This theory was developed and much modified through studies of magnoliifloran flowers in genera of Winteraceae, Degeneriaceae and Austrobaileyaceae (see EYDE 1976). Support for it is obtained from the early fossil appearance of sulcate or sulcate-derived pollen grains (in the Barremian of the Cretaceous) similar to those of plants in Magnoliiflorae and from the primitive vessels, or lack of vessels, in some of them.

We have noted in various connections above that there are very obvious links between representatives of Magnoliiflorae, such as Annonaceae, Aristolochiaceae (Fig. 19), Magnoliaceae, Degeneriaceae (Fig. 18) and Lactoridaceae (Fig. 20), and certain taxa in Dioscoreales of the Liliiflorae. The linking characters in Dioscoreales are: petiolate leaves with a small, scarcely sheathing leaf base and a reticulately-veined lamina, stamens with microsporangia below the apex, seeds with a minute embryo, the occasional occurrence of ruminate endosperm, an almost terminal plumule and, rarely, one ring of vascular strands in the young shoot. Those in the Magnoliiflorae are the monocotyledonous type of sieve tube plastids ($PII_c$-type sensu BEHNKE 1981, in *Saruma* and *Asarum* of Aristolochiaceae), a dorsal prophyll, frequently trimerous flowers, sulcate pollen grains, simultaneous micro-sporogenesis (rare) and cyanogenic compounds (tyrosine pathway).

A few of these similarities may be the result of convergence but they seem mainly to be based on a common ancestry. Given the connection, the question of direction remains to be decided. There are two possibilities:

1. The ancestors were dicotyledonous plants from which the earliest monocotyledons evolved by loss of cambial activity in the stem, fusion of cotyledons, and various other changes. This is the general view, compatible with the Ranalian or Strobilus Hypothesis.

2. The ancestors were little differentiated plants of roughly monocotyledonous type, from some groups of which the dicotyledons evolved (BURGER, 1981). The ancestors that could have given rise to dicotyledons are most logically to be sought among the Dioscoreales, but an alternative is the Nymphaeales. We would then assume that small plants (Fig. 22) with herbaceous atactostelic stems, linear, sheathing, petiole-less leaves with parallel veins, trimerous flowers, successive micro-sporogenesis and sulcate pollen grains gave rise to the dicotyledons via advanced monocotyledonous forms with petiolate leaves and reticulate venation. Further, that the acquisition of a cambium to form secondary vascular tissue and the development of two cotyledons evolved as novelties in some early but advanced monocotyledon derivatives which thus became the first dicotyledons.

We shall consider briefly each of these two alternatives.

## The Ranalean/Magnoliifloran Hypothesis

The monocotyledons obviously first appeared very early in the Cretaceous (see DAHLGREN 1983a), presumably in Aptian-Albian times (ca. 110 million years ago), when the ancestors of the Magnoliiflorae must already have acquired some of the present attributes of that group, but were less dif-

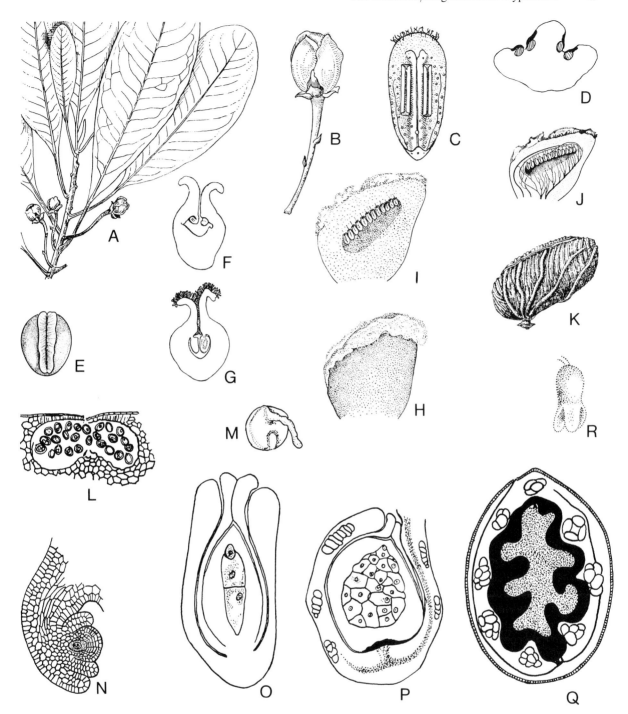

**Fig. 18.** Degeneriaceae, taken as an example of the woody Magnoliiflorae considered to be derivative of the primitive dicotyledon stock that has a shared origin with the monocotyledons. The single species *Degeneria vitiensis*. **A** Branch. **B** Flower, this has a tendency to trimery. **C** Stamen. **D** Stamen (extrorse), transverse section. **E** The single carpel, shown from above. **F–G** The carpel in different stages of development in transverse section. **H–I** The carpel in flowering stage, entire and in longitudinal section. **J** the carpel, with vascular system inserted. **K** Fruit. **L** Anther locule, at stage of pollen dispersal. **M** Germinating pollen grain (the pollen grain is sulcate). **N** Young ovule, note the very crassinucellate state. **O** Ovule, later stage, just after formation of the first wall of the cellular endosperm, the zygote on top. **P** Still later stage, the cellular endosperm well developed but zygote still undivided. **Q** Nearly mature seed showing ruminate endosperm, the minute embryo hardly visible. **R** Embryo; this is tricotyledonous, a highly unusual and undoubtedly derived condition. (**A–D** and **F–I** and **K** BUCHHEIM in MELCHIOR 1964; **E** and **J** WEBERLING 1981; **L–R** BHANDARI 1971)

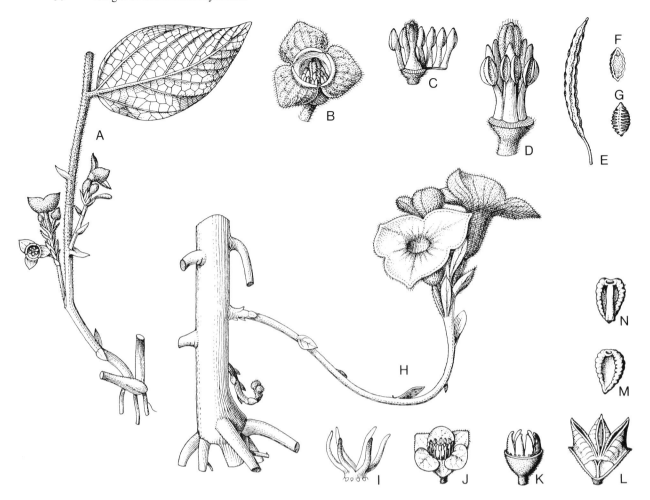

**Fig. 19.** Aristolochiaceae is a family which presumably evolved from arborescent, woody magnoliifloran ancestors with trimerous flowers. The primitive genera *Saruma* and *Asarum* have PIIc type sieve tube plastids, the stems are frequently herbaceous and the stele is "loosening up". Here is shown *Apama tomentosa* (**A–G**), *Thottea rhizantha* (**H**), *Thottea dependens* (**I**), and *Saruma henryi* (**J–N**). **A** and **H** Parts of branches with inflorescences. **B** and **J** Flowers. **C** Staminal whorl, opened up, and pistil. **D** Flower, perianth removed. **E** Fruit. **F–G** Seed, longitudinal section and unsectioned. **I** Stigmatic lobes. **K–L** Gynoecium in flowering and fruiting stage. **M–N** Seed in different views. (SCHMIDT 1935)

ferentiated, and when some other dicotyledonous groups had already branched off from the ancestral stock.

If we are to consider a derivation of monocotyledons from dicotyledons in a fair and realistic way, we must do away with certain misconceptions. The Ranalian Hypothesis does not necessarily require us to assume that the primitive angiosperms lacked vessels in the stem, or (as in *Magnolia*) had large flowers, a long floral axis, numerous perianth leaves and stamens and conspicuously flat stamens, or that they had carpels that were more or less "open" at some stage in their evolution (as in *Degeneria*). It seems unlikely that lack of vessels, which is met with in a number of extant dicotyledonous woody plants, is a primitive state in all of them (YOUNG 1981), because if this were true, we should have to argue that vessels evolved independently in a number of different lines, which is improbable. Furthermore, the first angiosperms probably had relatively modest flowers with inconspicuous perianth members and their stamens and carpels may not have been very numerous either. In the evolutionary line that evolved into the common ancestor of extant monocotyledons *and* part of Magnoliiflorae-Nymphaeiflorae, a clearly trimerous floral type (at least in the perianth) must have become established. Increase as well as decrease in number of stamens and carpels obviously happens easily, as is evident from the great variation in these parts within genera (cf. the variation

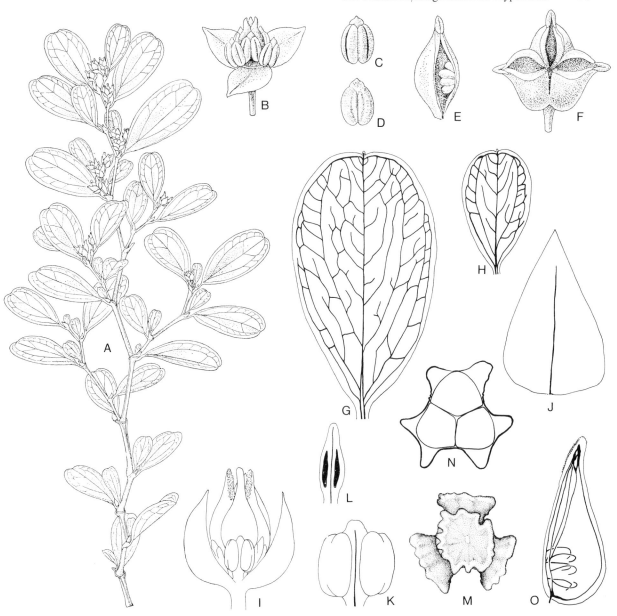

in carpel number in species of *Ranunculus*). It may be supposed, then, that large flowers such as those in most extant water-lilies (Nymphaeaceae) are secondary in relation to the condition in *Brasenia* (Cabombaceae) or *Barclaya* (Nymphaeaceae), and that the *Magnolia* or *Michelia* type of flower is a secondarily pleiomerous type in which the floral axis has elongated as a secondary adaptation, as can be assumed to have been the case in *Myosurus* (Ranunculaceae) and *Schisandra* (Schisandraceae).

On the basis of the distribution of character states in monocotyledons and dicotyledons, adherents of the Ranalean Hypothesis would probably characterize the dicotyledons that were the common an-

**Fig. 20.** Lactoridaceae. This family, with the single woody species *Lactoris fernandeziana,* has a number of the features characteristic of monocotyledons: leaves with converging main veins, flowers trimerous in all whorls (although there is only one perianth whorl) and pollen grains with a distal aperture. The pollen grains cohere in tetrads and the inner staminal whorl is staminodial. **A** Branch. **B** Flower. **C–D** Stamen (the filament very short, the anther extrorse, with an apically protruding "connective tip"). **E** Carpel, opened to show placentation. **F** Trifolliculus. **G–H** Large and small leaf from shoot (cf. **A**). **I** Flower, note the decurrent inwardly directed stigma. **J** Tepal. **K** Stamen, seen from outside. **L** Staminode. **M–N** Pollen tetrad (**N** in optical section). **O** Pistil, with venation depicted. (**A–F** Takhtajan 1969: **G–O** Carlquist 1964)

cestors of the monocotyledons and the Magnolii-florae-Nymphaeiflorae as below.

They were woody plants with a eustelic stem, long, narrow vessels in the xylem and scalariform perforation plates with many bars in the vessels. The leaves, which were presumably not deciduous, had a small base, but were petiolate and had a reticulate, somewhat irregular venation pattern with brochidodromous main venation tending towards an arching-parallel venation with anastomosing vein tips (HICKEY and WOLFE 1975). The stomata were paracytic or anomocytic. The flowers were borne in small panicles and their parts were spirally inserted but with a tendency for at least the perianth to be cyclic and disposed in two to three trimerous whorls (the outer being somewhat smaller than the inner, as in certain extant Annonaceae). The stamens tended to be more or less flat with the microsporangia situated well below their apex. The anther wall was of the Basic Type, the tapetum was glandular, microsporogenesis was successive, the pollen grains were sulcate and, when shed, bicellular. The carpels were mutually free, and may have tended to be three or multiples of three; whether they were foliar, style-less structures with a dispersed or long-decurrent stigma following the margins is uncertain. Probably the ovules were fairly numerous and anatropous, bitegmic (with rather thick integuments), crassinucellate, with a parietal cell (subsequently forming a parietal tissue) cut off from the archesporial cell, with *Polygonum* Type embryo sac formation, and with cellular endosperm formation (WUNDERLICH 1959). The gynoecium may have developed into a multifolliculus with primarily exarillate seeds with copious, ruminate endosperm containing mainly fat and aleurone as nutrients, and with a tiny embryo at the micropylar end. The embryo was dicotyledonous and lacked chlorophyll. As regards sieve tube plastids, these dicotyledons had protein crystalloids which may have had unstabilized shapes (cuneate-triangular, square, polygonal) and they might have had scarcely organized protein filaments. Ellagic acid and ellagitanins were lacking.

The derivation of an ancestral monocotyledon from this hypothetical taxon need only have involved a few steps: transition to a herbaceous state accompanied by the loss of the cambium and hence the capacity to produce secondary vascular tissue, change of the cotyledonary condition and, probably, changes in endosperm formation, stabilization of the floral formula as P3+3 A3+3 G3 (see floral morphology, p. 65), and fixation of a triangular shape for the protein crystalloids in the sieve tube plastids. It may seem that the change to the herbaceous state and the acquisition of an atactostele would be a drastic step, but this has obviously happened independently in both the Piperales and in the ancestors of Nymphaeales, as well as in one or more lines within the ancestors of extant Ranunculales. The nature of the change in the cotyledons is still not settled because there are diverse interpretations of the "cotyledon" in monocotyledons (see p. 44).

Supposing that the changes sketched above occurred in a line of protomonocotyledons, what would be the characteristics of the most ancestral monocotyledon group?

We conclude that it would be a group of herbs with a compact short stem (rhizome or multi-nodal corm), well-developed petiolate leaves with non-sheathing base, reticulate venation, hypogynous trimerous flowers, perhaps inconspicuous perianth, somewhat flattened stamens with the microsporangia below the apex, sulcate pollen grains, apocarpy, ruminate, copious endosperm, minute embryo and a subterminal plumule. Such a monocotyledon does not exist, but the closest extant plant is perhaps *Trichopus* (Trichopodaceae, Dioscoreales; Figs. 41, 42), differing from our hypothetical plant in its syncarpous gynoecium, which is, moreover, inferior. *Trichopus* is in our opinion not a direct link but it exhibits a number of features that are met with in Annonales. Some of the similarities could have evolved by convergence, but convergence is more probable where the plants have a similar habit and habitat (as with the aquatic Nymphaeales and Alismatales), and this can certainly not be said for the forest herbs of Dioscoreales and the trees of Annonales.

Another version of the Ranalian Hypothesis is favoured by TAKHTAJAN (1969, 1980) and CRONQUIST (1981); this supposes that at an early stage in the history of the woody Magnoliiflorae, an evolutionary line corresponding to or identical with the extant Nymphaeales was differentiated from them. As in the version detailed above, this differentiation involved transition to a herbaceous growth form which occurred in connection with adaptation to aquatic life. The resultant plants were thus provided with petiolate leaves with a floating lamina, they lost vessels completely and acquired a number of features met with in certain alismatiflorous monocotyledons, e.g. laticifers, stipules, atactosteles, extrorse anthers, laminar-dispersed placentation and helobial endosperm formation (Fig. 16). With the addition of a cotyledon-

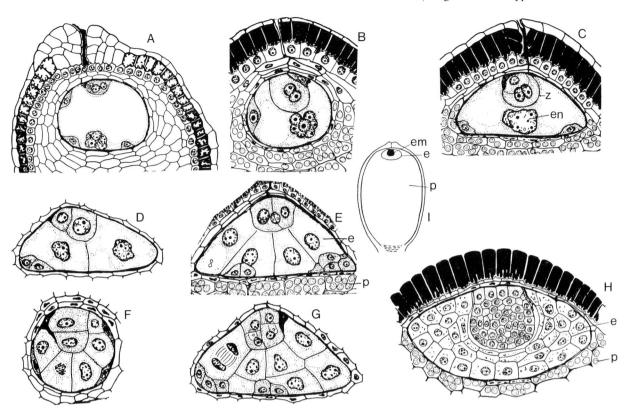

**Fig. 21.** Piperaceae. Details of embryology within the genus *Peperomia*. **A** Mature embryo sac with several polar nuclei. **B–C** Embryo sac shortly after fertilization; (*z* zygote; *en* primary endosperm nucleus). **D** First divisions of zygote and endosperm formation. **E** Somewhat later stage, with endosperm (*e*) being formed and the nucellar tissue already forming a starchy perisperm (*p*). **F** and **G** Stages somewhat later than **D** and **E**, respectively. **H** Part of mature seed showing thick-walled testa, a globose embryo, endosperm (*e*) and perisperm (*p*) comparable to the apical part of **I** (*em* embryo.) (All from NIKITICHEVA et al. 1981)

ary change, the plants would be classified as monocotyledonous. HAINES and LYE (1975) found that the two cotyledons of *Nymphaea* cohere slightly on one side and interpreted this as an indication of their monocotyledonous nature. Certainly the character states shared by some Nymphaeales (e.g. *Cabomba* of Cabombaceae which has a completely liliifloran floral diagram, P 3 + 3 A 3 + 3 G 3) with Butomaceae and Limnocharitaceae (Alismatiflorae) are impressive. However, under a more critical analysis this theory appears doubtful, because the seeds in all Nymphaeales have both a copious perisperm and a well-developed endosperm, a construction found elsewhere only in the Piperales. The seeds in the two orders are so similar and so peculiar that convergence

is a highly unlikely explanation of the resemblance. The seeds in Alismatiflorae, on the other hand, lack both perisperm and endosperm. Moreover, the sieve tube plastids in both Piperales and Nymphaeales completely lack protein crystalloids, whereas *all* monocotyledons have triangular protein crystalloids in the sieve tube plastids. Further, extant Nymphaeales possess ellagitannins, which are completely unknown in monocotyledons.

Therefore, in spite of the number of similarities between various Nymphaeiflorae and various Alismatiflorae, we assume that they are fairly distantly related and that the similarities can be divided into ancestral states (e.g. sulcate pollen grains) and convergences, whereas Piperales and Nymphaeales are sister-groups in the evolutionary sense.

Incidentally it may be noted that RIGGINS and FARRIS (1983), when criticizing a cladistic treatment of the magnoliifloran-nymphaeifloran families by YOUNG (1981), found that greater parsimony was obtained in a cladogram where the families of Piperales and Nymphaeales are placed close together as sister-groups.

It has also been argued (e.g. by LOTSY 1911, and EMBERGER 1960) that there are close relationships between the Piperales (dicotyledons) and the Arales (monocotyledons). This is a plausible conclu-

sion on the basis of a number of similarities between members of these two orders (DAHLGREN 1979; DAHLGREN and CLIFFORD 1981; 1982, p. 336). However, there are firm grounds for believing that the Piperales are derivatives of the same ancestors as the Annonales, Magnoliales, Laurales, etc.; these are the shared occurrence of essential oils, cellular endosperm formation, and the rare but now clear evidence for benzylisoquinoline alkaloids in *Piper* (HÄNSEL et al. 1975). Both the herbaceous and the climbing habit are found in both Piperales and Arales, and the orders are also similar in their leaf shapes (including occurrence of reticulate venation), the often random orientation of stomata, which moreover are often paracytic, the spadiciform inflorescence, the occasional occurrence, in Saururaceae, of white spathes, the similarly reduced flowers (in Piperales always naked, however), the occurrence of the Monocotyledonous Type of anther wall formation, the occurrence of sulcate pollen grains, the lack of a distinct style (so that the stigmatic lobes or papillae are placed directly on the top of the ovary), the occurrence of cellular endosperm formation, and the baccate fruit. These similarities between Piperales and Arales are quite as striking as those between some Nymphaeales and some Alismatales and if, as we have concluded (see Evolutionary Concepts, Classification), the relationship is nevertheless rather remote, the similarities must be mainly due to convergence. We believe that Arales are derived from primitive monocotyledons similar to Dioscoreales, and Piperales from primitive Magnoliiflorae similar to Annonales and Magnoliales, both groups independently adapted primarily to a life in tropical rainforests.

## The Hypothesis of Monocotyledons as Ancestral Angiosperms

Totally opposed to the above arguments and conclusions are those presented for the derivation of the monocotyledons by BURGER (1981). Earlier, BURGER (1977) had sketched an evolution of monocotyledons from Chloranthaceae-like ancestors (note that the first pollen grains of undoubtedly angiospermous nature, *Clavatipollenitis,* are nearly indistinguishable from those in the extant genus *Ascarina* of Chloranthaceae; MULLER 1981). BURGER's latest hypothesis is much more radical than that.

He now supposes that the first angiosperms were small monocotyledon-like plants, and that the monocotyledons, as a lineage, are more ancient than the dicotyledons. The most primitive angiosperms were small, simple, herbaceous plants with linear leaves with mainly parallel venation or at least venation of a few orders. They had scattered vascular bundles in the stem and simple leafy stems without aerial branching. The leaves clasped the stem and their base was continuous with the stem. These plants are supposed to have evolved from very primitive pteridosperms or paedomorphically reduced cycadophytes. Like many earlier morphologists (e.g. CAMPBELL 1930) BURGER notes the strong similarity in embryo shape between monocotyledons and those pteridophytes in which there is a solitary and terminal first leaf (or cotyledon).

BURGER found the new hypothesis to be more useful in interpreting early monocotyledon-like fossils (e.g. *Sanmiguelia*), even from the Triassic Period, as early angiosperms.

With regard to the reproductive structures BURGER (1981) has not developed his theory beyond that of his earlier work. However, we can undoubtedly apply the similarities between groups of monocotyledons and dicotyledons outlined above almost as well when deriving the dicotyledons from monocotyledons as the reverse. Then the assumptions of what is ancestral versus derived are generally totally reversed, too, but our out-group comparison (i.e. the comparison with dicotyledons) is no longer possible. Nonetheless some of the above remarks are still relevant, for example, the possible bases for considering Nymphaeales and Piperales as sister-groups and derived from magnoliifloran ancestors, even though it is also justifiable to discuss the chances of their being derived from the ancestors of the Alismatiflorae and the Ariflorae.

If one considers a possible transition from monocotyledons to dicotyledons, the "bridge" Dioscoreales-Magnoliiflorae is still relevant. In cases where there is evidence for an evolution of one group from another, it is a general feature that the derived group becomes constant in important features even where an explosion in diversification occurs, as in the Poaceae (Poales), Orchidaceae (Liliales) and the dicotyledon groups Fabales (Leguminosae) and Asteraceae (Compositae). With the dicotyledons as such, the new constant features would be the two lateral cotyledons, and the formation of a cambium between the phloem and xylem in a single ring of vascular strands.

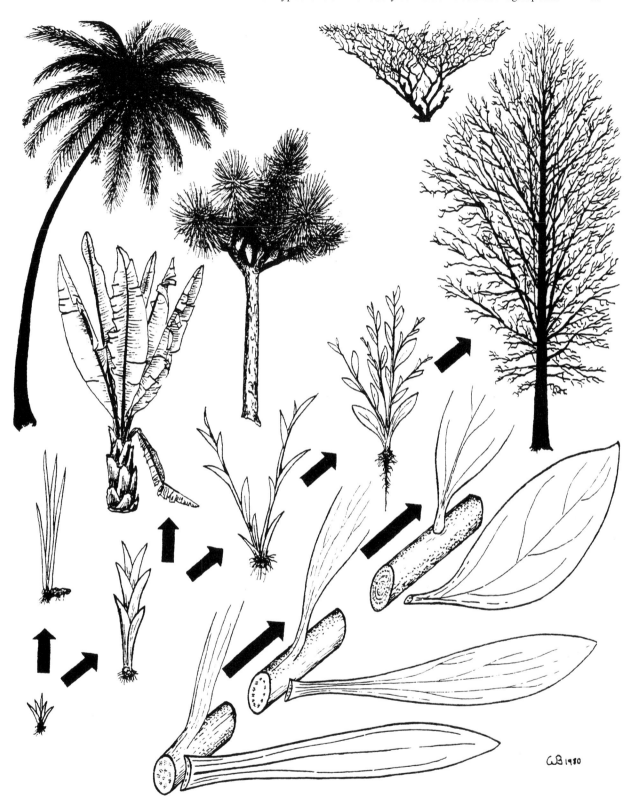

**Fig. 22.** A diagrammatic outline of the major trends in the evolution of the angiosperm plant body and leaf according to BURGER (1981). It is assumed by Burger that the first angiosperms were monocotyledonous short-stemmed plants with linear leaves, which gradually became more specialized and evolved a eustelar stem with secondary thickening growth at the time they evolved two cotyledons in place of one.

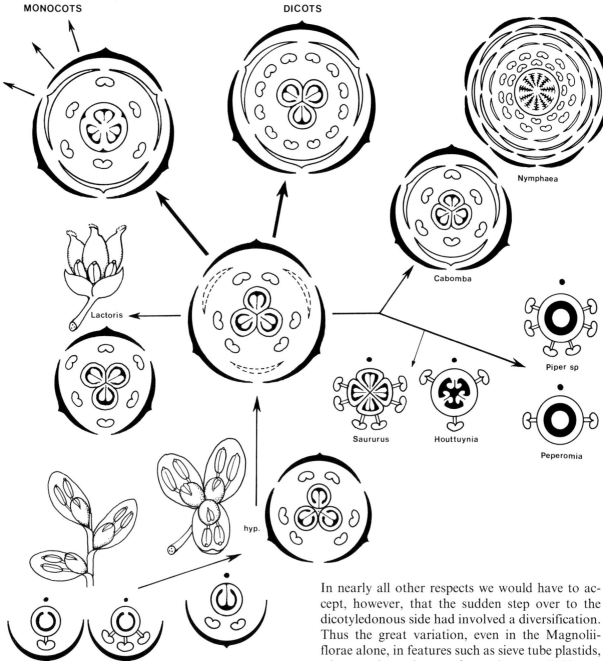

**Fig. 23.** Possible features of early angiosperm evolution when axes of the chloranthaceous type, with spirally set units (of one tepal, two stamens and a carpel), are taken as components which, when brought together by condensation would appear as a flower (*hyp*). Further differentiation of this type could have given rise to various magnoliifloran dicotyledonous groups (*dicots*) including *Lactoris,* to the common ancestor of the sister-groups Nymphaeales and Piperales, and to the monocotyledons. (Inspired by, but greatly changed from BURGER 1977)

In nearly all other respects we would have to accept, however, that the sudden step over to the dicotyledonous side had involved a diversification. Thus the great variation, even in the Magnoliiflorae alone, in features such as sieve tube plastids, where various shapes of protein crystalloids are represented, and also lack of them, contrasts sharply with their relatively uniform shape in the monocotyledons. As regards the vascular system there is likewise an enormous range of variation from lack of vessels to advanced vessels with simple perforation. The flowers vary in merism and may be apocarpous (frequent), syncarpous (Canellaceae, some Annonaceae) or monocarpellary.

The great variation in these characters is not what would be expected if the dicotyledons had evolved from already rather advanced monocotyledons. Rather the uniformity of floral symmetry, the constant lack of a cambium and the relative uniformity in sieve element plastid inclusions in the monocotyledons are comparable to the stable character conditions found in other large groups supposed to be monophyletic and evolved as secondary branches from ancestral angiosperms. In the examples within the dicotyledons and monocotyledons already cited, together with Ranunculiflorae and Caryophylliflorae (Caryophyllales only), the diversification is superimposed on a basic pattern which is fixed in important respects.

This plasticity (variability) of the dicotyledons, and even of the Magnoliiflorae, in combination with the weak evidence provided by the general type of pollen grains and the woody growth comparable to that in fossil gymnosperms that occur in the Magnoliiflorae justify our preference for regarding the monocotyledons as a derivative branch of early dicotyledonous ancestors.

# Fossils

Fossils contribute little to the determination of evolutionary courses of the monocotyledons.

The first palynomorph (pollen form type) that can almost certainly be ascribed to the monocotyledons has been described as *Liliacidites*. It possesses many characteristics of extant liliifloran pollen grains. *Liliacidites* appeared as early as the Albian (ca. 115 million years ago), fairly early in the Cretaceous. It is an elongate, monosulcate, coarsely reticulate pollen type with finer reticulation at the ends. Pollen types that are matched by those found in Restionaceae, Arecaceae and Pandanaceae are known from the end of the Cretaceous, in the Maestrichtian. The probably restionaceous pollen type is ulcerate with an irregular aperture margin and a scrobiculate surface, and the probably pandanaceous pollen grains are sulcate, with an echinate surface and with two indistinct layers (MULLER 1981). It is important to note that of the pollen ascribed to the palms three kinds occurred in the Maestrichtian. These correspond to those of extant *Nipa* (meridionosulcate), *Areca* (sulcate) and genera of Cocosoideae (trichotomosulcate).

From this we infer that the sulcate pollen type was the first to dominate in the monocotyledons. These were differentiated already in the Albian, ca. 115 million years ago, whereas the first pollen grains which almost certainly belonged to angiosperms, viz. *Clavatipollenitis,* are known from Barremian-Aptian, about 122 million years ago, a rather insignificant difference in time.

Leaf impressions from Aptian-Albian times are not of distinctly monocotyledonous type. HICKEY and DOYLE (1977) regard impressions classified as "*Acaciaephyllum*" as possibly monocotyledonous (Fig. 24), but if the first monocotyledons had broad, petiolate leaves of dioscoreaceous shape, other impressions could also be considered possibly monocotyledonous. Some plicate, palm-like leaves from the late Triassic Colorado clay, described as *Sanmiguelia* (see DAGHLIAN 1981) are here considered too dubious to be accepted as belonging to monocotyledons.

Of floral impressions from the early half of Cretaceous some six-lobate structures from the Dakota formation (Albian to Cenomanian) could be considered as possibly monocotyledonous, but it is more likely that they were of pre-Magnoliifloran affinity (see DILCHER 1979).

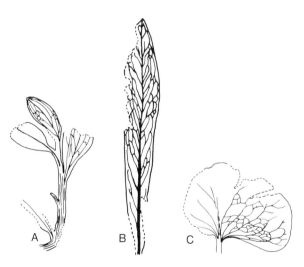

**Fig. 24.** Fossil angiosperm leaves from Zone I of the Potomac Group, dated to Barremian – early Albian of Cretaceous. **A** *Acaciaephyllum spatulatum,* branchlet with a few leaves. **B** *Rogersia angustifolia.* **C** *Proteaephyllum reniforme.* Some of these, at least **A**, may derive from monocotyledonous plants. (All from DOYLE and HICKEY 1976)

# Distribution of Character Conditions

## General

An extensive study of this topic in the monocotyledons has recently appeared (DAHLGREN and CLIFFORD 1982). Further, DAHLGREN and RASMUSSEN (1983) have made an assessment of which character states are ancestral and which derived in different groups of monocotyledons. Extensive comparative studies in the "Farinosae" (Zingiberiflorae; Commeliniflorae pro parte) were made earlier by HAMANN (1961, 1962b).

DAHLGREN and CLIFFORD (1982) displayed the distribution of each character condition or "attribute" on a "bubble-diagram" (cf. Figs. 25 and 26), representing the major groups of the mono-

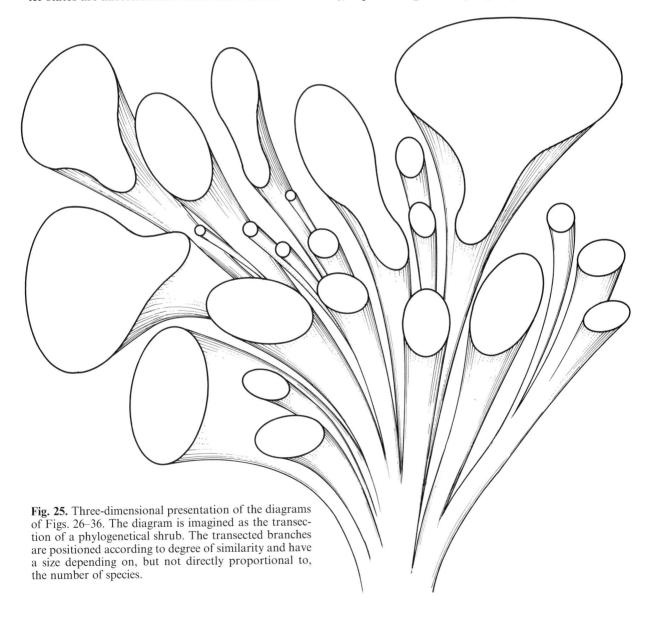

**Fig. 25.** Three-dimensional presentation of the diagrams of Figs. 26–36. The diagram is imagined as the transection of a phylogenetical shrub. The transected branches are positioned according to degree of similarity and have a size depending on, but not directly proportional to, the number of species.

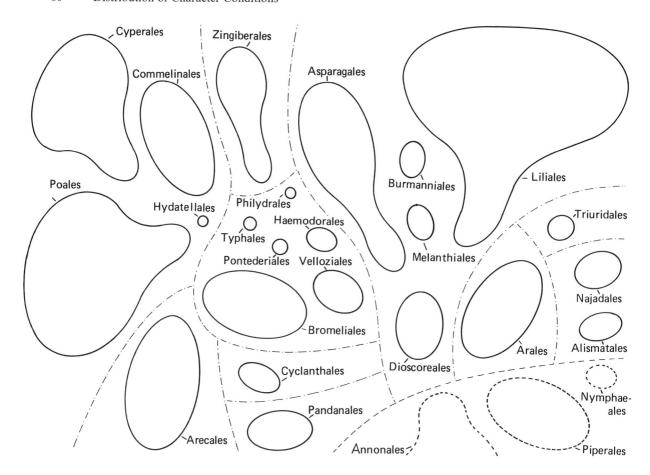

Fig. 26. A spatial diagram to illustrate the supposed phylogenetical relationships between the orders of the monocotyledons as circumscribed here. This diagram differs from those in DAHLGREN and CLIFFORD (1982) in various mostly unimportant respects. The relationships of Arecales, Pandanales and Cyclanthales are uncertain and the positions of their ordinal bubbles are provisional.

cotyledons. The assessment of the phylogenetic significance of the distributions of character conditions is always hampered by the possibility of convergence (see Evolutionary Concepts, Classification). The characters which appear to be of most use in phylogenetic studies are briefly surveyed below.

Explanations of terms are to be found in Morphological Concepts, unless otherwise indicated.

## Vegetative Morphology

*Bulbs* occur in most taxa of Hyacinthaceae, Alliaceae and Amaryllidaceae (Asparagales) and in Liliaceae and Calochortaceae and a few genera of Iridaceae (Liliales) as well as in some genera of Melanthiaceae (Melanthiales). *Corms* are the commonest underground organs in Iridaceae and Colchicaceae (Liliales) and also occur in orchids ("pseudobulbs"), Tecophilaeaceae, Cyanastraceae, Ixioliriaceae and some Alliaceae (Asparagales) and some Melanthiaceae (Melanthiales).

*Hypocotylar tubers,* finally, occur in many Dioscoreaceae and Taccaceae (Dioscoreales) and Araceae (Arales). (Rhizomes are widely distributed). Whereas most monocotyledons are herbaceous, quite a number of taxa are *shrubby or tree-like,* although it seems that a woody trunk has evolved independently in a number of evolutionary lines. Shrubby forms are, for example, common among the berry-fruited families of Asparagales, and large plants with a woody trunk are found in a number of other families in the same order, these families occurring in the Americas (Nolinaceae, Agavaceae), Africa (Asphodelaceae), Asia (Dracaenaceae), and Australia (Xanthorrhoeaceae, Dasypogonaceae). Similar forms also occur in Iridaceae (*Nivenia* etc.) of Liliales, Bromeliales (*Puya*), Vel-

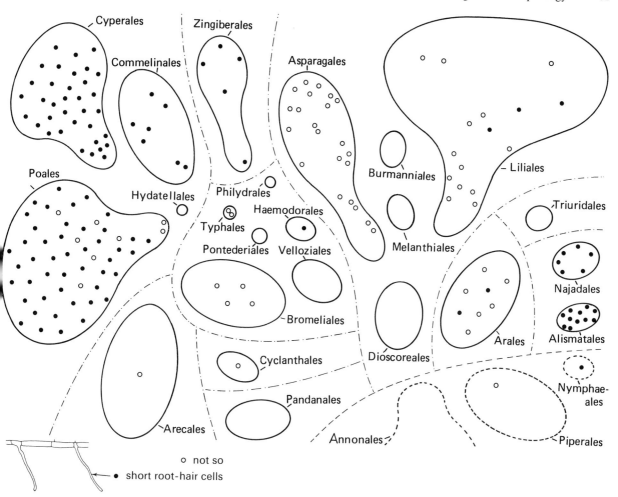

o not so

• short root-hair cells

**Fig. 27.** Distribution in the monocotyledons of root-hair epidermal cells which are much shorter than other epidermal cells and root-hair cells which are of the same size as other epidermal hairs. Each *dot* or *circle* represents an investigated species. (see DAHLGREN and CLIFFORD 1982)

loziales (*Vellozia*) and Cyperales (*Microdracoides*). In none of these orders does the woody trunk seem to be related to that of the palms (Arecales) or the bambusoid grasses (Poales). *Secondary thickening* growth in woody forms is restricted to some Asparagales and to a few genera of Iridaceae (Liliales); it takes a different form from that found in all dicotyledons.

*Phyllotaxy* is often of significance in monocotyledons. Thus nearly all Cyperales (including Juncales) have tristichous leaves and most grasses (and certain other Poales) have distichous leaves. Distichous leaves are also the commonest in most Bromeliiflorae other than Velloziales and Bromeliales, and are common in Zosterales in Alismatiflorae. In Arales and Arecales the leaves are only

rarely distichous. If seedlings are considered, distichy is even more widespread, many polystichous species having distichous juvenile foliage.

*Opposite and verticillate leaves* occur rarely in Dioscoreales (several families, Fig. 46A, Fig. 50A), Convallariaceae (Asparagales), Liliaceae (Liliales) and a few other families.

*Petiolate leaves with a broad leaf blade* are often associated with *reticulate venation*. Such leaves are common in the dioscorealean families, in Araceae (Arales), Smilacaceae, a few mainly berry-fruited families in Asparagales, Uvulariaceae (Liliales), a few Orchidaceae (Liliales) and Melanthiaceae (Melanthiales), etc.

*Digitately compound or palmate leaves* are restricted to Dioscoreaceae, Taccaceae (Dioscoreales) and Araceae (Arales).

*Ensiform leaves* occur in several presumably distantly related families. They are typical of Philydraceae (Philydrales), most Iridaceae (Liliales) and many Haemodoraceae (Haemodorales) and occur more rarely in Melanthiaceae, Juncaginaceae, Araceae, Restionaceae, Xyridaceae, Juncaceae and

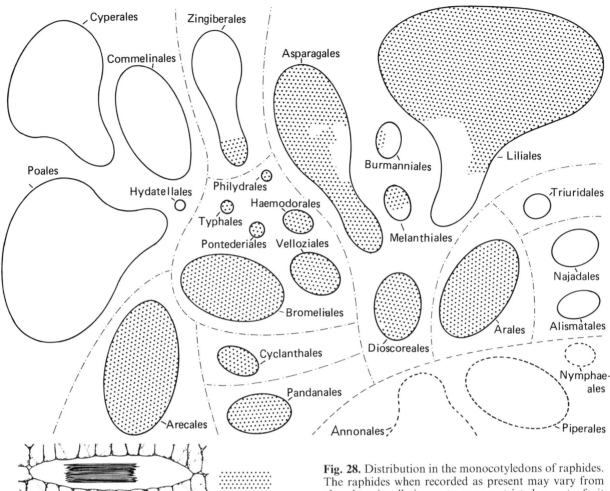

oxalate raphides

**Fig. 28.** Distribution in the monocotyledons of raphides. The raphides when recorded as present may vary from abundant in all tissues to very restricted, e.g. to fruit walls, tapetal layers, etc.

Orchidaceae. *Inverted leaf blades* are prevalent in the Southern Hemisphere families Alstroemeriaceae and Luzuriagaceae. They also occur in a few Poaceae with pseudopetiolate leaves.

*Stipules* occur in most Potamogetonaceae and some Hydrocharitaceae; in a number of families (Juncaginaceae, Joinvilleaceae, Juncaceae, Zingiberaceae, Pontederiaceae) there are membranous extensions ("stipules") from the leaf sheath which resemble ligules, the latter being typical of grasses. Similar structures occur in some palms (Arecales).

*Ptyxis* (CULLEN 1978) is highly variable in monocotyledons and is undoubtedly of phylogenetic significance. For example, *supervolute leaves* are found in Dioscoreales, Arales, broad-leaved Alismatales and certain Liliales and Zingiberales, while *plicate* and *conduplicate-plicate leaves* characterize Arecales, Cyclanthales and some Poales.

## Vegetative Anatomy

Lateral *roots* in monocotyledons generally arise *opposite the xylem strands* in the stem. However in many Commeliniflorae they may also arise *opposite the phloem* (VAN TIEGHEM 1887). *Root hairs* may develop from any root epidermal cells or from special short cells (Fig. 27). The latter condition is the commonest in Commeliniflorae and Alismatiflorae, and is also found in some orchids (Liliales) and Araceae (Arales), while otherwise the root hairs come from epidermal cells similar to other epidermal cells.

*Vessels* are absent from the stems in all Ariflorae, Triuridiflorae and Alismatiflorae and also in many families of Liliiflorae and Zingiberiflorae (e.g. Zingiberaceae) and all Cyclanthiflorae. Although vessels are generally present in roots they are lack-

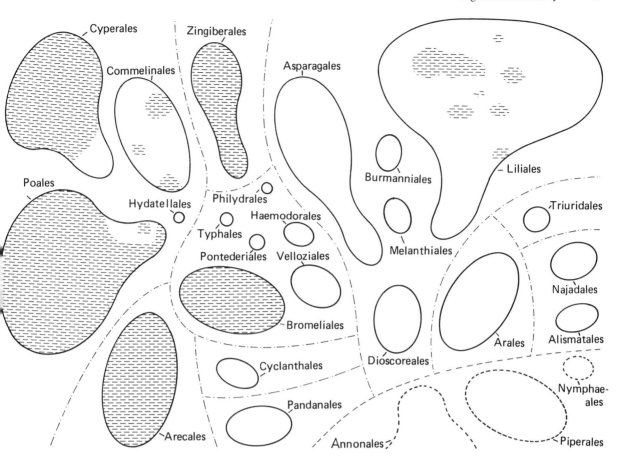

**Fig. 29.** Distribution in the monocotyledons of silica bodies (presence shaded). The silica bodies may occur in different tissues (epidermis, pericycle, etc.) and are of widely different size, shape and number. Presence in itself is therefore of limited importance. The presence of silica bodies is believed to have originated in several independent evolutionary lines.

ing at least in various Alismatiflorae. When vessels are present, the *perforation plates* are generally simple or simple and scalariform in Poales, Commelinales and many Cyperales of Commeliniflorae and in Areciflorae, but in Liliiflorae, Zingiberiflorae, Bromeliiflorae and certain Commeliniflorae the perforations tend to be scalariform (see WAGNER 1977; CHEADLE and KOSAKAI 1980, 1982).

*Laticifers* in monocotyledons are restricted to Aponogetonaceae (Alismatales), many Araceae (Arales), certain Alliaceae (Asparagales), Musaceae (Zingiberales) and Cyclanthaceae subfam. Cyclanthoideae (Cyclanthales). *Sieve tube plastids* (BEHNKE 1981) in monocotyledons are exclusively of the PIIc type containing cuneate protein bodies, a type that in dicotyledons is restricted to the gen-

era *Asarum* and *Saruma* of Aristolochiaceae (BEHNKE 1971; BEHNKE and BARTHLOTT 1983). This is of extraordinary significance as a piece of evidence that the monocotyledons are a monophyletic group. In Zingiberales the sieve tube plastids accumulate starch in addition to protein, a condition met with also in Dioscoreaceae and Araceae.

*Oxalate raphides* (Fig. 28) are widely distributed in monocotyledons but are absent from all Commeliniflorae except (nearly all) Commelinaceae, from all Triuridiflorae and Alismatiflorae, and from other groups such as the zingiberalean families with fewer than five functional stamens, most families of Liliales and certain families of Asparagales. *Silica bodies* (Fig. 29) occur in many families, but to judge from the diversity of shapes and their occurrence in the tissues have presumably evolved independently in several lines. Silica bodies are mostly conical in Cyperaceae (Cyperales), whereas other shapes are found in most Poales (they are lacking in Centrolepidaceae and many Restionaceae), Arecales, Zingiberales, Bromeliales and in scattered genera of Commelinales and Orchidaceae (Liliales).

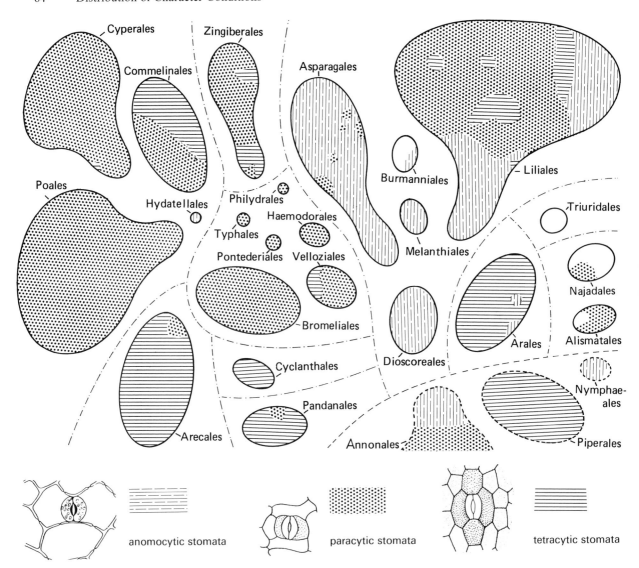

anomocytic stomata          paracytic stomata          tetracytic stomata

**Fig. 30.** Distribution in the moncotyledons of different types of mature stomatal complexes. (See DAHLGREN and CLIFFORD 1982). It should be noted that the significance of the type of mature stomatal types is not always phylogenetically great, because the same kind of ontogeny may lead to different "phenotypes" of mature stomata.

The number and arrangement of *subsidiary cells in stomatal complexes* (Fig. 30) shows certain regularities. Thus the families of Commeliniflorae, Bromeliiflorae and Zingiberiflorae generally have paracytic stomata (Commelinaceae and certain Zingiberiflorae having tetracytic stomata), most Arecales, Pandanales, Cyclanthales and Arales tetracytic stomata, and most Liliiflorae outside the variable Orchidaceae anomocytic stomata. However, H. RASMUSSEN (in preparation) has shown that for various Liliiflorae the mature stomatal complex has little correspondence with its ontogeny; for example perigenous stomatal formation may result in either anomocytic or paracytic stomata.

*Intravaginal squamules* (VON STAUDERMANN 1924) are restricted to the Alismatiflorae, where their presence is a general feature, and to certain Ariflorae.

*Epicuticular wax structures* (BARTHLOTT, in BEHNKE and BARTHLOTT 1983; BARTHLOTT and FRÖHLICH 1983) (Fig. 31) show some interesting patterns in monocotyledons. The epicuticular wax can be (1) unsculptured (2) sculptured, but the wax crystalloids without regular orientation (3) sculptured with scale-like crystalloids arranged in rows (*Convallaria* Type) or (4) sculptured, the wax crys-

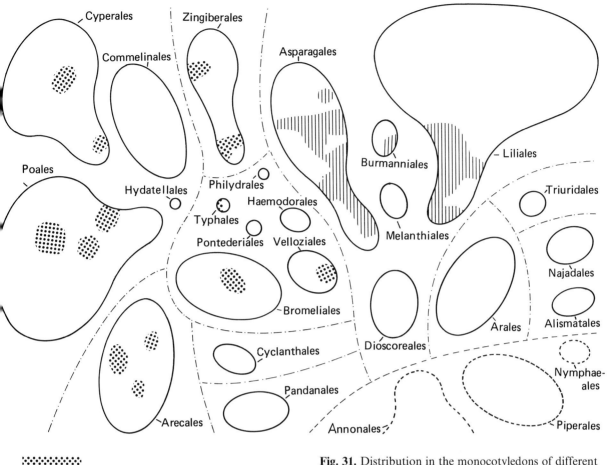

Strelitzia type

Convallaria type   epicuticular wax

**Fig. 31.** Distribution in the monocotyledons of different types of oriented epicuticular wax. Non-sculptured and sculptured but irregularly oriented wax is scattered in monocotyledons; when oriented the wax is of two types with very clear distributions. (Data after BARTHLOTT and FRÖHLICH 1983)

talloids aggregated into rod-like, often massive projections (*Strelitzia* Type). The *Convallaria* wax type is restricted to families (not all) in Asparagales, Liliales and Burmanniales, all in the Liliiflorae, whereas the *Strelitzia* wax type is known in members of Velloziales, Bromeliales and Typhales (Bromeliiflorae), in Zingiberales (Zingiberiflorae), in Juncales, Cyperales and Poales (Commeliniflorae) and in Arecales (Areciflorae).

## Floral Morphology

The *floral symmetry* in monocotyledons is nearly always derivable from a *basic actinomorphic type* with 3 + 3 tepals, 3 + 3 stamens and 3 carpels. *Zygomorphic flowers* are mainly concentrated in ani-mal-pollinated flowers with a showy perianth (or androecium, as in many Zingiberales). *Asymmetric flowers* occur in the zingiberalean Marantaceae and Cannaceae, where the tepals are less conspicuous than the petaloid staminodes and where there is only one functional theca. Zygomorphy often occurs in families and even genera where some members have actinomorphic flowers, and in these groups there is often a gradual transition between actinomorphic and zygomorphic flowers (*Aloë, Chlorophytum,* Fig. 84A). In some families, such as Apostasiaceae, Cypripediaceae, Orchidaceae, Corsiaceae, Philydraceae, Lowiaceae, Musaceae, Heliconiaceae, Strelitziaceae, Zingiberaceae and Costaceae the perianth is strongly zygomorphic, and in several of them this is combined with reduction in the number of stamens. Zygomorphy also occurs in many other families, such as Iridaceae and Commelinaceae and also in other families in Liliiflorae, Bromeliiflorae and Commeliniflorae apart from those already mentioned.

All six tepals are generally more or less petaloid in most Liliiflorae, Bromeliiflorae and Zingiberiflorae, although there are exceptions in each of these groups (as in Dasypogonaceae of Asparagales and Typhaceae of Typhales). A condition where the outer tepal whorl is *sepaloid* and the inner *petaloid* is met with in most members of Commelinales and in many Alismatales. In Liliiflorae, Bromeliiflorae and Zingiberiflorae it is quite common for the two perianth whorls to be petaloid but dissimilar; in certain Iridaceae (subfam. Iridoideae) and certain Orchidaceae, for example, they may differ conspicuously, and single perigone members such as the "labellum" in most orchids may be very different from others. *Fusion of* upper or lower *tepals* in a flower may also occur, as in Philydraceae. Fusion of tepals, either all six or only three, is also a condition that has appeared independently in various Liliiflorae and Zingiberiflorae in particular.

*Reduction of the tepals* to bract-like, often minute, scales, bristles or hairs, or total loss of tepals has occurred in various taxa in connection with adaptation to wind pollination or water pollination (e.g. Poales, Cyperales, Zosterales). As will be seen below, nectaries are also missing in flowers which are not animal-pollinated. In wind-pollinated flowers the filaments of the stamens are generally long and often flaccid, so that the anthers are pendulous, and in these flowers the stigmatic branches are often long and may have plumose stigmatic papillae.

In some of the insect-pollinated Liliiflorae, mainly in those that we place in the order Liliales, the tepals tend to *have a variegated or spotted colour pattern*. The pattern may be chequered, often in dark purple and white, or it may assume a "drop-like" or striate pattern which functions as a more obvious nectar guide to pollinators. With few exceptions the taxa with this type of pattern on the tepals *have nectaries in pockets at the base of the tepals*. The taxa with septal nectaries (see below) only very rarely have variegated tepals.

*Epigynous flowers* occur in groups in which animal pollination prevails. This condition has obviously evolved independently in a number of lines. The epigynous groups are the orders Zingiberales, Velloziales and Burmanniales, the families Orchidaceae, Cypripediaceae, Apostasiaceae, Iridaceae (except *Isophysis*), Geosiridaceae and Alstroemeriaceae in the Liliales, Campynemaceae in Melanthiales, the families Amaryllidaceae, Hypoxidaceae, Ixioliriaceae, Doryanthaceae, and (semiepigyny:) Tecophilaeaceae and Cyanastraceae in the Asparagales, the families Petermanniaceae, Dioscoreaceae and Taccaceae in Dioscoreales, and the Hydrocharitaceae in the Alismatales. Further taxa with epigynous flowers occur in other families where there are otherwise hypogynous or semiepigynous flowers.

It has been mentioned above that perigonal nectaries occur in most Liliales. An exception to this is Iridaceae subfam. Ixioideae with nectaries in the septa of the ovary, "septal nectaries". *Septal nectaries* (DAUMANN 1970) are quite common in insect-pollinated monocotyledons outside Liliales, and are recorded in the liliifloran orders Dioscoreales, Asparagales, Liliales, Melanthiales, Burmanniales, Arecales and in all bromeliifloran orders except Philydrales and Typhales and in most Zingiberiflorae, Zingiberaceae, however, having epigynous nectar glands which may or may not be gynoecial (Fig. 166I). In the apocarpous Alismatales nectaries on the sides of the separate carpels also seem to be homologous with septal nectaries.

A condition with *3 + 3 stamens* is found in at least some taxa of most orders of the monocotyledons, although a lower number is also frequent. Numerous stamens occur in many Alismatales, all Pandanales and Cyclanthales, several different groups of palms (Arecales) and in a few genera of other families (*Ochlandra,* Poaceae; *Vellozia,* Velloziaceae; *Gethyllis,* Amaryllidaceae); the condition seems to be secondary. Flowers with only three stamens are common in Poales and Cyperales (in each of which there are also taxa with 3 + 3 stamens), in many Dioscoreales, Pontederiales, Haemodorales, Liliales (e.g. all Iridaceae), Arales, etc. Further reduction, to two stamens or one, is typical of Cypripediaceae, Orchidaceae, Philydraceae, Hydatellaceae, Centrolepidaceae, Najadaceae, etc. In certain groups, e.g. in most Zingiberales, all but one of the stamens are missing or transformed into petaloid staminodes, and the one functional anther is sometimes monothecous.

The stamen morphology shows some interesting features. In several genera of the Dioscoreales, e.g. *Trichopus, Stenomeris* (Fig. 43C), *Stemona* (Fig. 46D), *Paris* (Fig. 50D) and *Tacca, the microsporangia are situated considerably below the apex of the stamen,* leaving the latter as a variously shaped "connective appendage", a feature common in the dicotyledonous Magnoliiflorae. This condition is also met with in various families, such as Najadaceae, Cyperaceae and Pandanaceae, belonging to other orders.

The microsporangia of the anthers are directed away from the centre of the flower, i.e. they are *extrorse,* in most Ariflorae, Alismatiflorae and Triuridiflorae. In many lilialean families they are directed tangentially, i.e. they are *latrorse,* as in most Poaceae (Poales), Arecaceae (Arecales), Cyclanthaceae (Cyclanthales) and Juncaceae (Cyperales). In most other monocotyledons the microsporangia face the centre of the flower, i.e. they are *introrse*. Sometimes there are fewer than four microsporangia per anther. Thus the monothecous anthers of Marantaceae, Cannaceae, Centrolepidaceae and most Restionaceae are bisporangiate. Bisporangiate anthers also occur in dithecous anthers (i.e. each theca is unisporangiate) of some Eriocaulaceae and in Smilacaceae, some Hydrocharitaceae and in isolated genera of a few other families. Although in most monocotyledons the anthers dehisce by longitudinal slits, in some groups they open apically, as in many Araceae. Dehiscence by *apical pores* occurs also in Luzuriagaceae, Cyanastraceae, Calectasiaceae, Tecophilaeaceae, several Dianellaceae, all Mayacaceae and Rapateaceae, some Commelinaceae and a few Amaryllidaceae (*Leucojum, Galanthus*).

The *carpels* are usually three in number in the monocotyledons and are generally syncarpous, i.e. fused with each other, at least in the ovary region. A condition with *free carpels* (apocarpy) is met with in the Alismatiflorae (some Juncaginaceae excepted), the Triuridiflorae and a number of palm genera (Areciflorae). Our theory that this may not be the ancestral condition in monocotyledons is discussed on p. 83. Carpels which are also free in the ovary region occur in some Melanthiaceae. Free stylodial branches occur in most Colchicaceae, many Poaceae, some Dioscoreaceae, some Arecaceae, Cyperaceae, etc., though in these families they may also be basally fused to form a style. In certain groups, such as many Arecaceae, Cyclanthaceae, Pandanaceae and Araceae, the stigma rests almost directly on the ovary, the style or stylodia being virtually suppressed.

The *stigmatic surface* is very variable in some groups of monocotyledons, e.g. the Liliiflorae (HESLOP-HARRISON 1982). It is as yet insufficiently known but may be useful in phylogenetic considerations in the future. The stigmatic surface may or may not be covered with a sticky secretion during the receptive phase, and is accordingly classified as "Wet" or "Dry" respectively (HESLOP-HARRISON and SHIVANNA 1977), the dry condition being by far the more common in monocotyledons. However, Wet stigmas occur in all Zingiberales, many Bromeliales and some Haemodorales in the Zingiberiflorae, in all Orchidaceae and in scattered genera of other Liliales and of Asparagales in the Liliiflorae, and in various Commelinaceae. Two liliifloran families in which all examples studied have dry stigmas are Iridaceae and Convallariaceae.

## Microsporangia, Pollen

*Anther wall formation* in monocotyledons nearly always seems to proceed according to the *Monocotyledonous Type*. According to this only the cells of the inner of the two parietal layers in the anther wall divide further and give rise to an inner tapetal layer and a middle layer. *Tacca* is reported to be an exception (having "dicotyledonous" anther wall formation) but this needs to be verified. Further exceptions are *Lemna* and *Wolffia* of Lemnaceae and *Najas* of Najadaceae, which have the "Reduced" Type of anther wall formation in which the cells of neither parietal layer divide further, the inner one functioning directly as tapetum.

The *endothecial layer* of the anther wall has thickened cell walls. The variations in the thickenings may be reduced to two main types: with a *Spiral* or a *Girdle-like* configuration, the former being the commoner in monocotyledons. Girdle Type endothecial thickenings occur in most Poales, many Cyperaceae (Cyperales), most Commelinaceae (Commelinales) as well as in Pontederiaceae (Pontederiales), some orchids (Liliales), some Alismatales and in a few other groups.

The *tapetal layer* (Fig. 32) may be classified either as *glandular-secretory* or *amoeboid-plasmodial*. The former type is the more common in monocotyledons, the Amoeboid-Plasmodial Type being restricted to Ariflorae, Triuridiflorae, Alismatiflorae, some orders in Bromeliiflorae, namely Haemodorales, Pontederiales (pro parte) and Typhales, Zingiberiflorae (rare), and some Commeliniflorae, namely Commelinaceae and the genus *Abolboda* (Xyridaceae) in Commelinales.

*Microsporogenesis* is of the *Successive Type* in most monocotyledon families, though the *Simultaneous Type* is found in some large complexes (Fig. 33): Dioscoreaceae (Dioscoreales), the orchid complex and Iridaceae (Liliales), most Arecaceae (Arecales) and Cyperales, but also in various

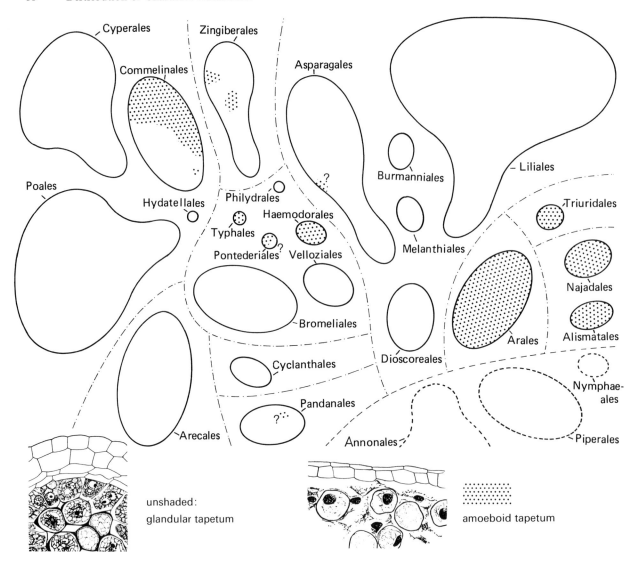

Fig. 32. Distribution in the monocotyledons of amoeboid-plasmodial and glandular-secretory types of tapetum. (see DAHLGREN and CLIFFORD 1982)

other, probably closely related families, namely Doryanthaceae, Phormiaceae, Asphodelaceae, Tecophilaeaceae, Cyanastraceae, Ixioliriaceae and Hemerocallidaceae (Asparagales) and probably Rapateaceae (Commelinales).

*The pollen grains are dispersed in the bicellular or the tricellular condition,* the bicellular being far commoner than the tricellular which, however, is common in Alismatiflorae and Triuridiflorae. It also occurs in Araceae (Ariflorae) and is widely distributed in Commeliniflorae except Commelinales, where both conditions can be found (BREWBAKER 1967).

The pollen grains are dispersed singly or in *tetrads,* the latter condition being restricted to Cyperales (all three families), Orchidaceae (Liliales) and a number of genera in each of the bromeliifloran families Velloziaceae, Typhaceae, Philydraceae and Bromeliaceae and in Araceae, in none of which pollen tetrads are universal.

*The aperture characters* (ERDTMAN 1952) are variable, the *sulcate* type being the most widely distributed, and probably the ancestral one in the monocotyledons. Sulcate pollen grains occur in most families of Liliiflorae and Bromeliiflorae, in many Commelinales and in Hydatellales of Commeliniflorae, in many Arecales (Areciflorae) and some Cyclanthales (Cyclanthiflorae) as well as in a few Arales (Ariflorae) and a few Alismatales (Alismatiflorae). *Trichotomosulcate* pollen grains occur in many palms (Arecaceae), in Phormiaceae and

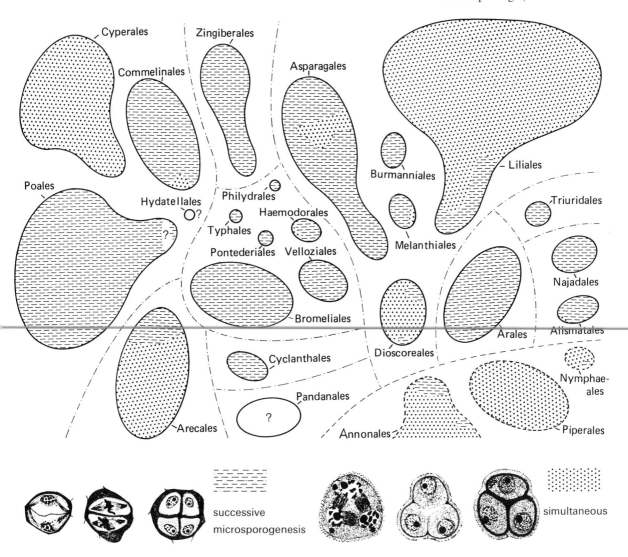

Fig. 33. Distribution in the monocotyledons of the successive and simultaneous types of microsporogenesis. (see DAHLGREN and CLIFFORD 1982)

some members of the tribe Johnsonieae of Anthericaceae, but are otherwise rare. *Ulcerate* pollen grains are found in all Poales and most Cyperales (where also foraminate pollen grains occur), and in all Pandanales and Typhales (all of which are wind-pollinated plants), but are otherwise rare in the monocotyledons. *Foraminate* pollen grains, with few to numerous circular apertures, are typical of Alismataceae (Alismatales) and are found in some members of Araceae, Dioscoreaceae (*Avetra*), Melanthiaceae, Haemodoraceae, Bromeliaceae and Cyperaceae, but are otherwise rare. *Sulculate* pollen grains occur in Dioscoreaceae, in some Dasypogonaceae etc., but are rare. The *Zonisulculate* Type, which is undoubtedly related to the Sulculate Type, is very rare and restricted to some genera of Rapateaceae and Dasypogonaceae. *Spiraperturate* pollen grains occur primarily

in the large family Eriocaulaceae and in Aphyllanthaceae. In some families the exine is very thin, and in some it is wholly absent or reduced to small spinules on the surface; this is the case in the *omniaperturate* and *inaperturate* types, either of which occurs in most orchids (Liliales), nearly all Zingiberales, all Najadales and Triuridales, many Arales as well as in a few other families (Smilacaceae, Velloziaceae pro parte, Hydrocharitaceae pro parte, etc.).

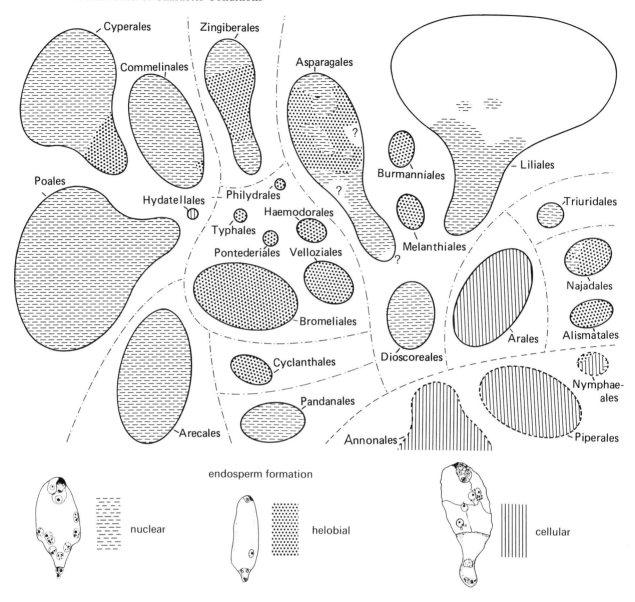

**Fig. 34.** Distribution in the monocotyledons of the Nuclear, Helobial and Cellular Types of endosperm formation. (see DAHLGREN and CLIFFORD 1982)

## Ovular Characters

In Alismatiflorae, where the carpels are generally solitary or separate from each other, the *placentation* is laminar-dispersed basal, apical or rarely lateral. *Laminar-dispersed* placentation occurs in the alismatalean Butomaceae, Limnocharitaceae and some Hydrocharitaceae (where the carpels are enclosed by the floral receptacle). Placentation is *basal* in Aponogetonaceae and Alismataceae (Alismatales) and in Scheuchzeriaceae, Juncaginaceae and Najadaceae (Najadales), as well as in the carpels of the Triuridiflorae. *Apical* placentation occurs in the other families of Najadales except for some Potamogetonaceae, where it is lateral. In Ariflorae the placentation varies between basal and apical.

Here the carpels vary from one to three in number; if there are two or three they are fused.

In the other superorders, which generally have syncarpous gynoecia, the ovules when numerous have *central* (axile) or *parietal* placentation. The latter condition prevails in relatively few families, namely Orchidaceae (Liliales), Burmanniaceae and Corsiaceae (Burmanniales), Taccaceae and Petermanniaceae (Dioscoreales), Philesiaceae (Asparagales), Mayacaceae (Commelinales), Juncaceae (Juncales) and Cyclanthaceae (Cyclanthales); in the last two families *axile* placentation is also common. Except in these groups and in Velloziales, where placentation is dispersed, and Typhales, where the single ovule is basal, central (axile) placentation occurs in nearly all families of the Liliiflorae, Bromeliiflorae and Zingiberiflorae.

*Central (axile)* placentation is also found in Commelinaceae and some Rapateaceae and Juncaceae of the Commeliniflorae. In the other Commeliniflorae the ovules are either *basal* (Cyperaceae, some Rapateaceae and Xyridaceae of Commelinales), *lateral* (some Poaceae and Thurniaceae) or *apical* (most Poales, Eriocaulaceae of Commelinales, Hydatellales). Basal placentation is found in Arecales and some Pandanales, parietal placentation in the remaining Pandanales and some Cyclanthales, and apical placentation in some Cyclanthales.

The *ovules* are generally *anatropous* in the monocotyledons except where placentation is apical, in which case the ovules are nearly always *orthotropous*. *Campylotropous* or *hemianatropous* ovules are, however, found in various families, such as most Anthericaceae, some Alliaceae and Convallariaceae and certain Poaceae.

*Tenuinucellate ovules* are common in monocotyledons and occur, for example, where the seeds are minute, in Orchidaceae, Burmanniaceae, etc., but the concept "tenuinucellate" is not well defined. *Absence of a parietal cell,* which is frequently associated with tenuinucellate ovules, is found in these groups and also in most Liliales except Iridaceae, in scattered taxa of Asparagales, in Velloziales, Cyclanthales, Hydatellales, in Eriocaulaceae and Xyridaceae (Commelinales), in nearly all Poales, and in many Arales, all Triuridales, and in Limnocharitaceae and Alismataceae (Alismatales). In some of these groups, such as most grasses, nearly all Cyclanthales, many Araceae and in Limnocharitaceae and Alismataceae, anticlinal divisions in the nucellar epidermis give rise to a nucellar cap, which makes the ovules appear crassinucellate (the term "pseudocrassinucellate" has been used for this condition).

*Embryo sac formation* in monocotyledons is mostly of the monosporic *Polygonum Type.* Disporic types, mostly the *Allium Type,* occur in Limnocharitaceae and Alismataceae (Alismatales), many Alliaceae (Asparagales), some Trilliaceae (Dioscoreales), Cypripediaceae and some Orchidaceae (Liliales) and a few scattered genera in other families. The *Fritillaria Type* perhaps occurs in Liliaceae (Liliales) only.

*Endosperm formation* (WUNDERLICH 1959) (Fig. 34) is *helobial* or *nuclear* in most monocotyledons, but *cellular* in Arales and Hydatellales. *Helobial* endosperm formation occurs in all Bromeliiflorae and in Zingiberaceae and Costaceae of the Zingiberiflorae, in many Liliiflorae, in Juncaceae (Cyperales) of Commeliniflorae, Cyclanthaceae (Cyclanthales) of Cyclanthiflorae and in Alismatales and some Najadales in Alismatiflorae. In Liliiflorae helobial endosperm is predominant in Burmanniales, Melanthiales, and many capsule-fruited Asparagales (most Agavaceae, Anthericaceae, Asphodelaceae, Aphyllanthaceae, Phormiaceae and Funkiaceae, and in many Hyacinthaceae, Alliaceae and Amaryllidaceae).

The helobial endosperm in Bromeliiflorae is characterized by the small, often starchless chalazal chamber.

*Embryogeny* in monocotyledons is generally of the *Asterad Type* although the *Onagrad Type* has also frequently been reported and seems to be the predominant type in Orchidaceae (Liliales) and Cyperales in particular. All Alismatiflorae and probably many (? most) Ariflorae have the *Caryophyllad Type* of embryogeny.

## Fruit and Seed Characters

The possible types of fruit derived from *apocarpous gynoecia* are different from those derived from syncarpous flowers. Individual (distinct) carpels may develop into *follicles,* as in some Alismatales (Butomaceae, Limnocharitaceae, Aponogetonaceae), into *achenes,* as in most Najadales, some Alismatales (Alismataceae) and most Triuridales, or into *monocarpellary drupes* (a few Najadales). Monocarpellary follicles may be secondarily derived from tricarpellary syncarpous ones, as is apparently the case with Typhaceae (Typhales), Centrolepidaceae (Poales) and one or both of the genera of the Hydatellaceae (Hydatellales), and monocarpellary berries or drupes probably also have such a derivation in some Arales and Arecales.

Of the fruits derived from *syncarpous* tricarpellary gynoecia the commonest in monocotyledons are *capsules,* these being predominant in the Liliiflorae, Bromeliiflorae, Zingiberiflorae and many Commeliniflorae (Commelinales, Juncaceae, Restionaceae). *Baccate fruits* are common in some orders, especially in Asparagales (a number of families) and Bromeliales (numerous genera) but have their main distribution in Arales, Arecales, Cyclanthales and part of Pandanales, while syncarpous *drupes* are common in Arecales and Pandanales (*Pandanus*). Syncarpous *nutlets* (including *caryopses*) are found in Poales (most Poaceae, some Restionaceae) and Cyperales (Cyperaceae) and in scattered genera of Liliiflorae, e.g. in some woody taxa of Asparagales, Bromeliiflorae and Zingiberiflorae. *Schizocarps* are rare and limited to *Triglochin* in Juncaginaceae (Najadales), Cyanastraceae (Asparagales) and, perhaps, some genus of Marantaceae (Zingiberales).

*Arils, strophioles and caruncles* are more or less fleshy appendages arising from various parts of the seed and they generally function as food for animals, which therefore collect and disperse the seeds. *Arils* occur in scattered groups of monocotyledons and are present in nearly all Zingiberales (except, for example, *Heliconia;* reduced in *Musa*) and are characteristic of the *Aloë* group of Asphodelaceae in Asparagales, Taccaceae in Dioscoreales, and some Commelinaceae. *Strophioles and caruncles* do not have extensive distributions in monocotyledons. They occur in several genera of Dioscoreales, for example, but there they are obviously largely independent, non-homologous structures in the different genera.

*Phytomelan* is a black, brittle, carbon-rich, resistant substance, which accumulates in the outer epidermis of the testa of the seeds in most capsulefruited Asparagales, but is not verified with certainty in any other order. Phytomelan is generally lacking in the seed coat of berry-fruited taxa of the order, but occurs, for example, in seeds of the berries of Luzuriagaceae, Asparagaceae (but not Ruscaceae) and some Phormiaceae. Some capsule- or nutlet-fruited taxa have phytomelan-less seed coats, among them several of the woody Australian genera in the Dasypogonaceae and Calectasiaceae and the African genus *Cyanastrum*. The hairy seeds of Blandfordiaceae and Eriospermaceae are, of course, also devoid of phytomelan, as are the seeds in some South African Amaryllidaceae-Amaryllideae.

Endosperm is the common storage tissue in the seeds of the monocotyledons. Perisperm and to a variable degree also chalazosperm replace the endosperm in the families of Zingiberales, chalazosperm being restricted largely to Costaceae, Cannaceae and Marantaceae; perisperm also replaces endosperm in the taxa of Hydatellaceae (Hydatellales) and *Yucca* spp. of Agavaceae (Asparagales) and a copious chalazosperm is developed in Cyanastraceae (Asparagales).

Although endosperm is generally present and copious in most monocotyledonous groups except those just mentioned, there are important exceptions: in Apostasiaceae, Cypripediaceae and Orchidaceae (Liliales) endosperm formation is arrested very early or does not take place at all, the diminutive seeds consisting of testa and embryo only. In all Alismatiflorae and many Araceae (Ariflorae), some Poaceae (Commeliniflorae) and certain Amaryllidaceae (Liliiflorae) the endosperm, although developed, is used up during the growth of the embryo and is absent or nearly absent in the ripe seed.

The *endosperm* may contain combinations of hemicellulose, aleurone, lipids and starch as storage substances. *Copious starch* (Fig. 35) is stored in the endosperm in nearly all taxa of Commeliniflorae and Bromeliiflorae (in most Zingiberiflorae, where there is more perisperm/chalazosperm, these tissues generally store starch) and in the endospermous Arales, but is more rarely found in other groups. Moderate to copious amounts of starch may, however, occur in the endosperm of certain Dioscoreales (*Stemona, Croomia, Paris, Trillium*), some Amaryllidaceae, etc., and starch is often present in unripe seeds of Burmanniales in particular but also in many Liliales.

*Ruminate endosperm* is of restricted occurrence in monocotyledons but occurs abundantly in the Arecaceae (Areciflorae, Fig. 219O and P) and Cyclanthaceae (Cyclanthiflorae). Elsewhere it occurs in a few genera of Dioscoreales, at least in *Avetra* and *Trichopus* (Fig. 42Q), and in some species of *Yucca* in Agavaceae (Asparagales), the last three genera being in the Liliiflorae.

The *shape and size of the embryo,* and its position in relation to the endosperm in the seed is very variable in monocotyledons (MARTIN 1946; DAHLGREN and CLIFFORD 1982; DAHLGREN and RASMUSSEN 1983). A small, basal, elongate embryo in copious endosperm (*Trillium* Type) is found in certain Dioscoreales (though not in *Dioscorea*), Smilacaceae (Asparagales), Colchicaceae (Liliales), etc. They are reminiscent of the embryos of the magnolialean families in the dicotyledons. A linear, terminal but larger embryo (*Urginea* Type) characterizes most other Liliiflorae, most Ariflorae, Alismatiflorae, many Zingiberiflorae, and also Areciflorae, Pandaniflorae and Cyclanthiflorae. A broad, often nearly lens-shaped embryo (*Xyris* Type), or a capitate or mushroom-shaped embryo (*Scirpus* Type), characterizes most Commeliniflorae except the grasses, where the embryo is excentric and situated at the side of the endosperm (*Avena* Type). A diminutive endospermless embryo (*Orchis* Type) is restricted to Apostasi-

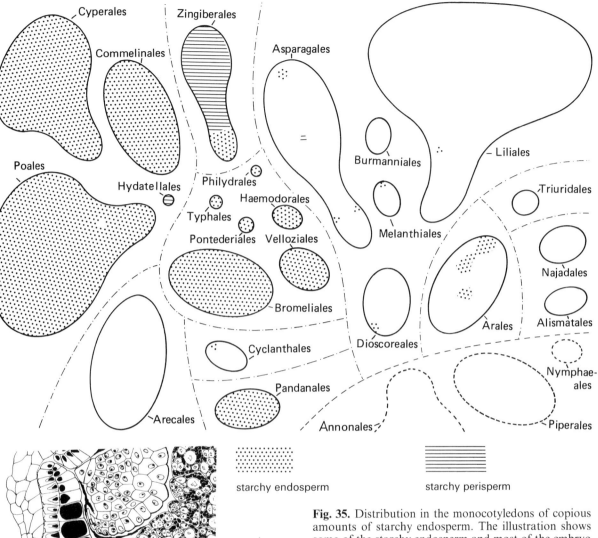

starchy endosperm

starchy perisperm

**Fig. 35.** Distribution in the monocotyledons of copious amounts of starchy endosperm. The illustration shows some of the starchy endosperm and most of the embryo in *Abolboda* (Xyridaceae, Commelinales)

aceae, Cypripediaceae and Orchidaceae (Orchidaceae s. lat.) in Liliales, and a diminutive embryo in an endospermous seed (*Burmannia* Type) to Burmanniales.

The *embryo* of some Dioscoreales has a *subterminal plumule,* whereas in other monocotyledons the cotyledon is terminal and the plumule emerges laterally. *Macropodous embryos* (i.e. with a thick and swollen hypocotylar part) characterize many aquatic Alismatiflorae and certain Ariflorae. A strongly *curved, horseshoe-shaped* embryo is characteristic of Limnocharitaceae and Alismataceae (Alismatales) and curved embryos are also common in aquatic Najadales, but occur also in various other families, e.g. Araceae (Arales), Alliaceae (Asparagales), Taccaceae (Dioscoreales) and Mar-

antaceae (Zingiberales). In the endospermless seeds of monocotyledons the embryo is often chlorophyllous (YAKOVLEV and ZHUKOVA 1980).

## Chromosomal Characters

We have not attempted here to survey the chromosome numbers in monocotyledons. In most families the numbers are too variable to be of much use in phylogenetic evaluations. The chromosome complement of 5 large and 25 small chromosomes, however, has helped in the decision to place the epigynous *Agave* group with the hypogynous *Yucca* group (Fig. 70 G–J) in the family Agavaceae (Asparagales) (MCKELVEY and SAX 1933).

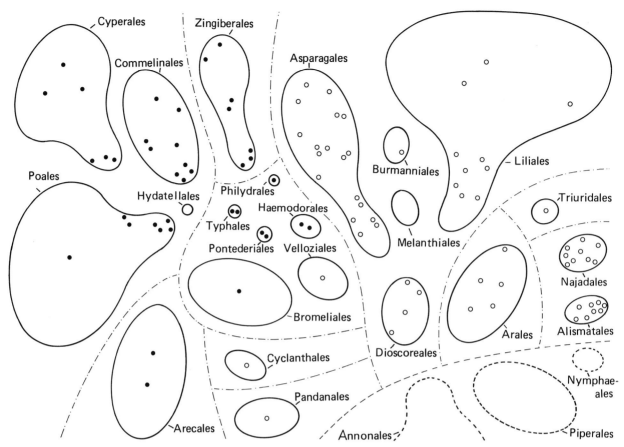

• UV-fluorescent compounds in cell walls

○ lack thereof

**Fig. 36.** Distribution in the monocotyledons of chemical substances (see further in the text) in the cell walls giving UV-fluorescence. (Data from HARRIS and HARTLEY 1980)

A similar complement is also present in *Hosta* (Funkiaceae), the affinity of which to Agavaceae is questionable, however. The chromosome numbers of the individual monocotyledon families will be briefly mentioned in the taxonomic section. In some families, such as Poaceae and Commelinaceae, the size and numbers of chromosomes have proved very important for estimating *internal* phylogenetic relationships. RAVEN (1975) concluded that "less can be said about the evolution of chromosome numbers in monocots than in dicots", but he recognized that x = 7, which has been claimed to be the basic number for angiosperms, is also much in evidence in the monocotyledons, and could easily be the original basic number in Arales, Alismatales, Poales and Zingiberales.

"Diffuse" or "polycentric" centromeres are present in at least some Juncaceae and Cyperaceae, where they are associated with post-reductional meiosis, a state that is extraordinary enough to indicate a close relationship between the two families.

## Phytochemical Characters

Raphides and silica (SiO$_2$·H$_2$O) have already been mentioned in this chapter under Vegetative Anatomy, as have phytomelan in seed coats and starch as a storage substance in seeds under Fruit and Seed Characters. Apart from these there are a number of chemical compounds that show interesting distributions in the monocotyledons.

*Aluminium* (CHENERY 1950) is accumulated in great concentrations in Rapateaceae and Xyridaceae (Commelinales), in *Aletris,* Melanthiaceae (Melanthiales) and *Centrolepis,* Centrolepidaceae (Poales).

*Steroidal saponins* occur especially in the orders Dioscoreales, Asparagales, Melanthiales and Liliales, but are also known from Arales, Zingiberales, Arecales and even Poales. The occurrence of chelidonic acid has a similar concentration, being reported in particular from Dioscoreales, Asparagales, Melanthiales, Liliales and Haemodorales.

Cyanogenic compounds, which in monocotyledons are nearly always tyrosine-derived (HEGNAUER 1977), are common in particular in Poales, Arecales, Arales and Zosterales (Juncaginaceae), but are also known in many other orders, e.g. Zingiberales, Pontederiales, Typhales, Asparagales, Pandanales, Cyclanthales and Commelinales (HEGNAUER 1963; GIBBS 1974).

*Alkaloids* of different kinds are known in several families of monocotyledons (SEIGLER 1977), the most conspicuous being perhaps Amaryllidaceae and Colchicaceae.

*Flavonoids* have been studied by BATE-SMITH (1968), HARBORNE (1973), GORNALL et al. (1979) and others. Of the flavonoids, tricin and sulphurated flavonoids are concentrated in particular in Poales, Cyperales and Arecales. C-glycoflavones, and luteolin are also common in these orders and a few others. 5-O-Me-flavones are mainly found in Cyperales (Cyperaceae and Juncaceae).

*Isoenzymes of the dehydroquinate hydrolyase* (DHQ-ase), which is activated by shikimic acid, were found by BOUDET et al. (1975, 1977) to be present only in monocotyledons. They were found to be present in the taxa of Poales and Cyperales investigated but only in 5 out of the 20 taxa of Asparagales and Liliales. In other monocotyledons studied (only one in Commelinales) they were lacking.

HARRIS and HARTLEY (1980) have shown that *ultraviolet fluorescence of cell walls,* linked with the presence of bound p-coumaric acid, ferulic acid and diferulic acid in trans forms, is concentrated in taxa of Zingiberiflorae, Bromeliiflorae (? except Velloziales), Commeliniflorae and Areciflorae (Fig. 36). This character, which is largely linked with epicuticular wax types, starchy endosperm (exceptions), and to some extent with stoma types, seems to have great phylogenetic significance.

Tuliposides were found by SLOB et al. (1975) to occur in rich quantities in Alstroemeriaceae and Liliaceae s.str. (both Liliales), but very rarely, and in small quantities, in other Liliiflorae.

## Parasites

*Host specificity of parasitic organisms,* mainly *rust and smut fungi* and herbivorous insects, is believed in many cases to indicate chemical similarities of the host plants that have a background in genetic relatedness. HOLM and NANNFELDT (in DAHLGREN and CLIFFORD 1982) and SAVILE (1979) give interesting fungal examples of host specificity. Thus the related *Puccinia atropuncta* and *Uromyces veratri* attack a number of genera of Melanthiaceae but no other plants, and *Uromyces muscari* attacks only genera of Hyacinthaceae. Several smut genera occur only on grasses, whereas *Entorrhiza* attacks only taxa of Juncaceae and Cyperaceae (both Cyperales).

Examples of *host specificity of insects* are given by SLATER (1976) for chinch-bugs of the family Lygaeidae subfam. Blissinae, and for plant lice (Hemiptera, family Aphididae) by EASTOP (1979). The chinch-bug genus *Ischnodemus* occurs on members of several families of Commeliniflorae and Zingiberiflorae, and related genera occur on taxa of Commeliniflorae, several of them on Poaceae only. Certain plant-louse groups feed only on Cyperaceae and Juncaceae (Cyperales), while others are restricted to grasses (Poaceae) and still others occur on Arecaceae and Pandanaceae.

Such data are sometimes of phylogenetic significance but one must always beware of over-emphasizing this kind of information because lack of observations on tropical groups, successful "jumps" to unrelated hosts, and other circumstances may distort the evidence.

# Ancestral and Derived Character States in Monocotyledons in General and in Various of the Major Groups

## General

The evolution of a group proceeds in such a way that if we currently conceive of it as a rather diversified order, it should correspond to a family a number of millions years ago and, then, to a genus some millions of years earlier than that. Still earlier it should be represented by a species. To calculate the general rate of evolution is not possible, but there is evidence that it has been vastly different in different complexes of plants. The fossil evidence is insufficient for estimating this rate. Fossil pollen is often used in estimating the occurrence and frequency of plant groups, but in such an evaluation pollen grains with thin exine will be lacking, and wind-pollinated groups will mostly be over-represented in relation to those that are animal pollinated.

The fossil data are far too incomplete for conclusions to be drawn concerning evolutionary courses in the monocotyledons. Hence hypotheses must be based on the distribution of character conditions in the groups of extant monocotyledons.

In this chapter we shall consider character conditions in the monocotyledons as a whole, and in certain major groups within them, and attempt to determine, for a number of characters, which condition is primitive and which derived. Having done this we shall, in the following chapter, be able to develop hypotheses about courses of evolution within the monocotyledons.

The principal procedure to be used is outgroup comparison (see Evolutionary Concepts). When we are considering the monocotyledons as a whole the outgroups to be looked at will be the group of dicotyledons which we regard as most closely related to monocotyledons, namely the Magnoliiflorae together with Nymphaeiflorae (see Origin of the Monocotyledons). This alliance, or some major complex within it, is therefore to be regarded as the sister-group of the monocotyledons. We are aware that the evidence in support of this viewpoint is not conclusive, but it is our working hypothesis.

If we had accepted BURGER's theory (1981) (see Origin of the Monocotyledons), the results of outgroup comparison on the status of character conditions in the monocotyledons as a whole would be reversed, for this theory treats the dicotyledons as derived from certain early but "advanced" monocotyledons.

In applying cladistic terminology, one must recognize that the determination of a character condition as plesiomorphic or apomorphic depends on the breadth of the group in which the occurrence of that condition is being studied. Thus the presence of perigonal nectaries is synapomorphic for the order Liliales, but for the family Orchidaceae within that order it is symplesiomorphic, and its substitution by osmophores is apomorphic for certain genera in Orchidaceae.

It is an axiom of cladistics that only synapomorphic states have significance in indicating groups as derived from the same ancestor, i.e. are phylogenetically related. The sharing of a common primitive character (symplesiomorphy) is phylogenetically uninformative.

The problem of convergence has already been discussed in Evolutionary Concepts, one of the most prevalent cases being the reduction of ovule number in the course of adaptation to wind pollination. To determine whether two instances of reduced ovule number are synapomorphic, the combinations of associated attributes in the two plant groups concerned must be analyzed. When a range of characters is examined in this way it is found that for some conditions convergence is frequent, whereas for others it is highly unlikely that they evolved more than once.

Sometimes a condition is very common in restricted groups without being universal in them. These are sometimes called "apomorphic tendencies". These cases may be the result of convergence, but perhaps more often there is an apomorphy that has been lost in a number of lines ("reversals"). In the case of chemical properties, established synthetic pathways can easily be broken in certain descendents of the group in which they are evolved. In other cases further mutations may affect morphological details so that a previously established property is concealed or eliminated (as in the case of the nectaries of orchids, mentioned above). Again, analysis of character combinations may make it possible to identify the synapomorphies involved.

Finally, by studying the distribution of supposedly derived conditions, the taxonomist is able to establish the sequence in which they have appeared in the course of evolution, the more widely distributed synapomorphies having evolved earlier than those with more restricted distributions. Through such an analysis showing the sequence of changes that are assumed to have occurred in the ancestors of particular units we may succeed in constructing a model of the actual evolution for the characters chosen. This evolution may, of course, be more or less intimately associated with shifts in important circumstances such as adaptation to the aquatic habitat or adaptation to specific kinds of pollination, in which cases it is likely that sequences of interdependent changes have occurred.

# Vegetative Morphology

As virtually all monocotyledons lack a cambium in the vascular strands giving rise to vascular cylinders, it is assumed that this was also the condition in the monocotyledon ancestor. An outgroup comparison, including the dicotyledons, however, indicates that the woody plants with such a cambium and thickening growth were probably present among the angiosperm ancestors (see Origin of the Monocotyledons). This is so, provided the monocotyledons did not (as proposed by BURGER 1981) appear before the dicotyledons or separately from them. Disregarding this possibility, then it must be presumed that even though the monocotyledon ancestor lacked the dicotyledonous type of cambium and thickening growth, it nevertheless evolved from a more distant ancestor which had such a cambium. On this assumption the scattered vascular strands and the herbaceous growth in monocotyledons are considered to be derived conditions.

Derivation of herbaceous monocotyledons from a hypothetical woody ancestor can easily be conceived, because herbaceous groups in Piperaceae and Aristolochiaceae show such an apparent shift to a secondarily herbaceous growth. In Piperaceae

we find in addition a presumably secondary condition with scattered vascular strands (an atactostele) while in Nymphaeales this condition is universal.

The loss of secondary thickening growth would also result in the restricted growth of the first root of the embryo, the radicle. In all monocotyledons the radicle persists only for a short period of time, being substituted for by adventitious roots emerging from the stem, mostly from its underground parts, but in many taxa, chiefly those of the rain forest, also from the aerial stem.

It is presumed that with the loss of the woody habit the promonocotyledon stem retained the power of perennial growth through its possession of well-developed underground parts, i.e. the plants were rhizomatous. We regard the rhizomatous habit as symplesiomorphic for the monocotyledons. Subsequently several adaptations (apomorphic conditions) evolved in response to varied soil conditions and climate. For example, the hypocotyl became tuberous, as in many Dioscoreales (Dioscoreaceae, Taccaceae) and Arales. In other groups stem units of *several* internodes became tuberous (corms); sometimes these were covered by sheathing leaf bases or scale-like leaves: "tunicated corms", a condition which is presumed to have arisen in several lines of evolution. The subterranean stem also became short and plate-like in certain groups, with the stored nutrients and water contained in fleshy scales or leaf bases, as is the condition in bulbs. These obviously developed in several lines of evolution. One such line may include Hyacinthaceae, Alliaceae and Amaryllidaceae, another the Liliaceae and Calochortaceae, a third part of the genus *Iris* (Iridaceae), and a further one certain species of *Triglochin* (Juncaginaceae). There is reason to believe that the acquisition of bulbs is not irreversible; in Amaryllidaceae they have, for example, probably been lost in the rhizomatous genus *Scadoxus* (cf. BJÖRNSTAD and FRIIS 1972) and in tropical species of *Crinum*. Problems remain as to whether the presence of rhizomes in *Agapanthus* or *Tulbaghia* is an ancestral condition or whether it represents a reversal in plants whose ancestors had bulbs like other Alliaceae. In Melanthiaceae bulbs may have evolved in more than one line.

Further, the rhizomatous herbs are likely to have evolved along several lines, giving rise either to perennials storing their nutrients and water in fusiform, swollen roots, as in many Alstroemeriaceae, Anthericaceae, Asphodelaceae and Orchidaceae (e.g. *Orchis* and *Dactylorhiza*), or to annuals, which die back at the end of the vegetative season and depend on seeds for surviving the unfavourable season.

In yet other plants the aerial stems became woody, either with a slender, shrubby habit, as in many Liliiflorae, especially the baccate-fruited Asparagales, or with a thick, stout trunk, as in other Asparagales and in various other groups. A thick trunk has evolved in a number of rosette plants or plants with a dense leaf tuft, which are shown by their differences in other characters to represent independent lines. In some of these groups the woody habit has become universal or dominant (Arecaceae, Pandanaceae, Velloziaceae, Xanthorrhoeaceae, Agavaceae, Nolinaceae), in others it is sporadic (Bromeliaceae, Cyperaceae). A special kind of cambium has evolved in the stems of a few of these woody plants, permitting new scattered strands to be formed in the pericycle long after elongation growth has been completed.

Thus, we consider both the annual herbs and the shrubby and tree-like perennials to be derived in the monocotyledons. It is not yet clear, at least in the woody and shrubby Asparagales, how many independent lines are involved. There are, for example, great problems as to the relationships of the woody asparagalean families presently distributed in different continents, the Agavaceae and Nolinaceae in America, the Asphodelaceae-Alooideae in Africa, the Dracaenaceae in Africa and Eurasia and, especially, the Xanthorrhoeaceae and Dasypogonaceae in Australia.

In the estimation of primitive and derived conditions of leaf characters, out-group comparisons with monocotyledon-like groups of dicotyledons in the Magnoliiflorae and Nymphaeiflorae imply that the primitive leaf type is wholly different from the one figured by BURGER (1981), who postulated that small herbaceous monocotyledons preceded all dicotyledons.

We shall here restrict ourselves to the former alternative. Accordingly we regard a leaf differentiated into a relatively short non-sheathing leaf base, a distinct petiole and a flat dorsiventral lamina with reticulate venation as the ancestral state even though this type of leaf is not particularly common in monocotyledons.

There is a transition between this type of leaf and leaves without a clear differentiation into petiole and lamina, and in some water plants both kinds of leaf occur on different parts of the same plant. The linear leaves in this case have a different ontogeny from the petiolate and laminar ones (KAPLAN 1973), just as do the cataphylls at the base of the

shoot and the bracts in the inflorescence in most monocotyledonous groups.

The main veins in monocotyledons generally converge towards the apex of the leaf, a condition which is, however, dubiously plesiomorphic in monocotyledons. In the leaves differentiated into petiole and lamina a reticulate minor venation is generally present between the main veins. This type of venation is also found in some early fossil leaves (e.g. *Alismaphyllum*) that have been referred to as monocotyledons, and is considered here as primitive in the monocotyledons in general. It is likely that the reticulate venation has become more prominent in some monocotyledonous groups as a consequence of secondary amplification of the leaf blade.

Compound leaves are common in Araceae, Taccaceae and Dioscoreaceae, where the leaf lamina is broad. They are likely to have evolved, secondarily, in several lines, and they have taxonomic significance only at subfamilial levels.

As regards leaf morphology, some character conditions are significant. Ensiform leaves, i.e. with isobilateral lamina, are undoubtedly derived in the families where they occur (Iridaceae, Melanthiaceae, Orchidaceae, Philydraceae, Haemodoraceae and Anarthriaceae) but there are cases (in Iridaceae) where one has reasons to postulate a reversal from ensiform to dorsiventral leaves. Reversed leaf blades occur in some Southern Hemisphere families or generic groups (Alstroemeriaceae and Luzuriagaceae) and are undoubtedly derived, although we do not know in how many lines this took place.

Continuations from the leaf sheaths in the form of stipules are rare in monocotyledons and we assume that they are apomorphies in each family where they occur (certain Hydrocharitaceae, Potamogetonaceae, Pontederiaceae). Quite often continuations from the leaf sheaths form hyaline so-called stipules, which are undoubtedly evolved independently in several lines. This is probably the case also with hyaline extensions from the *adaxial* side of the leaf sheaths (ligules) which are, for example, found in Zingiberaceae, Pontederiaceae and Poaceae.

Phyllotaxy is rarely constant within large families but one type is often dominant. Thus distichy occurs exclusively in several families in Zingiberiflorae and Bromeliiflorae, in nearly all Poaceae, all Iridaceae, and in various other groups of monocotyledons. Even in those families where the leaves are not distichous in the adult plants, the first leaves of the juvenile plants are generally distichous, as in Araceae (TOMLINSON 1970), Juncaceae (PEISL 1957) and Nolinaceae. An outgroup comparison does not give sufficient evidence for any assumption as to what phyllotaxy represents the ancestral state in monocotyledons, but it is likely that distichy of seedlings is plesiomorphic in them.

Tristichous leaves are characteristic of nearly all Cyperales and should be regarded as a synapomorphy for the members of this order.

## Vegetative Anatomy

The roots have been briefly commented on above. VAN TIEGHEM (1887) and VAN TIEGHEM and DULIOT (1888) observed that the lateral roots in many Commeliniflorae may originate opposite the phloem strands as well as opposite the xylem strands in the stem; in all other monocotyledons the lateral roots emerge opposite the xylem only, which is assumed to be the plesiomorphic state in monocotyledons.

The occurrence in roots of one or more outer layers of cells with unusual wall thickenings ("velamen", generally with a capacity for storing water) is concentrated in Liliiflorae, certain Ariflorae, etc. Velamen is here considered to be a specialization and apomorphic in the families (or orders) where it occurs. It must either have evolved in many evolutionary lines or on fewer occasions and been lost in some groups.

The differentiation of those cells of the root epidermis that bear the root hairs – as root hair short cells (LEAVITT 1904) – exhibits some regularity of distribution. Thus it is found in many Commeliniflorae, Alismatiflorae and Ariflorae and in some Bromeliiflorae and Zingiberiflorae, which indicates that it has a certain phylogenetic importance. We infer that the root hair short cells represent the derived condition in monocotyledons, but that reversals have taken place. In orchids the condition varies, and here seems to be of only minor importance.

Vessel conditions are very variable in monocotyledons and there is controversy as to what is ancestral and what derived (CHEADLE 1953; WAGNER 1977). Opinion here will depend on whether the Magnoliiflorae-Nymphaeiflorae are considered to be the sister-group of the monocotyledons and, if so, whether lack of vessels in certain members of this complex is regarded as the ancestral condition. YOUNG (1981), in a cladistic evaluation,

found that lack of vessels was probably the derived condition in various primitive dicotyledons (primarily of the Magnoliiflorae-Nymphaeiflorae complex). In that case it would be logical to regard the presence of long, slender vessels with oblique scalariform perforation plates as ancestral, at least in dicotyledons. We further conclude that the presence of such vessels not only in roots but also in the stems is the plesiomorphic state in the monocotyledons, and that lack of vessels in the stems (as in many Liliiflorae, many Zingiberiflorae, some Bromeliiflorae and all Cyclanthiflorae, Ariflorae, Triuridiflorae and Alismatiflorae) is derived, and even more so the lack of vessels in the roots of some Alismatiflorae.

Further, in agreement with CHEADLE (1943a, b, 1955a), CHEADLE and KOSAKAI (1980, 1982) and WAGNER (1977), we regard the shorter vessel elements with fewer bars and finally those with simple perforations as successively more derived. There are also functional aspects connected with growth form and life span to be considered in connection with vessels, and thus generalizations must be avoided.

Sieve tube plastids always, as far as is indicated by the ample material studied by BEHNKE (1981a) and BEHNKE and BARTHLOTT (1983), accumulate protein crystalloids of triangular shape (the $PIIc$ type sensu BEHNKE 1981). This type should therefore be regarded as having been already established in the ancestor of the monocotyledons. In dicotyledons this type is known only in two genera (*Saruma, Asarum*) of Aristolochiaceae (Aristolochiales, Magnoliiflorae) (BEHNKE 1971; BEHNKE and BARTHLOTT 1983), and it is uncertain whether these have evolved separately or are a relict condition in this group. We shall discuss this interesting character in other connections below.

Some subtypes of the $PIIc$ type plastids (Fig. 1) are to be regarded as apomorphies independently acquired in each of the monocotyledon groups: thus the $PIIcs$ subtype, with starch grains as well as with cuneate protein crystalloids, in all Zingiberales (=Zingiberiflorae) studied, in all Arales (=Ariflorae) and in Dioscoreaceae (five out of six genera) of Dioscoreales (Liliiflorae); the $PIIcf$ subtype, having protein filaments as well as cuneate protein crystalloids, is represented in a few genera that we refer to Asphodelaceae of Asparagales (Liliiflorae) and in *Musa* of Musaceae (Zingiberiflorae).

Oxalate crystals in the form of bundles of thin needles (raphides) are extraordinarily common and widely dispersed in monocotyledons. As they are lacking in some supposedly rather derived superorders (Alismatiflorae, Triuridiflorae, and most Commeliniflorae) but present in taxa of Liliiflorae, Bromeliiflorae, Zingiberiflorae, Areciflorae, Ariflorae and other superorders it seems most likely that they were present in the earliest monocotyledons, and thus constitute a symplesiomorphy for them, and that the absence of raphides is best regarded as derived. The value of raphides as a protective device against herbivores must vary according to circumstances, which would explain their loss in some groups.

Silica is present as separate bodies within the cells in a number of monocotyledonous groups. As these silica bodies are deposited in different cell layers and vary much in number and shape, we conclude that they arose independently in a number of lines.

Stomatal patterns can be classified both on the basis of the ontogeny and on the basis of the appearance of the mature stomatal complex, the former being phylogenetically most important, but still rather incompletely known except in the Liliiflorae (H. Rasmussen, in preparation).

Generally the stomata of monocotyledons are parallel to the axis of the leaf, but in most Ariflorae (Keating, personal communication) and various Dioscoreales they are randomly scattered. Agenous stomata (H. Rasmussen 1982) in which the surrounding cells do not divide and the meristemoid divides only to form the guard cells, are common in Liliiflorae and often result in anomocytic stomata. Neighbouring cells may divide to form daughter (perigene)-cells adjacent the stoma, which is then termed perigenous. If the resultant perigene cells are structurally distinct from other epidermal cells (which is by no means always the case), are two in number, and are parallel to the guard cells, which is often the case, the stomata become paracytic, which is extremely common in monocotyledons.

It is impossible to judge which kinds of stoma represent the plesiomorphic state in monocotyledons, the agenous or perigenous. Both types are common in certain orders of Liliiflorae, and occur sometimes in the same families, but in other superorders there is often great constancy in stomatal type.

Trichomes of various types occur in many groups of monocotyledons; certain kinds of hair, such as the peltate-stellate, multicellular hairs in Bromeliaceae, the multicellular many-rayed hairs in Hypoxidaceae and the two- to three-celled microhairs in Poaceae and Commelinaceae are typical of restricted groups and are undoubtedly apomorphic within the orders concerned.

There is argument as to whether the "intravaginal squamules" present in nearly all Alismatiflorae and in some Ariflorae are to be regarded as hairs. Their function is (or was originally?) protective, secreting tannins and compounds repelling unwanted organisms from the leaf axils. They are undoubtedly apomorphic in the Ariflorae-Alismatiflorae complex, and the crucial problem is whether, being such peculiar structures, they are likely to have evolved independently in both these superorders, or whether they evolved in their common ancestor, and have subsequently been lost in most Ariflorae.

Epicuticular wax, when sculptured and regularly oriented, is divisible into the *Convallaria* Type with scale-like rows of crystalloids, and the *Strelitzia* Type with rod-like crystalloids arranged as a rule as massive wax projections (see pp. 64–65). Both of these types should be regarded as derived and presumably can be regarded as synapomorphies for two groups of orders or superorders: the *Convallaria* Type for Liliales-Asparagales-Burmanniales of the Liliiflorae, and the *Strelitzia* Type for Bromeliiflorae-Zingiberiflorae-Commeliniflorae-Areciflorae.

## Floral Morphology

The floral symmetry in monocotyledons is generally derivable from an actinomorphic, trimerous, pentacyclic type with 3 + 3 tepals, 3 + 3 stamens and 3 carpels. As with dicotyledons, the inflorescence in primitive groups with actinomorphic flowers is a closed (determinate) one, in which the main inflorescence axis bears a terminal flower below which lateral flowers, or lateral branches, each with a terminal flower, are developed, a type that would be classified as a panicle.

Cases where the terminal flower is not developed and the laterals develop in acropetal succession (a thyrse, raceme or spike in the strict sense) should be regarded as derived. There is a tendency for lateral flowers in such racemose inflorescences to be zygomorphic, as in *Lilium* and in Iridaceae, where subfam. Ixioideae has racemose inflorescences and often more or less pronouncedly zygomorphic flowers, and the other subfamilies (e.g. Iridoideae) have inflorescences that are determinate and paniculate or monochasial (rhipidial) and always have actinomorphic flowers.

Various umbel-like inflorescences, such as those in Alliaceae, Hyacinthaceae and Taccaceae, are probably not racemose but usually represent condensed cymose types derived from paniculate inflorescences. Direct transitions between verticillate panicles and "pseudo-umbels" can be found in Alismatales.

Complex panicles and thyrses are found in various families, such as Iridaceae (panicles, see WEIMARCK 1939) and Marantaceae (thyrses, see ANDERSSON 1976). In dicotyledons zygomorphic flowers are more or less restricted to racemes (and related types of heads, spikes or umbels) and thyrses, i.e. to inflorescences where the main axis lacks a terminal flower, and the same relationship is evident in the monocotyledons. Interesting transitions can be found between cymose (most Commelinaceae) and obviously racemose inflorescences (*Cartonema*) (BRENAN 1966).

The actinomorphic, trimerous flower type is undoubtedly the plesiomorphic state in the monocotyledons. Zygomorphic flowers have arisen in many groups, often in single genera within families where actinomorphic flowers prevail (Anthericaceae, Asphodelaceae, Melanthiaceae, Amaryllidaceae). Sometimes, however, they characterize whole families (Corsiaceae, Philydraceae) or groups of families (Apostasiaceae-Cypripediaceae-Orchidaceae; Lowiaceae-Musaceae-Heliconiaceae-Strelitziaceae-Zingiberaceae-Costaceae). Families of which the flowers are asymmetric are Cannaceae and Marantaceae. Zygomorphy and asymmetry here are considered to be apomorphies.

The trimerous plan is surprisingly constant in the monocotyledonous flower. Deviations from it are mainly to be found in the Cyclanthiflorae and Pandaniflorae, which may be related. Apart from this the plan may be obscured as a result of reduction, unisexual monomerous or dimerous flowers having evolved independently in several families, e.g. Hydatellaceae, Centrolepidaceae, Cyperaceae (Mapanieae), Typhaceae, Araceae, Lemnaceae, etc. Isolated deviants from the trimerous plan are *Maianthemum* (Convallariaceae), species of *Paris*, *Kinugasa* and *Daiswa* (Trilliaceae) and the genera of Stemonaceae (where *Pentastemona* is, in fact, 5-merous). All these non-trimerous conditions are presumably derived. Polymerous androecia and gynoecia are considered later.

The trimerous pattern has its counterpart in various groups within the dicotyledonous superorders Magnoliiflorae and Nymphaeiflorae (see p. 46), and we assume that it had been evolved in the common ancestors of these groups and the monocotyledons long before the monocotyledons became a separate evolutionary line. In fact, trimerous flowers may have occurred in a large proportion of early angiosperms, the widespread occurrence of trimerous gyneocia being perhaps a relict state retained in various dicotyledons.

Petaloid, showy tepals predominate in the Liliiflorae, Bromeliiflorae and Zingiberiflorae and should be regarded as a specialization for animal pollination. The greenish yellow (see WILLEMSTEIN 1983), small or moderate-sized tepals found in most Dioscoreales could then be the ancestral state in the monocotyledons. The same condition is found in various Magnoliiflorae (e.g. Lactoridaceae, many Annonaceae, some Aristolochiaceae). In the Liliales the conspicuously petaloid condition of the tepals, their variegation with spots, chequering or striations and their possession of perigonal nectaries (in place of the septal nectaries prevalent in Asparagales etc.) are regarded as important synapomorphies.

In many Alismatales (Alismatiflorae), in nearly all Commelinales (Commeliniflorae), and in most Bromeliales (Bromeliiflorae) the tepals are differentiated into a sepaloid outer whorl and a petaloid inner whorl. The three orders are undoubtedly very distantly related, and the differentiation of tepals into sepaloid and petaloid is assumed to have evolved independently in each of these ordinal complexes.

Tepals which are scarious and bracteal or reduced to hairs or bristles are regarded as derived. The absence of tepals is also derived. In most cases reduction in size or loss of the perianth is associated with wind pollination, and is particularly common in the Commeliniflorae, but has also occurred in Typhales, certain Najadales and the Pandanales.

In Ariflorae and Cyclanthiflorae the flowers are densely crowded on an often thick, fleshy inflorescence axis to form a spadix, which is subtended and sometimes enveloped by one or more spathal bracts. The flowers in such groups have inconspicuous, often thick tepals or may lack tepals altogether; nectar may be secreted from the stigmas or elsewhere and odorous substances and heat are often emitted. All these specializations are here considered to be derived, and we assume, provisionally, that they evolved independently in the two superorders mentioned.

The semi-epigynous and epigynous conditions are restricted to flowers with a petaloid type of perianth, occurring in groups of Liliiflorae, Bromeliiflorae, Zingiberiflorae, Alismatiflorae (Hydrocharitaceae) and Cyclanthiflorae. Epigyny may be of selective advantage in giving greater stability to the tepals, when these are large, and in giving better protection to the ovules against predation. Epigyny is considered derived and has undoubtedly arisen in numerous independent lines of evolution. A number of families exhibit both hypogynous and epigynous flowers. In each of these epigyny has probably arisen separately. There are also pairs of families that agree in many features and are obviously phylogenetically closely related, where one has hypogynous and the other epigynous flowers (Alliaceae-Amaryllidaceae, Liliaceae-Alstroemeriaceae).

It has already been mentioned above that perigonal nectaries are characteristic of the families in Liliales, whereas septal nectaries have a much wider distribution. The occurrence of septal nectaries is so widespread in monocotyledons as to suggest that they occurred in the common monocotyledon ancestor, or at least evolved in some of the earliest lines. In the dicotyledonous groups Magnoliiflorae and Nymphaeiflorae nectar is only rarely secreted from the sides of the carpels (in some Nymphaeaceae).

Septal nectaries are thus assumed to have been lost in various groups. In a few of these they are thought to have been substituted by perigonal nectaries (Liliales) or epigynous nectar glands (Zingiberaceae). The occurrence of septal nectaries in Iridaceae subfam. Ixioideae of the Liliales is still unexplained. Presumably septal nectaries could have arisen secondarily here, the ancestors having had perigonal nectaries (which, again, are secondary in relation to septal nectaries in previous ancestors).

A complement of 3 + 3 stamens is distributed over most superorders of the monocotyledons, and as trimerous whorls are also common in those dicotyledons which are similar to the monocotyledons, we presume that this is the plesiomorphic state in the monocotyledons. This view is opposed to that of most contemporary taxonomists who regard the pleiomerous androecium, as in the Alismatiflorae, as the ancestral state in the monocotyledons. Conditions with numerous stamens that can easily be accepted as secondary are those in *Ochlandra* (Poaceae), many species of *Vellozia* (Velloziaceae), *Gethyllis* (Amaryllidaceae) and in a number of palm genera of different subfamilies (Arecaceae; see UHL and MOORE 1980). Numerous stamens are also found in Cyclanthaceae and Pandanaceae, but the ancestries of these are unknown. One reason for regarding numerous stamens as

ancestral in the Alismatales is that this condition is here combined with an apocarpous gynoecium which, in a traditional view, has been generally accepted as primitive.

SINGH and SATTLER's (1977, etc.) studies of the floral ontogeny of *Sagittaria* and other genera show, however, that 3 + 3 staminal and 3 carpellary primordia appear first; new initials superimposed on those first differentiated have here resulted in the high numbers of stamens and carpels. One can thus infer that the condition with numerous stamens is secondary in perhaps all monocotyledonous groups where it occurs and that high numbers have arisen in several independent evolutionary lines at different periods in the history of the monocotyledons. The modes of "dédoublement" are probably different between the groups where they have arisen.

Many Magnoliiflorae have numerous stamens but they are spirally arranged, whereas the stamens in monocotyledons, if numerous, are in whorls or in one or two whorls of fascicles.

The stamens in those members of Magnoliiflorae (dicotyledons) which show a number of conspicuous similarities to the monocotyledons are frequently flat, with the microsporangial pairs (thecae) situated at some distance below the apices. Flat stamens occur in a number of monocotyledons, though in these cases the stamens are not very similar to those of the Magnoliiflorae. However, there are a number of cases where the thecae are situated far below the stamen apex, and in various Dioscoreales (*Trichopus, Stenomeris, Stemona, Tacca, Paris*) this is a conspicuously frequent and striking characteristic (this order also shows other links with the Magnoliiflorae). In Pandanales, Cyperales, some Zingiberales and other groups this condition is also common, but is here not coupled with so many other features found in the Magnoliiflorae.

The anthers are introrse, latrorse or extrorse. In monocotyledons introrse stamens are commoner than extrorse, which prevail only in Ariflorae, Triuridiflorae and Alismatiflorae and in Liliales of the Liliiflorae. In the dicotyledonous Magnoliiflorae and Nymphaeiflorae both conditions occur. There is no strong basis for considering either type as ancestral in the monocotyledons, although we suspect that the introrse type is the original one.

As regards the attachment of the anther, the basifixed type or – more precisely – the "undifferentiated anther" (sensu SCHAEPPI, 1939), which is diffusely delimited from the filament, is assumed to be the plesiomorphic state in the monocotyledons. Dorsifixed-versatile anthers are considered to be derived, whether they are epipeltate or hypopeltate, both being divergent advanced types. Also derived is the condition where the connective forms a tube around the filament tip (as in some Asphodelaceae, Liliaceae, etc.).

Longitudinal anther dehiscence is considered to be the plesiomorphic state in the monocotyledons, whereas poricidal dehiscence which is rather rare and occurs in a number of scattered families, is derived. The condition of fewer than four microsporangia is uncommon in monocotyledons (various cases in Arales, Poales, Asparagales, Zingiberales and Zosterales) and is considered to be derived; then either the anthers are monothecous (and the theca bisporangiate) or they are dithecous (and the thecae unisporangiate).

The gynoecia in most monocotyledons consist of three carpels, although deviating numbers occur in various groups, being both lower (which we assume to be reductions) and higher (through "dédoublement", see above). The reason for regarding the trimerous gynoecium as the ancestral state in the monocotyledons is that it is found in various members of Magnoliiflorae-Nymphaeiflorae, which we think are related to the monocotyledons. It is also represented in nearly all monocotyledonous superorders and is generally the dominant state.

It is traditionally assumed that the condition of mutually free carpels represents a more primitive level than syncarpy in both the monocotyledons and the dicotyledons. Free carpels (or a solitary carpel) are found in Alismatiflorae and Triuridiflorae and in various Areciflorae, and nearly free carpels occur in a few Melanthiaceae (Melanthiales, Liliiflorae). As the first two groups are advanced in various respects and as there are indications that the high number of carpels could be derived, we also challenge the view that apocarpy is here necessarily ancestral in monocotyledons, although apocarpy is probably ancestral in angiosperms. However, the evidence is insufficient for a conclusive decision on this matter.

The standpoint that is taken on the question whether apocarpy is ancestral in monocotyledons also affects one's judgement as to whether pistils with free stylodial branches represent a more ancestral condition than pistils in which the apical, stylar parts of the carpels are fused to form a single style, and whether an apically tribrachiate style with separate stigmas is more ancestral than one with a single apical stigmatic head. The latter conditions are generally regarded as being more

derived, but we have no evidence that fusion of carpels proceeded gradually in evolution.

The stigmatic papillae may be smooth, "Dry", or may be covered by a sticky secretion, being then described as "Wet". The Dry state is regarded as the plesiomorphic, the Wet as derived, and it is assumed that Wet stigmas have evolved independently in a number of lines, judging by their occurrence in disparate groups.

## Microsporangia, Pollen

Anther wall formation in most monocotyledons and in a few orders of dicotyledons is of the so-called Monocotyledonous Type, which is generally considered to be derived from the Basic Type. An example of the so-called Dicotyledonous Type is reported for *Tacca* but needs verification. The Reduced Type, found in a few water plants (in Lemnaceae and Najadaceae), is probably derived in these groups from the Monocotyledonous Type. The Monocotyledonous Type is verified as nearly universal in monocotyledons and may represent a synapomorphy for them.

The cells of the endothecial layer of the anther wall have characteristic types of wall thickenings. These are generally spiral in the monocotyledons. Exceptions are most Commeliniflorae, the Alismatales of Alismatiflorae, and some other groups, where the thickenings are girdle-like. We regard the Spiral Type thickenings as the plesiomorphic state in the monocotyledons, and the Girdle Type as derived and polyphyletic.

The tapetum is generally glandular-secretory in the monocotyledons as it is in the dicotyledons, and this is here assumed to be the ancestral state. Amoeboid-plasmodial tapetum is then considered to be the derived condition. Its occurrence in some major groups which differ in many respects suggests that it has arisen in more than one evolutionary line.

Microsporogenesis can be successive or simultaneous, and both these types are well represented in the monocotyledons, the Successive Type being present in most families but the Simultaneous in several of the larger ones. In the dicotyledons, microsporogenesis is generally simultaneous, but the Successive Type occurs in a number of families in Magnoliiflorae and Nymphaeiflorae, where monocotyledonous attributes are most concentrated. In other dicotyledons successive microsporogenesis is rare and undoubtedly derived. It is uncertain that the successive microsporogenesis is the ancestral condition in the angiosperms, but on the basis of the evidence available we conclude that this is the case, and we draw the same conclusion for the monocotyledons.

The pollen tetrads formed with the Successive Type of microsporogenesis normally lie in one plane, forming a square or, more rarely, a row or a T-shape, whereas the tetrads formed with the Simultaneous Type of microsporogenesis are tetrahedral. In some groups the pollen grains of the tetrads remain coherent, which is here considered a specialization. This is the case in all Cyperales and in one or a few genera in each of several separate families elsewhere.

The pollen grains when ready for dispersal are either two-celled or three-celled. The three-celled pollen grains have their greatest concentration in Alismatiflorae, Commeliniflorae and Ariflorae, but are otherwise rare. In dicotyledons three-celled pollen grains are likewise concentrated in some supposedly specialized larger complexes, while two-celled grains have a much wider distribution. We consider the two-celled pollen grains to represent the plesiomorphic state in angiosperms in general (it is ubiquitous in the Magnoliiflorae) as well as in monocotyledons, and assume that the three-celled condition has evolved repeatedly in both monocotyledons and dicotyledons. Both kinds of pollen grain occur together in a number of families in, for example, the Asparagales and Liliales, but also in other orders, whereas in certain groups one or other type is predominant or universal.

As regards the aperture conditions in monocotyledons, we conclude that the Sulcate Type represents the ancestral state, as we assume it does also in the angiosperms in general. The earliest recorded pollen grains that undoubtedly belong to angiosperms (from the Barremian, $120 \times 10^{-6}$ BP) are sulcate. Such pollen is also typical of the gymnosperms and those dicotyledons that most closely resemble monocotyledons. The trichotomosulcate, sulculate, zonisulculate, ulcerate, forminate, spiraperturate and "inaperturate" (=partly, at least, exineless or omniaperturate) types of pollen grains are, accordingly, the derived types.

## Ovular Characters

The number of ovules per carpel is intimately connected with the mode of pollination, but it is possible to estimate some major trends of evolution.

The conclusion that a condition with several or numerous ovules is ancestral in the monocotyledons can be drawn from the fact that this is prevalent in the dicotyledons considered related to the monocotyledons, and from acceptance of the hypothesis that extreme adaptation to wind pollination was not present in ancestral angiosperms. Reduction of ovule number, however, must not be regarded as an irreversible change.

The possession of one or very few ovules per carpel or per locule of the ovary or of a single ovule in a unilocular pistil is generally a feature of the syndromes of both wind pollination (Poaceae, Cyperaceae, Scheuchzeriaceae, Juncaginaceae and most Potamogetonaceae) and water pollination (most other Najadales). The conditions described above presumably represent reductions which appear to have occurred in several lines of evolution, as there are grades in the level of reduction within, for example, each of the orders Poales (incl. Restionales) and Cyperales (incl. Juncales). Whether reduction has occurred in more than one evolutionary line in the ancestors of uniovulate water-pollinated Najadales is uncertain.

On the other hand, it is quite obvious that an *increase* in ovule number has occurred in certain insect-pollinated groups, the most extreme condition being found in Orchidaceae and the related Apostasiaceae and Cypripediaceae and, undoubtedly independently, in Burmanniales. The first three families, the "orchids", possess a syndrome of character states related to successful insect pollination on one or a few occasions per individual (see DAHLGREN and CLIFFORD 1982, pp. 40–41). This syndrome involves, together with the millions of ovules, also a transfer of a comparable number of pollen grains collected together in a pollinarium unit. When numbers of ovules are at their greatest, as in the orchids and Burmanniales, placentation is parietal. The states in these two groups of families are undoubtedly derived, as they are in other families (e.g. Philydraceae; Iridaceae-Geosiridaceae) where there are trends involving increase in ovule number.

Another case where we infer that an increase in ovule number per carpel has occurred late in evolutionary history is in the family pair Butomaceae-Limnocharitaceae with laminar-dispersed placentation. The basal-axile placentation type, which is present, for example, in Aponogetonaceae in the same order (Alismatales), is associated with few ovules and seems to have preceded the laminar-dispersed type. A comparable condition in dicotyledons is found only in Nymphaeales, where the laminar-dispersed placentation may also be derived from submarginal in more or less free carpels.

Thus both reduction and increase in number of ovules have obviously occurred repeatedly in monocotyledons and the ovule number is always dependent on conditions connected with pollination and dispersal, the above examples being only the most extreme ones. We have no direct evidence on the number of the ovules present in the ancestral monocotyledonous flower, and inference is difficult because we do not even know whether the latter possessed free or fused carpels. It is, however, likely that the ovules in these ancestors had rather thick integuments, that they were crassinucellate and that the seeds had copious endosperm. This permits us to guess that the ovule number per placenta was relatively low.

Of the ovular types, the anatropous type is the most widespread one not only in monocotyledons, but also in dicotyledons. There is, in the monocotyledons, a certain connection between placentation and ovular morphology. Thus, when the ovules are solitary and apical they tend to be orthotropous (as in many Poales, several marine families of Najadales, Smilacaceae of Asparagales, some Araceae of Arales, certain genera of Haemodoraceae in Haemodorales, and Eriocaulaceae of Commelinales). This connection is evidently derived and has arisen in a number of different evolutionary lines. Campylotropous and hemianatropous ovules, which both occur in sporadic groups, are, like the orthotropous, here considered to be derived. Anatropous ovules, generally found where placentation is axile or basal, are here considered plesiomorphic.

The ovules in the monocotyledons are nearly always bitegmic, although in a few genera in a number of families they are unitegmic, which condition is regarded as derived. In Amaryllidaceae unitegmic ovules characterize several genera and may prove to be of phylogenetic significance.

The ovules are also classifiable into crassinucellate and tenuinucellate according to the number of cells surrounding the archesporial cell or embryo sac, the crassinucellate state being regarded as ancestral in monocotyledons, as in the dicotyledons. The crassinucellate state is generally combined with the formation of a parietal cell which is cut off from the primary archesporial cell before it functions as a megaspore mother cell. The parietal cell generally divides, forming a variable number of cells above the embryo sac, but it may also remain undivided or divide only one or a few

times. The widespread occurrence of a parietal cell in dicotyledons as well as monocotyledons indicates that its presence represents the ancestral state. Absence (i.e. loss) of a parietal cell is thus derived; the scattered distribution of this state indicates that the formation of the cell is easily eliminated, in which case the primary archesporial cell functions directly to produce the megaspore tetrad. This is the case in various genera belonging to the families where a parietal cell is also encountered, but in other cases lack of a parietal cell is characteristic of entire families and groups of families (i.e. a number of families in each of Liliales and Commelinales, all Velloziales, Cyclanthales and Burmanniales, and at least nearly all families in Poales). In these cases the absence of the parietal cell can be of phylogenetic significance as an apomorphic condition.

Like anatropous ovule morphology the *Polygonum* Type of embryo sac has general and widespread occurrence in both monocotyledons and dicotyledons (DAVIS 1966), and there is ample evidence that this type was present both in the angiosperm ancestor and in the ancestor of the monocotyledons. Each of the other embryo sac types thus represents an apomorphic condition. The scattered distribution of the deviating types (the *Allium* Type occurring, for example, in Alismataceae-Limnocharitaceae, some Alliaceae, some Lemnaceae, most Trilliaceae, and the Cypripediaceae) suggests that they have arisen polyphyletically. The *Fritillaria* Type, which is characteristic of Liliaceae sensu stricto, is, however, an important autapomorphy for this family.

It has not been possible for us to determine which type of endosperm formation should be regarded as ancestral in the monocotyledons (see DAHLGREN and RASMUSSEN 1983). WUNDERLICH (1959), who surveyed the endosperm types in angiosperms, assumed on the basis of the pattern of endosperm formation in dicotyledons that the Cellular Type, found in many Magnoliiflorae, and in some families thought to be related to the Magnoliiflorae, is the ancestral type in the angiosperms. In monocotyledons this type is found in perhaps all Ariflorae and also in the small, specialized Hydatellales (Commeliniflorae), but is otherwise lacking. Most Ariflorae have an endosperm where the basal of the two cells formed after the first division does not divide further but enlarges and undergoes endomitosis, increasing the size of the nucleus. However, other Araceae (e.g. *Acorus*) and the Lemnaceae have a more normal cellular endosperm. To estimate which endosperm type is ancestral in Arales, and then whether this is likely to be ancestral in monocotyledons or derived from helobial or nuclear endosperm, needs profound study.

We assume here that cellular endosperm formation (as found in Arales) is a derived state in monocotyledons leaving us with the alternative of the Helobial or the Nuclear Types as ancestral. Even a choice between these would rest on a knife edge. However, it seems highly unlikely to us that the Helobial Type, which is a rather special type, has arisen repeatedly in the monocotyledons, as would be the case if this type were the derived one. It is much easier to accept that the very general Nuclear Type has arisen repeatedly in different monocotyledonous lines.

The helobial endosperm formation is variable in the monocotyledons and some "modifications" of it characterize particular groups, e.g. a type with a very small chalazal chamber where divisions are cellular is characteristic of the Bromeliiflorae, and we consider this type synapomorphic in this superorder, irrespective of whether the Helobial Type as such is ancestral or not in the monocotyledons.

Embryogeny in the monocotyledons is generally of the Asterad, or slightly less often of the Onagrad Type. These two types are closely related and either is likely to have occurred in the monocotyledon ancestor. Restriction to a choice between these two is supported by the fact that these types are very common in the dicotyledons (e.g. in Magnoliiflorae and Nymphaeiflorae). All Alismatiflorae and at least many Ariflorae have the Caryophyllad Type. We consider this to be an important synapomorphy for Ariflorae and Alismatiflorae (Triuridiflorae are unknown in this respect), and the exceptions in Arales are not unlikely to be reversals. In the rest of the monocotyledons types other than the Asterad or Onagrad (i.e. Solanad, Caryophyllad, Chenopodiad) Types are rarely encountered, some reports perhaps being in need of verification. These cases are so sporadic that we have not ascribed any significance to them. They are likely to represent derived conditions.

## Fruit and Seed Characters

We have already discussed the problems associated with judging whether apocarpy is necessarily the primary condition in monocotyledons. This is also fundamental for the estimation of the ancestral fruit type in the monocotyledons.

If the monocotyledon ancestor had an apocarpous gynoecium it is likely that the fruit was trifollicular with the follicles dehiscing adaxially. Such follicles are found, for example, in Aponogetonaceae (Alismatales). If we accept that increase in carpel number is derived, the multifollicular fruit in many Alismatales and a few Triuridales is derived from the trifollicular condition; also the multiovulate follicles in Butomaceae and Limnocharitaceae are probably derived from follicles with fewer ovules (see above). Dry, indehiscent monocarpellary fruits (achenes) are probably derived from the follicles; they are common in Triuridales (where they are numerous), and in Najadales (where they are solitary or few). There are also monocarpellary drupe-like fruits, as in Potamogetonaceae (Najadales). The fruit in Hydrocharitaceae is superficially very different from that in other Alismatales but we presume that it arose from a principally apocarpous gynoecium enveloped by the floral receptacle (reminiscent of that in *Nymphaea*, Nymphaeales, among the dicotyledons), a state which can be considered a synapomorphy for the genera of Hydrocharitaceae.

If the monocotyledon ancestor had a syncarpous gynoecium (as in Dioscoreaceae, most Araceae, some Juncaginaceae) and apocarpy is interpreted as a derived condition, we have to assume that a capsule of some sort was the ancestral fruit type, and that this gave rise to a trifollicular or multifollicular fruit in the ancestor of the triuridifloran-alismatifloran branch. This hypothesis is very rarely considered in phylogenetic speculations concerning the monocotyledons, but we regard it as a realistic alternative, mainly because the monocotyledons with apocarpous gynoecia are specialized in so many other attributes. Some of the monocotyledons with syncarpous gynoecia, on the other hand, exhibit numerous characters that are also present in the Magnoliiflorae with which they probably shared their ancestors.

From trilocular capsules, possibly with free stylodial branches, tricarpellary berries and drupes can be derived, which are indehiscent and fleshy, and also nutlets or nuts, which are dry and indehiscent and few- or one-seeded. There is ample evidence that such indehiscent fruits have arisen from capsules in many independent lines; sometimes in whole families or groups of families, sometimes only in some species of a genus.

A postulated evolutionary sequence of fruit types in the ancestors of Asparagales, in the Liliiflorae, represents an interesting cyclic process. In this order the seeds in the capsule-fruited taxa (some mainly Australian genera excepted) generally have a thin collapsed tegmen (seed coat layer formed from the inner integument) while the cell walls of the outer epidermis of the testa (formed by the outer integument) are provided with a layer of phytomelan which – as far as we know – is lacking in all other monocotyledonous orders. The phylogenetic significance of phytomelan has been stressed by HUBER (1969; personal communication), who believes that it has evolved as a compensatory novelty to protect the seeds after their seed coat had declined during an ancestral baccate stage (like that in some Dioscoreales). Thus the capsules of many Asparagales might be secondary to berries which are probably, in their turn, secondary to capsules (*Dioscorea* has capsules). Further, in some genera of Amaryllidaceae (*Clivia, Haemanthus*) there are berries which are presumably derived from capsules with phytomelan-coated seeds. If these surmises are true, the *Haemanthus* berry could have been derived via the sequence: trifollicular fruits → capsules → berries → capsules → berries! Such a sequence may be compared with the evolution sketched by STEBBINS (1974) for the insect-pollinated *Ficus,* which has presumably been derived from urticalean ancestors that were wind-pollinated like many extant Moraceae which, in turn, are assumed to have evolved from insect-pollinated forms comparable with certain Sterculiaceae in ancestral Malviflorae. These again may lead back to the gymnospermous ancestor of the angiosperms considered by Stebbins probably to be wind-pollinated. Such sequences of evolution resulting in repeated reversals must be expected to occur occasionally.

The fruit type partly reflects the pollination biology of the plant as outlined above (Orchidaceae: many tenuinucellate ovules, many pollen grains in pollinia, rare-event insect pollination, many-seeded capsule; – Poaceae: one (pseudo-) crassinucellate ovule, singly dispersed wind-borne pollen grains, one-seeded caryopsis); but it is also dependent on habitat and associated with dispersal, which can be brought about by mechanical scattering of seeds, by wind, by animals, by water, etc. Mutations and selection have resulted in a variety of fruits, and phylogenetically close taxa, sometimes even in the same genus, may have different fruits (see for example *Trillium; Yucca; Hypselodelphys,* Fig. 170).

The same can be said of arils and other fleshy seed appendages that have evolved from different parts of the ovule (funicle, chalaza, raphe, hilum, or the integument near the micropyle; see HUBER 1969, p. 232; DAHLGREN and CLIFFORD 1982, pp. 226–227). We are of the opinion that these structures are more often derived than ancestral (cf. CORNER 1954). Before regarding such a structure as a synapomorphy for supposedly related taxa, we must carefully check that they are histologically comparable and thus likely to be homologies. The arils in the Zingiberales, for example, probably have a common ancestry, and are re-

garded by us as an important synapomorphy for this order. The presence of an aril is also undoubtedly a synapomorphy for a number of genera in Asphodelaceae.

The significance of the presence of the black pigment phytomelan in the outermost epidermis of the seed coat in most Asparagales has been mentioned above, under fruit types. Admittedly, its presence has contributed much in defining the order Asparagales, but as will be seen in the discussion for Asparagales (pp. 130–131), this character is associated with other attributes, particularly characters of the seeds, and we consider presence of phytomelan to be an important synapomorphy for the Asparagales. Phytomelan is even found in the seeds of certain members of Asparagales with baccate fruits (as in *Asparagus, Geitonoplesium, Dianella,* each in a different family), although in the majority of the berry-fruited taxa the phytomelan layer is lacking, which we consider to be a secondary loss, although there may be doubts on this point. Phytomelan is lacking, presumably as a result of loss, in a few Hyacinthaceae and several Amaryllidaceae, its loss in the latter being apparently correlated with loss of one integument (the outer?). In each of the unigeneric families Eriospermaceae and Blandfordiaceae the seeds are covered with hairs (in *Walleria* with short pegs) which is not compatible with a phytomelan layer.

The minute seeds of Apostasiaceae (?), Cypripediaceae and Orchidaceae (Liliales) lack voluminous nutrient tissue. There is generally no endosperm formation at all in these groups. Only exceptionally is there an endosperm which develops beyond a four-nuclear stage. This condition is derived and represents an important synapomorphy for the members of the three families. In Alismatales, Najadales and a great many of the Arales, endosperm is formed, but is used up as the seed ripens. This state is a synapomorphy for Alismatales and Najadales (our Alismatiflorae), but we do not know whether the similar condition in many Araceae constitutes a convergence (or perhaps two or more convergences).

In some families seed nutrients, including starch, are stored in a perisperm or a chalazosperm or both. In these cases the endosperm is correspondingly reduced in amount and generally does not store starch. This is the condition in most families of Zingiberales. A chalazosperm is developed in Costaceae, Cannaceae and Marantaceae. The occurrence of copious perisperm and of perisperm and chalazosperm is obviously a specialization compared with the occurrence of a well-developed endosperm, where nucellar and chalazal tissues are insignificant, and thus is an important synapomorphy for most or all families of Zingiberales. If the development of chalazosperm has arisen only once and there has not been a reversal in Zingiberaceae (which have only perisperm), the presence of chalazosperm can be claimed as synapomorphic for Costaceae, Cannaceae and Marantaceae.

Other taxa where the seeds may contain perisperm and chalazosperm respectively, are the families Agavaceae and Cyanastraceae, which differ in many respects from the Zingiberales. These nutrient tissues have obviously arisen independently.

The bigeneric family Hydatellaceae, has been shown by HAMANN (1975, 1976) to have seeds with perisperm, their endosperm being reduced to a thin layer. The many differences between this and the perispermous Zingiberales indicate that the perisperm has arisen independently in Hydatellaceae (see Taxonomy).

The endosperm, when present in the ripe seed, may contain protein, fat, starch or cellulose in varying proportions. The presence of copious starch is here considered to represent a derived condition in monocotyledons. Endosperm with copious starch is only rarely present in those dicotyledons which have any considerable number of attributes in common with the monocotyledons. Furthermore, starchy endosperm is coupled in the monocotyledons with the presence of the *Strelitzia* Type of epicuticular wax (see p. 64) and of UV-fluorescent organic acids in the cell walls (see p. 75), both of which are regarded as derived attributes. Certain amounts of starch in the endosperm may, however, also occur outside the groups where starch is copious and constant, and in, for example, Liliales, the young seeds often contain starch grains that have disappeared when the seeds become ripe.

The endosperm is ruminate in certain monocotyledons, mostly in Arecales and Cyclanthales but also in some genera of Dioscoreales, e.g. *Trichopus* and *Stenomeris*. Outside the monocotyledons ruminate endosperm occurs particularly in taxa of the dicotyledonous superorder Magnoliiflorae. The condition is a peculiar one, and in some groups it may very well be ancestral, e.g. in Dioscoreales, where it is combined with a small (*Trillium* Type) embryo (see below) and some other attributes found in various Magnoliiflorae. Ruminate endosperm was, however, considered a derived condition in palms by MOORE and UHL (1982).

Embryo size and shape are to some degree typical of the superorders and orders of monocotyledons, although the variation may be great in some orders (e.g. Dioscoreales). Several types are distinguished (see pp. 72–73). A small, more or less linear embryo (*Trillium* Type) is the one that corresponds most closely with that found in those dicotyledons which have a considerable number of attributes in common with monocotyledons, i.e. in Annonales, Magnoliales and some related orders. With increasing size the *Trillium* Type embryo merges into the *Urginea* Type, which we consider more derived. Presumably parallel in the evolutionary sense to the *Urginea* Type, and likewise derived, are some types where the embryo is broad, lens-shaped (*Xyris* Type) or capitate or mushroom-shaped (*Scirpus* Type). Here the embryo is generally rather short in relation to the endosperm but much broader than in the *Urginea* Type. The former is present in some Commelinales, some Cyperales and Poales (excluding grasses) and the latter in other Commelinales, especially Commelinaceae, and most Cyperales. The *Avena* Type embryo may be derivable from a *Xyris*-like Type; it is lateral to the endosperm and asymmetrical. The embryos in the very small seeds of Apostasiaceae, Cypripediaceae and Orchidaceae, without endosperm, and the families of Burmanniales, with endosperm, are undoubtedly derived from the *Trillium* or *Urginea* Types.

A terminal cotyledon is characteristic of most monocotyledons, whereas in dicotyledons the two cotyledons are lateral and the plumule terminal. If we assume that the ancestors of the monocotyledons were dicotyledonous it is logical to conclude that a more or less laterally situated cotyledon, such as is found in Dioscoreaceae, represents a transitional condition and is thus ancestral in the monocotyledons.

## Chromosomal Characters

Surveys of chromosome numbers and their taxonomic implications in monocotyledons have been given by, for example, RAVEN (1975) and GOLDBLATT (1980), the latter account with particular emphasis on polyploidy. Whereas, in some families, the amplitude in chromosome numbers is great, there is in other families little variation. For several small families there are still only single records or, as for example in the families Ecdeiocoleaceae, Geosiridaceae, Mayacaceae, Thurniaceae, Hanguanaceae, Joinvilleaceae and Posidoniaceae, none at all. Well known on the other hand are the Poaceae, Hyacinthaceae, Alliaceae, Amaryllidaceae, Commelinaceae and Iridaceae.

Numerical data are very unevenly scattered in the monocotyledons. Conclusions or speculations on similarity or difference in numbers are somewhat risky if they are not accompanied by careful study of the chromosomes. Extensive studies of this kind exist, for example, for genera of Hyacinthaceae, Commelinaceae, Poaceae and other families, and have proven useful in settling infra- and inter-generic problems, but rarely problems at interfamily level.

One classical case of this kind forms an exception, namely that of Agavaceae-Funkiaceae. McKelvey and Sax (1933) found that members of *Agave* (with epignyous flowers) and of *Yucca* (with hypogynous flowers) have a chromosome complement consisting of 5 large and 25 small chromosomes ($x = 30$). The agreement was found to be so great that a close phylogenetic relationship between the genera was considered probable. At that time *Agave* and other Agavaceae with epignyous flowers were generally placed in Amaryllidaceae s. lat. and *Yucca* and related genera with hypogynous flowers in Liliaceae s. lat. The recognition of a unit, Agavaceae, neglecting to some extent the overestimated hypogyny/epigyny borderline, was a first step to a refined classification of the Liliiflorae. It has also been shown that *Hosta* (Funkiaceae) has a very similar chromosome complement with 5 large and 25 small chromosomes, and may be very closely related, but the significance of the similarity is not yet settled. Similar complements, with a few large and many smaller chromosomes occur in genera of Hyacinthaceae.

Relations between chromosome complement and phylogeny in Commelinaceae have been studied by K. Jones and Jopling (1972) and many similar studies have contributed to taxonomic understanding in the grasses (see Stebbins 1956).

The base number $x = 7$, which is common among those Magnoliiflorae considered to be primitive, and which is sometimes considered the probable base number in the angiosperm ancestor, can be compared with the number $n = 14$ in *Trichopus* (Trichopodaceae), Stemonaceae and Taccaceae, all these in Dioscoreales. We do not attach any considerable significance to this circumstance, but mention it as a matter of coincidence. The same number, $x = 7$, has been considered a likely original base number for the orders Arales, Alismatales, Poales and Zingiberales (Raven 1975).

"Diffuse" or "polycentric" centromeres are present in genera of Juncaceae and Cyperaceae, where this condition is associated with post-reductional meiosis. The two conditions are derived and indicate a close relationship between these two families, which are treated in our order Cyperales.

## Phytochemical Characters

Many problems that arise in the interpretation of chemical characters have been discussed in the chapter on Chemical Characters, but in addition to these there is the possibility that a compound may be restricted to a particular organ or a certain phase of the life cycle or may appear only under certain conditions of light, humidity, temperature, soil, etc.

The biosynthetic chain may be broken in derived plant groups so that less complex compounds occur in them than in their ancestors ("reversals" in an evolutionary sense). Convergence, loss and other phenomena are as relevant here as with morphological criteria, so the different kinds of characters need not be judged on entirely different grounds.

The cyanogenic compounds occurring in the monocotyledons are, with only scattered exceptions, formed along the tyrosine pathway. This is also true of the cyanogenic compounds in nearly all Magnoliiflorae, many Ranunculiflorae and certain other groups of dicotyledons (Hegnauer 1977). Although cyanogenic compounds are rather scattered in the monocotyledons, their uniformity of synthesis seems to support the theory that they occurred in the monocotyledon ancestors, and were then largely lost. We hold reservations about this hypothesis, which is based mainly on the lesser likelihood that cyanogenesis arose repeatedly, and always with the same biosynthetic pathway. The assumption that the cyanogenesis was present in monocotyledon ancestors and remains as a relict in scattered families leads to the conclusion that their presence is not of any considerable significance.

Steroidal saponins occur in various Liliiflorae, especially in Dioscoreales, Asparagales, Melanthiales, and Liliales (Iridaceae, Liliaceae), but are otherwise reported only sporadically in monocotyledons. Steroidal saponins may provide chemical protection, which is consistent with the fact that they are lacking in Amaryllidaceae and Colchicaceae where alkaloids are present (different in the two families). In the dicotyledonous Magnoliiflorae steroidal saponins are absent (or at least rare). We assume that they arose in the monocotyledons either at an early stage and were lost in some evolutionary lines or arose at an early stage in the ancestors of one or more lines leading to the orders mentioned.

Chelidonic acid has approximately the same distribution in the monocotyledons as have the steroidal saponins, being common in Dioscoreales, Asparagales, Melanthiales, Liliales and Haemodorales. A few reports of chelidonic acid come from taxa of other orders, and so this compound may have arisen repeatedly. Hence its shared occurrence is a dubious synapomorphy.

HARRIS and HARTLEY (1980) discovered that some organic acids, viz. bound ferulic acid, p-coumaric acid and diferulic acid in trans forms occur in the unlignified cell walls of many monocotyledons and show blue fluorescence in UV light. The distribution of fluorescing components in monocotyledons has a highly interesting and consistent pattern; they occur in all members of Bromeliiflorae, Zingiberiflorae, Commeliniflorae and Areciflorae investigated, but are lacking in all taxa of the Liliiflorae, Ariflorae, Alismatiflorae, Pandaniflorae and Cyclanthiflorae that have been studied, except for the single member of Velloziaceae (Bromeliiflorae) studied, which was UV-negative. We interpret the occurrence of these substances as a derived condition, possibly even a single synapomorphy for all the groups having them. As regards the presence of the individual acids, some were present in *Carludovica* (Cyclanthales) and *Lomandra* (Asparagales), where no fluorescence was observed, indicating that caution should be exercised in interpreting the results. (Note the coincidence in distribution between the UV-fluorescent compounds mentioned, the *Strelitzia* Type of epicuticular wax, and, except in Areciflorae, of a starchy endosperm.)

Another possibly significant chemical character is the occurrence of the iso-enzymes of dehydroquinate hydrolyase (DHQ-ase) (BOUDET et al. 1975), which represents a free form specifically activated by shikimic acid. This DHQ-ase occurs in fewer orders than do the UV-fluorescent compounds just mentioned, viz. in certain Commeliniflorae, especially Cyperales and Poales. Only rarely does it occur in genera of Liliiflorae (so far reported in only four genera). Although less significant than the above character and probably not wholly monophyletic (as indicated by the occurrence in a few Liliiflorae), it may represent a synapomorphy either for Poales-Cyperales or one for each of these orders.

Among the flavonoids (see HARBORNE in DAHLGREN and CLIFFORD 1982; GORNALL et al. 1979), some have phylogenetic significance. Thus it is highly significant that several kinds of the flavonoids, viz. tricin, sulphonated flavonoids, proanthocyanins, C-glycoflavones, and luteolin/apigenin are all found in Poales, Cyperales and Arecales, although the last three also occur in other orders. HARBORNE (1973) lays much stress on the distribution of the first two and we consider the presence of at least tricin and sulphonated flavonoids as derived, although it is unknown whether each of them has arisen only once or independently in two or more lines in the three orders.

## Parasites

Most plants are subject to attack by a bewildering array of parasites (both plants and animals), although the total host range of many parasitic organisms is imperfectly known. However, there are few higher plants that are not eaten or used as a site for oviposition, do not supply sap to suctorial insects or are not invaded as a source of shelter. Likewise plants serve as hosts for a variety of fungi (rusts, smuts, mildews) and bacterial and viral pathogens as well as parasitic angiosperms such as Orobanchaceae, Scrophulariaceae, Loranthaceae and others.

Such attacks do not appear to be random and in many cases particular parasites are restricted to one host or a limited set of hosts. This is probably because the host provides suitable nutrients in the form of proteins or secondary metabolites, essential for the completion of the life history of the parasite. The joint sharing of parasites may therefore reflect similarities in the metabolism of the host and so comprise a potential synapomorphy. This is especially so when the taxonomic range of hosts is not broad. Parasites do sometimes extend their ranges to include distantly related taxa, in which case shared parasites must be regarded as convergent.

# Evolution Within the Monocotyledons

Theories of evolution within the monocotyledons must be based on the character states considered to be ancestral versus derived in the group. Earlier, we presented two contrasted hypotheses on the origin of the monocotyledons, one regarding as ancestral certain states that are found in particular dicotyledonous groups (Magnoliiflorae-Nymphaeiflorae), the other regarding the same states as more or less derived in the monocotyledons but ancestral in the dicotyledons (BURGER 1981).

We have stated that we favour the first hypothesis, and our model for the evolution of the monocotyledons will be described in some detail below, largely in accordance with DAHLGREN and RASMUSSEN (1983).

First, however, we shall consider the alternative model.

## The Hypothesis of Monocotyledons as Ancestral Angiosperms

If the first angiosperms were small, herbaceous, monocotyledonous plants with simple stems, as proposed by BURGER (1981), evolution must have proceeded mainly in the direction of diversification of the stem to become increasingly branched, of differentiation of the leaves into a petiole and a broad lamina and of a narrowing of the leaf base and reduction of its sheath. We would assume, then, that the vessels of these leaves – although in extant monocotyledons they are generally associated with simple or mixed simple and scalariform perforation vessels, as in many Commelinales and most Poales and Cyperales – had scalariform perforation plates, that the stomata were paracytic, the flowers small, trimerous, wind-pollinated (and consequently without floral nectaries) and the pollen grains smooth and monoaperturate (in wind-pollinated taxa they are often ulcerate, but this is not likely to be ancestral, we presume, even with this theory). Further, to judge from the extant distribution and combination of character states in monocotyledons, the trimerous gynoecia probably became syncarpous at an early stage and had few crassinucellate ovules per carpel. Endosperm formation would have been nuclear, the seeds would have had copious endosperm, possibly filled with starch grains, and the embryo would have been a small, lens-shaped body close to the micropyle.

Evolution would then have proceeded towards the formation of various underground organs as adaptations to dry periods, mainly by the development of thick underground stems in the form of rhizomes, tubers or the short plate-like stems of bulbs. We cannot judge the chances that these are derived from the annual condition; in angiosperms such structures are usually considered primary in relation to the annual habit. The more differentiated vegetative structures and the woody forms in many Asparagales would, on this hypothesis (as with the magnoliifloran hypothesis), be secondary. The insect-pollinated types, in particular the Dioscoreales, would be highly advanced, and plants of Dioscoreales-like appearance (pre-Dioscoreales) would be those most likely to have taken the step over to the dicotyledonous state (cf. the several dicotyledonous attributes in *Trichopus* and other dioscorealean monocotyledons). One could, alternatively, assume that monocotyledons of alismatifloran appearance, ancestral to the Aponogetonaceae, Butomaceae and Limnocharitaceae, gave rise to Nymphaeales-like dicotyledons. To us this possibility appears even less likely, however, in the light of the very specialized characters of the Alismatales and Nymphaeales. Finally, an origin of Piperales-like dicotyledons directly from presumed monocotyledons such as the precursors of Dioscoreales and Arales is highly unlikely.

## The Ranalean/Magnoliifloran Hypothesis

The evolution of the monocotyledons becomes very different if we choose the magnoliifloran hypothesis, according to which the monocotyledons are derived from pre-Magnoliiflorae.

## The Liliiflorae

### The Dioscoreales

The Dioscoreales in this view becomes the most central order in the monocotyledons, even though extant representatives do not correspond precisely to the presumed ancestors. Thick vertical stems or rhizomes, herbaceous stems, petiolate leaves with a non-sheathing base and a broad lamina with reticulate venation are attributes that are assumed to have been present in the earliest monocotyledons. The venation of the primitive monocotyledons was probably neither as regular nor as dense as in most extant Dioscoreales, but rather like that in the fossil *Acaciaephyllum* (Fig. 24). A climbing stem, as in most extant Dioscoreales, may or may not have been developed in the earliest monocotyledons.

Extant Dioscoreales exhibit many specializations: in the climbing stems and their anatomy, the wide leaves with finely and regularly reticulate veins, the lamina which is sometimes compound (in species of *Dioscorea* and *Tacca*), and in floral characters. Many members of Dioscoreales (all Dioscoreaceae, Trichopodaceae and Taccaceae and some Stemonaceae) have epigynous flowers, which are undoubtedly secondary to hypogynous. The numeric condition in certain genera (*Stichoneuron, Croomia, Stemona* and *Pentastemona* of Stemonaceae and *Paris* of Trilliaceae) deviates from the trimerous state, and the stamens with the microsporangia (possibly primitively) situated below the apex are highly variable, and may be connected with specialized pollination mechanisms, although on the whole the members of the two perianth whorls are similar and rather inconspicuous. Arillar structures of different kinds are frequently present on the seeds of dehiscent fruits but the seed coat is unspecialized. In other cases the fruits are berries. The rather large and even T-shaped embryos in some Dioscoreaceae, presumably represent a derived condition, but minute embryos also occur. The endosperm may be ruminate (Trichopodaceae, Dioscoreaceae), which is possibly ancestral, as it occurs also in many taxa of Annonales (dicotyledons), which exhibit some of the same features as Dioscoreales. The same argument applies to the lateral cotyledon in some species of *Dioscorea*.

Thus, there are in extant Dioscoreales some attributes that we suspect are ancestral and others that are obviously derived. The general view that the Dioscoreales exhibit a concentration of features found in ancestral monocotyledons has been stressed in particular by HUBER (1969), although the evaluations here may deviate from his in some respects.

The further evolution of monocotyledons from the first, somewhat Dioscoreales-like, ancestors is likely to have proceeded in several directions. Some branches undoubtedly gave rise to other liliifloran complexes, among which the orders Asparagales, Liliales, Burmanniales and, perhaps separately, the Melanthiales may represent four parallel evolutionary branches (see DAHLGREN and RASMUSSEN 1983). Another main evolutionary branch has probably given rise to the Ariflorae, Triuridiflorae and Alismatiflorae. Further, the Bromeliiflorae, Zingiberiflorae and Commeliniflorae, which we assume to have a common ancestry, probably evolved from the first monocotyledons. More problematic are the three very dubiously closely related orders Arecales, Cyclanthales and Pandanales. In fact, the basis for assuming that these are close to each other is very tenuous and we shall here treat each as a separate superorder (DAHLGREN and RASMUSSEN 1983) as also does THORNE (1983). The Arecales (Areciflorae) turn out to have many similarities with the Poales, but this may be due to convergence. Cyclanthales may be an early derivative on the pre-Dioscoreales-pre-Arales line, but at the same time they show similarities to the palms in leaves and other structures; like certain Dioscoreales (Stemonaceae) they tend to have tetramerous flowers. Pandanales, again, has a highly uncertain ancestry; it might be shared with that of Areciflorae, Bromeliiflorae or Commeliniflorae.

Below, we comment in some detail on possibilities for the evolution of these complexes. In doing so, we shall start with the liliifloran orders mainly because the Dioscoreales are here referred to the Liliiflorae, and *not* because we consider the lily-like monocotyledons with large petaloid perianths as ancestral.

### The Asparagales

This order, first recognized with approximately the present circumscription and characterization by HUBER (1969), is believed to form a monophyletic group mainly on the basis of fruit and seed characters. As a rule the inner integument (tegmen) of the seed coat in the ripe seeds has completely collapsed, forming a thin (brown or colourless) membrane. The outer wall of the outer epidermis of the testa in nearly all capsule-fruited members, and also in some berry-fruited genera (*Geitonoplesium,*

*Asparagus, Dianella*) is provided with a more or less thick layer of phytomelan (see p. 132). The inner epidermis of the outer integument (testa) of the seed coat lacks the droplets of fat sometimes found in other Liliiflorae. In the berry-fruited taxa, the outer integument of the seed coat may partially disintegrate.

In this order (some berry-fruited families excepted) the flowers generally have septal nectaries, the tepals are non-patterned, and the style is usually simple. Of these attributes the last appears to be derived. We consider it highly unlikely that the phytomelan-coated seeds have arisen in several evolutionary lines in the monocotyledons. It is true that there are also blackish brown seeds in a few monocotyledon genera outside Asparagales, but the pigment in these groups does not seem to be phytomelan.

As regards the berry-fruited taxa, the loss of phytomelan is expected, for this highly resistant substance which gives the seeds protection is no longer required. Thus, in Asparagales berries are logically regarded as derived from capsules.

We have reservations on this, however, because one explanation of the evolution of phytomelan (HUBER, personal communication) is that a degeneration of the seed coat in the course of a stage with baccate fruits has later been compensated for by the formation of the phytomelan when the fruits, secondarily, became dehiscent.

Within the Asparagales nuclear endosperm formation and baccate fruits may be synapomorphic for one group of families including Philesiaceae (?), Luzuriagaceae, Convallariaceae, Asparagaceae, Ruscaceae, Herreriaceae, Dracaenaceae and Nolinaceae, which possibly form a monophyletic unit worthy of subordinal rank. The Philesiaceae show some obvious similarities to Alstroemeriaceae of the Liliales, but whether they serve as a "link" between Asparagales and Liliales and/or Dioscoreales is not fully clear. There is a great vegetative similarity between some genera of Convallariaceae and a few genera of the lilialean Uvulariaceae (incl. *Disporum* and *Clintonia*) with baccate fruits. The uvulariaceous affinity of these genera is seen from their lack of oxalate raphides and septal nectaries (both present in true Convallariaceae), the presence of perigonal nectaries (absent in true Convallariaceae) and even spurs (as in *Tricyrtis*), as well as some embryological characters such as nucellus type and lack of a parietal cell (present in Convallariaceae) (BJÖRNSTAD 1970).

A second probably monophyletic group of families within Asparagales is made up of Asphodelaceae,

Hemerocallidaceae, Doryanthaceae, Phormiaceae, Ixioliriaceae, Tecophilaeaceae and Cyanastraceae, which are all characterized by simultaneous microsporogenesis, a character state which we do not expect would appear very easily. Within this group the last three families also have corms, which may be superficially bulb-like, but which lack thick bulb scales.

Also probably closely allied to each other are Hyacinthaceae, Alliaceae and Amaryllidaceae, most taxa of which have bulbs, and all of which have a leafless, frequently fleshy, scape that carries an originally determinate inflorescence which varies in appearance from spicate (Hyacinthaceae) to umbel-like (Alliaceae, Amaryllidaceae).

A few taxa with capsular fruits in Asparagales do lack phytomelan. These include *Eriospermum* (Eriospermaceae) and *Blandfordia* (Blandfordiaceae), which have seeds with hairs, *Walleria* (Tecophilaeaceae), which has seeds with short, peg-like papillae, and the genera of Dasypogonaceae and Cyanastraceae (*Cyanastrum*). These are considered here to represent separately derived lines; their position in Asparagales, though sometimes doubted, is made probable by close similarities to taxa with phytomelaniferous seeds.

### The Liliales (incl. Orchidales) (Fig. 37)

This order also is here considered monophyletic on the basis of some character states shared by nearly all of its members, namely nuclear endosperm formation, perigonal nectaries (and lack of septal nectaries), and a variegated pattern on the tepals. Whereas the nuclear endosperm formation, which with some misgivings we consider derived from the Helobial Type, is found in all Liliales, the nectaries are septal in one group (Iridaceae subfam. Ixioideae). This exception is a puzzling one and possibly represents a specialization within the family. As regards the variegation of the tepals this occurs in all the families but is obviously an easily lost attribute. It is, however, considered significant because variegation of tepals in monocotyledons other than Liliales is rare.

Within the order further differentiation may have gone in the direction of loss of a parietal cell for one group of families and loss of oxalate raphides, loss of vessels in the stem, and acquisition of extrorse stamens in another group.

The group lacking a parietal cell consists of Alstroemeriaceae, Apostasiaceae, Cypripediaceae and Orchidaceae, a group that – as an alternative – may be supplemented with Uvulariaceae, Calo-

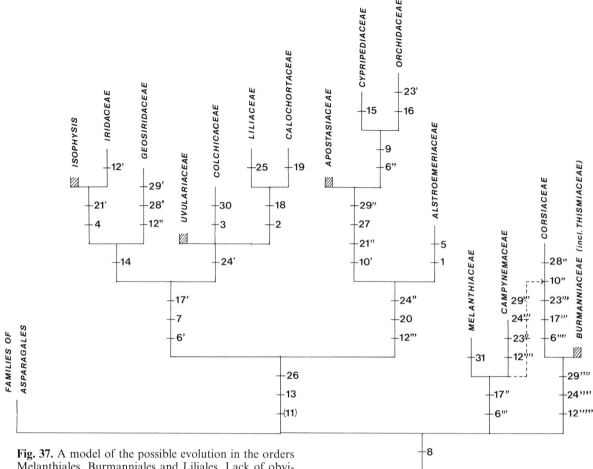

**Fig. 37.** A model of the possible evolution in the orders Melanthiales, Burmanniales and Liliales. Lack of obvious autapomorphies is indicated by *shading*. (Orig.) Note that apomorphies are here mentioned *before* the plesiomorphies.

 1  Storage roots – roots not swollen storage organs
 2  Bulb – bulb lacking
 3  Corm – rhizome
 4  Leaves ensiform – leaves dorsiventral
 5  Leaf lamina inverted – leaf lamina not reversed
 6  Vessels lacking in stems – vessels present (and with scalariform perforation plates)
 7  Oxalate raphides lacking – oxalate raphides present
 8  Perianth petaloid – perianth inconspicuous
 9  Median tepal of outer whorl specialized to form a labellum – all tepals of outer whorl more or less similar
10  Flowers zygomorphic – flowers actinomorphic
11  Tepals variegated, often with checkered, drop-like, or striated pattern – tepals with more or less uniform colour
12  Flowers epigynous – flowers hypogynous
13  Perigonal nectaries present – perigonal nectaries lacking (septal nectaries then generally present)
14  Flowers with three stamens only (the outer staminal whorl) – flowers with 3 + 3 stamens
15  Flowers with two functional stamens only – flowers with three or more functional stamens representing both staminal whorls
16  Flowers with one functional stamen only – flowers with three or more functional stamens representing both staminal whorls

17  Anthers extrorse – anthers introrse
18  Anther connective "pseudo-basifixed", in reality forming a tube over the filament tip thus appearing basifixed – filament tip not enveloped by connective
19  Style infinitely short, stigma crests on top of ovary – style distinct
20  Stigma wet – stigma dry
21  Microsporogenesis simultaneous – microsporogenesis successive
22  Pollen grains inaperturate – pollen grains with a distinct sulcus
23  Placentation parietal in a unilocular ovary – placentation axile in connection with a trilocular ovary
24  Parietal cell not cut off from primary archesporial cell – parietal cell cut off, and forming parietal tissue
25  *Fritillaria* Type embryo sac formation – embryo sac formation of *Polygonum* Type
26  Endosperm formation nuclear – endosperm formation helobial
27  Endosperm formation arrested at a very early stage (not more than 16 nuclei formed) or wholly inhibited – endosperm formation far beyond the 16-nucleate stage
28  Plants achlorophyllous – plants chlorophyllous
29  Seeds minute – seeds not minute
30  Colchicine alkaloids present – colchicine alkaloids lacking
31  Plants attacked by the related fungal parasites *Puccinia atropuncta* and *Uromyces veratri* – plants not attacked by these fungi

chortaceae and Liliaceae, also without parietal cell, which could also have evolved along the other branch (and, perhaps, Philesiaceae, which we place in Asparagales).

The three families comprising the "orchids" share a strong reduction in the androecium, so that *at least* two anthers are missing in the outer staminal whorl (the non-medial ones) and one in the inner staminal whorl (the medial one). In addition to this loss which makes the androecium zygomorphic, the perianth is also generally zygomorphic. A labellum is shared in common by the two parallel families Cypripediaceae (with the two anthers of the inner staminal whorl functional) and Orchidaceae s. str. (with only one anther of the outer staminal whorl functional). Although information is still somewhat incomplete, it seems that the three families also have as synapomorphies a Wet Type of stigma, simultaneous microsporogenesis, and a (nuclear) endosperm formation that is either arrested at a very early stage (with a maximum 16 nuclei in the endosperm) or fails to develop at all. These families, on the basis of their obvious similarities are often treated as one, Orchidaceae, which is frequently placed in its own order, Orchidales.

Alternative interpretations of orchids have been proposed. A theory proposed long ago is that they are closely related to the likewise small-seeded Burmanniales, which also have epigynous flowers and frequently parietal placentation (see further in GARAY 1972). Others consider that the closest relationships are with families that we place in our Asparagales (DRESSLER, personal communication and 1981).

We shall here mention another alternative hitherto neglected completely by current botanists. The presence of "hat-shaped", roughly conical silica bodies in the epidermal cells of Apostasiaceae, Cypripediaceae and many Orchidaceae has its counterpart in Zingiberales (H. RASMUSSEN, personal communication). It is further to be noted that Apostasiaceae is reported to have vessels not only in the roots but also in stems and leaves, which is only rarely the case in our Liliales. Further, the stomata often have two to four subsidiary cells, as is the case in Zingiberiflorae and Bromeliiflorae. Epicuticular wax of any of the *Convallaria* or the *Strelitzia* Types has not been found, so there is no indication in that character either in favour of a liliifloran position (Fig. 31). Root-hair epidermal cells may or may not be shorter than other epidermal cells (Fig. 27). As the formation of endosperm (nuclear when present) is hampered at a very early stage one cannot conclude whether the orchid ancestors had a starchy endosperm or not. It is true that features such as the simultaneous type of microsporogenesis (Fig. 33), the lack of UV-fluorescent compounds in the cell walls (Fig. 36), the spotted tepals and the occurrence of spurs argue for a liliifloran-lilialen affinity, but a profound future re-evaluation of the ancestry of orchids would not come as a total surprise to the authors.

The second group of families in Liliales (lacking oxalate raphides etc.) includes Uvulariaceae, Colchicaceae, Calochortaceae and Liliaceae which, like the first group, also have ovules in which a parietal cell is not formed. The other families in the same group are Iridaceae and Geosiridaceae. The latter of these are obviously closely allied, and might form a single family (Iridaceae in a wider sense). The Uvulariaceae are rhizomatous and the Colchicaceae have a corm, whereas the Calochortaceae and Liliaceae s.str. are bulbous plants, the bulbs being probably here a synapomorphy. A possible autapomorphy for the genera of Colchicaceae is their possession of colchicine-type alkaloids, but according to current information they are not always present. Also most Iridaceae have a corm, but this is probably not plesiomorphic in the family. Rhizomes occur in a number of supposedly primitive Iridaceae, e.g. the hypogynous *Isophysis,* and also in Geosiridaceae (*Geosiris*). The ancestors of Iridaceae and Geosiridaceae obviously lost one staminal whorl. Most Iridaceae (except *Isophysis*) and the Geosiridaceae (*Geosiris*) having epigynous flowers.

The ancestral Iridaceae as judged from *Isophysis* had hypogynous flowers (unless *Isophysis* represents a reversal in this respect). The subfamily Ixioideae of Iridaceae is anomalous in having septal nectaries (otherwise absent in Liliales) and racemose inflorescences (cymose-paniculate in other Iridaceae). Provisionally, we regard these attributes in Ixioideae as derived states.

**The Burmanniales**

The families of this order (Burmanniaceae, Thismiaceae and Corsiaceae) differ from the Liliales in having helobial endosperm formation, unvariegated tepals, and production of nectar in septal nectaries or on the surface of the ovary walls, the latter condition derivable from septal nectaries (RÜBSAMEN 1983). Thus they could only be included in Liliales if they have undergone reversals in these attributes.

In the past they have been closely associated with the Apostasiaceae, Cypripediaceae and Orchidaceae, having in common with these families parietal placentation, minute seeds, lack of a parietal cell (a consequence of having minute ovules) and epigynous flowers. However, the Burmanniales differ in having successive, not simultaneous, microsporogenesis, normal, not rapidly arrested, and helobial, endosperm formation, and copious endosperm in the ripe seeds. Primitive taxa of both groups have axile placentation. Thus the most conspicuous similarities are probably convergences.

The general seed structure, the helobial endosperm formation, which is of a type similar to that in Bromeliiflorae, the fact that the nearly ripe seeds are sometimes rich in starch (HAMANN, personal communication), the occurrence of vessels in the stems of *Burmannia* and some other details are reminiscent of conditions in Philydraceae (Philydrales, Bromeliiflorae) (HAMANN, personal communication). Obstacles to the association of Burmanniales with Bromeliiflorae, however, are the anomocytic stomata and lack of UV-fluorescence in the cell walls. As these are liliifloran characters we tentatively suggest that Burmanniales evolved from primitive ("liliifloran") monocotyledons as a separate evolutionary line, and regard an origin in early Liliales as unlikely.

### The Melanthiales

The Melanthiaceae also lack the synapomorphies that are used to define the Liliales, but are obviously liliifloran. The family contains some genera with septal nectaries and others with perigonal nectaries; the tepals are generally not variegated and the endosperm formation, unlike that in all other Liliales, is helobial (a plesiomorphy here?). The variability in attributes that are of significance in other groups makes the treatment difficult. AMBROSE (1980), who made a numerical study of the similarity of the genera, arrived at a grouping which is largely followed in the taxonomy of the family. The Campynemaceae form a separate line in Melanthiales, having epigynous flowers and parietal placentation.

### The Ariflorae-Triuridiflorae-Alismatiflorae Complexes (Fig. 38)

It is concluded here that this complex is monophyletic on the basis of a number of presumed synapomorphies, namely extrorse anthers, amoeboid tapetum, Caryophyllad Type of embryogeny and, perhaps, mainly *Urginea* Type embryos. We assume that the ancestors were somewhat Dioscoreales-like plants. Intravaginal squamules in the leaf axils are found in a relatively low proportion of the Ariflorae and in all Alismatiflorae; their disappearance in most Ariflorae and in the Triuridiflorae is not unlikely. Also laticifers are present in many Arales and about half of the Alismatales,

but convergent evolution of these is not improbable.

### The Ariflorae

The Ariflorae (=Arales) are presumed to have deviated from this main line at an early stage; they exhibit a number of autapomorphies including sieve tube plastids with starch (PIIcs, sensu BEHNKE 1981), tetracytic stomata (few exceptions), a spadiciform inflorescence subtended by a spathe, lack of septal nectaries (if these were present in the ancestors), cellular endosperm formation, and starchy endosperm in the ripe seeds (more often the endosperm is used up as the seed ripens, a further stage of advancement). Some dioscorealean features are still seen in the group, such as the broad, petiolate leaves, frequently with reticulate venation, and randomly oriented stomata, and the inflorescence type in the Potheae, which is not very different from the one in some small-flowered Dioscoreaceae.

Within the Araceae, differentiation proceeded further in the reduction of the perianth, in the loss of either sex (or both sexes) in the flowers and the segregation of the male and female (and often of neuter) flowers within the spathe, which may be constricted and sometimes even provided with a diaphragm, which are specializations for particular types of pollination. Production of odorous substances, the colouring of the spathe and the heating of the spadix appendix are involved in the specialized pollination syndrome. Two parallel lines of evolution in the Araceous complex have obviously led to the floating Araceae subfam. Pistioideae and the Lemnaceae, the former less reduced in vegetative respect but with endospermless seeds, the latter more reduced but with endospermous (and sometimes starchy) seeds.

### The Triuridiflorae

The Triuridiflorae, consisting of Triuridaceae only, and the Alismatiflorae may have a shared ancestry, their ancestors having acquired the following attributes that are synapomorphic for the two superorders (and in which they differ from the Ariflorae): lack of oxalate raphides, occurrence of three-celled pollen grains and apocarpy. It is not customary to consider apocarpy as a derived attribute, but the case for doing so in Alismatales is commented on above under Ancestral and

Derived States. Evolution of apocarpy in connection with reduced size and increased number of carpels is for us a conceivable possibility.

The Triuridiflorae are characterized by the following autapomorphic character states: lack of nectaries on the carpel sides (septal nectary homologues); lack of pollen grain apertures, or rather of a continuous ektexine; basal placentation (if that is not also ancestral in the Alismatiflorae); one ovule per carpel; lack of a parietal cell; nuclear endosperm formation; and lack of chlorophyll (the Triuridales being small saprophytes). In addition, the flowers are usually unisexual in Triuridiflorae and the fruits usually indehiscent (nutlets), but there are single exceptions to these states, one with bisexual flowers and one with follicles, which are assumed to be ancestral states.

## The Alismatiflorae

The Alismatiflorae are considered to form a monophyletic group with the following autapomorphies: root hair cells shorter than other epidermal cells on the roots, presence of intravaginal squamules and lack of endosperm.

Of all the Alismatiflorae, the unigeneric family Aponogetonaceae shows the fewest additional autapomorphies, having spicate inflorescences and three or fewer tepals. The family is placed in Alismatales (one of the two orders making up the superorder). Other families in Alismatales tend to have the sepals differentiated into an outer sepaloid and an inner petaloid whorl.

Members of the Aponogetonaceae may have six stamens and three carpels, but higher numbers are known for both types or organ. This tendency is stronger in other families of Alismatales and the higher numbers are considered to be derived (see Ancestral and Derived Character States).

The families fall into two groups in respect of placentation, which is basal in Aponogetonaceae and Alismataceae, and laminar-dispersed in Butomaceae, Limnocharitaceae and Hydrocharitaceae. Possibly basal placentation is ancestral in the superorder but, if not, both states may be derived from axile placentation.

Limnocharitaceae and Alismataceae represent a secondary monophyletic branch within the Alismatales, as is indicated by their shared possession of secretory ducts (TOMLINSON 1982), the Girdle Type thickenings of the endothecial cells, the pollen grains with two to numerous foramina, the horseshoe-shaped embryo, the lack of parietal cell,

and the *Allium* Type embryo sac. The Hydrocharitaceae form another clade characterized by the epigynous flowers.

Whereas the Alismatales frequently have numerous stamens and carpels, this is not the case in the remaining families which make up our order Najadales. These do not produce floral nectar and the pollen grains have a thin exine (without distinct apertures) or none. Scheuchzeriaceae and Juncaginaceae, which could perhaps be treated as one family, have basal placentation and are strongly cyanogenic, which attributes are both somewhat doubtful apomorphies. In Juncaginaceae the carpels each contain only one ovule and tend to be fused centrally, whereas in Scheuchzeriaceae the ovules are two per carpel and the carpels are more or less free, forming follicular fruits. The latter attribute is here considered to be ancestral. Both families are terrestrial marsh plants (rarely largely inundated), whereas the other members of Najadales are aquatic.

In the aquatic families the perianth is further reduced or lacking altogether, the carpels contain one ovule only, which is basal (Najadaceae) or lateral to apical (other families), and the fruits are always indehiscent and monocarpellary. They are probably closely related, and the further relationships are sometimes uncertain.

Among them, the Potamogetonaceae, Posidoniaceae and Zosteraceae have spicate inflorescences: their tepal-like structures are regarded as modified connectives of the stamens though by some, e.g. HUTCHINSON (1973), they are considered to be true tepals. The placentation is lateral or apical, the ovules campylotropous to orthotropous, and the embryo is more or less curved. Whereas the pollen grains in Potamogetonaceae are globose or (in *Ruppia*) elongate, in Posidoniaceae and Zosteraceae they are exineless and become filiform when dispersed, which occurs in the water (and representing then obviously pollen grains with pollen tubes). Both of the families grow in salt water. They have three or fewer stamens, a monocarpellary gynoecium and orthotropous ovules; the Zosteraceae, further, have flat inflorescence axes, male flowers with one stamen only, a nucellus lacking a parietal cell and nuclear endosperm formation.

Zannichelliaceae, Cymodoceaceae and Najadaceae have unisexual flowers clustered in the leaf axils and male flowers with three or fewer stamens. It is uncertain whether they form a clade. Among them, the Najadaceae are aberrant in having a solitary stamen, Reduced Type anther wall formation, a solitary carpel with a basal, anatropous ovule and nuclear endosperm formation, and may form a separate line of evolution. Zannichelliaceae and Cymodoceaceae are obviously more closely related, having apical placentation with an orthotropous ovule, and lacking the apomorphies of Najadaceae. Whereas Cymodoceaceae are marine, and have pollen grains dispersed as pollen tubes, Zannichelliaceae grow in brackish water or salt lakes, and do not have pollen grains dispersed as filiform pollen tubes.

## The Bromeliiflorae-Zingiberiflorae-Commeliniflorae Complexes (Fig. 39)

These large complexes are linked by some similarities, viz. starchy endosperm, epicuticular wax of the *Strelitzia* Type (BARTHLOTT and FRÖHLICH 1983), and the occurrence in the cell walls of UV-fluorescent organic acids (HARRIS and HARTLEY 1980) that we assume to be synapomorphic, as also, perhaps, is the occurrence of two or four (or six) subsidiary cells in the stomatal complexes.

---

**Fig. 38.** A model of the possible evolution of the Ariflorae, Triuridiflorae and Alismatiflorae. The Limnocharitaceae are here included in the Alismataceae. The position here of Triuridiflorae is uncertain (see the taxonomic text for this superorder). (DAHLGREN and RASMUSSEN 1983). Note that apomorphies are here mentioned *before* the plesiomorphies.

1 Shoot reduced to plate-like bodies (Lemnaceae) – shoot differentiated into stem and leaves
2 Stipules present – stipules lacking
3 Root-hair cells shorter than other epidermal cells – root hair cells equalling other epidermal cells
4 Vessels lacking in stem – vessels present in stem
5 Laticifers present – laticifers lacking
6 Sieve tube plastids with starch grains – lacking starch grains
7 Oxalate raphides lacking – oxalate raphides present
8 Stomata tetracytic – stomata not tetracytic
9 Intravaginal squamules present – lacking
10 Flowers in small dense clusters in leaf axils – flowers in panicles
11 Flowers in spikes or spike-like inflorescences – flowers in panicles
12 Flowers on a spadix (with a single spathe) – flowers in panicles
13 Inflorescence axis flat (with male and female flowers alternating) – inflorescence axis terete
14 Perianth hyaline or lacking – perianth fairly well-developed
15 Perianth differentiated into a more or less sepaloid outer and a petaloid inner whorl – perianth whorls similar or nearly so
16 Tepals lost – tepals present

17 Stamens with flat petaloid appendage – stamens without such an appendage
18 Flowers unisexual – flowers bisexual
19 Flowers epigynous – flowers hypogynous
20 Nectar production on carpel sides lacking – nectar secretion on carpel sides (incl. septal nectaries) present
21 Stamens more than six and/or carpels more than three by "dédoublement" – stamens six and carpels three
22 Stamens reduced to three or fewer – stamens six
23 Stamens solitary – stamens two or more
24 Anthers extrorse – anthers introrse or latrorse
25 Anthers latrorse – anthers extrorse
26 Anther wall formation of reduced type – anther wall of Monocotyledonous type
27 Endothecial thickenings of Girdle Type – endothecial thickenings of Spiral Type
28 Tapetum amoeboid – tapetum glandular
29 Pollen grains dispersed in dyads – pollen dispersed as separate grains
30 Pollen grains foraminate – pollen grains sulcate
31 Pollen grains inaperturate (entirely or almost lacking exine) – pollen grains sulcate, with continuous exine
32 Pollen grains germinating in water, dispersed as pollen tubes – pollen grains dispersed in their original form
33 Pollen grains prevailingly trinucleate when dispersed – pollen grains prevailingly binucleate when dispersed
34 Gynoecium apocarpous (here regarded as derived) – gynoecium syncarpous
35 Carpel solitary – carpels 2 or 3 per flower
36 Placentation basal – placentation marginal on carpel
37 Placentation laminar-dispersed – placentation marginal (or basal)
38 Placentation apical – placentation marginal (or basal)
39 A single ovule per carpel – two or more ovules per carpel
40 Ovules orthotropous – ovules anatropous
41 Parietal cell lacking – parietal cell (and parietal tissue) present
42 *Allium* Type embryo sac formation – *Polygonum* Type embryo sac formation
43 Cellular endosperm formation – helobial endosperm formation
44 Nuclear endosperm formation – helobial endosperm formation
45 Caryophyllad Type of embryo – Asterad Type of embryo
46 Fruits follicular – fruit a syncarpous capsule
47 Fruits indehiscent achenes or drupes – fruits follicular
48 Fruit (fleshy) enveloped by receptacle – fruits follicular not enveloped by receptacle
49 Fruit baccate – fruit a syncarpous capsule
50 Fruit a schizocarp of three achenes – fruits follicular
51 Endosperm lacking (used up) in the ripe seed – seeds endospermous
52 Endosperm starchy – endosperm not starchy
53 Embryo curved – embryo straight
54 Embryo of *Urginea* Type – embryo of *Trillium* Type
55 Plants lacking chlorophyll – plants chlorophyllous
56 Cyanogenic compounds abundant – cyanogenic compounds sparse

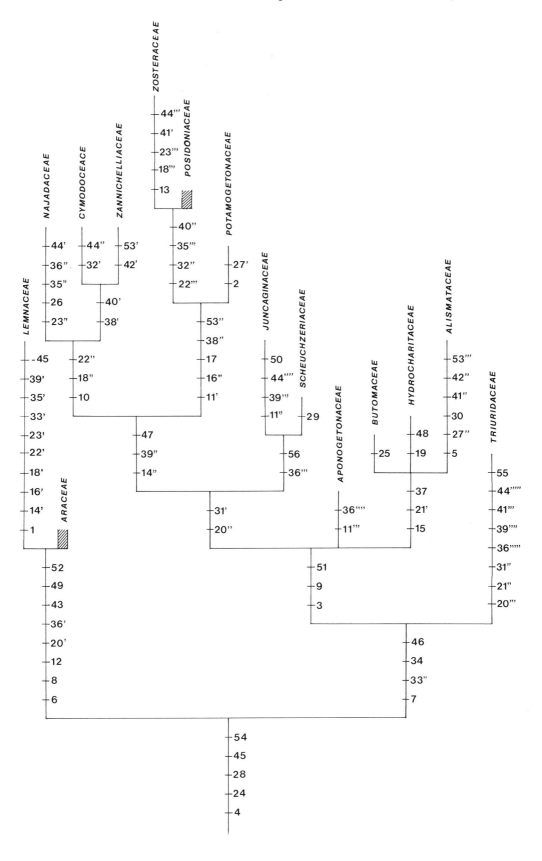

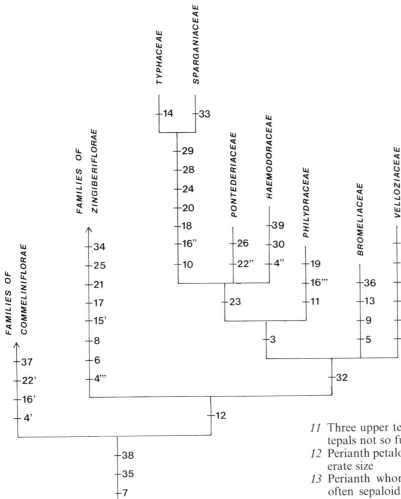

**Fig. 39.** A model of the possible evolution of the Brome-liiflorae and its relations to the Zingiberiflorae and Commeliniflorae. (DAHLGREN and RASMUSSEN 1983). Note that apomorphies are here mentioned *before* the plesio-morphies.

1  Plants woody at least at the base – plants herbaceous
2  Stem covered with a coat of roots and fibres – stem not so covered
3  Leaves distichous – leaves with other phyllotaxy
4  Root-hair cells shorter than other epidermal cells – root-hair cells equalling other epidermal cells
5  Complex, stellate or scale-like multicellular hairs present – hairs of simpler construction
6  Sieve tube plastids with starch grains – sieve tube plastids with protein bodies only
7  Epicuticular wax type of *Strelitzia* Type – epicuticular wax without sculpturing or with non-oriented sculpturing
8  Silica bodies present, often hat-shaped, druse-like, or rectangular – silica bodies lacking
9  Silica bodies large, solitary, present in epidermal cells – silica bodies lacking
10  Flowers small, crowded in head or spadix – flowers larger, not in dense heads or spadices

11  Three upper tepals fused to form a lip – upper three tepals not so fused
12  Perianth petaloid, showy – perianth discrete, of moderate size
13  Perianth whorls differing conspicuously, the outer often sepaloid and shorter than the inner petaloid – perianth whorls similar
14  Tepals hair-like – tepals not hair-like
15  Flowers epigynous – flowers hypogynous
16  Flowers lacking septal nectaries – flowers with septal nectaries
17  Stamens reduced to 5 – stamens 6
18  Stamens reduced to 3 – stamens 6 (evidence supports that this is independently from *17*)
19  Stamens reduced to one – stamens 6 (evidence supports that this is independently from *17* and *18*)
20  Pistil monocarpellary and as a result fruit probably follicular – pistil tricarpellary, fruit capsular
21  Stigma surface wet – stigma surface dry
22  Endothecial thickenings of Girdle Type – endothecial thickenings of Spiral Type
23  Tapetum amoeboid – tapetum glandular-secretory
24  Pollen grains ulcerate – pollen grains sulcate
25  Pollen grains without distinct apertures – pollen grains distinctly sulcate
26  Pollen grains bi- or trisulculate – pollen grains sulcate
27  Placentation laminar-dispersed – placentation central, axile
28  Placentation apical – placentation axile
29  Locule with a single ovule only – locule with more than one ovule
30  Ovules orthotropous (or in transitional cases campylotropous) – ovules anatropous

31 Parietal cell lacking – parietal cell (and parietal tissue) present
32 Helobial endosperm of characteristic subtype, with small, starch-free, sometimes haustorial chalazal chamber – helobial endosperm normal, with larger chalazal chamber
33 Fruit a drupe – fruit follicular (see 19)
34 Seeds arillate – seeds lacking aril
35 Endosperm with copious starch – endosperm without or with little starch
36 Starch of endosperm mealy – endosperm fleshy
37 Embryo of *Xyris* Type or *Scirpus* Type – embryo of *Trillium* Type or of a transitional *Trillium-Urginea* Type
38 Cell walls with UV-fluorescent compounds – cell walls lacking UV-fluorescent compounds
39 Chelidonic acid present – chelidonic acid lacking

## The Bromeliiflorae

The Bromeliiflorae complex as circumscribed here (Fig. 39), is somewhat difficult to define further, but most members of the group have showy petals, and all seem to have a particular type of helobial endosperm formation where the chalazal chamber is small and has an early wall formation, a type that is rare in other monocotyledons (but found in Burmanniales; HAMANN, personal communication). Further, with the exception of the Typhales, nearly all Bromeliiflorae are Southern Hemisphere groups which were presumably first differentiated when there were still possibilities for dispersal between some of the continents that were formed by the splitting up of Gondwanaland.

The Bromeliales (Bromeliaceae) and Velloziales (Velloziaceae), which have spiral phyllotaxy, form two separate derivatives within Bromeliiflorae, both centred in South America, although Velloziales are also common in Africa.

The Bromeliaceae have some autapomorphies in the form of complex peltate hairs, the silica bodies of the epidermal cells, a perianth where the outer tepals are more or less sepaloid and often stiff, and a mealy type of endosperm. This large family is restricted to America and concentrated in Central and South America (apart from a possibly late introduction on the coast of Africa). Part of the family has evolved epigyny.

The Velloziaceae have evolved a woody stem which is covered with a layer of adventitious roots mingled with the fibres of the old leaf sheaths; the flowers are always epigynous, the placentation is laminar-dispersed, and the ovules lack a parietal cell; all these characteristics are here regarded as apomorphies. In a group of *Vellozia* species each stamen has split into three or more, which is ob-

viously a specialization (and not ancestral as assumed by DE MENEZES 1980). As regards UV-fluorescence in the walls (see above, under the Bromeliiflorae-Zingiberiflorae-Commeliniflorae complexes) the only representative studied by HARRIS and HARTLEY (1980) did not fluoresce. As this species is the only exception for this attribute in the superorders Bromeliiflorae, Zingiberiflorae, Commeliniflorae and Areciflorae, the condition may represent a reversal affecting it alone or the whole family, and other species will have to be investigated to determine which is the case.

The orders Haemodorales, Pontederiales, Typhales and Philydrales generally have distichous leaves, which may represent a synapomorphic attribute for them. The tapetum is amoeboid in taxa of all of them except Philydraceae, which has a glandular tapetum, the amoeboid type being the derived state and a probable synapomorphy for the first three orders.

Philydrales (= Philydraceae) make up a homogeneous group which is derived in having ensiform leaves, zygomorphic flowers with the three upper tepals fused into an upper lip and strong reduction in the androecium so that only one anther is functional, and in lacking septal nectaries.

The Haemodorales (= Haemodoraceae), like the Philydrales, have distichous and frequently ensiform leaves, and are characterized by short root hair cells and a richness in chelidonic acid. The family has differentiated in pollen grain morphology (one derived type being biforminate), in flowers which are epigynous in some genera, and in the ovules which are mainly hemianatropous or orthotropous.

Of the three orders with an amoeboid tapetum (Haemodorales, Pontederiales and Typhales) the Pontederiales (= Pontederiaceae) stand out by having 2-3-sulculate pollen grains and Girdle Type wall thickenings in the endothecial cells, both states which are considered to be derived. The taxa are common in moist habitats and in water (some are floating) both in the Old and the New World. The leaves in, e.g., *Pontederia* approach in shape and venation those found in some Zingiberales, but the superficial similarities perhaps depend on convergence.

The Typhales (Sparganiaceae, Typhaceae), finally, are atypical in the order because of the small, wind-pollinated, unisexual flowers. The wind-pollination syndrome accounts for some of the synapomorphies of the two families, the inconspicuous perianth, ulcerate pollen grains, reduction in carpel number and within the carpel in ovule number (to one), the decurrent stigma of the pistil, etc. (see DAHLGREN and RASMUSSEN 1983).

## The Zingiberiflorae

The Zingiberiflorae ( = Zingiberales) are autapomorphic in the possession of the root-hair short cells, the sieve tube plastids that contain starch grains (BEHNKE 1981), the occurrence of silica bodies, the epigynous flowers, the lack of distinct apertures on the pollen grains (Costaceae being perhaps a case of reversal) and the occurrence of arils (lacking only in a few members).

Within this complex four families, Lowiaceae, Musaceae, Heliconiaceae and Strelitziaceae, have five to six functional stamens and their vegetative parts possess oxalate raphides. The Lowiaceae, still very incompletely known in embryological respects, seem to make up a very distinct unit, whereas Musaceae, Heliconiaceae and Strelitziaceae are more closely similar and with justification could, alternatively, be treated together as Musaceae sensu lato. In this complex the perisperm is already well developed, and in the remaining four families, Zingiberaceae, Costaceae, Cannaceae and Marantaceae, the perisperm, and in the last three families also a chalazosperm, make up nearly all the nutrient tissue of the seed.

In these last four families the androecium is largely transformed into petaloid staminodes, some of which are larger than the tepals, although some are generally small and some lacking, and there is only one functional stamen, the median member of the inner whorl. Zingiberaceae and Costaceae are different in many respects, the former having acquired a richness of ethereal oils contained in particular cells, and a strictly distichous phyllotaxy. There are also anatomical and embryological differences between the two families, the Costaceae having, for example, a copious chalazosperm lacking in Zingiberaceae. It is therefore, in spite of the obvious similarity in floral construction, appropriate to treat Costaceae as separate from Zingiberaceae.

A further derived condition, possibly a synapomorphy, for Cannaceae and Marantaceae is the loss of one of the two thecae of the single functional stamen. This is also associated with asymmetry in the other staminodes of the inner whorl. These two families have a conspicuous chalazosperm in which they agree with Costaceae. The relations of Costaceae to Zingiberaceae and to Cannaceae-Marantaceae are complex and not yet wholly clear.

## The Commeliniflorae (Fig. 40)

The Commeliniflorae complex is here assumed to be monophyletic, having as autapomorphies root hair cells shorter than the other epidermal cells, lack of septal nectaries, Girdle Type endothecial cell wall thickenings and a broad, capitate or lens-shaped embryo. The superorder consists of both insect- and wind-pollinated groups.

---

**Fig. 40.** A model of the possible evolution of the Commeliniflorae. Ecdeiocoleaceae and Anarthriaceae are here included in the Restionaceae, which explains the lack of autapomorphies of the latter family. A division of Restionaceae into these families makes each of the three families more distinct, Restionaceae being then characterized by having culms with a closed cylinder of sclerenchymatous tissue. Recent studies by FERGUSON and LINDER (unpubl.) have revealed further peculiarities of the pollen wall structure in Centrolepidaceae which add to the distinctness of this family. Lack of obvious autapomorphies is indicated by shading. (DAHLGREN and RASMUSSEN 1983). Note that apomorphies are here mentioned *before* the plesiomorphies.

1 Phyllotaxy tristichous – phyllotaxy distichous or otherwise
2 Leaf tip cirrhose, functioning as a tendril – leaf tip not cirrhose
3 Leaf base asymmetrical – leaf base symmetrical
4 Root-hair cells shorter than other epidermal cells – root-hair cells the same size as other epidermal cells
5 Silica bodies present – silica bodies lacking
6 Multicellular hairs with a single basal cell row – no such hairs
7 Oxalate raphides lacking – oxalate raphides present
8 Stomata only on abaxial side of leaf surface – stomata on both sides of the leaves
9 Stomata tetracytic – stomata not tetracytic
10 Perianth differentiated into a sepaloid outer and a petaloid inner whorl – perianth whorls similar or nearly so
11 (Inner) perianth reduced to lodicules – inner perianth not reduced to lodicules
12 Flowers unisexual – flowers bisexual
13 Septal nectaries lacking – septal nectaries present
14 Stamens 3 – stamens 3 + 3
15 Stamens solitary – stamens 3 or 3 + 3
16 Anthers extrorse – anthers introrse (or latrorse)
17 Anthers apically poricidal – anthers longitudinally dehiscent
18 Anthers with unisporangiate thecae – anthers with bisporangiate thecae
19 Endothecium with Girdle Type wall thickenings – endothecium with Spiral wall thickenings
20 Tapetum plasmodial – tapetum glandular-secretory
21 Microsporogenesis simultaneous – microsporogenesis successive
22 Pollen grains dispersed in tetrads – pollen grains separate

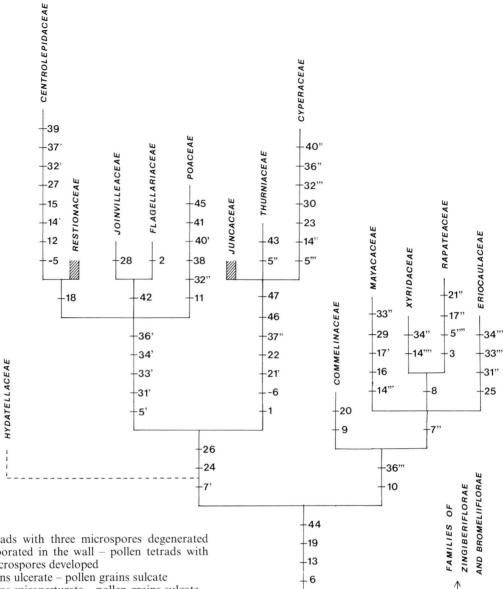

23 Pollen tetrads with three microspores degenerated and incorporated in the wall – pollen tetrads with all four microspores developed

24 Pollen grains ulcerate – pollen grains sulcate

25 Pollen grains spiraperturate – pollen grains sulcate

26 Pollen grains dispersed in the trinucleate state – pollen grains dispersed in the binucleate state

27 Gynoecium monocarpellary – gynoecium tricarpellary

28 Stigmas sessile, situated directly on the ovary top – stigmatic surfaces on more or less distinct stylar branches

29 Ovary unilocular with parietal placentation – ovary trilocular with axile placentation

30 Placentation basal – placentation axile

31 Placentation apical or subapical – placentation axile

32 One ovule per pistil – two or more ovules per pistil

33 Ovule orthotropous – ovule anatropous

34 Parietal cell not formed – parietal cell formed

35 Perianth whorls petaloid, showy – perianth discrete, of moderate size

36 Endosperm formation nuclear – endosperm formation helobial

37 Embryogeny of *Onagrad* Type – embryogeny of *Asterad* Type

38 Embryogeny of Grass Type – embryogeny of ordinary, *Asterad* Type

39 Fruit follicular – fruit capsular

40 Fruit an indehiscent nutlet – fruit capsular

41 Fruit a caryopsis with pericarp and testa fused – fruit a nutlet

42 Fruit at least with fleshy mesocarp – fruit dry

43 Seeds fusiform with subulate ends – seeds not with subulate ends

44 Embryo broad, lens-shaped or capitate, of *Xyris* or *Scirpus* Type – embryo of *Urginea* or *Trillium* Type

45 Embryo lateral, of Grass Type – embryo of *Xyris* or *Scirpus* Type (?)

46 Centromere activity not localized to one point, "diffuse" – centromere activity localized to one point

47 Meiosis of postreductional type – meiosis normal

One group that may have deviated early within the Commeliniflorae is Commelinales (Commelinaceae, Mayacaceae, Xyridaceae, Rapateaceae and Eriocaulaceae). Possible synapomorphies for this complex are the occurrence of multicellular hairs with a single-row base, a perianth differentiated into an outer sepaloid (or hyaline) and an inner petaloid whorl and nuclear endosperm formation. There are some doubtful points in these assumptions, since multicellular one-row hairs may be plesiomorphic in the group, and the evolutionary status of nuclear endosperm formation has been called into question by the report of the helobial for *Abolboda* in Xyridaceae (HAMANN, personal communication), which may either represent a "relict" feature or a reversal.

In having oxalate raphides the Commelinaceae are presumably more ancestral than the other families of the order, but the tetracytic stomata and the amoeboid tapetum in Commelinaceae are assumed to be specializations.

The relationships of the four families Mayacaceae, Xyridaceae, Rapateaceae and Eriocaulaceae are somewhat doubtful, and pairwise similarities in derived features (possible synapomorphies) are mutually exclusive, indicating no clear picture of their evolution. Thus in Xyridaceae and Rapateaceae stomata are restricted to the lower leaf surface; in Mayacaceae and Rapateaceae, the anthers are poricidal; in Mayacaceae and Eriocaulaceae the ovules are orthotropous; in Mayacaceae and Xyridaceae the flowers have three stamens only; in Xyridaceae and Eriocaulaceae the ovules lack a parietal cell. An evaluation of this situation is difficult, and we can only assume here, on strong evidence, that the families are rather closely related and probably represent parallel evolutionary lines; they are centred in South and Central America.

The remaining two orders Poales and Cyperales (incl. Juncales) are considered to have a common origin, their synapomorphies being lack of oxalate raphides and presence of ulcerate pollen grains which are dispersed in the three-celled stage.

The small bigeneric family Hydatellaceae, which was only recently described as distinct from Centrolepidaceae (HAMANN 1976), may have deviated from the common ancestors of Poales and Cyperales. Like these orders, it lacks raphides, but it has sulcate pollen grains dispersed in the two-celled stage (Fig. 40). In addition it has anomocytic stomata, which is the main feature making it aberrant in the Commeliniflorae. The position of the family is still highly debatable and its affinity could be with the Zingiberiflorae or Liliiflorae (see DAHLGREN 1980, and THORNE 1983). Its members are small grass-like plants with reduced unisexual flowers with one stamen or a uniovulate pistil. The ovule is anatropous, endosperm formation is cellular (undoubtedly a recent specialization) and the seeds, which are provided with a lid, have developed a copious starchy perisperm (another novelty), whereas the endosperm forms a unicellular layer only.

One of the two remaining main evolutionary branches comprises the three families Juncaceae, Thurniaceae and Cyperaceae which make up our Cyperales. Synapomorphies for these are: tristichous leaves, pollen grains dispersed as tetrads, Onagrad Type of embryo formation, non-localized centromere activity and post-reductional meiosis and, in addition, susceptibility to certain genera of parasitic fungi (such as *Entorrhiza*) and plant lice (e.g. the psyllid subfamily Liviinae and the aphid tribe Saltusaphidinae).

Juncaceae is basic in this order and it is difficult to find autapomorphies for this family. Its centre of variation is in South America, where Thurniaceae is endemic. Thurniaceae is a unigeneric family with dense globose inflorescences. In contrast to Juncaceae its tissues contain small silica bodies and its seeds are subulate at both ends. The third family, Cyperaceae, is further specialized in having epidermal cells with silica bodies which are as a rule conical, in having flowers with three stamens only, pollen tetrads in which three microspores degenerate and become incorporated in the wall of the fourth one, basal placentation, one ovule, nuclear endosperm formation (helobial in Juncaceae) and an indehiscent fruit.

Finally, the Joinvilleaceae, Flagellariaceae, Poaceae, Restionaceae, Anarthriaceae, Ecdeiocoleaceae and Centrolepidaceae form the terminal units on the last main branch of the Commeliniflorae. They constitute our Poales. Synapomorphies for this group are: silica bodies in epidermal cells (lacking in several of the taxa with monothecous anthers); lateral to apical placentation; orthotropous (or hemianatropous or campylotropous) ovules; lack of a parietal cell; and nuclear endosperm formation. Most taxa in this complex also have distichous leaves. The greatest concentration of ancestral conditions in this complex is met with in the unigeneric Joinvilleaceae and Flagellariaceae, which both have bisexual flowers with 3 + 3 bract-like tepals, trilocular pistils and tetrasporangiate anthers. The fleshy pericarp may be a synapomorphy for the two families.

The grasses (Poaceae) are more advanced in the reduction of the perianth to two or three lodicules representing the inner whorl (the hypothesis that the palea represents two outer tepals is highly speculative), in the unilocular pistils with a single ovule, in the grass type embryogeny (oblique cell walls in the early embryo), and in the peripheral, obliquely situated embryo with its peculiar structure. Further evolution in the grasses has affected the construction of the spikelets and led to the reduction of stamen number and the differentiation of the lodicules, silica bodies, chromosome sets, etc.

The remaining four families, of which Anarthriaceae and Ecdeiocoleaceae are often included in the Restionaceae, have evolved in the direction of unisexuality of the flowers, of reduction of the number of microsporangia to two per anther (Centrolepidaceae, most Restionaceae) etc. The Restionaceae are probably monophyletic, as is indicated by the consistent presence of a closed sclerenchymatous tube in the culms. Among the autapomorphies of the Centrolepidaceae are the lack of silica bodies, the male flowers reduced to a single stamen and the female flowers reduced to one carpel, and some pollen-morphological features. The flowers in Centrolepidaceae are grouped into condensed spikelets.

Several of the minor families, in particular the four last-mentioned, are pronouncedly Southern Hemisphere elements; Restionaceae is disjunctive, with a number of genera in each of Southern Africa and Southern Australia; Ecdeiocoleaceae and Anarthriaceae are confined to the South-West of Australia.

## The Areciflorae-Cyclanthiflorae-Pandaniflorae Complexes

The significance of the resemblances among Areciflorae (Arecales), Pandaniflorae (Pandanales) and Cyclanthiflorae (Cyclanthales), which have traditionally been treated as related, and the discernment of links between each of these groups and other monocotyledons are undoubtedly two of the greatest problems in monocotyledon evolution. The synapomorphies, if any, for these three complexes together are conspicuously few and include mainly the arborescent or liana form of growth (exception *Freycinetia* and some Cyclanthaceae), the tetracytic stomata (a few have paracytic), the indehiscent and more or less fleshy fruits, and the *Urginea* Type embryo.

The fact that the evidence for a close relationship between them is so minimal has led to the recognition of the three unifamilial orders as separate superorders (see also THORNE 1983).

One conspicuous similarity between the Areciflorae and the Cyclanthiflorae is provided by the similar, plicate leaves which split in a similar way. The inflorescence of unisexual flowers in Cyclanthales, which are regularly ordered on a thick axis, is also comparable with conditions in certain palm tribes. Provisionally we regard it as most likely that these two orders have a common origin, although possibly far back in the history of the monocotyledons. The lack of a parietal cell, the presence of laticifers, and certain other features of *Cyclanthus* itself (Cyclanthaceae subfam. Cyclanthoideae) are somewhat similar to conditions in certain Araceae. Cyclanthales have helobial endosperm formation, which is considered to be primitive, if not ancestral, in monocotyledons. The tetramerous flowers have their counterpart in some Dioscoreales, namely in most Stemonaceae, but this may well be by convergence. We thus regard the relationships of the Cyclanthiflorae as unresolved.

The Areciflorae, apart from a resemblance to the Cyclanthiflorae of highly questionable significance, have a number of features shared with the supposedly primitive Poales, in particular the Joinvilleaceae but also Flagellariaceae and Poaceae. Among the conspicuous similarities are the occurrence of vessels in the stems, the often very distinct nodes, the leaves having a sheathing base and a plicate ptyxis (probably primitive but not so common in grasses), the pronounced development of ligules and ligule-like structures, the occurrence of silica bodies, the same type of epicuticular wax (*Strelitzia* Type), the occurrence of organic acids in the cell walls giving UV-fluorescence, the conspicuously similar flavonoid compounds, such as tricin and sulphonated flavonoids, as well as various other flavonoid compounds, and abundant cyanogenesis.

The flowers in both complexes have 3+3 inconspicuous tepals in many of their representatives and the anthers in both groups tend to be latrorse, but in contrast to the grasses septal nectaries are common in palms, which are frequently insect-pollinated. Fleshy fruits occur in both groups (*Melocanna, Joinvillea, Flagellaria* among the Poales), but may be secondary in them. It should be noted that several pollen types represented in palms are known from the Maestrichtian, in the late Cretaceous; they correspond to types in extant *Nipa* (meridionosulcate), *Areca* (sulcate) and Cocosoideae (trichotomosulcate). At the same time the fossil record also shows ulcerate pollen grains of the type found in certain extant Restionaceae, whereas definite grass pollen from that time has not yet been discovered.

The Pandaniflorae, finally, are as yet wholly unsettled as regards their probable origin, although it may be presumed that it was amongst the ancestors of either the Arecales-Poales or the Cyperales-Poales.

# Taxonomy

## General

The classification followed here is mainly in line with the conclusions drawn by DAHLGREN and RASMUSSEN (1983). It deviates in various respects from the alternative system presented by DAHLGREN and CLIFFORD (1982), but is based to a large extent on the data presented in that work. More base data have been taken into consideration in the present work, and the approach has been more strictly evolutionary.

A historical account of monocotyledon classification, from LINDLEY (1853) onwards, has been given by DAHLGREN and CLIFFORD (1982) and is omitted here. Usually, only sporadic comments on classification of orders or families are given in the following text. We refer to the classifications of TAKHTAJAN (1980), CRONQUIST (1981), THORNE (1983) and DAHLGREN (1983a), to mention a few of the current classifications covering all the angiosperms and hence all the monocotyledons and those dicotyledons with more or less prominent monocotyledonous traits.

# Superorder Liliiflorae

*Five Orders:* Dioscoreales, Asparagales, Melanthiales, Burmanniales and Liliales.

This superorder is extraordinarily variable and contains some groups which, in our estimation, are likely to have retained many features from the ancestral monocotyledons. The wide range of variation makes a definition difficult and in an evolutionary sense this unit is undoubtedly paraphyletic rather than monophyletic. Some of its orders, on the other hand, we presume to be monophyletic.

The Liliiflorae are annual to perennial plants only rarely aquatic, but ranging from small herbs to shrubs or trees of considerable size and occasionally having a thick, tall trunk. Most families consist of terrestrial plants, but Orchidaceae is rich in epiphytes. This and some other families consist largely of mycorrhizal plants. Achlorophyllous taxa occur in several families, but are otherwise found only in Triuridiflorae.

The roots appear opposite xylem strands only. Except in some orchids the root hair cells are the same size as the other epidermal cells. Several-layered velamen is common in several of the groups, a feature otherwise known, among the monocotyledons, only in Ariflorae. Vessels are probably almost constantly present in the roots, where they often have scalariform perforation plates, although simple or simple as well as scalariform perforation plates occur in a great many families of Asparagales and Liliales.

The underground stem is frequently developed as a rhizome, but in many groups as a corm or tuber, which – except in the Ariflorae – is very rare in monocotyledons. Bulbs, i.e. a short, generally underground stem with fleshy leaves, occur in several families of this superorder but very rarely in others.

The aerial stem varies from slender and herbaceous, often twining, to slender and woody or thick and woody, and this is the only superorder among the monocotyledons where secondary thickening growth occurs (of the Monocotyledonous Type; see p. 9). Rarely the stem is provided with spines or prickles (*Petermannia*, species of *Smilax* and *Herreria*), and the branchlets may be flat and developed as phyllocladia (Ruscaceae, some Asparagaceae), which is never the case in other monocotyledons. Branching is monopodial or sympodial. Laticifers are rarely present (in some Alliaceae). Vessels are often lacking in the underground as well as aerial stem; where present they are nearly always provided with scalariform perforation plates. Hairs of various kinds occur; these and the stomatal complexes are mentioned in connection with the leaves.

The leaves vary much in their position and differentiation. They may be developed as simple cataphylls, they may be linear, with a sheathing base, or they may have a non-sheathing base, a distinct petiole and a broad leaf lamina, the latter type of leaf being more or less reminiscent of that in most dicotyledonous leaves. Rarely, the leaves are opposite or verticillate, which among the monocotyledons is otherwise the case only in some Alismatiflorae. Petiolate leaves in some families have a lobate or truly compound lamina, the latter being otherwise only known in Ariflorae among the monocotyledons. Where the leaves are distichous and linear they may be laterally compressed and ensiform, as in most Iridaceae. In some families the lamina is inverted basally, so that the morphological upper side faces downwards. The ptyxis is highly variable but generally supervolute, conduplicate or flat. Stipules are lacking or atypical. Tendrils from the leaf base occur in most Smilacaceae. Ligules are very rare. The venation in linear leaves is parallel, while in petiolate leaves with a broad lamina the main veins are separate from the base of the lamina and converge at the apex. In several families there is reticulate venation intercalated between the main veins, which in monocotyledons is otherwise the case only in (numerous) Ariflorae and (a few) Alismatiflorae. The stomata are parallel or rarely (Smilacaceae) randomly oriented in relation to the leaf axis. The stomatal complexes are agenous or perigenous (very rarely mesogenous, e.g. in some orchids) in the ontogenetic sense and become anomocytic or more rarely paracytic (very rarely tetracytic) when mature (note that there is *not* strict correlation between ontogenetic type and type of mature stomatal complex). Trichomes are very variable, being unicellular, uniseriate, tufted or otherwise, but often almost lacking. Intravaginal squamules are lacking. Epicuticular wax is unsculptured, irregularly sculptured or sculptured according to the *Convallaria* Type, never of the *Strelitzia* Type. Raphides are commonly present, except in a great part of Liliales and Burmanniales. Silica bodies are restricted to certain taxa in Apostasiaceae, Cypripediaceae and Orchidaceae. Vessels are generally absent in the leaves.

The inflorescences include both determinate and indeterminate kinds (see further under orders and families). The flowers are mostly trimerous but dimerous in some Dioscoreales, rarely on other numeric bases. It is characteristic of the Liliiflorae that the tepals are largely petaloid, although they are frequently inconspicuous and generally yellowish white or greenish white in Dioscoreales, Melanthiales, many Burmanniales and some primarily baccate Asparagales. In other groups they are generally brightly coloured. Scarious or "bract-like" tepals are rare. Perigonal nectaries occur in certain groups (they occur in very few monocotyledons outside this superorder), and in these groups (mainly Liliales) the tepals are frequently provided with a variegated colour pattern in the form of a checkered pattern, spots or streaks radiating from the centre of the flower, or as coloured sections, all probably important in attracting pollinators, and found in very few taxa outside this superorder. The flowers are hypogynous, hemiepigynous or epigynous. Differentiation in size and shape between the outer and the inner whorl of tepals is common, although the tepals in both whorls are petaloid. Rarely, the outer tepals are shorter, different in colour and tending to be sepaloid, although never typically so.

The stamens are generally 3 + 3 or 3 in number, but in some families reduced to two or one (e.g. two in Cypripediaceae and one in Orchidaceae). The filaments are free or more or less connate but in Orchidaceae and related families they are fused to the style, forming a gynostemium. The filaments are provided with spreading "food" hairs in some taxa in Asparagales and Melanthiales. In certain families of Asparagales they are flat and extended into a pointed lobe on each side of the anther. They may also be fused into a staminal tube. The anther is basifixed (impeltate) or dorsifixed (peltate) and, especially in some Dioscoreales, the connective extends beyond the microsporangia into a tip or lobe.

When dorsifixed the anthers may be epipeltate or hypopeltate. Sometimes the connective base forms a tube around the filament tip (making it appear basifixed). The anthers are usually introrse but extrorse mainly in some Liliales; they dehisce by longitudinal slits or rarely by apical pores.

The endothecial thickenings are generally of the Spiral Type (the Girdle Type has been reported in some orchids and in *Tacca*). The tapetum is secretory except, perhaps, in Hypoxidaceae, where it approaches the amoeboid type. Microsporogenesis is either successive or simultaneous. The pollen grains are generally dispersed singly or, in most orchids, as tetrads which generally cohere in massulae or pollinia. Their aperture condition is generally sulcate, sulculate or inaperturate (rarely trichotomosulcate, spiraperturate, zonisulculate or foraminate). They are nearly always dispersed in the bicellular state, having quite often a fusiform to crescent-shaped generative cell.

The gynoecium is generally tricarpellary with the three carpels fused at least basally in the ovary region, sometimes in the ovary region only, but in most taxa also in part of the stylar region. The stylar part of the gynoecium is generally long, rarely (as in Asteliaceae) very short or lacking. The placentation is axile or (especially in Orchidaceae of Liliales, most Burmanniales, some other families) parietal (but never laminar-dispersed); in a few genera with one ovule per locule it is basal or apical.

The ovules are generally anatropous, rarely campylotropous and very rarely hemianatropous or orthotropous, crassinucellate or tenuinucellate, and are with or without a parietal cell. Comparatively rarely, as in some Amaryllidaceae, the epidermal cells of the nucellus divide periclinally to form a nucellar cap. Embryo sac formation is generally of the *Polygonum* Type, but the *Allium, Endymion* and *Fritillaria* Types and others occur in scattered families and genera. Endosperm formation is helobial in Melanthiales, many Asparagales, and a few Dioscoreales, but is nuclear in all others and never cellular. In Orchidaceae and Cypripediaceae (and Apostasiaceae?) endosperm formation is soon arrested or (generally) does not take place at all, and endosperm is consequently lacking. In the other groups it becomes copious and stores primarily hemicellulose (in the cell walls), protein and fatty oil but only occasionally starch grains. Embryogeny is generally either of the Asterad or Onagrad Types.

The fruits are capsules or berries, very rarely trifolliculi (some Melanthiaceae) or nutlets. Arils are rare and strophioles or caruncles of different kinds occur in scattered groups. A black phytomelaniferous testal epidermis occurs in nearly all capsule-fruited and some berry-fruited Asparagales. The embryo is generally linear in shape but is sometimes mushroom-shaped (as in *Dioscorea* species) and rarely curved; it is central, varying from very small (*Trillium* Type) to fairly large (*Urginea* Type) and not macropodous. Perisperm occurs only in

*Yucca* (Agavaceae) and chalazosperm only in Cyanastraceae. The endosperm is ruminate only in some genera of Dioscoreaceae and (the perisperm) in *Yucca* (Agavaceae, Asparagales).

The embryo has a terminal cotyledon except in some Dioscoreales, where the plumule may be subterminal.

The chromosomal conditions are mentioned under the orders and families. The centromere is always localized.

*Chemistry:* Of chemical characteristics should be mentioned the common accumulation of calcium oxalate in the form of raphides, the very common occurrence of steroidal saponins in most taxa lacking alkaloids, and the common presence of chelidonic acid. Cyanogenesis is evidently not very common, but is known for example in Amaryllidaceae and Alliaceae. Allyl sulphides ("onion oils") occur in some genera of Alliaceae. Alkaloids of different types occur in Amaryllidaceae and Colchicaceae in particular (see under those families), but also in Orchidaceae (several types), Dioscoreaceae (quinuolidine type), Stemonaceae (tuberostemonine type). Of the flavonoids various kinds are known in the Liliiflorae, luteolin/apigenin especially in Alliaceae and Colchicaceae, and sulphated flavonoids in Alliaceae. UV-fluorescent compounds are not known in the cell walls of any Liliiflorae.

*Parasites.* Some fungal parasites are typical of isolated groups in the Liliiflorae but none are ubiquitous or typical of the superorder.

*Distribution.* The Liliiflorae do not show a restricted geographical pattern as do some of the other superorders. Many families, especially of Asparagales and Liliales (excluding the orchid group) are concentrated in the Southern Hemisphere and may have differentiated there. Other groups (most Dioscoreales, Burmanniales, orchids) are tropical to subtropical, whereas others again (Convallariaceae, Liliaceae, Melanthiaceae) are largely north-temperate.

*Relationships.* The Liliiflorae include those extant forms among the monocotyledons that combine the majority of the features here considered to be ancestral. A number of these features are also found in Magnoliiflorae, and we postulate that the monocotyledons and the Magnoliiflorae are sister-groups in the evolutionary sense. Most of the presumably ancestral attributes under this assumption are concentrated in Dioscoreales, in which we include Taccaceae.

In fact, the Ariflorae (e.g. *Pothos* group of Araceae) and the Alismatiflorae (e.g. the Aponogetonaceae), as well as the Asparagales, Liliales and Melanthiales, have obvious connections with members of Dioscoreales, and there are some problems in the delimitation of this order (Smilacaceae could, for example, be referred either to Asparagales or Dioscoreales; *Medeola* and *Scoliopus* connect Trilliaceae with Liliaceae and Uvulariaceae).

Taxa with a concentration of the following features are referred to Dioscoreales: twiners or herbs of forest floor habitats, leaves differentiated into petiole, often with a non-sheathing base, and a broad lamina with reticulate venation, which is often lobate or compound, flowers of moderate size, with inconspicuous perianth, without tepal variegation, microsporangia often below the stamen apex, and thus basifixed-undifferentiated anthers, nuclear endosperm formation, sometimes ruminate endosperm, non-phytomelaniferous, well-differentiated seed coat formed by the inner as well as the outer integument, and sometimes a non-terminal cotyledon. There are some supposedly derived character states in this order, too, e.g. epigynous flowers, non-sulcate pollen grains, nuclear endosperm and simultaneous microsporogenesis, as well as specialization in vegetative respects (tubers, compound leaves, anatomical details) in some taxa.

The Asparagales, we assume, comprise a probable clade, where the seed coat has been specialized, the capsule-fruited taxa having a phytomelan layer. Some family clusters here possibly make up separate clades, such as some shrubby, baccate-fruited families with nuclear endosperm formation and frequently without septal nectaries, others with a bulb and a scapose inflorescence, and yet others with a corm in combination with successive microsporogenesis.

Another probably monophyletic complex is the Liliales, with nuclear endosperm formation, non-phytomelaniferous seed coat, perigonal nectaries and often spotted tepals. In this group we have included Orchidaceae, Cypripediaceae and Apostasiaceae, but not the Burmanniales (Burmanniaceae, Thismiaceae and Corsiaceae); we do not think that these two groups of families are closely related to each other.

The Melanthiales and Burmanniales may each have evolved as a separate clade, but we are uncertain that this is necessarily so. Both have helobial endosperm, and both lack a phytomelaniferous seed coat and perigonal nectaries. A number of genera of Melanthiales share related uredinalean parasites, *Puccinia atropuncta* and *Uromyces vera-*

*tri.* The Burmanniales in their epigyny, minute seeds, and other features superficially agree with the orchids, but are probably not related with them.

The constellation of orders proposed here is largely based on HUBER (1969). It has gained some recent support from the serological results of CHUPOV and KUTIAVINA (1978, 1981), which are discussed by DAHLGREN (1983a, b). These studies show that there is generally little or no serological reaction between taxa that we refer here to different orders, whereas there is rather strong reaction between at least some families in each order.

We have here excluded from the Liliiflorae the families (orders) with the combination of copious starch in the endosperm, paracytic or tetracytic stomata, *Strelitzia* Type of epicuticular wax, and UV-fluorescent compounds in the cell walls, even where the groups with such attributes may have petaloid and even showy tepals, septal nectaries, raphides, and helobial endosperm, which are all widely distributed in the Liliiflorae.

## Order Dioscoreales

*Seven Families:* Trichopodaceae, Dioscoreaceae (incl. Stenomeridaceae), Taccaceae, Stemonaceae, Trilliaceae, Smilacaceae and Petermanniaceae.

Almost exclusively terrestrial, herbaceous or woody perennial plants with a rhizome or tuber which stores starch. The tuber, at least in *Tacca,* is apparently hypocotylar in character. The tuber in *Dioscorea* may be conspicuously thickened and covered with cork. Secondary thickening growth occurs sometimes in the tuber of *Dioscorea* sect. *Testudinaria.* The roots, as in species of *Dioscorea* (sect. *Testudinaria*) and *Tacca,* may have a single-layered velamen. The aerial stem is often long, twining or trailing, and it usually withers down annually. In Petermanniaceae and some Smilacaceae it is covered with hooks or spines. It has one, two or more rings of vascular bundles. Vessels with scalariform perforations in the end walls are always present in the roots and sometimes in the stems and (in Smilacaceae) also in the leaves. Laticifers are lacking. The leaves are variably arranged on the stem, being dispersed, opposite or verticillate. In most taxa the leaves are differentiated into a petiole and a flat, dorsiventral lamina; there is no petiole in Trilliaceae, however. The leaf base is simple or sometimes sheathing. Stipule-like, often filamentous, appendages are often present on the base, and in Smilacaceae the petiole bears two tendrils. The lamina is usually simple and entire, but may be pinnatisect, pinnatifid or sometimes palmately compound. When simple the lamina usually varies from lanceolate to cordate, with supervolute ptyxis. The primary venation is campylodromous or rarely pinnate and the veinlets, unlike those of most other monocotyledons, generally form a reticulate pattern (CONOVER, 1983). Ligules are lacking. Oxalate raphides seem to be universally distributed in the order. Trichomes are variable and occur as simple or branched nonglandular hairs, at least in *Dioscorea,* while *Tacca* has characteristic multicellular hairs (see below). The stomata are anomocytic and parallel to the leaf axis, or often, as in Dioscoreaceae, Trichopodaceae and Smilacaceae are randomly oriented.

The inflorescences are very variable in form, and most frequently axillary in position; they are sometimes terminal (and then occasionally one-flowered) on a leafless or leafy branch ("stem") arising directly from the subterranean rhizome or tuber. Flower construction is variable. Generally the flowers are trimerous, but deviations from this condition occur in Stemonaceae and Trilliaceae. The flowers are hypogynous, hemiepigynous or epigynous, actinomorphic, and bisexual or unisexual. The tepals are often inconspicuous, mostly pale yellow, green or dull-coloured (rarely bright white or purple), not spotted, those of the two whorls equal or different from each other, and sometimes fused into a campanulate structure. Nectaries, when present, are perigonal (e.g. species of *Trillium* and *Smilax*) or septal (e.g., in species of *Dioscorea, Trillium* and *Smilax*).

The stamens, like the tepals, occur in two isomerous whorls, one of which is staminodial in some Dioscoreaceae. The filaments may be adnate to or free from the tepals and are often relatively short; in Dioscoreaceae and Smilacaceae they may be fused with each other into a tube. The anthers are basifixed and dehisce introrsely or rarely (some Dioscoreaceae, *Scoliopus* in Trilliaceae) extrorsely. Frequently the stamens project apically beyond the microsporangia. The anthers are tetrasporangiate and anther wall formation proceeds according to the Monocotyledonous or, in *Tacca* (?), according to the Dicotyledonous Type. The endothecial thickenings may be of the Spiral or the Girdle Type. Anther dehiscence is by means of longitudinal slits. The tapetum is glandular and microsporogenesis is simultaneous (e.g. in *Dioscorea* and *Tacca*) or successive (e.g. in *Trichopus, Stemona, Trillium, Smilax*). The pollen grains are single, and may be sulcate, 4-sulcate, 4-foraminate (various Dioscoreaceae), or inaperturate (most Smilacaceae). They are binucleate when dispersed.

The gynoecium is syncarpous and has free stylodia or a single style or (in Stemonaceae) a nearly sessile stigma. The stigma is Dry or Wet. Except in Taccaceae, Stemonaceae and Petermanniaceae the ovary is trilocular with axile placentation. There are several to many ovules in each locule and these are anatropous except in *Stemona* and *Smilax,* where they are orthotropous. Further the ovules are bitegmic, usually crassinucellate and with a parietal cell cut off from the archesporium (except in *Trichopus*). Embryo sac formation is of the *Polygonum* Type or, in most Trilliaceae, of the *Allium* Type. The endosperm is mostly of the Nuclear Type, rarely helobial (*Trillium*). Embryogeny is probably of the Asterad Type although the Solanad Type has been reported for *Trichopus*.

The fruit is either capsular or, rarely, indehiscent and is then baccate (*Tamus, Trichopus*) or dry (*Rajania*). The seeds of capsular fruits frequently have an aril or caruncle, which may function as elaiosome; these are apparently diverse and non-homologous structures in the different families. The seeds vary in shape, and lack phytomelan, but may possess phlobaphene. The testa consists of both integuments. The endosperm is copious and contains aleurone and fat, while cellulose in variable amounts is deposited in the cell walls. In some genera (*Croomia, Stemona, Paris, Trillium, Ripogonum*) starch grains may also occur in the ripe seeds. Ruminate endosperm is found in *Trichopus* and *Avetra*. The embryo is basically linear and is frequently very small in size; in *Tacca*, however, it is curved. The plumule is often subterminal.

*Chemistry.* Dioscoreales are rich in steroidal saponins, and chelidonic acid is also common. Cyanogenic compounds are known in *Dioscorea.* Alkaloids occur in at least Stemonaceae and some Dioscoreaceae. Starch is generally deposited in the rhizomes and tubers, and these in *Tacca* contain also ceryl alcohol. The flavonoids are of the commoner types, including proanthocyanins, acylated anthocyanidin glycosides, cyanidin and pelargonidin; myricetin is known in *Tacca.*

*Distribution.* The Dioscoreales comprise a tropical group of families scattered over all the chief continents with a possible centre in southern Asia and Indonesia. Trilliaceae deviate in having their concentration in regions with a temperate climate in the Northern Hemisphere. The plants of the order are terrestrial and range from shade herbs on the forest floor to climbers in scrubby or rocky habitats. The latter are sometimes strongly adapted to arid conditions (e.g. *Dioscorea* sect. *Testudinaria*). The shade plants are often myrmecochorous.

*Relationships.* The present circumscription of the order Dioscoreales is unusual in including Taccaceae, as well as Trilliaceae, Smilacaceae and Petermanniaceae. Taccaceae have sometimes, possibly because of their pseudo-umbels, been associated with Amaryllidaceae, with which they have little in common. They have sometimes been treated as a separate order (e.g. by HUBER 1969 and DAHLGREN 1975) or even as a separate super-order, as by HUBER (1977). However, their position in (or at least in the immediate vicinity of) the Dioscoreales (-Stemonales) complex is supported by the whole spectrum of characters, *Trichopus* being perhaps the most apparent connecting link.

The Stemonaceae and Trilliaceae, which comprised a separate order, the Stemonales, in HUBER (1969), approach the Dioscoreaceae and Trichopodaceae, although Trilliaceae with *Scoliopus* has obvious affinities to *Medeola* and other Liliaceae s.str. in the order Liliales.

Smilacaceae and Petermanniaceae, referred previously (DAHLGREN and CLIFFORD 1981; DAHLGREN 1983a) to Asparagales, seem better placed in Dioscoreales on the basis of leaf morphology and floral appearance, but they form a bridge between the two orders.

The most interesting features of the Dioscoreales are certainly those indicating connections with the dicotyledons. These include vegetative features such as the subterminal plumule of the embryo, the few whorls of vascular bundles of the young plants tending to resemble the vascular systems of a dicotyledon seedling, the frequently opposite leaves, the petiolate, well-differentiated leaves with a non-sheathing base and with reticulate veins, and the anomocytic stomata having sometimes a scattered (i.e. not parallel) orientation. The frequently non-terminal microsporangia of the stamens provide a similarity with magnoliifloran families, while the dicotyledonous anther wall formation in *Tacca* may be a casual similarity (if not reported in error). Further, the endosperm is ruminate in *Trichopus* and *Avetra* as it is in numerous magnoliifloran families, and the embryo is generally very small in relation to the endosperm as in these families. Taking account of further considerations, including the frequently unspecialized nature of the perianth and the instability of the type of nectary,

HUBER (1969) strongly favoured the view that the Dioscorealean-Stemonalean families (he did not include Taccaceae, however) are the ones which most strongly approach the ancestors of the monocotyledons.

In this connection one should consider the variation spectrum and trends present in the Magnoliiflorae, the trends to herbaceous habit (independently) found in Aristolochiaceae and Chloranthaceaè, the monocotyledon-like flowers with obvious trimery in several dicotyledonous families, in particular Lactoridaceae, Aristolochiaceae (*Saruma*) and Annonaceae, and the basic, primitive types of ovules and pollen grains.

On the other hand, the Dioscoreales exhibit, in various of their members, a number of advanced vegetative structures and, with their epigynous flower, variable hair types, their often two- or more-aperturate pollen grains and their complex chemistry, are far from primitive. Similarly the Taccaceae, although apparently advanced in their bizarre inflorescences and complicated, epigynous flowers with unilocular ovary and parietal placentae, are probably more nearly primary in the monocotyledons in other character states. One interesting feature met with in several families of Dioscoreales is the common occurrence of both capsular and baccate fruits and sometimes a gradual transition between these fruit types. This flexibility may be considered as rather typical in the order but it also occurs in the Asparagales and Liliales.

Even though there may be most support for considering the basic pattern of the Dioscoreales as quite primitive in monocotyledons, there is not yet substantial fossil evidence for accepting this view, and certain features which are basically considered primitive here, such as the broad, reticulately veined leaf lamina, may in fact represent an adaptation related to the climbing habit in combination with a need for increased assimilating surface (cf. the Smilacaceae).

We claim, however, that the order Dioscoreales shows a number of features ancient for the monocotyledons and several which associate them with dicotyledonous families (even though similarities may have evolved by convergence). Some trends of affinity with primitive Arales (such as Pothoideae), where the flower size is reduced and attraction by a spathe has become general, can be discerned, but the common ancestry of these orders probably lies far back in time.

*Taxonomy.* The family concepts of the entities of the Dioscoreales are not yet stabilized. Dioscore-

aceae are generally acknowledged as a separate family, within which *Stenomeris* could be ranked as a tribe or subfamily or even be elevated to family rank. Trichopodaceae are often included in Dioscoreaceae. Stemonaceae (syn. Roxburghiaceae) are here also treated as a distinct family with Croomiaceae as a synonym. Trilliaceae are often treated as synonymous with Liliaceae s.lat., but in morphology they approach the families mentioned. Taccaceae have generally been regarded as a separate, and isolated, family with other affinities. *Ripogonum* of Smilacaceae is a very distinct genus and is in several respects (sulcate pollen grains, bisexual flowers, lack of tendrils) presumably more primitive than other Smilacaceae. Petermanniaceae, on the contrary, shows some specializations.

*Key to the Families*

1. Flowers 2-merous (rarely 5-merous) . . . . . . . . . . . . . . . **Stemonaceae**
1. Flowers 3-merous . . . . . . . . . . . . . . 2
2. Ovary inferior . . . . . . . . . . . . . . 3
2. Ovary superior . . . . . . . . . . . . . . 6
3. Erect, terrestrial herbs, either with a basal leaf rosette (Taccaceae) or with one or several leafy herbaceous stems (Trichopodaceae) . . . . . . . . . . . . 4
3. Twining woody or herbaceous vines . . . . . . . . . . . . . . . . 5
4. Flowers solitary or in groups opposite a large cauline leaf; placentation axile . . . . . . . . . **Trichopodaceae**
4. Flowers in umbel-like inflorescences, inserted between large leafy bracts and subtended by filiform, drooping phyllaries; placentation parietal . . . . . . . . . **Taccaceae**
5. Placentation central, axile; anthers introrse . . . . . . . . . .**Dioscoreaceae**
5. Placentation parietal; anthers extrorse . . . . . . . . . . . **Petermanniaceae**
6. Erect mesophytic herbs, never climbers . . . . . . . . . . . . . **Trilliaceae**
6. Woody or (partly) herbaceous vines . . . . . . . . . . . . . **Smilacaceae**

**Trichopodaceae** Hutchinson (1934)   1:1
(Figs. 41, 42)

Herbs with erect stems and a short rhizome. The leaves emerge in a rosette from the rhizome. They are simple, entire and petiolate and have three to five main veins, but lack stipule-like appendages (AYENSU 1966). The stomata are anomocytic and randomly directed on the leaf surface. The flowers

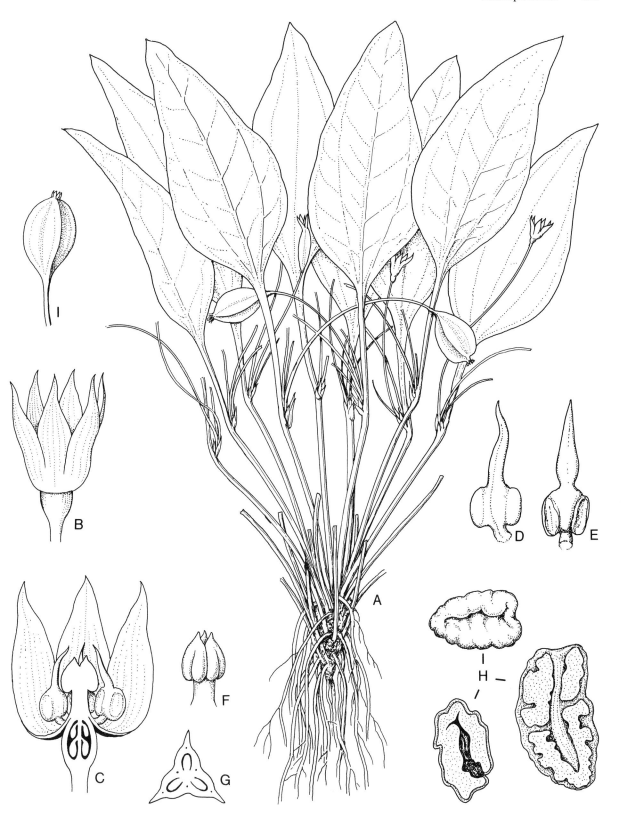

**Fig. 41.** Trichopodaceae. *Trichopus zeylanicus.* **A** Plant. **B** Flower. **C** Flower, longitudinal section. **D–E** Stamens, introrse, in different views. **F** Style and stigma lobes. **G** Ovary, transverse section. **H** Seed, below in transverse and longitudinal sections, showing the ruminate endosperm. (Redrawn from KNUTH 1924)

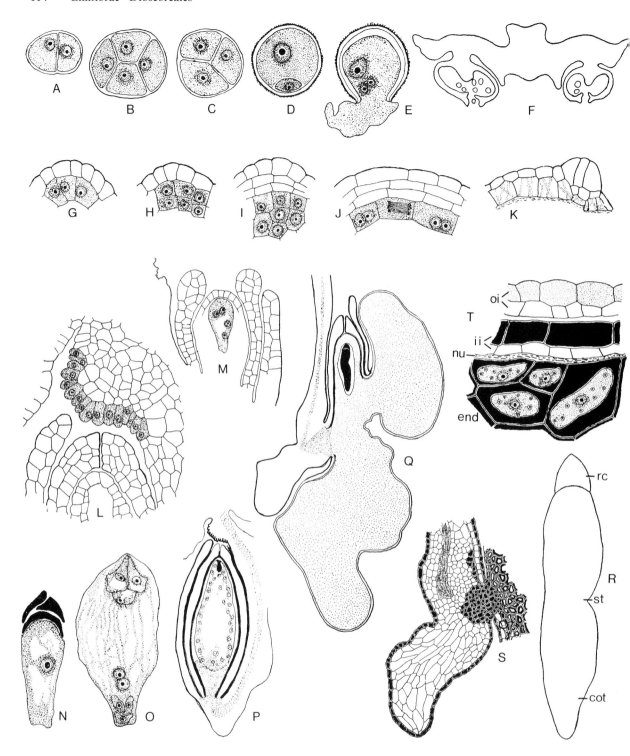

**Fig. 42.** Trichopodaceae. *Trichopus zeylanicus,* embryological details. **A–E** Successive microsporogenesis, mature pollen and pollen germination. **F** Anther in transverse section, showing the flat, leaf-like structure. **G–K** Pollen wall formation (Monocotyledonous type), tapetum (glandular type) and endothecium with spiral wall thickenings. **L–M** Details of the anatropous ovule. **N–O** Embryo sac mother cell and mature embryo sac (follows the *Polygonum* Type embryo sac formation). **P** Endosperm formation (nuclear type). **Q** Unripe seed showing the small elongate embryo and the extensive endosperm with marked rumination. **R** Embryo (*rc* radicle; *st* stem tip; *cot* cotyledon). **S** Chalazal portion of mature seed. **T** Seed coat and endosperm of mature seed (*oi* testal part; *ii* tegminal part; *nu* nucellar remnants; *end* endosperm cells). (NAGARAJA RAO 1955)

are solitary or paired and have long pedicels. They are bisexual and have 3 + 3 tepals forming a more or less campanulate perigone. There are 3 + 3 stamens which have short filaments and which are apically protracted into a long pointed tip far beyond the introrse thecae. Microsporogenesis (unlike that in the Dioscoreaceae) is successive. The pollen grains are sulcate and dispersed in the binucleate stage (NAGARAJA RAO 1955). The ovary is inferior and trilocular, as in Dioscoreaceae, and is covered with stalked glandular hairs. The three locules each bear one or two ovules. The inner integument becomes three- to four-layered at the micropyle. Unlike in the Dioscoreaceae a parietal cell is *not* cut off. Embryo-sac formation is according to the *Polygonum* Type and endosperm formation is nuclear. The fruit is three-winged, indehiscent and slightly fleshy. The seeds are ovoid and folded and have ruminate, almost starchless endosperm; the outer wall layers are hard and thick owing to heavy deposition of hemicellulose. The embryo in the mature seed is small, linear and straight, with a terminal cotyledon.

The family consists of *Trichopus* (1) *zeylanicus,* which is distributed in South and South-East Asia. It grows on the floor of forests.

Trichopodaceae combines features of Dioscoreaceae (in particular *Stenomeris*), Taccaceae, Stemonaceae and Trilliaceae. In several respects it exhibits supposedly primary character states, such as bisexual flowers, unspecialized rhizome, non-climbing growth, exserted connective tips, sulcate pollen grains and ruminate endosperm, while in other respects it is derived, as in the epigynous, winged fruit which is not readily dehiscent.

## Dioscoreaceae R. Brown (1810)   5:625
(Figs. 43, 44)

Herbaceous or rarely shrubby plants, mostly vines with a thick rhizome or tuber, more rarely (as in subfam. Stenomerioideae) with an elongate swollen rhizome. The tubers in subfam. Dioscoreoideae are of different morphological nature: i.e. the hypocotyl, the internode above it, or both. The tuber in some species is covered by a thick coat of cork. The rhizomes and tubers contain starch and sometimes have a continuous secondary growth. The aerial stem is usually long, twining in the vegetation or trailing on the ground. In stem twiners the stems may be rough through the possession of longitudinal ridges or prickles. The vascular strands are generally ordered in one or two rings.

At least in *Dioscorea* the stems as well as roots contain vessels with scalariform perforation plates.

The leaves are alternate, with variable phyllotaxis, or rarely opposite. They are petiolate and have a simple lamina (entire or lobate) or are digitately compound with from three to more than six leaflets. Venation is primarily palmate (campylodromous), with 3–13 main converging veins and reticulate veinlets of higher orders. The leaf tips often have a distinct water pore. Some species have stipule-like appendages at the leaf base. Hairs occur in about half of the taxa and are simple and sometimes peg-like, or stellate or bibrachiate, but glandular hairs are lacking. The stomata are anomocytic and often randomly directed on the leaf surface. Glands may occur, for example, on the forerunner tips, and extra-floral nectaries and mucilaginous pits are found in some species. Likewise there may be secretory cells with resinous contents and tanniniferous cells (idioblasts). Bundles of raphides are common.

Bulbils occur in the leaf axils of numerous species of *Dioscorea*; they are of cauline nature and contribute greatly to vegetative propagation.

The inflorescences are axillary panicles, cymes, spikes or racemes, which consist of many or few flowers. The flowers are epigynous, trimerous, and generally unisexual but bisexual in *Avetra* and *Stenomeris;* they are usually fairly small and inconspicuous, but rather conspicuous in *Avetra*. They are subtended by a small to relatively large bract and bear one bracteole or rarely two transverse bracteoles. The 3 + 3 tepals are sometimes fused into a tubular or campanulate structure. At least *Dioscorea* possesses septal nectaries (DAUMANN 1970). In *Tamus* a nectar-secreting tissue is stated to be present on the top of the ovary at the base of the tepal tube. There are 3 + 3 stamens, 3 of which may be staminodial. A disc is said to be present in a Bolivian species of *Dioscorea* (EMBERGER 1960). In *Stenomeris* and *Avetra* the stamens end in a long, subulate or filiform process. Sometimes the filaments may be fused into a tube. The anthers are introrse or extrorse and dehisce longitudinally. The endothecial thickenings are of Spiral Type. Microsporogenesis is simultaneous. The pollen grains are extraordinarily variable in aperture conditions: sulcate in *Stenomeris* and a few species of *Dioscorea*, 2-4-sulculate (*Dioscorea* and other genera) or 4(-5)-foraminate in *Avetra*, where the pollen grains are spinulose.

The inferior ovary is trilocular and the style is apically trilobate or tribrachiate. The stigma is Dry

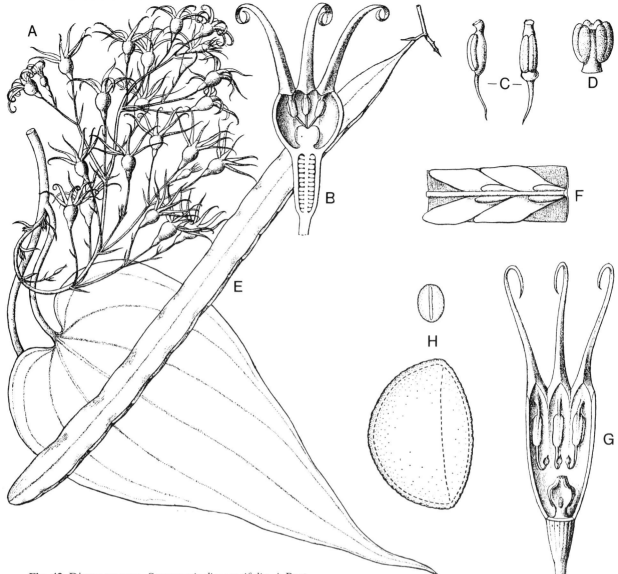

**Fig. 43.** Dioscoreaceae. *Stenomeris dioscoreifolia.* **A** Part of branch with inflorescence. **B** Flower, longitudinal section. **C** Stamens in the position they have in the flower, filament upwards. **D** Style and stigma lobes. **E** Fruit. **F** Seeds as located in the fruit. (KNUTH 1924) **G–H** *Stenomeris cumingiana.* **G** Flower, longitudinal section. **H** Pollen grains (sulcate). (ERDTMAN 1952)

at least in *Tamus*. Each locule has an axile placenta bearing from two to many ovules. A parietal cell is cut off from the archesporial cell. Embryo sac formation proceeds according to the *Polygonum* Type.

The fruits are usually capsules, which are generally triangular or three-winged, rarely a samara (*Rajania*) or a berry (*Tamus, Avetra*). The seeds, which are solitary to several, are often flattened; they are winged in *Dioscorea* and *Stenomeris,* while *Tamus, Avetra* and *Epipetrum* have globose seeds.

The testa of the seed is unusual in having several or sometimes many layers of cells, all with red to yellowish brown pigment (phlobaphene); all cells of the innermost layer of the outer integument in *Dioscorea* contain a crystal of calcium oxalate. The endosperm contains plenty of aleurone and lipids as well as hemicellulose which is deposited in its thick cell walls (pitted in *Epipetrum*). The embryo is small but well differentiated; it has an almost terminal plumule and a more or less lateral, broad, flat cotyledon. $x = 9, 10, 12, 14$.

*Chemistry.* Chemically the family is characterized by the rich occurrence of steroidal saponins (the aglycone of which is often diosgenin), and in the common occurrence of chelidonic acid. Species of *Dioscorea* contain the alkaloid dioscorrine.

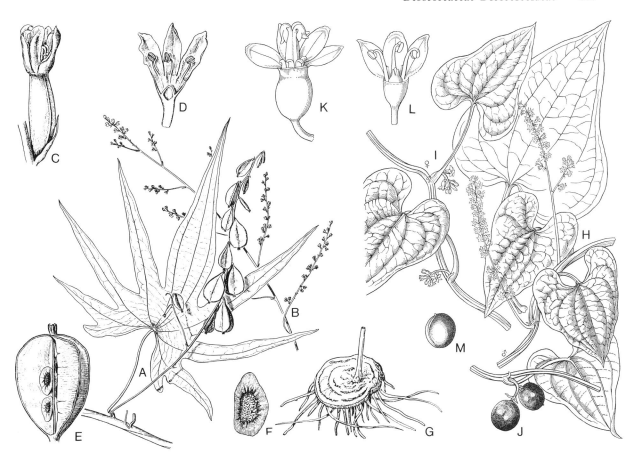

**Fig. 44.** Dioscoreaceae. **A–F** *Dioscorea brachybotrya.* **A** Branch with leaf and infructescence. **B** Male inflorescence. **C** Female flower. **D** Male flower. **E** Dehiscing capsule. **F** Seed. (CORREA 1969) **G** *Dioscorea sinuata,* tuber. (CABRERA 1968) **H–M** *Tamus communis.* **H** Piece of branch with male inflorescences. **I** Piece of branch with female inflorescences. **J** Fruiting branchlet. **K** Female flower, two tepals removed. **L** Male flower, three tepals and three stamens removed. **M** Seed. (ROSS-CRAIG 1973)

*Distribution.* Dioscoreaceae are widely distributed in the tropics and warm-temperate regions. While *Dioscorea* is widely distributed, *Stenomeris* occurs only in South-East Asia, *Tamus* is confined to Macaronesia and the Mediterranean, *Rajania* and *Epipetrum* occur in the West Indies and *Avetra* occurs on Madagascar.

## Subfamily Stenomeridoideae

Subfamily Stenomeridoideae consists of the (perhaps not very closely related) genera *Stenomeris* (5) (Fig. 43) in Indomalesia and the Philippine Islands, and *Avetra,* on Madagascar. They are vines with elongate swollen rhizomes and a twining stem with alternate, cordate or lanceolate leaves. The flowers, solitary or in axillary panicles, are bisexual. *Stenomeris* has an urn-shaped receptacle and six stamens inserted near the top of the tube on short basally directed filaments. The apices of the exserted connectives normally meet above the stigma. The pollen grains in *Stenomeris* are sulcate (a primitive feature in this context) but in *Avetra* four-(to five-)foraminate. The fruit is one-seeded and indehiscent (*Avetra*) or is a many-seeded loculicidal capsule (*Stenomeris*). In *Stenomeris* the fruit is long and linear and has a membranous wing, a similarity to subfamily Dioscoreoideae. This subfamily should perhaps be treated as two, (Stenomeridoideae and Avetroideae) and may deserve family rank.

## Subfamily Dioscoreoideae

Subfamily Dioscoreoideae makes up the main part of Dioscoreaceae and consists mainly of vines or trailing plants with an underground tuber which varies in shape and nature, but may be large and woody with a thick cork layer, as in *Dioscorea elephantipes* (sect. *Testudinaria*). The leaves vary from simple to digitately 6-7-foliolate and there is great variation in vegetative features. The flowers are invariably unisexual. In the male flowers the stamens are attached to the base of the perianth; three may be staminodial. The anthers lack apical continuation. The pollen grains are sulcate or bi-(to tri-)sulculate. In the female flowers there are often rudiments of the stamens; the inferior ovary develops into a trivalvular capsule, a berry (*Tamus*) or a samara (*Rajania*).

The subfamily exhibits great variation in various respects, such as tuber morphology, vascular anatomy, leaf morphology, gland morphology, pollen morphology and chemical contents.

*Tamus* (5) (Fig. 44 H–M) is a Macaronesian-Mediterranean genus of vines with red berries, containing globose seeds. The leaves and shoots of *T. communis* may be used as a vegetable. *Dioscorea* (ca. 600) is divisible into numerous sections according to tuber shape, inflorescences, seed wings, etc. *D. cayenensis, D. esculenta* and *D. batatas,* which are all from tropical Asia and Malaysia, and other species cultivated more locally, are used as Yams.

Yams are grown in the wet regions throughout the tropics for the starch-rich tubers and are of great importance as a starchy food. The plants are propagated vegetatively from tubers. The starch cannot be extracted as the grains are held together by mucilage; therefore the use of yams is somewhat restricted. The tubers contain the poisonous alkaloid dioscorine, but the concentration is low in all cultivars and the alkaloid is destroyed by boiling. They also contain steroidal saponins used for producing cortisone. *D. bulbifera,* "Air Potato", likewise from Asia, produces large tubers on the aerial stem as well as underground, and is cultivated in the U.S.A. In some species the tubers are very poisonous; the juice of tubers of *D. hispida* (tropical Asia), when mixed with Antiaria poison, can be used as arrow poison. The tubers of *D. rhipogonoides,* in China, can be used for tanning and dyeing fish nets. *D. villosa* and other species are exploited for the steroidal sapogenins which are used in the production of contraceptive pills and other medical products.

## Taccaceae Dumortier (1829)   1:10   (Fig. 45)

Terrestrial acaulescent perennials with a more or less thick rhizome or tuber containing large amounts of starch. The leaves are situated in a basal rosette and are interesting in being more or less long-petiolate and having a lamina varying from entire and lanceolate to broad and pinnatifid, bifid or even deeply palmately partite or palmatisect. The primary venation is pinnate or palmate, the secondary reticulate. The leaf base is more or less widened. From the growth apex of the tuber may be formed runners, which grow downwards and form new tubers. Vessels are confined to the roots and have scalariform perforation plates. Cells containing raphide bundles are concentrated mainly along the veins. Peculiar trichomes, consisting of a short row of cells bearing a multicellular body on which is another cell row at the apex, occur on both sides of the leaf. The stomata are anomocytic or in *Tacca* sensu stricto surrounded by one cell only ("axillocytic") and are randomly distributed on the leaf surface (LING PING-PING 1981).

The inflorescence is borne on a leafless, herbaceous peduncle directly from the rhizome or tuber. It is umbel-like, but apparently cymose, with the flowers inserted between foliose leaves. Long, filiform, drooping bracts subtend the flowers. These are epigynous and bisexual and have 3+3 more or less campanulately fused, dull, dark, brown-purple to greenish perianth members. The stamens, 3+3 in number, are inserted in the perianth tube. The filaments are short, flattened and adnate to the perianth except for the inflexed margins and the helmet-like apical continuation. The thecae are introrse. The anther wall formation is reported to agree with that in most dicotyledons (Dicotyledonous Type), as the outer secondary wall layer forms the endothecium as well as the middle layer, the inner wall layer forming the tapetum only. In all other monocots the middle layer is normally formed from the inner wall layer. The endothecial thickenings are of the Spiral Type. Microsporogenesis is simultaneous and the pollen grains are sulcate, verrucate to striate (ZAVADA 1983) and binucleate when dispersed.

**Fig. 45.** Taccaceae. **A–B** *Tacca plantaginea.* **A** Plant with inflorescence. **B** Flower, longitudinal section. (LIMPRICHT 1928) **C–M** *Tacca leontopetaloides.* **C** Plant. **D** Flower in centre of inflorescence. **E** Tepal and attached stamen excised from the syntepalous perianth. **F** Stamen from inside the "hood". **G–H** ovary and style, **H** in longitudinal section. **I** Ovary, transverse section. **J–K** Mature fruit, longitudinal and transverse sections. **L** Seed enclosed in aril. **M** Seed with aril removed. (HEPPER 1968) **N–O** *Tacca laevis;* pollen grain, **O** in transverse section. (ERDTMAN 1952)

The ovary is tricarpellary but unilocular, with three intrusive parietal placentas, each bearing numerous pendulous, ana- and apotropous ovules. The archesporial cell cuts off a parietal cell. Sometimes glandular cells or a "disc" may surround the stylar base but nectar is not secreted here; septal nectaries have been reported in some species. The style is simple, three-winged, and apically trilobate. The stigma is Dry, with a papillate surface. Embryo sac formation conforms to the *Polygonum* Type and endosperm formation is nuclear.

The fruits have a fleshy pericarp and seem to be dehiscent or indehiscent, i.e. capsules or berries. The seeds are prismatic, bean-shaped or almost horseshoe-shaped. They are provided with longitudinal ridges formed by the unusually high inner cell layer of the outer integument. Endosperm is copious and contains aleurone and fatty oils, but no starch. The embryo is small and ovoid. It has

a terminal plumule and a lateral cotyledon (CRONQUIST 1981). $n = 15$.

*Chemistry*. The chemistry is almost unknown. The tubers, which are very rich in starch, contain ceryl alcohol and a bitter principle.

*Distribution*. The family is pantropical, but is concentrated in Indomalesia, South-East Asia and the Solomons, where nine out of the ten species occur. The species grow on the ground in shady or open habitats: in open forests, forest margins and on savannahs and crop fields.

*Tacca* (incl. *Schizocapsa*; 10) is the only genus. The flowers are pollinated mostly by flies (*Diptera*). As in species of Aristolochiaceae the flowers exhibit the syndrome of sapromyophily and may function as "traps", so that the flies cannot leave the tubular chamber formed by the perigone without effecting pollination. Species of *Tacca* are used for their tubers, which contain starch; *T. pinnati-*

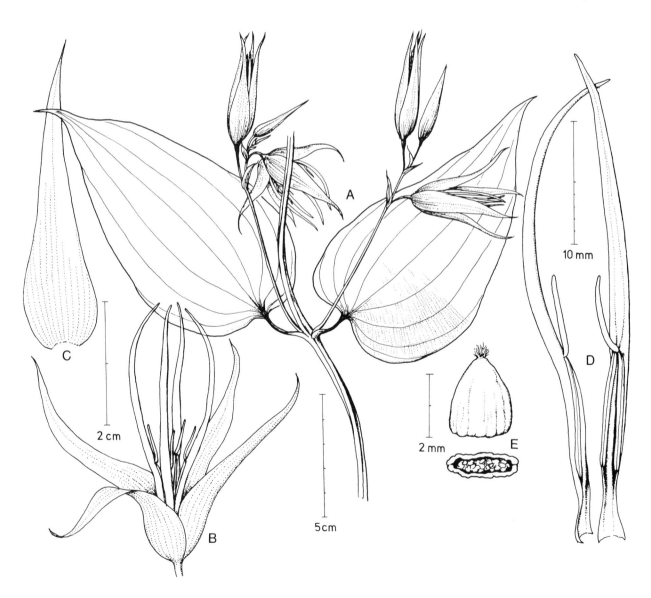

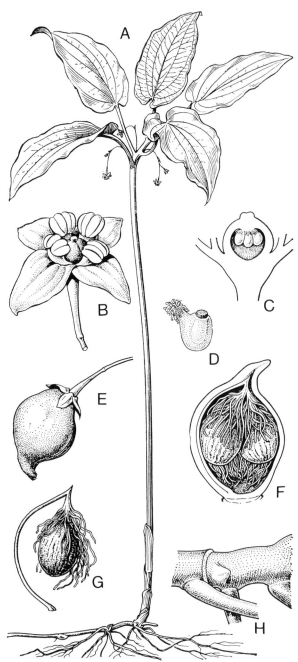

Fig. 47. Stemonaceae. *Croomia pauciflora.* **A** Plant.
**B** Flower. **C** Ovary, longitudinal section. **D** Ovule,
with funicular aril. **E** Fruit. **F** Same, longitudinal section.
**G** Seed; note the aril and the long funicle. **H** Rhizome
details. (Takhtajan 1982)

◁ **Fig. 46.** Stemonaceae. *Stemona tuberosa.* **A** Part of flowering branch. **B** Flower. **C** Tepal (outer whorl). **D** Stamens; note the long, exserted stamen, the distinctly separated bisporangiate thecae and the suprathecal, horn-
like process. **E** Ovary, lateral view and transverse section; note that the ovary is unilocular and has numerous
ovules on a basal placenta (a very rare condition in
monocotyledons). (Orig. B. Johnsen)

*fida* is used as "East Indian Arrowroot", but bitter
principles must be removed. The tubers can be
eaten as a vegetable.

The taxonomic position of Taccaceae has been
the subject of much speculation. The family has
been considered to be related to, for example,
Dioscoreaceae and Amaryllidaceae, and sometimes also to Aristolochiaceae, Velloziaceae, Apostasiaceae and Philydraceae. Drenth (1972), who
revised the group, considered it likely that it had
the closest connection with Amaryllidaceae, with
which it has the umbel-like inflorescence in common. However, the whole spectrum of characters
indicates a rather close relationship to genera like
*Stenomeris* (Dioscoreaceae), *Trichopus* (Trichopodaceae) or *Croomia* and *Pentastemona* (Stemonaceae).

**Stemonaceae** Engler in Engler & Prantl (1887)
4:30   (Figs. 46–48)

Herbs, herbaceous vines or shrublets with subterranean rhizomes or tubers and erect, short or long
and climbing stems. The vascular anatomy differs
from that in Dioscoreaceae, in that the bundles
are borne in one or two rings. The leaves are alternate or frequently opposite or verticillate. Further
they are petiolate, non-sheathing at the base, and
have an entire, often cordate lamina with 5–15
arching, convergent main veins and numerous secondary transverse cross-veins. Raphides are present or lacking. Vessels with scalariform perforations are known to occur in roots, stems and even
leaves (in *Croomia* in roots only). The stomata
are transversely oriented on the leaves in *Stemona*
and *Stichoneuron,* and are anomocytic. The flowers are solitary and axillary or occur in few-flowered cymes or racemes. They are bisexual or rarely
unisexual, hypogynous (*Stemona, Croomia*), half-
epigynous (*Stichoneuron*) or epigynous (*Pentastemona*), dimerous, with 2+2 tepals, or pentamerous, with five tepals in one whorl. The tepals are
petaloid or sepaloid and may be fused basally.
There are 2+2 or five stamens with short filaments, which are free from each other or more
often basally connate and sometimes adnate to the
perianth. The anthers are introrse, flat, and in *Stemona* produced into long, linear or lanceolate,
sometimes flattened connective tips. Their thecae
may be distant from each other. Microsporogenesis, as far as is known, is successive. The pollen
grains are sulcate and finely reticulate.

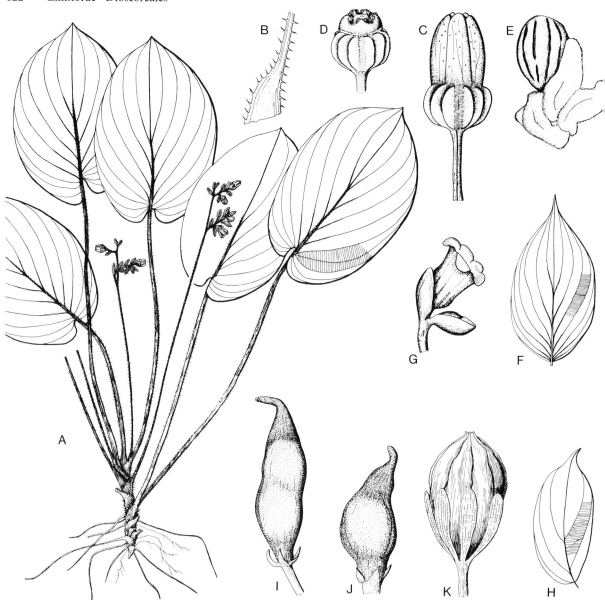

**Fig. 48.** Stemonaceae. **A–E** *Pentastemona sumatrana.* **A** Habit. **B** Base of leaf sheath. **C** Flower. **D** Flower with tepals removed. **E** Seed with funicular aril. **F–G** *Pentastemona egregia.* **F** Leaf. **G** Flower. **H–K** *Stichoneuron caudatum.* **H** Leaf. **I–J** Fruits. **K** Ribbed seed with lobed aril. (All from VAN STEENIS 1982)

The bicarpellary pistil has a unilocular ovary with sessile or subsessile stigma(s). The ovules are generally few and either basally inserted (*Stemona*) or apical and pendulous (*Croomia, Stichoneuron*). They are anatropous or (in *Stemona*) orthotropous. A parietal cell is cut off from the archesporial cell. Endosperm formation is nuclear. The fruit is a bivalvular, sometimes berry-like capsule with one or few longitudinally ribbed seeds, which bear an elaiosome or caruncle, formed from the hilum and sometimes also from the raphe. The elaiosome is surprisingly variable in the family. It generally contains fat. The endosperm is not ruminate. It contains aleurone, fat and cellulose, but in *Croomia* also copious starch (only a little starch in *Stemona*). $x = 7$.

*Chemistry.* Lactone alkaloids of a characteristic type (tuberostemonin, etc.) occur in at least the underground parts. They represent the effective substances in the insecticides sometimes produced from *Stemona sessilifolia* and *S. tuberosa* in tropical Asia.

*Distribution. Stemona* is distributed in southern Asia and Malaysia to northern Australia, *Croomia* in Japan and Atlantic North America, and *Stichoneuron* in eastern Asia. Some are shade plants and are obviously dispersed by ants.

Within the family, *Stemona* (*Roxburghia*) (25) (Fig. 46) has rather large, often pointed tepals, basal placentation and orthotropous ovules. It has the habit of some *Smilax* species. The outer integument of the seed has three to several cell layers, as in species of *Dioscorea*. The elaiosomes are very variable in the genus.

*Croomia* (3) (TOMLINSON and AYENSU 1968) and *Stichoneuron* (2) deviate in their smaller rounded tepals and apical, pendulous, and anatropous ovules. In their richness in starch the seeds of *Croomia* resemble those of some species of Trilliaceae.

*Pentastemona* (2) is a genus described by VAN STEENIS as late as 1982 (see also MEIJER and BOGNER 1983). The two species, *P. sumatrana* and *P. egregia* (Fig. 48), both from Sumatra, are extraordinary in monocotyledons by having actinomorphic pentamerous, tricyclic flowers (T5 A5 G3). They are rather small herbs with long-petiolate leaves with an ovate, mesomorphic leaf blade, cordate or rounded at the base and acute or subacuminate at the apex, pinnately veined with transverse secondary veinlets, tetracytic stomata and raphides as well as crystal styloids. The petioles and the widened sheathing leaf bases are ciliate. The leaves are rosulate, and the peduncles emerge from the ground level and are shorter than the leaves, with small flowers in a thyrse or raceme with one or two lateral components which may represent cincinni. The flowers are bracteate and supplied with a rather large dorsal bracteole, and are bisexual and epigynous with an urceolate-tubular perianth with five rounded lobes. The filaments are fused into a short fleshy tube, on the margin of which the basally widely separated thecae are situated. The pollen grains are inaperturate. The ovary is unilocular, with intrusive, longitudinal placentae, each with numerous ovules. The ovary and fruit are provided with ten longitudinal flanges or ribs. The style is short and thick; five (? nectar-) pouches are present at its base inside the staminal tube. The fruit is a membranous capsule with subglobose seeds which have a transparent sarcotesta, and a vesicular aril. The endosperm is copious and the embryo minute. We find this genus highly distinctive, and although VAN STEENIS (1982) is keen to stress that Stemonaceae should not be divided further, one may con-

sider regarding *Pentastemona* as about as distinct as *Trichopus* or even *Tacca,* and thus worthy of family rank (the Latin description of *Pentastemona* is given in VAN STEENIS 1982, p. 160).

## Trilliaceae Lindley (1846)   5:50   (Figs. 49, 50)

Rhizomatous herbs, generally with an erect aerial stem. The rhizome varies from long to short, thick and tuber-like. The stem is unbranched and generally bears a verticil of 3–6 or more leaves (generally the same number as each of the floral whorls) in its upper parts, except in *Scoliopus,* where the two leaves are basal. In the vascular strands, which are frequently disposed in three circles, vessels seem to be lacking; they are confined to the roots and have scalariform perforation plates. The leaf blades are entire, sessile and ovate to lanceolate in outline, and parallel-veined with pinnate secondary venation. Oxalate raphides are found in stems and leaves. The stomata are not always parallel to the main veins.

The flowers are solitary and terminal on the erect stem or (in *Scoliopus*) emerge from the rhizome. They are hypogynous, bisexual and usually 3–8-(-10)-merous. Tepals and stamens are persistent and remain withered around the fruit. The tepals occur in two whorls; the outer (especially in *Trillium*) are sometimes sepaloid, and are purplish or greenish; the inner are often petaloid and white, yellow, purple or of some other colour; they are either broader or, frequently, much narrower than the outer, sometimes even rudimentary (*Paris tetraphylla*). Aestivation is imbricate or contorted. The shape of the tepals varies from narrowly linear or filiform to ovate. Nectar is secreted from the tepal bases in at least some species of *Trillium,* in which genus septal nectaries have also been reported.

The stamens are isomerous with the tepals. They have a distinct filament and an elongate, basifixed, longitudinally dehiscing, introrse or (in *Scoliopus*) extrorse anther. In some genera, as in Stemonaceae, the connective may continue as a narrow apical process. Microsporogenesis is successive. The pollen grains are sulcate or inaperturate (*Trillium* s. lat.).

The ovary is usually 3-6(-8)-carpellary. It is roundish or angular, and sometimes depressed at the top and either has separate locules or is partly unilocular, and then with strongly intrusive placentae. The style is branched low down or there are free stylodia, which are slender or thick

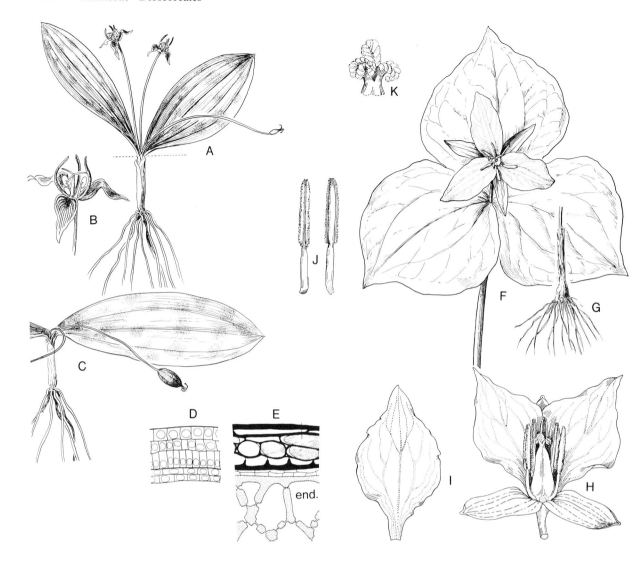

**Fig. 49.** Trilliaceae. **A–C** *Scoliopus hallii*. **A** whole, flowering plant. **B** Flower. **C** Fruiting plant. (HITCHCOCK et al. 1969) **D–E** *Scoliopus bigelovii*, seed coat in unripe and ripe seed, respectively. The seed coat does not have phytomelan. (BERG 1959) **F–G** *Trillium ovatum*. (HITCHCOCK et al. 1969) **H–K** *Trillium rivale*. **H** Flower with one of the inner petaloid tepals and the opposite stamen removed. **I** Tepal of inner whorl. **J** Stamen, the left seen from centre of flower (the anthers are introrse). **K** stylodial branches. (SEALY 1964)

(*Daiswa*), and then separated from the ovary by a transverse rim; the stigmas are Dry. The ovules are anatropous and a parietal cell is normally cut off from the archesporial cell. The embryo sac is generally of the *Allium* Type, or rarely (in *Scoliopus*) of the *Polygonum* Type. Endosperm formation is helobial in *Trillium* but nuclear in at least *Paris* and *Scoliopus*.

The fruit is a berry (*Paris, Kinugasa,* species of *Trillium*) or a fleshy capsule (*Daiswa,* species of *Trillium*). According to BERG (1958) the capsular fruits, which are the most common in *Trillium* (and also occur in *Daiswa*), seem to be derived from berries – a supposedly unusual evolutionary course. Seeds in *Trillium* may bear an oil-rich elaiosome developed from the raphe and hilum. The seeds have a testa formed mainly by the outer integument, which is 3–5 cell layers thick; in *Trillium* the innermost layer is strongly pigmented. A scarlet and juicy sarcotesta is present in *Daiswa* (TAKHTAJAN 1983). The copious endosperm contains aleurone, lipids and also plenty of starch grains. The embryo is small, undifferentiated, and globose to ovoid.

*Chemistry.* Chelidonic acid and steroidal saponins occur in Trilliaceae; in *Paris,* at least the latter are strongly poisonous. It is noteworthy that

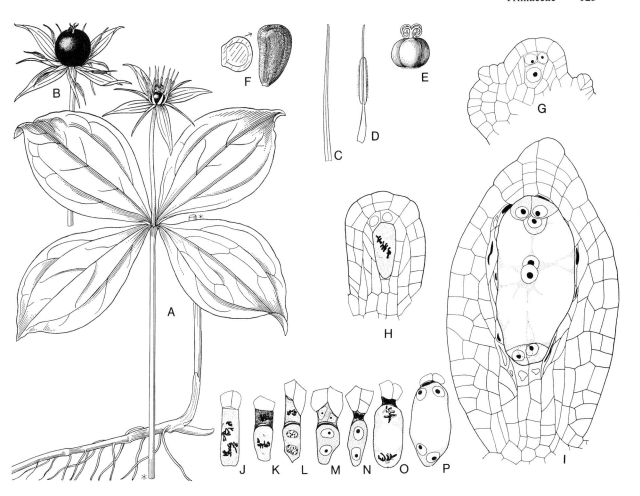

**Fig. 50.** Trilliaceae. **A–F** *Paris quadrifolia*. **A** Plant. **B** Fruit. **C** Tepal of inner whorl. **D** Stamen. **E** Ovary. **F** Seed, to the *left* in transverse section. (ROSS-CRAIG 1972) **G–I** *Paris quadrifolia*, embryology. **G–H** Early stage of ovule; parietal cells cut off. **I** Mature embryo sac. **J–P** Ontogeny of embryo sac; note that the four nuclei at the stage in **M** are the megaspore nuclei, and that the embryo sac is bisporic, of the *Allium* Type. (BERG 1962b)

*Paris* has poisonous, but *Kinugasa* non-poisonous and edible berries (TAKHTAJAN 1983). Several types of sapogenin are known in the family.

*Distribution.* Trilliaceae is mainly a Northern Hemisphere family with its widest distribution in temperate regions of Asia and North America. Most species are herbs growing in shady woods.

*Trillium* (30) (BERG 1958; 1962b) has trimerous flowers and three cauline leaves. The tepals are broad and often conspicuous; the inner are petaloid and white to reddish. The fruit is generally a three-valvular capsule, rarely a berry. Its seeds are provided with an elaiosome. The leaves of *T.*

*grandiflorum* may be eaten as a vegetable, while the rhizomes of *T. erectum* have medicinal use. Some species are used in ornamental horticulture.

In *Paris* (4), *Kinugasa* (1) and *Daiswa* (15) the tepals are narrow and linear, sometimes rudimentary. The connectives may be prolonged apically. *Paris* and *Kinugasa* have a berry, black and poisonous in *Paris*, but dark purple and edible in *Kinugasa*. The two genera also differ in that *Paris* has a slender and *Kinugasa* a thick rhizome. *P. quadrifolia* is common in European woods. Its leaves have medical use. The species of *Daiswa*, which have a thick rhizome, are characterized by having a fleshy loculicidal capsule and seeds with a sarcotesta. The flowers in this genus vary from being 4- to 8- (or even 10-)merous. These genera have a palaeotemperate distribution. The above division of *Paris* s.lat. was proposed by Takhtajan as recently as 1983.

The differences between *Scoliopus* (2) (Fig. 49) in North America, and other Trilliaceae in the embryo sac and other characters (BERG 1962b) sug-

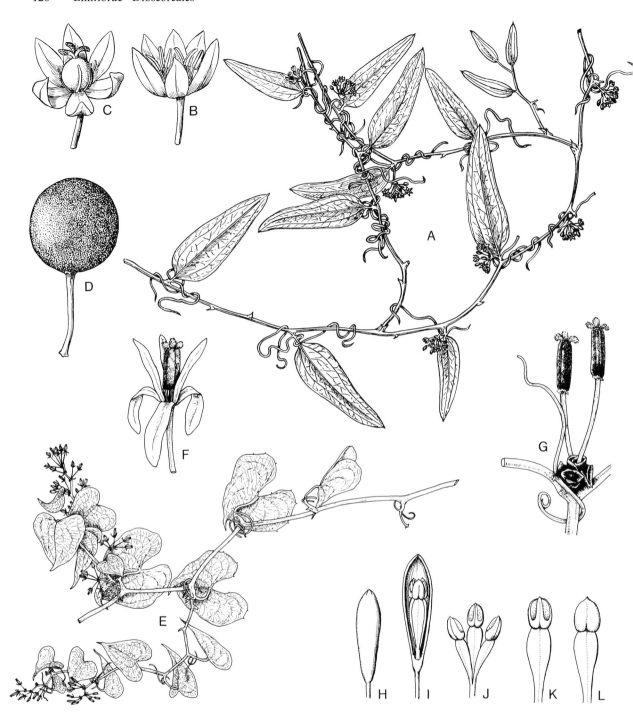

**Fig. 51.** Smilacaceae. **A–D** *Smilax campestris*. **A** Flowering branch. **B** Male flower. **C** Female flower. **D** Fruit. (CABRERA 1968). **E–G** *Smilax aspera*. **E** Flowering branch. **F** Flower. **G** Young fruits. (JAFRA and EL-GADI 1978) **H–L** *Heterosmilax japonica*. **H** Male flower bud. **I** The same in longitudinal section. **J** Androecium. **K–L** Stamen, front and back. (Redrawn from KOYAMA 1978)

gest that this genus might be better placed in Uvulariaceae. *Medeola,* which has also been referred to Trilliaceae, may have a similar affinity.

In its largely boreal-temperate distribution Trilliaceae deviates from the other families in Dioscoreales and joins up with Liliales (Melanthiaceae, Liliaceae) and Asparagales (Convallariaceae).

**Smilacaceae** Ventenat (1799)  4:310  (Fig. 51)

Herbs, shrubs or partially woody vines or twiners arising from thick, starch-rich rhizomes or tubers. The roots are mycorrhizal and lack root hairs. Secondary thickening growth is lacking. The branches climb or straggle with the help of tendrillar appendages from the base of the petiole (see below). In many species the stems are provided with prickly thorns (trichome homologues). The leaves are alternate or (in *Ripogonum*) opposite and are petiolate or subsessile. The lamina is lanceolate, broadly cordate or hastate, and is usually stiff and coriaceous; it has campylodromous primary venation, pinnate and marginally brochidodromous secondary venation, and a reticulate pattern of finer venation between the main veins. In all taxa except *Ripogonum* the petiole, at its junction with the only rarely broad and sheathing base, is provided with a pair of tendrils (which have sometimes been considered homologous with midveins of two lateral leaflets).

Laticifers are lacking. The vascular strands of the roots and stems, and at least sometimes also the leaves, contain vessels with simple or scalariform perforation plates. The stomata are anomocytic and often randomly oriented or transverse. The stomata in *Ripogonum* are mesoperigenous (CONOVER, 1983). Oxalate raphides contained in mucilage cells occur frequently, for example in the saponin-rich rhizomes or tubers.

The inflorescences are axillary or situated on the tips of branchlets. They often consist of one or more (superimposed) verticils of flowers (appearing like umbels or racemes). The flowers are actinomorphic, hypogynous and mostly unisexual (the plants dioecious) or rarely (*Ripogonum*) bisexual, and possess 3 + 3 more or less equal tepals, inconspicuous but petaloid, and sometimes fused into a short or (in *Heterosmilax*) long tube. The male (or bisexual) flowers have 3 + 3 (rarely 3 or 3 + 3 + 3 or up to 18) stamens. Their filaments are free from each other or sometimes more or less fused into a tube or column. The anthers are basifixed, introrse or latrorse, lack apical appendages, and dehisce longitudinally; they are tetrasporangiate and dithecous, but the thecae may become confluent (CRONQUIST 1981). The tapetum is glandular and microsporogenesis successive. The pollen grains are globose, inaperturate (or indistinctly aperturate) and spinulose in most genera, but sulcate and reticulate in *Ripogonum*. They are free and dispersed in the two-celled stage. Nectaries are often present at the base of the tepals or stamens.

The functional pistil is syncarpous and trilocular (rarely, in *Smilax pumila,* monocarpellary); it has three stylodia which are free or rarely fused basally for a short distance. Septal nectaries are present. The stigmas are papillate and Dry. The locules generally contain one or two pendulous ovules. These are incompletely anatropous, hemianatropous, campylotropous or (as a rule) orthotropous. A parietal cell is cut off from the archesporial cell. Embryo sac formation proceeds according to the *Polygonum* Type and endosperm formation is nuclear. The fruits are globose berries with one or three seeds. These are globose or ovoid and very hard; when ripe the outer epidermis is obliterated, and loosens from the inner shiny part of the seed, and the inner integument is comparatively well developed. Phytomelan is lacking. The hardness of the seeds depends on the endosperm which consists of radially elongated thick-walled cells (with scalariform-pitted walls). In addition to the cellulose, the endosperm contains aleurone and fatty oils (except *Ripogonum* where these products partly seem to be substituted by starch grains). The embryo is straight, linear and very small.

*Distribution.* Smilacaceae with its few genera is rich in species, the largest genus, *Smilax,* being pantropical-subtropical with the centre in America; in Europe, Eastern Asia and North America it reaches into temperate regions. The other genera have smaller East-Asiatic distributions, except *Ripogonum,* which occurs from Australia to New Guinea and New Zealand.

**Subfamily Ripogonoideae**

*Ripogonum* (7) (MACMILLAN 1972), deviates from the other genera in several characters. These include opposite leaves, lack of tendrils, bisexual flowers in spike-like inflorescences, non-confluent thecae, sulcate, reticulate pollen grains, anatropous ovules and starch-rich endosperm. It is arguable whether this genus belongs in Smilacaceae at all; HUBER is also inclined to treat it as a separate, closely related family.

**Subfamily Smilacoideae**

*Smilax* (300) and the smaller genera, *Heterosmilax* (6–15) in India–Japan with three stamens fused into a column, and *Pseudosmilax* (2), on Formosa with nine or more stamens, are all more or less similar in having leaves with variably long paired tendrils on the petiole and in having unisexual flowers (the plants are dioecious). *Smilax* is exceedingly variable in habit, leaf shape, number of seeds per berry and curvature of the tepals. A few species occur in Europe: *S. aspera* in a great part of the Mediterranean area and *S. excelsa* in the Balkan Peninsula. *S. canariensis* occurs in the Azores. The tubers of *S. china,* in Eastern Asia, are used in a decoction against gout. The roots of the Mexican *S. aristolochiifolia* and other species are important in yielding various types of "radix sarsaparillae" used in medicine as a tonic against digestive disturbances. The active substances are steroidal saponins, such as parillin, smilasaponin, and the poisonous sarsasaponin. The starch-rich tuber-like rhizomes of some species are used as potatoes, and the leaves of some species are used for tea. The berries of *S. megacarpa* in South-Eastern Asia (mostly Java) are eaten, mainly as preserves.

### Petermanniaceae Hutchinson (1934)    1:1
(Fig. 52)

A woody vine with more or less prickly stem (cf. *Smilax*) arising from a swollen, woody subterranean rhizome with irregularly distributed adventitious roots (TOMLINSON and AYENSU 1969). The leaves are spirally set, subsessile, broadly lanceolate and acuminate. Stipules, leaf tendrils and similar structures are lacking, and the primary venation is pinnate-campylodromous with reticulate finer venation. Vessels are restricted to the roots (cf. Smilacaceae), and have oblique end-walls with scalariform perforations. Raphides are present in all parts of the plant, and the stomata are anomocytic (TOMLINSON and AYENSU 1969). Branching is sympodial and some of the raceme-like paniculate inflorescences, principally terminal in position (but "leaf-opposed"), are transformed into tendrils. The flowers are bisexual, actinomorphic and epigynous. The 3+3 tepals are subequal and spreading to deflexed. There are 3+3 stamens with free, erect filaments and extrorse, tetrasporangiate

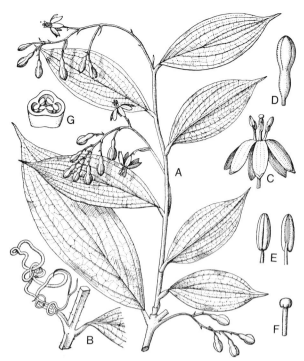

Fig. 52. Petermanniaceae. *Petermannia cirrosa.* **A** Flowering branch; branching is sympodial. **B** Tendril, which represents a transformed inflorescence. **C** Flower. **D** Flower bud. **E** Stamens (the anthers are extrorse). **F** Style apex and stigma. **G** Ovary, transverse section; note that it is unilocular and has parietal placentae. (KNUTH 1924)

anthers with contiguous loculi. The connective is not produced beyond the thecae, and the pollen grains are sulcate. The pistil has a unilocular, tricarpellary ovary. Its style is simple and stigma capitate. The parietal placentae bear fairly numerous ovules. The fruit is a berry with several seeds.

*Petermannia* (1) *cirrosa* is restricted to New South Wales and Queensland in Australia.

This genus, like *Ripogonum* and other Smilacaceae, combines features of Dioscoreales and Asparagales (e.g. the Philesiaceae). With some uncertainty, it seems to merit family rank separate from Smilacaceae.

# Order Asparagales

*The Order includes 31 families:* Philesiaceae, Luzuriagaceae, Convallariaceae, Asparagaceae, Ruscaceae, Herreriaceae, Dracaenaceae, Nolinaceae, Asteliaceae, Hanguanaceae, Dasypogonaceae, Calectasiaceae, Blandfordiaceae, Xanthorrhoeaceae, Agavaceae, Hypoxidaceae, Tecophilaeaceae, Cyanastraceae, Eriospermaceae, Ixioliriaceae, Phormiaceae (incl. Dianellaceae), Doryanthaceae, Asphodelaceae, Anthericaceae, Aphyllanthaceae, Hemerocallidaceae, Funkiaceae, Hyacinthaceae, Alliaceae (incl. Gilliesiaceae), Amaryllidaceae.

Perennial herbs as well as subshrubs, shrubs and sparingly branched trees usually with terminal leaf rosettes. In some families the stems are woody and in certain of these there is secondary thickening growth which is dissimilar to that in dicotyledons. The roots in rare instances are fusiform and nutrient-storing; one- to several-layered velamen is reported in several taxa. Most members have rhizomes or bulbs, bulbs being in particular characteristic of Amaryllidaceae, Alliaceae and Hyacinthaceae; corms occur in, for example, Tecophilaeaceae and some Alliaceae and Hypoxidaceae, and tubers in Eriospermaceae.

Vessels are present in the roots, where they have simple or scalariform perforation plates or both types; vessels with scalariform perforation plates also occur in the stems of several families, such as Philesiaceae and Luzuriagaceae and, in tree-like taxa, e.g. Agavaceae. Laticifers are restricted to some Alliaceae.

The leaves are alternate and distichous or more often with other phyllotaxies; rarely they are opposite or verticillate. They are always simple and generally narrow, with filiform or linear to lanceolate or oblong laminae; further they are mostly sessile and sheathing at the base with parallel or arching veins, but in some groups the leaves are differentiated into a petiole, which is not sheathing at the base, and a broad lamina which in a few families has reticulate venation. Succulent leaves occur in some families, sometimes in association with the presence of tough fibres. Ligules occur in at least some species of *Sowerbaea* and *Allium*. Peculiar appendages to the leaves occur in species of *Eriospermum,* Eriospermaceae.

The stomata are generally parallel to the leaf axis. They are agenous or perigenous in ontogenetic terms, and most often of the anomocytic type (lacking subsidiary cells) when mature, but paracytic stomata occur, for example in Hypoxidaceae, Xanthorrhoeaceae, Doryanthaceae and *Astelia* of Asteliaceae, and tetracytic stomata are frequent in Asphodelaceae subfam. Alooideae. Hairs are frequently lacking and when present are mostly simple and unicellular or uniseriate. Oxalate raphides contained in mucilage-filled cells or sacs are common throughout the order, while silica bodies are lacking.

The inflorescences are borne terminally or laterally and are often long-pedunculate (scapose), especially in groups with bulbs. Determinate as well as indeterminate inflorescences occur in the order, in the former case sometimes assuming an umbel-like appearance.

The flowers are actinomorphic or more rarely zygomorphic, hypogynous or epigynous, and generally pentacyclic, with two, mostly trimerous whorls of more or less petaloid tepals, two whorls of stamens and a pistil consisting of one whorl of carpels. In most cases the outer and inner tepals are similar to each other. They may be free or fused into a campanulate or tubular structure. The tepals are generally neither spotted nor variegated; they are provided with basal nectaries only in Philesiaceae and some Luzuriagaceae.

The stamens generally have basifixed or dorsifixed-hypopeltate anthers (see Fig. 3) and narrow or broad filaments, sometimes (as in many genera of Hyacinthaceae, Alliaceae and Amaryllidaceae) with lateral processes on each side of the attachment point of the anther; rarely, they are fused basally or high up into a staminal ring or tube. Hairy filaments occur in some genera of Asphodelaceae and Anthericaceae and in *Yucca* of Agavaceae. The anthers are tetrasporangiate and generally introrse but extrorse in, for example, Ruscaceae. They dehisce by lateral slits or, rarely, by pores, as in Tecophilaeaceae, Cyanastraceae, Calectasiaceae, some Phormiaceae and a few genera in other families. The endothecial thickenings are of the Spiral Type (except in *Lomandra* of Dasypogonaceae where they are of the Girdle Type). The tapetum is mostly secretory, but stated to be amoeboid at least in some Hypoxidaceae. Microsporogenesis is successive in most families but simultaneous in, for example, most or all Asphodelaceae, Tecophilaeaceae, Cyanastraceae, Phormiaceae, Doryanthaceae and Hemerocallidaceae. The pollen grains are mostly sulcate, but bisulcate in some genera of Dasypogonaceae and Amaryllidaceae, spiraperturate in Aphyllanthaceae and *Lomandra* of Dasypogonaceae, trichotomosulcate in some Luzuriagaceae and Anthericaceae and in

Phormiaceae and inaperturate in, for example, Philesiaceae. They are dispersed in the two-celled or rarely in the three-celled state (see WUNDERLICH 1936). The pistil is syncarpous and generally trilocular but rarely unilocular, as for example Philesiaceae and Calectasiaceae and in some Asparagaceae. Inferior ovaries (epigynous flowers) occur in several groups, the most important being the Amaryllidaceae, Hypoxidaceae and part of the Agavaceae, although single genera with epigyny occur in other families. Semi-inferior ovaries occur in particular in Tecophilaeaceae and Cyanastraceae. Septal nectaries occur in most members of the order but are absent at least from some Luzuriagaceae and Philesiaceae (with perigonal nectaries) and from Hypoxidaceae and Tecophilaeaceae. The septal nectaries are of external or internal types and quite variable; in groups with inferior ovaries they discharge at the style base. The style is simple or apically three-armed or three-lobed, free stylodia being absent from the order. In the Ruscaceae and Hanguanaceae and in some Asteliaceae the style is obsolete and the stigma(s) sessile or subsessile. The stigma is Dry or more rarely Wet. The placentation is mostly axile in trilocular ovaries with several to numerous ovules, and rarely basal or apical. A single ovule or two ovules per carpel are found in Ruscaceae, Dracaenaceae, Dasypogonaceae and a few genera of other families.

The ovules are mostly anatropous or, less often, hemitropous, campylotropous (frequent, for example, in Anthericaceae and Alliaceae) or orthotropous (occurring in several families). A parietal cell is usually cut off, forming a variably extensive parietal tissue, but is lacking in several groups, including some genera of Nolinaceae, Hypoxidaceae, Phormiaceae and Alliaceae, and also, for example, in Dracaenaceae and certain Amaryllidaceae, where the nucellar epidermis may divide periclinally to form a "cap". Embryo-sac formation generally conforms to the *Polygonum* Type, less often to the *Allium* Type, and very rarely to the *Scilla, Clintonia* or *Drusa* Types. Endosperm formation proceeds almost equally often according to the Nuclear and the Helobial Types (see under the separate families), and embryo formation according to the Asterad or Onagrad (rarely the Solanad) Types.

The fruit is generally a loculicidal capsule but in several families it is baccate (these families being treated first below); rarely it is dry and indehiscent. Other fruit types are the septicidal capsule (*Herreria, Blandfordia, Yucca, Excremis*) and capsules opening with a circumcissile slit (some Hypoxidaceae) or irregularly (*Ophiopogon*), the schizocarp (*Tricoryne, Cyanastrum*), and nutlets (genera of Nolinaceae and Dasypogonaceae).

The seeds are variable in size and shape, but are often ovoid or angular. The outer epidermis is obliterated in the seeds of most taxa with baccate fruits. In most taxa with capsules and a few of those with berries the outer integument is encrusted with phytomelan, an opaque, brittle charcoal-like substance which is chemically very inert. Rarely, the seeds have a brightly coloured testa (*Ophiopogon*). The inner integument in the seed coat is usually compressed into a colourless or red-brown membrane. Endosperm is usually present, but rarely substituted by chalazosperm (Cyanastraceae) or perisperm (*Yucca,* Agavaceae). The endosperm contains fatty oils, aleurone and cellulose, while starch grains are only occasionally present (then mostly in genera of Amaryllidaceae). The embryo is linear and usually straight, but is sometimes curved, e.g. in some Anthericaceae and Alliaceae. The plumule is lateral.

*Chemistry.* Chemically the order is characterized by the common presence of steroidal saponins, but these are lacking or rare in certain families, in particular in Amaryllidaceae, where there is instead a richness of particular alkaloids. Anthraquinone derivatives occur at least in Asphodelaceae and Xanthorrhoeaceae, bufodienolide in Hyacinthaceae and cardenolide in Asparagaceae. Fructan and glucomannan are frequently deposited in subterranean storage organs (rhizomes, corms, etc.). Chelidonic acid is widespread in the order, while cyanogenic compounds are rarer (certain Amaryllidaceae, Alliaceae, etc.).

*Distribution.* The order Asparagales is widely distributed and various groups of families show different patterns of distribution. RAVEN and AXELROD (1974) consider the group to be most likely of West Gondwanaland origin from which it could have invaded the Northern Hemisphere before the close of the Cretaceous. The rich representation of families such as Asparagaceae, Tecophilaeaceae, Asphodelaceae, Hypoxidaceae and Amaryllidaceae in Southern Africa suggests that their ancestors occurred on the Southern Hemisphere continents while these were still adjacent; the distributions of Xanthorrhoeaceae, Dasypogonaceae and Anthericaceae suggest a differentiation in continents derived from eastern Gondwanaland.

*Relationships.* The present circumscription of the Asparagales agrees largely with that proposed by HUBER (1969), who reinstated many of the families recognized earlier (e.g. by LOTSY 1911) but more recently submerged in a widely circumscribed Liliaceae s. lat. These families fall mainly under the Asparagales, the remainder into Liliales and Dioscoreales.

The differences between Asparagales and Liliales, can be summarized in Table 1 extracted largely from HUBER (1969:510–512).

It is obvious that taken singly none of the above characters is sufficient for distinguishing the families into different orders, but in conjunction they seem to be of great significance. Among the families which bridge the orders are Philesiaceae, with perigonal nectaries and tepals with variegated pattern. Philesiaceae could alternatively be placed in Liliales (as in DAHLGREN and RASMUSSEN 1983). Some families of Dioscoreales, especially Smilacaceae and Petermanniaceae, also show similarities to Asparagales, with which they could perhaps be treated with some justification.

A great problem is the circumscription of the families in Asparagales. While most taxonomists seem to be inclined to use a broad family circumscription, others restrict the circumscription and recognize many families. We have preferred not to give them a broader circumscription than is compatible with a probability that they are monophyletic. Exceptions are probably the Anthericaceae and perhaps the Alliaceae, which may turn out to be heterogeneous.

From an evolutionary point of view the Asparagales form a large fairly homogeneous complex of families which seem to have evolved parallel to the Liliales and Dioscoreales. Of primary importance for their recognition are the seed coat characters, especially the strongly deteriorated inner integument and, in most capsule-fruited taxa, *the phytomelan crust* of the outer epidermis. Phytomelan-encrusted seeds in the monocotyledons are probably restricted to Asparagales. Most berry-

**Table 1.** Comparison between the orders Asparagales and Liliales

|  | Asparagales | Liliales |
|---|---|---|
| Raphides | Common | Often lacking |
| Roots | Often fusiform, thickened | Rarely fusiform |
| Habit | Variable, rosette trees, shrubs, etc. rather common, many bulb plants. | No rosette trees, rarely shrubs, most often herbs with thick corms |
| Succulence | In several families | None |
| Inflorescence | If delimited from the vegetative region, terminal or axillary, often scapose | If delimited from the vegetative region, usually terminal, rarely scapose |
| Tepals | Generally not variegated or with drop-like colour pattern | Quite often variegated with drop-like colour pattern |
| Nectaries | Usually in the septa of the ovary | Mostly on the base of the tepals or filaments (except in the Ixioideae in Iridaceae) |
| Anthers | Basifixed or dorsifixed-hypopeltate | Basifixed, dorsifixed-epipeltate or tubularly arched over the filament tips ("pseudo-basifixed") |
| Anther dehiscence | Introrse (except in Ruscaceae) | Introrse, latrorse or often extrorse |
| Style (stylodia) | Usually simple style | Simple style or three stylodia |
| Ovules per locule | Numerous to two or one | Numerous |
| Fruit | Berries or mostly loculicidal capsules | Septicidal or loculicidal capsules (or separate follicles) |
| Unripe testa | Free from starch (except, for example, in some Asparagaceae and Convallariaceae) | Probably always with starch, disappearing at maturity |
| Testa | Never sarcotesta | Occasionally sarcotesta |
| Outer epidermis of testa | Obliterated in most baccate fruits; present and encrusted with phytomelan in capsular fruits | Always present and well developed, free from phytomelan |
| Tegminal part of seed coat | Usually completely collapsed to form a reddish brown or colourless membrane | The cellular structure usually retained |

fruited taxa (exceptions occur in at least *Dianella, Asparagus* and *Geitonoplesium*) lack the phytomelan crust. The seeds are also devoid of phytomelan in a few genera with capsules, such as all Doryanthaceae, Dasypogonaceae, Calectasiaceae and Cyanastraceae and a few species of Hyacinthaceae and Amaryllidaceae. In *Eriospermum* and *Blandfordia* the testa is hairy and in *Walleria* it has wart-like epidermal projections (bearing small hairs); phytomelan is lacking also here, but the taxonomic position of these genera is still uncertain.

Other characteristics mentioned in Table 1 are less consistent and can only be used to support the seed characters. For a few taxa there is great doubt as to their correct position, for superficial gross-morphological characters are opposed, for example, to embryological characters. Thus BJÖRNSTAD (1970) points out that *Disporum* and *Clintonia*, previously referred to Convallariaceae, show better agreement with Uvulariaceae, Colchicaceae and Liliaceae, and the genera are here transferred to Uvulariaceae of Liliales. As they have berries with phytomelanless seeds and are also neutral in most floral characters distinguishing Asparagales from Liliales, their previous position is understandable.

A similar case is Philesiaceae-Luzuriagaceae of Asparagales: the asparagalean affinity of the Luzuriagaceae being supported, for example, by their phytomelan-encrusted seeds. This family group is dubiously homogeneous; several members show affinity in various details with the Alstroemeriaceae, placed in Liliales with some hesitation. The most obvious similarity between the three families is perhaps provided by the inverted leaf blades, for which convergent evolution is a possible explanation, but in view of the similar geographical distribution of the families not indisputable. Philesiaceae (*Philesia, Lapageria*) show some further lilialean attributes, such as spotted tepal pattern and presence of perigonal nectaries.

CONOVER (1983) notes that the Philesiaceae (in our circumscription) show great similarity in venation and other leaf attributes with the Smilacaceae. Thus the Philesiaceae seem to form a link between the liliifloran orders Dioscoreales, Asparagales and Liliales, although this may well be due to derived states. In DAHLGREN and RASMUSSEN (1983) they are treated in Liliales near Alstroemeriaceae.

These cases provide insufficient reason, however, for lumping together indiscriminately the families here placed in Asparagales and Liliales, although if we were utilizing the category of suborder, these orders might appropriately be reduced to that rank and combined in a single order.

The specialized seed coat in the capsule-fruited Asparagales could, according to HUBER (personal communication), be explained by the theory that the common ancestors of the order, like many extant families, had baccate fruits and that in these the seed coat evolved to the point where it lost its protective ability. In a reversion to a second capsular stage the innovation of a phytomelan-encrusted outer epidermis has compensated for the degeneration of cell layers in the testa. This theory of HUBER's receives some support from the findings by BERG (1958), that the capsular fruit in *Trillium*, within the otherwise baccate-fruited Trilliaceae, most likely represents a secondarily capsular stage. Such a reversion represents a case parallel to that proposed for the Asparagales.

The following key has been prepared as a guide to the kinds of character that may be useful for distinguishing between families. It is meant to reflect the structure of the order rather than to be reliable key to identifying the families.

*Key to Families of Asparagales*

1. Ovary inferior or semi-inferior . . . . . . . . 2
1. Ovary superior  . . . . . . . . . . . . . . 8
2. Underground parts a (sometimes bulb-like) corm . . . . . . . . . . . 3
2. Underground part a rhizome or bulb . . . . . . . . . . . . . . . . . 5
3. Ovary semi-inferior  . . . . . . . . . . . 4
3. Ovary wholly inferior  . . . . . . **Ixioliriaceae**
4. Seeds with chalazosperm, lacking phytomelan, leaves pseudo-petiolate . . . . . . . . . . **Cyanastraceae**
4. Seeds lacking chalazosperm, with phytomelan, leaves not constricted into a pseudopetiole . **Tecophilaeaceae**
5. Inflorescence scapose, umbel-like, plants bulbous (seeds generally with phytomelan) . . . . . **Amaryllidaceae**
5. Inflorescence paniculate or racemose, plants rhizomatous . . . . . . . . . 6
6. Seeds lacking phytomelan (huge rosette plants with bright red bird-pollinated flowers) . . . **Doryanthaceae**
6. Seeds with phytomelan (plants variable, incl. huge rosette plants, but flowers then not bright red) . . . . . . . . . . . . 7
7. Massive, often fleshy rosette plants. . . . . . **Agavaceae** (subfam. **Agavoideae**)
7. Herbaceous non-fleshy plants of small or medium size (hairs often branched) . . . . . . . .**Hypoxidaceae**
8. Fruit a fleshy berry or drupe . . . . . . . . 9
8. Fruit hard or leathery, generally capsular, rarely a nutlet . . . . . . . .17
9. Placentation parietal; ovary unilocular . . . . . . . . . . . **Philesiaceae**
9. Placentation not parietal; ovary trilocular . . . . . . . . . . . . .10

10. Seeds with phytomelan . . . . . . . . . .11
10. Seeds without phytomelan . . . . . . . .13
11. Leaves small or rudimentary (not inverted), stems often green, sometimes cladodial . . . . **Asparagaceae**
11. Leaves large and well-developed; no phylloclades . . . . . . . . . .12
12. Leaf blades non-sheathing, usually with inverted blades . . . . **Luzuriagaceae**
12. Leaf blades sheathing, usually basally compressed . . . . . . . **Phormiaceae**
13. Branchlets developed into phylloclades, anthers extrorse . . . . . **Ruscaceae**
13. Branchlets not phyllocladial, anthers introrse . . . . . . . . . .14
14. Flowers small, unisexual, pistil ovate-globose with sessile stigma . . . . . . . . . . **Hanguanaceae**
14. Flowers variable in size, bisexual, pistil with distinct style . . . . . . .15
15. Leaves non-sheathing, blades usually inverted . . . . . . . **Luzuriagaceae**
15. Leaves sheathing, blades not inverted . . . . . . . . . . . . . . . .16
16. Plants with a thick aerial stem (with secondary growth), or short, basal, but leaves then very thick and fleshy . . . . . . **Dracaenaceae**
16. Plants lacking a thick aerial stem, leaves not particularly fleshy . . . . . . . . . . . **Convallariaceae**
17. Seeds without phytomelan . . . . . . .18
17. Seeds with phytomelan . . . . . . . . .23
18. Seeds invested with long hairs . . . . . .19
18. Seeds glabrous . . . . . . . . . . . . .20
19. Perianth syntepalous, campanulate, flowers yellow, orange and red; plant rhizomatous (Australia) . . . . . . . . . **Blandfordiaceae**
19. Perianth members free or almost free to the base, flowers more or less open; generally whitish; plants with a tuber (Southern Africa) . . . . . . . **Eriospermaceae**
20. Seeds with chalazosperm . . . . **Cyanastraceae**
20. Seeds without chalazosperm . . . . . . . .21
21. Dwarf shrubs, flowers with papery blue tepals, anthers poricidal . . . . . . . . . **Calectasiaceae**
21. Herbs, dwarf shrubs or large rosette plants, tepals not blue, anthers longitudinally dehiscent . . . . . . . . . . . . . . . .22
22. Guard cells rich in oil (New World) . . . . . . . . . . . . . **Nolinaceae**
22. Guard cells lacking oil (Australia) . . . . . . . . . . . **Dasypogonaceae**
23. Pollen grains spiraperturate (small blue-flowered herbs with reduced leaf blades) . . . . . **Aphyllanthaceae**
23. Pollen grains otherwise . . . . . . . . . .24
24. Pollen grains trichotomosulcate (medium-sized to large herbs with distichous, basally compressed leaves) . . . . . . . . **Phormiaceae**

24. Pollen grains not trichotomosulcate . . . . . . . . . . . . . . . . . . . . .25
25. Twiner, leaves in clusters, sometimes opposite (seeds winged) . . . . . . . . . . **Herreriaceae**
25. Non-twiners, or if twiners leaves not in clusters . . . . . . . . . . .26
26. Plants bulbous (inflorescence scapose) . . . . . . . . . . . . . . . .27
26. Plants non-bulbous, generally rhizomatous (inflorescence generally not scapose, rarely spicate) . . . . . . . . . . . . . . . . . .28
27. Inflorescence generally umbel-like . . . . . . . . . . . . . . . **Alliaceae**
27. Inflorescence racemose . . . . . **Hyacinthaceae**
28. Chromosome complement strongly dimorphic ($x = 30$; 5 large and 25 small) . . . . . . . . .29
28. Chromosomes more uniform in size and not as above . . . . . . . . .30
29. Scapose herbs; leaves strongly veined, pseudo-petiolate; flowers lilac to white, in a one-sided raceme . . . . . . . . . . **Funkiaceae**
29. Generally large, with short or tall woody trunk; leaves linear, flowers white to yellowish, in panicles or racemes **Agavaceae** (subfam. **Yuccoideae**)
30. Large rosette plants, often with a tall woody trunk, stomata paracytic, (flowers small, numerous, white, densely crowded in a spike; aril lacking) . . . . **Xanthorrhoeaceae**
30. Habit different, if woody generally rosette plants, the flowers then not white; stomata generally anomocytic, sometimes tetracytic . . . . . . . . . . . . . . . .31
31. Plants generally producing anthraquinones instead of steroidal saponins; ovules generally hemianatropous or orthotropous; seeds generally arillate (microsporogenesis simultaneous) . . . . . . . . . . . **Asphodelaceae**
31. Plants producing steroidal saponins; ovules anatropous or (especially in Anthericaceae) campylotropous; seeds not arillate but sometimes with a caruncle (microsporogenesis generally successive, simultaneous in Hemerocallidaceae) . . . . . . . . . .32
32. Inflorescence a (scorpioid) cyme; perigone yellow to ferruginous, campanulate-infundibular, similar to that in *Lilium*; microsporogenesis simultaneous . . . . . . . . **Hemerocallidaceae**
32. Inflorescence otherwise; flowers generally white, blue or violet; microsporogenesis successive . . . . . . . . . . . . . . . . . . . .33
33. Pollen grains spinulose . . . . . . . . . **Asteliaceae**
33. Pollen grains not spinulose . . . . **Anthericaceae**

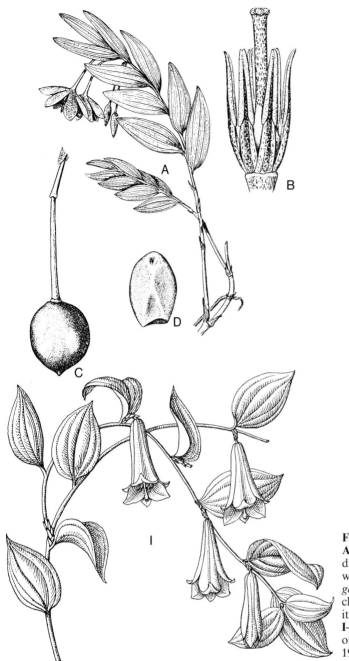

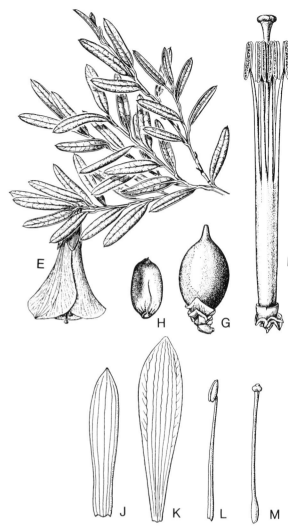

**Fig. 53.** Luzuriagaceae (**A–D**) and Philesiaceae (**E–M**). **A–D** *Luzuriaga radicans*. **A** Flowering branch. **B** Androecium and gynoecium. **C** Fruit (pendulous). **D** Seed without phytomelan. (CORREA 1969) **E–H** *Philesia magellanica*. **E** Flowering branch. **F** Staminal tube and (enclosed) pistil. **G** Berry (note the several bracteoles at its base). **H** Seed (without phytomelan). (CORREA 1969) **I–M** *Lapageria rosea*. **I** Flowering branch. **J–K** Tepals of outer and inner whorl. **L** Stamen. **M** Ovary. (KRAUSE 1930)

**Philesiaceae** Dumortier (1829)   2:2   (Fig. 53)

Erect undershrubs (*Philesia*) or vines (*Lapageria*) with woody branches, sometimes ascending from a branched rhizome. Spines are lacking, but crystals may be secreted on the stem surface. The leaves are alternate, sessile or petiolate, sometimes sheathing at the base, entire, ovate to lanceolate or broadly linear and twisted at the base, but the blade is not inverted. They have a few prominent parallel main veins, and between them a distinctly net-like or transverse venation. Vessels and vessel-tracheids occur in roots and stem, but not in the leaves. The stomata are anomocytic, and in *Lapageria* transversely oriented in relation to the leaf axis (CONOVER, 1983).

The flowers are solitary, situated in leaf axils or on the tips of short branchlets. Their pedicels characteristically bear a number of medial, dorsal and ventral scales (in a way superficially resembling that of some Epacridaceae and Ericaceae). The flowers are bisexual, actinomorphic and hypogynous. Their colour is variable, being at least partly rose or purple and more or less spotted on the inside. The two tepal whorls are very different in shape and size in *Philesia,* the outer short and nearly sepaloid. The tepals are free but form a funnel-shaped or campanulate structure. Nectary pouches occur on the base of the tepals.

The stamens are 3 + 3. Their filaments are free or fused half-way up. The anthers are subbasifixed, extrorse or introrse and dehisce longitudinally. Microsporogenesis is successive (CAVE 1966). The pollen grains are inaperturate and spinulose.

The style is erect and linear and the stigma is capitate or trilobate, and at least in *Lapageria* of the Wet Type (HESLOP-HARRISON and SHIVANNA 1977). The ovary is unilocular with intrusive parietal placentae. The ovules are anatropous and crassinucellate; a parietal cell is cut off before meiosis sets in; the embryo sac is formed according to the *Polygonum* Type; during meiosis the nucellar epidermis divides anti- and periclinally (CAVE 1966). The type of endosperm formation is unknown.

The fruit is a berry with few to numerous seeds, which are globose to ovoid. They sometimes shed the outer epidermis of the testa, which lacks phytomelan and collapses during its development (as with *Smilax*). The endosperm is massive and consists of cells with rather thin non-pitted walls. It contains aleurone and fatty oils. $n = 15, 19$.

*Chemistry.* Steroidal saponins (diosgenin) occur in *Philesia.*

*Distribution.* The family consists of the genera *Philesia* (1) and *Lapageria* (1), both in South America (Chile).

*Philesia magellanica* is a sclerophyllous, pink-flowered shrub, with linear leaves. It grows up to one metre in *Nothofagus* forests. *Lapageria rosea* is a vine with elliptic leaves and large campanulate flowers having a drop-like pattern within. The two genera hybridize in cultivation ( × *Philageria*).

**Luzuriagaceae** J. Dostal (1857)   4–5:6–8
(Figs. 53–54)

Slender shrublets or vines with thin, generally branched, woody aerial shoots. The roots in *Geitonoplesium* are fusiform and swollen. The leaves are alternate and distichously inserted, sessile or slightly petiolate, lanceolate to ovate or sometimes linear and nearly grass-like (the habit then approaching that of small bamboos), with the lamina inverted. Their parallel veins are closer and more numerous than in Philesiaceae and in contrast to this family they have only very slight reticulate or transverse venation between the parallel veins or none. Stipules and similar structures are lacking. Vessels are present in the roots and stems and also sometimes in the leaves; in *Geitonoplesium* the leaves may have vessel tracheids (WAGNER 1977). Oxalate raphides occur in most genera.

The flowers, (solitary or) few to several in axillary or terminal cymose inflorescences, are small and hypogynous. The tepals are white to pale violet, free almost to the base or, in *Behnia,* ± fused and campanulate. In *Eustrephus* the inner tepals are fimbriate. As in Philesiaceae the pedicel may bear a number of prophylls. Perigonal nectaries are present at least in *Luzuriaga.*

The stamens are 3 + 3 in number; their filaments are usually free, and the basifixed or dorsifixed, introrse or extrorse anthers are sometimes poricidal. The pollen grains are sulcate or trichotomosulcate (*Geitonoplesium*).

The style is erect, linear and sometimes short and the stigma is capitate to trilobate and, at least in *Geitonoplesium,* of the Dry Type. The ovary is trilocular with few anatropous or (in *Luzuriaga*) campylotropous, crassinucellate ovules. The fruit is a berry or, in *Eustrephus,* rather a fleshy capsule, as in that genus it opens and exposes the seeds. The seeds in *Geitonoplesium, Eustrephus* and, perhaps, *Behnia* have a testa encrusted with phytomelan, and otherwise agree in having a straight embryo and a massive endosperm, storing fat and aleurone. The seeds are arillate in *Eustrephus.* $x = 10$ in *Luzuriaga* (CAVE 1966).

*Chemistry.* Saponins are lacking in *Luzuriaga,* at least.

*Distribution.* The family occurs in southern South America, South Africa, Australia and adjacent regions as far as Java, New Guinea, New Zealand and New Caledonia. *Luzuriaga* occurs in South America, New Zealand and the Falkland Islands, and *Behnia* in Southern Africa. *Eustrephus* and *Geitonoplesium* occur in Australasia (and *Drymophila,* which is a dubious member of the family, is Australian).

*Luzuriaga* (3) (Fig. 53 A–D) consists of shrublets, on which the leaves are horizontally directed. The flowers are rather small, with free white, entire tepals. *Behnia* (1) (Fig. 54) is characterized by having a campanulate syntepalous perianth, to which the filaments are attached. The berries are white

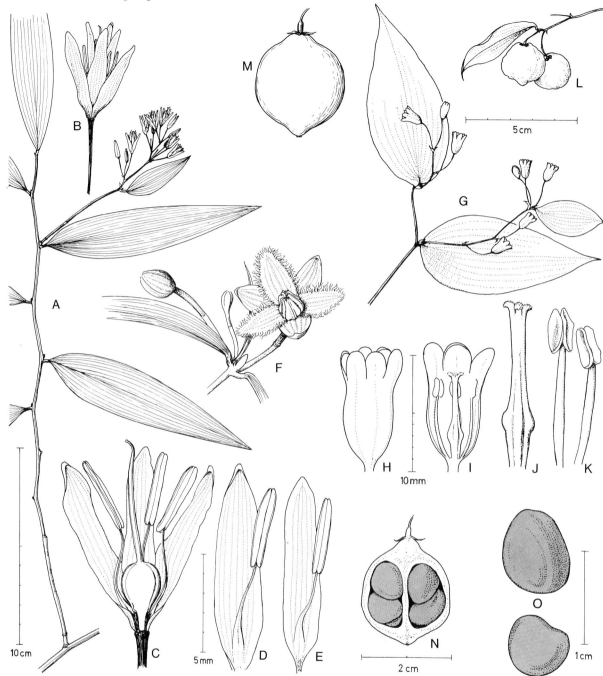

**Fig. 54.** Luzuriagaceae. **A–E** *Geitonoplesium cymosum.* **A** Flowering branch. **B** Flower. **C** Flower, late stage, with three tepals and three stamens removed. **D** A tepal and a stamen, each of the outer whorl. **E** A tepal and a stamen, each of the inner whorl. **F** *Eustrephus latifolus.* Inflorescence. **G–O** *Behnia reticulata.* **G** Flowering branchlet. **H** Flower. **I** Flower, half of perianth and stamens removed. **J** Pistil. **K** Stamens. **L** Fruiting branchlet. **M–N** Fruit, N in longitudinal section. **O** Seeds. The seeds of all these genera finally become dark, even where the fruit is a berry. (All orig. B. JOHNSEN)

and the large seeds remain almost white until the fruit is quite ripe, when they become dark.

*Eustrephus* (1) *latifolius* (Fig. 54), with fimbriate inner tepals and flowers basally prolonged into a "pericladium" and yellow fleshy capsules, occurs in Eastern Australia, Java, New Guinea and New Caledonia. *Geitonoplesium* (2; Fig. 54), with black berries, has a similar distribution and habit but extends to Fiji. These two genera have poricidal anthers and black, phytomelaniferous seeds.

*Drymophila* (2), in eastern Australia, with inverted leaf blades, is usually referred to Convallariaceae, but may be best placed in Luzuriagaceae or possibly in Uvulariaceae. This genus is at present being closely analyzed (CONRAN, personal communication). The genus *Elachanthera,* sometimes referred to this group of genera, is a synonym of *Myrsiphyllum asparagoides,* as shown by BURBIDGE (1963).

## Convallariaceae P. Horaninow (1834)   19:110
(Figs. 55–56)

Perennial herbs with sympodially or monopodially branching rhizomes, which are often thick and nutrient-storing. A multilayered velamen occurs in the roots of several genera, including *Aspidistra, Tupistra, Ophiopogon* and *Polygonatum.*

The vegetative leaves may be restricted to the base of the aerial stem (as in *Convallaria*) or may be spirally distributed (distichous or non-distichous), opposite or verticillate on an aerial stem. They are sessile or sometimes (as in *Aspidistra* or *Maianthemum*) petiolate and are linear, lanceolate, ovate or cordate in shape. The venation is parallel; stipules and similar structures are lacking. The stomata are anomocytic. Crystal raphides are present and contained in cells or sacs filled with mucilage. Vessels with scalariform or simple perforation plates occur in the roots; very rarely vessels with scalariform perforations occur also in the stem.

The flowers are distributed in various kinds of inflorescences. They are bisexual, actinomorphic, generally hypogynous (hypogynous to epigynous in *Peliosanthes* and *Ophiopogon*), trimerous, except in *Maianthemum* and *Aspidistra,* and with the tepals of the two whorls more or less similar. The tepals are generally connate to form an urceolate or campanulate perigone and are only rarely almost free from each other. They are usually white or greenish, only rarely purplish or of other colour.

The stamens are usually $3+3$ (rarely $2+2$ or $4+4$) in number and sometimes inserted high up in the perianth tube. In *Peliosanthes* the filaments are connate basally. The stamens are basifixed or rarely peltate (*Reineckea*). Anther dehiscence is longitudinal and introrse, and the connective is not protracted beyond the microsporangia, which are not confluent. Microsporogenesis is successive. The pollen grains are sulcate or inaperturate (at least in *Aspidistra*). They are dispersed in the two-celled or (at least in *Polygonatum multiflorum*) in the three-celled stage.

The pistil generally has an erect, simple style and a capitate or lobate stigma which generally or perhaps constantly has a Dry, papillate surface. The locules are separate and contain two or a few ovules each. Septal nectaries are present in most genera. The ovules are anatropous, campylotropous or nearly orthotropous and generally crassinucellate. A primary parietal cell is generally cut off from the archesporial cell. Embryo sac formation is unusually variable in Convallariaceae; generally it conforms to the *Allium* Type, but in certain genera to the *Scilla* (*Smilacina*) or *Drusa* Types (*Maianthemum, Smilacina*). Endosperm formation is of the Nuclear Type.

The fruits in Convallariaceae are normally berries, although those in *Liriope* and *Ophiopogon* are capsules and may rupture to expose the fleshy seeds. The berries vary from red to black or blue or may even be spotted (*Maianthemum*). Where the seeds are fleshy the testa is often bright blue, mimicking a fruit (*Ophiopogon*). In the seeds, the outer integument, which is several-layered, is almost completely obliterated during maturation, and a phytomelan crust is never formed. The inner integument, at least in the European genera, collapses during development (HUBER 1969).

The endosperm cells are often thick-walled (with pitted walls) and store aleurone and lipids, but not starch. The embryo is usually more than half as long as the endosperm.

*Chemistry.* Like the related families, Convallariaceae is characterized by rich contents of steroidal saponins and sapogenins, such as diosgenin and gentrogenin. Poisonous cardenolide glucosides (used as heart poisons and medicines) occur in the berries of *Polygonatum* and *Convallaria,* being especially rich in the latter genus. Chelidonic acid is also found in the family, e.g. in *Convallaria.*

*Distribution.* Convallariaceae are found chiefly in the Northern Hemisphere. Many taxa occur in the Himalayas and eastern Asia, but also some in North America and Europe. All species are terrestrial, and many grow in montane forests.

The family is divisible into the following tribes:

### Tribus Polygonateae

The rhizomes in this tribe are sympodial, the branch generations each ending in an erect stem with a variable number of cauline leaves. The inflorescences are terminal panicles or racemes; less commonly the flowers are solitary or few in the leaf axils. The basic chromosome number is $x = 23$.

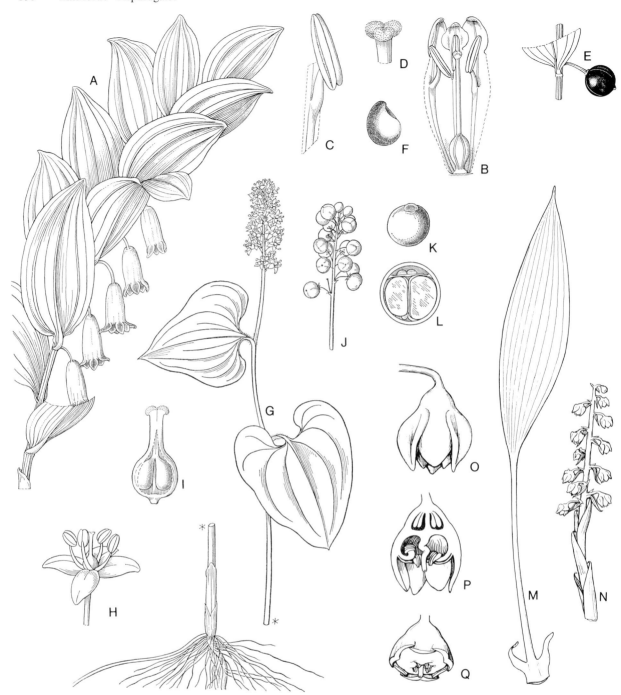

**Fig. 55.** Convallariaceae. **A–F** *Polygonatum odoratum*. **A** Flowering, aerial shoot. **B** Flower, half of perianth removed to show pistil. **C** Stamen, lateral view. **D** Style apex and stigma. **E** Berry. **F** Seed. **G–L** *Maianthemum bifolium*. **G** Plant. **H** Flower (note the dimerous condition). **I** Pistil, the wall opened to show placentation. **J** Infructescence. **K** Seed. **L** Berry, transverse section. (From Ross-Craig 1972) **M–Q** *Neolourya thailandica*. **M** Leaf. **O** Flower. **P** Same, longitudinal section. **Q** Fruit. (**M–Q** from Larsen 1966)

*Smilacina* (20) in Asia, particularly the Himalayas, and North and Central America, has terminal, often multiflorous paniculate inflorescences of flowers with almost free, spreading, white tepals. Some species are grown as ornamentals, e.g. *S. racemosa* with compound "racemes" and edible berries. *Polygonatum* (30) has erect leafy stems with solitary or few, axillary, pendulous, subcylindric flowers. The genus is subcosmopolitan, but chiefly developed in the Northern Hemisphere. The thick,

**Fig. 56.** Convallariaceae. **A–C** *Smilacina stellata*. **A** Aerial shoot with inflorescence. **B** Flower. **C** Berries. (HITCHCOCK et al. 1969) **D–G** *Peliosanthes cumberlegii*. **D** Leaf rosette. **E** Aerial shoot with inflorescence. **F** Flower. **G** Same, longitudinal section. (LARSEN 1966) **H–K** *Streptopus streptopoides*. **H** Rhizome and fruiting aerial shoot. **I** Part of flowering aerial shoot. **J** Berry. **K** Flower. (HITCHCOCK et al. 1969). *Streptopus* is a dubious member of Convallariaceae

starch-rich, sympodial rhizomes of some species, mainly in Japan, are eaten as vegetables or preserved in syrup. *Maianthemum* (3) consists of small forest herbs with dimerous flowers.

### Tribus Convallarieae

Tribus Convallarieae consists of rhizomatous geophytes. In this feature it agrees with Polygonateae, but while that tribe has a sympodial rhizome, the taxa of Convallarieae have monopodial, continuously growing rhizomes; the aerial stems are thus lateral. Generally there are only a few green assimilating leaves which are inserted at the base. However, in *Reineckea* the stem bears numerous leaves. The inflorescences in this tribe consist of axillary, scapose racemes or spikes with campanulate flowers. *Convallaria* (1) *majalis,* "Lily of the Valley", is a common Eurasian herb. It has usually two cauline leaves and white, campanulate, strongly scented flowers followed by red poisonous berries despite which it is often grown as an ornamental. *Reineckea* (1) *carnea,* in China and Japan, has pinkish, reflexed tepal lobes and large, one-seeded berries.

### Tribus Aspidistreae

Tribus Aspidistreae agrees with the Convallarieae in having monopodial rhizomes and lateral aerial stems.

*Aspidistra* (3) and *Tupistra* (3) are centred in eastern Asia. The former has solitary, rather large, dingily coloured, fleshy flowers, which usually have eight perigone lobes and a large stylar head. It is often grown as an indoor ornamental, but the flowers are seldom observed as they are produced at soil level.

### Tribus Ophiopogoneae

Tribus Ophiopogoneae is characterized by a short, thick rhizome, on which the fibrous roots are borne. These are sometimes developed as runners, sometimes thickened into storage organs. The leaves are narrow, lanceolate or linear, sometimes grass-like; the plants may be conspicuously tufted. The inflorescences are panicles, spikes or racemes with bracteate, often small flowers with white or violet tepals. *Liriope* has hypogynous and *Ophiopogon* and *Peliosanthes* hemi-epigynous flowers. In *Liriope* and *Ophiopogon* the tepals are free, in *Peliosanthes* fused into a campanulate perigone; in the last genus the filaments are also united into a ring. The fruit is a berry (*Peliosanthes*) or (the two other genera) a leathery irregularly rupturing capsule, and in these the seeds have a juicy, often bright blue sarcotesta.

This tribe consists of three genera, all Asiatic. The distribution ranges from the Himalayas to Japan and Malesia.

*Liriope* (5) occurs in China, Japan, and the Philippines. The roots of *L. spicata* are aromatic and have medical use. *L. platyphylla,* with pale violet pearl-like flowers, is grown as an ornamental. *Ophiopogon* (= *Mondo,* 3), from the Himalayas, eastwards, is also often grown as ornamentals ("Black Dragon"), forming tufts.

*Peliosanthes* (incl. *Lourya;* 10) is also distributed from the Himalayas to South-Eastern Asia.

There are strong indications that some genera normally placed in Convallariaceae, including at least the three southern Asiatic-North American genera *Disporum, Streptopus* and *Clintonia,* are misplaced in this family. As the bacciferous Asparagales generally lack phytomelan pigment in the epidermis of the seed coat, this criterion cannot be used to assign them definitely to the order. A secondarily bacciferous member of the Liliales, of the colchicaceous or uvulariaceous stock, might simulate the convallariaceous phenotype. The combination of attributes in *Disporum* and the other genera mentioned, namely absence of oxalate raphides, presence of perigonal rather than septal nectaries, lack of a parietal cell, and other embryological attributes (BJÖRNSTAD 1970), suggests that they should be transferred to the Liliales (Uvulariaceae). Also the Australian genus *Drymophila* with coriaceous, inverted leaves may be better placed in Uvulariaceae or in Luzuriagaceae (Asparagales).

### Asparagaceae A.L. Jussieu (1789)
3 : 305   (Fig. 57)

Shrubs, subshrubs or vines with woody or often partly herbaceous, persistent (evergreen) or annually withering branches growing from a short, sympodial rhizome. The roots are often swollen and fusiform, and sometimes provided with multiple velamen. Many species have green assimilating branchlets and a few have branchlets transformed into flat, leaf-like cladodes (phylloclades), as in *Myrsiphyllum* (sometimes included in *Asparagus*). The leaves on the long-shoots are normally reduced and more or less scale-like, as also are those of the short-shoots, if present. The fascicled, green, slender and needle-like to filiform assimilatory structures (as in *Asparagus officinalis*) were shown by ARBER (1925) to be of axial (stem) nature. In some species, e.g. *Asparagus densiflorus,* they may bear minute, reduced scale leaves. The vascular tissue of the roots contains vessels with simple or scalariform perforations, and that of the stems usually has vessels with scalariform perforations. Raphide cells are widely distributed in the family. Hairs are mostly lacking.

The flowers are small and solitary or assembled in umbel-like or raceme-like inflorescences which are probably mainly determinate in nature. The flowers are inconspicuous, hypogynous, and either bisexual or unisexual. The 3 + 3 tepals are all similar in shape; they are free and spreading or fused basally to form a campanulate perigone. Their colour is usually white, yellow or green. In the male or bisexual flowers the stamens are 3 + 3 in number

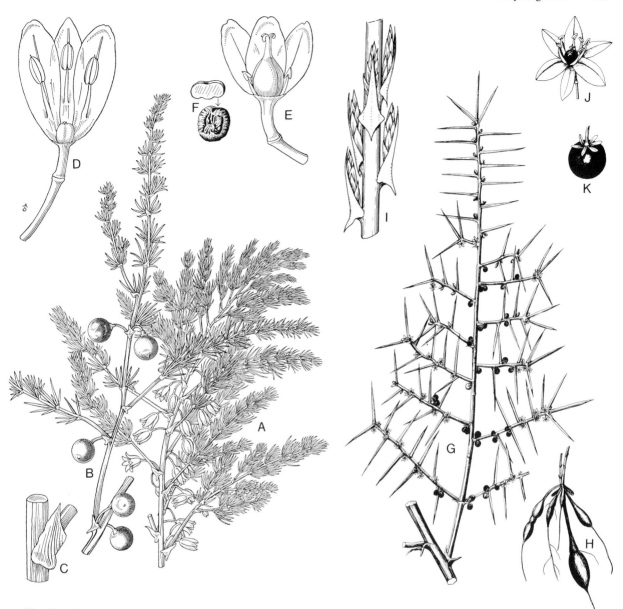

**Fig. 57.** Asparagaceae. **A–F** *Asparagus officinalis.* **A** Flowering branch. **B** Fruiting branch. **C** Scale leaf subtending lateral branch. **D** Male flower, longitudinal section. **E** Female flower, longitudinal section. **F** Seed. (Ross-Craig 1972) **G–K** *Asparagus stipularis.* **G** Flowering and fruiting branch. **H** Roots. **I** Scale leaves and axillary branchlets. **J** Flower. **K** Berry. (Jafri and El-Gadi 1978)

with the filaments free from each other, and their anthers are introrse-dorsifixed. Microsporogenesis is successive. The pollen grains are sulcate and dispersed in the two-cellular stage.

Where the flowers are unisexual stamens are always present but in the female flowers are non-functional (i.e. they do not produce pollen). The ovary is tricarpellary and trilocular and has a rather short style. The stigma is capitate or lobate and is either of the Dry or Wet Type. The placentation is axile and the ovules 2–12 per locule. The ovules are hemianatropous (or anatropous or almost orthotropous) and crassinucellate; a parietal cell is formed. Embryo sac formation is of the *Polygonum* Type and endosperm formation of the Nuclear Type.

The fruit is a globose, red, blue or otherwise coloured berry, the seeds of which have a deep black outer epidermis. The endosperm is starch-free and contains aleurone and lipids. The embryo in *Asparagus* may be slightly curved; it is from two thirds to almost as long as the endosperm.

*Chemistry.* Oxalate raphides are widespread in the family and chelidonic acid, acetidine carbonic acid and steroidal saponins are recorded in various species. Especially the rhizomes and swollen root tubers of *Asparagus* have been found to be rich in saponins, which can be obtained from, for example, rhizomes in *A. officinalis* and *A. acutifolius* or from roots of *A. thunbergianus*.

*Distribution.* Asparagaceae are widely distributed in the Old World; most species are found in regions with arid to Mediterranean climates, and extreme xeromorphic adaptations are common. Many species are practically leafless shrubs with thick underground stems or roots, which store water and nutrients. Assimilation is carried out by the green branch ends and branchlets.

Three genera may be distinguished: *Asparagus, Protasparagus* and *Myrsiphyllum,* the last two of which are perhaps best treated as subgenera of the first.

*Asparagus* and *Protasparagus* together comprise almost 300 species. *Asparagus* is widely distributed in the Old World, with many species in the Mediterranean Region and in Africa and part of Asia. It consists mainly of xeromorphic shrubs or shrublets but also of herbs. Common xerophytic species in the Mediterranean area are, for example, *A. stipularis, A. aphyllus* and *A. thunbergianus.* The young shoots of some species are used as vegetables, in particular those of *A. officinalis,* "Garden Asparagus", growing wild in temperate Eurasia. This species is widely cultivated and of great economic importance. Its seeds can be used as coffee substitute. Roots or tubers of some other species are edible, e.g. of *A. abyssinicus* in Northern Africa, of *A. lucidus* in China and Japan, and of *A. densiflorus* in Southern Africa. The more primitive genus, *Protasparagus,* with at least nearly 70 species, differs from *Asparagus* in several features. In *Protasparagus* the flowers are bisexual (whereas in *Asparagus* the flowers are unisexual and the plants dioecious), there are 4–12 ovules per locule (in *Asparagus* only 2), and the seeds are globose (in *Asparagus* they are dorsally convex and ventrally angular) (OBERMEYER 1983). – Flat cladodes occur in the genus *Myrsiphyllum* (12) in Southern Africa.

Asparagaceae, when given family status, is often circumscribed so as to include also the Ruscaceae, but these two groups are widely different, as will be shown below, and we consider them distinct here. Both of these are probably most closely related to the Convallariaceae, Herreriaceae and Dracaenaceae.

## Ruscaceae Hutchinson (1934)   3:8   (Fig. 58)

Subshrubs, often forming low thickets, or vines. The basal parts are woody and rather slender, and the branchlets are developed as flat assimilatory phylloclades which are ovate to lanceolate and pointed, (they are highly reminiscent of the leathery leaves in Luzuriagaceae). The phylloclades are alternate, opposite or even verticillate. The roots may have multiple velamen (known in *Semele*). The leaves are reduced and scale-like. The vascular tissues of the roots have vessels with scalariform and simple perforation plates, and vessels are either present or lacking in the stems, if present having scalariform perforation plates. Raphides occur throughout the plant. The stomata are anomocytic. Hairs are lacking.

The flowers are small and inconspicuous and occur in raceme-like clusters on the upper or the lower surface of the phylloclades (*Ruscus*), in umbel-like clusters on their margins (*Semele*), or in terminal racemes quite separately from the phylloclades (*Danaë*). In *Ruscus* and *Semele* the inflorescence is subtended by a single, scale-like or herbaceous leaf. The flowers are articulated with their pedicels. They are actinomorphic, hypogynous and bisexual or, in *Ruscus,* unisexual (bisexual and male occasionally in one species). The tepals are inconspicuous, greenish to pale yellowish-white, all more or less similar in size and shape; in *Danaë* they form an urceolate perigone. The stamens are fused by their filaments into a column and have extrorse, tetrasporangiate anthers. Microsporogenesis is successive and the tapetum glandular. The pollen grains are dissimilar to those in Asparagaceae in being inaperturate, at least in some taxa. They are dispersed in the two-celled stage.

The pistil is tricarpellary, and lacks or has only a very short style. The sessile or subsessile stigma has a Wet surface (at least in *Ruscus*). The ovary has one or three locules, each with (one or) two hemianatropous or orthotropous ovules. The embryo sac formation is reported to be of the *Allium* Type. The fruit is a red berry with one to four seeds. These are pale (not black as in Asparagaceae) and have a collapsed testa and a thick tegmen (unusual in Asparagales). The embryo is less than half as long as the endosperm. When germinated the seedling, except in some species of *Ruscus,* produces petiolate assimilating leaves with a flat lamina.

*Chemistry.* The chemistry of Ruscaceae is obviously similar to that in Asparagaceae, containing raphides, steroidal saponins, and chelidonic acid.

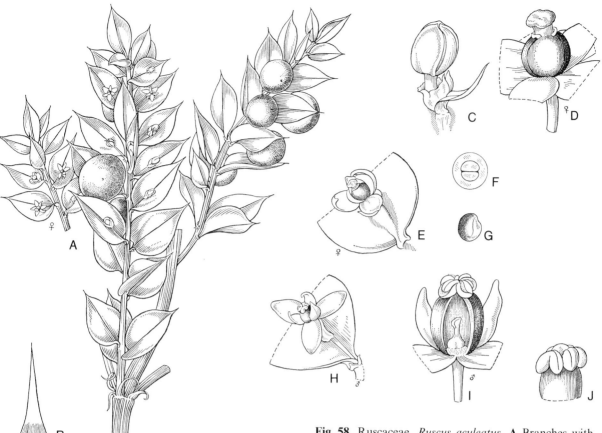

**Fig. 58.** Ruscaceae. *Ruscus aculeatus.* **A** Branches with male and female flowers and berries. **B** Apex of phylloclade. **C** Bract, bracteole and bud of female flower. **D** Pistil of female flower protruding from the staminal tube which has a rim formed by vestigial anthers. **E** Female flower in position on phylloclade. **F** Ovary, transverse section. **G** Seed. **H** Male flower in position on phylloclade. **I** Male flower, part of perianth cut away. **J** Staminal tube and anthers before dehiscence. (Ross-Craig 1972)

Rutin is found in the phylloclades of *Ruscus* and *Danaë.*

*Distribution.* Ruscaceae have a pronounced Mediterranean-Macaronesian distribution. The family consists of the following three genera.

*Danaë* (1), in forests from Syria to Iran, is a subshrub with terminal, raceme-like inflorescences with bisexual flowers free from the phylloclades, which are rather soft and mesomorphic. The flowers have six anthers. – *Semele* (1), a tall climber, in Macaronesia, has its inflorescences marginally on the phylloclades. The flowers, as in *Danaë,* are bisexual, and the anthers six in number.
*Ruscus* (6) (partly revised by YEO 1968), which is mainly Mediterranean in distribution, has stems that are either unbranched or branched once (disregarding the cladodes). The inflorescences are generally situated centrally on the adaxial or abaxial side of the cladodes. The flowers are generally unisexual, their green tepals often minutely dotted with violet. The staminal tube is fleshy, papillate and violet. It is equally well developed in female and male flowers. There are three anthers, the 12 microsporangia of which are fused to the rim of

the staminal tube. *R. aculeatus* is the commonest and most widely distributed species. It has stiff, ovate, pointed phylloclades. The young shoots, as in some *Asparagus* species, can be used as a vegetable.

## Herreriaceae S.L. Endlicher (1836)  2:9
(Fig. 59)

Climbing or twining subshrubs with subterranean tuberous rhizomes and with stems sometimes armed with prickles (as in *Smilax*). Vessels with scalariform perforation plates are recorded in the stem of *Herreria.* The leaves are concentrated in lateral clusters. They are linear or linear-lanceolate, sessile and somewhat coriaceous (cladode-

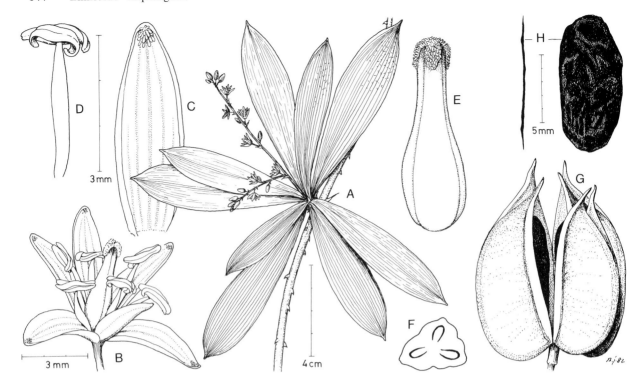

**Fig. 59.** Herreriaceae. *Herreria sarsaparilla.* **A** Part of branch with short shoot and inflorescence. **B** Flower. **C** Tepal. **D** Stamen. **E** Pistil. **F** Ovary, transverse section. **G** Capsule. **H** Seed in different views. (Orig. B. JOHNSEN)

like) with numerous parallel veins. Paniculate inflorescences, which are basally provided with several small, scale-like leaves, are situated in the leaf axils. The flowers have an articulate pedicel and are actinomorphic, hypogynous and bisexual. They have 3 + 3, spreading, free tepals, all of the same size and shape, 3 + 3 free stamens with basifixed, longitudinally dehiscent anthers, and a trilocular pistil with erect style and a small capitate stigma. The pollen grains are sulcate (KUPRIANOVA 1948). Septal nectaries are present in the ovary. The ovules, one to many in each locule, develop into flattened, helically winged seeds, encrusted with phytomelan. Endosperm is copious and the embryo is small. The fruit is a trilobate, septicidal capsule.

*Herreria* (7) occurs in temperate to subtropical South America and the other genus, *Herreriopsis* (2), occurs on Madagascar. *Herreria sarsaparilla* is common in moist forests in eastern Brazil.

The family, though having capsules, seems to approach the previous ones, but the exact affinities are not clear.

**Dracaenaceae** R.A. Salisbury (1866)   1–2:130
(Fig. 60)

Plants with a more or less woody trunk (*Dracaena*) or mostly without trunk (*Sansevieria*), the stem in the latter case short and partly subterranean. Trunk when present very variable in height and thickness, sometimes scandent, rarely considerable as with the trees of *Dracaena draco*, which may become many metres high and more than 2 m thick. In *Dracaena*, as well as *Sansevieria*, as in several genera of the Nolinaceae, there is extensive secondary thickening growth in the stem, caused by a meristematic zone continuously forming additional tissue in which new bundles are differentiated. The result is a great number of scattered vascular strands in which there are no real vessels, only tracheids. However, vessels are found in the roots and leaves, those in the roots mainly with simple and those in the leaves with scalariform perforation plates.

The leaves are narrowly linear to ovate and sessile, and have parallel venation. They may be of considerable size and in *Sansevieria* they are conspicuously succulent and sometimes terete or tubiform. They are frequently concentrated in rosettes either on the ends of aerial branches or (*Sansevieria*) from the apex of a mostly subterranean rhizome. The stomata are anomocytic; their guard

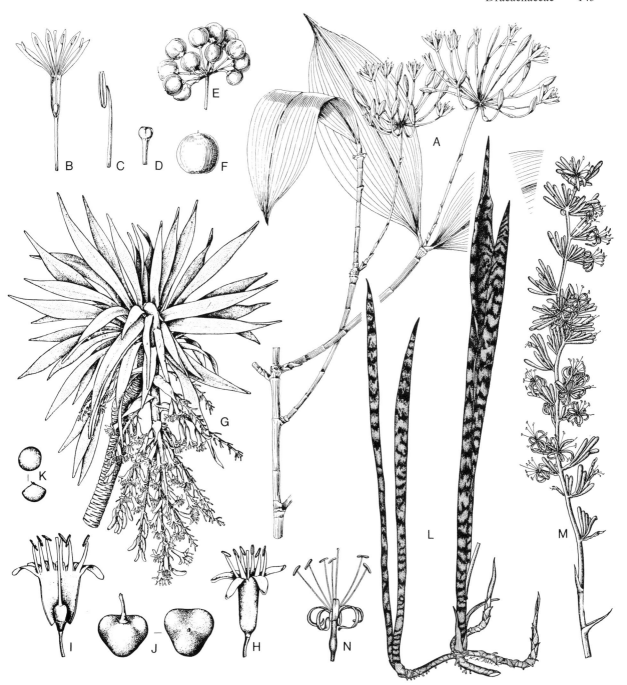

cells, unlike those in Nolinaceae, do not contain oil. Non-suberized, mucilage-filled cells with crystal raphides are present in the vegetative parts, whereas in Nolinaceae there are suberized cells with solitary, needle-like pseudo-raphides.

The inflorescences are axillary and pedunculate, emerging either from the leaf rosettes, i.e. from near the ground, in *Sansevieria,* or on the ends of the branches, in *Dracaena.* They consist of racemes or panicles, which may be elongate or may form pseudo-umbels. The flowers are articulated

**Fig. 60.** Dracaenaceae. **A–F** *Dracaena surculosa.* **A** Branches with inflorescences. **B** Flower, longitudinal section. **C** Stamen. **D** Style apex with stigma. **E** Infructescence. **F** Berry. (HEPPER 1968) **G–K** *Dracaena fernaldii.* **G** Branch with inflorescence. **H** Flower. **I** Same opened to show pistil. **J** Berry. **K** Seeds. **L–N** *Sansevieria guineensis.* **L** Rhizome with leaves. **M** Inflorescence. **N** Flower. (**G–N** from DEGENER and GREENWELL 1956)

on the pedicel, and are bisexual and hypogynous, with the elongate perianth members all of the same shape and size and connate at the base. The 3 + 3 stamens are inserted at the base of the lobes, and have dorsifixed, epipeltate, introrse anthers which dehisce longitudinally. Microsporogenesis is successive. The pollen grains are sulcate and two-celled.

The pistil is trilocular with an erect, simple, rather slender style and a trilobate or capitate stigma, which has been reported to be of the Wet Type in *Dracaena*. The ovary has septal nectaries. There is only one anatropous ovule in each locule (two in the Nolinaceae). No parietal cell is cut off from the archesporial cell but periclinal divisions occur in the nucellar epidermis, giving rise to a nucellar cap over the embryo sac (STENAR 1942). Embryo sac formation is of the *Polygonum* Type and endosperm formation is probably nuclear.

The fruit is generally a red or orange-coloured berry with up to three globose or elongate seeds, but sometimes it is hard and woody. The seed coat has a three to four-layered outer integument, the outer epidermis of which is thick-walled and lacks phytomelan, while the cells of the inner integument have collapsed and form a reddish brown membrane. The endosperm consists of cells with pitted walls and it stores aleurone and lipids but no starch.

*Chemistry.* Steroidal saponins occur in both genera; chelidonic acid is known at least in *Dracaena*. Resins are often secreted in the bark of species of *Dracaena*; by oxidation with nitric acid they yield the polycarbonic dracaenic and draco acids.

*Distribution.* The Dracaenaceae occur in subtropical-tropical regions of the Old World, both in rain forests and savannah.

Although the two genera are fairly different in appearance, they are very closely related and even dubiously distinct from each other.

*Dracaena*, with perhaps 80 species, but sometimes considered to contain up to 150 species, ranges from Macaronesia and subtropical Africa through Asia to northern Australia. It has a slender or occasionally very thick stem which may be of considerable height and the leaves vary from linear (almost grass-like) to broadly lanceolate. – *D. draco*, the "Dragon Blood Tree", on the Canary Islands may reach a height of 15 m or more and finally develops a thick trunk. Resins extracted from the stems can be used for varnishes, paper pigment and medicines. Other tree-like species are *D. reflexa* and *D. arborea* on Mauritius. Many

species are slender subshrubs or shrubs. Other resin-producing species are *D. cinnabari* on Socotra and *D. schizantha* in Arabia and tropical East Africa. The shoots of *D. mannii* in eastern Africa can be used as a vegetable. *D. draco* is often grown as an ornamental tree and *D. fragrans, D. marginata* and other species are ornamental foliage plants.

*Sansevieria* (50), ranging from Africa through Asia to Burma, consists of acaulescent rosette plants with more or less stiff, often variegated, leaves which contain tough fibres. The panicles are scapose and have pale yellow or cream-white flowers. Several species (*S. trifasciata, S. zeylanica, S. senegambica,* etc.) yield hemp, and various forms of *S. trifasciata* are common and tolerant foliar ornamentals.

## Nolinaceae T. Nakai (1943)   3–4:50   (Fig. 61)

Woody, generally large, arborescent plants with a stout, simple or sparingly branched trunk up to a few metres tall, with terminal leaf rosettes. Stem with secondary thickening growth of the kind described for Dracaenaceae. Vessels have not been recorded in the stems, but may occur in the leaves, where they have scalariform perforation plates (*Nolina*). The leaves are generally linear, sessile, parallel-veined, and have sunken stomata with oily contents. Suberized cells with pseudo-raphides (but not bundles of true raphides) occur in the family.

The inflorescences are panicles, often considerable in size and profusely flowered. The flowers are articulated on their pedicels, actinomorphic, hypogynous and polygamodioecious or dioecious. Their tepals are all equal, and are free from each other. There are 3 + 3 stamens, which have epipeltate anthers. The pollen grains are sulcate.

The pistil is generally trilocular, as in Dracaenaceae, rarely unilocular (*Dasylirion*). The ovary has septal nectaries. The style is relatively short. Each locule has two axially inserted ovules; in the unilocular ovary of *Dasylirion* there are three to six basally situated ovules. The ovules are anatropous; no parietal cell is cut off from the archesporial cell, nor are there any periclinal divisions in the nucellar epidermis; the embryo sac therefore becomes situated immediately below the epidermis. Nuclear endosperm formation has been recorded in *Nolina*. The fruit, unlike that in Dracaenaceae, is more or less dry and indehiscent, functioning rather as a nutlet than a berry (to

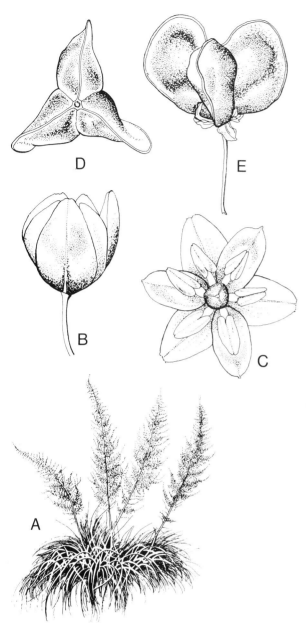

**Fig. 61.** Nolinaceae. *Nolina microcarpa.* **A** Habit, the inflorescences are to 2 m or more. **B–C** Flower in side and front view. **D–E** Fruit, top and side views. (CRONQUIST et al. 1977)

which, however, it closely corresponds). The seeds lack phytomelan and are similar to those in Dracaenaceae (HUBER 1969).

*Chemistry.* At least *Nolina* contains rich amounts of diosgenin, a steroidal saponin.

*Distribution.* Nolinaceae comprise three (to four) genera in the warm parts of America.

*Nolina* (25) in Texas, California and Central America has linear, tough leaves. Its flowers have

trilocular ovaries which develop into triangular or three-winged fruits. The leaves are used for thatching, mats, baskets, hats, etc. – *Dasylirion* (20–25), in the same regions, has inflorescences which are often of considerable size. The pulp from the shoot apex of *D. texanum* contains sugar and was used by the Indians as food or for preparing a beverage ("sotol"). The leaves of *Dasylirion* may also be used for basket work.

**Asteliaceae** Dumortier (1829)   4:50   (Fig. 62)

Dwarf to quite large herbs or woody rosette trees or shrubs, some species of *Cordyline* (e.g. *C. australis*) forming veritable trees up to 10 m high. Secondary thickening growth occurs at least in *Cordyline.* The herbs have a thick, tuberous underground rhizome. The leaves are spirally set or sometimes (in *Cordyline*) distichous. They are dorsiventral, mesomorphic to leathery and stiff, parallel-veined, linear to broadly lanceolate or elliptic, and occasionally constricted between sheath and lamina, or even pseudo-petiolate, e.g. in *Cordyline fruticosa.* The leaves at least in *Astelia* have paracytic stomata. Raphides are present in most taxa, e.g. *Cordyline* (they are absent, however, in *Astelia banksii*). In *Cordyline* the axis of the shoot may contain suberized cells with pseudo-raphides, but raphide cells may be present as well (HUBER 1969). Vessels seem to be lacking both in the stems and leaves.

The inflorescence consists of a system of bracteate racemes (or spikes), situated on the ends of leafy or leafless shoots. The flowers are hypogynous, trimerous and actinomorphic and have 3 + 3 tepals which are free or basally connate, and all of the same size and shape. They may be brownish or greenish (*Astelia*) or violet or white (e.g. *Cordyline*). The stamens are 3 + 3 in number and have basifixed or nearly basifixed, introrse anthers, which dehisce by longitudinal slits. The tapetum is secretory, and microsporogenesis, as far as known, is successive. The pollen grains are sulcate, spinulose in at least *Astelia* and *Milligania,* but non-spinulose in *Cordyline.*

The pistil is usually tricarpellary, either trilocular with axile placentation or, rarely, in some species of *Astelia* (where there may be four carpels), unilocular with parietal placentation. The style, although relatively long and simple in *Cordyline* and in some species of *Milligania,* is generally short or is lacking altogether in the other taxa, so that the stigma is subsessile or sessile. The stigma is

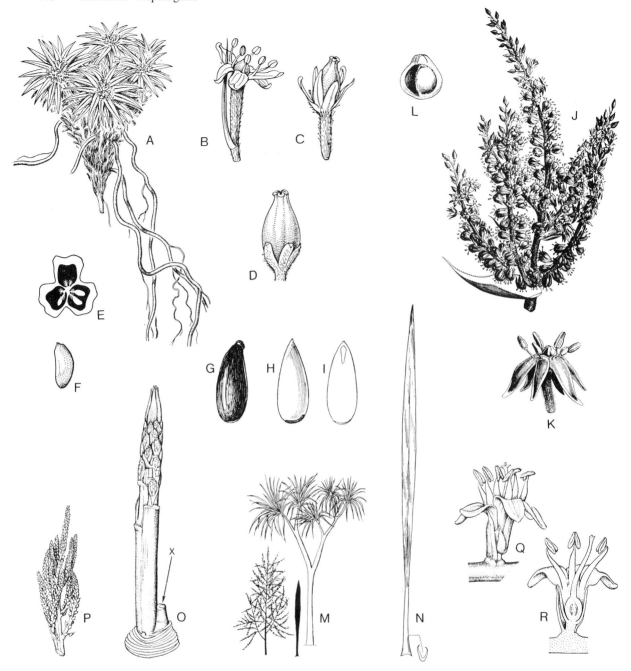

**Fig. 62.** Asteliaceae. **A–I** *Astelia pumila.* **A** Habit. **B** Flower. **C** Flower in late stage. **D** Fruit. **E** Ovary, transverse section. **F** and **G** Seeds. **H** Same, testa removed. **I** Seed, longitudinal section. (A from CORREA 1969; **B–D** and **F** from TAKHTAJAN 1982; **E** and **G–I** from CABRERA 1968) **J–L** *Astelia cunninghamii.* **J** Part of inflorescence. **K** Flower. **L** Fruit. (KRAUSE 1930) **M–N** and **Q–R** *Cordyline banksii.* **M** Tree, inflorescence and leaf, schematic. **N** Leaf. **Q** flower. **R** Same, longitudinal section. **O–P** *Cordyline australis.* **O** Young inflorescence (*x* lateral renewal shoot). **P** Inflorescence in early bud stage. (**M–R** from TOMLINSON and FISHER 1971)

Dry (at least in *Astelia* and *Cordyline*). Each locule contains 4 to ca. 15, anatropous or (in *Cordyline*) campylotropous ovules. The embryological information is incomplete.

The fruit is mostly more or less baccate, varying from fleshy to fairly dry (species of *Cordyline*), but in *Milligania* it is a globose capsule splitting from the apex. The seeds are ovate and often angular. Their testa is encrusted with phytomelan (even in the baccate forms). The embryo is straight and only a third to half the length of the endosperm.

*Chemistry*. Steroidal saponins occur in *Cordyline* and saponins have also been recorded in *Astelia*.

*Distribution*. Asteliaceae are distributed mainly in the Southern Hemisphere, but are absent from South Africa. *Astelia* itself has a highly disjunct distribution including Chile, Hawaii, the Marquesas, Tahiti, Samoa, Fiji, New Zealand, New Guinea, Australia and Reunion. *Cordyline* is even more widely distributed, occurring in Africa and South America as well as around the Pacific.

*Astelia* (25) consists of large to diminutive rhizomatous herbs with somewhat grass-like leaves. The stomata in the genus are reported to be paracytic, which is of particular interest. The colour of the tepals is green, dull red or fawn. All species are normally dioecious. Some species are large-growing herbs with leaves up to more than 2 m long, while others, like *A. pumila* (Fig. 62A–J) in Chile, are small and cushion-like with leaves only about 2–5 cm long. *A. grandis* in New Zealand is used for its tough fibres; another species, *A. nervosa,* in the same country, has edible berries, once used as food by the Maoris. *A. cunninghamii* is sometimes cultivated as an ornamental. *Milligania* (4), in Tasmania, differs from *Astelia* in having bisexual flowers and capsular fruits.

*Cordyline* (20) is distributed over Indomalaysia, New Zealand, Polynesia, Hawaii, Australia and even Africa, with one species, *C. dracaenoides,* in Brazil. The genus consists of shrubs and arborescent forms with leathery or stiff leaves, concentrated towards the ends of the branches. The flowers are yellow, greenish, blue or lilac. Their tepals are basally fused, with the lobes often recurved. Unlike *Astelia* the style is rather long and the ovules are campylotropous, but the genera agree (convergence?) in having baccate fruits with black seeds. *C. roxburghiana* and *C. australis* (Fig. 62O–P) are used for their fibres, and the underground parts of *C. terminalis* are fermented and used for a beverage, and the leaves to wrap fish, etc. The last two species among others are commonly grown as ornamentals.

It may seem that the Asteliaceae form a somewhat heterogeneous assemblage, but a small genus, *Cohnia* (2), in the Mascarenes and New Caledonia, may connect *Cordyline* with *Astelia* and *Milligania*. The basic chromosome numbers, $x=8$ in *Astelia* and $x=19$ in *Cordyline,* do not indicate a close connection. The former and the latter pairs of genera doubtless form two easily defined tribes within the family.

The family Asteliaceae shows some connection with the Asphodelaceae, and in the baccate fruit with black seeds it also resembles the Phormiaceae, although the similarity is most probably due to convergence. The two families differ markedly in pollen morphology.

**Hanguanaceae** Airy Shaw (1965)   1 : > 2
(Fig. 63)

Somewhat grass-like perennial herbs with thick creeping rhizomes, having a distinct endodermis and a loose central tissue with scattered, closed fibrovascular bundles. Aerial stems erect, with the linear or lanceolate leaves mostly concentrated towards the base. The leaf base is sheathing and pseudopetiolate and the lamina has supervolute ptyxis and pinnate-parallel venation with the distinct main veins diverging at intervals, and at a narrow angle, from the midvein (CRONQUIST 1981). Vessels are lacking in stems and leaves, but present in the roots, where they have scalariform perforation plates (TOMLINSON 1969). The aerial parts have short, uniseriate but branching hairs. Tanniniferous cells are common. The stomata are tetracytic (TOMLINSON 1969). Scattered cells with granular silica bodies occur in the endodermal cells around the vascular bundles, and larger silica bodies in hypodermal and mesophyll cells on the abaxial side of the leaf.

The inflorescence is a much-branched, bracteate panicle with small, unisexual flowers terminating the branchlets. The plants are dioecious, and the flowers actinomorphic. There are 3 + 3 tepals, which are small, rounded and pale, the outer being somewhat smaller than the inner. The male flowers have 3 + 3 stamens with long slender filaments (wind pollination) and basifixed, tetrasporangiate anthers which dehisce longitudinally. The axis in male flowers has conspicuous fleshy bodies. The pollen grains are inaperturate and spinulose. The female flowers have 3 + 3 staminodes, and a tricarpellary ovoid to globose pistil with a sessile, broad, triangular stigma (similar to that in *Astelia*). The ovary is trilocular and has one axile, nearly orthotropous ovule in each locule. The ovule has a funicular obturator and develops into a bowl-shaped seed with extensive endosperm and a small embryo.

The genus *Hanguana* (syn. *Susum*) (>2), occurs in Ceylon and Eastern Asia (Thailand to Australia). Its position is not yet settled. The presence of silica bodies, the tetracytic stomata, and the

**Fig. 63.** Hanguanaceae. *Hanguana malayana.* **A** Erect shoot and part of rhizome, the latter in longitudinal section. **B** Part of male inflorescence. **C** Male flower, note the basal callosities. **D** Part of female inflorescence. **E–F** Buds of male flowers. **G** Fruit. **H** Fruit, longitudinal section showing the seed. **I** Seed. **J** Seed, longitudinal section. (Orig. B. JOHNSEN)

seed shape do not support a position in Asparagales, but lack of vessels in the stem, lack of starch in the endosperm and other features indicate that *Hanguana* is after all closest to Asteliaceae (or Dasypogonaceae).

On Ceylon the species grows on the borders of pools and in swamps, where the rhizomes may form a veritable mat. Flowering of this form is rare. The forest floor *Hanguana* found in South-Eastern Asia is rather different and often flowers. These are probably not conspecific.

**Dasypogonaceae** Dumortier (1829)    7:53
(Figs. 64–65 and 68)

Mainly xeromorphic, rhizomatous perennial herbs or, more often, woody plants, which rarely may have a trunk up to 5 (−9!) m (the rosette tree *Kingia*) or which may form copiously branched shrublets, the branches of which are covered with short, stiff leaves or with leaf scars (see STAFF and WATERHOUSE, 1981). Stout, generally unbranched trunks with a terminal leaf rosette occur in *Kingia* and *Dasypogon* (*D. hookeri*), in at least the former of which there is a mantle of concealed aerial roots ramifying amongst the persistent leaf bases (reminiscent of the condition in Velloziaceae). Secondary thickening growth in the stem has been reported in *Lomandra,* where the stem is short.

The leaves are tough and xeromorphic, resembling those in some Cyperaceae. They are linear, dorsiventral, sheathing at the base, parallel-veined, and vary from one or a few centimetres (in *Acanthocarpus*) to more than 1 m long (*Kingia, Dasypogon*). They are generally tapering, but in *Chamaexeros* rounded-obtuse at the apex. The stomata are anomocytic. Raphides are reported in the *Lomandra* group (see below), but seem to be absent in the other genera. Vessels are always absent in the stem, but occur in the roots, where their perforation plates are usually simple (scalariform in *Kingia* and *Baxteria*). In *Acanthocarpus* vessels, with scalariform perforations, occur in the leaves (FAHN 1954).

The inflorescences are generally borne on separate branches with sparsely or often very densely set (imbricate) bracteal leaves. They may be elongate and loosely branched or consist of globose partial inflorescences (*Lomandra*) or they may constitute dense, multiflorous, often globose heads, which are obviously of paniculate or thyrsoid character, the flowering starting from the equator and proceeding towards both the base and apex (as often seen in *Dasypogon bromeliifolius*).

The flowers are generally small and inconspicuous (but ca. 6 cm long in *Baxteria*), hypogynous, actinomorphic, and either bisexual or unisexual; if they are unisexual the plants are dioecious. The flowers are not articulate on the pedicel. They have 3 + 3 free or basally connate tepals, which are white, cream or yellowish, and sometimes dry and scarious, but more often hyaline or petaloid to fleshy (as in several species of *Lomandra*). The two whorls of tepals may, as in *Dasypogon,* be conspicuously different in appearance, the inner much narrower and different in shape. A constriction may be present between a distal and a basal por-

tion. The large flowers of *Baxteria* are tubular. Hairs or bristles may be present on the outer surface of the tepals. Flowers with fleshy tepals are beetle-pollinated. Septal nectaries are present (*Dasypogon*) or absent (*Lomandra*).

The stamens are 3 + 3 in number, the 3 inner sometimes fused basally to the inner tepals. The filaments are generally rather long and filiform, and the anthers ovate to elongate, basifixed or slightly dorsifixed, introrse, and longitudinally dehiscent. Microsporogenesis seems to be unknown.

Pollen morphology (CHANDA and GHOSH 1976) is extraordinarily variable, varying from sulcate (*Kingia, Dasypogon*) to bisulcate (-circumsulculate) (*Acanthocarpus, Chamaexeros*) or circumsulculate to more or less spiraperturate, multiaperturate or irregular, spinulose (*Lomandra*).

The pistil is tricarpellary and trilocular (unilocular in *Dasypogon*), containing one ovule per carpel. The style is simple or apically tribrachiate, and in *Lomandra* very short. The stigma (in *Lomandra*) is of the Wet Type. Placentation is axile (in *Dasypogon* basal) and the ovules anatropous or half-campylotropous and crassinucellate. A parietal cell is formed at least in *Lomandra* and in this genus embryo sac formation is of the *Polygonum* Type (SCHNARF and WUNDERLICH 1939). Endosperm formation does not seem to have been investigated.

The fruits are capsules or nutlets: in *Acanthocarpus* and *Lomandra,* for example, there are loculicidal one- to three-seeded capsules, in *Baxteria* there are explosively opening capsules, while *Dasypogon* and *Kingia* have nutlets. The seeds, unlike those in *Xanthorrhoea,* lack phytomelan on the testal epidermis and are pale in colour. They are also smaller than in *Xanthorrhoea,* only 2–4.5 mm, elliptic to bean-shaped and circular in transverse section. Their outer epidermis consists of thick-walled, pitted cells, storing hemicellulose, aleurone and fat. The embryo is often relatively small (HUBER 1969). The basic chromosome number is $x = 7, 8, 9$.

*Chemistry*. Chemical data seem to be lacking.

*Distribution*. The Dasypogonaceae are almost restricted to Australia, where they are richly differentiated and adapted to arid habitats, mainly in a Mediterranean type of climate. *Lomandra* extends to New Guinea and New Caledonia.

The heterogeneity within Dasypogonaceae has long been recognized and several of the genera or groups of genera have been considered for family status (note that *Calectasia* often referred here is treated by us under Calectasiaceae). A tentative classification follows below.

**Fig. 64.** Dasypogonaceae. *Dasypogon bromeliifolius.*
**A–B** Plant. **C** Inflorescence. **D** One of the longer bracts
from the inflorescence. **E** A short bract (subtending a
flower). **F** Flower. **G** Stamen with dorsifixed hypopeltate
anther. **H** Pistil, longitudinal section. **I** Ovary, transverse
section (more compressed than in the natural state).
(Orig. B. JOHNSEN)

*Kingia* (1) and *Dasypogon* (3), both in the south
of Western Australia have tough, long leaves, and
both genera include veritable trees (*K. australis*,
*D. hookeri*) with a single (rarely branched) trunk.
Both genera have thick branches densely covered
with imbricate bract-like leaves, and globose heads
with small flowers. The tepals of both genera are
hairy on the outer surface, the fruit is indehiscent,
and the pollen grains are sulcate. – *Kingia* (1) *aus-*
*tralis* (Fig. 68) has linear, glossy, arching leaves,
and is similar in habit to arborescent species of

*Xanthorrhoea. – Dasypogon* (2) includes one species without an aerial trunk, *D. bromeliifolius* (Fig. 64), and one of trees, *D. hookeri.*

*Baxteria* (1), in the south of Western Australia, is a rosette plant, with a woody rhizome and linear leaves. The large, tubular flowers are situated at ground level in the rosette. The fruit is a capsule that opens explosively.

The remaining genera, *Lomandra, Romnalda, Acanthocarpus* and *Chamaexeros* form another possibly natural complex. They are shrublets or

**Fig. 65.** Dasypogonaceae. **A–E** *Chamaexeros serra.* **A** Plant. **B** Leaf margin; note the hyaline, dentate appendage. **C** Leaf, transverse section. **D** Flower. **E** Same, in detail, two tepals and stamens bent to show pistil; note the thick torus of the flower. **F–L** *Acanthocarpus preissii.* **F** Part of branch with fruits. **G** Fruit, seen from the base, with the six persistent tepals. **H** Same, lateral view. **I** Same, longitudinal section. **J** Seed (not with a black phytomelan layer). **K** Branchlet. **L** Flower. (Orig. B. JOHNSEN)

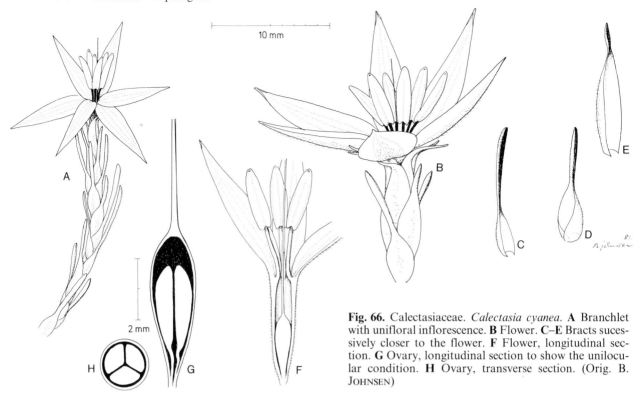

**Fig. 66.** Calectasiaceae. *Calectasia cyanea*. **A** Branchlet with unifloral inflorescence. **B** Flower. **C–E** Bracts sucessively closer to the flower. **F** Flower, longitudinal section. **G** Ovary, longitudinal section to show the unilocular condition. **H** Ovary, transverse section. (Orig. B. JOHNSEN)

herbs with a woody rhizome, and are all of moderate size, with a graminoid habit and the leaves concentrated basally or dispersed along the woody branches (*Acanthocarpus*). The inflorescences are generally dense, simple or complex heads or spikes but are sometimes looser, paniculate assemblages of flowers, which have generally yellow, brownish, or purple, hyaline or fleshy tepals. Raphides occur in the genera of this group. The pollen grains in this group are bisulcate, zonisulculate, spiraperturate or irregular. The fruits are capsular.

*Lomandra* (42) has the widest distribution in the family (see above) and is very variable in the appearance of the leaves as well as the inflorescences. The plants are dioecious and often have fleshy tepals eaten by beetles. Wind pollination may also occur. The anthers are dorsifixed and the pollen grains zonisulculate or spiraperturate.

*Acanthocarpus* (3) (Fig. 65) in the south of Western Australia is a shrublet with rather short, stiff, prickly leaves and a relatively large globose capsule covered with a scaly surface. *Chamaexeros* (3), (Fig. 65) in the same region, has a basal rosette of linear leaves with rounded apex and a cluster of small flowers.

The great variation in Dasypogonaceae may possibly justify a division into three families sketched above, although subfamilial status seems more appropriate. Family names such as Lomandraceae Lotsy and Kingiaceae Endl. are thus regarded as synonymous with Dasypogonaceae Dum.

## Calectasiaceae S.L. Endlicher (1836)    1:1
(Fig. 66)

Xeromorphic, rhizomatous, branched shrublets up to ca. 50 cm high, with slender, stiff branches covered by old leaf sheaths or leaves. Branchlets pubescent, with small, stiff, flat or concave, lanceolate, parallel-veined leaves with pointed apices and laminae disarticulating from the open sheaths. The stomata are anomocytic and raphides are lacking. Vessels are lacking in the stem but present in the roots and have simple perforation plates.

The flowers are solitary on the ends of short branchlets. They are subtended by a great number of small, imbricate leaves. The flowers are bisexual, hypogynous, actinomorphic and trimerous. They have 3 + 3 lanceolate, lilac-blue to purple tepals which are fused basally into a short tube but are otherwise more or less spreading to form a funnel-shaped or nearly open, stellate perigone, which has a glossy metallic sheen. The stamens, 3 + 3 in number, are inserted in the perigone tube and have rather short, flat filaments. The anthers are erect, basifixed, tetrasporangiate, introrse and stand close together to form a tube around the style. They each dehisce by two apical pores. The pollen grains are sulcate.

The pistil is tricarpellary. It has an erect filiform style and a unilocular ovary with three basal, erect, anatropous ovules. The embryology is incompletely known.

The fruit is dry, indehiscent and one-seeded (a nutlet), and the seed is elongate with a thin membranous testa lacking phytomelan. Unlike the Dasypogonaceae it has a basic chromosome number of $x = 9$.

The chemistry is unknown.

Calectasiaceae consists of the monotypic genus *Calectasia* only. *Calectasia cyanea* is restricted to Australia, where it grows in heath vegetation in the states of Western and South Australia and Victoria.

Calectasiaceae is evidently most closely allied to the Dasypogonaceae, in which its habit is matched by the genus *Acanthocarpus*. *Calectasia* deviates in several conspicuous floral characters and in the chromosome number. It is treated here as separate from Dasypogonaceae with some reservations.

## Blandfordiaceae Dahlgren & Clifford, fam. nov. 1:4   (Fig. 67)

Herbae perennes, rhizomatosae, erectae, ad 1.5 m altae. Folia in parte basali caulis aggregata, disticha, tenacia, linearia, dorsiventralia, ad 1 m longa et 8 mm lata, nervo centrali prominenti marginibusque asperis. Stomata anomocytica. Caulis rigidus foliis caulinis bracteatis instructus. Flores 1–20 ampliores, pendentes, campanulati, bracteati, pedicellati in racemo terminali dispositi. Pedicelli bibracteolati. Perianthium peristens. Tepala 6 in tubo aurantiaco, ca. 4/5 longitudinis coalita, partes libera lutea saltem ad margines. Stamina 6, filamentis in tubo inclusis, antheris dorsifixis, latrorsis, longitudinaliter dehiscentibus. Grana pollinis subglobosa, sulcata, bicellularia. Pistillum triloculare, breviter stipitatum; loculi ovula 40–50, anatropa, crassinucellata continentes. Saccus embryonis typi Polygoni formans; endospermium helobiale. Capsulae erectae, elongatae, septicidales. Semina 5–6 mm longa, fusco-pubescentia, sine phytomelano. Embryo linearis, longitudine seminis dimidius; endospermium inamyloideum. Plantulae Allii similes.

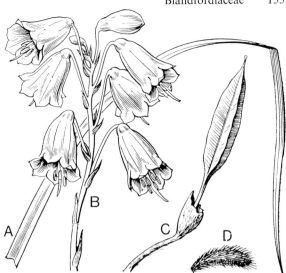

**Fig. 67.** Blandfordiaceae. *Blandfordia grandiflora*. **A** Leaf. **B** Inflorescence. **C** Winged capsule. **D** Seed, note the indumentum. (TAKHTAJAN 1982)

Genus typicus *Blandfordia* Sm.

Erect, herbaceous perennials to about 1.5 m tall, shortly rhizomatous with thick fibrous roots. The leaves are distichous and usually crowded at the base of the stem. They are linear, dorsiventral, glabrous, 30–100 cm long and 4–8 mm wide and have a prominent midrib and rough edges. The stomata are anomocytic.

The stem is rigid and its cauline leaves are bracteal and generally anthocyanic in colour. It ends in a raceme with 1–20 large, pendulous and more or less campanulate flowers, which are each subtended by a bract and (generally) two smaller bracteoles on their pedicels. Rarely, the flowers occur singly in the leaf axils. The perianth is persistent with the 3 + 3 tepals fused into a tube for about four-fifths of their length. The perianth tube is usually orange-red and the lobes yellow, at least on the margins.

There are 3 + 3 stamens, the filaments of which are inserted in the perianth tube about one third of the distance from the base. The anthers are dorsifixed and open latrorsely by longitudinal slits. The pollen grains are subsphaeroidal, sulcate and dispersed in the two-celled stage.

The pistil is tricarpellary and shortly stipitate, the stipe lengthening in fruit. The ovary is trilocular and surmounted by a short style that terminates in a small three-grooved stigma. Septal nectaries are lacking.

The ovules are 40–50 in each locule, inserted in two rows on axile placentas. They are anatropous and crassinucellate. The embryo sac is of the *Polygonum* Type and endosperm formation helobial (DI FULVIO and CAVE 1964).

The fruit is an erect, elongate, septicidal capsule 4–10 cm long (including the stipe). The seeds are 5–6 mm long with a soft felted surface of short brown hairs. Phytomelan is lacking. The embryo is about half the length of the seed and is embedded in a starchless endosperm. The seedlings are like those of *Allium*.

The single genus *Blandfordia* (4) is restricted to Eastern Australia and Tasmania, and generally grows in sandy, damp places or acidic moorland.

We consider this genus so distinct that familial recognition is justified. The habit is somewhat reminiscent of that of some Haemodoraceae, e.g. *Conostylis* (incl. *Blancoa*), but several details, such as the anomocytic stomata and starchless seeds, of *Blandfordia* argue against a position in Haemodoraceae. Possibly it is most closely allied to the Dasypogonaceae.

## Xanthorrhoeaceae Dumortier (1829)   1:15
(Fig. 68)

Xeromorphic plants, either rhizomatous with a woody trunk or nearly acaulescent, with a woody rhizome, the former comprising "Rosette Trees" with a trunk up to ca. 2 m tall. Where a trunk is developed this has secondary thickening growth of the kind described under Dracaenaceae. The trunk is covered with old leaf scars and leaf bases, and carries green leaves on the apical parts. The stems secrete large quantities of yellow, red and brown acaroid resins, which contain small amounts of essential oils. The leaves are spirally inserted on the stem, linear, dorsiventral, parallel-veined and tough, and may be up to more than 1 m long, but are sometimes only 3–5 cm, as in *Xanthorrhoea pumilio*. The stomata are paracytic. Vessels are present in roots and leaves and sometimes also in the stems; perforation plates are simple in the roots and scalariform in the leaves and stems. Crystal raphide cells are abundant.

The inflorescence is complex, dense, spike-like and multiflorous, with infinitely short secondary branches, and is situated on the end of a long, leafless peduncle. The flowers are small, actinomorphic, hypogynous and bisexual and are supplied with a relatively large bract and bracteole. The 3 + 3 perianth members are free from each other, and are more or less bract-like and partly

hyaline, the outer somewhat stiffer and shorter than the inner and generally hairy. There are 3 + 3 stamens with narrow filaments, which are free from each other and longer than the tepals. They bear peltate, elongate and introrse anthers dehiscing longitudinally. Microsporogenesis and tapetum do not seem to have been described. The pollen grains are sulcate.

Copious nectar exudes from septal nectaries, attracting trigonid bees (reported from Queensland).

The ovary is trilocular and has a long, erect, simple style and a trilobate or capitate to punctiform stigma. Each locule contains a few anatropous ovules, the embryological features of which have not yet been fully investigated. The fruit is an ovate, woody or cartilaginous capsule. Each capsule contains one or two ovoid-elongate seeds, rounded at one end, pointed at the other, and slightly compressed and trigonous in transection. The outer integument consists of several cell layers and its epidermis has a thick crust of phytomelan. The endosperm consists of rather thin-walled, isodiametric cells containing aleurone and fatty oils but no starch. The embryo is transversely situated in the centre of the endosperm; it is fusiform, straight or curved and about as long as the breadth of the seed (HUBER 1969). $x = 22$.

*Chemistry.* The family is well known for its contents of acaroid resins, which may contain p-cumaric acid, p-oxybenzoic aldehyde and essential oils like citronellol, paeonol, etc.

*Distribution.* The genus is restricted to Australia, where it is widely distributed.

*Xanthorrhoea* (15) shows a considerable variation in size and habit. Some species form veritable trees, which are characteristic components in the Australian chaparral. *X. preissii*, "Black Boy" (Fig. 68 A–H) has a stem up to more than 2 m high, leaves more than 1 m, and an inflorescence of 3–3.5 m. The wood of this species can be used for bowls and other wooden vessels. *X. australis* and *X. hastilis* yield a gum, "Grass Tree Gum" or "Red Acaroid Gum", which is used for varnishes. In *X. minus* the short trunk is mainly subterranean; *X. pumilio* has leaves only about 3 cm long and a scape of only ca. 30 cm.

The Xanthorrhoeaceae in terms of seed shape recall *Agapanthus* or *Phormium*, but are probably not most closely related to either of these, nor, perhaps, to the Dasypogonaceae, among which they resemble *Kingia* in habit and flower characters. The phylogenetic connections are still unsettled.

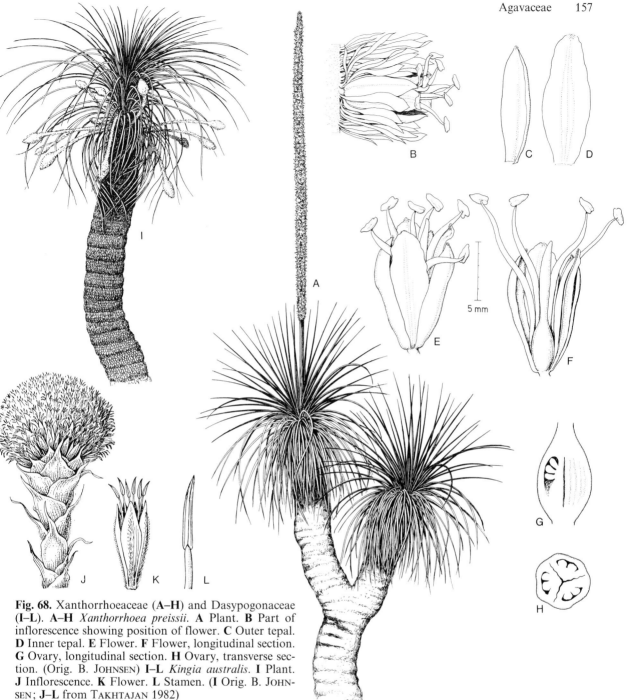

**Fig. 68.** Xanthorrhoeaceae (**A–H**) and Dasypogonaceae (**I–L**). **A–H** *Xanthorrhoea preissii.* **A** Plant. **B** Part of inflorescence showing position of flower. **C** Outer tepal. **D** Inner tepal. **E** Flower. **F** Flower, longitudinal section. **G** Ovary, longitudinal section. **H** Ovary, transverse section. (Orig. B. JOHNSEN) **I–L** *Kingia australis.* **I** Plant. **J** Inflorescence. **K** Flower. **L** Stamen. (**I** Orig. B. JOHNSEN; **J–L** from TAKHTAJAN 1982)

## Agavaceae S.L. Endlicher (1841)   8:300
(Figs. 69–70)

Predominantly large rosette herbs to giant rosette trees, which may have a stout trunk with secondary thickening growth. The underground parts consist of rhizomes or bundles of roots. The leaves are spirally set, dorsiventral, thick and succulent or tough and fibrous. They are lanceolate, linear or subulate, often broadest near the base and gradually tapered to a sharp point; their margins often bear lateral spines or teeth. In *Agave* and related genera, the leaves are fleshy and thick, with stomata more frequent on the upper than on the lower surface (otherwise stomata are mostly confined to the lower surface). The stomata are ano-

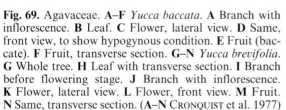

**Fig. 69.** Agavaceae. **A–F** *Yucca baccata*. **A** Branch with inflorescence. **B** Leaf. **C** Flower, lateral view. **D** Same, front view, to show hypogynous condition. **E** Fruit (baccate). **F** Fruit, transverse section. **G–N** *Yucca brevifolia*. **G** Whole tree. **H** Leaf with transverse section. **I** Branch before flowering stage. **J** Branch with inflorescence. **K** Flower, lateral view. **L** Flower, front view. **M** Fruit. **N** Same, transverse section. (**A–N** CRONQUIST et al. 1977) **O** The pollinator, a female of *Pronuba yuccasella*, collecting pollen on a *Yucca* stamen. (LOTSY 1911)

mocytic and generally deeply sunk. While crystal raphides are lacking, so-called pseudo-raphides contained in suberized cells and solitary crystals in small cells are typical of the family. The vascular bundles of the leaves are accompanied by thick and tough fibres, hence the use of several species as textile plants. The spines of the leaves in some species are filled with sclerenchymatous tissue.

The inflorescence-bearing stem is large and terminal, in some species 2 to several metres high and covered with few or numerous bracteal leaves. In

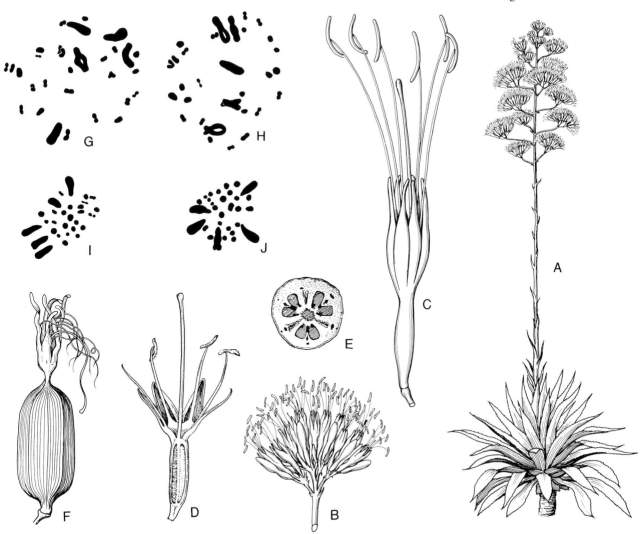

**Fig. 70.** Agavaceae. **A–C** and **F** *Agave* sp. **A** Plant. **B** Branch of inflorescence. **C** Flower. **F** Fruit. (LARSEN 1973a) **D–E** *Agave* sp. **D** Flower, longitudinal section. **E** Ovary, transverse section. (A.M. JOHNSON 1931) **G–J** Chromosome configurations of *Yucca flaccida* (**G** diakinesis), *Agave virginica* (**H** diakinesis), *Yucca filamentosa* (**I** metaphase), and *Agave virginica* (**J** metaphase). (MCKELVEY and SAX 1933)

most genera the inflorescence is a complex, much-branched panicle, the lateral components of which are cymose, consisting of monochasial units. In the hapaxanthic genera, after several years the plants produce a giant inflorescence and then die.

The flowers are mostly bisexual, trimerous, hypogynous or epigynous, actinomorphic or slighty zygomorphic and generally situated in the axils of well-developed bracts. The tepals, which are generally white or yellow, are free or more or less fused into a tubular or campanulate perigone; sometimes this is abruptly widened and urceolate in the outer part.

The stamens are $3+3$ in number and are inserted at the base of the tepals or on the inner side of the perigone tube. The filaments are sometimes basally widened; in *Yucca* they are relatively stout and short-hairy; in *Agave* they are filiform and often extend far beyond the perigone. The anthers are long or short, introrse, peltate ("epipeltate") or in *Yucca* nearly impeltate, and dehisce longitudinally. The tapetum is glandular (secretory) and microsporogenesis is successive. The pollen grains have a reticulate exine and a sulcate or (in *Polianthes* and *Prochnyanthes*) bisulculate aperture (ERDTMAN 1952). They are dispersed singly or rarely in tetrads (*Furcraea*; species of *Agave* and *Beschorneria*).

The ovary is trilocular, each locule having several to many anatropous ovules. The style is short (subfam. Yuccoideae) or rather long and simple (subfam. Agavoideae), with a punctiform or small and capitate to trilobate stigma with Dry or Wet stigma surface. A parietal cell is cut off from the archesporial cell. Embryo sac formation conforms to the *Polygonum* Type, and endosperm formation is helobial (rarely, as in *Furcraea,* nuclear). Haustoria in connection with the embryo sac and endosperm are lacking. In *Yucca* part of the nucellus divides actively before and after fertilization and forms a considerable perisperm tissue.

The fruit is a capsule or, in some *Yucca* species, rather a berry. It contains several or many seeds. These are sometimes flat and plate-like, crescentshaped or semi-circular, but sometimes less compressed (as in *Yucca*), and then with a furrowed endosperm. The outer integument of the testa consists of several to many (up to 20) cell layers; its epidermis has a thin crust of phytomelan. The endosperm contains aleurone and fatty oils.

The karyotype in all Agavaceae is uniform: $x = 30$, with 5 large and 25 small chromosomes (McKelvey and Sax 1933).

*Chemistry.* The family is rich in steroidal saponins, the sapogenins of which have 27 C-atoms. Of the simple sapogenins may be mentioned smilagenin, sarsasapogenin, tigogenin, neotigogenin, diosgenin and yamogenin. The saponins produced especially from species of *Agave* and *Yucca* are of great importance as hormonal substances (and form the active principle of contraceptive pills). Chelidonic acid is also widely distributed in the family, and the leaves may contain a certain amount of ascorbic acid. Essential oils can be obtained from the scented flowers of some taxa.

*Distribution.* The Agavaceae form an embryologically and cytologically homogeneous, but morphologically somewhat heterogeneous family. Both subfamilies, Yuccoideae and Agavoideae, are entirely American in origin, but *Agave* especially is now widely distributed in warm, arid areas all over the world. The members are prevailingly large prairie and steppe plants. The centre of distribution is in Mexico and adjacent warm and arid parts.

### Subfamily Yuccoideae

This consists of two genera with short to considerable trunks, sometimes several metres high, with secondary thickening growth. The leaves are not as succulent as in many Agavoideae, but are stiff or tough and fibrous, and are sometimes supplied with marginal fibres (Fig. 68 B). In the flowers,

which are hypogynous, the tepals are free or more or less connate basally and the anthers only weakly epipeltate.

Subfamily Yuccoideae is wholly American and consists of *Yucca* (incl. *Hesperoyucca, Clistoyucca* and *Samulea*) and *Hesperaloë* (2), which are both concentrated to the warmer parts of America.

*Yucca* (40) occurs in the plains and arid mountains of western North America. It consists of large or medium-sized plants which often have a woody stem: *Y. ("Clistoyucca") brevifolia,* "Joshua Tree" (Fig. 69 G–N) is a veritable tree with relatively short leaves covering considerable parts of the branches. In other species the stem is shorter and bears terminal leaf rosettes. The large flowers have almost free tepals. The stamens have rather stout, shortly hairy, filaments and small, arrowlike anthers. Some species have berry-like fruits; in others the fruits are dry or spongy. – The leaves of several species, especially *Y. australis,* yield fibres used locally for twine, cloth and baskets. The seeds or fruits of some species are eaten locally, and the young stems and leaves of *Y. australis* are fermented to an alcoholic beverage. – Several species are grown as ornamentals, in particular *Y. aloifolia, Y. brevifolia, Y. filamentosa, Y. glauca, Y. gloriosa* and *Y. smalliana.*

*Yucca* provides a classic case of interdependence between a plant and its pollinator, which in this case is the genus of small moths, *Tegeticula (Pronuba)*. One species of *Tegeticula* is specific to each of the two monotypic subgenera *Hesperoyucca* and *Clistoyucca* in the southwestern U.S.A., while a third pollinates the several *Yucca* species occurring here and further east. The moths, exhibiting a specific behaviour pattern, gather pollen into a ball, which is held under the head by the proboscis. They then fly to another flower and carry out alternate ovipositions (one egg at a time) and acts of pollination, in which they press pollen into the canal of the tubular style. Usually one egg is laid in each loculus of the ovary. Ovules in the neighbourhood of a moth's egg form gall tissue which nourishes the larva, while causing irregularities on the capsule (Fig. 69 E).

### Subfamily Agavoideae

Subfamily Agavoideae represents the epigynous genera of the family. Many taxa are giant herbs with secondary thickening growth and some, like *Furcraea longaeva,* have a considerable trunk. The leaves are often strongly succulent and contain tough fibres that are economically exploited. The plants, though often reaching considerable age, flower only once and then die, and so are hapaxanthic. They are generally propagated vegetatively. Their inflorescences are mostly profusely flowered and complex and may appear on tall axes with sparsely or closely set scale-like leaves. Some such species (in *Agave*) are adapted for bat pollination, while the syndrome of bird pollination is prominent in *Beschorneria*. The flowers are syntepalous with a very short to quite long perianth tube, and they mostly have long, slender, exserted filaments with epipeltate, versatile anthers. In some genera the pollen grains cohere in tetrads, and bisulculate pollen grains occur in other genera. The fruits are always capsular and the seeds mostly flattened.

Here belong *Agave, Furcraea, Beschorneria, Manfreda, Polianthes* and *Prochnyanthes*. Among other genera referred here at least *Littaea* is probably best treated as a subgenus of *Agave*.

*Agave* (300) in the southern U.S.A., central America and part of tropical South America consists of plants with a short, often woody stem, which at least in several species has secondary thickening growth. The fleshy leaves end in a spine and frequently (as in species of *Aloë*, Asphodelaceae) have lateral spines. The inflorescence in some species, like *A. americana*, is gigantic and reaches a height of 3 m or more. The flowers generally have yellow and greenish tepals, exceeded much in length by the stamens.

The leaves are covered with wax. The young inflorescence can be tapped for sugar-rich sap, which in large specimens may amount to hundreds of litres. When brewed this yields the alcohol-rich "Pulque", from which "Mescal" and "Tequil" are distilled. Many species are useful for their fibres.

*A. americana*, "Century Plant", from Mexico, is one of the commonest and largest species, cultivated in warm parts of the world as an ornamental or border plant and now widely naturalized. *A. sisalana* ("Sisal Agave") and *A. fourcroydes*, indigenous in the same region, are widely cultivated in the world for the sisal hemp fibres retted from the leaves and used for cordage and strings. Used in the same way are *A. funkiana, A. gracilispina*, and especially *A. cantala*, from Mexico, and *A. letonae* from Central America. Species used for "Pulque" are, for example, *A. complicata, A. atrovirens, A. mapisaga*, etc. The leaves of several species yield saponins.

Besides *A. americana*, many other species are grown as ornamentals, some being rather small in size.

*Furcraea* (20) and *Beschorneria* (7) from tropical America have flowers with tepals fused higher up than in *Agave*. The pollen in these genera is generally dispersed in tetrads. *Furcraea* includes some almost trunkless species, while others have a considerable trunk. The inflorescence of some species forms an enormous panicle. *Furcraea* yields fibres, which are of better quality than those from *Agave*; *F. gigantea*, "Piteira", cultivated mainly on Mauritius, *F. hexapetala* and *F. humboldtiana* are fibre plants. *Beschorneria* species are rosette plants without an aerial trunk. Their inflorescence is an erect panicle or raceme with pendent flowers.

## Hypoxidaceae R. Brown (1814)   10:150
(Figs. 71–72)

Herbs with a tuberous rhizome or a corm covered with a membranous coat of fibres. The leaves are radical and sometimes longer than the inflorescence. They are mostly tristichous and are dorsiventral, linear to lanceolate, sessile or, for example in some *Curculigo* species, constricted above the leaf base into a pseudopetiole. The lamina is prominently parallel-veined, and the ptyxis may be plicate. They are often clothed with fairly long, uniseriate hairs, but branched, multicellular hairs are also present in the family. Stomata, at least in most cases, are paracytic, rarely tetracytic. Raphide cells occur at least in *Curculigo*, and are probably widespread in the family. The inflorescence-bearing stem has only one circle of vascular bundles. Vessels are lacking in the stems and leaves, but are present in the roots where they have scalariform perforation plates.

The flowers are borne on a leafless, usually hairy scape arising from the rhizome or corm. The inflorescence consists of spikes, racemes or umbel-like clusters, or it may be reduced to a single flower.

The flowers are sometimes situated in the axils of fairly large bracts. They are actinomorphic, bisexual, trimerous and epigynous, with 3 + 3 tepals free or fused into a sometimes long and narrow tube above the ovary. The tepals of the two whorls are more or less equal and generally yellow or white, rarely red (in *Rhodohypoxis*); in most taxa they have attenuate apices and are hairy on the outside. Nectaries (of any kind) seem to be lacking.

There are 3 + 3 or rarely (*Pauridia*) 3 stamens inserted at the base of the perianth lobes; they have narrow filaments and basifixed or epipeltate, introrse anthers dehiscing longitudinally. The tapetum becomes periplasmodial at an early stage, and may be classified as amoeboid, which is exceptional in Asparagales. Microsporogenesis is successive.

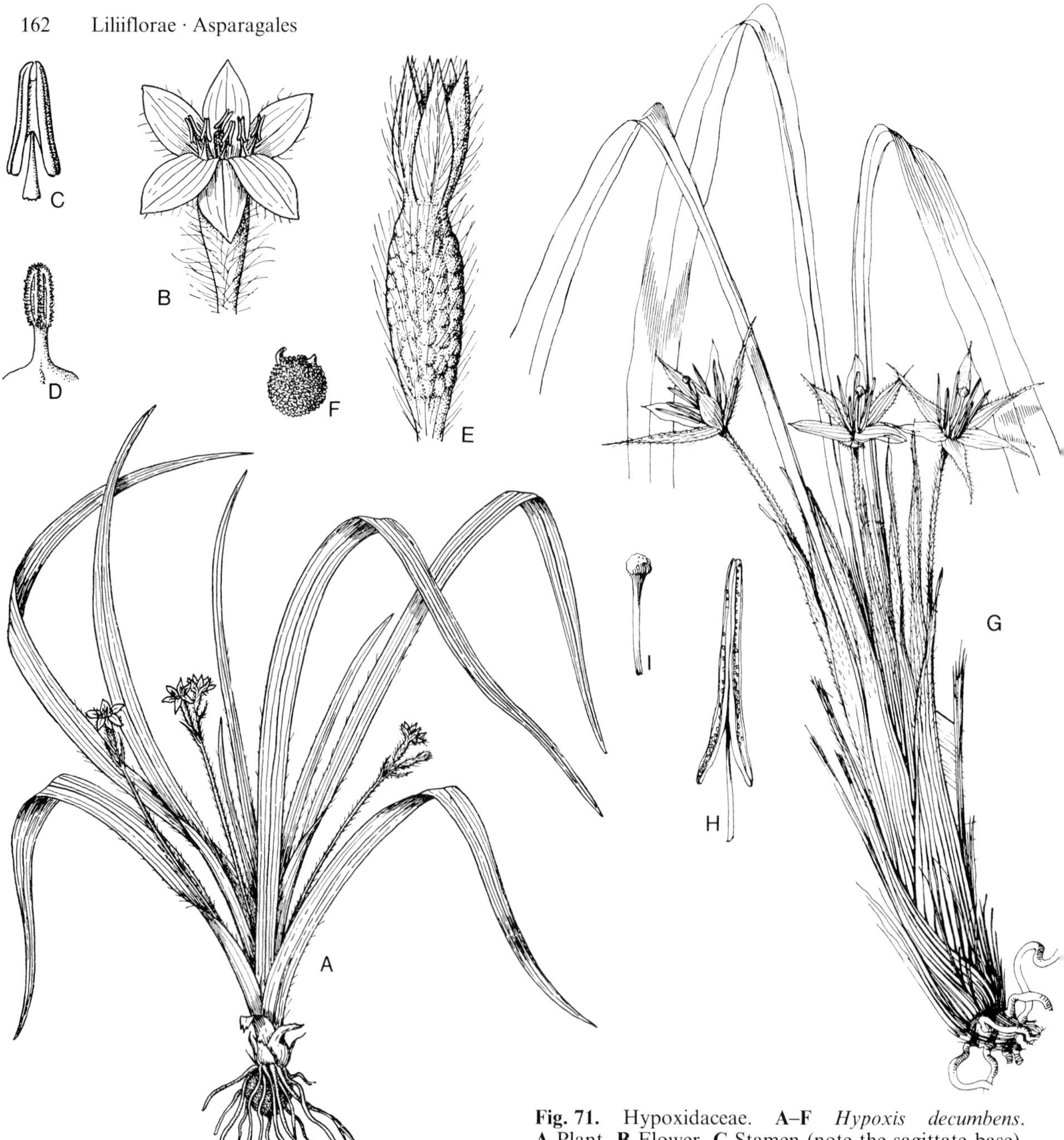

**Fig. 71.** Hypoxidaceae. **A–F** *Hypoxis decumbens*. **A** Plant. **B** Flower. **C** Stamen (note the sagittate base). **D** Style and stigmatic areas. **E** Fruit (with persistent tepals). **F** Seed. (CABRERA 1968) **G–I** *Curculigo pilosa*. **G** Plant (some leaves omitted). **H** Stamen. **I** Style apex and stigma. (HEPPER 1968)

The pollen grains are sulcate, with a finely reticulate exine, and are dispersed in the two-celled state.

The ovary is trilocular (unilocular in *Empodium*) and bears a short style which is apically divided into three branches. The stylodial branch tips are covered with a Dry stigmatic surface. The locules have several centrally (in *Empodium* parietally) inserted, anatropous or hemianatropous ovules. The archesporial cell functions directly as the megaspore mother cell and parietal tissue thus is lacking, but the epidermis often divides periclinally to form a nucellar cap. Embryo sac formation follows the *Polygonum* Type, more rarely the *Allium* Type. Endosperm formation is generally helobial but nuclear in *Pauridia,* where the chalazal part is somewhat haustorial.

The fruit is generally a capsule, which is mostly crowned by the remaining parts of the perianth,

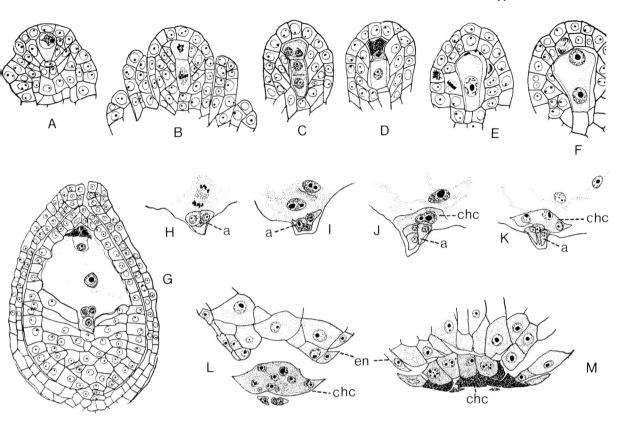

**Fig. 72.** Hypoxidaceae, embryology. **A–G** *Spiloxene aquatica*, embryo sac formation. A parietal cell is not formed here. **A–D** show the meiosis resulting in four megaspores, out of which the chalazal one continues to form an 8-nucleate embryo sac (**E–G**) according to the *Polygonum* Type. **H–M** *Spiloxene schlechteri,* helobial endosperm formation; **H–J** showing division of the primary endosperm nucleus with the formation of a wall between the micropylar and chalazal chambers; **K–M** showing the subsequent ab initio nuclear divisions in each chamber, with walls formed subsequently in the micropylar chamber (*en*) but not in the chalazal chamber (*chc*) (**L**), and then the degeneration of the chalazal chamber (**M**). (All after DE VOS 1948)

and opens by short vertical slits near the top; it may also be fleshy and indehiscent (*Curculigo* s. lat.). In all taxa the seeds are small and globose, with a prominent raphe. An elaiosome occurs at least in *Curculigo*. The seed coat consists of a few cell layers, the epidermis of which is palisade-like and has a thick, black phytomelan crust. The endosperm is composed of thin-walled, isodiametric cells storing aleurone and fatty oils, but no starch.

*Chemistry.* Alkaloids and steroidal saponins seem to be lacking, at least in *Hypoxis,* while chelidonic acid occurs in the family.

*Distribution.* The family is found mainly in the Southern Hemisphere, with centres in Southern (to Central) Africa, South America, Australia and the coastal regions of Asia. About 90 species in at least five genera occur in Southern Africa. The plants mostly grow in meadows, grassland or low macchia vegetation.

*Hypoxis* (90) is distributed in Africa, America, southern Asia and Australia. It has a vertical tuberous rhizome, and an often rather short inflorescence-bearing scape, the flowers of which have a yellow, stellate perianth and exserted stamens. Some species, e.g. the American *H. hirsuta* and the South African *H. stellata,* are occasionally cultivated as ornamentals. *Rhodohypoxis* (4) with red or white flowers and stamens hidden in the basal perianth tube occurs in the Drakensberg Mountains of Southern Africa. The South African genus *Spiloxene* (30) should probably also be distinguished from *Hypoxis,* differing in having a more bulb-like corm and in lacking pubescence on most parts. Besides, it generally has one or a few rather large flower(s) with spreading, orange, yellow or white (rarely partly green and black) tepals. – The small-sized species of *Pauridia* (2) in the same region have only three stamens and are doubtfully included in the family.

*Curculigo* (10) has about the same distribution as *Hypoxis*. The inflorescences are mostly more or less concealed by the far longer and larger leaves, and the indehiscent fleshy fruits are prolonged into a rostrum formed by the perianth tube. *Empodium* (10) in South Africa has a unilocular ovary with parietal placentae.

Further investigations are required to determine whether *Pauridia*, with nuclear endosperm formation and only three stamens, properly belongs in Hypoxidaceae.

The combination of attributes: paracytic stomata, lack of nectaries and tapetum which soon becomes periplasmodial is a peculiar one, the constancy of which needs further investigation. The closest affinities of the family are somewhat uncertain and also need further study. Sometimes Hypoxidaceae have been associated with orchids, an affinity that does not seem at all convincing to us.

The paracytic stomata, the fibrous leaf bases and the epigynous flowers may be taken to suggest an affinity with the Velloziaceae, but the phytomelan-coated, starchless seeds of the Hypoxidaceae show little similarity to the seeds of Velloziaceae, and the lack of vessels in stem and leaves indicates that Hypoxidaceae are best placed in the Asparagales, and the similarities with Velloziaceae are obviously due to convergent evolution.

## Tecophilaeaceae Leybold (1862)   5–7:15–20
## (Figs. 73–74)

Moderate-sized or fairly small, generally glabrous herbs with thick, often tunicated rhizomatous corms. The stem is erect and partly leafy, although the leaves may be concentrated at the base. The leaves are dorsiventral, linear to lanceolate, entire, parallel-veined and sheathing at the base. Vessels are lacking in the stems and leaves, but present in the roots; they have scalariform perforation plates (CHEADLE 1969). The stomata are anomocytic.

The inflorescences are simple or compound racemes or thyrses in which the flowers are subtended by bracts of variable size. The flowers are bisexual, trimerous, generally partly epigynous, and actinomorphic or, in particular with regard to the androecium, zygomorphic. The 3 + 3 tepals are free from each other or shortly connate at the base, generally lanceolate and in most taxa blue, violet, white or pale yellow. They are frequently somewhat reflexed. There are 3 + 3 stamens or sta-

men homologues, some stamens (often the upper) being often transformed into linear staminodes or being dwarfed and sterile (see the genera, below). The filaments are short and glabrous and the anthers, which are mostly basifixed (peltate in *Tecophilaea*), are sometimes connivent, with the connective sometimes produced at both ends. Anther dehiscence is by means of apical pores or by apical longitudinal slits. Microsporogenesis is simultaneous (at least in *Cyanella* and *Odontostomum*). The pollen grains are sulcate (with operculate sulcus) and two-celled when shed.

The ovary is generally half-inferior and the style erect and subulate to filiform, straight or slightly curved, ending in a small capitate stigma. There are three locules in the ovary, each with several to many ovules in two rows. The ovules are anatropous and crassinucellate, and a primary parietal cell is cut off from the archesporial cell. Embryo sac formation is of the *Polygonum* Type. In *Cyanella*, at least, the mature embryo sac forms a tubular chalazal haustorium. Endosperm formation is nuclear (recorded in *Cyanella*). Adventive embryony leading to polyembryony has been observed in *Cyanella*.

The fruit is an apically loculicidal capsule with rather numerous and small seeds. The outer integument of the testa consists of four or more cell layers, the outermost of which collapses and becomes encrusted with a thin layer of phytomelan. The endosperm cells contain lipids and aleurone.

*Chemistry.* The chemistry of Tecophilaeaceae seems to be practically unknown.

*Distribution.* As circumscribed here, Tecophilaeaceae are a mainly Southern Hemisphere group with centres in Chile, South America, (*Conanthera, Tecophilaea, Zephyra*) and in South Africa (*Cyanella* and perhaps *Lanaria,* see below). *Odontospermum* occurs in California.

*Conanthera* (5) (Fig. 73), in Chile, has flowers with six equal, functional stamens. The anthers are connate into a cone-like structure with projecting connectives. *Tecophilaea* (2), found in the same region, has blue flowers with three functional stamens and three linear staminodes. In contrast to the other genera, the stem is leafless in this genus. *Cyanella* (7) (Fig. 73) in South Africa are leafy plants with iridaceous habit and light blue or light yellow flowers in sparse racemes. The number of functional stamens is variable. Six similar functional stamens occur in *C. alba;* three are more common, as in *C. orchiformis,* and in *C. lutea* there is only one functional stamen, five being non-functional and reduced in size.

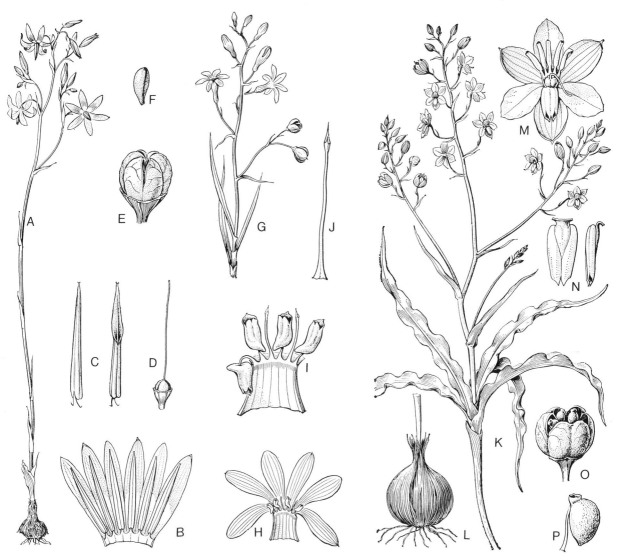

**Fig. 73.** Tecophilaeaceae. **A–F** *Conanthera bifolia*. **A** Plant with corm. **B** Perianth and stamens, spread open. **C** Stamen in different views; note the poricidal dehiscence. **D** Pistil (semiepigyny). **E** Capsule. **F** Seed. **G–J** *Zephyra amoena*. **G** Plant. **H** Perianth tube with attached stamens, spread open. **I** Androecium. **J** Staminode. **K–P** *Cyanella capensis*. **K** Plant. **L** Corm. **M** Flower; note that the upper five stamens are small and functionally sterile. **N** Functional stamen in different views. **O** Capsule. **P** Seed. (All from TAKHTAJAN 1982)

The South African genus *Lanaria* (1) has, according to DE VOS (1963), an embryology which coincides best with the Tecophilaeaceae, although it does not have poricidal anthers. *Lanaria plumosa* (Fig. 74) is a herb with distichous leaves with fairly small semi-epigynous flowers in a dense corymb-like panicle with monochasial components. The tepals are densely woolly. Of the six stamens, the three inner somewhat shorter than the outer and the anthers are peltate as in other Tecophilaeaceae, but open by longitudinal slits. The ovary is subinferior and trilocular, with biovulate locules, but develops into a one-loculed and mostly one-seeded capsule. Its seeds are similar to those of Hypoxidaceae and the stomata are also of the same type (paracytic). However, microsporogenesis is simultaneous and the tapetum secretory. The position of *Lanaria* may best be left open, although a position in Tecophilaeaceae may be considered. – *Lophiola* (1) *aurea*, occurring from New Jersey to Florida, with three stamens, may also belong to Tecophilaeaceae (see ZAVADA et al. 1983).

*Walleria* (1–5), in Southern Africa and Madagascar, is here included in Tecophilaeaceae with great reservations. It is a herb with a corm and with leafy stem and *Solanum*-like, hypogynous

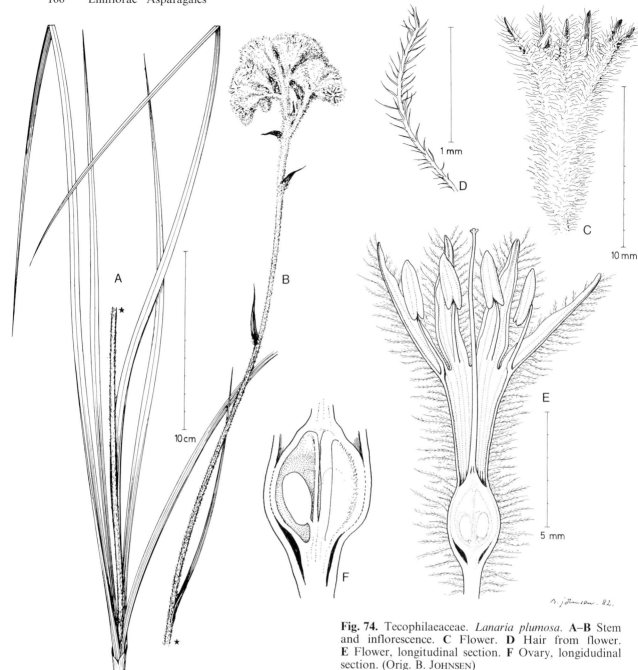

**Fig. 74.** Tecophilaeaceae. *Lanaria plumosa*. **A–B** Stem and inflorescence. **C** Flower. **D** Hair from flower. **E** Flower, longitudinal section. **F** Ovary, longidudinal section. (Orig. B. JOHNSEN)

flowers with somewhat recurved, violet tepals and large, coherent and apically poricidal anthers. The somewhat bean-like seeds are unique in being warty and lacking a phytomelan crust. The outer integument (testa) of the seed coat is multi-layered, and its warts are multicellular, hemispheric (although much longer at the chalazal end). The cell layers of the inner integument have collapsed completely in the seed into a thin membrane (HUBER 1969). A group of apical cells on each wart con-

tinue each into a thin hair, which might suggest affinity to *Eriospermum,* although this is highly uncertain.

*Cyanastrum,* here treated in a separate family, could also with some justification be included in Tecophilaeaceae (see further under Cyanastraceae).

The Tecophilaeaceae are probably most closely related to Cyanastraceae, Ixioliriaceae, Phormiaceae, Hemerocallidaceae and Asphodelaceae.

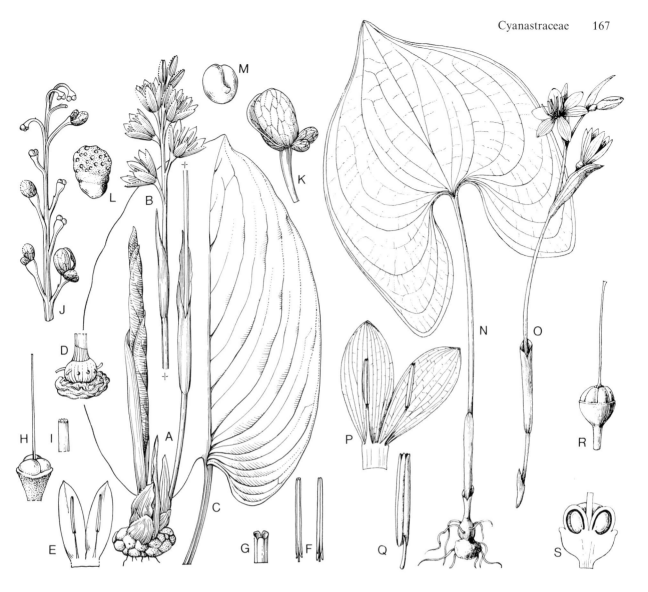

**Fig. 75.** Cyanastraceae. **A–M** *Cyanastrum johnstonii*. **A** Tuber, erect stem and leaf bud. **B** Inflorescence. **C** Leaf. **D** Tuber. **E** Two tepals and opposite stamens. **F** Anther in different view, note the poricidal dehiscence. **G** Apex of poricidal anthers. **H** Pistil. **I** Style apex with stigma. **J** Infructescence. **K** Fruit, this is a schizocarp where two mericarps have not come to development. (CARTER 1969) **N–S** *Cyanastrum cordifolium*. **N** Tuber and foliage leaf. **O** Inflorescence. **P** Two tepals and the opposite stamens. **Q** Stamen. **R** Pistil. **S** Young fruit, longitudinal section. (HEPPER 1968)

## Cyanastraceae Engler (1900)   1:6   (Fig. 75)

Herbs with a thick corm constricted at intervals. The leaves are few and concentrated basally, the lowest being usually bract-like, one or more being large, green and dorsiventral, sheathing the base and in some species with the base of the lamina constricted into a pseudopetiole. The lamina is lan-ceolate or ovate-cordate and in some species distinctly parallel-veined with fine transverse veinlets. Vessels are confined to the roots and have scalariform perforation plates. Raphides are lacking. The stomata are paracytic. The inflorescence is situated on an erect leafless scape which is basally subtended by a bract-like or spathe-like cataphyll. It represents a panicle, a raceme or a thyrse.

The flowers are bisexual, actinomorphic, trimerous and half-epigynous. The 3 + 3 tepals are similar in size and shape, generally blue, and often spreading but basally fused into a short tube, at the mouth of which the 3 + 3 stamens are inserted. These have short, often S-curved filaments and basifixed, elongate anthers which dehisce by an apical pore or a short slit. The tapetum is secretory and microsporogenesis is simultaneous. The pollen grains are finely reticulate and sulcate or sometimes trichotomosulcate (ERDTMAN 1952).

The pistil, which is only half-superior, is tricarpellary and trilocular, with a central simple, almost gynobasic style bearing a terminal punctiform stigma. The locules are separated from each other and from the style in a way reminiscent of the boraginaceous or ochnaceous pistil. Septal nectaries are present. Each of the locules contains two anatropous, basally inserted ovules with a funicular obturator. A primary parietal cell is cut off from the archesporial cell. Embryo sac formation is probably of the *Polygonum* Type. Endosperm formation is probably nuclear (helobial according to some reports). An extraordinary condition is that the tissue in the chalazal part of the ovule, opposite the micropyle, enlarges by cellular division to form a chalazosperm (FRIES 1919), in which the cells are swollen and filled with compound starch grains. The chalazosperm functionally but not histologically corresponds to a perisperm, which it resembles in being diploid.

In the three-lobed ovary only one seed normally ripens, the two remaining ones degenerating. The pericarp is membranous, and the fruit may be classified as a mericarp of a schizocarp. The testa consists of several layers of isodiammetric cells and completely lacks phytomelan. The embryo is depressedly globose and fills up about half of the seed. The chalazosperm is well-developed and starchy, whereas the endosperm is almost completely used up during seed development. The embryo has a large, terminal cotyledon and a sunken, lateral plumule. $x = 11, 12$.

*Chemistry.* The family seems to lack accumulation of calcium oxalate and tannins, but schizogenous spaces with an oil-like excretion are present in the leaves. They probably represent so-called cyanocysts (tissue masses enclosing anthocyanic contents).

*Distribution, Taxonomy.* The family consists of the genus *Cyanastrum* (6), in tropical Africa, which grows in shady forests. The species are quite variable in the shape of the leaves and the size, shape and number of flowers of the inflorescence.

The relationships of the family are somewhat uncertain. Some features, such as the corms, the petiolate leaves with distinct main veins and fine transverse veinlets discernible in between, and the simultaneous microsporogenesis, suggest an affinity with various Dioscoreales. The closest affinity is, however, no doubt with Tecophilaeaceae, with which *Cyanastrum* shares the corm, the general floral construction, the blue tepal colour, the basifixed poricidal anthers, the glandular tapetum, the simple style, and the simultaneous microsporogen-

esis. There is also some similarity to Eriospermaceae and Phormiaceae.

It is interesting to note that in *Walleria,* which is considered by HUBER (1969) to be related to *Cyanastrum* and the Tecophilaeaceae, the tissue of the raphe is rich in starch grains. In the Eriospermaceae, where chalazosperm is poorly developed, the endosperm, as in Cyanastraceae, is used up during the ripening of the seed, and the embryo is large and well developed. *Eriospermum* also shows some vegetative similarities to *Cyanastrum,* such as the corm or tuber and the solitary or few broad leaves.

## Eriospermaceae S.L. Endlicher (1836)    1:80 (Fig. 76)

Perennial herbs with a single, globose or sausage-like tuber or with a complex of tubers and stolons. The cut surface of the tubers is opaquely white, yellow, pink or red. In addition to some scale-like, reduced basal leaves there is a single leaf or there are up to three, rarely more, basally concentrated dorsiventral leaves each with a well-developed lamina. The leaf lamina varies from linear to ovate or cordate and sometimes lies flat on the ground; it is parallel-veined and either glabrous or clothed with simple or compound hairs. Of particular interest is the frequent presence of complex enations or a pubescent deeply dissected appendage at the base of the leaf blade. This appendage may have the shape of a "bottle brush", it may consist of a few filiform threads, or may resemble a bunch of feathers, the leaf itself being in these cases very small (MARLOTH 1915). In *Eriospermum dregei* and *E. paradoxum* the appendage is almost dichotomously branched, with linear, thick and fleshy segments (Fig. 76 K).

The inflorescence-bearing stem is an erect scape, leafless or with bracteal leaves, ending in an often sparse raceme with bracteate and sometimes long-pedicelled flowers. It develops in the summer after the green, assimilating (leaf or) leaves have withered. Fasciculate hairs, matched only in Haemodorales, are found on the inflorescence in some species (HUBER 1969). The flowers are hypogynous, actinomorphic, trimerous and bisexual, those uppermost in the raceme being sometimes minute and sterile. Their 3+3 tepals are free from each other, white, pink or yellow, and persistent, the outer being upright or spreading. The 3+3 stamens are adnate at the base to the tepals and have narrow or flat filaments and peltate, introrse, lon-

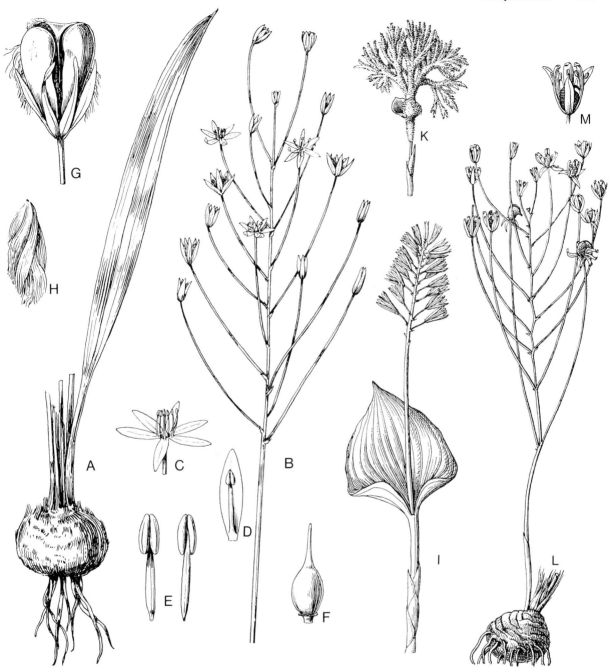

gitudinally dehiscent anthers. Tapetum is glandular and microsporogenesis is successive. The pollen grains likewise are poorly known but probably sulcate.

The tricarpellary, trilocular pistil has an erect simple style and a punctiform stigma. Each locule contains a few axile ovules, which are anatropous, bitegmic and weakly crassinucellate. A parietal cell is cut off from the archesporial cell. Embryo sac formation is of the *Polygonum* Type. Endosperm formation is nuclear(?), but the endosperm is used

**Fig. 76.** Eriospermaceae. **A–I** *Eriospermum abyssinicum.* **A** Corm with one leaf. **B** Inflorescence. **C** Flower. **D** Tepal and stamen attached to its base. **E** Stamens. **F** Pistil. **G** Capsule in state of dehiscence. **H** Seed with pubescence. (HEPPER 1968) **I** *E. majanthemifolium,* shoot. **H** *E. paradoxum,* young plant with one foliage leaf and, opposite this, an appendage of the unique kind found in part of this genus. **L–M** *E. abyssinicum.* **L** Plant, in flowering and fruiting stage. **M** Flower, in late stage. (**L–M** from KRAUSE 1930)

up in the course of seed ripening. The embryo becomes large and conical-cylindrical. Embryo formation follows the Solanad Type. Nucellar tissue envelops the chalazal parts of the embryo (LU, unpublished). The embryology is thus very unusual.

The fruit is a loculicidal, subglobose or often three- or six-lobed capsule with a few seeds. These are unique in the monocotyledons in being clothed with long, unicellular hairs consisting of epidermal cells which have grown up to a length of 8 mm, and form processes filled with air. The hairs are white or reddish brown. The seeds are pear-shaped to narrowly ovate and their testa lacks phytomelan. In addition to the hirsute epidermis the outer integument of the testa consists of two cell layers, which may be compressed and colourless or pigmented with phlobaphene (HUBER 1969). The endosperm of the ripe seed has thus been used up in connection with the extensive embryo, and nucellar tissue forms a cap around part of the embryo. This tissue accumulates fatty oils and aleurone, but no starch.

The chemical contents are otherwise not known.

*Eriospermum* (80) is confined to Africa with about 50 of its species concentrated in Southern Africa. Because the leaves and inflorescence are developed at different times, the taxonomy is somewhat difficult. The tubers of *Eriospermum* are edible. They contain mucilage but no starch or solid protein.

The genus is extraordinary in many respects, such as the occurrence of the peculiar leaf appendages, the epidermal hairs of the seed and the embryological attributes. HUBER (1969) associates *Eriospermum* with *Walleria* and *Cyanastrum* on the basis of seed characters, but the successive microsporogenesis and thin testa argue against this. The relations are still uncertain and *Eriospermum* is probably best treated as a separate family.

**Ixioliriaceae** T. Nakai (1943)   1:1–4   (Fig. 77)

A perennial, erect herb up to ca. 60 cm tall. The underground stem is developed as a bulb-shaped, tunicated corm, which seems to range over more than one internode (lateral corms being developed at some distance from its base; Fig. 77 F). The aerial stem is erect and leafy with the dispersed leaves flat and linear, sheathing at the base and supplied with a cylindrical-subulate apex. The tissues are rich in raphides (OGANEZOVA 1981; ARROYO 1982). The stomata are anomocytic.

The inflorescence is a panicle varying much in degree of branching and number of flowers; it is often few-flowered and quite frequently umbel-like. The flowers are bisexual, actinomorphic, trimerous and epigynous, superficially similar to those in Amaryllidaceae, in which *Ixiolirion* is often placed. The linear-oblanceolate tepals of the two whorls are similar in size and shape, the outer somewhat narrower than the inner, and with a more prominent point emerging slightly below and outside the apex. The tepals are blue, bluish violet, pale-blue or nearly white, and are free from each other. There are 3 + 3 stamens, with their narrow filaments attached basally to the tepals opposite which they are inserted. Their anthers are basifixed, tetrasporangiate, introrse-latrorse and longitudinally dehiscent. Microsporogenesis is simultaneous, and the pollen grains are sulcate.

The style is erect, slender and apically tribrachiate, with Dry stigmatic surfaces. The ovary is inferior and trilocular, with rather numerous axially inserted, anatropous, crassinucellate ovules. A parietal cell is cut off from the archesporial cell. Endosperm formation is helobial. The fruit is a capsule with numerous seeds. These are ovate to pear-shaped, reticulate and have a testal epidermis with phytomelan, making them black. Apart from the elongate, hexagonal epidermal cells, the testa consists of several cell layers with thin reddish brown walls; the tegmen is thin and membranous and the endosperm cells thin-walled and starch-free. The embryo is straight and fusiform and nearly as long as the endosperm (HUBER 1969). $x = 12$ (FEDOROV 1969).

*Chemistry*. Unlike the Amaryllidaceae, *Ixiolirion* lacks alkaloids. Chelidonic acid is present, but there seems to be no information on steroidal saponins.

The Ixioliriaceae consist of the single genus *Ixiolirion* (1–4) which is distributed in South-West Asia (WENDELBO 1970). Some taxonomists claim that the genus contains only one species.

Ixioliriaceae is to be separated from Amaryllidaceae on the basis of its corm, inflorescence structure and lack of alkaloids. Rather it could be most closely related to some Alliaceae (e.g. Brodiaeeae), to Tecophilaeaceae or Phormiaceae. The combination of simultaneous microsporogenesis and corms is important for drawing this conclusion. A close link with Liliaceae or Alstroemeriaceae is excluded because of the seeds which are coated with a phytomelan layer.

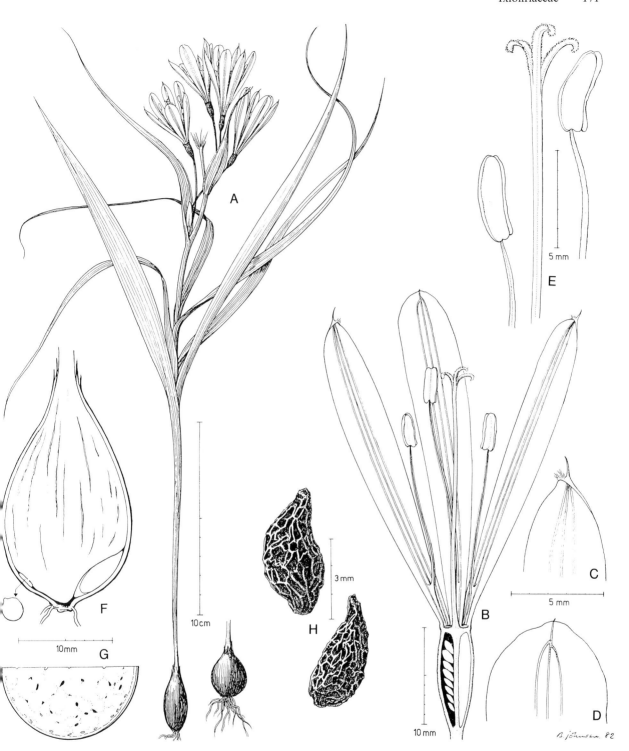

**Fig. 77.** Ixioliriaceae. *Ixiolirion montanum* s. lat. **A** Plant. **B** Flower, longitudinal section. **C** Apex of outer tepal. **D** Apex of inner tepal. **E** Style and two stamens, one of each whorl in the relative positions they have in the flower. **F** Corm; note the lateral new corms formed on both sides in the lower part. **G** Corm, transverse section. **H** Seeds, black from the phytomelan layer. (All orig. B. JOHNSEN)

**Phormiaceae** J.G. Agardh (1858)    7:32
(Figs. 78, 79)

Medium-sized to tall, rhizomatous herbs or sub-shrubs without secondary thickening growth. Rhizome often thick, with fascicled, fibrous roots. The leaves are distichously inserted and concentrated on the base and lower part of the stem. They are linear, often rigid and tough, and the lowest are developed as scale-like cataphylls. The foliar leaves are long, linear, parallel-veined, basally strongly compressed and nearly unifacial, otherwise dorsiventral and V-shaped or flat. Short antrorse teeth may be present on the leaf margins and midribs, making them rough or serrulate. Raphides are often present except in *Phormium,* where suberized cells with pseudo-raphides are present, a similarity to Nolinaceae. Stomata are of the anomocytic type and more or less sunken. The fibres of the leaves are sometimes conspicuous but not very strong (they are exploited in *Phormium,* as "New Zealand Flax"). Vessels seem to be lacking in the stem and leaves, but are present in the roots, and have simple and scalariform perforation plates.

The inflorescence, which is terminal on the stem, is a sparingly to densely branched panicle, which varies from rather few-flowered to very many-flowered. The pedicel is sharply demarcated from the perigone (Fig. 79 G). The flowers are bisexual, actinomorphic or slightly zygomorphic, hypogynous or nearly so (*Phormium*). There are 3 + 3 tepals, which are free or nearly free from the base, the outer often slightly smaller than, and sometimes different in texture and colour from the inner, which are more petaloid and often have rounded and/or recurved tips. The tepal colour varies from (greenish) yellow to red (*Phormium*), but is more often blue, violet or white.

The 3 + 3 stamens have narrow, basally somewhat thickened filaments. Rarely, they are fused at the base into a ring (*Excremis*). Sometimes they are of unequal length (*Phormium*). They may also have papillate or villous hairs (*Stypandra*). In *Dianella* and some other genera they are thickened apically near the attachment point of the anther to form a globose or ovoid structure covered with nectar-secreting papillae. The anthers are elongate, basifixed or dorsifixed-epipeltate, and dehisce with apical pores (as in *Dianella* and *Stypandra*) or longitudinal slits. The tapetum is glandular and microsporogenesis simultaneous. All the genera seem to have trichotomosulcate pollen grains.

The pistil is tricarpellary and trilocular and has an erect, narrow style, which may be slightly up-curved in lateral flowers of *Phormium*. The stigma is punctiform and has a Dry surface. The locules each contain from four to rather numerous anatropous (epitropous) ovules which are weakly crassinucellate. No parietal cell is formed as the archesporial cell functions directly as the megaspore mother cell; periclinal divisions of the nucellar epidermis may (*Stypandra, Phormium*) or may not (*Dianella*) occur, in the former case forming a nucellar cap three to four cells in thickness. Embryo sac formation follows the *Polygonum* Type, and endosperm formation is helobial (CAVE 1955, 1975).

The fruits are either capsules or berries, the capsules in *Phormium* being elongate and often curved, those in other genera shorter. The berries in *Dianella* are often blue. The seeds are ovoid in most genera but elongate, elliptic-oblong and surrounded by a wing in *Phormium*. They are always provided with a black layer of phytomelan (even where the fruit is baccate). The seed coat has a multilayered outer integument with compressed but not collapsed cells, and a tegmen which has collapsed to form a thin membrane. The endosperm cells are relatively thin-walled and contain aleurone and lipids. The embryo varies from only about a third of the length of the endosperm to almost the same length. The basic chromosome number is $x = 8$.

*Chemistry.* The chemical content seems to be little known. Steroidal saponins seem to be lacking in *Phormium;* the rhizome in this genus contains wax (ca. 1%) and its leaves contain "rubber", a hemicellulose-like polysaccharide.

*Distribution.* Phormiaceae occur mainly in South-East Asia and Australia and on the Pacific Islands, including New Zealand, *Dianella* extending, however, westwards to Africa, Madagascar and the Mascarene Islands. *Excremis* occurs in tropical South America.

*Dianella* (25) is widely distributed in the Old World tropics. It is a remarkable genus in its blue (to white) flowers and its often blue berries. Its branches are herbaceous or somewhat woody and the panicles are slender and often much branched. *D. ensifolia* (Fig. 78 A–D) reaches a height of up to 2 m. Another species with wide distribution is *D. nemorosa,* ranging from Indomalesia to Madagascar and East Africa. *Stypandra* (3) in Australia has woolly filaments and capsular fruits. Some species are toxic if grazed. – *Agrostocrinum* (1), in South-Western Australia, has a capsular fruit

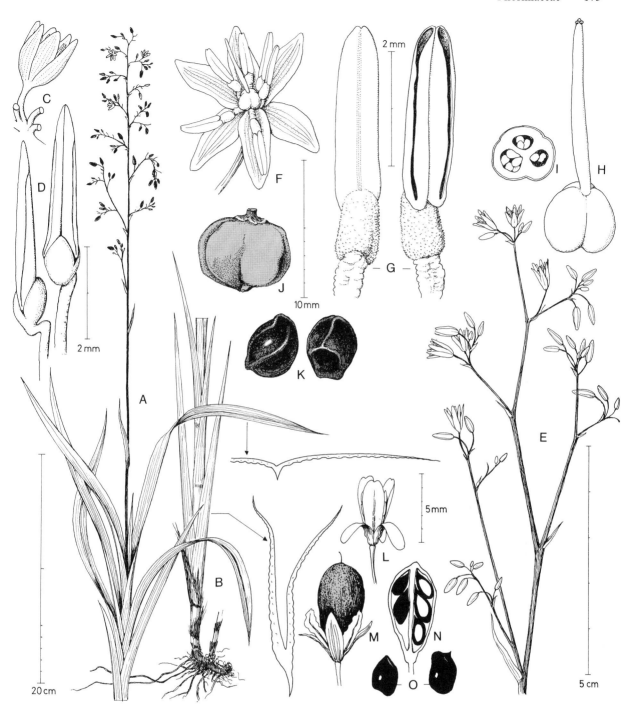

enclosed by the persistent base of the perianth. – *Excremis* (1) is a plant of the paramos from Peru to Colombia growing up to 1.5 m tall. – *Phormium* (2), occurs primarily in New Zealand but the best-known species, *P. tenax*, is also found on Norfolk Island. It is a large herb with leaves up to 2.5 m long and an inflorescence axis 2–4.5 m long; it grows mainly in lowland swamp habitats.

**Fig. 78.** Phormiaceae. **A–D** *Dianella ensifolia* (specimen from Kew). **A–B** Plant, transection of leaves shown to the *right*. **C** Flower. **D** Stamen in different views. **E–K** *Dianella* sp. **E** Inflorescence. **F** Flower. **G** Anther, different views. **H** Pistil. **I** Ovary, transverse section. **J** Berry. **K** Seeds, note the black colour caused by the phytomelan layer. **L–O** *Dianella nigra*. **L** Flower. **M** Berry. **N** Same in longitudinal section. **O** Seeds. (Orig. B. Johnsen)

**Fig. 79.** Phormiaceae. **A–B** *Phormium cookianum*. **A** Inflorescence and leaf. **B** Flower. (Redrawn from EVERARD and MORLEY 1970) **C–E** *Phormium tenax*. **C** Flower bud. **D** Stamen from flower bud. **E** Pistil, in same bud. (Orig. from Kew Gardens) **F–I** *Phormium cookianum*. **F** Capsule, transverse section. **G** Lateral branch of infructescence; note the sharp delimitation ("joint") between pedicel and flower. **H** Seeds. **I** Same, transverse section. (Orig. from Kew Gardens; B. JOHNSEN)

The species is known as "New Zealand Flax" and was used by the Maoris as a source of fibre for making cloth and cord. At present it is grown commercially on a moderate scale in New Zealand, the U.S.A., Central Africa, Mauritius, etc., and is used mostly for cordage and sacking material. A smaller species, *P. cookianum* (Fig. 79 A–B), with which it hybridizes, grows in open heath and montane scrub.

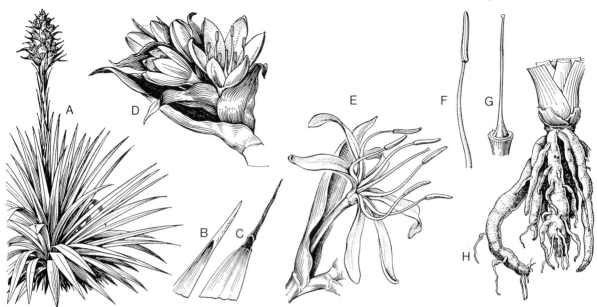

**Fig. 80.** Doryanthaceae. **A–D** *Doryanthes palmeri*. **A** Plant. **B–C** Apex of young and old leaf. **D** Branch of the thyrsoid inflorescence. **E–H** *Doryanthes excelsa*. **E** Flower. **F** Stamen. **G** Pistil. **H** Root system. (All from TAKHTAJAN 1982)

It is possible that the genus *Xeronema* (2) (EVERARD and MORLEY 1970), on New Zealand and New Caledonia, consisting of large herbs with one-sided panicles of reddish flowers, is closely related to *Phormium* and also belongs to Phormiaceae, although further studies on its embryology and pollen morphology are needed. Alternatively, a position in or near Asteliaceae or Doryanthaceae is to be considered.

Phormiaceae, which is here circumscribed as by TAKHTAJAN (1980), is probably rather closely related to Doryanthaceae in which, however, the leaves are spirally set and not basally compressed, and the pollen grains are not trichotomosulcate. Both families have developed large flowers with the bird-pollination syndrome and have the simultaneous microsporogenesis and nucellar cap in common. Although these details do not convincingly prove a close relationship the shared geographical distribution may be regarded as a support.

**Doryanthaceae** Dahlgren & Clifford, fam. nov. 1:3 (Fig. 80)

Plantae giganteae foliis in rosula maxima spiraliter aggregatis. Folia linearia, crassa, carnosa, dorsiventralia, ad 1.5 m longa vel longiora, apice tereti infuscato vel ferrugineo quasisuberoso munita. Contextus partium vegetativarum sine rhaphidibus pseudorhaphidibusque. Stomata paracytica. Caules floriferi usque ad 5 m alta, recti, rigidi, foliis reductis instructi et inflorescentiis thyrsoideis elongatis vel subcapitatis terminati. Flores diametro 10–15 cm, laete rubri, raro candidi, epigynici, actinomorphi vel parum zygomorphi, bisexuales. Tepala patentia, interiores ad basin leviter dilatati. Stamina 3 + 3, antheris elongatis pseudobasifixis, longitudinaliter dehiscentibus. Microsporogenesis simultanea. Grana pollinis sulcata. Ovarium triloculare; stylus simplex trisulcatus, in stigma triangulare terminans. Nectaria septalia praesentia. Ovuli quoqueloculo plures vel multi, anatropa, crassinucellata. Saccus embryonis typi Polygoni formans; endospermium helobiale. Fructus capsula loculicidalis. Semina aliquot compressa et alata vel conica, sectione transversali saepe triangularia, sine phytomelano. Endospermium olea pinguia et aleuronem continens; embryo obconicus.

Genus typicus: *Doryanthes* Corréa.

Giant rosette plants with a short basal stem. The leaves are spirally set, linear, thick, dorsiventral, up to more than 1.5 m long, with a terete, brownish tip, 1–2 cm long. The leaves lack raphides as well as pseudoraphides, but cells with twin crystals are common (raphides, however, are found in the tepals). The stomata are of the paracytic type (BLUNDEN and JEWERS 1973).

The inflorescence is borne on straight, stiff aerial branches up to 5 m long, with numerous short leaves; it is a thyrse, elongate or sometimes globose, with numerous, large flowers, which may be substituted by bulbils. The flowers are 10–15 cm long, with tepals varying in colour from nearly white to bright maroon-red. The flowers are epigynous, actinomorphic or slightly zygomorphic and bisexual. The six tepals are spreading, the inner being slightly dilated at the base.

The 3 + 3 stamens have linear-subulate filaments, broadening towards the base, and elongate anthers which are "pseudobasifixed", i.e. in reality peltate but with the apex of the filament enclosed in a tube formed by the connective (as in *Tulipa,* for example). Anther dehiscence is longitudinal. The tapetum is of the Secretory Type. Microsporogenesis is simultaneous (as in Phormiaceae; CAVE 1955). The pollen grains are sulcate.

The inferior ovary is trilocular and the style is simple, with three furrows, and bears a triangular stigma. Septal nectaries are present. There are several to many, anatropous, crassinucellate ovules in each locule. A primary parietal cell is cut off from the archesporial cell and, additionally, the nucellar epidermis divides periclinally to a nucellar cap (CAVE 1955). The embryo sac formation follows the *Polygonum* Type, and the endosperm formation is helobial (NEWMAN 1928–29).

The fruit is a loculicidal capsule with several seeds per locule. The seeds vary from somewhat compressed and winged (*D. palmeri*) to conical; they are often triangular in transection and elongated transversely. They lack phytomelan entirely. The testal layer of the seed coat consists of seven to many cell layers; its epidermis consists of isodiametric cells, reddish brown (with phlobaphene). The inner integument is thicker than in Agavaceae and has a distinct cuticle. The endosperm consists mainly of isodiametric cells storing aleurone and fatty oils, but not starch. The embryo varies in size and has a broadened obtriangular cotyledon (HUBER 1969). Saponins are now known to occur in the family.

Doryanthaceae consists of the single genus *Doryanthes* (3) in Queensland and New South Wales, Australia, impressive rosette plants with large, brightly coloured flowers. They are known as "Spear Lilies", and are sometimes grown as ornamentals.

The affinity of Doryanthaceae is somewhat uncertain. Possibly it comes closest to Phormiaceae. Doryanthaceae was first suggested as a family by HUBER (1969) and was also acknowledged by TAKHTAJAN (1980), but has not formally been published.

## Hemerocallidaceae R. Brown (1810)   1:16
(Fig. 81)

"Lily-like", glabrous herbs to ca. 1 m tall with short rhizomes and fleshy, swollen, sometimes nearly fusiform roots. The larger leaves are concentrated at the base, and are dorsiventral, linear, mesomorphic, sheathing at the base and parallel-veined. The stomata are anomocytic. Crystal raphides are present. Vessels are lacking in stem and leaves; those of the roots have scalariform perforation plates (CHEADLE and KOSAKAI 1971).

The inflorescence is borne on a bracteate, but otherwise largely leafless scape, and is considered to represent one (or two) double helicoid cymes (KRAUSE 1930). The flowers are fairly large and few (mostly 5–12) and are not delimited from the pedicel by a joint. They are hypogynous, trimerous and funnel-shaped (*Lilium*-like) in outline, having 3 + 3 petaloid, yellow to orange or brick-red tepals, which are basally connate into a tube. They may be striped (but not variegated with a drop-like pattern) and may be recurved apically.

The stamens, which are inserted in the tubular part of the perigone, are all slightly upcurved, making the flowers zygomorphic. Their filaments are long, glabrous and free from each other; the anthers are epipeltate-versatile, often twisted, and dehisce longitudinally. The tapetum is secretory. Microsporogenesis is simultaneous (CAVE 1955). The pollen grains are sulcate, two-celled and rather similar to those in Funkiaceae.

The pistil is trilocular and has a long, slender, slightly upcurved style and a punctiform-capitate stigma with a Wet surface. Septal nectaries are present in the rather triangular ovary. Each of the locules contains numerous anatropous ovules, in which the archesporial cell functions directly as the megaspore mother cell. There are no periclinal divisions in the nucellar epidermis. The embryo sac formation follows the *Polygonum* Type, and the endosperm formation is nuclear.

The fruit is a loculicidal capsule opening from the apex. The seeds are subglobose to prismatic or slightly elongate (not flat) and have a smooth, shiny black epidermal layer encrusted with phytomelan; the inner layers of the outer integument are compressed and rust-coloured, while the inner integument is collapsed. The endosperm contains aleurone and fat, but not starch, and the embryo is of about the same length as the endosperm.

The chemistry seems to be unknown.

It is interesting to note that the chromosomes ($x =$ 11) are not dimorphic, as they are in *Hosta,* with which *Hemerocallis* is sometimes associated.

The family Hemorocallidaceae ("Day Lilies") consists of the single genus *Hemerocallis* (16), distributed mainly in temperate regions of Asia but extending into southern Europe. *H. fulva* and several garden hybrids with different perigone colours are grown as ornamentals and are reminiscent of

**Fig. 81.** Hemerocallidaceae. **A–E** *Hemerocallis fulva.* **A** Roots and young shoot. **B** Inflorescence. **C** Flower bud, longitudinal section. **D** Anther in different views. **E** Ovary, transverse section, note the septal nectary cavities. **F–H** *H. middendorfii.* **F** Infructescence. **G** Opening capsule, one valve removed. **H** Seeds. (Orig. B. JOHNSEN)

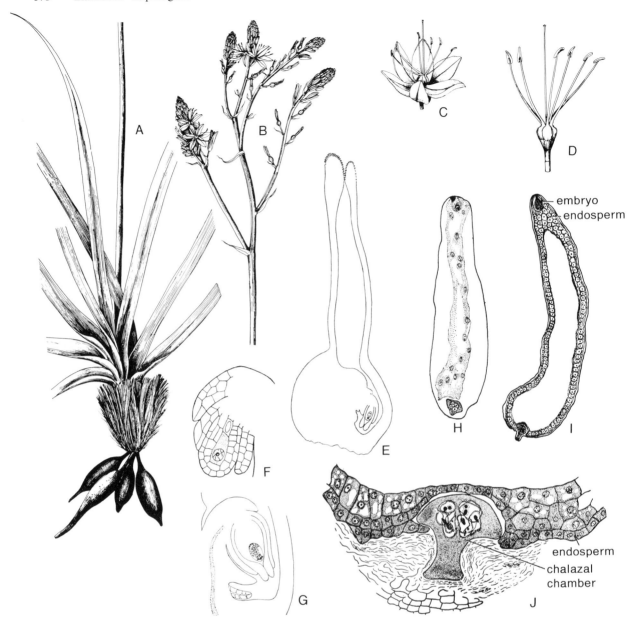

certain lilies (*Lilium*), from which they are easily distinguished by their rhizome and the black, prismatic to rounded (not pale brown and flat) seeds.

The closest affinities of *Hemerocallis* are uncertain. Although in seed coat structure the genus agrees with the Alliaceae (HUBER 1969) its simultaneous microsporogenesis, the lack of parietal tissue, and its inflorescence type exclude it from that family. Comparisons with the Asphodelaceae, Anthericaceae and Funkiaceae suggest no valid reasons for including *Hemerocallis* in any of these families, most features being perhaps concordant with Asphodelaceae subfam. Asphodeloideae.

**Fig. 82.** Asphodelaceae. **A–D** *Asphodelus aestivus.* **A** Lower part of plant. **B** Inflorescence. **C** Flower. **D** Androecium and gynoecium, the filament bases widened to conceal ovary. (JAFRI and EL-GADI, 1978). **E** *Asphodelus fistulosus,* carpel in early stage, showing ovule with beginning of aril formation. (STENAR 1928a). **F** *Echeandia ternifolia.* Ovule. (SCHNARF and WUNDERLICH 1939). **G** *Bulbine annua,* anatropous ovule with aril. (STENAR 1928a). **H–J** *Asphodelus tenuifolius,* endosperm formation. **H** Helobial endosperm, the lower dark portion being the chalazal chamber. **I** Later stage; wall formation has taken place in the micropylar chamber. **J** Details of same section, showing the chalazal chamber. (EUNUS 1952)

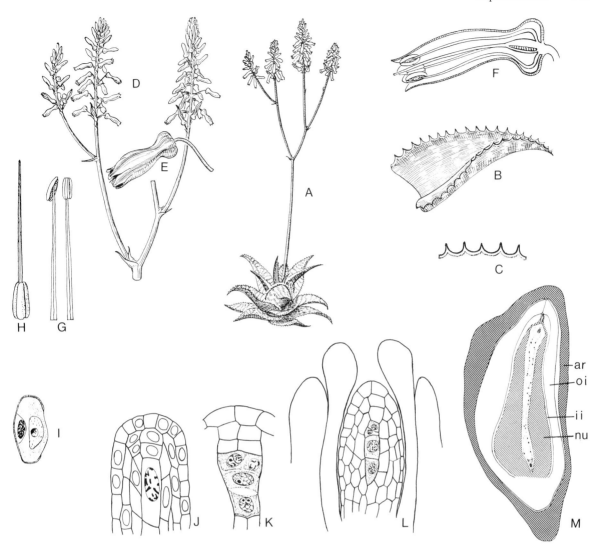

**Fig. 83.** Asphodelaceae. **A–H** *Aloë macrocarpa.* **A** Plant. **B** Apical part of leaf. **C** Leaf margin. **D** Part of inflorescence. **E** Flower. **F** Same, longitudinal section. **G** Stamens. **H** Pistil. (HEPPER 1968). **I** *Aloë humilis,* pollen grain (sulcate, two-celled). **J** *A. ciliaris,* nucellus with parietal cells. **K** *A. variegata,* T-shaped megaspore tetrad. **L** *A. brownii,* linear megaspore tetrad. **M** *Kniphofia praecox,* ripe seed (*ar* aril; *oi* outer integument; *ii* inner integument; *nu* nucellus; the endosperm, with a small basal chamber, is seen in the *centre*). (**I–M** from SCHNARF and WUNDERLICH 1939)

**Asphodelaceae** A.L. Jussieu (1789)    18:750–800
(Figs. 5, 82, 83, 84 G–L, 89 Q–T)

Mostly herbs, but also woody forms provided with a trunk up to several metres high. The latter forms bear rosettes of succulent leaves at the ends of the branches, whereas the leaves are otherwise normally radical. Secondary thickening growth occurs in species of *Aloë* and *Kniphofia.* The roots sometimes (e.g. in *Asphodelus*) are inflated and fusiform; multi-layered velamen is known in some genera. The leaves are generally dorsiventral and often thick and succulent; they may be terete but are never ensiform. They are usually spirally set (occasionally distichous, as in several species of subfam. Alooideae), from linear or subulate to thickly conical or even elliptic, sheathing at the base and, in the succulent genera, often serrate-dentate and apically spiny. The veins are longitudi-

nal but are indistinct or invisible when the leaves are succulent. The stomata are generally anomocytic or tetracytic, rarely paracytic. Vessels, with scalariform perforation plates, are only occasionally present in the stem (species of *Asphodelus*). Raphide cells are abundant. So-called aloine cells, which are parenchymatous cells arranged as a cap at the phloem pole of most vascular bundles of the leaf, occur in many taxa, and are characteristic of subfam. Alooideae. These cells secrete a range of anthraquinones and other substances. In subfamily Alooideae the aloine cells generally contain coloured secretions. The parenchyma cap may be replaced by sclerenchyma. In other genera of the family (i.e. taxa of subfam. Asphodeloideae) typical caps of aloine cells are lacking, but there may be a large parenchymatous region at the xylem pole of the vascular strands (e.g. in *Kniphofia*). Tannin cells then replace cells with anthraquinones (CUTLER, personal communication).

The inflorescences are simple or compound racemes or spikes. They are situated on the ends of terminal peduncles, which are sharply demarcated from the vegetative part of the shoot and are either nearly leafless or beset with bracteal leaves (densely so in *Jodriella,* for example). The growth of the shoot is often sympodial, so that the inflorescence-bearing axes, which are terminal, are pushed aside and appear lateral. As in the Anthericaceae, the flowers may have a basal "pericladium".

The flowers are hypogynous, trimerous and normally bisexual. In certain genera (e.g. *Asphodelus*) the tepals are free or nearly free from each other, while in subfam. Alooideae they are often more or less fused with each other into a tube. The perigone may then be conspicuously zygomorphic, with the apical parts (together with the filaments) somewhat curved (*Gasteria*) or slightly two-lipped (*Haworthia*). The tepals vary much in colour, from white or rose-coloured to bright red and/or yellow or pale purple and greenish (but not blue or blue-violet); in large-flowered, ornithogamous forms they may be rather stiff. The inner tepals may be more or less different from the outer.

The stamens are $3+3$ in number and generally inserted at the base of the ovary. The filaments are linear and free from each other, and rarely (*Bulbine, Jodriella, Simethis*) provided with hairs. The anthers are dorsifixed-epipeltate and introrse and dehisce longitudinally. Microsporogenesis is simultaneous (a noteworthy difference from Anthericaceae, where it is successive). The pollen grains are sulcate and two-celled when dispersed.

The pistil is syncarpous, tricarpellary and trilocular and has a simple, long style with a small stigma of the Dry or, more rarely, of the Wet Type. Septal nectaries are present in the ovary walls. Placentation is axile. The ovules (Fig. 5) are two to rather numerous per locule, anatropous, hemianatropous or (as in species of *Aloë* and *Asphodelus*) nearly orthotropous, and crassinucellate, with a parietal cell usually cut off from the archesporial cell. Embryo sac formation conforms to the *Polygonum* Type, and endosperm formation is of the Helobial Type (Fig. 89 Q–T). An aril frequently develops as an annular invagination at the distal part of the funicle and envelops the seed to a variable extent during its growth, as if it were a "third integument" (SCHNARF and WUNDERLICH 1939). The aril is two or more cells thick; it is well developed in *Bulbine, Kniphofia* and the genera of subfam. Alooideae; in *Kniphofia* it is supplied with purple pigment; an aril, although thinner, is also present in *Asphodelus, Verine, Bulbine* and *Bulbinopsis*.

The fruit is a loculicidal capsule, which is carnose in *Lomatophyllum*. The generally arillate seeds (see above) are elongate and ovoid, winged in some species of *Eremurus*. The outer epidermis of the outer integument develops a black phytomelan crust while the layers of the inner integument become more or less collapsed. The endosperm cells store lipids and aleurone, and the embryo is straight, linear and generally at least three-quarters the length of the endosperm. $x=7$ (in *Kniphofia* 6).

*Chemistry.* Chemically the Asphodelaceae differ from the Anthericaceae in lacking steroidal saponins and in synthesizing, quite frequently, anthraquinones, which are accumulated in the aloine cells mentioned above. Chelidonic acid is often present. The underground organs accumulate fructans and other carbohydrates but not starch.

*Distribution.* The family consists of ca. 17 genera and has a wide Old World distribution with a clear centre in Southern Africa, in particular subfamily Alooideae.

## Subfamily Asphodeloideae

This consists of a number of herbaceous genera with mesomorphic to succulent or otherwise xeromorphic, rarely subterete, leaves and a central, branched or unbranched, pedunculate inflorescence. The plants are rhizomatous and may have thickened storage roots (as in species of *Asphodelus,* Fig. 82 A). Typical caps of aloine cells are obviously not formed at the phloem poles of the vascular strands of the leaves, not even in *Kniphofia*, which has previously been referred to the *Aloë* group of genera (see below). The stomata are generally anomocytic. The flowers are usually white or pink, but are yellow in, for example *Bulbine,* and yellow and red in most species of *Kniphofia*.

The locules of the ovary have two or, in *Kniphofia, Eremurus* and *Trachyandra,* more ovules. An aril is present in some genera, but it is often thin and devoid of lipids. Although the information is incomplete, anthraquinones seem to be present in some taxa of the subfamily, e.g. in *Simethis* (HEGNAUER 1963; GIBBS 1974); aloine reactions are also reported in *Bulbine* (HEGNAUER 1963). Tannin contents rather than anthraquinones are, however, found in the mesenchymatous cells at the vein endings of the leaves in *Kniphofia* (CUTLER, personal communication).

The Asphodeloideae form a varied group of genera, often with a less African concentration than in the Alooideae. *Eremurus* is Asiatic with most species in the U.S.S.R. and Afghanistan and *Asphodelus* is Mediterranean and West Asiatic; *Trachyandra, Bulbine* and *Kniphofia* are, however, centred in Southern Africa.

*Asphodelus* (12) (Fig. 82) is chiefly Mediterranean-West Asiatic and includes small to fairly tall herbs with white or pink flowers. The roots are swollen and fusiform; those of *A. albus* are used for fermenting alcohol, those of *A. ramosus* for gum. *A. aestivus* is a common plant on disturbed ground in the Mediterranean Region; it can be used for dyeing carpets. – Related is *Asphodeline* (20) in the same region. – *Eremurus* (35) is concentrated in the steppes of the high plateaus in central Asia, up to 6,000 m in the Himalayas. It consists of herbs with scapose racemes up to 3 or more metres high with white, pinkish or yellow flowers on rather slender pedicels. The seeds in some species have broad wings. – *Trachyandra* (50), mainly in South Africa, has scabrid filaments but otherwise superficially resembles *Anthericum,* in which it was long included.

*Simethis* (1) (Fig. 84G–L), in western Europe and the western Mediterranean, has often been placed in the *Anthericum* group of genera (our Anthericaceae), but agrees with Asphodelaceae in several respects, on the basis of which we place it here. The white, somewhat *Anthericum*-like flowers have hairy filaments (Fig. 84I) and the thickly melaniferous seeds (HUBER 1969) have an elaiosome formed from the hilum. The locules, as in most Asphodeloideae, contain two ovules each. Anthraquinones (e.g. emodin) are reported from the leaves (HEGNAUER 1963; GIBBS 1974). Certain embryological details, however, agree better with the Anthericaceae.

*Bulbine* and related genera are likewise herbaceous with a single central inflorescence; their leaves are usually not succulent (but sometimes strongly so).

The tepals are free from each other and yellow or rarely white. The inner layer of the aril contains lipids. *Bulbine* (c. 50) and *Bulbinella* (c. 15) occur in Southern Africa, the latter also in New Zealand, and the related, recently described *Jodriella* (3) occurs in tropical Africa (BAIJNATH 1978).

*Kniphofia* (70) in Africa, mostly in the southern parts, are acaulescent or rarely caulescent herbs of medium size, with rosulate, rarely distichous, linear leaves, sometimes with serrate margins. The inflorescences are more or less dense spikes, within which the flowers may vary in size and colour, the upper being often red and spreading or upright, the lower yellow and pendulous (hence the popular name, "Red-Hot Poker"). Rarely, the flowers are white or purple (see CODD 1968) and they may all be short and spreading, *K. breviflora* being conspicuously *Muscari*-like. Some species are grown as ornamentals, e.g. *K. triangularis, K. ensifolia* and various hybrids. *Kniphofia,* with its tubular, usually bright-coloured flowers largely pollinated by birds, is peripheral in subfamily Asphodeloideae and has generally been placed with *Aloë* and other genera of our Alooideae. As shown by BAIJNATH (1980) and CUTLER (personal communication), there are great differences in the anatomical construction of the leaves between *Kniphofia* and the Alooideae, and it seems most appropriate to place *Kniphofia* outside the latter subfamily.

### Subfamily Alooideae

The members of this subfamily vary from minute plants with a rosette of succulent leaves, as in *Haworthia,* to giant trees branching pseudo-dichotomously, as with some species of *Aloë* (*A. dichotoma, A. bainesii*); the variable genus *Aloë* also contains small rosette herbs. Each plant often develops a number of inflorescences. As the shoot continues its growth sympodially the terminal inflorescences are pushed aside to a pseudo-lateral position.

The taxa of this subfamily are all conspicuously succulent-leaved. The leaves are usually spirally set, but distichous in species of *Aloë* as well as *Haworthia.* They vary in shape from linear or lanceolate to elliptic or triangular-conical and often possess lateral spines or teeth and frequently end as a sharp spine. The vascular bundles are present in a ring around the central ground parenchyma, which is not the case in the genera of the Asphodeloideae. A cap of aloine cells is present at the phloem pole of most vascular bundles of the leaf.

The aloine cells contain coloured secretions. In *Poelnizia,* several species of *Gasteria* and *Haworthia* and single species of *Aloë* the parenchyma cap (aloine cells) are replaced by sclerenchyma. The stomata are generally tetracytic.

The leaves are often perennial, dying off gradually. The flowers are more or less tubular and vary in colour from red, red and green or red and orange to white (as in *Haworthia*). The large-sized flowers are generally fleshy and bird-pollinated (which is also the case in *Kniphofia*). The perigone in *Haworthia* is two-lipped and thus distinctly zygomorphic. There are several to many ovules in each locule. The seeds are arillate.

The members of Alooideae produce anthraquinones.

The subfamily consists of six to seven genera with a markedly Southern African concentration.

Subfamily Alooideae is rather distinct, especially in its anatomical characters, and with the family concept adopted here may deserve that status. We assume that it is a monophyletic group of genera derived from some ancestral Asphodelaceae. The remaining Asphodelaceae (subfamily Asphodeloideae) are more varied and probably make up a paraphyletic group.

*Aloë* (330), ranges along the whole of Africa, including Madagascar, but is most concentrated in South Africa. They vary from modest herbs to veritable trees ("Kokerboom"). Species like *A. dichotoma* and *A. bainesii* in Namaqualand, western Cape, may attain a height of up to 7 m or more, having thick and seemingly dichotomously branched trunks. Other species have a trunk 1–5 m tall, like the common *A. ferox*. The leaves, which are very thick and fleshy, are sometimes distichous, but more often spirally set and may bear lateral spines. The carnose, often bright red outer tepals are fused into a tube. This is one of the genera with "aloine cells", mentioned above. The contents, rich in anthraquinones, are used against intestinal worms and as a cathartic. *A. ferox* is used for this purpose in South Africa and *A. vera* in the Mediterranean; *A. saponaria* in South Africa has a similar medical use. Many species are grown as ornamental leaf succulents, e.g. the small-sized *A. picta* and *A. variegata*.

Related to *Aloë* are the genera *Chamaealoë* (1), *Poelnizia* (1) and *Astroloba* (12), all in Southern Africa. – The species of *Lomatophyllum* (14), on Madagascar and the Mascarenes, have fleshy fruits.

*Gasteria* (50), likewise mainly South African, has a curved perigone with subglobose base; its leaves are strongly succulent and sometimes facetted. Several species are grown as greenhouse succulents. – *Haworthia* (ca. 150), differs from *Gasteria* in having a straight, apically two-lipped, rarely actinomorphic, and generally whitish, pink or red, perianth tube. The seeds are small. Many species are grown as ornamentals because of their imbricate, succulent or leathery leaves with smooth, denticulate or ciliate margins.

The common presence of anthraquinones, the lack of steroidal saponins, the simultaneous microsporogenesis, the different ovular morphology and the common presence of an aril are typical attributes of the Asphodelaceae. We have largely adopted the circumscription of SCHULZE (1975a).

The great phenetic similarity between members of Asphodelaceae subfamily Asphodeloideae and certain Anthericaceae may cause practical problems. However, it is likely that the Asphodelaceae could be more closely related to other families which have simultaneous microsporogenesis, e.g. Hemerocallidaceae, Phormiaceae and Cyanastraceae.

## Anthericaceae J.G. Agardh (1858)    33:620
(Figs. 84–86)

Rhizomatous herbs with most leaves concentrated in a basal rosette and with an erect, largely leafless aerial stem bearing a terminal inflorescence.

The leaves are reduced in certain Australian genera, where the scapes are the main assimilating parts. A woody trunk is never developed, and secondary thickening growth is lacking. The leaves are spirally set or rarely (for example, in *Caesia* and *Sowerbaea*) distichous. They are flat and dorsiventral, sometimes terete or triangular, generally linear, sheathing at the base, mostly inconspicuously parallel-veined and rarely scleromorphic, rigid or succulent but then not to the same extent as in Asphodelaceae. Ligule-like structures are present at the top of the leaf sheath in *Sowerbaea*. The stomata are anomocytic. Oxalate raphides are widely distributed in the family. Vessels are surprisingly often present in the stems (e.g. in taxa of *Anthericum, Arthropodium, Caesia, Johnsonia* and *Tricoryne*), where they have scalariform or rarely simple perforation plates.

The inflorescences are simple or compound racemes, spikes or panicles, sometimes condensed into dense heads or clusters (e.g. in some "Johnsonieae").

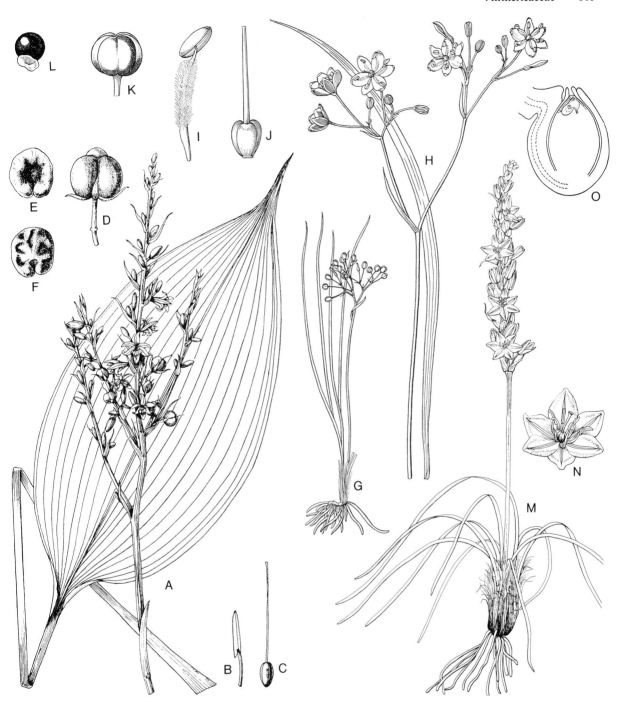

The flowers are hypogynous, trimerous and generally actinomorphic and bisexual. They are often jointed to a "pericladium". The tepals are 3 + 3, free from each other or basally connate into a tube; the two whorls may be similar or slightly different in size and shape. The tepal colour is white (to rose-coloured), yellow (*Tricoryne*), or blue or bluish violet (the last-mentioned colours being absent in Asphodelaceae). Zygomorphy af-

**Fig. 84.** Anthericaceae (**A–F, M–O**) and Asphodelaceae (**G–L**). **A–F** *Chlorophytum orchidastrum.* **A** Inflorescence and leaf. **B** Stamen. **C** Pistil. **D** Capsule. **E–F** seed in different views. (HEPPER 1968). **G–L** *Simethis planifolia.* **G** Plant. **H** Leaf and inflorescence. **I** Stamen. **J** Pistil. **K** Capsule. **L** Seed with caruncle. (ROSS-CRAIG 1972). **M–N** *Eremocrinum albomarginatum.* **M** Plant. **N** Flower. (CRONQUIST et al. 1977). **O** *Anthericum ramosum,* campylotropous ovule. (STENAR 1925)

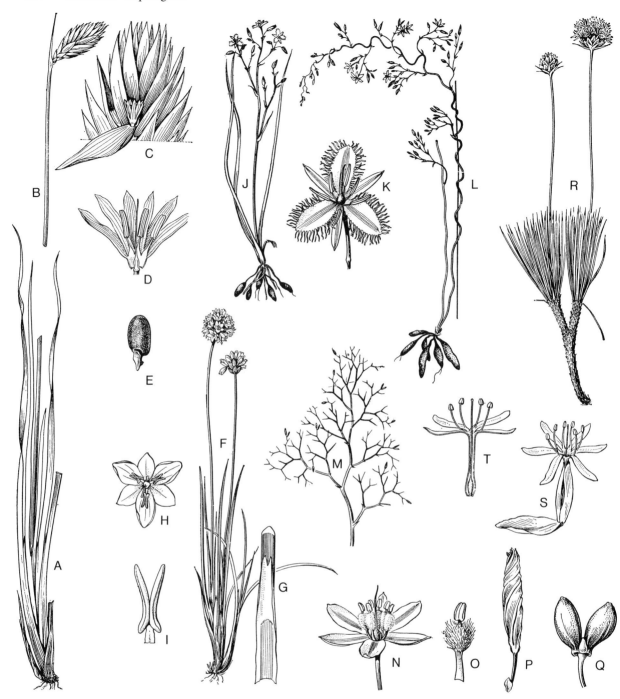

**Fig. 85.** Anthericaceae. Australian genera of the Johnsonia-Caesia groups. **A–E** *Johnsonia lupulina.* **A** Base of the plant. **B** Inflorescence, the flowers concealed between the bracts. **C** Bract bent away to show flower. **D** Flower, tepal tube spread open to show the three attached stamens and the pistil. **E** Seed. **F–I** *Sowerbaea juncea.* **F** Plant. **G** Ligule (ligules are very rare in the family). **H** Flower. **I** Anther, sagittate. **J–K** *Thysanotus tuberosus.* **J** Plant, note the swollen ends of some roots. **K** Flower. **L** *Thysanotus patersonii,* plant; the whole plant, as with *Bowiea* of Hyacinthaceae, consists mainly of a branched, scandent inflorescence. **M** *Thysanotus spiniger,* inflorescence, here much-branched, with rigid and straight branches. **N–Q** *Tricoryne elatior.* **N** Flower. **O** Stamen with hairy filament. **P** Spiralled withered flower. **Q** Fruit, here a schizocarp with two locules developed. **R–T** *Borya septentrionalis,* a somewhat dubious member of the family. **R** Branches, the leaves stiff and prickly. **S** Flower and bract. **T** Flower, longitudinal section. (All from TAKHTAJAN 1982)

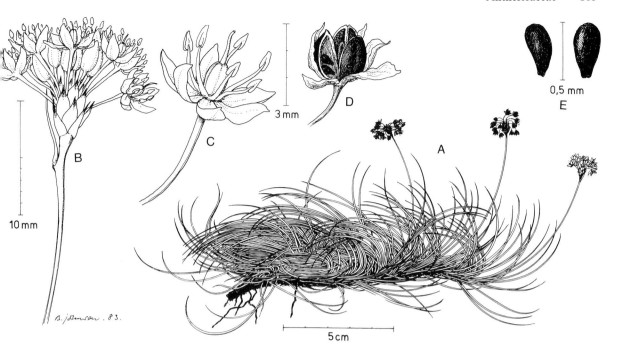

**Fig. 86.** Anthericaceae. *Alania endlicheri,* a member of the heterogeneous Johnsonia group. **A** Part of plant. **B** Inflorescence. **C** Flower. **D** Opening capsule. **E** Seeds; note that in this genus they are black, covered with a phytomelan layer. (Orig. B. JOHNSEN)

fecting shape and size of the tepals occurs rarely in, for example, *Chlorophytum.*

The stamens are usually 3 + 3 in number, but are reduced to 3 in several genera (*Anemarrhena, Arnocrinum, Hensmania, Johnsonia, Sowerbaea, Stawellia,* and some species of *Thysanotus*). The filaments are free or basally connate (*Echeandia*), and are glabrous or sometimes hairy (*Arthropodium, Glyphosperma*), and the anthers are introrse and dorsifixed-epipeltate or more rarely basifixed, and dehisce longitudinally. Microsporogenesis, in contrast to the Asphodelaceae, at least in the taxa known, is usually successive (but simultaneous in *Tricoryne*), and the pollen grains are sulcate or rarely (*Arnocrinum, Johnsonia*) trichotomosulcate or tetrachotomosulcate, i.e. provided with a three- or four-lobed distal aperture, or otherwise, and are normally dispersed in the two-celled state.

The pistil is syncarpous, tricarpellary and trilocular, with septal nectaries at least in many of the genera. The style is simple, erect and apically trilobate or punctiform, and the stigma is usually of the Dry Type. The placentation is axile. Each locule contains from two to many, generally campylotropous, crassinucellate ovules, in which a parietal cell is cut off from the archesporial cell. Embryo sac formation conforms to the *Polygonum* Type, and endosperm formation is generally helobial.

The fruit is a loculicidal capsule. The seeds occasionally have a conspicuous elaiosome structure (in *Caesia, Hensmania, Johnsonia* and *Stawellia*).

They are elongate, ovoid and often very sharply angled, and their outer epidermis is generally encrusted with phytomelan. The endosperm contains lipids and aleurone, and has a linear embryo which is usually shorter than the endosperm. The embryo is slightly curved in several genera (*Anthericum, Arthropodium, Dichopogon* and *Thysanotus*).

*Chemistry.* Steroidal saponins seem to be typical, while anthraquinones are lacking, in which respect this family differs from the Asphodelaceae. Chelidonic acid is widespread in the family. Cyanogenic compounds are known in *Chlorophytum.*

*Distribution.* The Anthericaceae are distributed over most parts of the world with the tribe Caesieae and the probably highly heterogeneous "tribe" Johnsonieae concentrated in Australia. Also in Australia are *Thysanotus* and a number of smaller genera, while the large genera *Chlorophytum* and *Anthericum* have wide, mainly tropical distributions.

*Anthericum* (excl. *Trachyandra;* ca. 65) has its centre in Africa but is represented also on other continents. It is chiefly tropical and characterized by linear leaves, three- to seven-nerved tepals, nearly basifixed anthers, and a rounded capsule with several seeds. *A. liliago* grows in sandy calcareous habitats in Europe. – *Chlorophytum* (ca. 80–100) is an even larger genus with a wide tropical distribution. It differs from *Anthericum* in having distinctly trilobate, triangular or three-winged capsules. *C. comosum* from Southern Africa propagates itself effectively by producing plantlets in the inflorescence. A form with variegated leaves is a common foliage ornamental.

*Thysanotus* (47), which is mainly Australian, is interesting in having unequal outer and inner tepals, which are generally violet, the inner ones conspicuously fringed with hairs. A similar genus, *Bottinaea* (1), occurs in Chile. *Caesia* (9) and some other genera, mainly in Australia, are probably rather closely related to *Thysanotus*. They have blue or violet tepals, which twist spirally after anthesis. They also have anatropous ovules, and rarely nutlets (*Corynotheca*) instead of capsules.

*Johnsonia* (3) and a few other genera (*Arnocrinum, Hensmania, Stawellia*) are rhizomatous perennials with *Juncus*-like scapes. They have compact, spicate inflorescences with the flowers subtended by bracts. The leaves are sometimes reduced and assimilation is then restricted to the scapes. The stamens may be reduced to three only. The seeds are black and reniform and have a white caruncle. Some genera, for example *Arnocrinum* (3), resemble *Caesia* in having blue, finally twisted tepals. Other Australian genera, *Laxmannia* and *Sowerbaea,* have umbellate inflorescences, filiform leaves that are triangular in transection, and black prismatic seeds without an elaiosome. They are dubiously closely related to the former (KEIGHERY, personal communication).

The great diversity of the genera hitherto referred to a single tribe, the Johnsonieae, has recently been pointed out by KEIGHERY (unpublished). Besides the genera mentioned above, a few more are frequently placed here, viz. the two, likewise mainly Australian, genera *Borya* (7) (Fig. 85 R–T) and *Alania* (3) (Fig. 86), both subshrubs with narrow, scleromorphic, either short and prickly or almost grass-like leaves, tubular, syntepalous flowers, spicate inflorescences, small, rounded to ovate seeds, and mycorrhizal roots (KEIGHERY, personal communication). These are somewhat woody at the base, and we suspect that they are better placed in or near the Dasypogonaceae. The genus *Tricoryne* (6), in Australia, which has grass-like leaves, tepals that are spirally twisted after flowering, umbel-like inflorescences, simultaneous microsporogenesis, a schizocarp rather than a capsule, and pale, globose seeds, may

be misplaced in Anthericaceae and needs further investigation.

Thus the Anthericaceae comprise a complex which is probably heterogeneous. Decisions must await further gross-morphological, anatomical, embryological and pollen-morphological studies, in which the different generic groups should be considered separately. None of the groups seems to come particularly close to the Asphodelaceae, although there are great similarities between species of *Anthericum* or *Chlorophytum* (Anthericaceae) and *Simethis* and species of *Asphodelus* (Asphodelaceae).

**Aphyllanthaceae** G.T. Burnett (1835)    1:1
(Fig. 87)

A relatively small herb with a short rhizome and leaves reduced to their sheaths only. Secondary thickening occurs in the rhizome (CHAKROUN and HÉBANT 1983). The inflorescences are terminal on green, assimilating scapes which are much longer than the leaf sheaths (cataphylls). Vessels are lacking in the stem but are present in the roots, where they have simple and scalariform perforation plates (TOMLINSON 1965b). Unsuberized rather large cells with raphide bundles occur in the cortex of the stem. The stomata are deeply sunken and of the anomocytic type. The outer and lateral walls of the epidermis of the scapes are thick and comprise most of the mechanical tissue and the scapes also have a thick cuticle.

The inflorescence is a small, compressed, spike-like cluster with 1 or 2 (-3) flowers and their imbricate, hyaline bracts. Each flower is enclosed by one or two free and five basally fused scales, indicating that the inflorescence is in fact compound and branched. It appears to possess one terminal and a few lateral flowers (= branches), thus representing a reduced panicle.

The flowers are hypogynous and bisexual. The tepals are 3 + 3 in number and are basally fused into a tube; the lobes are narrowly elliptic-oblong and blue. There are 3 + 3 stamens inserted in the mouth of the perigone tube. They have narrow, glabrous filaments and epipeltate, longitudinally dehiscent anthers. Microsporogenesis is successive. The pollen grains are spiraperturate and similar to those of *Lomandra endlicheri* (Dasypogonaceae) (TAKHTAJAN 1980) and have a finely spinulose surface.

The pistil is trilocular and has an erect style which is apically tribrachiate with Dry papillate stigmatic surfaces. The ovary has septal nectaries and is tri-

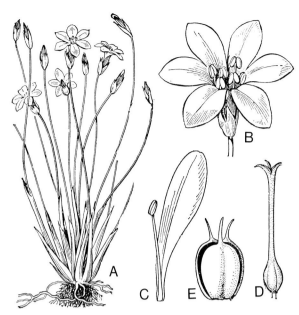

**Fig. 87.** Aphyllanthaceae. *Aphyllanthes monspeliensis.* **A** Plant. **B** Flower. **C** Tepal and attached stamen. **D** Pistil. **E** Capsule. (TAKHTAJAN 1982)

locular with each locule containing one anatropous ovule. Endosperm formation is helobial. The fruit is a loculicidal capsule, the seeds of which are slightly flattened and have a fairly thin outer epidermis thinly encrusted with phytomelan. The endosperm contains aleurone and lipids, and the embryo is straight and as long as the endosperm.

*Chemistry.* The rhizome contains steroidal saponins and the rhizome and leaves also contain wax alcohol.

*Distribution.* The single species, *Aphyllanthes monspeliensis,* is a west Mediterranean xerophyte growing on rocky hills. The reduction of the leaves and the scleromorphic, assimilating scapes are adaptations to the extreme climatic and edaphic conditions.

The relationships of *Aphyllanthes* are uncertain. The extraordinarily variable Anthericaceae in Australia, almost certainly a heterogeneous assemblage, include forms (*Caesia, Thysanotus*) with violet or blue flowers where the inflorescence is the assimilating part, but these genera deviate in various, different respects from *Aphyllanthes,* and the vast geographical distance makes any attempt to associate *Aphyllanthes* with any of them highly speculative and improbable. Lacking a better alternative we follow some previous taxonomists in recognizing Aphyllanthaceae as a distinct family. On the basis of the spiraperturate pollen grains Aphyllanthaceae is sometimes associated with Da-

sypogonaceae (see TAKHTAJAN 1980), but we consider this similarity the result of convergence.

## Funkiaceae P. Horaninow (1834)    3:12
(Fig. 88)

Rhizomatous herbs or (*Hesperocallis*) basally woody plants with a short stem and bulb-like corm. The roots are often fleshy, although not fusiform. The leaves are spirally set, concentrated at the base or in *Hesperocallis* on the lower part of the stem, dorsiventral, linear, lanceolate or (within *Hosta*) even ovate to subcordate. The leaves are sheathing at the base and in some *Hosta* species the section between sheath and lamina is narrowed into a pseudopetiole. Further, the leaves are parallel-veined. The stomata are anomocytic. Oxalate raphides are known at least in the fruit wall of *Hosta*. Vessels are absent from the stem and leaves but are present in the roots (where they have scalariform perforation plates), at least in *Hosta*.

The inflorescence in all three genera is a simple raceme on a bracteate scape. The flowers are hypogynous and bisexual and may or may not be jointed to the pedicel. The tepals are fused into a cylindrical, campanulate, or funnel-shaped perigone with the six lobes of variable length and sometimes recurved. The perianth is actinomorphic or nearly so and blue, violet or white in colour.

There are 3+3 stamens inserted in the perigone tube, with mutually free, glabrous filaments, introrse anthers, the connective of which forms a tube over the filament tip. Anther dehiscence is longitudinal. The tapetum is secretory and microsporogenesis successive. The pollen grains are sulcate.

The pistil is syncarpous and three-locular, with a simple, filiform style with a minute, capitate or three-lobed stigma (*Hosta* having a Wet stigmatic surface). Placentation is axile and the ovules numerous and anatropous. The nucellus has a parietal tissue formed by a parietal cell cut off from the archesporial cell. Periclinal divisions occur in the nucellar epidermis except in *Leucocrinum*. A hypostase is formed (lacking in Hemerocallidaceae). Embryo sac formation conforms to the *Polygonum* Type, and endosperm formation, at least in *Hosta*, is helobial.

The fruit is an elongate to subglobose, loculicidal capsule with numerous seeds, which are fairly small, flattened or compressed and often elliptic.

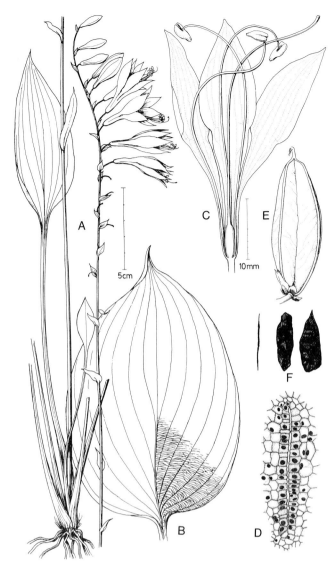

**Fig. 88.** Funkiaceae. **A** and **C.** *Hosta japonica*. **A** Plant, the basal parts and the inflorescence. **C** Flower, longitudinal section. **D** *Hosta* sp., septal nectary. (WEBERLING 1981). **B, E, F** *Hosta ventricosa*. **B** Leaf; note the minute, transverse and partly reticulate venation. **E** Capsule. **F** Seeds in different views. (Orig. B. JOHNSEN)

They have a thick, black coat of phytomelan. The outer integument is several layers thick, but all layers except the epidermis are strongly compressed, as is also the inner integument. The endosperm consists mainly of isodiametric thin-walled cells containing protein and fat. $n = 24.30$.

*Distribution.* While the largest genus, *Hosta*, is centred in China and Japan, *Hesperocallis* and *Leucocrinum* occur in Colorado, the former extending to southern California.

*Hosta* (ca. 10) has a scape with or without basal scale leaves. The leaves vary much in breadth and colour, and the flowers are situated in a one-sided, slightly nodding raceme. *Hosta* consists of shade plants. Several species and hybrids are grown as garden ornamentals, e.g. *H. caerulea, H. glauca, H. plantaginea* and *H. sieboldiana*. *Leucocrinum* (1) and *Hesperocallis* (1) occur in sandy valleys and in deserts, respectively. They are not particularly similar to *Hosta* nor to each other, but various shared features suggest relationships. *Leucocrinum* may prove more closely allied to Hemerocallidaceae (CAVE 1964).

The position of Funkiaceae, like that of Hemerocallidaceae, is uncertain. Some details indicate that the family may be related to Agavaceae, this being supported most strongly by the karyotype of *Hosta* which agrees almost totally with that in Agavaceae. The chromosome number in *Hosta,* as in Agavaceae, is $n = 30$, but that in *Hesperocallis* $n = 24$. The serological investigations by CHUPOV and KUTIAVINA (1981) show affinity between *Hosta* and both the Agavaceae and *Camassia* of Hyacinthaceae.

## Hyacinthaceae Batsch (1802)    40:900
(Figs. 89–91)

Mostly glabrous scapose perennial herbs with bulbs (except *Schoenolirion* and *Chlorogalum,* which have a rhizome). The bulbs generally have a membranous tunic and a number of free or coalescent bulb scales. The roots are sometimes thick and generally contractile. The leaves, concentrated at the base, are solitary to numerous, generally spirally set, flat and dorsiventral, and generally linear to linear-lanceolate, rarely elliptic to orbicular. Further, they are normally mesomorphic, sheathing at the base, non-petiolate and parallel-veined. The stomata are anomocytic. Crystal raphides contained in mucilage cells or canals are widely distributed in the family. Vessels are present in the roots only and have scalariform and/or simple perforation plates.

The leafless scape usually bears a simple or, more rarely, branched raceme or spike, its axis being elongate (in contrast to most Alliaceae). Rarely, however, as in *Massonia,* there is a head-like cluster of flowers. The inflorescence has few to many flowers and is generally bracteate, at least in the lower part.

The flowers are generally bisexual, hypogynous, trimerous and actinomorphic. The 3 + 3 tepals are

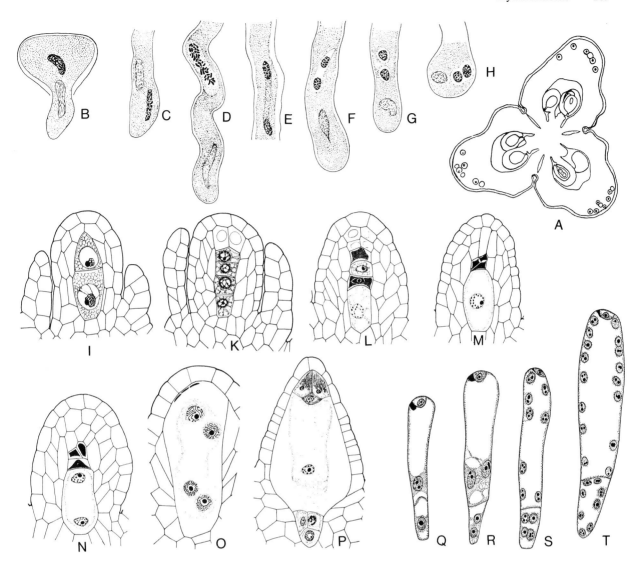

free or more often connate, forming then a campanulate, urceolate or tubular perigone. The tepal colour varies considerably, being white, blue or violet, or more rarely yellow, red, brown or even nearly black. The tepals of the two whorls are generally similar in appearance, but may differ in size, shape and position (as in *Albuca*). In some cases the upper flowers are sterile and of different colour from the fertile, having only the function of attracting insects.

There are 3 + 3 stamens, inserted either at the base of the tepals or in the tepal tube. Their filaments, as in Alliaceae, are often broad and flat and in some genera each is produced into two lobes, one on either side of the anther (e.g. in species of *Chionodoxa, Ornithogalum, Eucomis*). In *Puschkinia* they even have appendages forming a "paracorolla". The anthers are introrse, "epipeltate" and dehisce longitudinally. The tapetum is glandu-

**Fig. 89.** Hyacinthaceae (**M–P**) and Asphodelaceae (**Q–T**). **A** *Albuca fastigiata,* ovary, transverse section. **B–H** *Muscari atlanticum* (=*racemosum*). **B** Pollen grain in the state of germination. **C–H** Pollen tube, with successive stages in division of the generative cell into two sperm cells, in **H** oriented at the tip of the pollen tube. (WUNDERLICH 1937). **I–P** *Urginea indica,* meiosis and formation of embryo sac (*Polygonum* Type), **I** after the first meiotic division, **K** with a linear megaspore tetrad, and showing at the top two cells of a parietal layer, **M–P** showing the chalazal megaspore developing into an 8-nucleate embryo sac. (CAPOOR 1937). **Q–T** *Eremurus himalaicus,* development of helobial endosperm. (STENAR 1928a)

lar and microsporogenesis is of the Successive Type. The pollen grains are sulcate and two-celled, the generative nucleus being enclosed in an elongate, thin-walled cell often centrally located in the pollen grain (WUNDERLICH 1937).

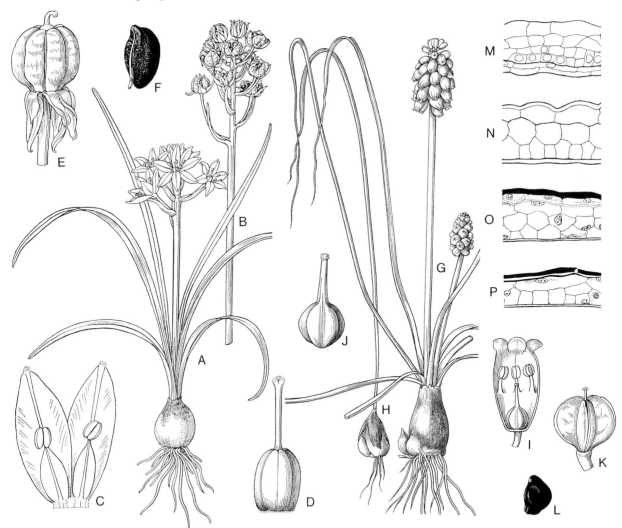

The pistil is tricarpellary and trilocular and has septal nectaries. Its style is simple, terminating in a punctiform or sometimes distinctly trilobate stigma with either a Wet or Dry papillate surface. The ovules are two to numerous in each locule, anatropous and with the axis straight or almost straight or, in a few genera, curved. They are crassinucellate, and a primary parietal cell is cut off from the archesporial cell and in addition the nucellar epidermis may divide periclinally to form further cell layers of a nucellar cap. Embryo sac formation conforms to the *Polygonum* or (rarely) *Scilla* or *Allium* Types, and endosperm formation is helobial or, more rarely, nuclear (nuclear in species of *Camassia, Hyacinthus, Scilla* and *Urginea*).

The fruit is a loculicidal capsule with, as a rule, two or more seeds per locule. The seeds are ovoid to pear-shaped and vary from rounded to strongly angular in transection; they may be small

**Fig. 90.** Hyacinthaceae. **A–F** *Scilla verna.* **A** Flowering plant. **B** Infructescence. **C** Two tepals and opposite stamens. **D** Pistil. **E** Capsule. **F** Seed. **G–L** *Muscari atlanticum.* **G** Plant. **H** Young bulb with leaf. **I** Flower, longitudinal section. **J** Pistil. **K** Capsule. **L** Seed. (ROSS-CRAIG 1972). **M–P** *Muscari atlanticum,* development of seed coat, **M–N** before and **O–P** after the development of the black phytomelan layer in the outer epidermis; the inner integument, which is two-layered in **M,** collapses and becomes a thin membrane. (WUNDERLICH 1937)

(1.2–2.0 mm), as in *Lachenalia* and *Massonia,* or large (5–8 mm), as in *Eucomis* or *Veltheimia.* The outer epidermis of the testa consists of several cell layers, and normally has a phytomelan crust, which is rather thick in *Massonia, Hyacinthoides* (*Endymion*) and *Hyacinthus,* but more often rather thin, and may even be lacking in *Chionodoxa, Puschkinia* and a few species of *Scilla,* where the outer epidermis collapses and forms an almost un-sculptured yellowish brown layer. The innter inte-gument forms a thin membrane, while the endo-sperm consists of cells with thin to rather thick and pitted walls and with contents of aleurone and fatty oils but usually no starch. Only exceptionally, as in one species of *Eucomis* and in *Scilla bifolia,* does the endosperm contain starch grains. The em-bryo is cylindrical or almost so, and straight or, in *Chlorogalum,* slightly curved.

*Chemistry.* Hyacinthaceae are characterized by producing steroidal saponins (often abundantly, as in the bulbs of *Albuca*) and chelidonic acid, which is known to occur in a number of genera. Salicylic acid occurs in the scape and bulb scales of *Hyacinthus.* The bulbs of *Urginea maritima* con-tain cardiotoxic glucosides (belonging to the so-called bufodienolides) and are therefore used as poison and in medicine. The bulbs of Hyacintha-ceae contain fructans and also starch.

*Parasites.* Within the rust genus *Uromyces* one species, *U. muscari,* attacks species of *Muscari, Scilla, Dipcadi, Hyacinthus, Ornithogalum* and *Ur-ginea,* all in Hyacinthaceae, but not members of other families. Another example of restricted fun-gal parasitism involves the smut genus *Ustilago.* SAVILE (1979) reports the following host genera in Asparagales: *Albuca, Bellevalia, Eucomis, Hya-cinthus, Muscari, Ornithogalum, Urginea* and *Al-lium.* Of these genera all except *Allium* belong to Hyacinthaceae.

*Distribution.* The Hyacinthaceae are widely dis-tributed, but most richly represented in Southern Africa and in a region from the Mediterranean to South-West Asia. They are apparently best adapted to a fluctuating moist-arid climate, with-ering down to the bulbs in the arid period.

Many genera provide widely cultivated mainly spring-flowering ornamentals.

There seem to be great difficulties in systema-tizing the genera. Some genera are pronouncedly peripheral in Hyacinthaceae and may merit the rank of one or two separate families. Included here are *Bowiea, Schizobasis, Chlorogalum* and *Schoeno-lirion,* which are also referred to Hyacinthaceae by HUBER (1969) and seem to fit better here than with any other family. Some of these genera will briefly be discussed below.

Among the genera having *free or only basally fused tepals* are the following. *Ornithogalum* (100) is an Old World genus richly represented in dry habitats, such as steppes and grassland from Southern Africa to Asia. The flow-ers are mostly white to greenish and as a rule have spreading tepals, flat filaments and rounded-ovoid seeds. The habit is most variable, from large herbs with racemose inflorescences to small, shortly sca-pose plants with few-flowered sometimes nearly umbel-like inflorescences. Several species are orna-mentals, such as the commercially important *O. thyrsoides* from the Western Cape, South Africa, the "Chinkerinchee". Among the European spe-cies cultivated since the Middle Ages are *O. umbel-latum* and *O. nutans.* Whereas some species are strongly poisonous, others such as the European *O. narbonensis* and *O. umbellatum* (bulbs) and the West European *O. pyrenaicum* (young shoots) are edible. – *Albuca* (50), centred in Southern Africa, has sparse racemes of *Galanthus*-like, white to yel-low flowers, the inner tepals being erect to in-curved and different from the outer which are more spreading. Some species are cultivated as or-namentals. – *Urginea* (50–100) is a variable genus of small and few-flowered to very large and multi-florous herbs, which ranges from South Africa and the Mediterranean eastwards to India. The free tepals of this genus are shed after flowering, and the seeds are angular. Best-known is *U. maritima,* a common species on coastal mountain slopes and shores in the Mediterranean. It has extremely poi-sonous bulbs containing the glucosides scillarin-A and -B. The red-bulbed varieties are used as rat poison; other forms provide heart medicines. – *Scilla* (80), in the Old World, is characterized by free white, blue or violet tepals and rather narrow filaments. The seeds are ovoid to globose and not strongly angular. Several species are grown as or-namentals. *S. sibirica,* indigenous in the Caucasus and southern Russia, is spring-flowering with few, nodding, campanulate bright blue flowers. *S. peru-viana* from the Mediterranean, has a many-flow-ered raceme of long-pedicelled, violet flowers, with spreading tepals. The *S. bifolia* complex in south-ern and central Europe and western Asia has rath-er few-flowered unilateral racemes. – A related genus is *Hyacinthoides* (=*Endymion*) with three or four western European species (*H. hispanica* of-ten cultivated), with lax racemes of nodding, light blue, campanulate flowers and bulbs with coales-cent bulb scales (free in *Scilla*). – *Camassia* (4), indigenous to North America, is characterized by racemes of flowers with spreading, blue-violet or pale yellow tepals; some are cultivated for orna-ment. – To this group may also be referred the Southern African genus *Eucomis* (10), which is pe-culiar in having the inflorescence crowned by a fascicle of green leaves. The racemose inf-lores-cence has greenish yellow to brown flowers.

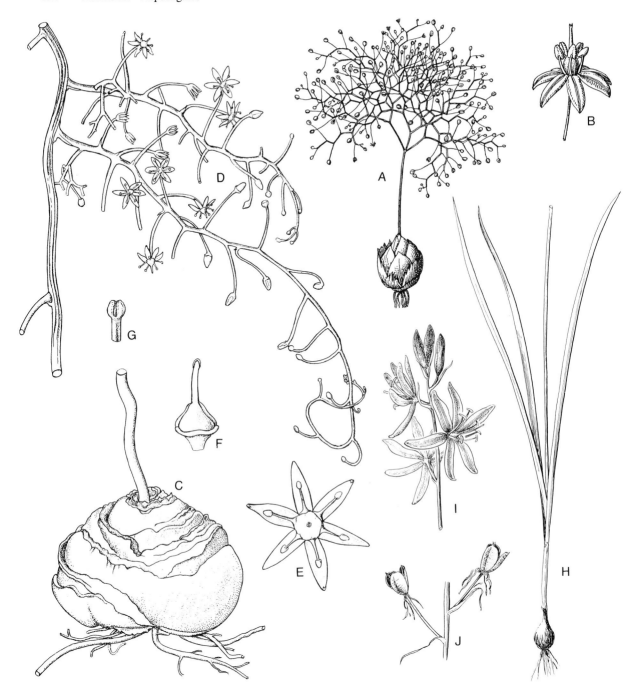

**Fig. 91.** Hyacinthaceae. **A–B** *Schizobasis intricata.* **A** Plant, the leaves in the adult plant, as in *Bowiea,* are restricted to the bulb scales. **B** Flower. (KRAUSE 1930). **C–G** *Bowiea volubilis.* **C** Bulb. **D** Part of the intricately branched inflorescence. **E** Flower. **F** Pistil. **G** Style apex with stigma. (DYER 1941). **H–J** *Camassia quamash.* **H** Base of plant. **I** Upper part of inflorescence. **J** Two capsules. (HITCHCOCK et al. 1969)

Among the genera with *more or less connate tepals* are the following.
*Chionodoxa* (8), mostly in Asia Minor and *Puschkinia* (2), in the same region and eastwards, are *Scilla*-like genera with two basal leaves and bluish to white flowers with the tepals fused only basally. The stamens are joined to the mouth of the perigone tube; in the former genus they have free but flat filaments, while in *Puschkinia* they are fused into a tube. Both genera contain popular garden

ornamentals like *C. sardensis* and *P. scilloides*. Detailed cytological studies in these genera and in *Scilla* by SPETA (1976) and GREILHUBER and SPETA (1976) show that morphological and cytological evidence is often contradictory and the conventional generic borderlines artificial. – *Hyacinthus* (3) is indigenous to South-Eastern Europe and Asia Minor. *H. orientalis,* with heavily fragrant flowers, is one of the most popular spring-time ornamentals. Various cultivars with white, yellowish, pink, crimson or blue flowers are cultivated commercially. Closely related to *Hyacinthus* are a number of genera such as *Hyacinthella* (11), *Brimeura* (2), and *Alrawia* (2). – *Bellevalia* (50) with a wide Old World distribution, and *Muscari* (incl. *Leopoldia,* 55), range from the Mediterranean as far as Caucasus and likewise contain a number of ornamentals. In *Muscari* the flowers are globose-urceolate with short tepal lobes; *M. botryoides* from the central Mediterranean and *M. armeniacum* from Asia Minor are frequently cultivated species. In *M. comosum* and other species, often treated in the genus *Leopoldia,* the uppermost flowers are sterile, closed and brightly coloured, while the fertile are dull in colour.

*Dipcadi* (55) ranges from South Africa to the Mediterranean. – *Lachenalia* (90) in South Africa is extremely variable in floral colour and often has a variegated scape and (one to five) variegated or spotted leaves. – Large mesomorphic leaves and a variegated scape characterize the likewise African genus *Veltheimia* (6). *Galtonia* (3) in South Africa are herbs with yellowish white, campanulate, pendulous flowers in an elongate raceme. *Galtonia candicans* and *Veltheimia viridifolia* are grown as ornamentals.

The South African genus *Schizobasis* (1) has a compound, twining inflorescence (Fig. 91 A–B).

*Massonia* (30) and some other Southern African genera have very short scapes and a densely compressed capitate or umbel-like inflorescence situated between two broad, sometimes orbicular leaves pressed against the ground. The habit of *Massonia* resembles that of some species of *Haemanthus* (Amaryllidaceae). The flowers have a tubular perianth, the lobes of which are sometimes reflexed. Their stamens have long, narrow filaments inserted in the floral tube.

*Bowiea* (1–3), in Southern to Central Africa, is a most peculiar plant which has a large bulb of thick green scales and filiform, quickly withering leaves (Fig. 91 C–G). The inflorescence is developed as a richly branched, herbaceous vine, with part of the branches transformed into tendrils. The

flowers have six, equal, free, more or less reflexed, greenish tepals. The bulbs are poisonous.

The above grouping of the more important, "typical" genera of Hyacinthaceae is certainly artificial. Fusion of the tepals has undoubtedly occurred in several evolutionary lines within the family; in fact, *Chionodoxa* species can give rise to hybrids when crossed with members of the *Scilla bifolia* complex, and it has been proposed to include them in *Scilla,* whereas *Puschkinia* evidently has other affinities (SPETA 1976). Other divisions are based on whether the seeds are ovoid and rounded or sharply angular and flattened, and on the seed size. Chromosome structure can also be used in constructing a more phylogenetic classification in the family. Serological studies, finally, indicate other constellations, and suggest that neither *Camassia* nor *Bowiea* is very closely allied to the majority of the genera (CHUPOV and KUTIAVINA 1978, 1981).

Two genera that fall somewhat outside the ordinary variation pattern of the family and need to be reconsidered as regards their most appropriate position are *Schoenolirion* (4) and *Chlorogalum* (3) in North America, the former with rhizome, the latter species with a narrow bulb.

**Alliaceae** J.G. Agardh (1858)    30:ca. 720
(Figs. 92–94)

Perennial herbs with a bulb or a bulb-like corm, which has membranous or fibrous outer scales, or more rarely a rhizome (*Agapanthus* and *Tulbaghia*). The leaves are basally concentrated, but sometimes sheathing the scape for a considerable distance and therefore appearing cauline (as in *Allium scorodoprasum*). They are spirally set or distichous (at least in *Agapanthus*) and filiform-linear, lanceolate or rarely ovate. They may be flat, terete, fistulose or angular and are sheathing at the base, parallel-veined and rarely (as in *Allium ursinum*) constricted into a pseudopetiole between the sheath and a broad lamina. The stomata are anomocytic. Raphides are present in several genera but not, for example, in *Allium, Milla* and *Tulbaghia* (which contain allylic sulphides and other compounds, which are the source of the onion smell; see below).

The scape is terete (sometimes conspicuously fistulose) or angular and bears an umbel-like inflorescence of short- or long-pedicelled flowers. The inflorescence represents, at least in most cases, one or more contracted helicoid cymes. Rarely (as in

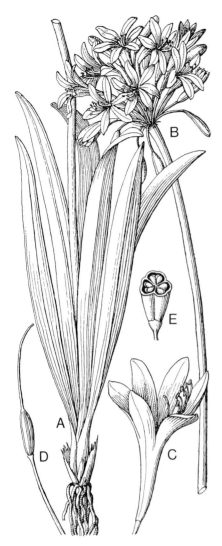

**Fig. 92.** Alliaceae. *Agapanthus umbellatus.* **A** Leaf rosette and the short rhizome, showing absence of bulb. **B** Scape with inflorescence. **C** Flower. **D** Young capsule. **E** Capsule, transverse section. (KRAUSE 1930)

*Milula*) the inflorescence is cylindrical and spike-like.

The inflorescence is subtended by an involucre of (one,) two or more membranous spathal bracts, which may be free from each other and then usually spreading, or united at the base and then mostly erect and enclosing the buds, like a calyptra.

Articulation of the flowers with the pedicels occurs rarely. The flowers are trimerous, hypogynous, generally bisexual, and actinomorphic or (in most genera of the Gilliesioideae, for example) zygomorphic. The 3 + 3 tepals are free or often connate to form a campanulate or tubular perianth with erect, spreading or sometimes recurved lobes. The

tepals are generally similar in the two whorls; they vary in colour from white to blue, violet, purple or even yellow. The functional stamens are 3 + 3 or sometimes 3 or 2, in the latter case (e.g. *Leucocoryne* and genera of the *Gilliesia* group) there are several staminodes. The filaments are inserted at the base of the tepals or in the perigone tube; they are more or less flat and those of the inner staminal whorl sometimes end with a tip on each side of the anther. The anthers are elongate (rarely short), epipeltate and introrse, and dehisce with longitudinal slits. The tapetum is secretory and microsporogenesis successive. The pollen grains are sulcate and two-celled when dispersed.

The pistil is tricarpellary and trilocular and has a single erect style with a trilobate or capitate stigma, which has a Dry or sometimes (*Bloomeria, Leucocoryne*) Wet papillate surface. Septal nectary grooves are present on the ovary. Each of the locules contains two to several ovules, which are campylotropous or less often anatropous, with a straight or curved axis. A parietal cell is generally *not* cut off from the archesporial cell, although this has been observed at least in *Agapanthus,* but the nucellar epidermis may divide periclinally to form a nucellar cap. Embryo sac formation seems to conform to the *Allium* Type, at least in *Allium* and *Leucocoryne,* but according to the *Polygonum* Type in *Nothoscordum, Muilla* and *Brodiaea.* Endosperm formation has been found to be nuclear in several species of *Allium* and *Brodiaea* but helobial in *Nothoscordum, Tulbaghia* and perhaps *Triteleia.*

The fruit is a loculicidal capsule with few to numerous seeds. These are sometimes rather small and ovoid or ellipsoidal to subglobose and rounded in transection (*Brodiaea, Triteleia* and related genera, having anatropous ovules), but more often triangular in transection and half-ovoid, half-globose or tetrahedral in shape. They are often larger where the ovules are campylotropous and the ovule axis curved 120 degrees (in *Allium, Nothoscordum, Muilla* and related genera) than when the ovules are anatropous. The testal epidermis is usually covered by a rather thick crust of phytomelan and consists of subisodiametric or slightly elongate cells. The inner layers of the testa, which is several-layered, are compressed or collapsed, as is the tegmen. Fatty oils and aleurone, but not starch, are deposited in the endosperm cells, which usually have rather thick, pitted walls. The embryo varies in length; it is usually more than half the length of the endosperm and becomes straight in the anatropous ovules, but more or less

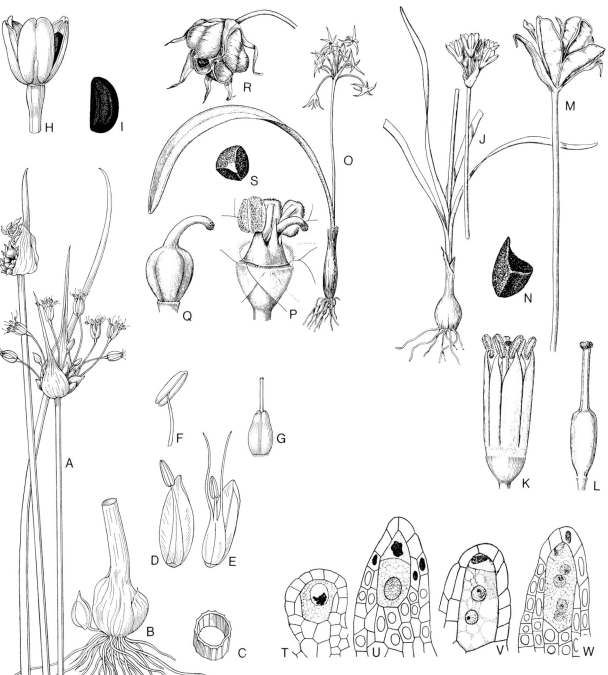

**Fig. 93.** Alliaceae. **A–I** *Allium vineale*. **A** Scape with inflorescences; note the bulbils at the base of the flowers. **B** Bulb. **C** Leaf, transverse section. **D** Tepal of outer whorl and the opposite stamen. **E** Tepal of inner whorl and the opposite stamen; note the long filamental lobes. **F** Anther and top of filament. **G** Pistil. **H** Capsule. **I** Seed. (Ross-Craig 1972). **J–N** *Nothoscordon inodorum*. **J** Plant. **K** Stamens and pistil. **L** Pistil. **M** Scape with capsule. **N** Seed. (Correa 1969). **O–S** *Solaria attenuata*. **O** Plant. **P** Androecium and gynoecium. **Q** Pistil. **R** Capsule. **S** Seed. (Correa 1969). **T–W** *Allium cepa*, early stages in development of embryo sac (*Allium* Type); note that the chalazal dyad (**V**) develops into the embryo sac. (After H.A. Jones and Emsweller 1936)

curved when the ovules are semi-campylotropous or campylotropous.

*Chemistry.* The family contains steroidal saponins with, for example, aigogenin (*Allium*), yuccagenin (*Agapanthus*) and agapanthagenin (*Agapanthus*) as sapogenins. Chelidonic acid is known to be present in several genera. While raphides are lacking in *Allium,* this genus possesses oxalate crystals in various other shapes (which may be characteristic of the species). Most or all species of *Allium* and *Tulbaghia,* and also, at least, species of *Ipheion, Androstephium, Leucocoryne, Milula* and *Tristagma,* contain sulphur compounds, such as allyl sulphides, propionaldehyde, propionthiol and vinyl disulphide, which participate in the essential oils causing the onion odour. The typical scent is released only when the tissues are wounded or during decay. The mechanism involved can be demonstrated by adding an enzyme, alliinase, to an amino acid, alliin, which is converted into the strongly "onion-scented", water-soluble compound allicin and other compounds. The bulb scales in *Allium* contain flavonoid compounds, e.g. quercetin, which is the reason for using onion scales in dyeing eggs. Finally, the thick bulb scales in *Allium* and other genera contain carbohydrates, in the form of inulin-like fructans but, at least in *Allium,* not starch.

*Distribution.* Alliaceae are widely distributed, especially the genus *Allium* itself. The other genera show characteristic distribution patterns: the subfamily Agapanthoideae is South African and, within the subfamily Allioideae, many smaller genera of the tribe Brodiaeeae (e.g. *Brodiaea* and *Milla*) are American (California, Mexico, Chile, etc.), while the subfamily Gilliesioideae has its centre in Chile in South America.

The subdivision of the family may be based on various characters, such as the underground parts (bulbs, tunicated corms, rhizomes), tepal characters (free or connate), stamen characters (filaments free or fused into a tube, narrow or broad and flat, with or without lateral tips), and ovule shape (anatropous or variably strongly campylotropous) which is also clearly connected with size and shape of the seeds. Besides, presence or absence of raphides and presence or absence of allyl compounds can be used, as well as chromosome characters and other details. A strong emphasis on embryology and seed characters may be the most acceptable approach, but for practical reasons the division here will follow more conventional lines.

## Subfamily Agapanthoideae

This subfamily consists of only two genera. Both are rhizomatous, sometimes large, herbs with flat leaves and umbel-like inflorescences with two spathe-bracts, enclosing the floral buds. The pedicels are not articulate, the tepals are basally connate, and the ovules are half-campylotropous and develop into relatively large (4–11 mm), angular seeds provided with a wing at the funicular end. Their phytomelan crust is thinner than in most other Alliaceae. The two genera may not be particularly closely allied.

*Agapanthus* (7) (Fig. 92), in Southern Africa, are large, flat-leaved, saponin-rich herbs with many-flowered inflorescences of deep to light blue (rarely white) flowers without corona or stamen appendages. Embryo sac formation seems to be of the *Polygonum* Type. Crystal raphides occur in this genus, while allyl sulphides and similar compounds are lacking. Some species, e.g. *A. praecox, A. africanus* and *A. campanulatus,* are grown as ornamentals.

*Tulbaghia* (24, VOSA 1975) has its centre in South Africa and is characterized by the fewer-flowered "umbels" with violet (*T. fragrans*) or more often dull-coloured, green, brown or white tepals. These are alike in shape and fused basally into an urceolate or cylindrical tube. Three or six conspicuous, fleshy filamental "corona lobes" are present in the mouth of the perianth tube opposite the inner tepal lobes. The filaments are adnate for most of their length to the tepal tube, the subsessile anthers being inserted at different levels of the corona. Crystal raphides are lacking, and "onion-smelling" sulphur compounds are present in this genus. *T. fragrans* is grown for ornament.

## Subfamily Allioideae

Subfamily Allioideae consists of herbs with a bulb covered by membranous or fibrous-coated as well as fleshy scales or with a corm covered by a membranous or fibrous coat. In most features the subfamily conforms to the family description. It is the largest and most important part of the family and apart from *Allium* has a clearly American concentration. It may be divided into two tribes, the Brodiaeeae and the Allieae. These are rather distinct and may alternatively be treated as subfamilies.

### *Tribus Brodiaeeae*

This consists of some ten genera with a corm covered by membranous or fibrous scales (MOORE 1953). The scape bears an inflorescence subtended by three or more separate, spreading spathal bracts, which do not enclose the flower buds. The flowers are often articulate on their pedicel. They are actinomorphic and have 3 + 3 functional stamens or 3 stamens and 3 staminodes. The ovules are anatropous or rarely (*Muilla*) campylotropous,

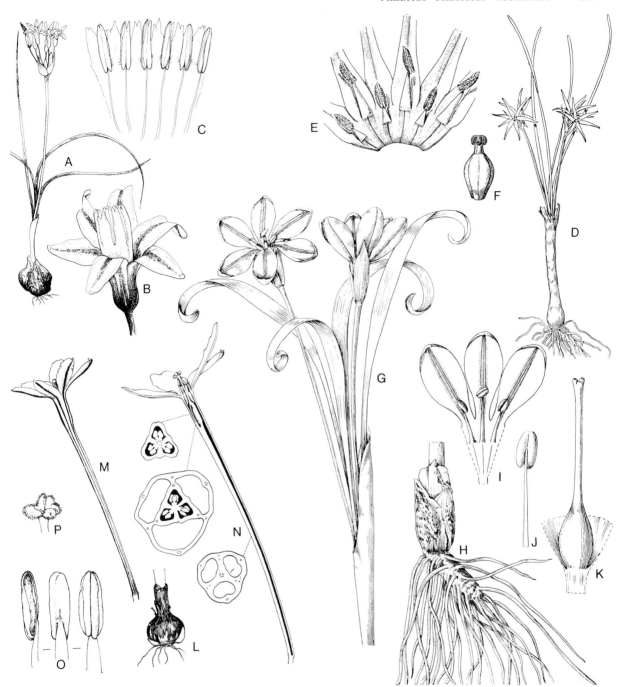

having generally a straight axis, and developing into rather small, ovoid to subglobose rounded seeds. The phytomelan crust is thick. All the genera occur in America and most of them in North America.

*Brodiaea* (10) (Fig. 94G–K), *Dichellostemma* (6) and *Triteleia* (16) have tubular flowers. They contain some ornamentals, e.g. *Triteleia laxa* with blue flowers and *Dichellostemma ida-maia* with tubular, cigar-shaped, red and yellow flowers. – *Leucocoryne* (14), mainly in Chile, and two other gen-

**Fig. 94.** Alliaceae. **A–C** *Androstephium breviflorum.* **A** Plant. **B** Flower. **C** Androecium, forming here a staminal tube. (Cronquist et al. 1977). **D–F** *Tristagma anemophilum.* **D** Plant. **E** Stamens, attached at their base to the perianth. **F** Pistil. (Correa 1969). **G–K** *Brodiaea circinnata.* **G** Inflorescences. **H** Rhizome, note that the bulb is here dissolved into an elongate rhizome; cf. *Scadoxus* in Fig. 95. **I** Half of the perianth with attached stamens. **J** Stamen. **K** Pistil. (Sandwith 1937). **L–P** *Milla magnifica.* **L** Corm. **M** Flower. **N** Flower in longitudinal section, showing also transverse sections at three levels. **O** Stamen in different views. **P** Style apex with stigmatic lobes. (Moore 1953)

era have three functional stamens only. *L. alliacea* is often cultivated as an ornamental.

Some related genera, among them *Milla* (16), have parallel tunic fibres, covering a corm (in most genera the fibres are reticulate), a long-stipitate ovary, and a narrow-tubular perianth. The stipe may be connate to the tube by three lateral flanges. With respect to ovules and seeds these genera, like *Leucocoryne,* seem to be intermediate between the Brodiaeeae and Allieae. *Milla,* with its geographic centre in Mexico, has white or blue tepals.

### Tribus Allieae

Tribus Allieae consists of seven genera which are truly bulbous plants with membranous or fibrous tunics covering a number of thick bulb scales. The leaves are very variable, being flat or terete, and sometimes falsely cauline (see the family description above). Two hyaline bracts subtend the inflorescence. These are normally more or less fused and enclose the young inflorescence. The flowers are not articulate on the pedicel. Their tepals are free or connate and variously coloured, but often violet. The stamens are $3+3$ in number. The ovules are campylotropous and have a strongly curved axis. The unwinged seeds in this group of genera deviate conspicuously from those in the typical Brodiaeeae in being broad and triangular, but as in that tribe they have a thick phytomelan crust. The embryo is more or less curved. While *Allium* is widely distributed, the other, smaller genera are all centred in South America.

Among them is *Nothoscordum* (18), mainly in America. It has connate tepals and helobial endosperm formation. *N. fragrans* is a white-flowered species introduced in the Old World. – *Tristagma* (5) in Chile has flowers superficially reminiscent of those in *Tulbaghia*. – *Ipheion* (25) *uniflorum,* with a uniflorous scape, is indigenous in Argentina and Uruguay, but is widely cultivated.

*Allium* (incl. *Nectaroscordum*; ca. 550) is the largest genus in the family and is mainly distributed in the Northern Hemisphere. More than 110 species occur in Europe, most of them around the Mediterranean; other centres are in Asia and North America. The genus is characterized by having bulbs enclosed in membranous rather than fibrous scales, free or almost free tepals and often a subgynobasic style and by producing allylic sulphides and similar sulphur-containing compounds. Some species of *Allium* have true ligules, and another peculiarity is the occurrence in the genus of laticifers. The genus is richly represented in regions that are seasonally dry. It is divisible into many sections according to colour of the perianth and stamens, leaf characters (position and shape, whether terete, fistulose, flat or keeled, etc.) transection and thickness of the scape, and occurrence of nectariferous pores. In many species bulbils are developed in the inflorescence between the flowers. These serve as an effective means of vegetative propagation, especially in taxa with poor seed-setting.

Some species are cultivated as ornamentals, e.g. *A. christophii, A. karataviense, A. rosenbachianum* and *A. aflatunense,* all with stellate, violet flowers in globose heads, the former three from Turkestan, the last mentioned from Central Asia. *A. flavum* and *A. moly* from Turkey and South-Western Europe respectively are yellow-flowered, and *A. narcissiflorum* and *A. siculum* from the Alps and Southern Europe respectively, have pendulous, campanulate, reddish to brown flowers. *A. carinatum* with smaller, nodding flowers, is native in South-Eastern Europe. *A. cyathophorum* var. *farreri,* with purple flowers, and *A. cyaneum,* with blue, both from China, are dwarf ornamentals.

Several species are economically important. *A. sativum,* "Garlic", indigenous from Southern Europe to Central Asia, is grown for its white bulbs that are used to flavour meat and other food. The oil extracted from them, oil of garlic, is used in medicine and for flavouring food. It is rich in allyl-propyl disulphide and diallyl disulphide which with the enzymes released on bruising the tissues give a strong scent. – More widely used and milder are the diverse sorts of *A. cepa,* "Onion", which are cultivated throughout the world and are eaten raw in salads or roasted, boiled or fried with various kinds of food. The juice is mildly antiseptic. The species is grown in various varieties and forms. Var. *bulbiferum,* "Top Onion", produces rather large bulblets in the inflorescence. – *A. ascalonicum,* "Shallot", probably indigenous in western Asia, is grown as a garden vegetable and mostly used pickled. – *A. schoenoprasum,* "Chive", growing wild in Eurasia, for example on rocky habitats in the Baltic, is grown for its terete leaves used as sandwich salad. – *A. porrum,* "Leek", indigenous in the Mediterranean, is widely cultivated for the mildly flavoured bulb and leaf bases used in cooking. – Also other species, such as *A. ampeloprasum, A. chinense, A. fistulosum, A. ledebourianum* and *A. odorum* are cultivated for their bulbs or leaves.

### Subfamily Gilliesioideae

Subfamily Gilliesioideae perhaps represents the most "advanced" of the subfamilies. It consists of about nine small genera, all in South America and most of them confined to Chile. They are small, bulbous herbs with a few basal, linear leaves and mostly few-flowered inflorescences subtended by two unequal, spathal bracts. The tepals are free or slightly fused and more or less unequal, sometimes by fusion appearing five or three in number. Adaxial appendages are often present inside and at the base of the tepals. The filaments are basally widened and fused into a staminal tube or cup. Three or four of the stamens may be transformed into staminodes, only the upper two or three then being fertile.

In a few genera, each with but one or two species, the flowers lack extrastaminal scales, e.g. in *Solaria* (2; Fig. 93 O–S) in Chile. Such scales are present in some other genera, among them *Gilliesia* (3), which has unequal tepals and a synandrium with only three functional anthers. *G. graminifolia* is sometimes grown as an ornamental.

The taxonomic position of Alliaceae has been variously interpreted. Some botanists include the Alliaceae in Amaryllidaceae, (HUTCHINSON 1934, 1973) or regard them as closely related to this family on the basis of similarities in many morphological characters, but this is not unequivocally supported by chemical characters, as the Amaryllidaceae lack saponins and possess unique alkaloids. Another family which may be more closely allied to Alliaceae than Amaryllidaceae is Hyacinthaceae. Most characters found in Alliaceae are met with in this family although its inflorescence is racemose and a parietal cell is cut off in the nucellus, a condition rare in Alliaceae. Finally, as is pointed out in particular by HUBER (1969), there are several interesting similarities between the Alliaceae and some generic groups in the Anthericaceae, such as the occurrence (at least in certain cases) in both families of articulated pedicels, of campylotropous ovules, of a thick phytomelan crust on the seed, of helobial endosperm formation and of similar chemistry (as regards saponins and crystal raphides).

Alliaceae may not be such a homogeneous family as is generally assumed. Important features such as endosperm formation, raphides, underground parts and articulated or non-articulated pedicel are quite variable. An analysis of the variation in the family is obviously needed.

**Amaryllidaceae** Jaume St. Hilaire (1805)  50:860 (Figs. 95–97)

Perennial or biennial herbs with subterranean bulbs with thick, fleshy bulb scales, only rarely without a typical bulb and with a rhizome, as in *Scadoxus* and *Clivia*. The roots are contractile and in all cases known have vessels with scalariform perforation plates (in contrast to the frequently more advanced, simply perforated vessels in the roots of Alliaceae). The basal stem is herbaceous, short, sympodially branching, generally with disti-

chous, basally concentrated leaves. The inflorescence-bearing scapes are terminal (as in Hemerocallidaceae, Hyacinthaceae and Alliaceae), but through sympodial branching may appear to be lateral.

The leaves are generally flat and dorsiventral, linear to almost orbicular, sheathing at the base, parallel-veined (veins often indistinct), mostly glabrous, and provided with anomocytic stomata and mucilage-filled cells or elongate sacs with raphides.

The usually glabrous scapes bear an umbel-like inflorescence comprising one to several helicoid cymes, the axes of which are suppressed. The inflorescence is subtended by from two to eight involucral scales which are free or basally connate. The flowers are not articulate on the pedicel. They are always epigynous, trimerous and bisexual and in most cases actinomorphic or weakly zygomorphic (strongly so in *Sprekelia*). The tepals are generally similar in the two whorls and vary from free to connate into a longer or shorter tube. Their colour may be white, yellow, purple or red, but not blue (violet in *Grittonia*). A perigonal corona structure, "paraperigone", forming a ring or tube is present in *Narcissus* (incl. *Tapeinanthus*; see under tribus Narcisseae).

The stamens are 3+3 in number (rarely 3 in *Zephyra* and up to 18 in *Gethyllis*). Their filaments are narrow or flat and inserted at the base of the tepals or in the tepal tube. In several genera a lateral, stipule-like, subulate appendage is present distally on the filament on each side of the anther (as in some genera of Alliaceae and Hyacinthaceae). The filaments may even be expanded and connate at the base to form a staminal "corona" structure, as in *Pancratium*, *Hymenocallis* and related genera. The anthers are epipeltate or perhaps basifixed in some genera (*Hessea*, *Leucocrinum*, *Galanthus*). They are generally elongate, longitudinally dehiscent or rarely (in *Galanthus*, *Leucojum* and *Lapiedra*) opening by apical pores. The tapetum is secretory (except, perhaps, in *Galanthus*, where it is reported to become amoeboid at an early stage); microsporogenesis is successive. The pollen grains are sulcate in most genera, but bisulculate in the tribe Amaryllideae, and are normally dispersed in the two-celled stage.

The ovary is inferior, tricarpellary, trilocular and provided with more or less distinct septal nectar grooves (except in Galantheae, where the nectar may be secreted from the distal part of the inner tepals). The style is simple with a punctiform, capitate or trilobate stigmatic apex, generally with a

Dry, papillate surface (more rarely Wet, e.g. in *Sprekelia*). The locules generally contain several to many, centrally inserted, anatropous or nearly anatropous (but never campylotropous), ovules, with two integuments or more rarely one and occasionally none. The ovules are crassinucellate and a primary parietal cell is probably cut off from the archesporial cell in most genera, but not in taxa studied of *Crinum, Eucharis, Narcissus* and some other genera, where the archesporial cell functions directly as the megaspore mother cell. Embryo sac formation generally conforms to the *Polygonum* Type but to the *Allium* Type at least in species of *Crinum* and, perhaps, *Pancratium*. The endosperm formation is nuclear or helobial, each type being found in several genera, and possibly of taxonomic interest within the family (see below).

The fruit in most genera is a capsule, but baccate, indehiscent fruits occur for example in *Chloanthus, Clivia, Cryptostephanus, Gethyllis* and *Haemanthus*. The seeds are highly variable: in European-centred genera, like *Galanthus, Leucojum* and *Narcissus*, they are mostly globose, ellipsoidal or ovoid, while in most of the primarily extra-European genera they tend to be more or less flattened. Exceptions are the species where the seeds do not enter a dehydrated stage but remain water-rich (75–92%; other seeds contain but 12–25% water) and germinate directly upon maturation. Such taxa are found, for example, in *Amaryllis, Boophane, Clivia, Haemanthus, Hymenocallis* and *Nerine*; the seeds in these cases are disseminated when the embryo is still quite small or are sometimes "viviparous".

The raphe sometimes develops a crest or wing, and an elaiosome is formed in the chalazal region in *Galanthus* and *Leucojum*. In the dehydrated seeds, the seed coat largely agrees with that in Alliaceae and Hyacinthaceae: the outer integument consists of several cell layers and its epidermis is covered with a mostly thin phytomelan crust. This, however, is lacking more often in Amaryllidaceae than in Hyacinthaceae, for example in species of *Leucojum, Galanthus* and *Sternbergia*, and in particular in water-rich seeds of some tropical and South African genera, especially in the tribe Amaryllideae. The subepidermal layers of the outer integument are mostly strongly compressed and those of the inner integument are collapsed into a thin film. The endosperm cells contain aleurone and fatty oils and in the terete seeds also cellulose (stored in the cell walls). Starch is frequently present, but in small amounts. The embryo is straight or slightly curved, cylindrical or in the flat seeds somewhat compressed, and generally more than half as long as the endosperm.

The water-rich seeds, on the other hand, are large, globose to ovoid, often green and lacking a phytomelan crust. The endosperm is richly developed and starch is the most important nutrient compound; it is also present in the integument(s), especially in the single integument of the seeds in some Amaryllideae (*Amaryllis, Boophane, Brunsvigia, Nerine*). In the ategmic (i.e. naked) seeds of some species of *Crinum* the nucellar remnants, endosperm and embryo make up the whole seed; the embryo is then large and green and may contain as much as 92% water.

*Chemistry*. Amaryllidaceae are easily characterized chemically. Steroidal saponins as well as allyl sulphides and similar compounds are probably lacking throughout, whereas particular alkaloids, not known to occur in other plants, are constantly present. Among the specific ("amaryllis") alkaloids are lycorin, belladin, haemanthanin, homolyrin, lycorenin, galanthanin, crinin and tazettin (WILDMAN 1968). About a hundred such bases are known which are considered to be biogenetically related. According to HEGNAUER (1963) an over-production of the amino acid tyrosine, which is the chief component in these alkaloids, has led to their biosynthesis. Chelidonic acid is widely distributed in Amaryllidaceae. The bulbs are rich in carbohydrates and also contain organic acids and soluble nitrogenous compounds. Calcium oxalate is found mainly in the form of raphide bundles, which are concentrated in mucilage-rich cells or sacs.

*Distribution*. The Amaryllidaceae are widely distributed. They are richly represented in the tropics and have pronounced centres in South Africa and, but less so, in Andean South America. Other groups have their centre in the Mediterranean. Groups of genera supposed to be phylogenetically related often have a particular geographic concentration.

*Taxonomy*. The family may be divided according to whether
1. the filaments are filiform or strongly modified, sometimes fused to form a coronal tube,
2. the pollen grains are sulcate or bisulculate,
3. the spathe bracts are two or more in number,
4. the style is strumose or not,
5. the integuments are single or double,
6. the fruit is a capsule or berry,
7. the anthers dehisce longitudinally or by apical pores,

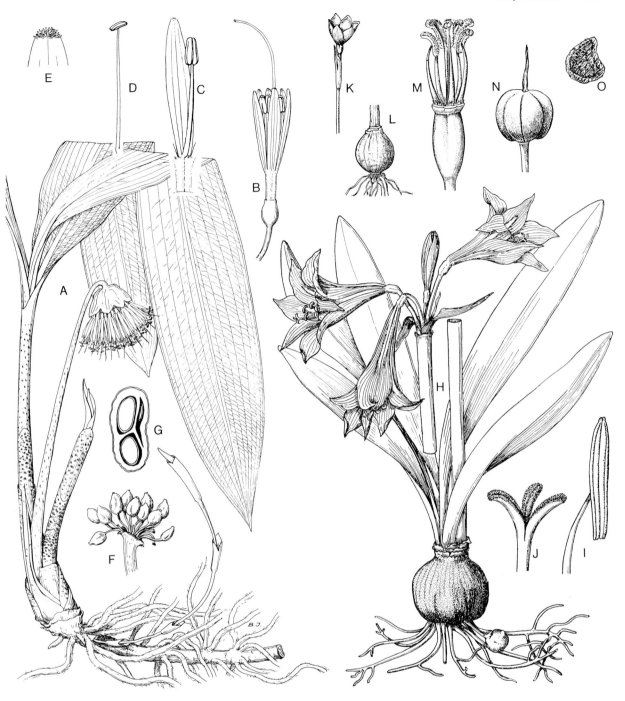

8. the bulb coat is fibrous or not when torn,
9. the scape is hollow or not, and
10. the flowers are actinomorphic or zygomorphic.

There have been various attempts to divide the family into natural generic groups. The present division is a simplified version of some of the recent ones (by PAX and HOFFMAN 1930; TRAUB 1957, 1970; D. and U. MÜLLER-DOBLIES 1978).

**Fig. 95.** Amaryllidaceae. **A–G** *Scadoxus nutans.* **A** Flowering plant; note the rhizome. **B** Flower. **C** Perianth segment and opposite stamen. **D** Stamen. **E** Apex of perianth segment. **F** Infructescence (berries). **G** Berry, transverse section. (FRIIS and BJÖRNSTAD 1971). **H–J** *Hippeastrum rutilum.* **H** Plant. **I** Anther. **J** Style apex with stigmatic lobes. (CABRERA, 1968). **K–O** *Zephyranthes filifolia.* **K** Unifloral inflorescence. **L** Bulb. **M** Androecium and gynoecium. **N** Capsule. **O** Seed. (CORREA 1969)

### Tribus Haemantheae

Tribus Haemantheae consists of about six genera and has its centre of variation in Southern Africa. They usually have quite a big bulb, but this is replaced by a rhizome and thick sheathing leaf bases in *Scadoxus* (10), which has a more tropical distribution. The scape is solid or sometimes hollow (variable within *Cyrtanthus*), and bears an inflorescence of from one to many flowers. The spathal bracts vary from two to eight and are free or basally fused. The perigone is generally long and tubular, funnel-shaped or cylindrical and sometimes slightly zygomorphic. The six filaments generally lack distal tips or appendages and are not fused into a cup. They are usually inserted in the perigonal tube. The pollen grains are consistently sulcate.

Within this group there is a variation from loculicidal capsules with black, phytomelan-encrusted seeds (as in *Cyrtanthus*), to baccate fruits with few or even solitary seeds, which are globose, water-rich and lack phytomelan (as in *Haemanthus*).

*Haemanthus* (40) is often considered to be a large genus but many of the species seem to be local colour variants. The leaves are distichous and vary from rather narrow and erect to elliptic or circular and pressed against the ground (as in *Massonia* of Hyacinthaceae). The scape is generally broad and flattened, and often spotted. The spathal bracts are wide, often more than two in number, and sometimes fleshy. In some species they are bright red and serve as the main attraction, while the flowers are small, numerous and often very densely clustered. The locules are one- or two-ovuled and the fruit is a berry with globose, water-rich seeds. The genus is conspicuous in the Cape flora, and several species appear abundantly after veld fires. Especially forms of *H. coccineus* are often grown as ornamentals. The genus is well-known by cytologists for its large chromosomes (BJÖRNSTAD and FRIIS 1972). – *Scadoxus* (10) (Fig. 95 A–E), from Arabia to Southern Africa, grows in shady forest habitats; the species have a loose bulb, with a distinct rhizomatous part or, more often, an elongate rhizome, and leaves that are not distichous.

*Cyrtanthus* (incl. *Vallota* and *Anoiganthus;* 48) is mainly South African. Most species have narrow red, yellow, pink or white tubular flowers. The ovary contains rather numerous ovules and the fruit is a loculical capsule with flat, ovate to oblong seeds with a wing at the micropylar end. *C. purpureus* (the former genus *Vallota*), has *Amaryllis*-like flowers. This and several other species of *Cyrtanthus* ("Kaffir Lilies") are grown as ornamentals.

*Clivia* (4) consists of rather large herbs without a typical bulb. They have several distichously arranged leaves and their tepals are fused only basally to form a funnel-shaped perigone, orange-red in colour. There are five to six ovules in each locule. The fruit, unlike that in *Cyrtanthus*, is a globose berry with one to few globose, water-rich seeds. *Clivia miniata* is commonly cultivated.

Possibly related to the Haemantheae, or to the Amaryllideae (below), is the genus (*Gethyllis*) (10) which has a short scape with a single spathal bract, a single flower with up to 18 stamens and a succulent fruit. Another possibly related genus is *Apodolirion* (6). Both of these genera are South African.

### Tribus Amaryllideae

Tribus Amaryllideae consists of about ten genera and has a pronounced centre in the winter rainfall area of Southern Africa, although the genus *Crinum* has a pantropical distribution (see below). The tribe consists of small to large-sized herbs with well-developed bulbs. When bulb coats and leaves are torn apart minute fibres become visible (except in secondarily cartilaginous leaf coats). The leaves vary from narrow and linear (sometimes coiled or undulate) to broad, and the scape is always solid. The inflorescence consists of one to many flowers and is always subtended by two free spathal bracts. The flowers are actinomorphic or, the larger funnel-shaped ones in particular, slightly zygomorphic. The perianth tube varies from short to long and from broad and funnel-shaped or campanulate to narrowly tubular. The filaments generally lack appendages and never cohere into a tube; nor is there a perigonal corona. The most distinctive characteristic of this tribe is that the pollen grains are *bisulculate*. The style is slender or in some genera (e.g. *Strumaria*) swollen and triquetrous at the base; the stigma is trilobate or capitate. An embryological peculiarity is that the ovules, at least in several of the genera, are unitegmic, and in at least one species of *Crinum* there is no integument at all. The fruit has a dry pericarp and is either indehiscent or bursts open irregularly or loculicidally. The single integument in the seed is generally green and, like the endosperm, is rich in starch grains. In *Nerine* its epidermis even has stomata! These seeds are rich in water and do not have a true resting period.

*Brunsvigia* (20) consists of rather large herbs, the flowering scape of which is often developed much

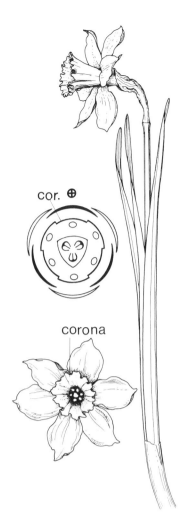

cor. ⊕

corona

**Fig. 96.** Amaryllidaceae. *Narcissus pseudonarcissus.* Plant, floral diagram and flower with coronal structure. (LARSEN 1973a)

earlier than the two to numerous distichous leaves. The locules of the ovary contain rather few ovules. The development of the globose infructescence is characteristic; its pedicels enlarge, then it dries up, separates from the stem and rolls around driven by the wind. The capsules are inflated and turbinate and the seeds are green and may germinate in situ. *B. josephinae* is sometimes grown as an ornamental. – While *Brunsvigia* has up to ca. 30-flowered "umbels", *Boophane* (5) may have a hundred flowers, but otherwise resembles the former genus in most features. – *Nerine* (ca. 30) has almost free tepals which are often undulate and usually pink to crimson. *N. sarniensis,* the "Guernsey Lily", *N. bowdenii, N. curvifolia* and other species are important in horticulture. – A few related genera are *Strumaria* (10), *Hessea* (10) and *Carpolyza* (1), all restricted to the dry regions of South Africa. They are small herbs with bulbs

covered by fibrous tunics and have small, white to pink flowers with short or no perianth tube; the seeds are soft and fleshy as in the other genera of this group. *Strumaria* and *Hessea* are closely related; both having a strumose style and a globose fruit (GOLDBLATT 1977a).

*Amaryllis (1) belladonna,* the "Cape Belladonna", is a somewhat *Crinum*-like or *Hippeastrum*-like, fairly large herb with large, pink, campanulate flowers. It is indigenous in the south-western part of the Cape Province, but is now a widely cultivated ornamental. There has long been a nomenclatural confusion between this genus and the American genus *Hippeastrum,* the latter having frequently been called "Amaryllis" (TRAUB and MOLDENKE 1949; TJADEN 1981). The nomenclature seemed to be settled (DANDY and FOSBERG 1954; etc.), but TRAUB recently (1983) has considered *Amaryllis* to be typified by a species of *Hippeastrum.*

*Crinum* (130) consists of herbs which may reach considerable dimensions and may have a falsely leafy stem (resulting from the scape being enclosed by leaf sheaths as in *Musa*). In some species prolonged (basal) growth of the leaves results in their tips trailing on the ground. The flowers are generally large and have a long white, rose or crimson perigone tube in which the filaments are inserted, and the fruits are capsules with fleshy seeds. The ovules are generally unitegmic or occasionally ategmic, in which case the seeds consist mainly of the endosperm and embryo. *Crinum* is widely distributed in the tropics and warm temperate regions. It contains many tall species, a number of which are cultivated, e.g. the Indomalaysian *C. amabile,* the tropical Asiatic *C. latifolium* and *C. asiaticum,* the Himalayan *C. amoenum,* the American *C. americanum* and *C. erubescens,* the South African *C. bulbispermum* and *C. campanulatum,* the West African *C. laurentii* and *C. nutans* and the Australian *C. pedunculatum.*

### Tribus Hippeastreae

Tribus Hippeastreae consists of about ten genera with a pronouncedly American distribution. They represent herbs of various dimensions with tunicated bulbs which do not expose any fibres when torn apart. The leaves are linear and flat, and the scape generally hollow. The inflorescence is subtended by an involucre of two or four spathal bracts which may be free from each other (as in *Hippeastrum* and some related genera) or, more often, fused into a tube (as in *Zephyranthes* and several related genera). The flowers may be solitary or rather numerous in the inflorescence, and are actinomorphic or often zygomorphic (as, in particular, in *Sprekelia*). The tepals are almost free or more often fused into a tube, while a "paraperigone" is mostly lacking or, when present, inconspicuous. The filaments are often of different lengths. The anthers dehisce longitudinally and the pollen grains are sulcate. The style is not thickened and the locules have few to several, bitegmic ovules. The fruit is a loculicidal capsule with seeds that are generally flat, dry, encrusted with phytomelan, and often winged.

*Hippeastrum* (55–75) (Fig. 95H–J) is American, ranging from the West Indies and Mexico to Argentina. It consists of often large herbs with distichous flat leaves and scapes with two free spathal bracts. The large, mostly purple or red flowers are funnel-shaped and slightly zygomorphic. The tepals are free or basally connate, and the filaments are unequally long and more or less upcurved. The application of the name "*Amaryllis*" to this genus, here (dubiously) considered incorrect, persists in horticultural circles. Numerous species are grown as ornamentals, e.g. *H. advenum, H. bifidum, H. elegans, H. puniceum, H. reginae* and *H. striatum*. – *Rhodophiala* (30), in South America, consists of species having slender scapes with a few-flowered pseudo-umbel of funnel-shaped, variously coloured, flowers. The flowers may have a "paraperigone" of scales or bristles at the base of the tepals.

*Zephyranthes* (ca. 60) is distributed from Chile to far up into North America. It has narrow, linear leaves and a slender, short scape with a unifloral inflorescence subtended by a tubular involucre of spathal bracts. The flower is actinomorphic, erect or suberect and funnel-shaped, with a short tube. A "paraperigone" of bristles or scales may be present. The capsules have locules with few, flat, black seeds. Ornamental species include *Z. atamasco, Z. bifolia, Z. brasiliensis, Z. candida, Z. smallii* and *Z. tubiflora. Sprekelia* (1) *formosissima*, in Mexico to Central South America, is extraordinary in having a single-flowered scape with a strongly zygomorphic, firm, red flower with spreading tepals, the lower three of which enclose the long filaments and style. The flower is strongly adapted to bird pollination. The seeds are D-shaped, narrowly winged and black.

### Tribus Lycorideae

Tribus Lycorideae differs from the preceding mainly in having a compact (not hollow) scape, and in being Asiatic in distribution. – *Lycoris* (17), distributed from Burma to Korea and Japan, has a somewhat *Nerine*-like appearance, but the capsular fruits have round, smooth, black seeds. It may represent a rather primitive element in the family. A "paraperigone" is sometimes present in the form of small scales at the throat of the tepal tube. – Related is *Ungernia* (8) in central Asia.

### Tribus Stenomesseae

Tribus Stenomesseae (incl. Eustephieae), like the following tribe, is South American. It includes ca. 14 genera. In this group the leaves, which are sometimes petiolate, vary from thin to fleshy, as in *Rauhia*. The scape is solid and the flowers mostly yellow or red and often pendulous (in many cases exhibiting the syndrome of ornithogamy). Their stamens are often fused basally into a staminal cup; in other genera they are free but variously appendiculate. The filaments may be callous at the base and their appendices may even be petaloid. The filaments are sometimes long and declinate. The fruit is capsular and the seeds are black, and often flat, either D-shaped and winged or oblique and unwinged.

*Stenomesson* (20), ranging between northern Chile and Ecuador, generally has pendulous flowers and a well-developed staminal cup. – *Chlidanthus* (2), from Mexico to Bolivia, has flowers with a long tepal tube. The filaments of the outer staminal whorl are short and inserted in the mouth of the perigone tube, while those of the inner whorl are longer and inserted basally in the tube.

*Phaedranassa* (6), which ranges from Andean Peru to Costa Rica, has petiolate leaves. – The tubular flowers of *Eustephia* (4), in Peru and Argentina, have free filaments with tooth-like appendages inserted in the perigone tube.

### Tribus Eucharideae

Tribus Eucharideae, with about five genera, is likewise mainly American (*Eurycles* excepted). The staminal filaments in this group, as in some of the following, are either laterally dilated and fused into a basal cup or supplied with basal appendages. In this tribe the seeds are generally large, globose and fleshy, but in contrast to the Amaryllideae, with similar seeds, the fruit is a loculicidal capsule. The flowers vary in colour, from being white and *Pancratium*-like to red-and-yellow as in the Narcisseae.

*Eucharis* (20) in Central America and northern South America, has a short perigone tube dilated at some distance from the base. The filaments, inserted on the margin of the perigone tube, are basally indistinctly appendiculate and narrow. The flowers are white or slightly yellowish, as in the related genus *Hymenocallis* (55), which ranges from the temperate South-Eastern U.S.A. to Northern South America. In *Hymenocallis*, the filaments are basally fused into a cup of varying appearance. *Eucharis* and *Hymenocallis* include several species widely cultivated as ornamentals, e.g. *Eucharis grandiflora* and *E. candida*, and *Hymenocallis narcissiflora, H. macleana, H. vargasii, H. longipetala* and *H. expansa*.

### Tribus Pancratieae

Tribus Pancratieae consists of four Old World genera only, ranging from South Africa to Macaronesia and the Mediterranean Region and further eastwards into tropical Asia. They are large to small herbs with linear or linear-lorate leaves and inflorescences with one to several generally white flowers. The perigone is basally tubular and funnel-shaped with spreading segments. The six stamens are either fused basally into a staminal cup with narrow filaments continuing above, or free and with lateral basal appendages. A capitate stigma terminates the style. The fruit is capsular and the seeds dry, black, turgid and angled or spherical.

*Pancratium* (20) ranges from the Canary Islands through the Mediterranean Region to Asia. The Mediterranean *P. maritimum* and *P. illyricum* are grown as ornamentals. – *Vagaria* (1), in Asia Minor, has free stamens with basal appendages.

### Tribus Narcisseae

Tribus Narcisseae consists of small to medium-sized herbs with linear leaves and a solid scape bearing an inflorescence of one to several (rarely numerous) flowers. Its spathal bracts are basally fused into a tube. The flowers are actinomorphic and have six equal tepals. Inside this a "paraperigone" or "corona" is generally present, which varies from a dentate, rudimentary cup to a large, broadening and sometimes undulate tube. The anthers dehisce longitudinally and the pollen

grains are sulcate. The ovary is trilocular, with several bitegmic ovules in each locule. The fruit is a capsule with globose to angular, dry and black seeds.

*Narcissus* (27) has a typically West-Mediterranean distribution and is here taken to include *Tapeinanthus* (= *N. humilis*) in Spain-Morocco, which has a reduced paraperigone. The colours of the tepals usually vary in the genus from white to bright yellow and the paraperigone from white or light yellow to orange or orange-red.

*Narcissus* is extremely variable in Spain and Portugal, where most species occur. Most are spring-flowering, but a few, namely *N. viridiflorus* (which is indeed green-flowered), *N. humilis* (see above), *N. serotinus* and *N. elegans* flower in the autumn. Several of the spring-flowering taxa have been subject to intensive breeding and are important garden ornamentals. They include the variable *N. tazetta* ("Tazettas"), indigenous in central Mediterranean, *N. poeticus* ("Poet's Narcissus"), indigenous in regions from Spain to Greece, and *N. pseudonarcissus* ("Daffodil") (Fig. 96). *N. poeticus* and *N. tazetta* have a short paraperigone, coloured reddish orange and yellow respectively, while in *N. pseudonarcissus* the paraperigone is large, yellow and undulate. A variable small-sized species complex is centred on the West-Mediterranean species *N. bulbocodium* which, relative to the perigone, has a large paraperigone. This and the likewise small-sized, western *N. cyclamineus*, with reflexed tepals, are often cultivated in gardens.

*Sternbergia* (5), ranging from the Mediterranean to Iran, though lacking a paraperigone, perhaps belongs here.

### Tribus Galantheae

Tribus Galantheae might be united with the Narcisseae, but may be distinguished in terms of the anthers dehiscing by pores instead of slits. The tribe includes two or three genera with a Mediterranean-Western Asiatic distribution.

These consist of small to moderate-sized herbs with tunicated bulbs and sessile, lorate leaves, which are from broadly linear to almost filiform. The scape is cylindrical or compressed, solid or hollow, and bears one or few, usually pendent, flowers subtended by one or two partly green spathal leaves. The tepals are free or basally fused and are white, often with green or yellow spots. Usually the anthers open by apical pores. The pollen grains are sulcate. The pistil is trilocular and has several bitegmic ovules in each locule. It devel-

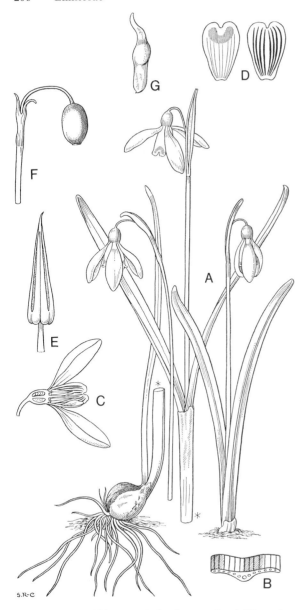

ops into a capsule with turgid, subglobose or ellipsoidal, pale or black seeds.

*Galanthus* (5) ranges from Mediterranean Europe to Asia Minor and the Caucasus. Some species are polymorphic and up to 13 species are often recognized. The scape is normally one-flowered and the bract single but two-ribbed. The tepals are unequal, the inner being short, truncate, notched and convolute. The flower is white but the inner tepals have one or two green blotches externally and green striation within. All the species are early-spring flowering, but the widespread *G. nivalis* has autumn-flowering forms. *G. nivalis* and other species, "Snowdrops" (Fig. 97), are very common garden ornamentals in cool-temperate regions.

*Leucojum* (8), ranging from Portugal to Crimea and Armenia, has one- to five-flowered inflorescences with one or two spathal leaves. The flowers are campanulate. The tepals are almost equal and have green or yellow tips. The subglobose, black or brown seeds may or may not have a crest-like appendage. This genus also includes mostly spring-flowering ornamentals, such as *L. vernum* and *L. aestivum* but *L. autumnale,* a slender herb, flowers in the autumn.

The Amaryllidaceae form one of the climax groups in the Asparagales. They are probably most closely allied to Alliaceae and Hyacinthaceae and share with them the bulb, the leafless scape, the normally phytomelan-coated seeds and various other attributes. They are definitely not closely related to the Hypoxidaceae, the Agavaceae (subfam. Agavoideae), the Haemodoraceae or the Alstroemeriaceae, with which they have formerly been united on the basis of the shared epigynous flowers.

**Fig. 97.** Amaryllidaceae. *Galanthus nivalis.* **A** Flowering plants and bulb. **B** Leaf, transverse section. **C** Flower, longitudinal section. **D** Tepals of the inner whorl, outer and inner surface. **E** Stamen, longitudinally poricidal dehiscence. **F** Fruit. **G** Seed, with terminal appendages. (Ross-Craig 1972)

# Order Melanthiales

*Two Families:* Melanthiaceae and Campynemaceae.

Perennial or rarely annual herbs with short or long rhizomes, rarely corm-like, and exceptionally with a bulb-like base although without nutrient bulb-scales. The plants are rarely woody in the basal part. The roots are fibrous and generally provided with vessels that have scalariform perforation plates, whereas vessels are absent in stems and leaves. Vessels are completely lacking in one or two wholly chlorophyll-deficient genera. The aerial stem varies from being leafy to comprising a leafless scape and a basal rosette.

The leaves are usually spirally set but sometimes distichous. They are dorsiventral, often more or less flat, in some genera ensiform and *Iris*-like. The leaves are sheathing at the base, linear, lanceolate or rarely ovate, except in the achlorophyllous taxa, where they are scale-like; occasionally tapering at the base and nearly pseudopetiolate. Ligules occur rarely (*Pleea*). Cells containing crystal raphides occur in most genera.

The inflorescences are simple or compound racemes or spikes or rarely panicles; the flowers are bracteate or ebracteate, bisexual or very rarely unisexual, hypogynous, half-epigynous or epigynous (Campynemaceae), trimerous, and generally actinomorphic, being zygomorphic only in *Chionographis*.

The perigone consists of 3+3 tepals, which are free from each other or sometimes slightly connate basally or occasionally forming a campanulate or tubular perigone (in the hemi-epigynous taxa). It is generally inconspicuous, white, pale yellow, brown to purple or, rarely, bright yellow (*Narthecium*). It is fused to the basal parts of the ovary in a few genera. The tepals are generally similar in the two whorls and in some genera have basal nectaries. Spotted patterns and spurs are usually lacking.

The stamens are usually 3+3 but in *Pleea* up to 9 or 12. The filaments are free from each other; in *Narthecium* they are provided with spreading hairs. The anthers are basifixed or dorsifixed-hypopeltate, generally extrorse but sometimes introrse (a combination thereof in *Campynemanthe*), and dehisce longitudinally. The tapetum is secretory (Fig. 98 W). Microsporogenesis is successive or rarely (*Tofieldia*) simultaneous. The pollen grains are generally sulcate, but are sometimes bisulculate (Tofieldieae) or tetraforaminate (Chionographi-

deae). The exine surface is generally reticulate, rarely spinulose. The pollen grains are dispersed in the two-celled state; the generative cell is small, narrow and peripheral in the pollen grain.

The pistil is tricarpellary and trilocular or rarely unilocular (at least in *Campynema*). In most genera the stylodia are separate. Sometimes the locules are free in their upper part, and in the Petrosavieae and in *Harperocallis* the three carpels are separate nearly to their base (EL-HAMIDI 1952). An apically tribrachiate style occurs in *Aletris,* but some genera, such as *Tofieldia,* have a single style and a capitate stigma. The stigmatic surface is Dry in *Veratrum.* Septal nectaries are often lacking but occur in several genera (e.g. *Petrosavia, Protolirion, Tofieldia, Zigadenus*). The locules contain two to numerous ovules, which are crassinucellate and anatropous or rarely (*Petrosavia*) campylotropous. A parietal cell is cut off from the archesporial cell (Fig. 98 C) and forms a parietal tissue. Embryo sac formation follows the *Polygonum* Type (Fig. 98 I–L), and endosperm formation where known is helobial (Fig. 98 N–O) (a difference from all Liliales) (STENAR 1925, 1928 b; EUNUS 1951).

The fruit is generally a capsule or rarely (where the carpels are almost separate) consists of three follicles which open up along their inner suture. The capsules are loculicidal (most Aletreae and Metanarthecieae) or septicidal (*Clara, Hewardia* and *Pleea*), but *Tofieldia* and the Veratreae have carpels free in the apical parts; their capsules open ventricidally from the apex towards the base and along the central parts. Arils and fleshy caruncles are lacking. The seeds are generally rounded and isodiametric in transection, always elliptic, elongated longitudinally but not conspicuously flattened, and rarely longitudinally striate (*Aletris*). However, they are generally provided with wings or terminal appendages. The seeds are winged all around in *Melanthium* and *Veratrum.* Terminal appendages are of considerable length in some genera, for example, in *Narthecium* and *Tofieldia,* and the chalazal one is usually longer than the micropylar. The testal part of the seed coat has few, sometimes only two, cell layers; the epidermal cells are elongate and colourless, either well-preserved or collapsed. The testa lacks phytomelan (a difference from nearly all capsular Asparagales) as well as phlobaphene. The tegminal part of the seed coat is thin and collapsed, red-brown or yellowish; only rarely are its cells retained, though flattened (*Narthecium*). The endosperm usually consists of thin-walled, frequently isodiametric cells containing aleurone and fatty oils and sometimes traces of starch, the latter in the form of rounded grains. More often than in Liliales is the embryo small; it is ovoid or globose ($^1/_7$–$^1/_9$ of the length of the endosperm though sometimes considerably longer).

*Chromosome Numbers.* There is no single clear basic chromosome number in the order: $x$ may be 8 or 11 (*Veratrum*), 12 (*Chionographis*), 13 (*Narthecium, Metanarthecium, Aletris*), 15 (*Tofieldia, Xerophyllum*), or 17 (*Helonias, Heloniopsis*), suggesting an aneuploid series.

*Chemistry.* Among chemical characteristics should be mentioned the common presence of raphides (in both families). Steroidal saponins are known to occur in many genera; those of *Narthecium* are thought to cause liver disease (icterus) in non-pigmented sheep (HEGNAUER 1963). Alkaloids derived from steroidal precursors such as cholestanol as well as those of the C-nor-D-homo Type (SEIGLER 1977) are found in several genera in the Veratreae. Thus, roots of *Veratrum* contain toxic alkaloids (veratramin, jervin and allied compounds). Also the alkaloids of the seeds of *Schoenocaulon* (= *Sabadilla*) have medical use; dissolved in acetic acid they are frequently used as insecticides.

*Parasites.* It is interesting to note that two related species of the rust genus *Puccinia* (section *Caricinae*) attack genera of this order only, viz. species of *Amianthium, Schoenocaulon, Veratrum, Xerophyllum* and *Zigadenus* (HOLM 1966), which further indicates that these genera form a biogenetically related group as regards proteins and/or other compounds.

*Distribution.* Melanthiaceae is a mainly Northern Hemisphere family distributed over Eurasia and Northern America with a few genera extending to South America. The members are mainly temperate to boreal (rarely alpine or tropical), and grow in forests, marshes, meadows, etc. Two genera (*Petrosavia* and *Protolirion*) are achlorophyllous, sometimes quite whitish, forest floor plants occurring in South-Eastern Asia and on Borneo, and on the Malay Peninsula, respectively. In contrast, Campynemaceae occur in the Southern Hemisphere, *Campynema* in Tasmania and *Campynemanthe* in New Caledonia.

*Taxonomy, Relationships.* The treatment of these families in a separate order is the logical consequence of their combination of attributes. They lack the features typical of each of the Asparagales, Liliales and Burmanniales.

Thus they differ from the Asparagales in that the seeds are not phytomelaniferous (although the fruits are capsular), the tepals are inconspicuous, the stylodial branches are generally separate and the embryo is small.

Unlike the Liliales endosperm formation is helobial and septal nectaries may be lacking; the tepals are less conspicuous and rarely spotted as is characteristic of the Liliales. Raphides are generally present.

The Melanthiales might seem to be more closely allied to the Burmanniales, but the tenuinucellate ovules (where also no parietal cell is formed) and diminutive seeds that characterize Burmanniales make a technical separation easy. Also, the vast majority of Burmanniales are achlorophyllous plants. The Burmanniales share some embryological and seed attributes with Philydrales (HAMANN, personal communication), and we hesitate to associate them intimately with Melanthiaceae.

AMBROSE (1980), who studied the Melanthiaceae from a cladistic point of view, questioned that the family is monophyletic. Its members are variable in a number of characters which are generally constant in other liliifloran families. Further evidence as regards the host specificity of particular *Puccinia* species such as pointed out by HOLM (1966) may become important in this connection.

*Key to the Families*

1. Flowers hypogynous or half-epigynous . . . . . . . . . . . **Melanthiaceae**
1. Flowers epigynous . . . . . . **Campynemaceae**

**Melanthiaceae** Batsch (1802)    25:150
(Figs. 98–100)

Description as for the order, above, but flowers not epigynous and placentation never parietal.

It is possible that a division of Melanthiaceae into three or four families would better reflect the variation in the group. For example, Petrosaviaceae has already received recognition. Division of the family into tribes has been proposed in earlier treatments (e.g. KRAUSE 1930; HUTCHINSON 1934, 1959, 1973).

The family has been dispersed among a number of tribes, generally placed in Liliaceae s. lat. in previous literature. In BENTHAM and HOOKER (1883) its genera were placed mainly in the tribes Narthecieae and Veratreae; in KRAUSE (1930) under the tribes Tofieldieae, Petrosavieae, Helonieae, Hewardieae, Veratreae and Aletreae, and in HUTCHINSON (1973) under the tribes Narthecieae,

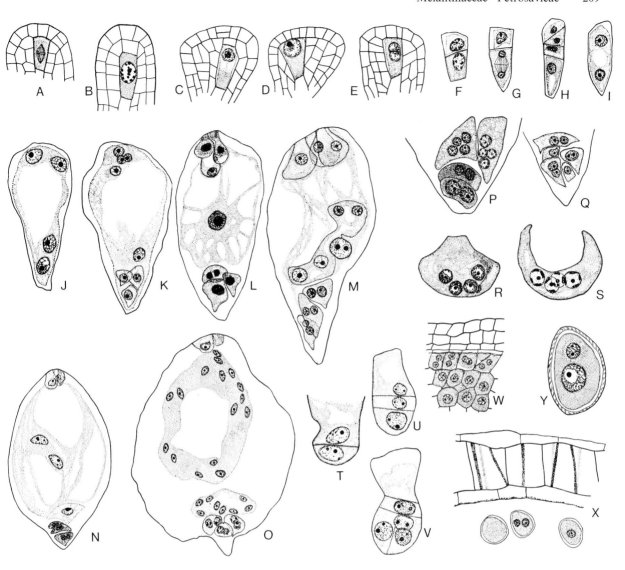

Helonieae and Veratreae and in the family Petro-saviaceae. Recently, AMBROSE (1975, 1980) has investigated their variation to test their possible homogeneity, in which he utilized numerical methods (AMBROSE 1980), finding rather different constellations of genera.

We shall try here to present a division of the family taking into account data from the available sources.

### Tribus Petrosavieae

Tribus Petrosavieae consists of two saprophytic, achlorophyllous genera, *Petrosavia* (1) and *Protolirion* (3) (Fig. 99), which grow in rain forests in Eastern Asia, Malaya and Borneo. They are small, vesselless herbs with thin rhizomes and scaly leaves. The flowers are almost hypogynous, having

**Fig. 98.** Melanthiaceae. *Amianthium muscaetoxicum,* embryological details. **A–L** development with (**B–C**) and without (**D–E**) formation of parietal cell, meiosis (**F–H**), and embryo sac formation (**I–L**), with the polar nuclei having fused in **L. M** Two embryo sacs in the same nucellus, one mature and one in the 4-nucleate stage. **N–O** Endosperm formation and development, according to the Helobial Type. **P–Q** Increase in nucleus number in antipodal cells and increase also in number of antipodal cells. **R–S** Antipodal cells assuming peculiar shapes (resembling chalazal chambers). **T–V** Development of the early embryo. **W** Anther, sectioned to show tapetum and archesporial cells (both dark). **X** Endothecium (with spiral wall thickenings) and nearly mature pollen grains. **Y** Pollen grain. (EUNUS 1951)

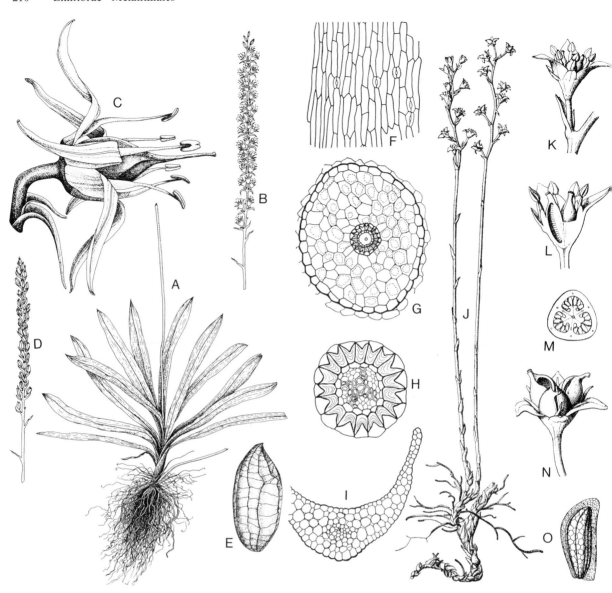

**Fig. 99.** Melanthiaceae. **A–E** *Metanarthecium luteoviride.* **A–B** Plant. **C** Flower. **D** Infructescence. **E** Seed; note that this is not black. (UTECH 1978 b). **F–I** *Petrosavia stellaris.* **F** Stomata. **G** Stem, transverse section. **H** Stele. **I** Scale-leaf, transverse section. (STANT 1970). **J–O** *Protolirion sakuraii.* **J** Plant. **K** Flower. **L** Same, longitudinal section. **M** Ovary, transverse section. **N** Trifolliculus. **O** Seed. (KRAUSE 1930)

only basally fused carpels which mature into follicles. Septal nectaries are present. The pollen grains are sulcate to inaperturate. – These genera have an anatomy very much like that in Triuridales. Their carpels, largely free, have been taken as an indication that these genera form a very primitive group of Liliiflorae.

### Tribus Melanthieae

This tribe consists of large or medium-sized herbs with bulb-like bases (*Amianthium, Stenanthium, Zigadenus* spp., *Schoenocaulon,* etc.) or often short, thick rhizomes (*Melanthium, Veratrum*). The leaves are variable, from narrow and grass-like to broad, lanceolate-ovate and prominently veined (as in *Veratrum*). The stems end in a spike, raceme or panicle which is sometimes compound and extensive. The flowers are mostly white to dark purple, and the stamens bear cordate to reniform anthers dehiscing extrorsely. The pollen grains are sulcate. Each carpel of the pistil ends in a separate stylodium. The carpels open ventricidally and their seeds are larger than in the following groups. In the seed coat the outer integument is retained (not collapsed as in most other genera). The plants are rich in alkaloids. Among the genera in this group may be mentioned *Schoenocaulon, Stenanthium, Amianthium* and *Melanthium,* restricted to America, and *Zigadenus* and *Veratrum* which occur in Eurasia as well.

*Schoenocaulon* ( = *Sabadilla;* 10) ranges from Florida to Peru. It consists of tough herbs with a bulb-like corm, grass-like leaves and dense-flowered spikes with partly unisexual flowers. The seeds of *S. officinale,* in mountain meadows of Guatemala and Venezuela, are used to produce an insecticide (see above). – *Zigadenus* (15) occurs from Siberia to North America and in Mexico. Its tepals are basally adnate to the ovary and bear basal nectar glands. *Z. elegans* is the most widespread species. Some species are grown as ornamentals.

*Melanthium* (5) is a North American genus. Like *Veratrum* it is rhizomatous and has a pubescent inflorescence. The leaves are, however, narrow and the tepals clawed. *M. virginicum* is grown as an ornamental. – *Veratrum* (45) is the largest genus in the family and has a wide distribution in the Northern Hemisphere. The plants grow in meadows. They are mostly large herbs, which may reach a height of 2 m or more and have prominently veined leaves. The flowers vary from white or pale greenish to dark purple. The pistil has separate stylodia; its ovary develops into a sometimes inflated septicidal capsule. Several species, like the Eurasiatic *V. album* and *V. nigrum,* the Japanese *V. mackenii,* and the American *V. californicum, V. insolitum* and *V. viride* are grown as ornamentals. *V. album* is grown for the subterranean parts, which contain alkaloids (see above) and are used in medicine as a sedative and cardio-vascular depressant. – *Stenanthium* (5) in North America, has polygamous flowers and tepals basally fused to the ovary. It is a bulbous plant with linear or lanceolate leaves.

## Tribus Narthecieae

Tribus Narthecieae consists of a number of genera with short, tuberculate rhizomes and erect herbaceous, leafy stems with terminal spikes or racemes. The stamens are mostly extrorse and the pollen grains are sulcate. Nectaries are found at the tepal bases, and septal nectaries are lacking. The carpels are either apically separate to form free stylodial branches or completely fused and forming a single style, as in *Heloniopsis, Nietneria, Narthecium* and *Metanarthecium.* In *Aletris* and *Nietneria* the perianth is adnate to the ovary for

about half its length. The capsules are mostly loculicidal, and the seeds small, generally with a collapsed outer integument; terminal appendages are present or absent. Saponins and oxalate raphides are present but no alkaloids are known in this tribe.

*Helonias* (1) *bullata* is a herb up to 80 cm, occurring in swampy regions in Atlantic North America and sometimes grown as an ornamental. The tepals are purplish and spreading, the anthers blue. – *Heloniopsis* (4), in Korea, Japan and Taiwan, has campanulate, carmine-red flowers; *H. orientalis* is often grown as an ornamental. – *Aletris* (10) in Eastern Asia, Indomalaysia and North America, has grass-like leaves and yellow or white, connate tepals. The dried underground parts of *A. farinosa,* "Colic Root", in eastern North America, are used by the Indians as a diuretic, and against colic. – *Metanarthecium* (5) (Fig. 99 A–E) occurs, in subalpine meadows in Japan and Formosa. – *Nietneria* (1), in the mountains of Guayana, has stiff somewhat falcate leaves. – *Narthecium* (8) in Eurasia as well as North America grows in marshy places. In AMBROSE's (1980) study it was shown to be the genus most closely similar to *Nietneria,* but it has a number of singular features. It is a herb with creeping rhizomes forming clones of plants with crowded, ensiform, *Iris*-like leaves. The erect stems are basally leafy and bear racemes of yellow, stellate flowers with hairy filaments and introrse anthers. The seeds are pale and have long terminal appendages. In contrast to the other genera, the tegmen does not collapse but consists of distinct cells with lemon-yellow contents. *Narthecium ossifragum* (Fig. 100 A–E) is a common subatlantic herb growing in acid bogs in western Europe. – Also *Lophiola* (2) in Atlantic North America seems to belong in this group of genera (AMBROSE 1980).

## Tribus Tofieldieae

This consists of two genera, *Tofieldia* (17) (Fig. 100 F–K), a North temperate and arctic genus growing in moist alpine to cold-temperate habitats. In this genus should perhaps be included *Pleea* (1) *tenuifolia* (UTECH 1978 b), a marsh plant in Southern U.S.A. Closely allied is also *Harperocallis* (1), in Florida. These genera are relatively small herbs with creeping rhizomes, distichous, often unifacial leaves and small, often whitish flowers with a non-caducous perigone. *Tofieldia* has 3 + 3 or (*Pleea*) 6 + 3, or even 6 + 6, stamens with introrse, subbasifixed anthers. Microsporogenesis

**Fig. 100.** Melanthiaceae. **A–E** *Narthecium ossifragum.* **A** Plant. **B** Dehiscing capsule. **C** Tepal and opposite stamen; note the pubescent filament. **D** Ovary. **E** Seed with long, terminal appendages. **F–K** *Tofieldia pusilla.* **F** Plant, to the *left* in fruiting stage. **G** Flower. **H** Stamen. **I** Pistil. **J** Capsule. **K** Seeds. (**A–K** ROSS-CRAIG 1972). **L–P** *Xerophyllum tenax.* **L–M** Plant. **N** Inflorescence. **O** Flower. **P** Fruit. (**L** and **P** HITCHCOCK et al. 1969; **M–O** TAKHTAJAN 1982)

at least in *Tofieldia* is simultaneous, and the pollen grains are bisulculate. Septal nectaries are known in *Tofieldia*. The carpels are usually free in the apical part, with free stylodia or in some species of *Tofieldia* with a tribrachiate style. In *Harperocallis* the carpels are free almost to the base. The capsules are ventricidal to septicidal, and the seeds are small and have terminal appendages. Oxalate raphides are lacking, at least in *Tofieldia*.

meriaceae the leaf lamina is generally inverted. The veins are parallel or arching and apically converging; when the leaves are broad (as in *Tricyrtis, Disporum,* certain orchids) a minor reticulate venation may be present between the main veins. Vessels are normally lacking in the leaves (present only in some orchids and in *Sisyrinchium* of Iridaceae and then having scalariform perforation plates). The stomata are anomocytic except in Orchidaceae and Cypripediaceae, where the stomatal conditions are highly variable and partly characteristic for the subgroups (see further under Orchidaceae). Silica bodies are lacking except in a number of genera of Apostasiaceae, Cypripediaceae and Orchidaceae. Raphides are lacking in all families except Alstroemeriaceae, Apostasiaceae, Cypripediaceae and Orchidaceae, where they are abundant and in Uvulariaceae, where they occur very rarely (GOLDBLATT, personal communication). Unicellular or uniseriate hairs are scattered in the order. Laticifers are lacking.

The inflorescence is terminal or (e.g. in some Orchidaceae) lateral. It is generally not situated on a leafless scape from ground level, except in some Cypripediaceae and many Orchidaceae. It constitutes a raceme or a thyrse or is composed of one or more generally monochasial cymes. There is no articulation between the flower and its pedicel, nor is a typical "pericladium" developed.

The flowers are bisexual or rarely unisexual, trimerous, hypogynous or frequently epigynous, actinomorphic or weakly to (often) very strongly zygomorphic, and vary from minute to large and conspicuous. They are pollinated by various kinds of insects or rarely by birds.

As in Asparagales, the tepals are generally 3 + 3 in number and may or may not be equal in the two whorls. Unlike most Asparagales and many Melanthiales, the tepals generally have perigonal nectaries, although their function may be substituted by osmophores. An exception is Iridaceae subfamily Ixioideae, which has septal nectaries (see the discussion under Iridaceae). Nectariferous spurs or sacs are sometimes present at the base of one, three or more of the tepals. In Cypripediaceae and Orchidaceae, where the flowers are strongly zygomorphic and generally resupinated, the inner median tepal, then directed downwards, is generally developed as a more or less conspicuous *labellum,* which may be provided with a spur (see further under these families). The tepals vary from inconspicuous to very large and showy and in some taxa of nearly all families have variegated, checkered, drop-like, radially striate or transverse-

ly figured colour patterns, which are otherwise very rare in monocotyledons.

The stamens are 3 + 3, 3, 2 or 1 in number; when they are three in number they usually represent the outer whorl (Iridaceae, Geosiridaceae), when two they belong to the inner whorl (Cypripediaceae, Apostasiaceae pro parte) and when solitary the stamen belongs to the outer whorl (Orchidaceae). The filaments are generally narrow and without lateral appendages or apical lobes; they are more or less connate in some Iridaceae, in Calochortaceae and rarely in Colchicaceae (*Sandersonia*). In Apostasiaceae, Cypripediaceae and Orchidaceae, there are only one to three stamens, but sometimes rudimentary staminodes, and these are more or less intimately fused with the style into a gynostemium, the details of which are described under each family. The anthers are basifixed or (as in many Liliaceae) dorsifixed-epipeltate or the connective may form a tube around the tip of the filament, making the anthers falsely basifixed. The anthers are extrorse in most families, but introrse in others, e.g. in Apostasiaceae, Cypripediaceae and Orchidaceae. Anther dehiscence is longitudinal. The endothecial thickenings are of the Spiral Type or, in certain Orchidaceae, of the Girdle Type. The tapetum is secretory, and microsporogenesis is simultaneous in Iridaceae, Apostasiaceae, Cypripediaceae and Orchidaceae but successive in other families investigated.

The pollen grains are dispersed separately or, in most Orchidaceae, in tetrads which in addition cohere in so-called massulae or in pollinia (which comprise all the pollen tetrads of each microsporangium or of each theca). The different specializations and mechanisms connected with pollen transfer in Orchidaceae are described under this family. The pollen grains are sulcate in most families, but in the majority of the Orchidaceae the exine is thin and the pollen grains thus have no distinct aperture. Bisulculate, ulcerate, foraminate and spiraperturate pollen grains are restricted to isolated genera or species. The pollen grains are dispersed in the two-celled stage.

The pistil is tricarpellary and trilocular in most families but unilocular in nearly all Orchidaceae and some Cypripediaceae and Alstroemeriaceae. The carpel apices sometimes extend into separate stylodia, as in most Colchicaceae, but are generally fused at least basally (STERLING 1975, 1977), forming one style which is often tribrachiate or trilobate at the apex (the complicated structures in Orchidaceae are described separately). The stigmatic surface is Dry in Iridaceae, Calochortaceae and part of Colchicaceae and Liliaceae, Wet in all Orchid-aceae studied and in Alstroemeriaceae, Trilliaceae and some Colchicaceae and Liliaceae. Placentation is axile in the taxa with trilocular ovaries and parietal in those with unilocular ovaries. The ovules are generally numerous and in Apostasiaceae, Cypripediaceae, Orchidaceae and Geosiridaceae extremely numerous. They are nearly always anatropous (weakly campylotropous in some Colchicaceae) and bitegmic; they are crassinucellate (Iridaceae), weakly crassinucellate or often tenuinucellate (Apostasiaceae, Cypripediaceae, Orchidaceae, etc.). A parietal cell (and parietal tissue) is developed only in the Iridaceae. Embryo sac formation generally conforms to the *Polygonum* Type or, in Liliaceae, to the *Fritillaria* Type, and in Cypripediaceae and a few Orchidaceae, to the *Allium* Type. Endosperm formation is constantly of the Nuclear Type, but a conspicuous feature in Cypripediaceae and Orchidaceae (and presumably also in the Apostasiaceae) is that endosperm formation is arrested in the very earliest stage and often does not take place at all; fusion of polar nuclei (or the central nucleus) with one sperm nucleus may not take place or, if it does, the primary endosperm nucleus may not divide or may divide only twice (in *Vanilla* up to four times). Embryogeny is of the Asterad or Onagrad Type.

The fruits in nearly all Liliales are loculicidal or septicidal capsules, or the rupture is between six ribs, as in many Orchidaceae, for example. Berries occur in some Alstroemeriaceae, in one or two genera referred here to Uvulariaceae (*Disporum, Clintonia*) and in a few orchids. Loculicidal capsules predominate in the Iridaceae, Liliaceae and Alstroemeriaceae, septicidal ones in Calochortaceae and Uvulariaceae and in most Colchicaceae.

The seeds are disc-shaped, angular, ellipsoidal or elongate, and often (in particular in Orchidaceae) have wings or flattened processes at both ends. In Orchidaceae and related families the seeds are diminutive and occur in enormous number (see under those families). Aril-like structures occur in some Iridaceae (*Iris* sect. *Regelia, Pseudoregelia* and *Oncocyclus*) and caruncles are found in certain other groups, developed from the chalazal region or from the raphe or hilum (*Erythronium, Gagea, Uvularia, Colchicum, Hermodactylus, Patersonia*). The testal layers of the seed coat in the unripe seeds often temporarily store starch, which has disappeared in the ripe seeds. The testal layers of the seed coat are always retained and sometimes form a sarcotesta, and the epidermis is never encrusted with phytomelan (though black in *Dietes* of Iridaceae and *Vanilla* of Orchidaceae); its inner epidermis is sometimes developed as a lipid layer. Also the tegmen is generally retained as a cell layer, and is not as completely collapsed as in Asparagales. The endosperm where it occurs (i.e. in all taxa except Apostasiaceae, Cypripediaceae and Orchidaceae) consists of cells with thick or thin pitted walls; the cells store aleurone and lipids. Starch grains are normally lacking, though rarely present in small amounts. The embryo is generally straight and rather well developed (about two thirds of the endosperm length or more) though sometimes short. It is curved in some scattered genera.

*Chemistry.* In chemical characters the Liliales are characterized by the rather uneven occurrence of steroidal saponins except, at least, in the alkaloid-bearing Colchicaceae. Whereas raphides are absent from several families (see above), calcium oxalate is often present in Iridaceae as long prismatic bodies. Chelidonic acid is common in taxa of Colchicaceae and is known in some taxa of Iridaceae and Alstroemeriaceae. Specific alkaloids, totally different from those found in Amaryllidaceae, occur in most or all Colchicaceae, Calochortaceae and a few genera of Liliaceae and Iridaceae, and are largely different in these groups. Alkaloids are probably absent in Alstroemeriaceae, Iridaceae, most Liliaceae and Uvulariaceae. In the Colchicaceae, one molecule of phenylalanine and one of tyrosine unite to form a series of compounds including colchicine and androcymbine. *Fritillaria* of the Liliaceae and some genera of Iridaceae have alkaloids of similar structure. C-glycoflavones are common in Orchidaceae, where also a number of other flavonoids occur, although Liliales are otherwise rather poor in flavonoids.

*Distribution.* Liliales has a world-wide distribution. The largest family, Orchidaceae is also subcosmopolitan, having a tropical concentration, whereas the Apostasiaceae are found in Indomalaysia and tropical Australia and the Cypripediaceae over a great part of the Northern Hemisphere and also in South America. Iridaceae, the next largest family, has a southern distribution with a chief centre in Southern Africa. This is also largely the case with Colchicaceae, while Alstroemeriaceae is South American. Geosiridaceae is known from Madagascar. The other families have a Northern Hemisphere distribution, Liliaceae with its main centre in western to central Asia and Calochortaceae in the western U.S.A.

*Relationships, Taxonomy.* The Liliales have here been circumscribed largely as by HUBER (1969) with the addition of what is generally known as Orchidales (Apostasiaceae, Cypripediaceae, Orchidaceae), which we may call the orchid families. These have been included here on the basis, firstly, of the reasons presented on p. 95, which indicate that these families arose out of early lilialean ancestors with some features in common with each of Alstroemeriaceae and Uvulariaceae, and thus that the Liliales would become a paraphyletic group without them. We are convinced, however, that the former Orchidales comprise a monophyletic group.

Minor problems remain as regards the delimitation of Liliales. Thus at least Philesiaceae, here placed in Asparagales, might be better placed in Liliales near Alstromemeriaceae, as in DAHLGREN and RASMUSSEN (1983), and *Medeola,* here placed in Liliaceae, shows some resemblance to *Scoliopus* of Trilliaceae (Dioscoreales). These examples illustrate that more research is needed before the taxonomy can be stabilized.

Burmanniales, which was formerly associated with Orchidaceae in the wide sense, has been considered by us to be rather distantly related to it (see further on p. 95 and under Burmanniales), the similarities like epigyny, small seeds and frequent lack of chlorophyll being presumably conditions that have evolved independently in the two groups. A third, presumably independent, group with these attributes is the Geosiridaceae, which shares most features with Iridaceae.

Within the Liliales the Alstroemeriaceae, and the three families Apostasiaceae, Cypripediaceae and Orchidaceae, form a couple of possibly related groups. Uvulariaceae, Calochortaceae and Liliaceae, which also approach the few rhizomatous Colchicaceae, form another group, Iridaceae and Geosiridaceae approaching, to some extent, Uvulariaceae and Colchicaceae.

The Alstroemeriaceae and the orchid families often have vessels in the stems, they contain raphides, the flowers are epigynous, the anthers are introrse and the ovules lack parietal tissue. The Alstroemeriaceae are, however, not clearly related to the orchid families; some similarities may depend on convergence. The orchid families are advanced in further characteristics, such as stronger zygomorphy, loss of some stamens, development of a labellum, simultaneous microsporogenesis, reduction of endosperm or total loss of endosperm formation. Their enormous number of small seeds and specialized pollen-clump pollination in combi-

nation with mycorrhizal life have obviously been very favourable, partly in combination with epiphytic life, and Orchidaceae is now one of the two largest families of flowering plants (the other is Asteraceae).

Some families of terrestrial plants, mainly of the Northern Hemisphere, that lack raphides and have rather large and showy tepals, weakly crassinucellate or tenuinucellate ovules without parietal tissue form one familial group. Here belong Liliaceae and Calochortaceae, both with bulbs, and the former with *Fritillaria* Type embryo sac formation, and Uvulariaceae, with rhizomes. A few genera which have tended to be placed in other families, *Disporum, Clintonia* and *Kreysigia,* are here placed in Uvulariaceae. Colchicaceae, with corms and usually with typical kinds of alkaloids, approach the Uvulariaceae. The seeds are more or less distinctive in shape: flat and platelike, or transversely or longitudinally prolonged, and often with the outer epidermis colourless or with a water-soluble yellow pigment. The cotyledon is not coleoptile-like. The stigmatic surface is Dry or Wet. Alkaloids of various kinds, mainly from steroidal precursors, are sporadically present.

Finally, Iridaceae and Geosiridaceae may form another group approaching the Colchicaceae. These seem to form a stock that consists mainly of corm geophytes with a strong concentration in the Southern Hemisphere. Their seeds are mostly isodiametric, and neither flat nor provided with wings or appendages. The outer epidermis of the seed coat has straight radial walls. The inner integument retains its cellular character and generally contains the brownish pigment phlobaphene. In addition, embryo sac formation conforms to the *Polygonum* Type, the cotyledon often has a coleoptile-like appearance, the stigmatic surface is mostly Dry and oxalate raphides are lacking. Styloids are often present in Iridaceae. Alkaloids when present are of the colchicine type (see above).

*Key to the Families*

1. Flowers hypogynous . . . . . . . . . . . . 2
1. Flowers epigynous . . . . . . . . . . . . . 6
2. Plants bulbous . . . . . . . . . . . . . . 3
2. Plants with rhizomes or corms . . . . . . . 4
3. Fruit septicidal (*Polygonum* Type embryo sac formation) . . . . . **Calochortaceae**
3. Fruit loculicidal (*Fritillaria* Type embryo sac formation) . . . . . . . **Liliaceae**
4. Plants rhizomatous . . . . . . . . . . . . . 5
4. Plants with a corm . . . . . . . . **Colchicaceae**
5. Leaves bifacial (dorsiventral) . . . . **Uvulariaceae**
5. Leaves unifacial-ensiform . . **Iridaceae**:*Isophysis*

6. Stamens 1 or 2, very rarely 3, filaments fused with the style into a gynostemium . . . . . . . . . . . . . . . . 7
6. Stamens 3 or 6, filaments not fused with the style . . . . . . . . . . . . 9
7. Anthers 2 or 3 . . . . . . . . . . . . . . . 8
7. Anther solitary . . . . . . . . . **Orchidaceae**
8. Labellum large, sac-shaped . . . **Cypripediaceae**
8. Labellum not conspicuously different from the other tepals . . . **Apostasiaceae**
9. Stamens 6 . . . . . . . . . . **Alstroemeriaceae**
9. Stamens 3 . . . . . . . . . . . . . . . . 10
10. Chlorophyllous plants generally with ensiform leaves . . . . . . . **Iridaceae**
10. Achlorophyllous plants with scale-like leaves . . . . . . . **Geosiridaceae**

**Alstroemeriaceae** Dumortier (1829)    4:160
(Fig. 105)

Erect or twining herbs with sympodial rhizomes. Certain of the roots are fusiform and modified to store nutrients and water. The roots have vessels with scalariform and sometimes simple perforation plates, and the stems have vessels with scalariform perforation plates. The stems of the vines may be up to 4 m long or more. The leaves are scattered on the stems and are linear to lanceolate, narrowing at the base and not sheathing. They are generally twisted at the base so that their morphologically lower side is turned upwards. The leaves are glabrous or in some species of *Bomarea* provided with unbranched simple hairs on the lower (morphologically upper) surface. Oxalate raphides are known in *Alstroemeria*.

The branches end in inflorescences which consist of helicoid cymes and are generally umbel-like; they are rarely unifloral. The flowers, which are generally subtended by relatively large, green and leaf-like bracts, are trimerous, epigynous, bisexual and actinomorphic or slightly zygomorphic. The mostly free tepals vary from nearly similar to conspicuously different in the two whorls, the outer being often shorter, of different colour and less variegated than the inner, which often bear a dotted or striate-dotted colour pattern. Their colour varies from orange to rose-coloured, purple or green. Nectaries are present near the base of two or all of the inner tepals. The ovary on the surface shows distinct decurrent borders of the outer tepals; likewise a circular "scar" around the top of the fruit is left after the shedding of the perigone.

There are 3 + 3 stamens which have narrow filaments and elongate pseudo-basifixed anthers. Dehiscence is introrse and longitudinal. Microsporogenesis is successive. The pollen grains are large and sulcate, usually plano-convex with the sulcus on the convex part of the grain. The pistil is trilocular or rarely unilocular and has a single style with three apical stigmatic branches, the stigmatic surface of which (at least in *Alstroemeria*) is of the Wet Type. DAUMANN (1970) reported septal nectaries for *Alstroemeria*, but this needs verification. The ovary contains numerous anatropous ovules on axile (*Alstroemeria, Bomarea*) or parietal (*Leontochir, Schickendantzia*) placentae. No parietal cell is cut off from the archesporial cell. Embryo sac formation conforms to the *Polygonum* Type.

The fruit is a capsule which is generally loculicidal; occasionally it is indehiscent or opens explosively. The seeds are globose or rounded-ellipsoidal. The outer integument of the seed coat in *Alstroemeria* is dry and consists of few to eight cell layers, but in *Bomarea* it is thick and multi-layered and forms a veritable sarcotesta with sparse starch grains. Cell differentiation in the outer integument is weak; the cells are colourless or contain a yellow pigment, and there are sparse subepidermal "bladder cells" in *Alstroemeria* containing oxalate raphides. The inner integument collapses into a thin membrane, while the endosperm cells have thick pitted walls and contain aleurone and fatty oils but no starch. The cylindrical embryo is about two thirds the length of the endosperm. When germinating, the seedling does not have a coleoptile-like cotyledon. The basic chromosome number is $x = 8$ (*Alstroemeria*) or $x = 9$ (*Bomarea*).

*Chemistry.* In chemical respects the family is characterized by its probable production of steroidal saponins, and its production of chelidonic acid, while alkaloids are lacking. The storage roots contain starch. In some of these respects Alstroemeriaceae is not very characteristic of the order. Alstroemeriaceae share with Liliaceae the occurrence of tuliposides (SLOB et al. 1975).

*Distribution.* Alstroemeriaceae is restricted to South and Central America and has its centre in Andean South America.

*Alstroemeria* (60) consists of erect herbs of variable size, usually with relatively large, yellow, orange or reddish flowers which are often more or less zygomorphic, generally with striated inner tepals (especially the upper two) and upcurved filaments. The leaf blades are inverted. *A. aurantiaca*, "Peruvian Lily", and other species, mainly from Chile, are frequently grown as ornamentals. Several hybrids and breeding products have been pro-

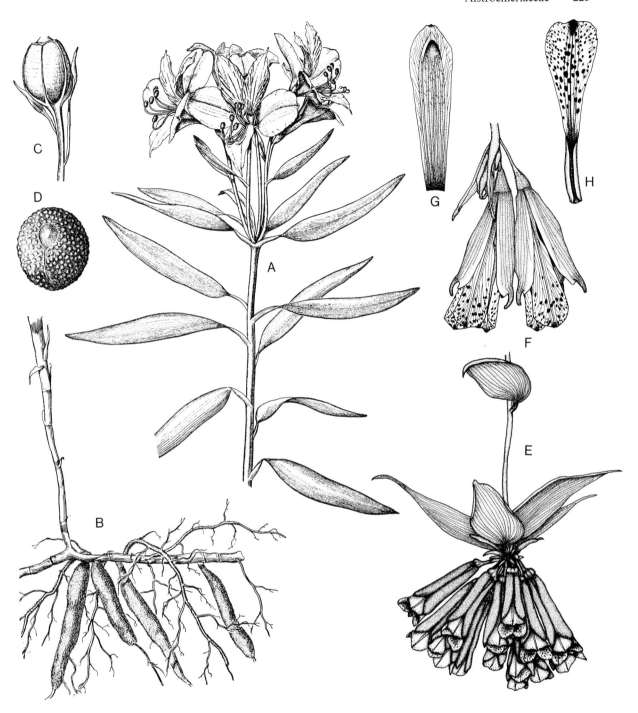

duced. *A. pulchella* in Brazil reaches a height of 90 cm and has striking red-and-green flowers.

*Bomarea* (100) species are mostly vines with actinomorphic, frequently more or less tubular flowers having straight filaments. The outer tepals are generally much smaller than the inner, sometimes appearing sepaloid although not being green. The perigone is often salmon-red, and is sometimes green on the apical parts; the inner side of the

**Fig. 105.** Alstroemeriaceae. **A–D** *Alstroemeria aurantiaca*. **A** Upper part of flowering plant. **B** Rhizome with swollen roots. **C** Capsule. **D** Seed. (CORREA 1969). **E–H** *Bomarea pardina*. **E** Pendulous inflorescence. **F** Two flowers. **G** Tepal of outer whorl. **H** Tepal of inner whorl. (NEUENDORF 1977)

inner tepals generally has a variegated or spotted pattern. The inflorescence in some species, such as *B. cordei,* has long-pedicelled flowers. The storage roots of some species are edible and taste slightly similar to "Jerusalem Artichoke" (*Helianthus tuberosus*). Some species are grown as ornamentals, e.g. the orange-flowered *B. caldasii* from Ecuador.

While these two genera have flowers with a trilocular ovary, two small genera, *Leontochir* (1) in Chile and *Schickedantzia* (1), in Andean Argentina, have unilocular ovaries with parietal placentae. The former is a vine, the latter a small, erect herb.

Alstroemeriaceae has sometimes been included in Amaryllidaceae because of its epigynous, hexastaminate flowers in an umbel-like inflorescence, but is only distantly related to that family.

The possession of oxalate raphides and, most probably, of steroidal saponins and of the rhizomatous underground parts provide differences from the Liliaceae, Uvulariaceae and Calochortaceae, but most evidence points to a close alliance with these families and the orchids. The presence of septal nectaries in Alstroemeriaceae may represent a relic property.

Another possible affinity should be mentioned, namely with the Philesiaceae, where such characters as the climbing habit, twisted leaf blades, differentiation of the perigonal whorls, variegated tepal colour, perigonal nectaries and phytomelan-free seeds are found. Whether most or all of these result from convergent evolution is still uncertain.

**Colchicaceae** A. P. de Candolle in Lamarck & de Candolle (1805)   17:170   (Figs. 106–107)

Low or tall, erect or sometimes twining herbs, with a subterranean, starch-rich corm. In some genera (*Gloriosa, Littonia, Sandersonia*) the corm is stoloniferous. Such stolons or runners may form apical tuberculate corms from which buds are developed. Vessels are present in the fibrous roots only and have scalariform perforation plates.

The leaves may be few and basally concentrated, several on a short leafy stem, or numerous and scattered on a long, erect or climbing, sometimes branched, stem. They are dorsiventral, mostly linear or lanceolate, sessile, or rarely constricted into a pseudopetiole, sheathing at the base, parallel-veined and, rarely, cirrhose, ending in a tendril (as in *Gloriosa*). In those species of *Colchicum* which flower in autumn the leaves are developed after the flowers. Raphides are lacking.

The aerial stem is herbaceous and usually stiff and erect, sometimes climbing, ending in racemes (or monochasial systems) or rarely in a single flower.

The flowers are trimerous, actinomorphic, hypogynous and generally bisexual. The 3 + 3 tepals are more or less similar in size and shape: they are free from each other or basally connate, in *Colchicum* forming a long narrow basal tube. They vary much in size and are basally provided with nectaries. (*Colchicum* and *Androcymbium* lack perigonal, but have androecial, nectaries.) Tepal colour varies from white to purple, red or yellow, and sometimes it is variegated (as in species of *Colchicum*) or, more often, two-coloured, the base with a different colour from the rest.

The stamens are free from each other and have narrow or basally broad, glabrous filaments and mostly short, ovate, longitudinally dehiscent anthers; these are usually extrorse, but rarely latrorse or nearly introrse. Microsporogenesis is successive. The pollen grains are generally sulcate, but 2-, 3- or 4-foraminate and spiraperturate pollen grains occur in species of *Colchium* and *Androcymbium* (RADULESCU 1973b).

The pistil is trilocular and either has three totally free stylodia or a style which is tribrachiate at least in its upper part (STERLING 1975, 1977). The stigma is Dry or Wet. Each locule contains several to many anatropous (to weakly campylotropous) ovules on axile placentae. A parietal cell is not cut off from the archesporial cell, but in some genera (at least some Iphigenieae) the epidermal cells of the nucellus divide periclinally to form two or more layers. Embryo sac formation conforms to the *Polygonum* Type. Nucellar embryony occurs at least in *Colchicum*.

The fruit is capsular and dehisces according to various modes, usually septicidally but in some genera loculicidally. The seeds are generally globose, but ovoid in *Onixotis*. Appendages arising from the chalazal region occur in *Colchicum*. The testal part of the seed coat consists of few to many cell layers, which in the unripe seed contain starch. The outer epidermis is retained or compressed (to collapsed) and often contains the red-brown pigment phlobaphene. The tegmen is variably compressed, with or without the pigment phlobaphene or other yellowish red pigments. The endosperm generally contains somewhat thick-walled, pitted cells with aleurone and fatty oils, but no starch. A straight, relatively short, elongate or rarely subglobose embryo is enclosed in the endosperm. An important feature of this family, like Iridaceae,

**Fig. 106.** Colchicaceae. **A–G** *Androcymbium striatum.* **A** Flowering plant. **B** Flower. **C** Tepal and attached stamen. **D** Pistil. **E** Carpel, longitudinal section. **F** Fruit, wall partly removed to show seeds. **G** Seed. (KRAUSE 1930). **H–K** *Ornithoglossum parviflorum.* **H** Habit. **I** Flower. **J** Tepal. **K** Stamen. (NORDENSTAM 1982). **L–P** *Iphigenia oliveri.* **L** Habit. **M** Flower. **N** Pistil, wall partly removed to show placentation. **O** Capsule. **P** Seed. (KRAUSE 1930)

is that the seedling has a conspicuously coleoptile-like cotyledon. $x = 5, 7, 10, 11$ or 12 (in Colchiceae 7, 8, 9, 10, 11, 12, 19), 11 being perhaps the basic number.

*Chemistry.* The family is characterized by the absence of steroidal saponins and oxalate raphides. Chelidonic acid and alkaloid bases are present in most or all genera. The alkaloids represent colchicine and related compounds (WILDMAN and PURSEY 1968) and are extremely poisonous (see further under the ordinal description). Starch is accumulated in the vegetative parts, e.g. in the corms.

*Distribution.* Colchicaceae has a pronounced centre in the summer rainfall regions of South Africa, but ranges through Africa to the Mediterranean end of western Asia, where the largest genus, *Colchium* (incl. *Merendera* and *Bulbocodium*), has its main distribution area. A few genera occur in Asia and a few others are found in Australia. *Iphigenia* and *Wurmbea* (incl. *Anguillaria*) are disjunct, occurring in South Africa and Australia (*Iphigenia* also in Tropical Asia and New Zealand).

The following division of Colchicaceae largely follows NORDENSTAM (1982).

### Tribus Anguillarieae

Tribus Anguillarieae consists of erect, non-climbing herbs with a bulb-like corm and with sessile flowers in spicate bractless inflorescences. The flowers are generally small or medium-sized with white, brownish, purplish or violet tepals. The tepals are free or connate and the anthers extrorse

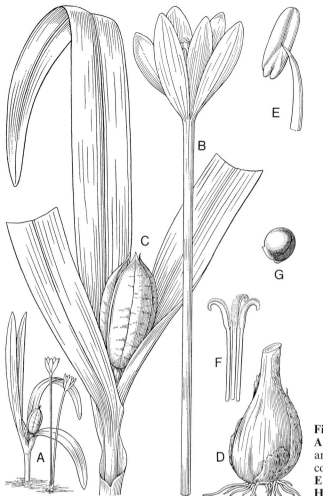

**Fig. 107.** Colchicaceae. **A–G** *Colchicum autumnale.*
**A** Habit; note that flowers appear in the autumn, leaves
and fruits in the winter and spring. **B** Floral tube (ovary
concealed underground). **C** Capsule and leaves. **D** Corm.
**E** Stamen. **F** Stylodia. **G** Seed. (ROSS-CRAIG 1972).
**H** *Gloriosa superba,* branch. (HERKLOTS 1976)

or latrorse. The pollen grains are sulcate. The pistil
has three fairly long stylodial branches. The capsules are septicidal or "endocidal" (*Neodregea,
Onixotis, Wurmbea* pro parte) or loculicidal
(*Wurmbea* pro parte). This group is largely South
African, but occurs also in Australia.
*Wurmbea* (incl. *Anguillaria;* ca. 35) is a South Afri-
can-Australian genus of hyacinthaceous habit with
white, yellow, pink or more or less purple or
brown-patterned, basally connate tepals (see MAC-
FARLANE 1980). – *Onixotis* (*Dipidax*) (ca. 4), has
white or rose-coloured to purple-violet flowers; *O.
triquetra,* a decorative meadow species, is sometimes grown as an ornamental.
Related to Anguillarieae is the genus *Baeometra*
(1), in the Cape Province of South Africa, with
a bracteate spike and with tepals tapering at the
base and coloured yellow inside and red outside.

### Tribus Iphigenieae

This consists of erect herbs, which either have a
short aerial stem and basally concentrated leaves
or an erect or twining stem on which the leaves
are dispersed. The leaves are sometimes pointed or
cirrhose. The underground part consists of a bulb-
like corm, from which runners may be formed.
The inflorescence is a bracteate raceme, and the
bracts may be foliose (as in *Sandersonia, Littonia*
and *Gloriosa*). The flowers vary from medium-
sized to rather large and conspicuous, and have
free, or more rarely fused tepals, which vary in
colour from white to bright red, yellow or dark-
violet or almost black. The style or stylodia are
often abruptly bent at right angles to the ovarian
axis. *Iphigenia* (6) (Fig. 106 L–P) ranging from
South Africa through India to Australia and New
Zealand, comprises erect, non-climbing herbs with
loculicidal capsules. The Southern African genera

*Ornithoglossum* (8) (Fig. 106 H–K) (NORDENSTAM 1982) and *Hexacyrtis* (1) (WILLIAMSON 1983) are similar.

*Gloriosa* (6) (Fig. 107 H) has a wide range from South Africa through Ethiopia to eastern Asia. It is a vine with branches up to 150 cm long, pendent flowers, yellow to bright or dark red or often bicoloured tepals and a style bent basally almost at right angles. *G. superba* and *G. rothschildiana* are frequently cultivated ornamentals. The corms are rich in colchicine and very poisonous. – Two other genera, *Littonia* (8) and *Sandersonia* (1), both African, belong to this tribe. They are twining or erect herbs and may have cirrhose leaves. In *Sandersonia* the tepals are fused to form a campanulate perigone.

### Tribus Colchiceae

Tribus Colchiceae consists of two genera, *Androcymbium* and *Colchicum,* the latter of which is sometimes divided into several genera. The subterranean parts are tunicated bulb-like corms, and the flowers, which are solitary or few, are situated on a very short stem (actually axillary in a few-flowered spike), basally in the centre of the shoot; they are frequently developed before the green leaves and have long-clawed tepals, the claws being free or fused into a long tube. The anthers are latrorse. The pollen grains are variable: usually 2–4-foraminate, sometimes spiraperturate. There are three stylodia which are free or fused basally into a style. The pistil develops into a septicidal capsule. – *Colchicum* (incl. *Merendera, Synsiphon* and *Bulbocodium;* 65) (Fig. 107 A–G) is centred in the Mediterranean Region and western Asia. It includes a number of chiefly autumn-flowering species, many superficially resembling species of *Crocus* (Iridaceae). The ovary is concealed above the corm in the underground part of the perigone, growing above ground at maturity. Many species are grown as ornamentals, including *C. autumnale, C. alpinum, C. cilicicum* and *C. speciosum.* The corms and seeds are rich in the toxic alkaloid colchicine, which is used in medicine. It is also used in plant breeding to induce polyploidy. The most frequently grown species for these purposes is *C. autumnale.* In the species sometimes treated as *Merendera* and *Bulbocodium,* the tepal claws are not fused as in the other species of *Colchicum.* – *Androcymbium* (40) (Fig. 106 A–G), which ranges from South Africa to the Mediterranean, has densely clustered flowers enveloped by bracteal leaves.

## Uvulariaceae C.S. Kunth (1843)   8:ca. 40
(Fig. 108)

Erect, medium-sized herbs up to ca. 1.5 m, with creeping rhizomes, and an erect aerial stem with few to numerous, well-developed lanceolate to ovate leaves scattered on the stem, the lower often not larger than those on the upper parts of the branches. Cataphylls are often present at the base of the aerial stem. The leaves are generally sessile (rarely shortly pseudopetiolate) acute to acuminate, basally cuneate, truncate, cordate or rarely amplexicaul, glabrous or hairy and with arching, converging main veins, and in some cases (*Tricyrtis,* some species of *Disporum*) with clearly reticulate venation between the main veins (Fig. 108 C, K). Vessels with scalariform perforation plates, at least in *Tricyrtis* but perhaps in all genera, are present in the stems as well as in the roots. Raphides are usually lacking, but have been observed in *Disporum* and *Streptopus* associated with crystal sand (GOLDBLATT and HENRICH, personal communication).

The inflorescence is a generally few-flowered and sparse panicle (or thyrse), with cymose components (Fig. 108 J) or a spike with solitary or paired axillary flowers. Green foliose leaves are generally present in the inflorescence. Recaulescence phenomena are often present in the inflorescence region.

The flowers are fairly small to medium-sized or rather conspicuous (*Tricyrtis*), bisexual, actinomorphic, hypogynous, and trimerous. They are white, pale yellow or more or less purple-spotted (*Tricyrtis*). The outer or all tepals often have a basal, nectariferous, swelling or sac, which may be globose and shiny, as in *Tricyrtis,* or a spur (in some species of *Disporum,* s.lat., Fig. 108 G). The tepals are often suberect to slightly spreading, pointed and in some genera narrow and tapering. The stamens are 3+3 in number and have erect glabrous filaments, which are narrow, erect and sometimes connate basally, the anthers are dorsifixed-hypopeltate and extrorse and dehisce longitudinally. The pollen grains are sulcate and dispersed in the two-celled state.

The ovary is globose to ovoid, tricarpellary and trilocular, and has a tribrachiate style (STERLING 1977). There is great variation, even within the genus *Disporum* s.lat., in the relative length of the style: in some taxa the stylodial branches are free almost from the base, in others there is a style with three rather short apical lobes (Fig. 108 D, E); sometimes the stylodial branches are spreading

**Fig. 108.** Uvulariaceae. **A–D** *Disporum* (*Prosartes*) *lanuginosa*. **A** Flowering branch. **B** Rhizome. **C** Leaf; note the reticulate venation. **D** Flower, 3 tepals and 3 stamens removed. **E–F** *Disporum* (s.str.) *smilacinum*. **E** Flower, 3 tepals and 3 stamens removed. **F** Tepal. **G–L** *Disporum* (s.str.) *calcaratum*. **G** Flower, 3 tepals and 3 stamens; note the extrorse anthers. **H** Tepal of outer whorl, with long nectariferous spur. **I** Stamen. **J–O** *Tricyrtis latifolia*. **J** Flowering branch; note the monochasial branches of the inflorescence. **K** Leaf, showing the reticulate venation. **L–M** Flower from one side and from above. **N** Tepal of outer whorl and the opposite stamen. **O** Capsule. **P** *Tricyrtis* sp., seed. (All orig. B. JOHNSEN)

and bifurcate (Fig. 108 M). The stigmatic surface is of the Wet (*Tricyrtis*) or the Dry (*Clintonia*) Type. The ovules are anatropous and weakly crassinucellate (nucellus of "*Scoliopus* Type"; BJÖRN-STAD 1970); no parietal cell is developed. Embryo sac formation is of the *Polygonum* (or rarely the *Clintonia*) Type (OGURA 1964). Endosperm formation is nuclear.

The fruit in *Tricyrtis, Kreysigia* and *Uvularia* is a globose to elongate cylindrical-triangular capsule which is septicidally or loculicidally dehiscent, in *Schelhammera* it is a somewhat fleshy capsule and in *Disporum* s.lat. and *Clintonia* it is a berry. The seeds are flat and disc-like (*Tricyrtis, Streptopus*) or generally globose (the other genera). The testal characters are probably variable; the endosperm consists of cells storing aleurone and fatty oils, and the embryo, at least in some cases, is rather small and little differentiated, for example only $^1/_5$–$^1/_{12}$ of the length of the endosperm in *Tricyrtis*. The basic chromosome numbers are, $x =$ 13 (*Tricyrtis*), 14 (*Uvularia, Schelhammera, Kreysigia*), and 16 (*Disporum* s.lat., *Streptopus*).

*Distribution.* The family as circumscribed here occurs mainly in the Northern Hemisphere, with *Tricyrtis* in eastern Asia, *Uvularia* in North America, and *Disporum* s.lat. (see below), *Clintonia* and *Streptopus* in both continents. Of the smaller genera, *Kreysigia* ranges from Malaysia to Eastern Australia, and *Schelhammera* from New Guinea to Australia.

It is very uncertain that Uvulariaceae with this circumscription is a homogeneous family but it should at least be more adequate than previous constellations. Further investigations into the correlation of morphological, embryological and chemical characters is needed. For instance, uncertainty as to the borderline between Uvulariaceae and Colchicaceae might possibly be cleared up by an investigation of alkaloids.

### Tribus Tricyrtideae

Within the family, the genus *Tricyrtis* (11) stands out as quite distinct. It occurs in Eastern Asia, and is characterized by its purple spotted, variegated flowers, provided with globose nectariferous pouches on the outer tepals. The capsules are elongate and septicidal, and in each locule have numerous flat pale seeds, ovate to orbicular in outline, which are piled upon each other as in several genera of Liliaceae (e.g. *Lilium, Tulipa, Fritillaria*). The outer epidermis of the testa may contain lignin, whereas the inner is totally collapsed. – *T. stolonifera,* from Formosa, *T. hirta,* from Japan, and other species, some with a rather *Lilium*-like appearance, are frequently grown as ornamentals.

### Tribus Uvularieae

The remaining genera generally have smaller and paler flowers, and the fruits, whether loculicidal capsules or berries, are more globose, and their seeds are globose rather than flat. The following three genera have baccate fruits.

*Disporum* (20) (UTECH and KAWANO 1976) occurs both in southern and eastern Asia and in North America. The American species are fairly similar to *Tricyrtis* in having leaves with reticulate venation and non-spurred tepals. The Asiatic species lack distinct reticulate venation, and the tepals often have distinct spurs, although they always have nectar grooves. The American species are probably quite distinct from the Asiatic and should be placed in a separate genus, *Prosartes* (CONOVER 1983), which has a greater similarity to *Tricyrtis* than has *Disporum* sensu stricto. – *Clintonia* (6) with a similar distribution, has the leaves more basally concentrated and the flowers are solitary or several together in umbel-like inflorescences. The leaves of *C. borealis* (Labrador to North Carolina) can be used as salad. There is some doubt as to whether this genus belongs here or in the Convallariaceae.

*Streptopus* (4) consists of *Polygonatum*-like herbs with the flowers situated solitarily or in pairs, on filiform pedicels, displaced by recaulescence. The seeds may contain starch.

*Uvularia* (5) has campanulate flowers with mutually free, mostly pale yellow perianth members and anthers with a prolonged connective. The genus is reminiscent of taxa of Stemonaceae-Trilliaceae, to which it may also be distantly related. *U. grandiflora* is grown as an ornamental, and the young shoots of *U. sessiliflora* are edible!

*Schelhammera* (3) in Eastern Australia and New Guinea and *Kreysigia* (1) in approximately the same regions, consist of small-sized plants.

*Scoliopus* and *Medeola,* often referred to Trilliaceae, have proved to be more or less out of place in that family (BERG 1962a, b), but they are also very different from each other. Either or both of these is perhaps best placed in Uvulariaceae.

### Calochortaceae Dumortier (1829)   1:60
(Fig. 109)

Erect, often slender herbs with a tunicated bulb having two thick, nutrient-storing leaves and a membranous or fibrous-reticulate coat. The stem bears only a few leaves, or sometimes there is only one basal leaf apart from the inflorescence bracts. The leaves are flat, dorsiventral and usually linear; there is normally a single basal leaf which is much larger than the cauline leaves. Bulbils may be formed in the axils of the lower leaves. Crystal raphides are lacking, as in Liliaceae.

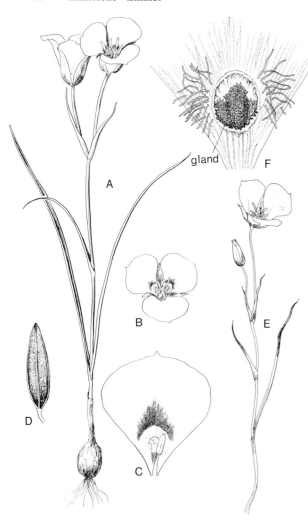

**Fig. 109.** Calochortaceae. **A–C** *Calochortus nuttallii.* **A** Plant. **B** Flower. **C** Tepal of inner whorl and opposite stamen. (HITCHCOCK et al. 1969). **D** *C. aureus,* fruit. **E–F** *C. excavatus.* **E** Plant. **F** Base of inner tepal. (**D–F** from CRONQUIST et al. 1977)

The inflorescence is a branched or unbranched raceme; the long-pedicelled flowers are situated in the axils of green, narrowly lanceolate or linear bracts, which are twisted into a position between the flower and the axis.

The flowers are hypogynous, trimerous, bisexual, and their tepals are free and conspicuously different in the two whorls. Those of the outer whorl are ovate or lanceolate, generally glabrous, and either green and sepaloid or of the same colour as the tepals of the inner whorl. The latter are generally broad, rounded or obtriangular to broadly obovate, petaloid, cuneate or clawed, frequently fringed or ciliate on the margins and sometimes pubescent or bearded on the inner face; their

colour is usually yellow, rose or crimson, and they are frequently marked with conspicuous large or small spots, streaks or other patterns (nectar guides). Conspicuous nectary glands are present on the tepal bases, and are sometimes associated with a knee-like or spur-like pocket. The stamens, 3 + 3 in number, are free from each other and inserted at the bases of the tepals; their filaments are usually dilated, and their elongate anthers pseudo-basifixed, i.e. surrounding the tip of the filament with a tubular part of the connective (as in *Tulipa*). Anther dehiscence is by longitudinal slits. The pollen grains are sulcate and dispersed in the two-celled stage.

The ovary is tricarpellary and trilocular and the style very short or obsolete. The three stigmatic branches are situated directly on the apex of the ovary. Their surface is of the Dry Type. Each locule contains numerous anatropous ovules in two rows. As in the Liliaceae, the ovules lack a parietal cell, but embryo sac formation, unlike that of Liliaceae, follows the *Polygonum* Type. Endosperm formation is nuclear (BERG 1960). The capsules are linear to globose, three-angled or three-winged, and septicidal (as in *Tricyrtis*). Their seeds are lanceolate to circular and usually strongly flattened, although not as regularly horizontal as in most Liliaceae; in *C. macrocarpus* they are narrowly winged. In the seed coat, the outer integument consists of fairly intact cells and lacks pigments; its outer epidermis shows a hexagonally reticulate pattern (and is effectively water-absorbing). The inner integument is flattened but with the cells retained, and containing a lemon-yellow pigment. The endosperm cells contain fat and aleurone and the slender embryo is from about half to nearly as long as the endosperm. The chromosome number varies much, and $x = 7$–20.

*Chemistry.* *Calochortus* lacks oxalate raphides and chelidonic acid. The seeds of *C. macrocarpus,* at least, contain alkaloids.

*Calochortus* (60), "Mariposa Lily", is distributed over the temperate western parts of North America from British Columbia to Central America. The species grow in grassland, chaparral and semi-desert vegetation, less commonly in woodlands. Many species are highly decorative and are cultivated as ornamentals, e.g. *C. venustus.* The bulbs of many species were formerly eaten by the Indians because of their high starch content.

*Calochortus* seems to form a late evolutionary branch comparable to the Old World Liliaceae, which *Calochortus* resembles in important respects, e.g. the bulb, the leafy stem, the nectariferous tepals, the anther construction, the absence of a parietal tissue in the nucellus, the flat seeds, and the chemical contents. The septicidal capsules, *Polygonum* Type of embryo sac, and the basic chromosome number are different; *Calochortus* can easily be distinguished from Liliaceae in that the inner and outer tepals are strongly different from each other. There are also differences between the families in leaf anatomical characters. As remarked by BERG (1960) *Calochortus* is distinct in the combination of embryological details but in these comes closest to genera placed here in Colchicaceae and Uvulariaceae.

## Liliaceae A. L. Jussieu (1789)   13:385
(Figs. 110–112)

Erect herbs, always with bulbs. The bulbs are of variable construction in the family, being tunicated or non-tunicated, and having from one (*Gagea*) to very numerous fleshy nutrient scales. The roots are contractile and contain vessels with scalariform perforations only.

The erect stem has from one to very many leaves, which may be basally concentrated (*Gagea, Tulipa*) or distributed all along the stem. The leaves are alternate or rarely verticillate (species of *Lilium* and *Fritillaria*). They are flat, dorsiventral, linear to ovate-lanceolate, bifacial, parallel-veined and often sheathing at the base. Further, the leaves may be dark-spotted, apically attenuate and sometimes petiolate. Bulbils are present in the leaf axils of several species of *Lilium*. Oxalate raphides are lacking.

The cauline stem ends in a determinate or rarely indeterminate inflorescence. It may form a raceme (*Lilium*) or be umbel-like (*Gagea*), or is sometimes restricted to a single terminal flower (*Tulipa* spp.).

The flowers are bisexual, hypogynous, trimerous and generally actinomorphic, but more rarely weakly zygomorphic. Their tepals are free from each other to the base and generally similar in the two whorls or, rarely, different, as in *Nomocharis,* where the inner three tepals, unlike the outer, are often fimbriate or pubescent. Nectaries are present on the tepal bases. The tepal colour is highly variable and patterns of striation, spotting or checkering are frequently present.

The stamens are 3 + 3 in number and in *Lilium, Notholirion,* and *Nomocharis* they have epipeltate anthers, and in the other genera pseudobasifixed anthers (the filament tip being surrounded by the

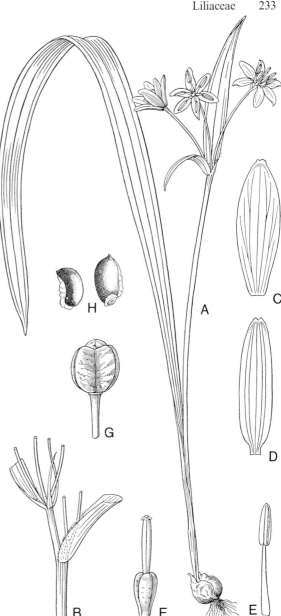

**Fig. 110.** Liliaceae. *Gagea lutea.* **A** Plant. **B** Part of branch system in the inflorescence region. **C** Tepal of outer whorl. **D** Tepal of inner whorl. **E** Stamen. **F** Pistil. **G** Capsule. **H** Seeds with appendage. (ROSS-CRAIG 1972)

tubular connective). The filaments are glabrous, free and narrow, sometimes dilated basally. Microsporogenesis is successive. The pollen grains are large, often coarsely reticulate and normally sulcate, only in *Tulipa* being occasionally operculate or with an irregular aperture. Further they are two-celled, the generative nucleus being enclosed in an easily stainable, rather wide cell almost as long as the diameter of the pollen grain lumen.

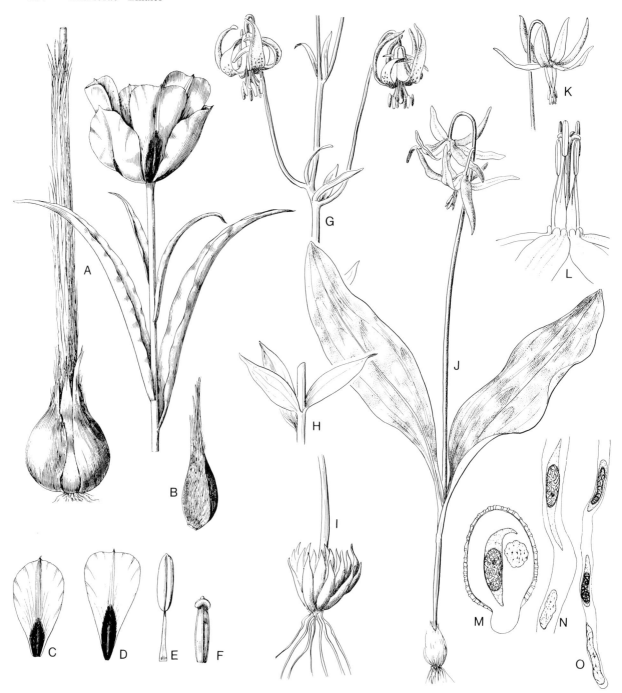

**Fig. 111.** Liliaceae. **A–F** *Tulipa borszczowii.* **A** Plant. **B** Lateral bulb. **C** Tepal of outer whorl. **D** Tepal of inner whorl. **E** Stamen. **F** Pistil. (SEALY 1938). **G–I** *Lilium columbianum.* **G** Part of inflorescence. **H** Leaves, in a pseudoverticel. **I** Bulb. (HITCHCOCK et al. 1969). **J–L** *Erythronium revolutum.* **J** Plant. **K** Flower. **L** Androecium and gynoecium. (HITCHCOCK et al. 1969). **M–O** *Lilium regale,* pollen and pollen tube development. **M** Two-celled pollen grain. **N** Part of pollen tube before division of generative cell. **O** Same after the division with two sperm cells. (COOPER 1936)

The pistil is tricarpellary, syncarpous and trilocular, and the style is short to long, apically trilobate or with three stigmatic crests. The stigmatic surfaces are papillate and Dry (species of *Erythronium* and *Tulipa*) or Wet (species of *Fritillaria* and *Lilium*). Each locule contains several to many anatropous ovules, which are rather tenuinucellate and in which a primary parietal cell is not cut off from the archesporial cell. Embryo sac formation normally conforms to the tetrasporic *Fritillaria* Type (see P. MAHESHWARI 1946; Fig. 112).

The fruits are loculicidal capsules, the seeds of which in typical cases are flat and disc-shaped, placed like piles of coins on top of each other, and often provided with a marginal rim. Sometimes, however, they are ellipsoidal rather than flattened or sharp-angled (species of *Gagea, Lloydia, Erythronium,* rarely *Tulipa*). They are not winged, but may have a lateral raphe-crest or a raphal elaiosome, the latter in *Gagea arvensis* and related species, while a chalazal elaiosome occurs in *Erythronium*. The outer integument of the seed coat is thin: sometimes only two-layered. Its cells are flat or even collapsed, and sometimes they have lignified walls, especially on the outer epidermis. The inner integument is strongly suppressed. In the endosperm of ripe seeds the cells mostly have rather thickened, pitted walls and contain aleurone and fatty oils but no starch. The embryo is generally little differentiated, ovoid or ellipsoidal, and only about $^{1}/_{5}$–$^{1}/_{12}$ the length of the endosperm. The seedling does not have a coleoptile-like cotyledon. The basic chromosome number is $x = 12$.

*Chemistry.* Raphides and chelidonic acid are probably lacking in the family. Steroidal saponins occur at least in several genera, including *Lilium* (with the sapogenins lilagenin and yuccagenin), and steroidal alkaloids occur at least in *Fritillaria,* in which the family agrees with the tribe *Veratreae* of Melanthiaceae. The accumulation of x-methylene-glutamic acid derivatives is common. *Erythronium* produces x-lactone. Tuliposides were found by SLOB et al. (1975) to be present in several genera.

*Distribution.* Liliaceae has a Northern Hemisphere distribution with a pronounced centre in South-Western and Himalayan Asia to China. The taxa are mostly spring-flowering plants growing in steppes and mountain meadows.

The present circumscription of Liliaceae is supported also by HUBER (1969) and SCHULTZE (1980); it is a rather homogeneous family, the closest relatives of which are undoubtedly the Calochortaceae, Uvulariaceae and Colchicaceae.

*Gagea* (90) (Fig. 110) is a widespread Eurasian genus of mostly small spring-flowered herbs with yellow (rarely white or purplish), externally green-banded flowers in (one- to) few-flowered, umbel-like inflorescences, which are usually subtended by two spathe-like upper leaves. The bulbs consist of two(–four) bulb scales, only one (rarely two or three) of which is a fleshy nutrient scale. *G. lutea, G. spathacea* and other species are common European species, growing in grassland and woods, *G. arvensis* in cultivated fields, and *G. pratensis* on dry, sandy ground. – *Lloydia* (18), widely distributed in the Northern Hemisphere, is a genus of small, white-flowered, *Gagea*-like herbs with bulbs having a single nutrient scale. *L. serotina* ranges from western Europe through the Alps and Himalaya to the western mountains of North America.

*Tulipa* (60–100) (Fig. 111 A–F), "Tulips", are distributed from western Europe (*T. sylvestris*) to Asia; the western and central parts of Asia being the centre of variation. The bulbs are few-leaved, and the erect stem has a limited number of basally concentrated leaves and mostly a single terminal flower. When there are two or more flowers, the lateral are ebracteate. Tepal colour is variable, often red or yellow, and often bicoloured (but white, violet and other colours also occur). Some species are polymorphic with yellow and red or orange phases. The anthers are pseudobasifixed (see above), and the stigmatic crests usually almost sessile on the top of the ovary. The floral parts are robust and it has been suggested that the flowers are adapted to beetle pollination. *T. sylvestris* is wild in central and southern Europe, a rather distinct subspecies, ssp. *australis* occurring in the western part of the Mediterranean. Many of the western Asiatic species have been brought into cultivation, often under the name *T. gesneriana.* Some of the commonly cultivated species are the mostly red *T. eichleri, T. fosteriana, T. greigii, T. kaufmanniana* (also yellow), *T. linifolia, T. praestans* and (violet) *T. violacea,* the yellow-flowered *T. kopalkowskiana, T. stellata, T. tarda* and *T. whittallii,* and the white-flowered *T. turkestanica.* Numerous crosses and aberrant forms have been produced, so that hundreds of cultivars are available on the market. In Holland, especially, but also in other countries, tulips have great economic importance. They are perhaps the commonest garden ornamentals in Europe.

Related to *Tulipa* is no doubt *Erythronium* (20) (Fig. 111 J–L) with the greatest concentration in North America and one species, *E. dens-canis,* occurring from southern Europe through Asia to Japan. In this genus the basal, green leaves are not more than two in number, often broad, subpetiolate and frequently spotted, and the nodding flowers are few or solitary. Their patent to reflexed tepals give them a *Cyclamen*-like habit. *E. dens-canis* and other species are grown as ornamentals.

*Fritillaria* (100) (Fig. 112) is restricted to the Northern Hemisphere with an Asiatic centre. It has few- or many-leaved stems with alternate or

**Fig. 112.** Liliaceae. **A–G** *Fritillaria meleagris.* **A** Plant.
**B** Bulb. **C** Tepal of outer whorl. **D** Tepal of inner whorl.
**E** Androecium and gynoecium, two stamens removed.
**F** Capsule. **G** Seed, sectioned above. (ROSS-CRAIG 1972).
**H–L** *Fritillaria davisii.* **H** Upper part of plant. **I** Tepal
of outer whorl. **J** Tepal of inner whorl; note the basal
nectar grooves. **K** Stamen. **L** Pistil. (TURRILL 1943).
**M–T** *Fritillaria persica,* embryo sac formation of the
*Fritillaria* Type, see further in the text on p. 14.
(BAMBACIONI 1929)

verticillate leaves, and the flowers are almost always nodding or pendulous, campanulate but with
free perianth members. The tepals are yellow, orange or pale to dark brownish, rarely white and
frequently variegated; they are supplied with a
nectariferous pit above the base. Many species are
grown as ornamentals. Best known is perhaps *F.
imperialis,* a large herb from Iran and Afghanistan,
with verticillate leaves and a whorl of nodding
flowers below an apical crown of leaves. Also conspicuous are *F. meleagris* (purplish with checkering
or white) and *F. aurea* (yellow with checkering).
These belong to the largest section of the genus,
which has unifloral stems. *F. meleagris* is Central

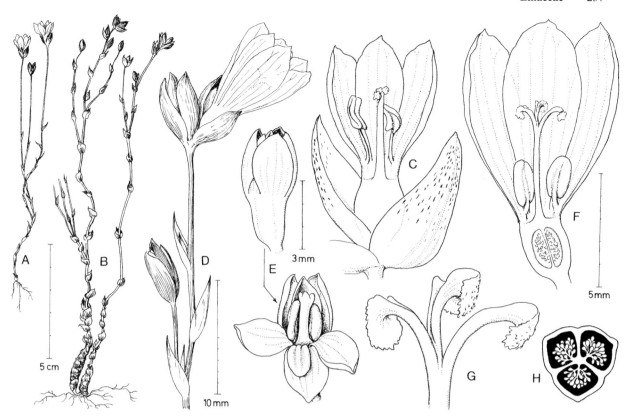

and South European, and *F. aurea* Turkish. The multiflorous *F. pallidiflora* and *Lilium*-like *F. camtschatcensis* are also cultivated. A number of the West-Asiatic species with sombre campanulate flowers appear to be pollinated by queen wasps (*Hymenoptera, Vespinae*).

*Lilium* (75), "Lilies", is a variable genus, widely distributed over the Northern Hemisphere. The bulbs generally have numerous imbricate fleshy scales, and the stem is often multifoliate with alternate or verticillate leaves at several levels. Sometimes, as in *L. bulbiferum,* dark-coloured axillary bulbils are developed. The flowers are few or numerous and are borne in racemose or sometimes umbellate inflorescences, or they may be terminal and solitary. They are mostly large and magnificent, funnel-shaped, often with more or less recurved tepals, which vary in colour from white to yellow, orange, bright red or dark purple, the inner side being mostly spotted and sometimes supplied with hairs. The anthers are epipeltate and their coarsely reticulate pollen grains often bright red. – Many species are grown as ornamentals. Among the white-flowered are *L. candidum* from Greece to Syria, *L. regale* and *L. langkongenese* from China and *L. auratum* from Japan. *L. speciosum,* likewise from Japan, has white to pink tepals

Fig. 113. Geosiridaceae. *Geosiris aphylla.* **A** and **B** Plants. **C** Flower with bract and bracteole; 3 tepals and 4 stamens removed. **D** Inflorescence; note the prophylls of the laterals. **E** Flower bud; *below* opened to show stamens and pistil. **F** Flower, longitudinal section; note the extrorse anthers. **G** Stylar branches. **H** Ovary, transverse section. (Orig. B. Johnsen)

with red spots. Yellow-flowered ornamentals are, for example, *L. canadense,* from eastern North America, *L. pyrenaicum,* from the Pyrenees, and various hybrids. *L. henryi,* from China, *L. bulbiferum* and *L. croceum* from Central Europe and other species have orange-coloured, *L. japonicum* rose-coloured and *L. pardinarium* and *L. martagon* purple flowers. Among the ornamentals many are hybrids and cultivars.

Related to *Lilium* is the Asiatic genus *Notholirion* (4) in Central Asia, mostly Himalaya, which differs from *Lilium* in forming subterranean bulbils from the main bulb, in having a tunicated (not scaly) bulb, and in having long basal leaves. – *Nomocharis* (13), in highland meadows in Burma, Tibet and inner China, generally has open, pink, more or less nodding flowers and short stamens. *N. pardanthina* and *N. saluenensis* are sometimes grown as ornamentals. – *Cardiocrinum* (3) is another Himalayan lily-like genus and has petiolate cordate leaves. It is a large plant with trumpet-shaped flowers and a bulb which is replaced annually.

Whether also the genus *Medeola* (2) should best be treated in Uvulariaceae or Liliaceae is not yet fully clear (see BERG 1962a, b; UTECH 1978a).

## Geosiridaceae Jonker (1939)   1:1   (Fig. 113)

A small, sparingly branched, achlorophyllous mycorrhizal saprophyte with small, scale-like, lanceolate to ovate, hyaline leaves more densely situated on the lower than on the upper part of the erect stem. Rhizome and corm are both obviously lacking, the basal part of the stem being supplied with thin mycorrhizal roots. The inflorescence has not been thoroughly analyzed. It is few-flowered and each lateral branch is supplied with a basally inserted prophyll, as in many Iridaceae. Apically it often bears two or more densely set flowers in the axils of relatively large bracts more than half the length of the flowers, and supplied with a prophyll, in the axil of which a floral bud may develop. No information on vegetative anatomy is available.

The flowers are bisexual, actinomorphic and epigynous with 3 + 3 subequal, obovate, petaloid, pale tepals, which are fused basally and inserted at the top of the ovary, the outer lobes slightly overlapping the inner marginally. The three stamens are opposite the outer tepal lobes (i.e. they represent the outer staminal whorl, as in the Iridaceae) and are inserted on these at their base. Their filaments are relatively short and their anthers subbasifixed, extrorse, tetrasporangiate and longitudinally dehiscent. The pollen grains are uniaperturate (CRONQUIST 1981). The ovary is trilocular, with branched intrusive placentae and numerous minute ovules (Fig. 113 H). There is a single style which is apically tribrachiate with flattened deflexed stigmatic ends. The fruit is a triangular-obconic pyxis or capsule (JONKER 1939), with an annulus at the truncate end. The seeds are numerous and minute.

The single species, *Geosiris aphylla* (Fig. 113), occurs on Madagascar and some islands of the Indian Ocean.

The strong similarity in appearance between *Geosiris* and the Burmanniaceae is probably due to convergence. In several features, such as the inflorescence, the floral construction (e.g. the three episepalous stamens) and the style it is in close agreement with Iridaceae. We assume that it is derived within some group of that family, possibly Iridaceae subfamily Iridoideae, and would logically be included in this, as it was originally by BAILLON (1894), but we keep it distinct until the embryological details have been fully investigated.

## Iridaceae A.L. Jussieu (1789)   70:1.400
(Figs. 114–118)

Mainly herbs, but a few genera woody at the base and sometimes shrubby at length. The underground parts consist of rhizomes or subterranean corms, which are usually tunicated (in *Ferraria* non-tunicated), rarely (in genera allied to *Iris*) bulbs. The tunic enclosing the corm in many genera is longitudinally or reticulately fibrous and characteristically sculptured. The roots usually have simple or, in particular in the Aristeoideae, scalariform perforation plates. Otherwise vessels are lacking in stems as well as in leaves (a notable exception is *Sisyrinchium,* where vessels with simple and scalariform perforation plates have been observed in the stem).

The stems are generally erect and leafy and often compressed. In a few genera (*Klattia, Nivenia, Witsenia*) the stem is basally woody and has secondary thickening growth, which is otherwise not known in Liliales. The densely leafy stem of, for example, *Witsenia* may be up to 2 m high. The leaves are generally distichous, narrow or broad, either flat and dorsiventral or terete or, most often, ensiform (unifacial), linear or rarely lanceolate or filiform, sheathing at the base and parallel-veined. They are glabrous or provided with simple hairs. Crystal raphides are invariably lacking, calcium oxalate being instead accumulated in the form of long, prismatic crystals contained in characteristic dead cells.

The inflorescences in subfamilies Iridoideae and Aristeoideae are panicles, thyrses or cymes composed of monochasial units (rhipidia), but in the large subfamily Ixioideae they are spikes. They frequently have a complicated construction and may be difficult to interpret, especially in the Aristeoideae; sometimes they are arranged in one plane because of the strictly distichous, unifacial leaves and the sometimes compressed stems.

The flowers are trimerous, epigynous (hypogynous in *Isophysis*), mostly bisexual, and actinomorphic or often weakly, but sometimes strongly, zygomorphic. The outer and inner whorls of tepals are both petaloid, but in a number of genera (e.g. in *Libertia* and *Iris*) they are conspicuously different

in shape and colour, the inner sometimes much smaller than the outer. They are blue, violet, white, yellow, red, purple, and even blue-green (*Ixia viridiflora*), and variegation and striation are rather common, in particular in the Tigrideae and Mariceae. Nectaries are present on the tepal bases except in subfamily Ixioideae, where the pistils have septal nectaries.

**Fig. 114.** Iridaceae. **A–E** *Nivenia binata*. **A** Flowering branch. **B** Pair of flowers enclosed by brown bracts. **C** Flower, longitudinal section. **D** Ovary, transverse section. **E** Base of flower, part of the tube removed to show lobate top of ovary. **F** *Nivenia corymbosa*, fruit. **G** *Nivenia dispar*, woody stem. **H–I** *Nivenia* sp., cluster of fruits and seeds. (All orig. B. JOHNSEN)

The stamens are generally three in number (two in *Diplarrhena*), representing the outer whorl of hexastaminate Liliiflorae. The filaments are narrow and free from each other, or sometimes connate basally or entirely; they are often inserted in the tepal tube (e.g. *Tigridia, Sisyrinchium, Patersonia*). The anthers are basifixed, often basifixed-sagittate, or rarely hypopeltate; they are always extrorse, and dehisce by longitudinal slits. Microsporogenesis, interestingly enough, is simultaneous in all cases studies. The pollen grains are generally sulcate, rarely bisulculate (as in *Tigridia*), spiraperturate or inaperturate (in forms of *Crocus* and *Syringodea*). (See SCHULZE 1971).

The pistil is tricarpellary and trilocular (in *Hermodactylus* unilocular), the style either tribrachiate and often with the branches dichotomous or further branched or differentiated, sometimes strongly petaloid (e.g. in *Iris, Dietes, Moraea*). The stigmatic surfaces are papillate and of the Dry Type (HESLOP-HARRISON and SHIVANNA 1977). Each locule usually contains numerous ovules, rarely few or even a single ovule. The ovules are anatropous, and the archesporial cell cuts off a parietal cell, which forms one or more parietal layers. Embryo sac formation conforms to the *Polygonum* Type. Endosperm formation is nuclear.

The fruit is a loculicidal capsule with a thin to leathery, rarely hard, wall. The seeds are semi-globose or angular, rarely round and flat (*Diplarrhena*) or elongate-ellipsoidal (e.g. *Cypella* and *Nivenia*). The raphe often forms a ridge or wing, especially in subfamily Ixioideae, and a wart may also be present on the micropylar side. Arils occur in some sections of *Iris*, elaiosomes derived from chalaza and raphe are found in *Patersonia*, and others derived only from the chalaza occur in *Hermodactylus*. The outer integument of the testa usually consists of several cell layers which are retained for the most part in the seed. The outer epidermis is retained or is rarely collapsed, and usually has brownish contents, while the inner epidermis is mostly well-developed; the inner integument is often partly collapsed. Aleurone and fatty oils are accumulated in the endosperm cells, which have thick, pitted walls (thus being also rich in cellulose). Starch is mostly not found in ripe seeds (an exceptional case, *Radinosiphon*, is reported by HUBER 1969). The embryo is straight and cylindrical and from about one third to two thirds as long as the endosperm. As in Colchicaceae, the cotyledon of the seedling has a coleoptile-like appearance.

*Chemistry.* Characteristic of Iridaceae are prismatic oxalate crystals (see above) and mucilage ducts. Chelidonic acid is rarely present (as in *Lapeirousia, Romulea* and *Trimezia*). Alkaloids largely seem to be lacking, although certain alkaloids or alkaloid-like compounds, like homeridin, which has a digitalis-like effect on heart function, occur, for example, in *Homeria*. Steroidal saponins are known to occur in corms, flowers or fruits of a number of genera (e.g. *Crocus* and *Gladiolus*). Several genera accumulate phenolic compounds such as leucoanthocyanins, anthraquinones and tannins. The flavonoid spectrum is more varied than in other families of Liliiflorae. The corms and rhizomes accumulate carbohydrates, such as saccharose, fructan and starch in variable proportions. The contents of the styles in *Crocus sativus*, "Saffron", include lycopin, $\beta$- and $\gamma$-carotin, zeaxanthin, the water-soluble orange pigment crocin, and the strongly fragrant safranal.

*Distribution.* The Iridaceae are most richly represented in the Southern Hemisphere, where the subfamily Ixioideae forms a climax group in the winter rainfall area of South Africa. Another comparable centre is in tropical and subtropical America. The genus *Isophysis*, which is probably a relict, having hypogynous flowers, occurs only in Tasmania. A few genera show markedly disjunct distributions. Among these are *Libertia*, occurring in New Zealand, New Guinea, Australia and Chile, *Diplarrhena* in Australia and South America, and *Orthrosanthus* in Australia and both North and South America.

**Subfamily Isophysoideae**

This is clearly a member of Iridaceae as judging from the vegetative characters. It consists of a single species, *Isophysis* (1) *tasmanica*, endemic in Tasmania, a herb with a thick rhizome, distichous, ensiform leaves and a solitary, actinomorphic and stellate flower. This has subequal, lanceolate, free tepals and a pistil with *superior* ovary, a short style and simple, recurved stigmatic branches. If the superior ovary does not represent a derived state, i.e. in this case a reversal (epigyny is generally assumed to be an irreversibly derived state), and the ensiform leaves have not arisen two or more times, then there is evidence that the ancestral Iridaceae resembled *Isophysis*, and that the dorsiventral leaves in many Iridaceae are secondary.

## Subfamily Aristeoideae

Subfamily Aristeoideae consists of evergreen rosette herbs or even shrublets, often with persistent leaf sheaths, as in the Velloziaceae. The roots have vessels with mostly scalariform perforations. Secondary thickening growth has been recorded in the stem and roots of a few genera. The leaves are distichously inserted and either unifacial-ensiform or terete (most species of *Bobartia*). The inflorescences are bracteate panicles or thyrses with lateral components of rhipidial monochasia. Sometimes they are strongly contracted. The flowers are actinomorphic, with frequently fugacious, violet, more rarely white or yellow tepals which are generally fused into a tube of variable length. Nectar secretion takes place at the tepal base. The stylar branches are short or obsolete and the filaments are free. The seeds lack wings and other appendages. Their seed coat, as in *Aristea*, may have a testal layer of only two unpigmented cell layers, one often carrying lipids, and a thick, several-layered tegminal layer. The testal layer may dissolve. In the endosperm cells the walls are thin and unpitted. The embryo is very small, only ca. $1/10$ to $1/5$ of the endosperm length. This subfamily is wholly African and concentrated mainly in the winter rainfall regions of South Africa. It consists of four to five genera.

*Aristea* (50) has its centre in the Cape but extends to Madagascar. It is naturalized in the New World tropics. The frequently branched inflorescences consist of rhipidia which are one- or few-flowered. The tepals are blue, violet or white and fugacious. – Related is *Nivenia* (8; Fig. 114A–I) in South-Western Cape. It is a shrub with branched woody stems bearing scars of old leaf sheaths. – *Witsenia* (1) in the same region is a slender shrub up to more than 2 m tall. It has bird-pollinated, tubular cylindrical flowers, that are green, black and (apically) yellow in colour, and hairy on the outer side. – *Bobartia* (15), likewise in the Cape, should possibly also be referred to this subfamily. It has yellow (rarely violet) flowers in dense and contracted clusters, rarely loose panicles. The stem is leafless and the leaves mostly terete and stiff, giving the plants a *Juncus*-like appearance. Alternatively, *Bobartia* may have its closest relatives in Iridoideae, near *Dietes,* with which it shares pollen shape and chromosome number ($x = 10$) (GOLDBLATT 1971).

## Subfamily Sisyrinchioideae

This consists of herbs with a rhizome. They rarely (*Patersonia*) have secondary thickening growth. The roots have vessels with scalariform and/or simple perforation plates, and vessels have been reported in stems and leaves of *Sisyrinchium*. The leaves are distichous but bifacial and flat or triangular in transection. As in the Aristeoideae, the inflorescence is principally a panicle with rhipidial components. The flowers are actinomorphic; the tepals are free or basally fused into a short tube and the two whorls are generally similar in shape and size. Nectaries are present on the perigone. The stamens have free or basally connate filaments. The ovary lacks septal nectaries, and the three stigmatic branches, which alternate with the stamens, are undivided or slightly divided only. The seeds are normally globose or ellipsoidal and only in *Diplarrhena* are they flat and disc-shaped. A lipid-rich elaiosome is present in *Patersonia*. The outer integument of the seed coat contains several layers of cells (as also in the following subfamilies); its outer epidermis tends to be completely collapsed and is either red-brown or colourless (HUBER 1969). The embryo varies in length, but is very short in *Patersonia*.

This somewhat dubiously homogeneous subfamily has great affinities to subfamily Aristeoideae.

All genera have rhizoms. *Patersonia* (20) in Australia, Borneo, the Philippines and New Guinea, stands out as remarkable. The leaves are basally concentrated, sometimes on a short, woody and linear stem, the aerial stem is leafless and ends, as a rule, in a double rhipidium with blue or yellow flowers. Of the six tepals the inner three are reduced to small teeth; the staminal filaments are connate into a tube around the style, and the three simple stigmatic lobes are deflexed, rather broad and ciliate. In addition, the seeds have terminal appendages.

The probably related *Libertia* (10) also has a Southern Hemisphere distribution. They are small or medium-sized rhizomatous herbs with narrow, flat leaves and flowers with white or blue inner tepals, the outer smaller than the inner and slightly "sepaloid" in nature. The stylar branches are simple and narrow-subulate. – *Orthrosanthus* (10), distributed in Australia and both North and South America, is doubtless related to *Libertia*. – *Diplarrhena* (2), in Australia, has totally zygomorphic, large flowers in a loose rhipidium. The stamens are unequal, the median one staminodial, and the

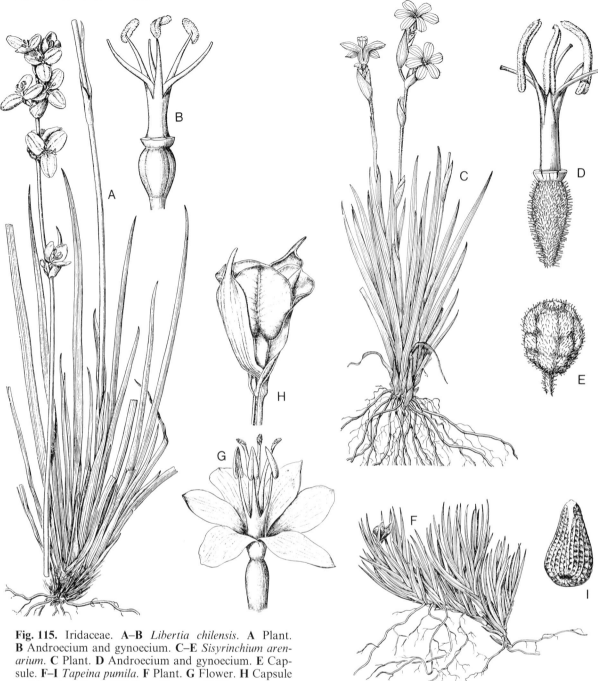

**Fig. 115.** Iridaceae. **A–B** *Libertia chilensis.* **A** Plant. **B** Androecium and gynoecium. **C–E** *Sisyrinchium arenarium.* **C** Plant. **D** Androecium and gynoecium. **E** Capsule. **F–I** *Tapeina pumila.* **F** Plant. **G** Flower. **H** Capsule and bracts. **I** Seed. (All from CORREA 1969)

stylodial branches, as is often the case in the following genera, are flattened.

More specialized than these genera is *Sisyrinchium* (80) in South America (principally in the Andes) and Central and North America, with an outlier in Ireland. It is a rhizomatous genus with ensiform leaves and paniculate or thyrsoid inflorescences with yellow, white or blue, actinomorphic flowers.

Superficially the genus resembles *Aristea,* but there is no perigone tube and the stylar branches are longer, simple as in the previous genera, and not flat as in most of the Iridoideae. The seeds lack a lipid layer. – Several species are grown as ornamentals, e.g. the southern North American *S. bermudiana* with violet flowers and the South American *S. striatum* with pale yellow flowers.

## Subfamily Iridoideae

Subfamily Iridoideae consists for the most part of herbs with leafy stems withering down each season. The subterranean parts consist of short or long, sometimes horizontal rhizomes or, quite often, of tunicated corms; rarely, in the tribe Irideae, are there true bulbs. The leaves vary from unifacial-ensiform to bifacial and flat or canaliculate; sometimes they are terete. The inflorescences are paniculate or simple and consist of one or more rhipidial cymes. The flowers are often large; they are epigynous and generally actinomorphic, and have free or connate tepals, which are frequently fugacious. There is often great difference in shape and size between the outer and the inner tepals, the former of which have basal nectar secretion, septal nectaries being absent. The stylar branches are generally apically bifurcate and in a number of genera flat and petaloid, more rarely narrow, showing rich differentiation in the subfamily. The seeds are half-ovoid, sometimes sharply angular, discoid or irregular-polyhedral, and always lack a wing. In the seed coat, the testal layer consists of several, sometimes many, persistent cell layers. If lipids are present they occur in one to several layers and contain amorphous masses of red-brown pigment. The testal cells are not radially elongated, as in the Ixioideae. The walls of the endosperm cells are pitted except in the periphery of the endosperm; the embryo in most genera is one third to two thirds as long as the endosperm.

Like the Ixioideae, this subfamily represents a climax group whose flowers have differentiated richly within a characteristic pattern of variation. One major complex of genera has radiated in the Old World, and makes up the tribe Irideae, while two smaller, probably more recent groups, the Tigrideae and Mariceae, have developed, respectively, almost entirely and wholly, in the New World.

## Tribus Irideae

Tribus Irideae is the most variable of the three tribes and covers most of the above variation. Typical of most genera and possibly basic in the tribe are the often large, flat and petaloid stylar branches sheltering the three stamens and their extrorse anthers. Besides the genera traditionally referred here are included *Ferraria* and *Homeria*, sometimes referred to the Tigrideae.

Within the Irideae, the genus *Dietes* (6) in South Africa and on Lord Howe Island possesses a number of features, some of which may be basic and primitive in this tribe, being evergreen herbs and having unifacial leaves, a woody rhizome, flowers with free stamens and broad petaloid stylar branches (GOLDBLATT, personal communication). Superficially the genus is fairly *Iris*-like.

The remaining genera may be divided into two groups, which may be called the *Iris* and the *Homeria* Groups. The former has differentiated chiefly in the Northern Hemisphere (mainly the Old World), and the latter in Southern Africa.

The Iris Group, apart from the largest genus, *Iris*, includes three genera with bulbs, *Iridodictyum*, *Xiphium* and *Juno*, and two small genera with rhizomes, *Hermodactylus* and *Belamcanda*.

*Iris* (200) is the largest genus of the family. It has a wide, Northern Hemisphere distribution with a centre in Asia. When the above bulb-bearing genera have been excluded, as done by RODIONENKO (1960), which is debatable and not yet accepted by most taxonomists and horticulturalists, the remaining *Iris* s.str. is more easily defined than before. It then consists of rhizomatous species with equitant, ensiform or linear leaves, the shoots appearing as a rule quite flat. Further classification depends on whether the outer perianth members are crested or bearded or unornamented, on the seed shape and whether or not the roots are fleshy (see DYKES 1913; LAWRENCE 1953).

Subgenus *Nepalensis* consists of two deviating species, including *I. nepalensis*, with a short rhizome bearing a fascicle of fleshy roots. This subgenus is centred in Yunnan (China).

The other groups have a conspicuous rhizome. Among them certain species lack a crest or bearded zone on the outer tepals. One of the species, *I. dichotoma*, in eastern Asia, with stout stem and compressed seeds with parchment-like testa, constitutes subgenus *Xyridion*.

Most other species without crest or beard comprise subgenus *Limniris* (= sect. *Apogon*). They have other seed shapes and often comprise large plants. They represent a large group of Eurasian as well as North American species, including *I. sibirica*, *I. orientalis*, *I. macrosiphon*, *I. foetidissima*, *I. ruthenica*, *I. spuria* and *I. unguicularis*. Certain interesting features occur among some of these, *I. sibirica* and *I. orientalis* having a hollow aerial stem; *I. macrosiphon* and other species being North American; *I. foetidissima* having orange-red seeds; *I. ruthenica* being low and grass-like; and *I. unguicularis* having a floral tube 10–20 cm long. Subgenus *Limniris* (subsect. *Laevigatae*) includes species from Europe (*I. pseudacorus*), Eastern Asia (*I. laevigata*) and North America (*I. versicolor*), having a thin-walled capsule which breaks up irregularly at maturity.

Subgenus *Crossiris* (= sect. *Evansia*) comprises relatively few, mainly small-growing species having a crest (but not a hairy ridge) on the outer tepals. *I. cristata* in Atlantic North America almost lacks an aerial stem. The

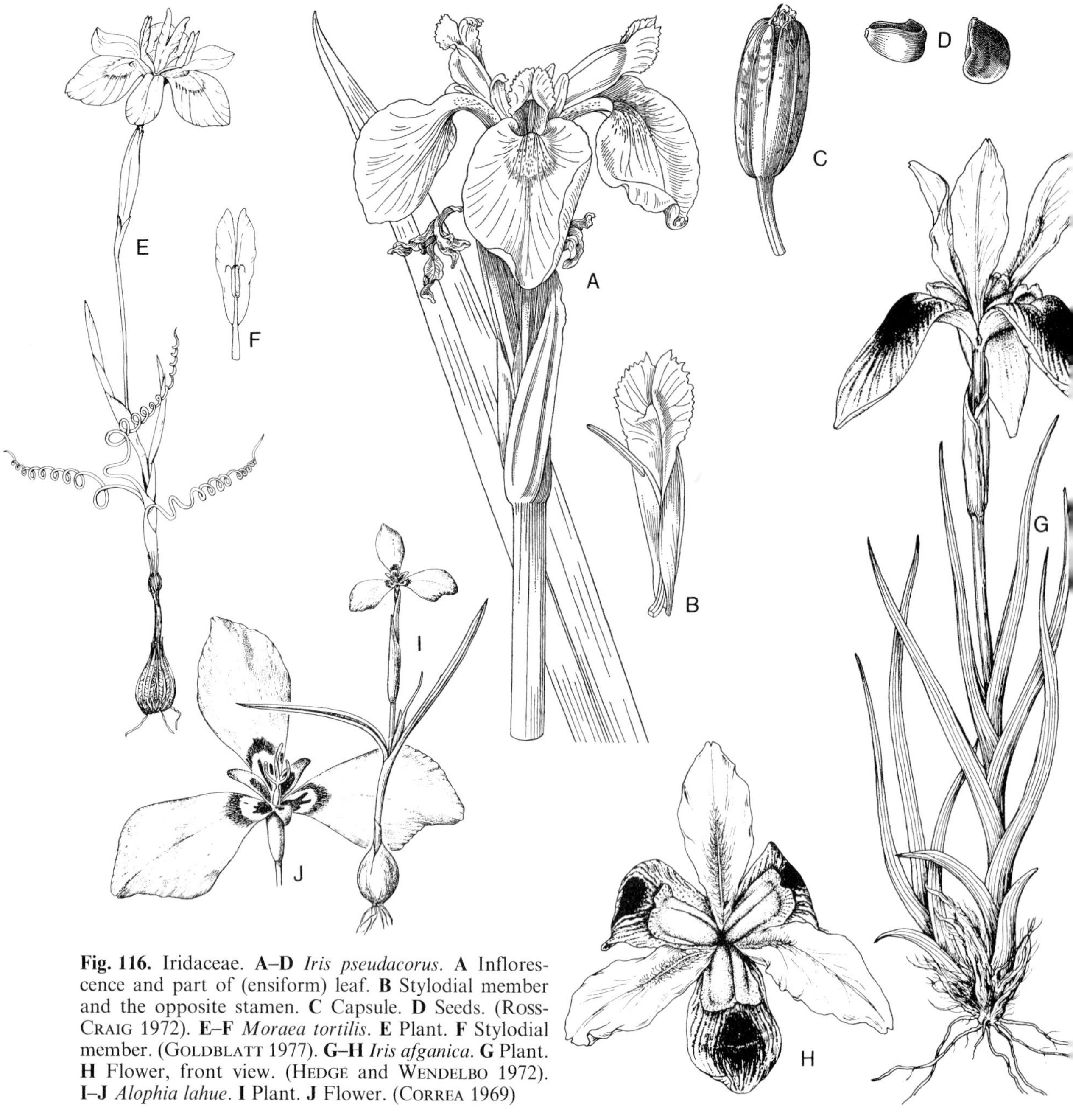

**Fig. 116.** Iridaceae. **A–D** *Iris pseudacorus*. **A** Inflorescence and part of (ensiform) leaf. **B** Stylodial member and the opposite stamen. **C** Capsule. **D** Seeds. (Ross-Craig 1972). **E–F** *Moraea tortilis*. **E** Plant. **F** Stylodial member. (Goldblatt 1977). **G–H** *Iris afganica*. **G** Plant. **H** Flower, front view. (Hedge and Wendelbo 1972). **I–J** *Alophia lahue*. **I** Plant. **J** Flower. (Correa 1969)

other species occur in Japan and China and other parts of eastern Asia, among them the violet and large-flowered *I. tectorum*.

Subgenus *Iris* (incl. the sections *Oncocylus, Regelia, Pseudoregelia,* all with arillate seeds, and sect. *Pogoniris,* without aril), is characterized by having a bearded ridge or crest in the centre of the outer tepals. This subgenus is large and variable and includes both large and small-sized species. The *Oncocyclus* species are often small-sized and have unifloral spathes and seeds with a yellowish white aril. They are concentrated in western Asia and include the small *I. paradoxa* and the larger *I. su-*siana. The related *Regelia* species are also West-Asiatic, but usually have two to three flowers per stem. Among these may be mentioned *I. falcifolia.* An aril smaller than the seed characterizes sect. *Pseudoregelia,* with a few species with twisted rhizomes. They range from Himalaya to Western China, and include *I. hookeriana.* The species of sect. *Pogoniris* are fairly numerous, at least 40, and occur in Asia and southern Europe. Here belong a number of unbranched, blue, violet or yellow-flowered species, including *I. pumila, I. tigridia* and *I. flavissima,* and species with branched stem, such as *I. aphylla, I. variegata,* and the frequently cultivated *I. germanica.*

The genus *Xiphidium* (= *Iris* sect. *Xiphium*) (7), in the western Mediterranean, are often slender herbs with dorsiventral, flat or canaliculate leaves and a bulb with entire, smooth and membranous tunics. The nominate species of these "bulb irises" is *X. xiphium,* used in horticultural breeding and represented in cultivation by white-, yellow-, violet- or brownish-flowered forms. Others are *X. filifolia* and *X. juncea.*

*Juno* (= *Iris* subgen. *Scorpiris* or sect. *Juno*) (20) with a centre in Western Asia (Iran and neighbouring countries), are less slender, often stout plants with dorsiventral, broad and broadly sheathing, flat leaves. The bulb has entire smooth tunics, and fleshy roots both of which persist during the dormant period. The inner tepals are small or even minute. Here belong *J. persica, J. bucharica,* and *J. aucheri,* sometimes grown as ornamentals.

*Iridodictyum* (= *Iris* subgen. *Scorpiris* or sect. *Reticulata*) (10), with a centre in western Asia, consists of erect herbs with short stem, one flower, narrow tubular leaves with four or eight ridges, and a basal bulb with fibrous tunic; as in *Xiphidium* the roots do *not* persist in the dormant period. These small plants, often less than 20 cm high, are often grown as ornamentals and blossom quite early in the spring. They include *I. reticulatum, I. histrioides* and *I. danfordiae.* – *Hermodactylus* (1) is a Mediterranean genus with thick rhizome, linear, four-angled leaves, and a long spathe with one flower and parietal placentation. The flowers are blackish brown. The perianth members in the two whorls are dissimilar, the outer being erect, the inner smaller and erect-spreading.

The South African group of genera, centred around *Homeria,* is characterized by having a tunicated corm rather than a rhizome or bulb. The leaves vary from equitant to terete or dorsiventral and canaliculate to flat.

*Homeria* (31), in Southern Africa, has mostly short stems, which bear one or more long, linear leaves which exceeds the stem in length. Generally the tepals, which vary much in colour, are of about equal size and shape, while the filaments are fused into a tube, and the stylar branches are linear or somewhat broader. – The related *Galaxia* (12) has similar, very ephemeral flowers; it consists of small, acaulescent plants, often with rather broad leaves. – *Moraea* (110) occurs in sub-Saharan Africa and has a pronounced centre in the winter rainfall regions of South Africa, where ca. 60 species are concentrated. Like *Homeria* and *Galaxia* it has a fibrous corm tunic and thus approaches these genera rather than the superficially more similar *Iris* (GOLDBLATT 1977b). The filaments are also fused for half of their length and like the anthers are pressed against the flat stylar branches. In contrast to *Iris* the leaves in *Moraea* are flat and bifacial (? a secondarily reversed state). Some species of *Moraea* are cultivated for their decorative flowers, e.g. *M. neopavonia, M. villosa* and *M. aristata* (= *glaucopsis*). A considerable number of species are very local and close to extinction. – *Gynandriris* (9) is another mostly South African genus with tunicated corms; it is closely allied to *Moraea. Gynandriris* also includes the common Mediterranean species *G. sisyrinchium,* often referred to *Iris.*

## Tribus Tigrideae

This is a wholly American group consisting of about six genera, of which *Tigridia, Cypella* and *Alophia* are mentioned here. These are characterized by "woody" bulbs and flat, ensiform, plicate leaves. The flowers are often conspicuous with their bright-coloured and variegated tepals, the inner of which are often much smaller than the outer. The stylar branches are variable, often being narrow and bifurcate.

*Tigridia* (15) ranges, mainly along the Cordilleras, from Mexico to Chile. *T. pavonia* and other species are frequently cultivated ornamentals. – *Cypella* (10) and *Alophia* (10) (Fig. 116I–J) are also Central and South American.

Other genera, sometimes treated in a separate tribe, Cipureae, have their proper place in Tigrideae, e.g. *Cipura* (1), with simple stylodial branches, and *Nemastylis* (21) and *Calydorea* (15), where the stylodial branches are divided.

## Tribus Mariceae

Tribus Mariceae exhibits a parallel trend, possibly more primitive, in its possession of rhizomes or bulbs, than the Tigrideae, and is mainly concentrated in the Americas. The leaves are flat, equitant and ensiform, and are not plicate. The flowers are reminiscent of those in the Tigrideae, the inner perianth segments being for example smaller than the outer and mostly variegated.

*Neomarica* (15) is concentrated in neotropical regions with its centre in Brazil. Some species have ornamental value. – *Trimezia* (6) ranges from Mexico through Central America to Southern Brazil. It is naturalized in the Old World tropics.

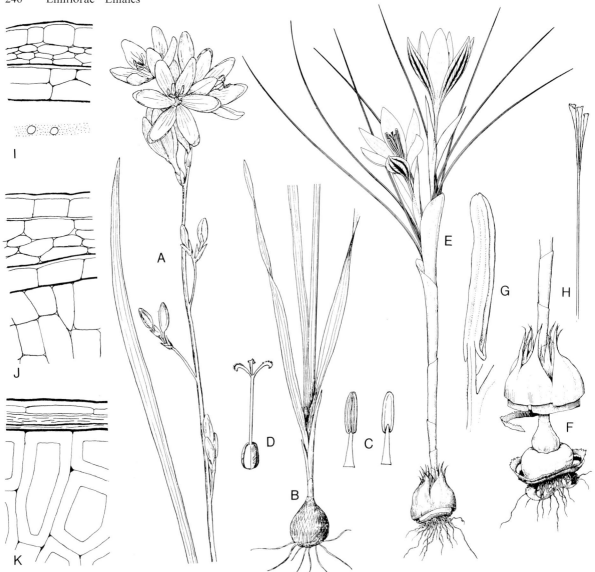

## Subfamily Ixioideae

Subfamily Ixioideae consists of herbs, always with a tunicated corm, the tunic structure of which often supplies good taxonomic characters. It may consist entirely of lignified layers or of soft reticulate fibres. The roots have vessels with mostly simple perforation plates. The leaves are sometimes bifacial, and flat or canaliculate, as in *Crocus* or *Syringodea,* but are more often unifacial (ensiform). The plants generally wither after each season. One of the most important characteristics is that the inflorescence is basically indeterminate: a simple or rarely compound (*Pillansia, Lapeirousia*) spike, which may become panicle-like. The tepals in all genera are connate into a tube, which

**Fig. 117.** Iridaceae. **A–D** *Ixia flexuosa.* **A** Inflorescence. **B** Corm and leaves. **C** Stamen in different views. **D** Pistil. (LEWIS 1962). **E–H** *Crocus stridii.* **E** Plant. **F** Corm, tunics splitting at the base. **G** Stamen. **H** Style apex. (PAPANICOLAOU and ZACHAROF 1980). **I–K** *Romulea* sp., development of seed coat, three successive stages from above. The seed coat is here rather thin; it does not contain a black layer on the outer epidermis. (STEYN 1973)

may, however, be very short, as in most species of *Romulea.* Nectar glands are not present on the base of the tepals, but in glandular cavities opening at the style base, thus representing septal nectaries. The seeds are globose or subglobose and, in some genera, winged, e.g. *Gladiolus, Homoglos-*

*sum, Oenostachys, Anomalesia* and *Watsonia;* they have a multi-layered outer integument of persistent cells containing a lipid layer, one cell thick, of sometimes palisade-like cells. The cell walls in the endosperm in most cases are largely or totally unpitted (rarely pitted except in the periphery). This very large and well-defined subfamily has its centre in Southern Africa, especially in the winter rainfall area of the Cape Province. Only *Crocus* is distributed entirely outside southern Africa.

The subfamily has been revised by GOLDBLATT (1971). The infrasubfamilial division will not be presented in detail. The following genera are worthy of mention.

*Watsonia* (45) in Southern Africa consists of often large herbs with fibrous corm tunics and ensiform leaves. The spicately borne flowers are often large, and are long-tubular and brightly coloured; the seeds are winged. *W. aletroides, W. pillansii* and other species are grown as ornamentals; *W. meriana* is naturalized in Macaronesia and southern Australia. – *Lapeirousia* (25) are mostly smaller herbs, with ensiform leaves and a "flat" appearance. The woody tunics of the corms show variable sculpture. The flowers are short- to long-tubed and variable in colour. The genus ranges from the Cape through Namaqualand to Ethiopia and to Nigeria. – *Schizostylis* (1) *coccinea* in Central to South Africa is a slender streamside plant with attenuated corm and rose-coloured, actinomorphic flowers. It is often cultivated. – Related genera are *Hesperantha* (50) and *Geissorhiza* (80), the former more widely distributed, the latter restricted to the Cape. – *Melaspherula* (1) *graminea,* in South-Western Africa, has a much-branched, thin and slender inflorescence with rather small, slightly zygomorphic, whitish flowers.

*Ixia* (44), *Sparaxis* (6) and *Dierama* (20), centred mainly in South Africa, are slender herbs with their corm covered by reticulate fibres. The flowers are unspecialized and actinomorphic or almost so, and highly variable in tepal colour. Several species are grown as ornamentals, e.g. *D. pulcherrima* with wiry inflorescence axes and pendent flowers.

*Gladiolus* (150) is a large, variable African to Mediterranean and western Asiatic genus with ca. 105 species in South Africa. It consists of small to fairly tall herbs. The corms are ovate or globose to flattened and generally clothed with entire, papery tunics. The bracts of the inflorescence may be rather large, and the flowers are tubular or funnel-shaped with syntepalous, actinomorphic or, more usually, weakly or often very strongly zygomorphic perianth. Some or all tepals may be conspicuously striate. A wide diversity of pollination syndromes is exhibited. The wild species are often fairly small in total size and have smaller flowers than the cultivated. *G. cardinalis* and *G. dalenii* (= *G. psittacinus*), besides various hybrids, are among the numerous cultivated forms originating in South Africa. Conspicuous South African species are *G. alatus, G. carinatus* and *G. carneus.* Common species in the Mediterranean region are the rose- to crimson-flowered, rather slender species *G. illyricus* (Fig. 118A–G) and *G. italicus,* the latter of which is a weed. – *Homoglossum* (10) and *Anomalesia* (3) in Southern Africa are closely allied to *Gladiolus* (there are fertile hybrids between *Gladiolus* and *Homoglossum*); in some species the narrowly tubular flowers have long protruding upper lobes. – *Freesia* (20), *Tritonia* (40) and *Crocosmia* (7) are additional South African genera with globose to depressed corms which are covered by tunics with fairly fine fibres. The tubular, often zygomorphic flowers are white, yellow, orange, red or violet and in *Freesia* mostly strongly scented. – While the wild forms of *Freesia* are usually yellow or white, the cultivated forms may be yellow, orange-red or lilac. They are ornamentals with great economic value. – *Tritonia* differs from *Freesia* in the entire or only apically (not deeply) bifid styles. The plants sometimes have rather short stems and membranous to scarious inflorescence bracts. The flowers usually have calluses on the lower perianth lobes. – *Crocosmia* has broader capsules than *Freesia* and slightly to strongly zygomorphic, yellow to orange-coloured flowers without tepal calluses, usually showing the syndrome of ornithogamy. *C. masonorum, C. aurea* and, especially, the artificial hybrid *C. × crocosmiiflora* with its different cultivars, are commonly grown ornamentals, and the last is naturalised in Western Europe and Australia.

*Anapalina* (7) has tubular, strongly zygomorphic, bright red flowers with hooded upper lobes. – *Babiana* (63) and the related *Antholyza* (2) are more or less hairy, low herbs with ensiform, plicate leaves. The former genus varies in flower colour, mostly in the white-violet range, while *Antholyza* has large, bright red tubular-zygomorphic flowers which are bird-pollinated. All these genera are concentrated in the Cape Province. Occasionally representatives of them are brought into cultivation. – *Romulea* (90) also has its centre in South Africa but ranges as far as the Mediterranean and the Canary Islands. Its species have a bulb-like corm and in many species the stem is short, and wholly subterranean. The few flowers (? in a spike)

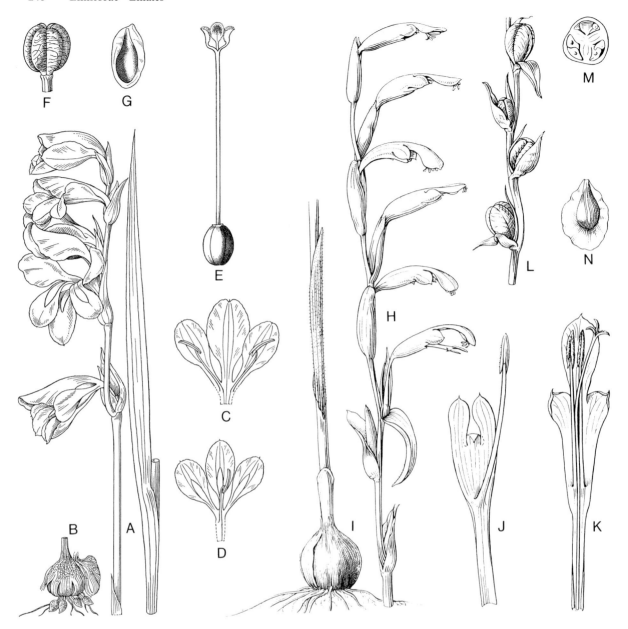

appear one at a time. The leaves are principally ensiform, but often have adaxial longitudinal ridges. The flowers are actinomorphic and somewhat *Crocus*-like, usually with a short tube, and the stylar branches are bifid or rarely multifid. Several species are grown as ornamentals, e.g. *R. bulbocodium* from the Mediterranean region and *R. macowanii* from South Africa. – Related is *Syringodea* (8) in South Africa.

It is uncertain whether *Crocus* (80), which is distributed in the Mediterranean and western Asia to Iran, is closely related to *Romulea* and *Syringodea,* or whether the similarity could have developed by convergence. The leaves of *Crocus* are bifacial, flat or canaliculate. Most species flower

**Fig. 118.** Iridaceae. **A–G** *Gladiolus illyricus.* **A** Inflorescence and leaf. **B** Tunicated corm. **C** Upper perianth segments and two stamens. **D** Lower perianth segments and one stamen. **E** Pistil, the perianth and stamens removed from the top of the ovary. **F** Capsule. **G** Seed. (ROSS-CRAIG 1972). **H–N** *Oenostachys vaginifer.* **H** Inflorescence. **I** Corm and basal leaves. **J** Lower perianth segments and one stamen. **K** Upper perianth segments, two stamens and style. **L** Capsules. **M** Ovary, transverse section. **N** Seed. (MILNE-REDHEAD 1950)

in early spring, generally before the leaves are fully expanded. The flowers, as in *Romulea,* are sessile and the ovary does not appear above ground level. The perianth is actinomorphic and yellow, violet or white, and often bicoloured or tricoloured. Within the genus there is an interestingly great variation in the style, the three main branches of which may be repeatedly branched, palmate or fan-shaped; it is usually yellow to red in colour and may be rich in crocin and safranol (see the chemical characters of the family). *C. sativus* is cultivated in France, Spain and other mainly Mediterranean countries for its styles, which yield the yellow "Saffron", a spice used since antiquity to flavour and dye bread and other food. About a hundred flowers are needed to produce a gram of saffron. It is also used medicinally. Numerous species of *Crocus* are also grown as ornamentals, e.g. the early spring-flowering *C. vernus, C. chrysanthus, C. angustifolius, C. flavus, C. etruscus, C. laevigatus, C. minimus, C. sieberi* and *C. tomasinianus.* Some autumn-flowered species are also grown as ornamentals, e.g. *C. kotschyanus, C. nudiflorus, C. pulchellus, C. speciosus* and also the "Saffron Crocus", *C. sativus.*

# Orchids

F.N. Rasmussen

The traditional order Orchidales is in this work included in Liliales sensu lato, expressing the view that the orchids have evolved within the Liliales. As the orchids are here treated as three distinct families (see the discussion below), and as no rank between order and family (suborder) is employed in this book, the orchids are left as a group without formal rank.

Another solution to the ranking problem, when the orchids are included in Liliales, would be to treat them as a single family, reduce the present families to subfamilies and create a new rank for the subfamilies of Orchidaceae, such as the contribes of Vermeulen (1966). However, the phylogenetic hierarchy of nature cannot be adequately expressed by the relatively few ranks employed in practical systematics. Thus, the ranking of the orchidaceous taxa in this treatment is only partly an expression of the supposed phylogenetic hierarchy (see Fig. 120).

Despite the lack of formal rank the orchids are here treated in the same way as the orders, with an introductory characterization of the group followed by the family treatments.

*Three Families:* Apostasiaceae, Cypripediaceae, Orchidaceae.

Perennial herbs, sometimes saprophytic. The Apostasiaceae and Cypripediaceae are usually terrestrial and the Orchidaceae prevailingly epiphytic but to a great extent also terrestrial. Orchid roots generally have a multi-layered velamen. In terrestrial species the roots are often swollen and tuberous; the epiphytic groups often have aerial roots.

The stems are sympodial or (in certain Orchidaceae) monopodial; they are often rhizomatous at the base and sometimes form corms. Internodes functioning as storage organs (*pseudobulbs*) are common in the Orchidaceae. The leaves are entire, linear to orbicular and have a tubular sheathing base; sometimes pseudopetiolate.

Hairs of different kinds are found on stems and leaves of most species, but few are distinctly hairy. Stomata are variable and include anomocytic and paracytic types. Silica bodies enclosed in special cells (stegmata) in stems and leaves are reported from both genera of Apostasiaceae, from *Cypripedium* in the Cypripediaceae and from numerous genera of the Orchidaceae (Solereder and Meyer

1930; MØLLER and RASMUSSEN, 1984). Calcium oxalate raphides are generally present.

The inflorescences are indeterminate and terminal or lateral (some Orchidaceae), rarely one-flowered.

The flowers are epigynous and zygomorphic with a trimerous perianth. The tepals of the outer whorl are usually referred to as sepals, although they are rarely truly sepaloid; the lateral tepals of the inner whorl are called petals, the median the *labellum*. The number of stamens is always reduced in relation to the two trimerous whorls of stamens of the complete monocotyledonous floral diagram. The median stamen of the outer whorl and the laterals of the inner are developed as either fertile stamens or as more or less distinct staminodes. Stamens and style are fused into a *gynostemium* which forms a "column" except in a few genera of Orchidaceae with almost sessile anther and stigma. The vascular strands of the adaxial stamens are always present in the gynostemium (RAUNKIÆR 1895–1899; SWAMY 1948). The anthers are basifixed to dorsifixed. The tapetum is glandular with uninucleate cells in all species investigated, except one species of *Paphiopedilum* (Cypripediaceae) and *Arundina* (Orchidaceae) with binucleate tapetal cells (SWAMY 1949a; A.N. RAO 1967). Microsporogenesis is simultaneous. The pollen grains are single in Apostasiaceae and Cypripediaceae, but cohere in tetrads in the majority of Orchidaceae. The individual pollen grains are two-celled at dispersal.

The ovary is trilocular or generally unilocular. There are widely different interpretations of what is seen in a section of an orchid ovary. It appears to consist of six parts: three fertile elements bearing double placentae, alternating with three sterile elements (Fig. 123). The classic view (BROWN 1833; SWAMY 1948) holds that the ovary is tricarpellary with marginal placentae. However, VERMEULEN (1966) and earlier workers cited by him have advanced the view that the ovary is indeed made up of six parts, the three sterile ones being either solid carpels or extensions of the receptacle. The orchidaceous ovary may be interpreted as tricarpellary with marginal placentation. The ridges seen on the outside of the ovary are the basal parts of the perianth-segments (Fig. 123, right), the sepal bases often splitting the carpels.

The style is more or less apically inflexed and terminated by a trilobate stigma with a Wet surface.

The numerous ovules are anatropous, tenuinucellate and usually bitegmic (unitegmic ovules are reported in *Paphiopedilum* of the Cypripediaceae and *Epipogium* and *Gastrodia* of the Orchidaceae). The development of the embryo sac is triggered by pollination. The actual fertilization sometimes takes place as late as 5–6 months after pollination (WIRTH and WITHNER 1959). The information on embryology cited here is based on observations from the Cypripediaceae and Orchidaceae only. Embryogenesis of the few species investigated has proved very variable. The embryo is always immature in the ripe seed. Endosperm formation is arrested very early (not later than in the 16-nucleate stage) or wholly omitted. When it does take place it is nuclear.

The seeds are characteristically minute and numerous, but may nevertheless vary considerably in shape and size. Typically, only the outer layer of the outer integument persists as a membranaceous seed coat, which may have an adaptive significance with regard to aerodynamic properties and the wettability of the seed (BARTHLOTT 1976b).

On germination the embryo forms a tubercle called a protocorm which is covered with rhizoids on most of its basal part. A radicle is not formed and usually no cotyledon; eventually several leaves develop from the apex (VEYRET 1974). Most orchids under natural conditions will germinate only after establishment of symbiosis with a fungal mycelium (endotrophic mycorrhiza). The fungus in many cases is referable to the imperfect genus *Rhizoctonia*, but mycelia of several perfect genera, e.g. *Corticium, Clitocybe, Marasmius, Xerotus* and *Fomes* have been isolated from saprophytic Orchidaceae. The adult plants when green are usually able to grow without the mycorrhiza, but achlorophyllous, mycotrophic holosaprophytes are known from the Orchidaceae. The relationship orchid-fungus is usually imagined as one of parasitism by the orchid on the root fungus, but it is still being discussed whether the orchid mycorrhiza may be of mutual benefit for the organisms involved (HIJNER and ARDITTI 1973; HADLEY 1982).

*Chemistry.* Very little is known about orchid chemistry. So far as is known, the orchids have no distinct chemical characters.

*Distribution.* The orchids are distributed all over the world, but the vast majority of taxa on all levels are tropical. The smallest family, Apostasiaceae, is restricted to Indomalaysia and tropical Australia , and the Cypripediaceae are not found in Africa.

*Relationships, Taxonomy.* The orchids as circumscribed here correspond to one of the classic orders of flowering plants, comprising about 40%

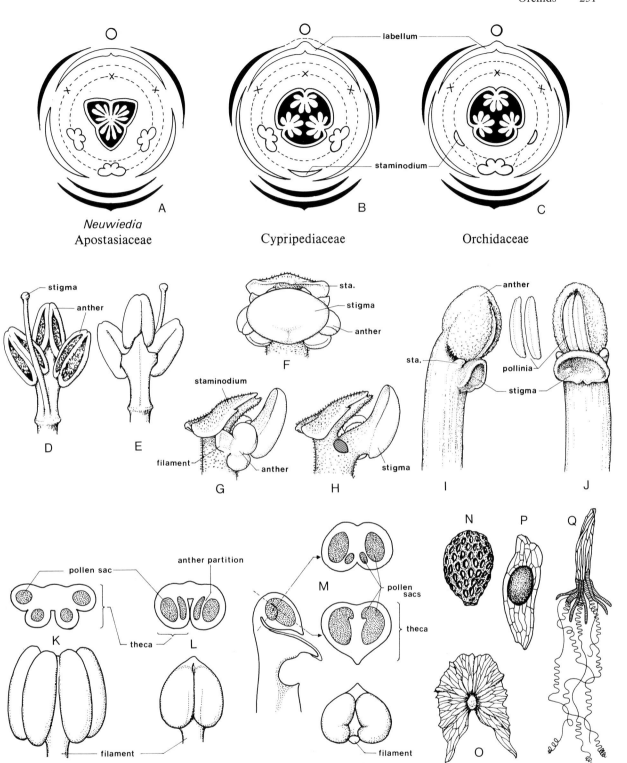

**Fig. 119.** Apostasiaceae, Cypripediaceae and Orchidaceae. **A–C** Floral diagrams of the three orchid families. **D–E** Apostasiaceae: *Neuwiedia inae,* androecium and style in different views. **F–H** Cypripediaceae: *Paphiopedilum appletonianum,* gynostemium in different views, one anther removed in **H. I–J** Orchidaceae: *Cephalanthera longifolia,* gynostemium in different views.

**K–L** Anther in Liliaceae (**K**) compared with anther in Orchidaceae subfamily Neottioideae (**L**). **M** Details of gynostemium and anther in Orchidaceae subfamily Vandoideae. **N–Q** Seeds in Orchidaceae. **N** *Vanilla.* **O** *Galeola.* **P** *Orchis.* **Q** *Chiloschista.* (Orig. and redrawn after BARTHLOTT, JOHNSEN and RASMUSSEN)

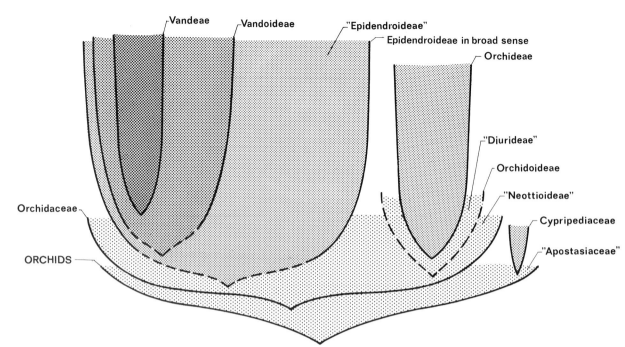

**Fig. 120.** Clades in the orchids. *Broken lines* indicate less certain clades. Names of clades are attached to the periphery of the respective groups, names of "remnant groups" are attached to the interior of the clades to which they belong and put in quotes in accordance with WILEY's conventions of phylogenetic classification (WILEY 1981). The clade Neottieae (in Orchidaceae) and many large lower-order clades, e.g. Angraecinae in Vandeae, are omitted for the sake of clarity in the scheme. The argumentation for each of the clades is found in the text. (Orig.)

of the monocotyledons. It is an almost certainly monophyletic group, distinct from all other orders in the reduction of the adaxial stamens plus the possession of numerous small, endospermless seeds (Fig. 119). The orchidalean gynostemium (Fig. 119) is a unique feature in the monocotyledons (analogous gynostemia are found in the Stylidiaceae and Asclepiadaceae among the dicotyledons).

The epigynous, zygomorphic flowers, the numerous small seeds and the mycotrophy are characteristics shared with the families here grouped as Burmanniales. This similarity has been stressed by several taxonomists when including both groups in one order, the "Microspermae". HUTCHINSON (1959, 1973) considered the Apostasiaceae "closely connected with the Hypoxidaceae, particularly with *Curculigo*", apparently mostly because of similarity in habit, and he excluded the family from the Orchidaceae. VERMEULEN (1966) argued for a relationship between the Orchidaceae, the Commelinaceae and the Pontederiaceae, pointing out a reduction of the number of stamens as a common character. It is, however, not the same stamens that are reduced and the suggestion of relationship is contradicted by numerous differences.

The Orchids are most probably derived from epigynous Liliales-like ancestors with six stamens, simultaneous microsporogenesis and nuclear endosperm formation, through three-staminate forms comparable to the recent genus *Neuwiedia*. For a quite different interpretation see p. 95. A probable extant phylogenetic sister-group cannot be pointed out with our present knowledge.

While there seems to be a general agreement among different workers on the major groups of Orchids regarding their circumscription, their ranking and the supposed relationships between these groups differ widely. Many modern treatments follow GARAY (1960, 1972) in considering the Orchids as consisting of one family, containing five or six subfamilies. The recognition of three families as suggested by VERMEULEN (1966) is accepted here in accordance with the family concept applied throughout this textbook. GARAY (1972) has argued that raising the Apostasiaceae and Cypripediaceae to family rank on morphological grounds would require similar recognition of the Orchidoideae s.str. as a distinct family. This statement was partly based on a later disproved interpretation of the rostellum in this group, and on incomplete information about the actual distribution of character states (see under Orchidaceae).

**Apostasiaceae** Lindley (1833) 2:15
(Figs. 119, 121)

Erect, perennial, sometimes (*Apostasia*) rhizomatous herbs. Roots penetrate the bases of the lower leaves; branching takes place in the subterranean part, which is sometimes tubercled, swollen or woody. Vessels with prevailingly simple perforation plates are recorded (WAGNER 1977).
The stems are woody at the base and sometimes also branched. Pseudobulbs are never formed. The leaves are spirally arranged, entire, plicate and herbaceous to papyraceous. Paracytic stomata were reported by SIEBE (1903). The inflorescences are terminal, sometimes branched racemes.
The flowers may be resupinate by twisting of the pedicel in *Neuwiedia* but not in *Apostasia*. They are slightly zygomorphic, white or yellow, and have 3 + 3 free tepals. The median inner tepal ("labellum") may be broader than the others. The stamens are two or three in number, representing the median of the outer whorl (sometimes missing) and the lateral ones of the inner whorl. The stamens are only partly fused with the style in the gynostemium. The anthers are dorsifixed or subbasifixed, introrse and dehiscent by longitudinal slits. The pollen grains are free, monosulcate with an operculum, and reticulately sculptured (NEWTON and N.H. WILLIAMS 1978; SCHILL 1978). Pollinia are not formed. The style is terminated by a two- to three-lobed stigma.
The ovary is trilocular with central placentation (Fig. 121 C). The fruit is a thin-walled or fleshy capsule, usually disintegrating (in *Neuwiedia veratrifolia* opening loculicidally by three valves). The seeds are ovoid to elliptic, 0.2–0.4 mm long, and usually alveolate to reticulate, rarely provided with long appendages.
Data concerning embryology and chemical contents are wanting.

*Distribution.* The Apostasiaceae are distributed in South-East Asia, from the Himalayas and Ceylon to New Guinea and Northern Queensland, Australia. The highest number of species is found on Borneo.

*Relationships, Taxonomy.* The Apostasiaceae, especially *Neuwiedia,* are very close to most botanists' concept of an ancestral orchid, and may be considered a phenetic link between the Liliales sensu stricto and the two other families of the orchids.
"Affinities" with Hypoxidaceae (HUTCHINSON 1973) seem to be based mainly on resemblance in habit to *Curculigo* (see p. 162). Most morphological evidence (DE VOGEL 1969) and also the anatomy of the flower (V.S. RAO 1974) support a close relationship with Orchidaceae.

It has been suggested by GARAY (1972) that the two genera of Apostasiaceae are not closely related. They do not share any unique, advanced characteristics which would suggest a close relationship, and they differ in some notable respects, such as the number of stamens and the occurrence of a rhizome.

*Neuwiedia* (7) (Fig. 121 A–D), Thailand to New Guinea, are non-rhizomatous terrestrial plants with three fertile stamens (Fig. 119 D–E). The flowers are resupinate by twisting of the pedicel. The inflorescences are more or less hairy. – *Apostasia* (incl. *Adactylus;* 11), Ceylon to Northern Queensland, Australia, are rhizomatous plants. The flowers have two fertile stamens; in *A.* sect. *Apostasia* the median stamen is present as a staminode. The inflorescence is glabrous and the flowers are not resupinate.

**Cypripediaceae** Lindley (1833) 4:100
(Figs. 119 G–H, 121 E–G)

Herbs, usually terrestrial, with fleshy or fibrous roots on short rhizomes. Vessels with simple perforation plates are reported in the roots of three of the four genera, but vessels are lacking in stems as well as leaves (ROSSO 1966).
The stems are sympodial with scattered or basally crowded, spirally arranged or distichous leaves. The leaves may be thin and plicate or fleshy and conduplicate; in the latter case they are usually strap-shaped. Stomata are anomocytic or paracytic, with agene or perigene ontogeny (H. RASMUSSEN 1982). The inflorescences are terminal racemes or sometimes one-flowered.
The flowers are resupinate, strongly zygomorphic and of various colours. The two lateral "sepals" are fused into a synsepalum. The lateral "petals" are often considerably longer and narrower than the other tepals, in some species of *Phragmipedium* reaching a length of 50 cm. The median petal (labellum) is characteristically slipper-shaped. The two lateral stamens of the inner whorl are always present, and the median stamen of the outer whorl is present as a characteristic, shield-like staminode. The filaments are fused with the style, forming a thick, inflexed gynostemium (Fig. 119 F–H). The anthers are subglobose and latrorse, dehiscing by longitudinal slits. The pollen grains are free, sulcate, ulcerate or porate, and reticulate, verrucate or scabrate or may lack distinct sculpturing (NEW-

TON and N.H. WILLIAMS 1978; ZAVADA 1983). The pollen is more or less viscid, but true pollinia are reported only from *Phragmipedium longifolium* and *Selenipedium chica.*

The ovary is trilocular and with axile placentation or unilocular with parietal placentation. Embryology is of the Asterad Type as well as the Onagrad Type (VEYRET 1974), and the endosperm may reach the four-nucleate stage. The fruit is a capsule with numerous small, membranaceous seeds.

*Chemistry.* Very little is known about the chemical contents of Cypripediaceae. Alkaloid-positive reactions were found in some species of *Paphiopedilum* by LÜNING (1974).

*Distribution.* The Cypripediaceae are widely distributed in the northern boreal and north temperate zone (*Cypripedium*) and in the tropics of the Old as well as New World, but are lacking in Africa. The largest number of species is found in Indo-Malaysia.

*Relationships.* The Cypripediaceae form a very distinct and homogeneous taxonomic entity. They differ from Orchidaceae in having two fertile stamens, neither of which corresponds to the single fertile one in Orchidaceae. The flower is strongly adapted to insect pollination: once an insect has entered the "slipper" it will head for one of the two openings between the base of the labellum and the staminode, where it must pass the stigma and one of the anthers (NILSSON 1981).

*Paphiopedilum* (50) (Fig. 121 E–F) from India to the Solomon Islands, has conduplicate leaves and unilocular ovaries. Ornamental species and artificial hybrids of this genus are widely cultivated and commercially important, e.g. *P. insigne* (North India and Nepal). It has a characteristic white upper margin of the dorsal outer tepal and has been used extensively in breeding. *P. barbatum* and *P. callosum* (Thailand and Indo-China) have ciliate calli on the sepals. *P. villosum* (North India to Thailand) is an epiphytic species with large, brownish flowers. – *Phragmipedium* (10), in Central and South America, has conduplicate leaves and a trilocular ovary. – *Cypripedium* (35), in boreal, temperate and subtropical zones of America, Europe and Asia (the highest number of species in Asia), has plicate leaves and a unilocular ovary. *C. calceolus* ("Lady's Slipper") occurs in boreal Eurasia from France to Northeast Asia. The closely related *C. pubescens* occurs in temperate North America. – *Selenipedium* (4), in the Central and South American tropics, has plicate leaves and trilocular ovary. Some species become as tall as 5 m (Fig. 121 G).

**Orchidaceae** A.L. Jussieu (1789)   ca. 730:20,000
(Figs. 119–126)

Perennial herbs with extremely variable morphology and size, terrestrial, litho- or epiphytic, sometimes climbers. Most of the "epiphytic" species are very well capable of growing on the ground when they are given the chance in forest clearings etc. Two completely subterranean species and several achlorophyllous saprophytic forms are known. The roots may be thin and wiry or sometimes form thickened, cord-like or tuberous storage organs. The roots of the epiphytic forms are usually thick and have a velamen; they emerge by breaking through the leaf sheaths. Chlorophyll-containing assimilating roots occur in some genera. Vessels have been observed in the roots of the few genera investigated for this character (WAGNER 1977).

The basal parts of the stems may be modified in various ways. Branching is usually sympodial and in terrestrial forms the branched part may be rhizomatous (Fig. 121 K), whereas in the epiphytes there may be branching leafy shoots which can be prostrate, erect or pendent. Many epiphytic orchids develop thickened storage organs called pseudobulbs from one or more internodes of the stem (Fig. 121 L–M).

In monopodial species the whole plant usually consists of a single continually growing shoot. As with sympodially branching orchids, roots may be borne on the leaf-bearing part (e.g. *Vanilla* and the tribe Vandeae, Fig. 121 N). Many species have a very short stem with crowded leaves. The texture of the stem varies from succulent to almost woody. Stems of very few species have been investigated anatomically; some have vessels with scalariform perforation plates (WAGNER 1977).

The leaves are alternate, rarely opposite or whorled, spirally or distichously arranged, sometimes closely imbricate. They are often jointed to their sheathing base, falling off when old. Their shape is generally linear to broadly ovate, more rarely circular, angular-cordate, hastate, etc., very rarely lobed. Their texture varies from membranaceous to coriaceous, in which case they may function as storage organs. The ptyxis of the mesomorphic leaves is often plicate or convolute, and the coriaceous often conduplicate. Terete and ensiform leaves occur in several groups. Some species have reduced, scale-like leaves only. Stomata with subsidiary cells have been found in most groups, but subsidiary cells are often lacking in subfamily Orchidoideae. The subsidiary cells, when present,

**Fig. 121.** Apostasiaceae (**A–D**), Cypripediaceae (**E–G**) and Orchidaceae (**H–N**). **A–B, D** *Neuwiedia inae*. **A** Habit. **B** Flower. **D** Fruit. **C** *N. veratrifolia*, ovary, transverse section. **E–F** *Paphiopedilum* sp. **F** A flower in lateral view. **G** *Selenipedium chica* habit. **H–N** Orchidaceae, roots, tubers and corms. **H–J** *Orchis*, tuberoid roots. **H** second year and first year tuberoid. **I–J** "Tuberoid", in longitudinal section and transverse sections, schematically. **K** *Epipactis*, rhizome with fibrous roots. **L–M** *Pholidota* (**L**) and *Bulbophyllum* (**M**), "pseudobulbs", which are swollen internodes. **N** *Vanda* sp., with monopodial growth, thick leaves and aerial roots. (Orig. and redrawn after AMES, PFITZER and VAN CREVEL)

generally have perigenous development, the perigene cells being formed by oblique divisions of surrounding cells; mesogene cells are reported from the genera of the tribes Cranichideae (N.H. WILLIAMS 1979) and Orchideae (H. RASMUSSEN 1981).

The inflorescences may be terminal or lateral; in some genera both terminal and lateral ones occur on the same shoot. Lateral inflorescences frequently break through the sheath of the subtending leaf or, if the leaves are closely set, those of several leaves. The inflorescence is always basically racemose, but many kinds of modification occur, from racemes or spikes to branched panicles, dense capitula, umbels, spadices, etc.

The flowers are usually resupinate by torsion of the pedicel and/or ovary. The flowers are generally bisexual, but unisexual flowers occur in some species. Their perianth may be almost actinomorphic, but is generally strongly zygomorphic. The shape of the tepals varies strongly. The median inner one, the labellum, is in the majority of species clearly different from the others. It is frequently spurred and often adorned with various keels or calli. Spurs and calli may occur also on the other tepals. Fusion of two or more tepals or of tepals and gynostemium is characteristic of many genera. The fused basal parts of the labellum and the lateral sepals (and, sometimes, also of the lateral petals) may appear as a ventral extension of the column carrying the labellum – "*column-foot*" as in, e.g., *Dendrobium* and *Bulbophyllum* (Fig. 125 M). The sepals and the column-foot may form a chin-like projection called a *mentum*.

The androecium normally includes only one fertile stamen, the median one of the outer whorl, and two more or less distinct staminodes representing the lateral stamens of the inner whorl. However, a specimen with two fertile anthers, the median one of both whorls, has recently been described as a distinct genus, *Diplandrorchis*, apparently closely related to *Cephalanthera* (CHEN 1979). The anther in Orchidaceae varies from erect with four distinct pollen sacs to incumbent with indistinct partitions between the pollen sacs (Fig. 119 L–M).

The connective is frequently dorsally and apically enlarged, enclosing the thecae as a tunic. The connective and the empty thecae are often then referred to as the "*anther cap*" or "*operculum*". The anther is introrse and dehisces by longitudinal slits. The pollen grains are generally united into tetrads, but are free in a few genera (*Cephalanthera; Vanilla* and allies). The pollen grains vary from prominently reticulate in most of the neottioid orchids to smooth and tectate, as in most epidendroid and vandoid orchids. The pollen grains generally appear as inaperturate, except for example, in some neottioid genera with ulcerate or porate pollen grains (SCHILL and PFEIFFER 1977; ACKERMANN and N.H. WILLIAMS 1980; ZAVADA 1983). The pollen grains of *Vanilla* and related genera have been reported to have three to four pores (ERDTMAN 1952). The pollen tetrads are agglutinated into *pollinia* by strands of sterile sporogenous material. The texture of the pollinia varies from very soft and easily disintegrating, to waxy or bony. The basic number of pollinia is four, one for each pollen sac. The interior strands of anther-tissue which are not differentiated into pollen will remain as septa between the pollinia and form the so-called anther partitions, which dry up together with the anther cap. By reduction of the partitions between the pollen sacs (microsporangia) (or, ontogenetic confluence of the sporogenous areas in the anther) the pollinia may be only two, and by formation of secondary partitions the number of pollinia may increase to up to twelve. In some orchids the pollinia are further divided into granular packets called *massulae*, in which case the pollinia are said to be *sectile* (Fig. 124 F). Each massula corresponds to one pollen mother cell (HOFMEISTER 1861).

The style is apically inflexed and terminated by a three-lobed stigma. The three stigmatic lobes – i.e. the apices of the three carpels – are seen in the early developmental stages of all orchids, but they usually become obscured during the growth of the gynostemium eventually forming a "stigmatic cavity", the margins of which represent the apices of the carpels. The median stigma lobe, as well as participating in the formation of the stigmatic cavity, is always more or less modified apically. Its modified apical part is often referred to as the *rostellum*. However, the term rostellum is also frequently defined as "the median stigma lobe", i.e. including the receptive basal part *and* the modified apical part. The latter usage gives the term a more precise morphological connotation and is strongly recommended. It is often stated in textbooks that the median stigma lobe is sterile (non-receptive). This error originates from confusion in the use of the term "rostellum", and the discussions in orchid literature as to whether the rostellum is fertile (receptive) or not are in fact caused merely by contradictory definitions. In the majority of orchid species most of the receptive stigmatic surface does in fact belong

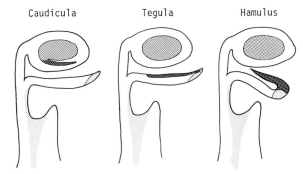

Caudicula     Tegula     Hamulus

**Fig. 122.** Orchidaceae. Different kinds of pollinium stalks, schematically. A caudicula is formed inside the anther whereas a "stipes" is derived from the rostellum. A tegula is an epidermal strap, and a hamulus is the recurved apical part of the rostellum. (Orig.)

to the median stigma lobe, as seen in the column ontogeny of *Doritis* (Fig. 126 F–K).

In its simplest form the rostellum is only slightly larger than the lateral stigma lobes (e.g., *Cephalanthera,* Fig. 124 A), but usually it is more elaborate, e.g. elongated into a prominent beak-like structure, the rostellar projection (as in many vandoid orchids, Fig. 126 K). It can also be lobed (Orchidoideae) or transversely folded (*Epipogium*). The rostellum plays an important role in the pollination of the species of Orchidaceae, as the pollinia are deposited on its abaxial side (the "*clinandrium*") after the beginning of anthesis, and are glued to the pollinator by viscid material originating from its apex. This principle is often developed into very intricate relationships of structures: in many orchids the viscid apex is detached from the rest of the rostellum and removed in its entirety together with the pollinia. The portion of the rostellum that produces the viscid matter is termed the *viscidium* ("retinaculum", "glandula", "viscid disc"). Two viscidia may be present if the rostellum is lobed or bifid.

The pollinia in most species are connected with the viscidium by some kind of *pollinium stalk* (Fig. 122). This may simply be an elongation of the pollinia (*caudicula*), consisting of sporogenous tissue with scattered pollen tetrads embedded in it, as in many epidendroid orchids (Fig. 125 L). The anther partitions may also take part in the formation of an elastic, hyaline pollinium stalk, e.g. in Orchidoideae (Fig. 125 C) and *Epipogium*. Alternatively pollinium stalks in some orchids can be differentiated from the rostellum in various ways. In one type a ridge or strap-like plate of thick-walled cells (a *tegula*, litt.: "small roof", Fig. 122) becomes freed from the abaxial side of

the rostellum by disintegration of the anticlinal cell walls; after anthesis this is glued to the pollinia by means of viscid sterile sporogenous material (most vandoid orchids, Fig. 126 K; some Goodyerinae). Another kind of pollinium stalk is the hook-like *hamulus* (Fig. 122), formed by apical growth of the rostellum; it is found in some neottioid orchids, the subtribe Prasophyllinae of the Diurideae (Fig. 124 K), a few species of *Bulbophyllum* and perhaps some related genera. The pollinium stalks provide diagnostic characters in orchid classification, and various intricate (and often confused) terms have been applied to them. The term *caudiculae* (sing. *caudicula*) is most frequently reserved for stalks derived from the anther itself, while stalks of rostellar origin are collectively termed *stipites* (sing. *stipes*). The pollinia are generally attached to their stalks by their apices ("acrotonic" orchids), but in forms with an erect anther and a relatively short rostellum the attachment may be lateral or basal ("basitonic" orchids, e.g. *Cephalanthera,* Fig. 119 I–J, and all Orchideae). The unit that is removed as a whole by the pollinator is termed a *pollinarium* [one or more pollinia plus stalk(s), if present, and a viscidium].

The ovary is unilocular except in very few genera. Embryo sac formation is occasionaly bisporic (*Allium* Type), but the majority of species studied have the monosporic "*Polygonum* Type", although often with a reduction in the number of nuclei. Polyembryony and apomixis have been observed in many orchids (VEYRET 1974). Embryogeny is either of the Asterad or the Onagrad Type, and also irregular types are reported. Proper endosperm is not formed, but formation of a few endosperm nuclei has been observed in ovules of some genera (up to 12 or 16 in species of *Vanilla* and *Galeola*; VEYRET 1974). Double fertilization generally does not take place at all.

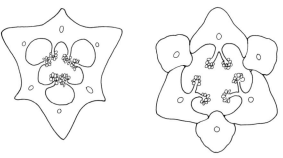

**Fig. 123.** Orchidaceae. Transverse sections of ovaries. *Left Cephalanthera rubra.* (Redrawn from RAUNKIÆR 1895–1899). *Right Nervilia fordii.* (Orig.)

The fruit is a capsule, opening by three or six longitudinal slits, except in the baccate fruits of certain species of *Vanilla* and *Galeola* and *Palmorchis* (all Vanilleae) and in *Rhizanthella* of the Diurideae.

The seeds vary considerably in shape and size (Fig. 119 N–Q), from almost filiform types where the membranaceous testa is exserted at each end, reaching a length of up to 5 mm, to very minute, oblong to subglobose and less than 0.1 mm long. It has been observed in a species of *Chiloschista* (Vandeae) that one end of the seed coat has specialized cells with strong spiral thickenings, sending out long threads (to 4 mm) when moistened (BARTHLOTT and ZIEGLER 1981). *Vanilla, Galeola* and related genera have a different kind of seed with opaque, sclerotic seed coats.

*Chemistry*. Biochemical data on Orchidaceae are very scattered. LÜNING (1974) and other investigators, who tested 2,000 species of 281 genera for presence of alkaloids, found that only few species are alkaloid-rich.

*Distribution*. Orchidaceae is a subcosmopolitan family, with representatives in all phytogeographical regions and climates, but the majority of taxa on all levels are tropical. Most genera are, however, restricted to only one continent; only 11 genera are pantropical, and only 21 more genera show a transoceanic distribution, 27 of these being terrestrial, which may indicate an early expansion of the family, before the epiphytic mode of life evolved (GARAY 1964).

*Pollination Biology*. The Orchidaceae are famous for their unique examples of complicated adaption to insect pollination. The components of the orchidaceous pollination-biology character syndrome may be summarised thus: (1) Zygomorphy of the flower imposes a unidirectional approach by the pollinator (as is usual in other groups). (2) Pollen is presented to pollinators in only a few packages per flower and it is glued to their bodies. (3) Floral attributes restrict the range of potential visitors to each species: this is achieved by special attractants on the one hand and structures with a selectively exclusive effect on the other. (4) The limited range of visitors makes possible precision in placing the pollen on their bodies; without this, the above-mentioned packaging of the pollen would not be effective. Special adaptations, by which insect visitors are positioned in the flower and the pollinia (pollinaria) are positioned upon the insect, are developed. (5) For the same reason, the shaping and positioning of the stigmatic cavity are adapted to the removal of the pollen from the correct part of the body of the right insect. A few orchids are bird-pollinated.

This syndrome has far-reaching effects on the biology and, hence, on the evolution of the family. It may lead to closer and closer specialization and more and more opportunities for speciation based on floral isolation [see, as general sources, VAN DER PIJL and DODSON (1966), PROCTOR and YEO (1973), NILSSON (1981),

DRESSLER (1981)]. This floral isolation may become effective not only by using different insects but by depositing pollinia on different parts of an insect which is attracted to more than one species of orchid.

In the course of progressive floral specialisation, arrangements for the positioning of the insect have evolved beyond simply shaping the flower to fit the insect closely when it has attained the reward (in fact, there may be no reward). There may be loss of control by slipping, temporary imprisonment or intoxication. A further step is active "handling" of the insect by motility which is passive (pivoted epichile in various genera, e.g. *Drakaea* and, especially, *Bulbophyllum*) or active (irritability of the epichile or whole labellum, e.g. *Pterostylis*). An intermediate step is the enforcement of a one-way passage through the flower as in *Coryanthes* (see also Cypripediaceae). Handling and intoxication also appear to have a rôle in promoting outcrossing by stimulating the insect to move to another plant and/or causing a delay sufficient to allow the pollinia to adjust their position for reception by a stigma.

The orchids display amazing diversity of form and colour of the flowers and even of inflorescences, the axes of which may function visually or mechanically in the pollination process. Some orchids offer nectar to potential pollinators in a conventional way, but never pollen. However, a specially produced pollen-like foodstuff ("pseudo-pollen") is offered by some species of *Polystachya* and some other genera, and nutritive oil is probably offered by some genera (VOGEL 1978 b). The food-gathering instincts of insects may also be exploited without the offer of any reward: deceit may be achieved by a normal-looking flower, perhaps with a spur, which is devoid of nectar (some Orchidoideae) or by visual imitation of anthers (*Arethusa, Calopogon* and *Calypso*) or pollen (*Cephalanthera*: DAFNI and IVRI 1981); alternatively the flower may mimic another flower in the habitat (some *Oncidium* spp., *Epidendrum ibaguense*). Also, the Orchidaceae exploit, to a degree shown by no other plant group, instincts other than those of food-gathering, namely mating (*Ophrys, Cryptostylis, Caladenia*, etc., attract male Hymenoptera; other genera attract male Diptera), egg-laying by imitation of substrates such as carrion (*Bulbophyllum* spp.), fungal fruit bodies (*Dracula*, and probably *Corybas* (VOGEL 1978 a) or aphids (*Epipactis consimilis*; IVRI and DAFNI 1977); gathering of sexual pheromones (chemical signalling compounds) (occurring especially in members of subtribes Catasetinae and Stanhopeinae, pollinated by male bees of the family Apidae, tribe Euglossini); territory-holding (some *Oncidium* spp., pollinated by male bees); and roosting (*Serapias*, pollinated by male bees; DAFNI et al. 1981, and earlier references therein).

Plant groups which rely on a restricted range of specialized pollinators tend to have weak post-pollinatory (physiological or genetic) sterility barriers between species. Orchids, the most specialized large family in this respect, are notorious for their interfertility, which readily transgresses the established generic boundaries. Such hybridization, while not common in nature, is the basis of a large and widespread horticultural industry. In nature it keeps open the possibilities of speciation or enhancement of variability by hybridization.

Although highly specialized pollination relationships are common in Orchidaceae, many orchids have a broader range of potential pollinators. Autogamy is not rare in the Orchidaceae; in some cases it may be a consequence of the breakdown of the pollination relationships. It is doubtful, however, whether any orchid species is pollinated exclusively by one insect species.

*Taxonomy.* The features chiefly emphasized when dividing the Orchidaceae into major groups (subfamilies, tribes) are:

1. Pollinia: texture soft and friable to waxy or cartilaginous; sectile or not.
2. Pollinium stalks: absent or present; when present: caudicula, tegula or hamulus.
3. Viscidium: absent or present; structure and shape.
4. Position of attachment between pollinia and viscidium: basal ("basitonic") or apical ("acrotonic").
5. Position and shape of the anther: erect or incumbent to deflexed; long and tapering or short and broad; overtopping the rostellum or not.
6. Persistence of anther wall: drying up and falling off after anthesis ("operculate anther") or not.
7. Number of pollinia.
8. Shape and length of the rostellar projection: broad, shelf-like to slender and beak-like; projecting between the thecae of the anther (as in Orchideae) or not.
9. Position of inflorescences: terminal on the shoots or axillary.
10. Vegetative habit and mode of growth: terrestrial with or without subterranean storage organs; epiphytic with or without pseudobulbs; sympodial growth or monopodial one-shoot plants.
11. Foliage: thin, pleated, convolute leaves or leathery, conduplicate leaves.

None of the larger groups (subfamilies, tribes) in any orchid classification is uniquely designated by any one of these characters (except, perhaps, the tribus Orchideae – see below). Attributes once believed to be confined to certain groups are often found in modern investigations to be more widely distributed, suggesting many parallelisms. Furthermore, many previously supposed homologies have proved to represent cases of convergence.

Among the most important characters on subtribal level are the arrangement of roots (fascicled or scattered on the rhizome); shape and structure of pseudo-bulbs; arrangement of leaves, and whether articulated or not; morphology of perianth and gynostemium. Recent research has revealed that micro-characters such as seed structure

**Table 2.** *Subdivision of Orchidaceae*

Subfamily Neottioideae
    Tribus Epipactieae (subtribus Limodorinae sensu DRESSLER, 1981, pro parte)
    Tribus Neottieae (subfamily Spiranthoideae DRESSLER plus subtribus Listerinae sensu DRESSLER)
        The *Listera* Group (subtribus Listerinae)
        The *Goodyera* Group (subtribus Goodyerinae)
        The *Spiranthes* Group (tribus Cranichideae)

Subfamily Orchidoideae
    Tribus Diurideae
        including subtribus Chloraeinae
        The *Diuris* Group { (subtribus Diuridinae) (subtribus Prasophyllinae)
    Tribus Orchideae
        The *Orchis* Group
        The *Habenaria* Group
        The *Disa* Group

Subfamily Epidendroideae
    Tribus Arethuseae
    Tribus Vanilleae
    Tribus Gastrodieae
    Tribus Epipogieae
    Tribus Coelogyneae
    Tribus Malaxideae
    Tribus Calypsoeae
    Tribus Epidendreae
        The *Epidendrum* Group (subtribus Laeliinae)
        The *Eria* Group (subtribus Eriinae)
        The *Pleurothallis* Group (subtribus Pleurothallidinae)
        The *Dendrobium* Group (subtribus Dendrobiinae)
        The *Bulbophyllum* Group (subtribus Bulbophyllinae)

Subfamily Vandoideae
    Tribus Polystachyeae
    Tribus Cymbidieae
        The *Cymbidium* Group (subtribus Cyrtopodiinae)
        The *Catasetum* Group (subtribus Catasetinae)
        The *Stanhopea* Group (subtribus Stanhopeinae)
        The *Oncidium* Group (subtribus Oncidiinae)

    Tribus Maxillarieae
        The *Corallorhiza* Group (subtribus Corallorhizinae)
        The *Zygopetalum* Group (subtribus Zygopetalinae)
        The *Maxillaria* Group (subtribus Maxillariinae)

    Tribus Vandeae
        The *Vanda* Group { (subtribus Aerangidinae, in a broad sense, including subtribus Sarcanthinae nom. illeg.)
        The *Angraecum* Group (subtribus Angraecinae)

and seed sculpture, pollinium surface, stomata and stomatal development etc. may also characterize groups.

The attempts of various authors to divide the family into "natural" groups have resulted in several highly contrasting systems of classification. Until now, the most influential has certainly been that of SCHLECHTER (1926). GARAY (1960, 1972),

DRESSLER and DODSON (1960), VERMEULEN (1966), and DRESSLER (1981) have suggested very different hierarchical arrangements of the larger groups and very different positions of the odd genera. The following division is modified from that of DRESSLER (1981). DRESSLER (1981) acknowledges 4 subfamilies, 21 tribes and 70 subtribes of monandrous orchids; 16 of these tribes are mentioned here (Table 2). A few names are different from the ones used by DRESSLER due to different positions of nominal genera. The "groups" mentioned here under some of the tribes in most cases correspond to DRESSLER's subtribes (names in brackets).

None of the classifications cited above has been claimed to be phylogenetic in the sense that all taxa are presumed to be strictly monophyletic groups (clades) with ranks reflecting the branching hierarchy of a hypothetical phylogenetic tree. Some groups are clearly paraphyletic "remnant groups", defined by shared primitive (ancestral) characters, other groups are given rank according to their degree of difference from other groups although they are supposed to have evolved within groups given equal or lower rank. At present far too little is known about the homology and actual distribution of characters in the Orchidaceae to construct a reliable phylogenetic scheme with a resolution of the "remnant groups", but the presumed status of the taxa mentioned here is noted under each taxon (see also Fig. 120).

## Subfamily Neottioideae

The subfamily Neottioideae in most classifications comprises a number of genera with mainly primitive characters. The neottioid orchids are usually terrestrial hygro- or mesophytes and very few are epiphytes. They never have pseudobulbs or any other of the vegetative specializations characterizing the epidendroid and vandoid orchids. The anther is erect or bent slightly forward. It contains four or two bipartite, soft, granular or sectile pollinia often without pollinium stalks (although all known kinds of pollinium stalk have now been found in the Neottieae; F.N. RASMUSSEN 1982). The attachment to the more or less well-differentiated viscidium is usually acrotonic.

Recent investigations have demonstrated that certain of the genera and generic groups of the traditional Neottioideae are actually better conceived of as the most primitive members of the subfamilies with which they share apparently homologous advanced characters. As here circumscribed, the Neottioideae consist of one small unresolved "remnant group", the tribe Epipactieae, and one larger tribe, Neottieae, which is supposed to be a monophyletic entity.

*Tribus Epipactieae* (Subtribus Limodorinae pro parte in DRESSLER 1981)   7:50

This contains a number of mainly northern temperate genera, the most primitive of the monandrous orchids. They are terrestrials, often saprophytic, with elongated rhizomes carrying more or less clustered, rather thin roots. The leaves are spirally dispersed on the erect stem, convolute or plicate and nonarticulate. The inflorescences are terminal and the flowers have free tepals. The labellum may be spurred or saccate and is often divided into hypo-, meso- and epichile. The anther is slightly inflexed (Fig. 124 A). The pollinia are very soft, and in *Cephalanthera* composed of single grains. The viscidium is not detachable. – *Cephalanthera* (12) ranges from between Europe and North Africa to Korea and Japan and has a single species in the Western U.S.A. The anther projects high beyond the hardly modified rostellum. The stigma is cup-shaped with somewhat revolute margins. *Epipactis* (30) is distributed over the Northern Hemisphere; most species are Asiatic but a single species extends in the south to Mozambique. *E. helleborine* is a common forest floor herb in central and North-Western Europe, naturalized in North America. Several autogamous species have been segregated from this species. *E. veratrifolia* (Asia Minor to China), *E. gigantea*, North America and Asia Minor to Himalaya, and *E. africana* are all stout herbs up to 2 m tall. *Limodorum* (1) *abortivum*, in Southern Europe, is a robust saprophyte with bracteal leaves and long-spurred, violet or bluish flowers.

*Tribus Neottieae* (Subfamily Spiranthoideae DRESSLER plus subtribe Listerinae)   85:1,000

This tribe has a worldwide distribution, but is centred mainly in the tropics of Asia and America. It is possibly a monophyletic group, which is characterized by the dorsal, erect anther and the corresponding elongated, tapering rostellum (Fig. 124 B). This basic arrangement is varied in a number of ways, distinguishing separate phylogenetic lines within the group. The majority of genera have two bipartite pollinia, tapering towards the apex, where they connect with a distinct, detachable viscidium. However, all above-mentioned kinds of pollinium stalks are found in this tribe,

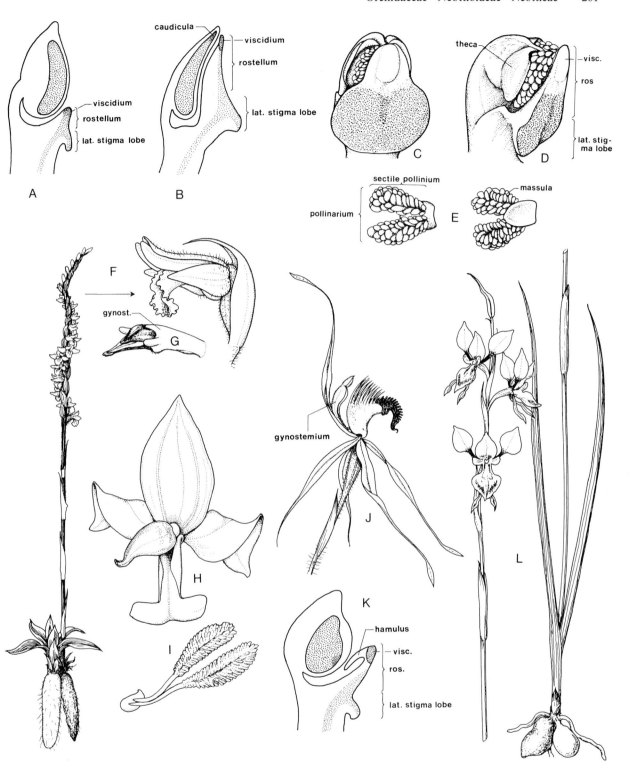

**Fig. 124.** Orchidaceae subfamily Neottioideae (**A–I**) and subfamily Orchidoideae tribus Diurideae (**J–L**). **A–B** Gynostemium in longitudinal section, *Cephalanthera* and *Spiranthes*, respectively. **C–E** Gynostemium in front (**C**) and side (**D**) views and pollinarium (**E**) in the genus *Goodyera*. **F–G** *Spiranthes spiralis* with flower and gynostemium (**G**). **H–I** *Ludisia discolor*, flower and pollinarium. **J** *Caladenia dilatata*, flower. **K** *Microtis*, gynostemium in longitudinal section. **L** *Diuris aurea*. (Orig. and redrawn after BRIEGER, NICHOLLS, RASMUSSEN and RUPP)

and two genera have a peculiar sensitive viscidium, which releases sticky matter when touched.

**The Listera Group** (Subtribus Listerinae) (3:40) in the Northern Hemisphere, is characterized by the sensitive rostellum. The green and assimilating leaves, when present, are only two in number. This group is often referred to the previous tribe, in which case the names of the tribes are different. DRESSLER (1981) refers the Listerinae and the Limodorinae to the subfamily Orchidoideae.

*Listera* (25) occurs in the temperate Northern Hemisphere. Characteristic are the two apparently opposite foliage leaves, hence "Twayblade". – *Neottia* (10), occurs in Northern and central Asia but with one species, *N. nidus-avis,* in Europe. It consists of brownish yellow saprophytes with bracteal leaves. The short subterranean rhizome carries a dense cluster of thick roots, hence the name "Bird's-nest Orchid".

**The Goodyera Group** (Subtribus Goodyerinae) (35:430). Pantropical and Northern Hemisphere. Forest floor plants with creeping, succulent rhizome and often more or less variegated leaves. The lateral petals are often fused with the dorsal sepal to form a hood over the gynostemium. The labellum is saccate or spurred, often with a two-lobed blade. Inside the spur there are sometimes conspicuous glands. The pollina are sectile.

*Goodyera* (50) occurs in temperate and tropical Eurasia. *G. repens,* which is circumboreal, is the only species in Europe. It is a small, creeping herb growing in the moss of coniferous forests. – *Erythrodes* (about 80) occurs in South-East Asia and tropical America. – *Ludisia* (1) *discolor* (Fig. 124) (syn.: *Haemaria d.*), in South-East Asia, has peculiar twisted flowers with a white perianth and a striking yellow anther. Its leaves are often vividly variegated. It is frequently cultivated. – *Anoectochilus* (40), in tropical Asia, is cultivated for the variegated leaves with a metallic lustre, "Jewel Orchids". – *Zeuxine* (50), in tropical Asia and Africa, has some species with tegula (see above). *Zeuxine* and *Anoectochilus* belong to a subgroup in which the stigma forms two dome-shaped areas, one on each side of the rostellum.

**The Spiranthes Group** (Tribus Cranichideae in DRESSLER 1981) (45:500). This is characterized by leaves in a basal rosette, and by clustered, often thick storage roots.

*Spiranthes* (40), is distributed in temperate zones of Eurasia and North America and has a few species in the tropics, including Eastern Australia; the small flowers are placed on a spirally twisted rachis. *S. spiralis* (Fig. 124F–G) ranges from North Africa to Denmark. *S. romanzoffiana,* in North America from California to Alaska, is also found in a few places in Ireland and western England. – *Stenorrhynchus* (40) occurs in temperate South America to South-East U.S.A.; it has a much elongated, bayonet-like rostellum and viscidium. Some species have fairly large, showy flowers. *S. speciosus,* Mexico, Northern South America and the West Indies, is cultivated for its bright red or purple flowers and inflorescence bracts. – *Cranichis* (50), in tropical America, especially the Andes, often has pseudopetiolate leaves and many small, non-resupinate flowers. – *Ponthieva* (25), related to *Cranichis,* includes some species with a pollinium stalk of the hamulus-type. The non-resupinate flowers with spreading sepals may be quite attractive. – *Cryptostylis* (10), Indo-Malaysia and Australasia, is probably also related to *Cranichis.* Pollination is by pseudo-copulation involving Ichneumonidae.

### Subfamily Orchidoideae

The subfamily Orchidoideae contains about 115 genera with approximately 2,500 species. It is distributed all over the world, but mainly in the temperate zones. As here conceived, the Orchidoideae consist of one clearly monophyletic group, the tribe Orchideae sensu lato and an orchidoid "remnant group", the tribe Diurideae, within which the Orchideae may have evolved. This hypothesis is based on the shared occurrence in the tribes of so-called root-stem "tuberoids" (DRESSLER 1981), and on the scattered occurrence in the Diurideae of orchidoid characters, such as basitony. The tuberoid is a root structure around a core of stem tissue with an apical bud, which in the next season will grow into a new shoot, with one of the axillary buds forming a new tuberoid. The tuberoid is usually polystelic. It should be noted that its anatomy is known only from a few genera of Orchideae, and very little is known about root and tuber anatomy of other terrestrial orchids. As the genera of Diurideae do not share distinct and probably uniquely derived attributes with any other orchid group, the classification followed here is chosen as a tentative phylogenetic approach. However, the inclusion also of the Epipactieae and the Listerinae of the Neottieae (DRESSLER 1981) seems unjustified. As pointed out by DRESSLER (1981), the primarily South African Orchideae, the Australian Diurideae and the temperate South American *Chloraea* Group show an interesting vicariance pattern.

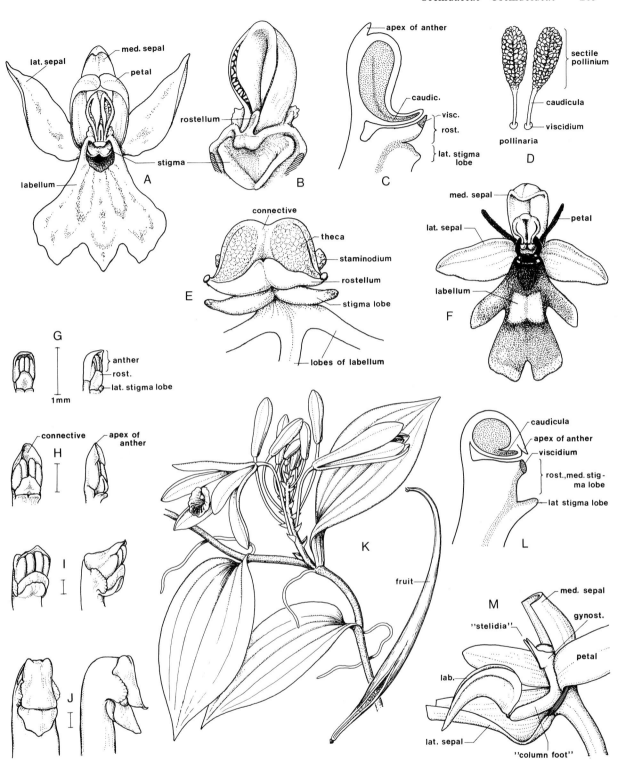

**Fig. 125.** Orchidaceae subfamily Orchidoideae (**A–F**) and subfamily Epidendroideae (**G–M**). **A–D** Flower and floral details in *Dactylorhiza maculata*. **A** Flower. **B** Gynostemium. **C** Gynostemium in longitudinal section. **D** Pollinaria (two for each flower). **E** *Habenaria reniformis*, gynostemium. **F** *Ophrys insectifera*, flower.

**G–J** Ontogeny of the gynostemium in *Vanilla roscheri*. **K** *Vanilla planifolia*, flowering branch and fruit. **L** *Epidendrum*, gynostemium in longitudinal section. **M** *Bulbophyllum reptans*, details of flower. (Orig. and redrawn after BERG and SCHMIDT, BRIEGER, JOHNSEN, RASMUSSEN and SEIDENFADEN)

All Orchidoideae are small to medium-sized terrestrial plants, or very rarely epiphytes, usually with a rather primitive foliage of thin, spirally arranged, basally clustered, convolute, plicate or conduplicate, nonarticulate leaves. The inflorescences are always terminal and simple, with few to many spirally arranged flowers. The gynostemium may be from very short and hardly forming a column (as in certain Diurideae) to much elongated. The anther is usually upright or, in some Orchideae, strongly reflexed, but never becomes strongly inflexed during ontogeny as in the Epidendroideae and Vandoideae. The apex of the rostellum of all Orchideae and some Diurideae protrudes between the thecae of the anther and is laterally expanded between their bases, where viscidia may be formed (Fig. 125C). Pollinium stalks of the hamulus type occur in a few genera of Diurideae, whereas the Orchideae always have caudicles. Viscidia may be diffuse or detachable; the attachment is acro-, meso- or basitonic. The pollinia are always rather soft, often composed of single pollen grains in the Diurideae, but are otherwise sectile.

### Tribus Diurideae    35:550

One group (subtribus Chloraeinae, 6:100) occurs in temperate South America and New Caledonia, but the tribe is primarily Australian. A few genera are represented also in tropical Asia. The general characters are as described for the subfamily, but usually without the attributes distinguishing the Orchideae. There are many odd specializations in the foliage and, especially, in the flowers, notably in the Australian endemics, in this tribe. – *Chloraea* (50) from Chile to northern Peru, and *Gavilea* (13), in Argentina and Chile, perhaps lack the tuberoid but otherwise strongly resemble the Australian genus *Caladenia* (80) (Fig. 124J) and its allies. – *Drakaea* (4), *Spiculaea* (3) and *Caleana* (5) have bizarre flowers with a hinged and mobile labellum, functioning in pseudo-copulation with wasps: Scolioidea: Thynnidae (STOUTAMIRE 1974, 1975). – In *Pterostylis* (70), mainly in Australia, the sepals form a hood inside which the narrow, sensitive labellum is hinged, moving actively when touched, trapping small pollinating flies, so that these cannot escape without touching the anther. They are occasionally cultivated as "Greenhoods". – *Rhizanthella* (1), in South-West Australia is a subterranean saprophyte with a dense, capitate inflorescence, which is a few centimetres across and surrounded by large bracts. It barely reaches the soil surface. The unusually large seeds are perhaps dispersed by animals feeding on the succulent, indehiscent fruits (GEORGE 1980).

**The Diuris Group** (Subtribus Diuridinae and Prasophyllinae) (8:110) comprises genera with a short erect gynostemium and almost no column; the staminodes are erect, elongated and wing-like. – *Diuris* (35), mainly in Australia, has grass-like leaves and peculiar spreading petals. – *Prasophyllum* (70) and *Microtis* (10) in Australia and New Zealand, have a solitary terete and hollow leaf ("Leek Orchids") and sectile or granular pollinia, often attached to a distinct hamulus (Fig. 122). – *Thelymitra* (50), mainly in Australia but with a few species extending to Indonesia and the Philippines, has almost regular flowers; the staminodia and column form a kind of hood, "mitra", enclosing the anther and stigma; the apices of the mitra may be adorned with tufts of variously coloured cilia. The flowers are often brightly coloured, e.g. *T. ixioides* with blue flowers. Occasionally cultivated, "Sun Orchids".

### Tribus Orchideae (Orchideae plus Diseae in DRESSLER 1981)    74:2,100

With world wide distribution, but mainly concentrated in temperate zones, especially Southern Africa. The Orchideae is one of the most distinct and indisputably monophyletic of all major tribes of Orchidaceae. It is characterized by the erect or (in part of the *Disa* group) reflexed anther, the base of which is completely united with the column, firmly fixing the anther in position relative to the much integrated rostellum. The anther contains two bipartite sectile pollinia, always on caudicles. The apex of the rostellum protrudes between the thecae, forming detachable viscidia at their base (basitonic attachment) (Fig. 125C). The interthecal part of the rostellum in some genera is conspicuously developed, in others rudimentary. The infratribal division followed here is rough and provisional.

**The Orchis Group** have parallel thecae, the bases of which are enclosed in two lateral pouches, "bursiculae", of the rostellum; there is however only one bursicula in genera with a reduced intrathecal rostellum part.

*Orchis* (35), is characterized by spheroidal tuberoids and *Dactylorhiza* (30) by palmate tuberoids; the two genera are distributed in subtropical, temperate and boreal Eurasia, the highest number of species being in Mediterranean Europe. Modern treatments of this intricate complex have segregated a number of smaller genera, e.g. *Galearis*

(7), in the Himalayas and East Siberia to North America, and *Amerorchis* (1), in northern U.S.A. to western Greenland. – Within *Ophrys* (25), a genus concentrated in Mediterranean Europe, North Africa and Asia Minor, all species are pollinated by bees or wasps through a process of pseudo-copulation, a pollination strategy first described for this genus. The flowers mimic female insects in odour, shape, colour and surface texture. The similarity is quite striking even to the human eye: thus for *O. insectifera* (Fig. 125 F) the petals represent the antennae, the lateral labellum lobes the wings and the median lobe the body of the mock insect. However, the scent and surface texture appear to be more important than the form, suggesting that pseudo-copulation could be more widespread in orchids. In *O. speculum,* "Mirror Ophrys" the centre of the labellum is glossy blue, bordered by conspicuous brown hairs; the pollinator here is another Scolioid wasp, in this case belonging to the Scoliidae (cf. Diurideae). – *Serapias* (10) is another Mediterranean genus. In this the sepals form a helmet hiding the diminutive lateral inner tepals: the labellum is large and tongue-shaped. The connective of the anther is distinctly elongated. Other genera in this group are *Anacamptis* (1); *Himantoglossum* (4), *Barlia* (1) and *Aceras* (1). *Aceras anthropophorum,* "L'homme pendu", has greenish yellow flowers which resemble miniature human bodies.

**The Habenaria Group** is characterized by thecae which are more or less spreading and a rostellum which does not form bursicles.

*Platanthera* (85) has a world wide distribution, but is absent from Africa south of Sahara and South America. It has about 50 endemic species in East Asia, 7 in Europe and about 10 in North America. The delimitation of this genus is much disputed. – *Gymnadenia* (7) occurs in temperate Eurasia from Spain to Japan. – *Habenaria* (about 800) is distributed all over the world, but concentrated mainly in the tropics. It is one of the largest and most widely distributed genera of orchids. The stigma is divided into two convex stigmatic areas which are borne on stalks, "stigmatophores", which may be rather long and are parallel to the (sometimes widely) spreading thecae (Fig. 125 E). The lateral petals and the spurred labellum are often deeply lobed, giving the flowers a spidery appearance. Some species have large showy flowers, e.g. *H. rhodocheila* in Malaysia and Indo-China, with scarlet or brick-red flowers.

**The Disa Group** generally has a reflexed anther, bending back from the column.

In *Disa* (150), which occurs in Africa, especially Southern Africa, the dorsal sepal has the shape of a hood or a spoon and is usually spurred. The flowers are often very conspicuous. *D. uniflora,* in the Cape Province of South Africa has large, brilliant scarlet to orange flowers, and is widely cultivated. – In *Satyrium* (115) in tropical and southern Asia, the Himalayas and western China, the anther is reflexed and positioned below the stigma. The hooded labellum of the non-resupinate flowers has two spurs, descending on each side of the ovary.

## Subfamily Epidendroideae

The subfamily Epidendroideae, with the circumscription here adopted, is the largest of the subfamilies. It includes about 220 genera and perhaps more than 10,000 species, most of which are tropical epiphytes.

The attributes distinguishing this group in classical orchid systematics are the incumbent anther with broad tunica-like connective and caducous anther cap (the anther is "operculate"), the usually rather hard pollinia which either have caudicles or no pollinium stalks at all, and generally the possession of one or more vegetative adaptations for epiphytic life, such as thick aerial roots with multiseriate velamen, aerial corms made up of one or more internodes, articulate, coriaceous conduplicate leaves, etc. The next subfamily in the present treatment, Vandoideae (often included as a subgroup of the Epidendroideae) is segregated by having, in addition, stipites of the tegula type and even harder pollinia. However, these characters have now been demonstrated to have a wider distribution than was earlier believed. Thus pollinium stalks of all kinds are reported from the Epidendroideae in the present sense, although true tegulae seem to be very rare (tribus Calypsoeae?). Therefore, the incumbent anther and the horizontal or deflexed rostellar projection produced by a strong inflexion of the anther during ontogeny (Fig. 125 G–J) appear to be the only features generally characterizing the Epidendroideae. In the Vandoideae, this inflexion begins much earlier in the ontogeny of the gynostemium and may continue to deflexion of both anther and rostellar projection through 180°, suggesting that this group represents a further development within the epidendroid orchids. It remains, however, an open question whether the anther inflexion may be regarded as a uniquely derived character, designating a monophyletic group. Some epidendroid orchids have erect anthers which, on the basis of this hypothesis, must be considered cases where this ontogenetic incumbence has been retarded. Recent investigations indicate that most probably the subfamily Vandoideae comprises several groups of relatively advanced orchids and forms an "evolutionary level" (DRESSLER 1974) rather than a single branch of evolution.

Despite the difficulties in reconstructing the phylogeny of the epidendroid orchids, it is a fairly easily recognized group according the present definition: incumbent anther and/or presence of one or more of the advanced characters mentioned above. When, in addition, very early ontogenetic inflexion of the anther occurs and a tegula is present, the orchid is referred to the Vandoideae.

The traditional orchid taxonomy encountered difficulties in delimiting the epidendroid orchids from the neottioid. This was because the subfamily Neottioideae was treated as a monophyletic taxonomic entity defined solely on primitive characters which were considered more important than the advanced ones. *Vanilla, Nervilia* and related genera have thus often been classified as being neottioid due to their soft pollinia.

### Tribus Arethuseae   35:540

This tribe, with a worldwide distribution, consists mainly of cormous terrestrial plants, which often have slender reed-like stems and distinct pseudobulbs. The leaves are plicate or less often conduplicate, and may or may not be articulate. The inflorescences are most often lateral. The apical part of the column may be broad and petaloid, the pollinia, which may be soft or relatively hard, are very often two in each pollen sac due to incomplete development of the sporogenous tissue. The two pairs of pollinia are connected by ventral caudicles. The stigma is often somewhat emergent.

This tribe consists primarily of "primitive" epidendroid genera, having no conspicuous attributes justifying their grouping in separate taxa. They share no particular, advanced characters with any of the "advanced" larger tribes. Any one of the other epidendroid or vandoid groups could have evolved within the Arethuseae as here defined.

*Arethusa* (1) *bulbosa* is a small North American bog plant with an inflorescence bearing one or two attractive pink flowers. – *Calopogon* (4), in North America, consists of small herbs growing in wet habitats, "Grass Pink". – *Sobralia* (95), in the Central and South American tropics, con-

sists of tall terrestrials with thin reed-like stems and broad, plicate leaves. The flowers are large and showy, but fugacious. – *Elleanthus* (70), in the American tropics and West Indies, has flowers in an inflorescence which is sometimes subcapitate and has hard conspicuous bracts, an adaptation to hummingbird pollination. – *Thunia* (5–6), in tropical Asia, and *Phaius* (30), in tropical Asia, Madagascar and Australia and one species in tropical Africa, are stout terrestrials with reed-like stems and large plicate leaves. *P. tancarvilleae*, in tropical Asia and Australia, may become up to 2 m tall; the many-flowered inflorescences carry very showy, white-and-brown flowers 10–12 cm across with a gullet-shaped lip. – *Calanthe* (ca. 200), in subtropical and tropical Asia, Australia and Africa, comprises stout terrestrial herbs with more or less pseudobulbous stems. Some species have leafy shoots alternating seasonally with inflorescences. The spurred flowers have a characteristic three-lobed, deeply bifid labellum, partly fused with the column. All parts of the plant become blue when injured. *C. triplicata,* which is widely distributed in tropical Asia and Australia, is a large species with showy white flowers with a red or yellow callus on the labellum; it is frequently cultivated. – Of the genus *Bletilla* (9), in East Asia, *B. striata,* from South China and Japan, is a popular, fairly hardy garden plant. It has several-flowered inflorescences with violet flowers 4 cm long.

### Tribus Vanilleae   13:240

In this tribe will be mentioned the *Vanilla* Group only.

**The Vanilla Group** (Subtribus Vanillinae) consists of 5–7 genera with ca. 180 species. *Vanilla* is pantropical, the other genera being palaeotropical or neotropical, two small ones endemic to New Caledonia. They are terrestrial or monopodial lianas. Some genera are saprophytic. The leaves are rigidly coriaceous or fleshy, convolute or conduplicate and non-articulate. The flower is gullet-shaped and often provided with a tuft of retrorse hairs or crests, forcing visiting insects, when retreating, to raise their body and thereby touch the versatile anther. The four pollinia are very soft and mealy and are composed of free pollen grains. The three-lobed stigma is somewhat emergent; the broad rostellum, which forms a diffuse viscidium at its apex, is inflexed, hiding the lateral stigma lobes. The fruit of *Vanilla* and *Galeola* is unusual for the family in being dehiscent or indehiscent (a capsule or berry), somewhat resembling a banana. It contains numerous small seeds with a thick, opaque, sometimes black seed coat, which may be winged. The seeds of *Vanilla* were earlier regarded as primitive as they are superficially similar to seeds in Apostasiaceae. It is more likely, however, that they are derived, adapted to a new mode of seed dispersal in connection with the secondarily fleshy, fragrant fruits. SEIDENFADEN (1978) supposes that the fruits are eaten and dispersed by animals. – *Vanilla* (100), a pantropical genus, has thick, fleshy leaves or no leaves at all, and consists of vines with assimilating stems, climbing by the twining aerial roots which are produced opposite the leaves or bracts. *V. planifolia* (Mexico) (Fig. 125) is grown commercially throughout the tropics, especially Madagascar, for producing "Vanilla Beans", which are the dried and fermented fruits. The characteristic flavour of the product comes from a mixture of oleoresins and an aromatic compound, vanillin, formed during the fermentation (or, less efficiently, by natural ripening of the fruit) through hydrolysis of glucovanillin. "Vanilla" was used by the Aztec Indians of Mexico for flavouring chocolate. A few other species of the genus are grown locally for vanilla production, and quite a few are grown as ornamentals, e.g. the impressive *V. imperialis* (tropical Africa). – *Galeola* (25), in tropical Asia and Australia, consists of saprophytes, terrestrial humus plants or vines. The terrestrial species have striking red *Vanilla*-like fruits and wingless seeds; the vines have dehiscent capsules with winged seeds. *G. altissima* is said to reach 40 m, and is perhaps the largest of all saprophytic plants.

### Tribus Gastrodieae   7:130

This occurs all over the tropics, but mainly in tropical Asia. The tribe consists of small terrestrial saprophytes (except the autotrophic genus *Nervilia*), frequently with subterranean corms. The tepals are often more or less fused and the pollinia are often sectile. – *Nervilia* (80), in tropical Asia, Australia and Africa has only one leaf which is almost circular or cordate, plicate and non-articulate. The leaf and inflorescence are produced at different seasons. – In *Gastrodia* (30), in tropical Asia and Australia ("Potato Orchid") the stigma is positioned at the base of the slender, erect gynostemium.

### Tribus Epipogieae   2:3

This tribe consists of small peculiar-looking, succulent, leafless terrestrial saprophytes of the Old World. They are rhizomatous or bear a corm. – *Epipogium* (2) has spurred, non-resupinate flowers. The lumpy, incumbent anther holds two bipartite sectile pollinia, each on a long recurved caudicle. The stigma is basal on the gynostemium as in *Gastrodia*, but the systematic position and possible phylogenetic history of this odd genus have been much disputed. *E. aphyllum*, in the temperate and boreal zones of Eurasia, has a well-developed transversely folded rostellar projection with a detachable viscidium. The autogamous *E. roseum* is found in the Old World tropics.

### Tribus Coelogyneae   42:440

These, which occur in tropical Asia and southern China, are epiphytes or lithophytes with pseudobulbs made up of a single internode. The flowers have four superposed pollinia with distinct caudicles. The column is often apically more or less petaloid. – *Coelogyne*, in tropical Asia, consists of over 100 species of epiphytes with distinct ovoid or conical pseudobulbs carrying one or two apical, plicate leaves; some have showy flowers and are cultivated. – *Pleione* (10) occurs in the Himalayas and the mountainous areas of Thailand, Laos, China and Taiwan; it consists of small but large-flowered epiphytic or lithophytic plants. They are frequently cultivated.

### Tribus Malaxideae   6:900

These genera have a worldwide distribution but are mainly found in the tropics. They are small terrestrial or epiphytic plants, usually with pseudobulbs of a few internodes. The leaves are plicate or (in *Oberonia*) laterally flattened. The terminal inflorescence usually carries very small to small greenish, non-spurred flowers. The anther contains four rather hard pollinia, which are "naked", i.e. lacking pollinium stalks of any kind. The Malaxideae in this respect resemble the Dendrobiinae and the Bulbophyllinae. – *Liparis* (250), is cosmopolitan, but occurs mainly in the tropical zone. It usually has resupinate flowers with an erect, slender column. *L. loeselii* occurs in temperate Europe and North America. – *Malaxis* (synonym: *Microstylis*) (300), is a subcosmopolitan genus. The ovary is often twisted through 360°, so that the flowers appear non-resupinate. The column is shorter than in *Liparis*. *M. (Hammarbya) paludosa*, from Europe through Siberia to Japan, has leaves which often produce bulbils at their tips. – *Oberonia* (300), mainly in Asia, is a genus of small epiphytes with fleshy, laterally much compressed leaves, giving the plant a peculiar, fan-like appearance. The flowers are very minute, often less than 1 mm across, situated in whorls on a dense raceme.

### Tribus Calypsoeae   2:3

Includes *Calypso* (1) *bulbosa,* a delicate, circumboreal, terrestrial herb with a single plicate leaf and a solitary, fairly large, pink flower. Its systematic position is uncertain.

### Tribus Epidendreae   115:8,000

This tribe, as here circumscribed, is by far the largest of the orchidaceous tribes. It is very diverse and difficult to define on the basis of shared characteristics. Most probably it represents a group of various, relatively advanced lines of epidendroid orchids. Most of them are epiphytes with slender stems or pseudobulbs of one or more internodes. The leaves are often conduplicate, rarely plicate, and distichous. In many of the subgroups the flowers have a well-developed column foot and a mentum formed from the lateral sepals. Only a few genera have spurred flowers. The pollinia vary in number from two to eight, and are usually of a rather hard texture. Most genera have distinct caudicles of the same kind as those in Coelogyneae, situated as "runners" underneath the pollinia (see *Epidendrum,* Fig. 125 L). The large subtribes Dendrobiinae and Bulbophyllinae have no pollinium stalks except for a hamulus found in a few species of *Bulbophyllum.* The small subtribe Sunipiinae, which is closely related to the Bulbophyllinae, has species with peculiar, tegula-like pollinium stalks.

**The Epidendrum Group** (Subtribus Laeliinae) consists of ca. 45 genera and 850 species, all neotropical. – *Epidendrum* (ca. 500) ranges from North Carolina to Argentina. Terrestrial and epiphytic species. They have slender, often branching stems and one to many leaves, which may be terete or flattened. The inflorescences are terminal (rarely lateral), racemose, sometimes almost umbellate, or paniculate. The flowers vary from minute to quite large and showy and have more or less spreading tepals. The column is adnate for its full length to the labellum and functions as a spur. The anther contains four waxy pollinia with runner-like caudicles. This genus comprises several popular greenhouse plants: *E. ibaguense* (incl. *E. radicans* and *E. schomburgkii*), which is common throughout tropical America, a variable terrestrial

(or epiphytic) orchid up to 1.5 m tall, having subumbellate inflorescences with medium-sized, white to scarlet, flowers. The flowers of *E. medusae* (Ecuador) have a striking maroon-coloured labellum up to 5 cm broad with deeply fimbriate margins. – *Encyclia* (150), mainly in Mexico, has been segregated from *Epidendrum,* having pseudobulbs and only partial adnation between labellum and column or none. *E. cochleata* and *E. fragrans,* from Florida to Venezuela, have non-resupinate flowers. They are widely cultivated. – *Laelia* (ca. 50), from Mexico to Brazil, is closely related to *Epidendrum* and *Cattleya.* It has distinct pseudobulbs, few leaves, and mostly large and showy flowers with eight pollinia in the anther. – *Cattleya* (30), in the tropics of Central and South America, consists of epiphytic or lithophytic plants with more or less thickened pseudobulbous stems and one or two thickly coriaceous apical leaves. The one- or few-flowered inflorescence is subtended by a large spathaceous bract. The flowers are often very large and spectacular and have an anther with four pollinia. This is one of the most important genera of cultivated orchids. Intergeneric hybrids between *Cattleya* and other genera of Laeliinae play an important role in commercial orchid breeding. *C. labiata,* in eastern Brazil, has flowers 12–15 cm across.

**The Eria Group** (Subtribus Eriinae), 8:500, occurs in tropical Asia (one genus in Africa). It consists of small to medium-sized epiphytes with slender stems or with pseudobulbs of several internodes. The anther has eight pollinia with caudicles and usually with a well-defined viscidium. – *Eria* (300–400), in tropical Asia, is a vegetatively rather variable genus, which includes very minute plants with densely crowded button-shaped pseudo-bulbs as well as larger forms without distinct pseudobulbs but with fleshy, sometimes terete leaves. The flowers often have a distinct column foot, resembling that in small *Dendrobium* flowers. Several species are covered with a tomentose or woolly indumentum.

**The Pleurothallis Group** (Subtribus Pleurothallidinae) contains about 26 genera with about 3,800 species (according to DRESSLER 1981), all in tropical America. They have characteristic, unifoliate, non-pseudobulbous stems. The leaf is conduplicate and very fleshy, functionally substituting a pseudobulb. The flowers have a joint between the pedicel and ovary and usually exhibit characteristic adaptations to fly pollination like the mainly Asiatic group Bulbophyllinae. There are usually two pollinia; the caudicles are often very short, if existent.

– *Pleurothallis* (at least 1000 species, perhaps many more), consists of small to medium-sized epiphytes or lithophytes. The lateral sepals of this genus are more or less united, and often the column foot is distinct. The pollinia are waxy. – *Stelis* (ca. 500) is another small-flowered genus. – In *Masdevallia* (ca. 250), in tropical America, the sepals are united into a tube. The free apices of the outer tepals are spreading, giving the flowers a characteristic trilobate appearance when seen in front view. Many species are cultivated. – *Restrepia* (30), differs in having four pollinia. – *Dracula* (ca. 60), which has recently been segregated from *Masdevallia*, has strange-looking flowers said to imitate fungi, with their odour and labellum structure, attracting fungus gnats (Diptera: Mycetophilidae) which serve as pollen-vectors (VOGEL 1978). *D. vampira* and related species are cultivated as curiosities (northern Ecuador).

**The Dendrobium Group** (Subtribus Dendrobiinae), which occurs mainly in tropical and subtropical Asia, contains about 1,700 species in one very large genus, *Dendrobium,* and six to seven small segregate genera. They include epiphytes and a few terrestrials. The vegetative morphology is highly variable. The inflorescences are only rarely truly terminal on the shoots. The flowers are characterized by having a pronounced column foot and four (sometimes two) ellipsoid pollinia, without pollinium stalks, in the versatile anther. The viscidium is diffuse and usually not detachable. This group seems to be further characterized by cells containing globular, muricate silica bodies. This type of silica body is restricted to *Dendrobium,* certain Eriinae and the Vandeae (MØLLER and H. RASMUSSEN, 1984).

*Dendrobium* (at least 1,400 species), exhibits an incredible variation of the vegetative parts. Yet the flowers, although quite variable as to size and form, conform to the characteristics described above. Some species have showy flowers and are of great horticultural importance. *D. nobile* (Himalaya to East Asia) has reed-like, slightly pseudobulbous stems, spreading whitish pink sepals and lateral petals, and a gullet-shaped labellum with maroon-coloured throat. It is one of the most commonly cultivated of all orchids. *D. bigibbum,* especially its variety *phalaenopsis,* (South-East Indonesia and tropical Australia) has durable white, purple or lilac flowers and is very common as a commercial cut flower. Commonly cultivated also are *D. chrysotoxum, D. lindleyanum* (syn. *D. aggregatum*) and *D. fimbriatum,* all with golden or yellow flowers. *D. senile* (Thailand and Indochina)

has bright yellow, lemon-scented flowers and is conspicuously hairy on the pseudobulbs and leaves. Smaller waxy pinkish purple flowers in a dense one-sided inflorescence characterize *D. secundum,* "The Tooth Brush", which is a common dry forest epiphyte in tropical Asia. *D. leonis* (Thailand to Indonesia), with fragrant, dirty yellow-brown flowers to 1.5 cm across, belongs to a section with fleshy, laterally compressed densely distichous leaves, a vegetative appearance found also in other groups of orchids, e.g. in *Oberonia* and in several genera of the subfamily Vandoideae. *D. unicum* (Thailand and Indo-China; frequently misidentified as "*D. arachnites*") has bizarre, spidery, purple-striped flowers to 6 cm across, and is widely grown as an ornamental. – The genus *Flickingeria* (60), in tropical Asia and Australia, consists of much branched bushy epiphytes with flowers lasting for only a few hours.

**The Bulbophyllum Group** (Subtribus Bulbophyllinae) consists of one large genus, *Bulbophyllum,* and five to seven small genera. They are all small to medium-sized epiphytes with more or less conical or ellipsoid, sometimes angular or winged pseudobulbs, which are made up of one internode and carry apically one or two fleshy, conduplicate leaves. The inflorescence is lateral and arises at the base of the pseudobulbs. Its flowers have a column foot and usually naked pollinia. This group may be imagined to have evolved within the preceding group as a further development of one of the lines with single-internode pseudobulbs, leading to the reduced number of leaves and standardized architecture observed in *Bulbophyllum* (Fig. 121 M). However, it appears that the Bulbophyllinae have no silica cells; these may have been lost in the course of evolution. – *Bulbophyllum* (ca. 1,000), is pantropical but concentrated mostly in the Old World tropics, especially in tropical Asia (Figs. 10–11). The usually rather small flowers often show adaptations to fly pollination with various mobile appendages, with dark (often deep purple or violet) flower colours and sometimes with a disagreeable smell. The column is short and stout, with two often distinctly elongated teeth ("stelidia") on each side of the anther (Fig. 125 M). The characteristic, fleshy, curved labellum is hinged to the apex of a prominent column foot, on which the lateral petals are inserted. The pollinia are usually without appendages and only few species have a detachable viscidium. Two species with a distinct hamulus are known.

The distinct vegetative mode and basic construction of the flower make this an easily identified genus. There is nevertheless a substantial variation in the overall appearance of the species, especially concerning the inflorescence and arrangement of flowers. A number of more or less distinct subdivisions are recognized, e.g. sect. *Megaclinium* (Africa and tropical America), with a flattened, leaf-like inflorescence rachis, and sect. *Cirrhopetalum* with often subumbellate inflorescence and apically connivent lateral sepals much longer than the other tepals. Only a few species of *Bulbophyllum* are regularly cultivated, e.g. *B. lobbii* (Thailand, Indonesia), which has solitary brownish yellow flowers 6–10 cm across. *B. wendlandianum* ( = *Cirrhopetalum collettii*), *B. fascinator* and related species have extraordinary flowers with caudate lateral sepals 20–30 cm long.

**Subfamily Vandoideae**

The Subfamily Vandoideae (tribus Vandeae in many older classifications) contains about 300 genera and 5,000 species, most being tropical epiphytes. In traditional orchid systematics, and as here accepted, it is an assemblage of the most advanced orchids, possessing one or more of several advanced attributes: lateral inflorescences, strongly inflexed anther and rostellum, indistinct anther partitions, superposed and hard, waxy to bony pollinia, a detachable viscidium, and a pollinium stalk of the tegula type. However, none of these characteristics is unique to the Vandoideae, nor is any of them shared by all members traditionally included, although only very few (some Cymbidieae and some Corallorhizinae) lack the tegula. The inflexion of the anther and rostellum apex begins very early in the ontogeny of the gynostemium of all vandoid genera of which the development has been investigated (Fig. 126 F–K). Inflexion at an early stage may be a unique character of the Vandoideae as here defined, but in view of the otherwise conflicting character distribution it appears rather unlikely that the Vandoideae form a monophyletic group – most probably they consist of several advanced evolutionary lines from "epidendroid" ancestors (see DRESSLER 1981). DRESSLER (1981) interprets the vandoid anther as erect and opening basally, literally "basitonic". This is, however, generally not the case – it remains to be shown whether it is true for some genera.

The subdivision of Vandoideae into tribes and subtribes varies greatly in different classifications. However, at least the tribe Vandeae appears to be a distinct, monophyletic group.

*Tribus Polystachyeae*    4:220

This pantropical tribe contains the large pantropical genus *Polystachya* and three small Afro-Madagascan genera. They are epiphytic or more rarely terrestrial herbs, usually with slender reed-like stems or multinodal pseudobulbs. Most species have rather small flowers in racemose or paniculate, usually terminal inflorescences. A prominent column foot makes the flowers resemble those of the epidendroid genera *Eria* and *Dendrobium,* but the short but distinct stipes places the tribe Polystachyeae among the vandoids.

*Polystachya* (210), occurs predominantly in tropical Africa. The flowers are not resupinate. *P. concreta* (syn. *P. flavescens, P. tesselata, P. luteola*) is a variable pantropical species, perhaps the most widespread of all orchids.

*Tribus Cymbidieae*    130:1800

As circumscribed by DRESSLER (1981) this is a very heterogeneous group. It includes all vandoid orchids with two pollinia except the Vandeae. It is only in this very broad sense that it comprises as many species as indicated here.

*The Cymbidium Group* (Subtribus Cyrtopodiinae) contains about 25 genera and 430 species. It is pantropical, but is mainly concentrated in the Old World tropics, and consists of terrestrial or epiphytic plants; a few species are saprophytes. This group is very diverse, including principally the vandoid orchids having only one or a few of the advanced characters mentioned for the subfamily. Quite often the members have plicate leaves and multinodal pseudobulbs. The flowers often have a broad and somewhat winged column and a prominent column foot. The pollinia in most genera are attached to a broad tegula (Fig. 126 D).

*Eulophia* (200), is a pantropical genus, but most species occur in Africa. They are mostly terrestrial herbs, often found in grasslands, but a few are saprophytes. Many species have large, showy flowers, for example *E. cucullata,* widespread in tropical Africa, which has purplish pink flowers. *E. petersii,* in dry habitats of East Africa, has thick, coriaceous, finely serrate leaves, resembling a small-sized *Aloë.* The branched flowering shoots may exceed 2 m. – *Ansellia* (1–2), widespread in tropical Africa, is epiphytic: it has showy yellow flowers with brown spots. *A. gigantea,* "Leopard Orchid", is frequently cultivated. – *Cymbidium* (ca. 50; Fig. 126A), mainly in tropical Asia, con-

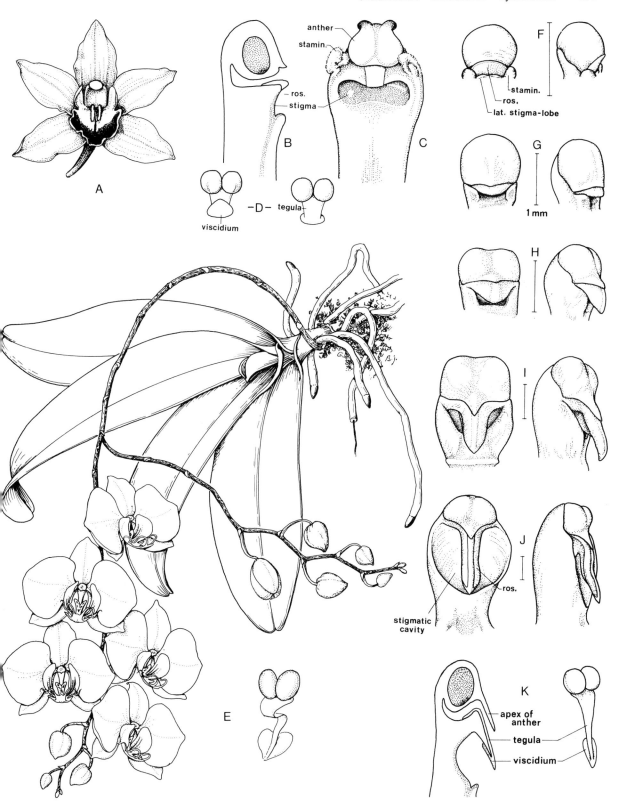

**Fig. 126.** Orchidaceae subfam. Vandoideae. **A** *Cymbidium lowianum*, flower. **B–D** *Eulophia andamanensis*. **B–C** Gynostemium, lateral and front views. **D** Pollinarium in different views. **E** *Phalaenopsis amabilis*, of the tribe Vandeae, with pollinarium. **F–K** *Doritis pulcherrima*. **F–J** Ontogeny of the gynostemium. **K** Gynostemium and pollinarium. (Orig.)

sists of epiphytic or rarely terrestrial herbs, usually with a short stem and long, narrow and tough distichous leaves. The often very showy flowers have a three-lobed labellum, borne on a short column foot. The labellum is usually conspicuously bicarinate at the base, and the midlobe is recurved. *Cymbidium* hybrids are among the most important commercially grown orchids. *C. viridioides* (syn.: *C. giganteum*), *C. lowianum* (tropical Himalaya) and a few closely related species (or varieties) and their hybrids are widely sold as cut flowers. In the flowers, up to 10 cm across, the labellum often contrasts in colour with the other, yellowish green tepals. *C. finlaysonianum,* in the lowlands of tropical Asia, is a robust epiphyte with dull yellow or brownish flowers up to 6 cm across. – *Grammatophyllum* (ca. 10), in tropical Asia, is apparently closely allied to *Cymbidium,* but has stalked pollinia. *G. speciosum* is a large epiphyte, with pseudobulbs up to 3 m long and 6 cm in diameter. The basal, erect inflorescence may be up to 2 m, carrying numerous, greenish yellow spotted flowers up to 10 cm across.

**The Catasetum Group** (subtribus Catasetinae), with 145 species in 5 genera, occurs in tropical America. It consists of epiphytes with multinodal pseudobulbs and scattered, distichous, plicate leaves. The inflorescences are lateral and racemose and the flowers often unisexual. The column usually bears two basal antennae, which function as levers releasing the large, elastic stipes when touched. The fairly large pollinarium is thus thrown off and the sticky viscidium may hit the scutellum of a visiting bee with considerable force. – *Catasetum* (50), found in tropical America, consists of epiphytes with unisexual flowers. Male and female flowers of the same species may look very different. Both monoecious and dioecious species are found. Some species also have facultatively hermaphroditic flowers, which may be different from either of the unisexual ones. The fleshy labellum is often helmet-shaped, and often with a fimbriate margin – especially in male flowers. Some species are widely grown.

**The Stanhopea Group** (subtribus Stanhopeinae), with 17 genera and ca. 190 species, occurs in tropical America. These are all epiphytes with uninodal pseudobulbs, terminally with plicate, usually petiolate leaves. Like many Catasetinae they are pollinated by male euglossine bees attracted by specific fragrances. The flowers do not provide any food but exude perfumes when scratched. In most genera the flowers are organized so as to take advantage of the retarded re-

flexes of the drugged insects which are manipulated to fall through a chute, down a slide, or (in *Coryanthes*) into the water-filled, bucket-shaped epichile of the labellum. When crawling out of this they have to pass the apex of the gynostemium. The flowers of Stanhopeinae are bisexual, self-pollination being prevented by dichogamy. – *Coryanthes* (c. 20), from Guatemala to Peru and Brazil, is the genus of the famous "Bucket Orchids", occasionally cultivated for their large, bizarre flowers. – *Stanhopea* (c. 25), widespread in tropical America, is epiphytic (occasionally terrestrial) consisting of medium-sized to large plants with one-leaved pseudobulbs and large fleshy flowers. The genus is frequently cultivated. – *Gongora* (25) is also frequently cultivated.

**The Oncidium Group** (subtribus Oncidiinae) comprises 60 genera and 950 species, predominantly epiphytes, in tropical America. Most taxa have uninodal pseudobulbs, but forms with slender stems, and even monopodial forms, occur in this group. The two pollinia are only slightly notched and attached to a rather long and slender stipes. The column top and pollinarium strongly resemble those in the Old World Vandeae, and it has sometimes been speculated that the Oncidiineae and the Vandeae represent respectively a neotropical and a palaeotropical evolutionary line from a common ancestral stock.

*Oncidium* (more than 400), in subtropical and tropical America, for the most part consists of pseudobulbous epiphytes. The labellum has a characteristic basal, warty callus, and the column top is mostly auriculate or winged. Infrageneric classification is difficult and the delimitation from several other genera of the subtribe is unclear. The flowers vary from minute to very large. Yellow and brownish flower colours are common. The flowers frequently have a characteristically three-lobed labellum with a bipartite spreading midlobe: "Dancing Girls". Many species are cultivated. *O. bicallosum* (Mexico and central America) has brownish yellow flowers up to 5 cm across. *O. kramerianum* and *O. papilio* also have large flowers with narrowly spathulate, upwardly directed sepals, 5–7 mm long and resembling the tentacles of an octopus, and with broad, wing-like, deflexed petals. *O. altissimum,* with flattened pseudobulbs, has scrambling inflorescences up to 3 m long with yellowish flowers. – *Odontoglossum* (ca. 100) occurs in subtropical and tropical central and South America, often at fairly high altitudes. The base of the labellum is parallel to and often partly adnate to the column. The genus contains many popular species. *O. grande,* a robust epiphyte, has showy flowers with purple-brown spots and yellow stripes; it is sometimes placed in a segregate genus, *Rossioglossum,* on account of the free labellum. *O. crispum,* in Columbia, has showy white or rose flowers, spotted with red or purple, but a yellow labellum. – *Miltonia* (20), in the Andean highlands (these often separated as *Miltoniopsis*) and South East Brazil, has a shorter column than that in the preceding genera. – *Brassia* (20), in tropical America, has linear-lanceolate sepals, and a short, wingless column.

### Tribus Maxillarieae    80:760

As circumscribed by DRESSLER (1981) this tribe comprises all Vandoideae with four pollinia except the tribes Vandeae and Polystachyeae. Apart from certain members of the Corallorhizinae this is an entirely American tribe. The Maxillarieae are vegetatively very diverse. They generally produce a corm and have plicate leaves, but the more "advanced" members may have uninodal pseudobulbs and conduplicate leaves. A few genera are monopodial. The flowers are typically vandoid. Spurs are usually lacking, and the gynostemium usually has a long slender rostellar projection and a correspondingly slender tegula.

**The Corallorhiza Group** (subtribus Corallorhizinae) in the classification followed here consists of 9 genera with about 60 species distributed mainly in North Temperate and Tropical America (*Corallorhiza* being found also in the Old World). This is a group of terrestrial, often saprophytic cormous herbs with plicate (or reduced) leaves and terminal inflorescences. The details of the gynostemium are surprisingly poorly known, but it is reported to have a distinct, detachable viscidium and a small stipe. The *Corallorhiza* group is differently circumscribed by most authors and not associated with the Maxillarieae with which it has only the four superposed pollinia in common.

*Corrallorhiza* (ca. 10), distributed over the Northern Hemisphere, are saprophytic bog or forest plants characterized by the much branched, coral-like rhizome, "Coral Root". *C. trifida,* circumboreal (including Greenland), is the only species also found in Europe.

**The Zygopetalum Group** (subtribe Zygopetalinae) (including Lycastinae) (20:160) is tropical American. It consists mainly of epiphytes of varying habit, all genera except one being reported as sympodial. There is often a well-developed column foot. The rostellar projection is prominent and the

tegula is often long and slender. – *Zygopetalum* (30–40), comprises tropical South American epiphytes which are usually recognized by the fleshy callus at the base of the labellum. The flowers are often showy. A popular, cultivated plant is, for example, *Z. intermedium.* – *Lycaste* (25), from Mexico to Peru, has uninodal, often somewhat flattened pseudobulbs. The attractive, often yellow flowers have large sepals. *L. virginalis* is a well-known greenhouse plant.

**The Maxillaria Group** (subtribus Maxillariinae) (9:400) likewise occurs in tropical America. It resembles the Zygopetalinae but has one-flowered inflorescences. Some species of *Maxillaria* are said to have no pollinium stalk.

*Maxillaria* (300), widely distributed in the American tropics, has coriaceous, non-plicate leaves and a prominent mentum formed by the bases of the lateral sepals and the column foot (*Maxillaria* = little jaw). The pollinarium has a rounded or crescent-shaped viscidium. There are two major groups in the genus, one with untufted plants with pseudobulbs borne on an elongated rhizome, the other with densely tufted plants. Cultivated are, e.g., *M. picta,* from Eastern Brazil, with clustered pseudobulbs and yellowish, purple-spotted flowers up to 6 cm across.

### Tribus Vandeae    135:1,700

This occurs essentially in the Old World tropics; it includes only a few New World genera. They range from dwarf plants to very large epiphytes with showy flowers. The tribe is characterized by the monopodial habit and the vandoid gynostemium with a distinct tegula. Some genera have four pollinia, but others have two, which may be more or less notched or entire (see also Fig. 119M). The stem apex grows continuously and the base gradually dies away; pseudobulbs are not developed. The distichous leaves have generally taken over the water-storing function and may be very succulent, in some genera even terete. The lateral inflorescences and the thick aerial roots penetrate the imbricate leaf sheaths. However, a few genera lack foliage leaves altogether and have green assimilating roots. The leaves may be spaced on an elongated erect or pendent shoot, or may be condensed on a short shoot so that the long leaves become arranged as a fan. The flowers vary from a few millimetres to very large and showy. Their labellum may be pointed and is often saccate or spurred. The column may have a prominent foot, and the rostellum has frequently a long and beak-like projection, pointing downwards across the stigmatic cavity.

The Vandeae appear to form a monophyletic group distingishable by its habit. A few other orchids are monopodial (e.g. *Vanilla*) but all of these show other relationships demonstrating that the monopodial habit has in these cases evolved independently.

*The Vanda Group* (subtribus Aërangidinae in broad sense, syn.: Sarcanthinae nom. illeg.) contains ca. 120 genera and 1,300 species, ca. 1,000 of which are found in tropical Asia. There are almost no genera in common between Africa-Madagascar and Asia so the genera in these regions are often treated in different subtribes. The taxonomy of genera and generic complexes in this group is highly controversial.

*Phalaenopsis* (50), in tropical Asia, has a characteristic callus on the labellum. This genus contains several very important ornamentals, e.g. the large, white-flowered *P. amabile* or *P. aphrodite* and their hybrids (Fig. 126E), often seen in wedding bouquets. *P. cornucervi* has smaller yellowish brown flowers with a complicated callus. – *Kingidium* (2–3) and *Doritis* (2) are genera segregated from *Phalaenopsis; D. pulcherrima,* from South East Asia, is a common greenhouse ornamental. – *Aerides* (ca. 20), in South Asia, are robust epiphytes, with dense racemes of flowers, having a distinct column foot continuous with the three-lobed labellum. The spur is often forwardly projecting. *A. odoratum* is a commonly cultivated species. – *Vanda* (30–40), widespread in tropical Asia, comprises small to very large epiphytic or terrestrial orchids, often with very showy flowers. The spreading perianth segments are tapered towards their bases. There is no column foot. The tegula is broad, almost square. *V. tricolor,* (Java), is a coarse, erect epiphytic or terrestrial plant with large, variably coloured flowers; *V. coerulea,* from North India to Thailand, is famous for its pale blue or lilac flowers; and *V. teres* (sometimes referred to *Papilionanthe*), from Burma to Indochina, is a scrambling epiphyte with terete leaves. Several important commercial cultivars are intergeneric hybrids between *Vanda* and the related genera *Arachnis* and *Renanthera*. – *Cleisostoma* (syn.: *Sarcanthus*; ca. 100), widespread in tropical Asia, has small, fleshy flowers. – *Aërangis* (ca. 15), widespread in tropical Africa and Madagascar, has very short stems and long-spurred, often whitish flowers, borne in lax racemes. The attractive *A. luteoalba* var. *rhodosticta,* in Cameroun to East Africa, has a bright scarlet column. – *Taeniophyllum* (ca. 90), widespread in tropical Asia, extending to Eastern Australia (one species in Africa), and *Microcoelia* (26), in tropical Africa and Madagascar, are small leafless epiphytes with very short stems and more or less dense clusters of branching, assimilating roots.

*The Angraecum Group* (subtribus Angraecinae), with ca. 16 genera and 1,600 species, is mainly an Afro-Madagascan group, but includes four American genera. It is characterized by a notched rostellum and a correspondingly very short tegula. There are only two pollinia. This group has probably evolved within the Aerangidinae.

*Angraecum* (25) has its centre on Madagascar. It is a vegetatively variable genus, ranging from tiny plants, sometimes with flattened, densely imbricate leaves (e.g. *A. distichum*) to large *Vanda*-like species (e.g. *A. eburneum*). The genus is recognized by the spurred, concave labellum, which more or less envelops the column. The flowers are most often white or greenish. *A. sesquipedale,* on Madagascar, has spurs 30–35 cm long. The existence of a hawk moth with a correspondingly long proboscis was predicted by Charles Darwin 40 years before this moth became known to zoologists. – Closely related to *Angraecum* is probably *Polyradicion* (2–4) in Florida and the West Indies: leafless root-assimilating epiphytes having surprisingly large, long-spurred flowers (up to 12 cm across).

# Superorder Ariflorae

*One Order:* Arales.

The two families of Arales are quite dissimilar. Whereas the Araceae as a rule comprise well-differentiated perennial herbs, the Lemnaceae are small, free-floating aquatics, the green shoot of which comprises a thallus-like assimilatory plate. The flowers in Araceae vary from bisexual trimerous and pentacyclic to unisexual, naked and consisting of either only one stamen or one carpel, whereas the flowers in Lemnaceae are always naked, unisexual, and monomerous. There is, however, evidence that the families are closely related (see below), and we shall here supply a broad superordinal description, that will mostly be relevant for Araceae only, in order to give a reference base for comparisons with other superorders.

Mainly perennial herbs with rhizomes or subterranean tubers. The well-differentiated Araceae vary from small plants without an aerial stem to branched, frequently climbing herbs; rarely the shoots form stolons, as e.g. in the floating *Pistia*. In the Lemnaceae the leaves and stems are not clearly differentiated, but the plate-like assimilatory bodies are connected sympodially.

The roots are either subterranean, or in various branched vines they may be green and climbing; there are thick pendent roots, as in *Monstera*, or adhesive roots appressed to the substratum. The vascular supply of the roots originates opposite the xylem strands. The root hairs are generally derived from unspecialized cells of the epidermis, but in certain genera they are reported to arise from cells shorter than ordinary epidermal cells. A multiple velamen is rare, but recorded occasionally. In the Lemnaceae the roots vary from few to one or are completely lacking.

The leaves are spirally set or rarely distichous. In most genera they are differentiated into a petiole, an extended sheathing base (not differentiated in *Symplocarpus* and *Lysichiton*), and a broad lamina, which may be simple or divided. Rarely they are linear and occasionally ensiform, as in *Acorus*, or broad and sessile, as in *Pistia*. The main venation is pinnate, and in many genera there is also a minor reticulate venation. A row of intravaginal squamules occurs in the axils of the leaf sheaths or in the axils of the cataphylls in some genera (see VELENOVSKÝ 1907; ENGLER 1920C). Ligule-like structures occur rarely, and the leaf sheath is sometimes widened into stipule-like "ears". Occasionally, as in *Pistia*, a membranous axillary stipule is present.

The Ariflorae are richly differentiated in vegetative anatomy. Laticifers are present in particular generic groups of the Araceae, and are usually simple and uniseriate. T- or H-shaped spicular cells occur in many of the genera lacking laticifers. Resin ducts are sometimes present, and rows of mucilage cells occur in some genera. Vessels are probably present in the roots of most Ariflorae, always with scalariform perforation plates; they are consistently lacking in stems and leaves. Silica bodies are lacking, whereas oxalate raphides occur, at least in the majority of the members, though lacking in some Lemnaceae. They are often plentiful and may be important in protecting the plants from predators.

The cells of the leaf epidermis are mostly polygonal with five to eight sides. Unicellular hairs occur occasionally (uniseriate hairs in *Pistia*). The stomata are generally hexacytic or tetracytic (WEBBER 1960), and are frequently randomly directed on the leaf surface. Rarely, but in all Lemnaceae, well-defined subsidiary cells are lacking; in other cases there may be a considerable number of subsidiary cells.

The inflorescence is situated on a leafless scape (peduncle) of variable length and is subtended by a spathe (rarely more than one), which is generally wide and dorsiventral (but ensiform and appearing as a continuation of the scape in *Acorus*, Fig. 127A). It is generally entire and ovate and has distinctive colours that may attract the pollen vectors. Attraction is often connected with emission of odorous substances and in some cases with nectar secreted from the stigmas of the flowers. Rarely, the spathe is green and only slightly different from the blades of the vegetative leaves. It encloses the spadix in the juvenile stage, and in

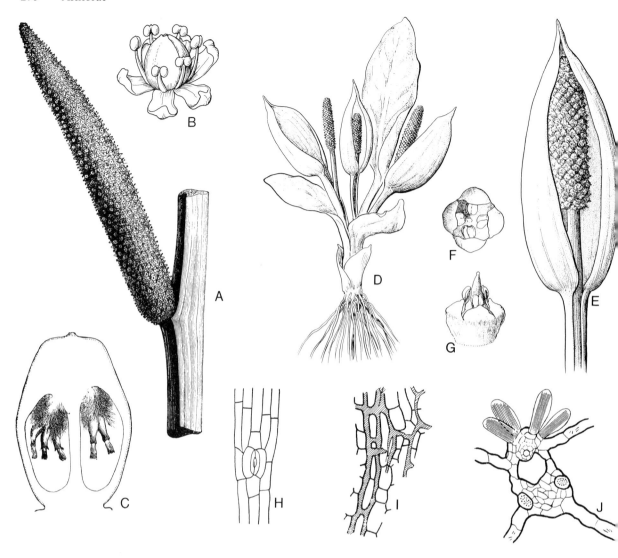

**Fig. 127.** Araceae. **A–C** *Acorus calamus.* **A** Spadix, distal part of scape and basal part of spathe. **B** Flower. **C** Ovary, longitudinal section. (**A** and **C** HITCHCOCK et al. 1969; **B** LARSEN 1973a). **D–G** *Lysichiton americanum.* **D** Plant. **E** Inflorescence; note the long internode between the spathe and the flowers of the spadix. **F–G** Flower in front and lateral view. (HITCHCOCK et al. 1969). **H** *Acorus calamus,* stoma. **I** *Xanthosoma sagittifolium,* laticifer. **J** *Typhonodorum madagascariense,* raphide cells. (**H–J** from SOLEREDER and MEYER 1928)

many taxa it encloses at least part of it during anthesis, and sometimes afterwards.

In the Lemnaceae the (very reduced) inflorescence is located in a groove in the assimilatory plate.

The spadix or spike is normally simple. Its spirally (rarely verticillately) arranged flowers are subtended neither by bracts nor by bracteoles (prophylls). Only in *Pedicellarum* do they have a short pedicel. In the Lemnaceae they are few and the inflorescence can hardly be identified as a spike. The differentiation of the spadix in Araceae is considerable. The original state is presumably one where the axis is uniformly covered with bisexual flowers to its apex. In more specialized inflorescences the flowers are unisexual, and there may also be neuter flowers with particular functions in pollination. In its most derived state the spadix may be apically flowerless and form a naked, often strongly coloured continuation, the appendix.

The flowers vary considerably, from having $3+3$ (or $2+2$) tepals, $3+3$ (or $2+2$) stamens, and a 3-(to 8-)carpellary, syncarpous pistil, to naked and unisexual, with either one to three stamens or a monocarpellary pistil. This reduction seems to have taken place in several evolutionary lines in Araceae. The flowers are always hypogynous and actinomorphic.

The tepals when present are small, often thick and prismatic, and free from each other. Perigonal nectaries are lacking. The stamens when several are either free from each other or often united. They have a short, glabrous filament and a basifixed, either bi- or tetrasporangiate anther, which dehisces abaxially or apically through slits or porelike openings. Anther wall formation is of the Monocotyledon Type (Araceae, some Lemnaceae) or of the Reduced Type (some Lemnaceae). The endothecial thickenings are of the Spiral Type, and the tapetum is amoeboid. Microsporogenesis is successive. The pollen grains are separate or very rarely coherent in tetrads. They are sulcate or sulcoidate, bi- or trisulculate, zonisulculate, bi- or tetraforaminate, trichotomosulcate, or, in many genera, inaperturate. They are dispersed either in the two- or in the three-celled stage.

The pistil has a very short style or none. The ovary is usually unilocular, with a basal, lateral, parietal, central or apical placentation. The stigmas are Dry or Wet. The ovules are generally few or solitary, rarely numerous. They are anatropous, orthotropous, or rarely hemianatropous and are bitegmic and weakly crassinucellate or tenuinucellate. A parietal cell may or may not be cut off from the archesporial cell; in the latter case a nucellar cap is generally formed by periclinal divisions in the nucellar epidermis (not so in *Pistia,* however, where the embryo sac lies directly under the nucellar epidermis). Embryo sac formation is generally of the *Polygonum* Type, but in most Lemnaceae (except *Spirodela*) of the *Allium* Type. Endosperm formation, in contrast to that of nearly all other monocotyledons, is of the Cellular Type. In most Araceae the chalazal chamber remains uninucleate, increases much in size, and shows haustorial activity. Embryogeny, in at least part of the Araceae, conforms to the Caryophyllad Type, but in, for example, *Pistia* and the Lemnaceae, to the Onagrad Type.

The fruits are nearly always berries. The seeds may have a sticky, fleshy exotesta or the exotesta may be dry and then tends to form longitudinal ridges. The inner layers of the testa contain thick-walled cells. The tegmen remains unspecialized and often collapses. In the ripe seeds the endosperm remains in the Lemnaceae and in some groups of Araceae, but in other groups it is absorbed by the growing embryo (as in all Alismatiflorae). When present, the endosperm cells in at least a number of cases studied, in both families, contain starch grains as well as protein.

The embryo is generally linear, but may be straight or curved, chlorophyllous or achlorophyllous, and sometimes macropodous (having a swollen hypocotyl or radicle). Seedlings arising from non-endospermous seeds usually have one or two cataphylls intercalated between the cotyledon and the foliage leaves, whereas such scales are usually lacking on seedlings from endospermous seeds.

*Chromosome Numbers.* The predominant chromosome numbers in the superorder are $n = 14$ and $n = 21$; other common numbers are $n = 13$ and $n = 12$. The original basic number is possibly $x = 7$, but this number is not known in recent taxa (e.g. Raven 1975).

*Chemistry.* Steroidal saponins occur at least in certain genera, but chelidonic acid does not seem to be reported at all. Cyanogenic compounds are common. Flavonols are the major components of the flavonoid profile, with C-glycoflavone as a minor one (Gornall et al. 1979; C.A. Williams et al. 1981). Tricin, which is common in palms, is lacking in the Ariflorae.

*Parasites.* The Ariflorae appear to have few fungal parasites, although *Acorus* is attacked by the rust *Uromyces sparganii* (see below), *Arum* by species of *Melampsora* and by *Puccinia sessilis,* which otherwise among monocotyledons attacks mainly liliiflorous hosts. *Lemna* is susceptible to the smut genus *Doassansia,* which is distributed on a number of diverse aquatic angiosperm genera.

*Distribution.* The Ariflorae are distributed over most of the world but have a marked tropical-subtropical distribution in the Old World as well as the New World. The centre is clearly in the Old World tropics. Many genera play a dominant role in the vegetation of rainforests where they are particularly common in moist, humid habitats. All Lemnaceae and *Pistia* of the Araceae are free-floating aquatics.

*Relationships.* The affinities of the Ariflorae have generally been considered to be with the Arecales, Cyclanthales, Pandanales and Typhales. There are indeed a number of similarities with these groups, and until recently the Typhales were placed by Thorne (1976, 1981) in the Arales. The fact that *Sparganium* (Typhales) and *Acorus* (Araceae) are both attacked by *Uromyces sparganii* (Parmelee and Savile 1954) has been considered to support the postulate of affinity between the two orders. They differ, however, in many respects: vessels are present in the stems of Typhales but are lacking in stems of the Arales, stomata are paracytic in Typhales but mostly tetracytic or hexacytic in Arales, the pollen grains in Typhales are ulcerate, whereas this type is virtually missing in Arales, the pistils are very different in shape, a parietal cell is formed in Typhales, which is only rarely the case in Arales, endosperm formation is helobial in Typhales but cellular in Arales, embryo formation is of the Onagrad Type in Typhales but generally of Caryophyllad Type in Arales, the fruits are drupes or follicles in Typhales but mostly berries in Arales, etc. However, similar features are found in the floral construction, amoeboid tapetum occurs in both groups, and there is great superficial similarity as regards the inflorescence.

More significant, in our view, are the similarities between the Ariflorae and the Alismatiflorae (DAHLGREN and CLIFFORD 1981). These include the lack of vessels in the stems, the well-differentiated leaves with petiole and lamina, the occurrence in both groups (though perhaps rarely in the Ariflorae) of intravaginal squamules, the occurrence in both complexes of a spathe and of spicate inflorescences (cf. Juncaginaceae, Aponogetonaceae), of laticifers and of extrorse anthers, amoeboid tapetum, Caryophyllad Type of embryogeny, etc. It may also be noted that the cyanogenic substance triglochinin occurs in Araceae as well as in certain Alismatiflorae, e.g. Juncaginaceae (NAHRSTEDT 1975). The fact that the multistaminate flowers in certain Alismatiflorae may represent a derived condition (SATTLER and SINGH 1978, etc.) should be considered in this context.

There are also obvious similarities between some Ariflorae and some Dioscoreales, notably the Dioscoreaceae, in the Liliiflorae. Thus some primitive Ariflorae (such as members of Araceae subfam. Pothoideae) in several features resemble the Dioscoreales: in the climbing growth, the well-differentiated leaves with reticulate venation, the spicate, unspecialized inflorescences, etc. However, the precise kind of venation found in Dioscoreaceae is not known in the Araceae, where the venation is more like that in some Alismatiflorae. In their hypogyny the Ariflorae are more primitive than the Dioscoreaceae. Thus a likely hypothesis is that the Ariflorae share their ancestors with those of the more primitive Liliiflorae. The Alismatiflorae may represent a sister-group.

# Order Arales

N. JACOBSEN

*Two Families:* Araceae and Lemnaceae.

The order consists of two families, Araceae and Lemnaceae.

It is the single order of the superorder Ariflorae; hence the description and comments under the superorder are valid for the order, too.

**Araceae** A.L. Jussieu (1789)    110:2,450 (Figs. 127–131)

Perennial herbs with an underground or aerial stem or both. Underground stems often form tubers (which may be of considerable size) or rhizomes. The aerial stems may be erect (*Dieffenbachia*), climbing (*Monstera, Raphidophora, Scindapsus*) or procumbent (*Anthurium radicans*). Some taxa have stolons, e.g. the floating plants of *Pistia*. Juvenile plants, at least in most cases, seem to have monopodial growth, shifting to sympodial with the first-developed inflorescence, the inflorescences being terminal. With some vines exhibiting sympodial growth, the greater part of the stem is made up of the first internodes of the successive lateral branch systems. The internode between the cataphyll (scale leaf) and the next foliage leaf does not elongate. Monopodial growth throughout is characteristic of some Pothoideae.

Rarely, adventitious bulbils are formed; they may occur on the leaves (e.g. in *Amorphophallus bulbifer*), on the stem (e.g. *Xanthosoma viviparum*) or on special branches developed from the tubers (*Remusatia vivipara*).

The leaves vary considerably in shape. They are sheathing at the base and usually petiolate. The lamina is typically flat, lanceolate to broadly ovate, and frequently with a cordate, sagittate or hastate base; sometimes it is palmately compound, palmatifid or pedate, but strap-shaped or linear leaves also occur. Unique in the family are *Pistia* with broadly cuneate, sessile leaves and *Acorus* with ensiform leaves. The leaf sheaths usually enclose the new leaves, but in sympodial systems they enclose the terminal inflorescences while the cataphylls protect the continuation axes. Ligule-like structures occur in a few genera, e.g. *Arisaema* and *Dieffenbachia*; a true ligule is developed in e.g. *Piptospatha*.

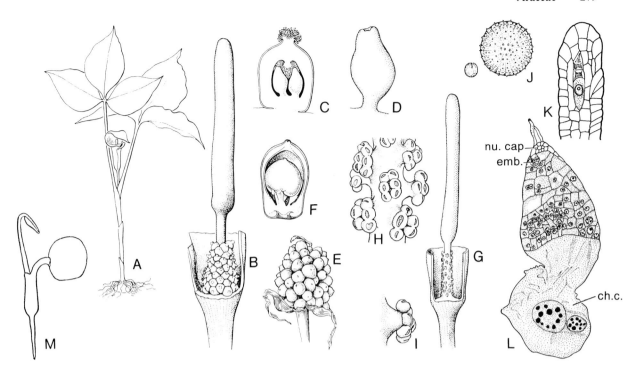

Genera in which intravaginal scales have been reported include *Philodendron, Cryptocoryne* and *Lagenandra*. Stipules occur in *Pistia*.

Heterophylly, as regards the shape of the leaf lamina, is found in many genera, the leaves of the young shoot being entire, although frequently cordate, and those on branches formed later lobed. Arrested growth in intercostal areas as in the developing leaf of *Monstera deliciosa* causes "holes" to form in the adult lamina.

The main veins of the leaf blades are generally converging. The veins are subparallel in the Philodendroideae, and in other groups an intercostal, reticulate venation is frequent, as in Pothoideae and Monsteroideae, but is also found within Calloideae, Lasioideae and Aroideae, in all groups of which the veins are directed towards the leaf margins.

The cells of the leaf epidermis are mostly polygonal in outline with five to eight sides, but are only four-sided in *Acorus*. Unicellular hairs are found in a few genera, e.g. *Xanthosoma* and *Homalomena*; uniseriate hairs occur in *Pistia*. Some species of *Philodendron* have filiform protuberances on the petioles.

The stomata are generally randomly oriented on the leaf blades, and not in parallel files as in most other monocotyledons. They are generally tetracytic or hexacytic, the guard cells being surrounded by one pair of (sometimes weakly defined) subsidiary cells at the ends of and one pair

**Fig. 128.** Araceae. **A–J** *Arisaema atrorubens*. **A** Plant. **B** Female spadix, most of the spathe removed. **C** Ovary, longitudinal section. **D** Ovule. **E** Infructescence. **F** Berry, longitudinal section. **G** Male spadix, most of the spathe removed. **H** Part of male spadix with naked flowers, the stamens fused into synangia. **I** Male flower (=synangium), lateral view. (WILSON 1960). **J** *Arisaema consanguineum*, pollen grain. (ERDTMAN 1952). **K–L** *Arisaema wallichianum*. **K** Nucellus with megaspore tetrad. **L** Endosperm, embryo, and nucellar cap. (S.C. MAHESHWARI and KHANNA 1956). **M** *Arisaema dracontium*, seedling. (LOTSY 1911)

parallel to the guard cells. However, in *Arisaema* well-defined subsidiary cells are lacking, while in other cases, as in *Dieffenbachia seguine*, there may be quite a number of subsidiary cells.

Laticifers (Fig. 127I), which are probably secondary in the Araceae, are absent from the presumably primitive genera (e.g. of subfamily Pothoideae). They are usually simple and uniseriate, but laterally branched and anastomosing laticifers occur in several genera (incl. *Caladium, Xanthosoma, Colocasia, Alocasia* and *Syngonium*). T- or H-shaped spicular cells occur in many of the genera lacking laticifers, e.g. *Spathiphyllum* and *Monstera*. Long resin-ducts occur in, e.g., *Philodendron*, shorter and thicker ones in *Homalomena*. Rows of mucilage cells occur in, e.g., genera of the Colocasioideae. Vessels are probably present in the roots of most Araceae and always have scalariform perforation plates; in stems and leaves vessels are constantly lacking.

Calcium oxalate crystals are of general occurrence in Araceae, both in the form of spherical or other crystals and as raphides (Fig. 127J).

The spadix is subtended by one (rarely more) spathe(s) of variable size and shape, which may be leaf-like and leave the spadix exposed, as in some Pothoideae; it may be strongly coloured while still leaving the spadix exposed as in *Anthurium,* or it may enclose the spadix basally (*Lasia, Monstera*) or totally (*Cryptocoryne*), being then sometimes coloured and even variegated, especially on the inner side.

In some groups in which the spathe encloses the base of the spadix the former is constricted between the closed chamber and the upper hooded or laminar limb. This constriction is often at the point of division between the male and female zones of the spadix. However, in the more advanced groups, which have a well-developed spadix appendix, the constriction is between this and the floral zone. In *Cryptocoryne* (already mentioned) and a few other genera the basal chamber is formed by fusion of the spathe margins, not merely their overlap. In these groups, in which the spathe is longitudinally differentiated, the basal part is commonly persistent and helps to protect the developing fruits, as in Philodendroideae, Colocasioideae, and Aroideae.

There is great variation in the form of the spadix. The most ancestral state is undoubtedly where the axis is uniformly covered to its apex with bisexual flowers as in *Pothos, Anthurium, Acorus* and *Spathiphyllum.* In other cases the flowers are unisexual and often reduced both in perianth and numerical condition, in which case the spikes may show considerable variation as regards the position of bisexual, female, male or neuter flowers. When the flowers are unisexual, the females are usually proximal and the males distal on the spadix (as in *Zantedeschia, Aglaonema,* and *Homalomena*). Staminodial, neuter flowers may also be intercalated between the female and the male flowers (as in *Philodendron, Xanthosoma* or *Arum*), or there may be a naked, sometimes constricted zone separating the female and male flowers (as in *Cryptocoryne*). Frequently the spadix ends in a flowerless and naked appendix, which may be strongly coloured and variously differentiated.

As in Cyclanthaceae, the spadix produces heat that is often accompanied by a strong odour which varies from sweet to noxious or pungent. The odorous substances are sometimes produced from other parts, e.g. from sterile flowers, from the appendix of the spathe, or from glands on the spathe. In the appendix of *Sauromatum* it has been found that indole production can be detected more than 10 hours before the first detectable rise in temperature. The function of the heating may therefore be to improve the diffusion of the scent (CHEN and MEEUSE 1972). The production of heat is generally most pronounced during the first day or days of anthesis.

The flowers are protogynous, as is also the spadix where flowers are unisexual. The flowers are variable; in *Spathiphyllum,* for example, they have $3+3$ tepals, $3+3$ stamens, and a tricarpellary, trilocular pistil, but reductions and specializations are common; flowers with $2+2$ tepals, $2+2$ or $3+3$ stamens, and a unilocular ovary occur, for example, in *Lasia;* in *Rhodospatha* they have $2+2$ stamens and a bilocular pistil. In *Cryptocoryne* the flowers are unisexual, the stamens are solitary, and each female flower develops into a monomerous pistil with secondarily parietal-"adaxial" placentation.

The tepals are mutually free, small, inconspicuous, generally of equal size and shape, often thick, sometimes prismatic as a result of spatial conditions.

The pollen grains (THANIKAIMONI 1969) are sulcate or sulculate, foraminate or inaperturate (see p. 277), with a smooth, striate, reticulate or scabrous exine pattern. There is reasonably good correlation between the grouping suggested by pollen morphology and the taxonomic groups proposed by ENGLER (1920a).

The ovary is often unilocular irrespective of whether the carpels are one, two or three in number; rarely it is trilocular. The placentation in trilocular ovaries may be central and axile, but otherwise it varies from basal to lateral, parietal or (rarely) apical. The ovules are numerous or more often few or solitary, the solitary usually being basal.

The fruits are generally berries and the seeds often have a fleshy testa, but in *Pistia* they are drupes, and in some other cases the pericarp dries so that the fruit becomes nut-like. They are generally animal-dispersed. The endosperm may be retained or used up by the developing embryo. Occurrence of endosperm in the mature seed is to some extent used in the infrafamilial classification.

*Chromosome Numbers.* There is a wide variation in chromosome number (e.g. MARCHANT 1973) ($2n=14$ to 140) and in chromosome size. Some genera have only one number, e.g. *Anubias*

**Table 3.** Araceae in major geographical areas (number of genera – number of species). Each genus is included in only one of the geographic categories

| | North Temperate | South America | Old World | Africa and Madagascar | Asia | Miscellaneous |
|---|---|---|---|---|---|---|
| **Pothoideae** | | | | | | |
| Pothoeae | | | 1–50 | | 3–11 | |
| Heteropsideae | | 1–12 | | | | |
| Anthurieae | | 1–700 | | | | |
| Acoreae | 1–2 | | | | | Australia 1–1 |
| **Monsteroideae** | | | | | | |
| Monstereae | | 3–57 | 1–60 | | 3–48 | |
| Spathiphylleae | | 1–45 | | | | New Guinea 1–3 |
| **Calloideae** | | | | | | |
| Symplocarpeae | 3–4 | | | | | |
| Calleae | 1–1 | | | | | |
| **Lasioideae** | | | | | | |
| Lasieae | | 4–36 | 1–19 | | 4–7 | |
| Pythonieae | | | 1–100 | 2–12 | 3–9 | |
| Nephthytideae | | | | 3–15 | | |
| Montrichardieae | | 1–2 | | | | |
| Culcasieae | | | | 1–20 | | |
| Zamioculcaseae | | | | 2–6 | | |
| Callopsideae | | | | 1–1 | | |
| **Philodendroideae** | | | | | | |
| Philodendreae | | 1–350 | | | 8–260 | New Guinea 1–1 |
| Anubiadeae | | | | 1–8 | | |
| Aglaonemateae | | | | | 2–22 | |
| Dieffenbachieae | | 1–30 | | | | |
| Zantedeschieae | | | | 1–6 | | |
| Typhonodoreae | | | | 1–1 | | |
| Peltandreae | 1–4 | | | | | |
| **Colocasioideae** | | | | | | |
| Colocasieae | | 6–67 | 1–2 | | 6–95 | |
| Syngonieae | | 2–34 | | | | |
| Ariopsideae | | | | | 1–1 | |
| Protareae | | | | | | Seychelles 1–1 |
| **Aroideae** | | | | | | |
| Stylochitoneae | | | | 1–21 | | |
| Arophyteae | | | | 3–11 | | |
| Asterostigmateae | | 8–23 | | | | |
| Zomicarpeae | | 5–7 | | | | |
| Areae | | | 8–38 | | 4–76 | Worldwide 1–150 Asia and Australia 1–25 |
| **Pistioideae** | | | | | | Worldwide 1–1 |
| Total: 110–2,456 | 6–11 | 34–ca. 1,360 | 13–ca. 270 | 16–ca. 100 | 34–ca. 530 | 8–ca. 180 |

with $2n = 48$; others may have both diploids and tetraploids, e.g. *Spathiphyllum* with $2n = 30$, 60, or more rarely a combination of euploid and aneuploid numbers, of which the most extreme example is *Cryptocoryne* with $2n = 20$, 22, 28, 30, 33, 34, 36, 42, 54, 66, 68, 72, 85, 88, 102, 112, and 132. The numbers $2n = 24$, 26, 28, 30, 32, 34, and 36 are common, which would suggest a primary base number of $n = 6$, 7 or 8; however, in several cases the numbers may represent an aneuploid series of secondary base numbers, e.g. $n = 12$, 13, 14, 15, 16, 17, and 18. The numbers contribute only moderate help in the classification of the genera in the family.

*Distribution.* The geographical distribution of the Araceae (ENGLER 1920b; CROAT 1981) is worldwide (see Table 3).

*Pollination*. The structural trends in the evolution of the Araceous inflorescence lead to functionally highly specialized pollination systems based on the "deception" of insects that are unspecialized with respect to flowers, such as various Diptera and Coleoptera. Often there is manipulation of a passive kind (slide zones) and imprisonment of insects for a definite term, as in *Arum,* followed by their release. However, it is not so easy to discern the functional significance of the most primitive inflorescences in the family. In some *Anthurium* species the insect visitors are Diptera and Coleoptera, but the type of inflorescence hardly seems to exemplify any of the recognized pollination syndromes, and the secretion of sweet stigmatic fluid which is consumed as "nectar" (MADISON 1979) is suggestive of improvisation. Nevertheless, it seems possible that this inflorescence form is an effective adaptation to mutually beneficial relationships with unspecialized insects, for it is closely paralleled in Piperaceae (SEMPLE 1974). These simple inflorescences have also become adapted to serve advanced insects. Thus, some *Anthurium* species are pollinated by *Trigona* (social bees of the family Apidae) while others, and apparently most *Spathiphyllum* species, exhibit the syndrome of "Euglossine pollination". That is, they receive visits from the males of the solitary bees of the tribe Euglossini (also of the family Apidae) that obtain not food, but aromatic oils which apparently become sex pheromones after they have been gathered (MADISON 1979; VOGEL 1963a). The supposedly primitive floral type found in *Magnolia* and nymphaeaceous plants attracting beetles to a large, fragrant white cup, often provided with food bodies, and operating on the "mess and soil" principle, corresponds to that in *Zantedeschia,* which, however, is not the most primitive of the Aroids in its construction, for it has unisexual flowers in separate zones. The primitively constructed Araceae apparently were pollinated by unspecialized insects, and the form of the inflorescence lent itself to evolution towards exploitation of these insects by deceptive syndromes, but not towards co-evolution with more advanced insects, which has taken place only along chemical lines.

*Classification*. The classification made by ENGLER (1905, 1908, 1911, 1912, 1913, 1915, 1920a–c) is still largely accepted. A revisional study of the interrelations in the family is likely to change this, and progress in that respect is being made in several fields, e.g. in pollen morphology (THANIKAIMONI 1969), chemistry, and anatomy (KEATING). Taxonomical studies are also being made in various generic groups.

Many genera have only few species each; thus about 15% of the species belong to 90% of the genera. Numerous genera are widespread (see Table 3).

ENGLER's classification of Araceae (1905–1920) has recently been updated and revised by BOGNER (1978), who proposed the division of the family into eight subfamilies and about 30 tribes.

## Subfamily Pothoideae

Subfamily Pothoideae (6 tribes, 11 genera) is marked by several characters assumed to be primitive. Laticifers and spicular cells are lacking. The leaves are spirally set or distichous. Lateral veins of the second and third orders are connected by veinlets to form a reticulum; rarely the veins are parallel. The flowers are bisexual.

*Pothos* (50), of the tribe Potheae, is found throughout Asia and has a few species in Australia and on Madagascar. It has distichously arranged leaves on long, climbing stems. The leaves usually have an ovate or lanceolate lamina and an expanded, flat ("leaf-like") petiole. The spadix is ovoid to ellipsoid and is subtended by a greenish, unspecialized spathe. The flowers have 3 + 3 tepals, 3 + 3 stamens, and a trilocular ovary, each locule with one ovule. Pollen morphology varies greatly. The berries are red, with non-endospermous seeds. *P. scandens* and other species are sometimes grown as ornamentals.

*Anthurium* (ca. 700) is the only genus of the tribe Anthurieae and is the largest in the family. It is distributed in the American tropics, ranging from northern Argentina to Mexico, and consists of terrestrial or epiphytic rain forest plants. The leaves vary from narrowly elliptic to ovate or obpyriform, hastate, tripartite, pedate-radiate or lobate. The inflorescence, often on a long peduncle, has a narrowly to broadly ovate, usually patent or reflexed spathe, coloured greenish to white or red, and an approximately cylindrical spadix. The flowers are bisexual and have 2 + 2 tepals. Unlike those of *Pothos* the seeds are endospermous. Commonly cultivated as leaf and flower ornamentals are *A. scherzerianum* and *A. andraeanum* with bright red spathes, *A. crystallinum* with spectacular, white-nerved leaves, and *A. veitchii* with pendent leaves up to 1 m long.

The tribe Acoreae differs from the previous tribes in having linear and ensiform leaves and apical, pendulous, orthotropous ovules. The two genera are perhaps rather isolated in the subfamily.

*Acorus* (2) (Fig. 127 A–C) has a creeping rhizome. The peduncle and the leaves are both laterally compressed and conspicuously alike. Also the spathe is ensiform and forms a continuation of the peduncle. The flowers are trimerous and bisexual. *A. calamus,* in temperate Asia, was formerly used as a medicinal plant and has therefore been introduced in many places. All parts are fragrant.

*Gymnostachys* (1) *anceps,* in Australia, also has ensiform leaves. Its two to six more or less pendulous spadices are situated in a cluster on the 1–2 m high, unifacial peduncle. The flowers are bisexual and dimerous.

## Subfamily Monsteroideae

Subfamily Monsteroideae (two tribes) is characterized by the lack of latex vessels, by the presence of spicular cells, and by the bisexual flowers. Most of the genera are twining herbs. The pollen grains are zonisulculate or inaperturate.

The tribe Monstereae (seven genera) has a deciduous spathe and the flowers are without a perigone. In *Raphidophora* (60), from Asia, the seeds have an endosperm; the leaves are large and entire or divided. *R. decursiva* and *R. aurea* (*Scindapsus aureus, Epipremnum aureum*) are widely used ornamentals.

*Monstera* (22), from South America, has seeds without endosperm; the leaves are entire or divided. *M. deliciosa* is widely cultivated as an ornamental because of its large, divided and perforated leaves and for its edible fruits ("Ceriman", "Cheese Plant").

The genus *Spathiphyllum* (45), with most species in the American tropics, is characterized by having bisexual flowers with a perigone and a persistent spathe. The rhizomes are erect or procumbent. Several species are grown as ornamentals, e.g. *S. commutatum* and *S. wallisii,* which have white spathes.

## Subfamily Calloideae

Subfamily Calloideae, with four genera, consists of herbs growing in wet habitats in the Northern Hemisphere. These genera have laticifers and bisexual flowers. The pollen grains are bisulculate.

Three of the genera have flowers with a perigone and seeds lacking endosperm. *Lysichiton* (2) (Fig. 127 D–G), with a yellow or white spathe, flowers before the large leaves are fully developed. The lower part of the spathe forms a sheath around the base of the inflorescence axis. *Symplocarpus* (1–2) *foetidus,* "Skunk Cabbage", flowers before the leaves appear, but has a shorter, more hooded spathe, which is dark red or green with brown spots. *Orontium* (1) *aquaticum,* "Golden Club", is also conspicuous on account of the swollen white tip of the inflorescence axis and the yellow spadix. It is an amphibious plant resembling

*Lysichiton* in the inflorescence axis ensheathed by the spathe, but differing in the very small spathe limb ending well below the spadix.

The genus *Calla* (1), in the temperate parts of the Northern Hemisphere, is a marsh plant with a creeping rhizome, white spathe, naked flowers, and endospermous seeds.

## Subfamily Lasioideae

Subfamily Lasioideae, with about 23 genera, is characterized by having simple latex vessels, bisexual or unisexual flowers, and generally non-endospermous seeds. It is a rather polymorphic subfamily which may turn out to be an unnatural assemblage of genera. The pollen grains are sulcate or inaperturate.

The genera with bisexual flowers and a long-persistent spathe include *Lasia* (3), *Cyrtosperma* (19), and *Dracontium* (13). *Lasia* is an Indomalesian genus with a rather slender, creeping rhizome, long, armed leaf petioles, sagittate or pinnatifid laminae, and a long, narrow spathe. *Cyrtosperma,* which has its centre in New Guinea, has a tuberous rhizome and leaves up to 4 m long with warty or spiny petioles and hastate, sagittate laminae. The spathe has a broad opening. *Dracontium,* in South and Central America, includes some of the tallest herbs of the family in the Americas, having a petiole up to 3 m long and a repeatedly divided blade.

Another group of genera centres around *Amorphophallus* (100) and includes plants with unisexual flowers lacking a perianth. The spadix bears the female flowers, which have a one- to four-locular pistil, in the lowermost part and above these the male flowers, that have three to four stamens, and it is terminated by a large, sterile appendix. The leaves are solitary and arise from a corm. They vary in size according to species, the largest having a petiole 3.5 m long and a deeply divided blade up to 4 m in diameter. Within *Amorphophallus,* which is a widespread genus in the Old World tropics, may be mentioned *A. paeoniifolius* (*A. campanulatus*), in Asia, with a spathe up to 50 cm long and equally broad, and a large, irregularly shaped spadix appendix. Even larger is *A. titanum,* in Sumatra, also with a campanulate spathe, but more than 1 m in width and with a spadix more than 1 m long; the tuber weighs up to 50–70 kg. Several much smaller species are commonly cultivated, e.g. *A. rivieri* and *A. bulbifer.*

The genus *Montrichardia* (2), in marshy places of tropical America and the West Indies, has unisexual flowers; the female, with a unilocular ovary, and the male, with three to six stamens, cover the terminal part of the spadix. Some species are herbs up to 4 m high with a stout scape and sagittate leaves.

## Subfamily Philodendroideae

Subfamily Philodendroideae, with several tribes and numerous genera, occurs both in the New and the Old World. It is characterized by having straight-walled latex vessels and naked, unisexual flowers. The leaves have convergent venation. The pollen grains are inaperturate.

One group of genera around *Philodendron* has male flowers with free stamens and endospermous seeds and consists of upright or climbing herbs. *Homalomena* (140), with most species in moist forests of tropical Asia, has female flowers with or without one or more staminodes; the lowest male flowers may or may not be sterile. They are procumbent to erect herbs with mostly ovate leaves and a persistent spathe without a constriction at the middle. About ten species occur in South America, including the frequently cultivated *H. wallisii* with white-spotted leaves. – *Schismatoglottis* (100) has most species in tropical Asia but some in America. It has a sterile spadix appendix patterned with rudimentary flowers and sometimes has sterile flowers forming a zone on the spadix between the female and the male flowers. Some of the female flowers also have staminodes. The lowest part of the spathe is more or less fused with the spadix; the distal part is usually shed after anthesis. The species are more or less erect herbs with mostly ovate leaf blades. They occur in moist places in tropical forests. Only a few species are cultivated.

The genus *Philodendron* (350) is the second largest genus of the family. It occurs in tropical South and Central America, both in the rain forests and in drier regions. It is characterized by having female flowers lacking staminodes and male flowers with two to six prismatic stamens. Some species have scale-like sterile flowers in a zone above the female ones. The spathe is persistent. Leaf shape is variable, from ovate to deeply divided. *Philodendron* is divisible into a number of sections based on growth habit, shape of the leaves, and details of inflorescence and flowers. Many species are ornamentals, and some are differentiated into a range of cultivars. Among the more notable are the vine *P. scandens,* with green, ovate leaves, and *P. erubescens,* with bright, shiny, more or less tri-angular leaves. *P. bipinnatifidum* and *P. selloum* have large, deeply divided leaves. In *P. speciosum* the (entire) leaf blade measures up to $1.5 \times 1$ m in size.

*Anubias* (8) forms another tribe, having stamens fused into a synandrium and non-endospermous seeds. The species are procumbent herbs in West Africa, growing in moist forests. The polymorphic *A. barteri* is widely used as an aquarium plant.

*Aglaonema* (21), in Asia, has tightly situated male flowers with solitary stamens, the anthers of which open with two apical pores; the seeds are non-endospermous. The erect or procumbent herbs have ovate leaves which are often variously white- or silver-spotted. Some species, e.g. *A. commutatum* and *A. pictum,* are grown as ornamentals.

*Dieffenbachia* (30), in tropical forests of Central and South America, has male flowers with a synandrium of three to four fused stamens, while the female flowers are provided with staminodes. The spathe is fused to the spadix with its basal dorsal part and has a constriction near the middle; after anthesis it closes tightly around the spadix. The plants are erect to procumbent and often have variously marked leaves. Several species are grown as ornamentals, for example *D. seguine* (incl. *D. maculata*), of which there are many cultivars.

*Zantedeschia* (6), in Southern Africa, forms its own tribe, having male flowers with two to three stamens closely pressed together and female flowers in some species provided with staminodes. The spathe is asymmetrical and infundibular, the leaves are sagittate or lanceolate, and the underground stem is tuberous. *Z. aethiopica,* "White Calla Lily", is common as a weed in South Africa ("Pig Lily").

## Subfamily Colocasioideae

Subfamily Colocasioideae is defined as having anastomosing (rarely straight) latex vessels, unisexual, naked flowers, stamens fused into synandria, and leaves with reticulate venation. The pollen grains are inaperturate.

A group of about 15 genera, centred around *Colocasia,* has mutually free synandria in the male section of the spadix and seeds provided with endosperm. Here belong several conspicuous genera.

*Remusatia* (2), Africa to Formosa, has female flowers, the ovaries of which have parietal placentation. The distal part of the spadix is sterile. *R. vivipara,* which is sometimes cultivated, is known for the bulbils developed in the axils of scale-like leaves covering upright stems arising from a tuber.

The bulbils are covered with hooked cataphylls. – In *Caladium* (7), in South America, the pistil of the female flowers has no style and has central placentation. The genus consists of tuberous herbs with ovate, sagittate or peltate leaf blades that are often variegated with a shiny surface. The spathe is constricted near the middle. *C. bicolor,* with red, white and green spotted leaves, is a widely cultivated ornamental.

*Xanthosoma* (45), of the Neotropics, differs from *Caladium* mainly in the conspicuous stylar discs which cohere between adjacent female flowers. The plants are terrestrial, almost "arborescent" or have subterranean stems or tubers. *X. sagittifolium,* in the West Indies, has an arborescent stem up to 4 m high and is used as a source of starch. Several other species are used as vegetables.

*Alocasia* (70), in Asia, has short, subterranean tubers or thick aerial stems and cordate, often peltate leaf blades. The pistils have basal placentation. *A. macrorrhiza* has a stem up to 5 m high and 20–30 cm in diameter. Both the stems and the leaves, which are rich in starch, are cooked and used for food (poisonous when fresh). Other species, such as *A. cuprea* and *A. sanderiana,* are grown as ornamentals because of their colourful leaves.

*Colocasia* (8), also from Asia, resembles *Alocasia,* but has a pistil with parietal placentation. The stem is subterranean and tuberous or upright and the leaf blades are more or less cordate. *C. esculenta* ("Taro") is widely cultivated for its starch-rich tuber. Many forms and cultivars are found throughout Asia and the Pacific, each adapted to its special habitat and way of use.

*Syngonium* (33), in Central and South America, consists of vines with hastate, ternate or pedate leaves and club-shaped spadices. The male flowers have synandria free from each other, and the seeds lack endosperm. *S. podophyllum* and other species are grown as ornamentals and may have variegated leaves.

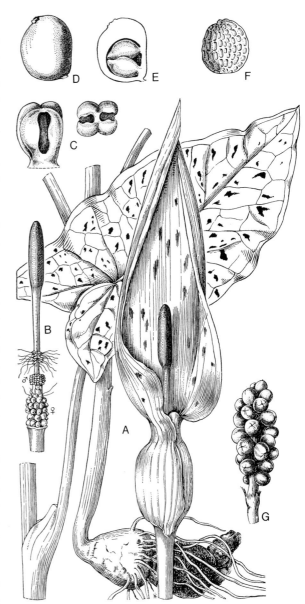

**Fig. 129.** Araceae. *Arum maculatum.* **A** Inflorescence, tuberous rhizome and leaf. **B** Spadix, showing from below female, male and neuter zones and appendix, spathe removed. **C** Male flower, consisting of two fused stamens. **D** Female flower, consisting of a unicarpellary ovary. **E** Same, longitudinal section. **F** Seed. **G** Infructescence. (Ross-Craig 1973)

## Subfamily Aroideae

Subfamily Aroideae, with numerous tribes, is perhaps the most homogeneous of all the large subfamilies. It is characterized by having straight laticiferous elements in stems and leaves, unisexual flowers, and generally endospermous seeds. Most genera have subterranean tubers or corms. The pollen grains are inaperturate.

Only one genus, *Stylochiton* (21), in tropical Africa, has flowers with a perigone (both on male and female flowers). The female flowers are situated in a whorl or in a spirally arranged row at the base of the spadix. The genus consists of herbs with more or less sagittate leaves.

A group of eight genera, all tuberous herbs in South America, is defined by having, as a rule, rudiments of stamens and pistils in the female and male flowers, respectively, and a three- to eight-locular fruit. Here belongs *Taccarum* (4), in which the spadix is free from the spathe, and in which the leaves are tripartite and much divided.

Another group, centred around *Arum,* is characterized by having naked female and male flowers without rudiments of stamens or pistils, respectively, and a unilocular ovary. In *Arum* (15; Fig. 129), in Europe and the Mediterranean, the spadix ends as a yellow or purple club emerging from the spathe, which encloses with its base the floriferous, basal part of the spadix. The female flowers, at the base of the spadix, consist only of a pistil, which has basal placentation; between these and the male flowers is inserted a zone consisting of several whorls of sterile flowers with filiform appendages; a further whorl is located above the male flowers in the constricted part of the spathe. The lower part of the spathe forms a trap for the pollen vectors, which may be flies or beetles, according to species. *A. maculatum,* "Lords and Ladies", occurs in central Europe and *A. italicum* in the Mediterranean; they are pollinated by Psychodidae (Owl Midges). In the related genus *Typhonium* (25), which consists of tuberous herbs in Asia, the leaves are sagittate, the spadix likewise extends out of the spathe, and the spadix appendix is separated from the floriferous part by a constriction. Also in this genus sterile flowers are inserted between the male and female ones, and the placentation is basal. The spathe has a large limb and there is a conspicuously thickened constriction above the level of the flowers.

*Arisaema* (150; Fig. 128), with its centre of diversity in Asia and a few species in North America, differs from the previous genera in lacking sterile flowers on the spadix. The spathe lacks a constriction and its limb is often shorter than the furled part. The distal appendix of the spadix is variable, in some species short and club-shaped, in others long and flagelliform. Paradioecious and/or monoecious spadices occur in various species, in some cases correlated with the size of the plant. The leaves are tripartite to palmately or pedately compound. Examples of Asiatic species are *A. griffithii* and *A. filiforme,* both with a long, whip-like spadix appendix; a well-known North American species is *A. triphyllum* ("Jack in the Pulpit").

In some dioecious species the spathe of the male inflorescence gapes at the base, and insects that have entered at the top emerge at the bottom. The upper part of the spathe often has stripes of white with remarkable refractive properties, making them appear very bright. Pollination is by various small Diptera (see under Pollination).

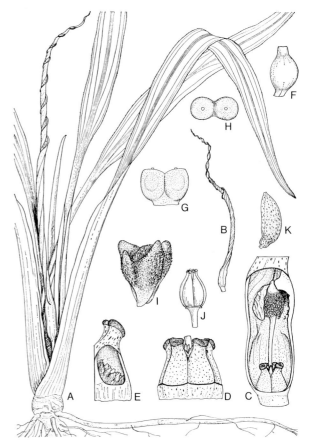

**Fig. 130.** Araceae. *Cryptocoryne retrospiralis.* **A** Main part of the partly submersed plant. **B** Spathe. **C** Chamber of spathe with diaphragm; note the partly fused female flowers in a basal whorl and the congested male flowers partly enveloped by the diaphragm. **D** Female flowers. **E** Two female flowers, one opened. **F** Ovule. **G** Stamen (male flower), lateral view. **H** Same seen from above. **I** "Scent body". **J** Ring of fused fruits. **K** Seed. (DE WIT 1971)

The spadix appendix of *Arisarum proboscideum* mimics the appearance and structure of a Basidiomycete "fruiting body" and has a damp epidermis. Fungus gnats (Mycetophilidae) lay their eggs in the spathe and while doing so apparently cause pollination (VOGEL 1978a).

*Cryptocoryne* (50) (Fig. 130) occurs throughout Asia and Malesia and consists of aquatic to amphibious herbs. The spadix is totally enclosed by the spathe and bears female flowers in a single basal whorl; between these and the male flowers it is stalk-like and naked apart from some scent-producing bodies. The spathe often forms a long tube leading to a basal chamber that contains the spadix. A flap developed from the spathe separates this basal chamber from the tube. The fruit is a

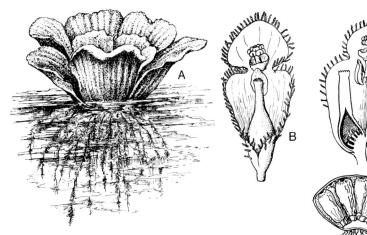
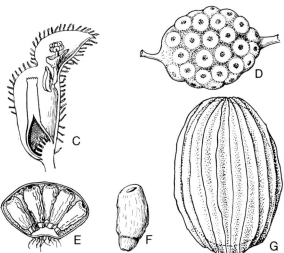

Fig. 131. Araceae. *Pistia stratiotes*. A Free-floating plant. B–C Spathe and inflorescence, front view and longitudinal section. D–E Fruit, note the thin, partly transparent pericarp through which the seeds are visible. F Seed. G Same, outer layer removed. (Cabrera 1968)

syncarpium. The plants are rhizomatous but propagate vegetatively by stolons; they have linear, lanceolate, ovate or cordate leaves which are often wholly submerged. Many species are used as aquarium plants, among them *C. wendtii* and related species, with ovate leaves, *C. cordata,* with cordate leaves, and *C. crispatula,* with linear, bullate to crispate leaves.

## Subfamily Pistioideae

Subfamily Pistioideae is monotypic, consisting of *Pistia stratiotes* ("Water Lettuce"; Fig. 131) which is a free-floating aquatic found in all tropical regions. Laticifers are lacking. The leaves, in a rosette, lack a distinct stalk and are obovate and hairy. Stolons are sent out from the rosettes and secure effective vegetative propagation. The spathe is small, short, and greenish. There is a single, naked female flower consisting of a unilocular pistil, partly attached to the spathe, with numerous basal-parietal ovules; above the female flower there is one naked male flower with four to six stamens united into a synandrium. The pollen grains are sulcate. The fruit is a dryish, juiceless "berry", the seeds of which are non-endospermous. It is often stated that this genus forms a link between the Araceae and the Lemnaceae, but this is unlikely, even though they may share a common ancestor not too far back in the phylogenetic history of the Araceae.

## Lemnaceae S.F. Gray (1821)    4:30
(Figs. 132–133)

Small to minute floating or submerged aquatics, where the entire plant consists of more or less plate-like bodies (fronds). These are solitary or cohere in groups due to vegetative reproduction, being connected by short to sometimes rather long "stipes". The roots are simple or may be lacking; when present, they are solitary or several per frond. The inflorescence consists of one female and one or two male flowers. These are naked or enclosed in a sheath. The male flowers consist of a single stamen. The anther wall formation proceeds according to the Monocotyledon Type (*Spirodela*) or the Reduced Type (*Lemna, Wolffia*). The tapetum is amoeboid. The pollen grains are ulcerate and, at least in *Lemna,* tricellular. The female flower consists of one sessile, unilocular ovary with one to four basal, anatropous (*Spirodela*), hemianatropous (*Lemna*) or orthotropous ovules. A parietal cell is cut off from the archesporial cell. Embryo sac formation follows the *Polygonum* Type (*Spirodela*) or the *Allium* Type (*Lemna, Wolffia*). Endosperm formation is cellular, and the chalazal cell first cut off is not enlarged, as in many Araceae, but divides like the micropylar one. The fruits are rather dry and not fleshy as in most Araceae. They contain one or few seeds. The endosperm fills up part of the ripe seed and at least in *Lemna* is rich in starch grains (S.C. Maheshwari 1956) (Fig. 133).

*Chromosome Number.* The chromosome numbers are $n = 5$, 20, 21, 22, 25, 28–30, 35, and 40 (Raven 1975). The original basic number may be $x = 5$.

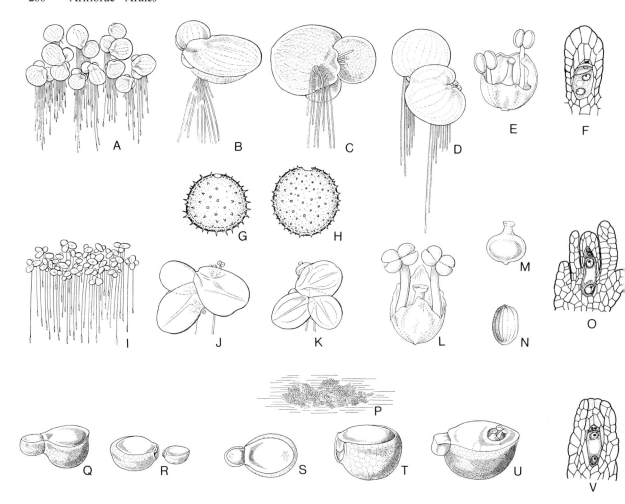

**Fig. 132.** Lemnaceae. **A–F** *Spirodela polyrhiza*. **A** Colony of specimens. **B–C** Budding fronds. **D** Flowering plant. **E** Spathe and male and female flowers. **F** Nucellus with embryo sac in binucleate stage. (**A–F** Ross-Craig 1973; **F** S. Maheshwari and N. Maheshwari 1963). **G** *Lemna gibba,* pollen grain. **H** *L. trisulca,* pollen grain. (**G** and **H** Erdtman 1952). **I–O** *Lemna minor*. **I** Colony of specimens. **J–K** Flowering specimen. **L** Spathe and male and female flowers. **M** Pistil. **N** Seed. **O** Ovule with megaspore tetrad, two megaspore nuclei taking part in the embryo sac formation (*Allium* Type). (**I–N** Ross-Craig 1973; **O** S.C. Maheshwari and Kapil 1963). **P–V** *Wolffia arrhiza*. **P** Colony on the water, note that this is the smallest of all angiosperms. **Q–R** Budding. **S** Budding seen from above. **T** Specimen, showing the budding pocket to the left. **U** Flowering specimen, spathe is here lacking. **V** Nucellus in the same stage as for *Lemna minor* in **O**. (**P–V** Ross-Craig 1973; **V** S.C. Maheshwari and Kapil 1963)

*Ecology and Distribution.* The members of Lemnaceae, "Duckweeds", (see den Hartog and von der Plas 1970) are usually gregarious and then form green, floating mats on the surface of stagnant or slowly running waters, often reaching "pest" proportions. *Wolffia* and *Wolffiella* are the smallest among the angiosperms. The distribution is subcosmopolitan.

*Relationships.* The relationships to the Araceae are not wholly obvious, but the cellular endosperm formation, the amoeboid tapetum, and other details support the general conclusion, drawn mainly from the appearance of the reduced flowers and their position (cf. *Pistia*), that the Lemnaceae are derivatives of the araceous stock. It is true they are in the cladistic sense extremely derived offshoots from araceous ancestors (the closest ones being perhaps not necessarily *Pistia*) and could logically be included within the Araceae (S.C. Maheshwari 1958). Their very distinct appearance and the uncertainty as regards the closest relatives, however, makes it practical to treat the Lemnaceae as a family.

## Subfamily Lemnoideae

Subfamily Lemnoideae is characterized by the presence of raphides and roots. Each frond has two flattened budding pouches, one basal and one lateral on either side of the axis. The inflorescence is developed in one of the budding pouches and consists of one female and two male flowers enclosed by a membranous spathe. The male flowers each consist of a single stamen with a bilocular anther, which dehisces transversely. The female flower is lateral to the male flowers.

*Spirodela* (4), which is subcosmopolitan, has a dorsal and a ventral scale on the 3–15-veined frond and has one to several roots. The fronds are larger than in *Lemna* and always float on the water surface. They often have red pigmentation. *S. polyrhiza* is subcosmopolitan.

*Lemna* (9) lacks dorsal and ventral scales on the fronds, which are one- to three-veined, and there is always one root. The fronds are always smaller than in *Spirodela,* except in the submerged *L. trisulca.* In most species the fronds float on the water surface, singly or in groups. *L. minor* is a common aquatic in the Northern Hemisphere; a related species, *L. gibba,* has thick, swollen fronds.

## Subfamily Wolffioideae

Subfamily Wolffioideae is characterized by lack of raphides and roots, and the fronds have only one, median budding pouch. The inflorescence is developed in a dorsal cavity (not in a budding pouch), it lacks a spathe and has one female and one male flower. The anthers are unilocular and dehisce apically.

*Wolffia* (7), which is subcosmopolitan, has globular or ellipsoidal fronds which are solitary or cohere two together, each rarely exceeding 1 mm. There is only one inflorescence. The budding pouch has a circular outline. Related is *Wolffiopsis*

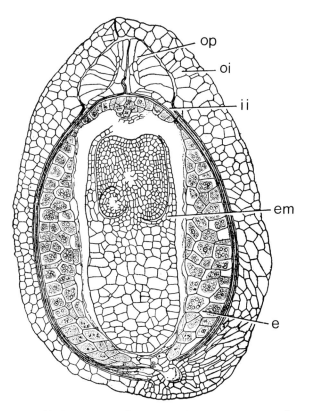

**Fig. 133.** Lemnaceae. *Lemna paucicostata,* mature seed, longitudinal section (*op* operculum; *oi* outer integument; *ii* inner integument; *em* embryo, frond with apical germ pocket; *e* endosperm, containing here copious amounts of starch). (S.C. MAHESHWARI and KAPIL 1963)

(1), in tropical America and Africa, with a submerged, pigmented frond like that of *Wolffia,* but symmetrical. *Wolffiella* (8), in tropical America and Africa, has an asymmetrical, linear, more or less bullate frond with an asymmetrical budding pouch, which opens by a transverse slit. It has one or two inflorescences. Some species are submerged while others are floating.

# Superorder Triuridiflorae

*One Order:* Triuridales.

Small, glabrous, achlorophyllous, usually white, yellowish, purplish or pale violet terrestrial, generally monoecious herbs. Bulbs, corms and stem-tubers are lacking. The underground stem forms a slender system of rhizomes often covered with scales. The roots are usually thin with a cortex of one to three mycelium-containing cell layers. The cauline leaves, which are scattered on the stem, are scale-like and have a single median vascular strand. Intravaginal scales have not been reported in the group, nor have stipules or ligules. The vascular tissue is weakly developed, with the vascular bundles in a single ring; vessels are lacking. Stomata, raphides and silica bodies are lacking (TOMLINSON 1982).

The inflorescence is determinate (sympodial) in *Triuris,* otherwise monopodial and racemose with simple, hyaline bracts. The flowers are small, actinomorphic, usually unisexual (rarely bisexual within *Sciaphila*) and hypogynous. There is generally a marked difference between the male and the female flowers. The perianth consists of three to ten (frequently six) equal perigone members, which seem to form one series. The tepals are more or less connate basally; they are often reflexed and may form a star-like configuration. Sometimes they are conspicuously attenuate apically, and may end in appendages or glands. The stamens are two to six, often three, as in *Triuris,* and seem to be situated in a single series. Their filaments are distinct or indistinct (lacking in *Triuris*); the anthers are extrorse, 2-, 3- or 4-sporangiate, and open by longitudinal or more often transverse slits. In *Andruris* and *Seychellaris* the connective extends into a long appendage. Staminodes in the form of filamentous or glandular structures are often present. The tapetum is probably glandular and microsporogenesis successive. The pollen grains are simple, inaperturate or sulcate (*Sciaphila*), globose, smooth, and three-celled when dispersed.

The carpels are free from each other and vary in number from 6 to ca. 50. The ovaries are globose or obconical-obovate, asymmetric, and have a smooth or often rugose surface. The stylodia are generally more or less lateral and gynobasic, rarely almost terminal. Each pistil contains a single, basal, anatropous, erect ovule, which is bitegmic and tenuinucellate. A parietal cell is not formed. The embryo sac is reported to develop according to the *Polygonum* Type and the egg cell may develop parthenogenetically. The dry fruitlets when mature may rupture longitudinally along the side opposite the stylodium and thus function as achene-like follicles; they are indehiscent in *Soridium*. In the mature seed the seed coat is formed by the outer integument and the endosperm is copious (the latter being a difference from all the Alismatiflorae). It contains protein and fat but not starch.

*Chemistry.* The group is almost totally unknown in regard to chemical contents and parasites.

*Distribution.* Triuridales is concentrated mainly in tropical regions in the Old as well as New World and consists of plants growing on mostly shady ground of rain forests, in forest litter, rotten wood or rarely on termite nests. Most of them are extremely small and slender.

*Relationships.* As a consequence of the apocarpy and high number of carpels the order is generally considered a derivative of primitive monocotyledons. It is no doubt strongly specialized, however. It is sometimes considered closely allied to, and often included in, Alismatiflorae (related to Alismataceae according to CHANT 1978). Superficial similarities to Alismataceae in the free, numerous carpels with gynobasic styles are obvious. Whether the more conspicuous similarities have developed by convergence or not is yet uncertain. The gynobasic styles, the extrorse stamens, the inaperturate and three-celled pollen grain, the vessel-less stem, and the slightly *Scheuchzeria*-like habit are further characters found among Alismatiflorae, but these features may very well have evolved by convergence. TOMLINSON (1982) remarks that there are sufficient differences in ovule organization, endosperm and embryo to exclude a close relationship with Alismatiflorae. He also remarks that as it has not been verified that

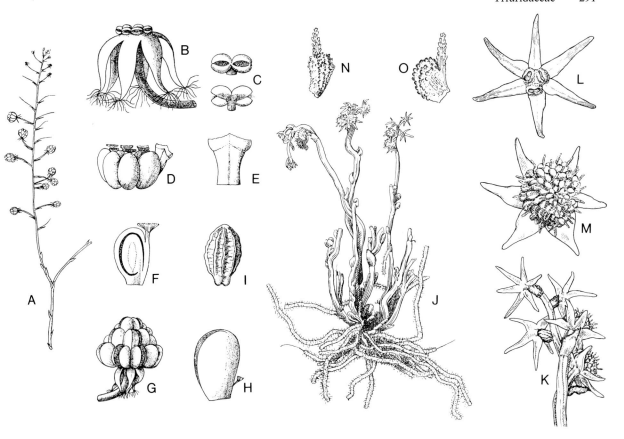

**Fig. 134.** Triuridaceae. **A–I** *Sciaphila ledermannii.* **A** Aerial part of shoot. **B** Male flower. **C** Stamen in different views; note the transverse dehiscence. **D** Carpels and staminode from female flower. **E** Staminode. **F** Carpel, longitudinal section. **G** Female flower, fruit stage, the carpels opening up. **H** Fruit. **I** Seed. (HEPPER 1968). **J–O** *Sciaphila thaidanica.* **J** Plant. **K** Inflorescence. **L** Male flower. **M** Female flower. **N–O** Carpel in different views. (LARSEN 1961a)

the Triuridaceae have a single cotyledon (the embryo is so undifferentiated that this cannot be decided) even their place in the monocotyledons is uncertain and rests mainly on the trimerous flowers.

Painstaking studies on embryology and germination may give a better basis for evaluating the relationships of the Triuridaceae.

## Order Triuridales

*One Family:* Triuridaceae.

**Triuridaceae** G. Gardner (1843)    7:70–80
(Fig. 134)

Description as for the superorder.

The family may be subdivided into the following two groups.

***The Sciaphila Group*** consists of monoecious plants or has bisexual flowers, the perigone segments of which are pointed and often supplied with hairs but hardly caudate; the anthers are uni- to tetrasporangiate. – The largest genus, *Sciaphila* (50–55) is concentrated mostly in the Old World tropics, but has four species in America. One species, *S. purpurea,* may exceed 1 m in height. – *Andruris* (16), which has an extended connective, is mainly Indonesian.

***The Triuris Group*** consists of dioecious plants with three or six caudate perigone segments and tetrasporangiate longitudinally dehiscent anthers. – *Triuris* (1) and *Hexuris* (2), both chiefly in Brazil, are herbs rarely more than 10 cm tall.

# Superorder Alismatiflorae

*Two Orders:* Alismatales and Najadales.

Herbaceous aquatic, semi-aquatic and marsh plants, some growing even in marine waters, which is unique in the angiosperms. The root hairs are developed from epidermal cells shorter than the ordinary epidermal cells (LEAVITT 1904). Starch-rich rhizomes occur in several families but bulbs, corms and tubers are rare. Hibernacula occur in certain groups. The leaves are alternate or opposite. They may be uniformly linear and ribbon-like or differentiated into petiole and lamina. The veins in the former kind of leaf are parallel, those in the latter are arching and converging at the apex. Sparsely reticulate or transverse veins may occur. The leaf base is usually sheathing and invests the stem loosely. Stipules are present in various Zosterales and a few Alismatales (Hydrocharitaceae), and form the so-called "ligular sheaths" in the Potamogetonaceae. Intravaginal scales (or "squamules") are almost diagnostic to the superorder (ARBER 1923; VON STAUDERMANN 1924; TOMLINSON 1982). Laticifer-like schizogenous lacunae occur in Alismataceae and Limnocharitaceae, and laticifer-like elements occur in Aponogetonaceae and *Lilaea* (Juncaginaceae) (TOMLINSON 1982). Silica bodies and raphides are lacking. Stomata are generally lacking in the obligately submerged water plants; when present they are paracytic (STANT 1964) or in certain Hydrocharitaceae (Limnobieae) also anomocytic.

Apart from the intravaginal scales (which are perhaps trichomes), hairs are mostly lacking on the vegetative parts. Vessels are lacking in the stems and leaves but are present in the roots in part of the superorder (CHEADLE 1944; WAGNER 1977).

Paniculate and spicate inflorescences prevail in the superorder; in Alismatales, in particular, the panicles may have tiered branches terminating in cymose components, but sometimes the inflorescence is umbel-like. Spikes occur in several families of Zosterales. The flowers vary from actinomorphic to more or less zygomorphic (as in Aponogetonaceae and Zosteraceae). They may be perfect and have two trimerous perianth whorls, 3 or $n \times 3$ or numerous stamens, and three to numerous pistils which are free or very rarely united. Reductions occur, however, and the flowers may be naked or oligotepalous with one or few stamens and one or two pistils. In a few families, viz. Juncaginaceae, Potamogetonaceae and Zosteraceae, tepals, according to current interpretations, are lacking but functionally substituted by laminar "connective appendages". Alternative interpretations are that there are true tepals in these families, or that the flowers are of synanthial nature. The flowers are hypogynous except in Hydrocharitaceae where the floral receptacle envelops the carpels. In the hypogynous taxa the gynoecium is constantly apocarpous or (in some Juncaginaceae) syncarpous but schizocarpic at maturity. The tepals, when present, tend to be hyaline and white, pinkish or reddish, rarely blue, brown, purple or yellow; in some groups (the Alismataceae and most Hydrocharitaceae) they may be differentiated into green "sepals" and white, pinkish or rarely yellow "petals". The petals are usually not variegated, but very rarely they are minutely spotted-striated or provided with a basal nectar guide spot. Perigonal nectaries are known rarely in *Echinodorus* (Alismataceae).

The stamens vary from one to many, in the latter case representing multiples of trimerous whorls (LEINS and STADLER 1973). The filaments are usually free and linear-filiform but sometimes broad and laminar or very short; rarely they are united in pairs, rings or groups. Androecial nectaries occur in some Hydrocharitaceae and in *Echinodorus*, Alismataceae (DAUMANN 1970). The anthers are basifixed, with apical or sometimes lateral thecae, which usually dehisce extrorsely with longitudinal slits. They are generally tetrasporangiate, but sometimes bisporangiate, and very rarely (*Najas minor,* Najadaceae) unisporangiate. The endothecial thickenings are of the Girdle or the Spiral Type (UNTAWALE and BHASIN 1973). The tapetum is amoeboid with uninucleate tapetal cells (WUNDERLICH 1959). Microsporogenesis is of the Successive Type. The pollen grains are generally free although they may cohere in threads, in tetrads or, in Scheuchzeriaceae, in dyads. They are generally sulcate or bi- to polyforaminate, but sometimes

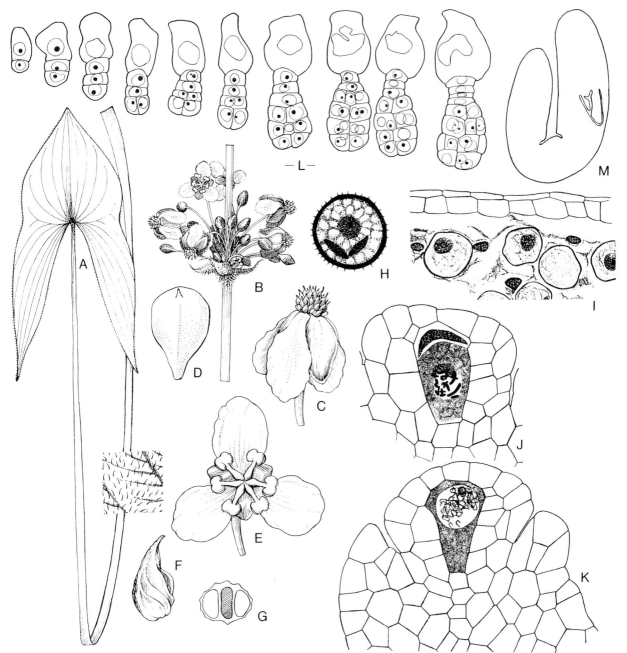

inaperturate (almost exineless) in Alismatales or almost or wholly exineless in Zosterales (ERDTMAN 1952). In many of the marine taxa pollination occurs in the water, and the pollen grains grow out to become filiform or if remaining globose may cohere in moniliform threads. The pollen grains are perhaps most often three-celled when shed (BREWBAKER 1967), but two-celled pollen grains are reported for several genera.

The carpels vary from solitary to numerous, in the latter case probably by secondary multiplication of initials and thus they are *not* spirally set (SATTLER and SINGH 1978). A significant degree

**Fig. 135.** Alismataceae. **A–G** *Limnophyton angolense.* **A** Leaf with detail showing pubescence. **B** Verticil of the paniculate inflorescence. **C** Female flower. **D** Tepal from the inner whorl. **E** Male flower. **F** Achene. **G** Same in transverse section. (CARTER 1960). **H–K** *Limnophytum obtusifolius.* **H** Three-celled pollen grain in the stage of dispersal. **I** Plasmodial tapetum in late stage, invading the anther loculus. **K–J** First meiotic division, the lower cell in **J** forming the embryo sac directly (*Allium* Type). (JOHRI 1935a). **L–M** *Sagittaria guayanensis.* **L** Embryogeny, showing that the basal cell formed at the first division remains undivided and forms a unicellular haustorium (*Caryophyllad* Type embryogeny). **M** Embryo in the ripe seed; this is characteristically curved in Alismataceae. (JOHRI 1935c)

of syncarpy is known in Juncaginaceae only (*Triglochin*). Nectar glands are sometimes formed laterally on the base of the carpels (in Alismataceae, Butomaceae and Aponogetonaceae) and correspond to septal nectaries in the Liliiflorae (DAUMANN 1970). The stigmas are generally of the Dry Type. Placentation varies from typically laminar (Butomaceae and Limnocharitaceae) to submarginal, apical or basal. The ovules vary from numerous (in many Alismatales) to few or solitary. They are orthotropous, anatropous or rarely (*Ruppia*) campylotropous, and generally weakly crassinucellate. A parietal cell is cut off from the archesporial cell, except in Alismataceae, Limnocharitaceae and *Zostera* (Zosteraceae). The embryo sac generally develops according to the *Polygonum* Type but according to the *Allium* Type in Alismataceae and Limnocharitaceae. Endosperm formation is prevailingly helobial, but often nuclear. Embryogeny conforms to the Caryophyllad Type (JOHRI 1935a, b, c; VON GUTTENBERG 1960; YAMASHITA 1972).

When the gynoecium is apocarpous the fruits are usually follicles or achenes; in the epigynous Hydrocharitaceae the fruit is fleshy or dry, splitting up stellately or irregularly or not at all. The endosperm is used up during seed maturation, and the mature seeds are exendospermous or have a very thin layer of endosperm only. Arils and elaiosomes are lacking. The testa is sometimes double and elaborately constructed. The embryo is linear, straight or (in, for example, all Alismataceae and Limnocharitaceae) curved, and often macropodous. It frequently contains chlorophyll already in the seed (NETOLITZKY 1926).

*Chemistry.* Information on chemical conditions within Alismatiflorae is fragmentary. Cyanogenic compounds are known in some genera and seem to be typical of Scheuchzeriaceae and Juncaginaceae. Steroidal saponins are absent or nearly so, and so also apparently are alkaloids, leucoanthocyanins and chelidonic acid, but certain types of tannins may be formed.

Anthocyanins and anthocyanin pseudobases occur within Aponogetonaceae, Butomaceae and Hydrocharitaceae. C-glycoflavones and luteolin/apigenin are also known in the superorder (GORNALL et al. 1979). In the carbohydrate metabolism part of the group resembles certain dicotyledons, starch and sugar being accumulated in rhizomes or other parts. Rust and smut fungi rarely attack members of this superorder, although some are hosts of the smut genus *Doassansia*.

*Distribution.* The members of the Alismatiflorae are subcosmopolitan as might be expected from their aquatic occurrence, and it is difficult from the distributional pattern to draw any conclusion about their earliest differentiation centre.

The Alismatiflorae have sometimes been considered to be derived from or to represent the ancestors of the monocotyledons as a whole. This is not very likely (DAHLGREN and CLIFFORD 1981, 1982) in the light of the number of obviously derived and specialized character conditions in this complex (see p. 43). The hypothesis that the family Alismataceae represents a particularly primitive group is strongly contradicted by its pollen morphology, embryo sac formation, laticifer-like vessels, curved embryo and other specializations.

*Relationships.* Members of Alismatiflorae have often been considered the most important connecting link between dicotyledons and monocotyledons, the Nymphaeiflorae (see p. 43) and even the Ranunculiflorae being then regarded as the groups of dicotyledons which most closely resemble their monocotyledonous ancestors. The herbaceous habit, the scattered vascular strands of the stem, the apocarpous, often multicarpellary gynoecium and the polyporate pollen grains are found in members of both Ranunculaceae and Alismataceae. However, these features are without any doubt the result of convergent evolution; it may be noted, for example, that in spite of the similarity in vasculature the Alismataceae completely lack vessels in the stem, whereas the Ranunculaceae have vessels of an advanced type, with simple perforation plates.

There are even more numerous and far-reaching similarities between various Alismatales and various Nymphaeales, which cannot so easily be dismissed. Some of these are mentioned on p. 43. Examples are the construction of the root cap (*Nuphar*–monocotyledons), sulcate pollen grains (Cabombaceae-Butomaceae), helobial endosperm formation (Cabombaceae–most Alismatiflorae) and laminar placentation (Nymphaeaceae-Butomaceae, etc.). As previously stressed, some of the similarities (such as the sulcate pollen grains) are presumably shared primitive attributes, whereas others probably are adaptations to a similar habitat. Of contrary significance are some striking differences in important characters, e.g. in sieve tube plastids, in the occurrence of endosperm and perisperm, of ellagic acid and in the number of cotyledons.

# Order Alismatales

*Five Families:* Aponogetonaceae, Butomaceae, Hydrocharitaceae, Limnocharitaceae and Alismataceae.

Perennial or to a lesser extent annual herbs, which are marsh plants, freshwater aquatics and (*Halophila* and *Thalassia*) marine aquatics, rooted in the bottom (or *Hydrocleys* often floating). The vegetative parts are strongly aerenchymous. A creeping rhizome is generally present, which may be rich in starch and sugar. The leaves are spirally set, not distichous, and generally concentrated in a basal rosette, rarely verticillate. They vary from linear and ribbon-like to petiolate and provided with a well-developed, lanceolate, oval or sagittate (rarely trilobate) lamina with supervolute ptyxis. The base is sheathing, and stipules are rarely developed (*Hydrocharis*). Intravaginal squamules are always present. Stomata are generally present and of the paracytic type, but are lacking in some Hydrocharitaceae. Laticifer-like schizogenous lacunae occur in Alismataceae and Limnocharitaceae. Vessels are present in the roots only or are lacking altogether; when present they have scalariform and/or simple perforation plates.

The inflorescences are paniculate but highly variable, sometimes being extensively branched and with verticillate branches, sometimes resembling umbels, racemes or spikes, or the flowers may be solitary. The flowers are hypogynous or (in Hydrocharitaceae) epigynous, bisexual or unisexual, actinomorphic (zygomorphic in most Aponogetonaceae) and trimerous. They are insect-pollinated or more rarely wind- or water-pollinated. They are generally subtended by a bract, and the floral clusters often by spathal bracts. The tepals are $3+3$ or more rarely 3, but sometimes (in Aponogetonaceae) fewer or 0; in the first case they are quite often differentiated into a sepaloid outer whorl and a petaloid inner whorl, but this differentiation is sometimes weak (as in Butomaceae). The sepaloid whorl is generally smaller than the petaloid, which is white, pink, yellow or lilac, and which may have red, purple or yellow blotches (nectar guides). Nectaries rarely occur at the base of tepals or filaments. Partial or complete reduction of one tepal whorl or both may occur in Hydrocharitaceae. In *Aponogeton* the perianth members, which when present are white or purplish, are reduced asymmetrically to 3, 2 or 1 or may be totally lacking.

The stamen number is variable, from 3 (*Wiesneria* of Alismataceae), $3+3$ or $3+6$ to very numerous. The filaments are sometimes united in pairs opposite the outer tepals (*Butomus*) or are arranged in groups or rings (Hydrocharitaceae). The anthers are tetrasporangiate, extrorse or rarely (*Limnophytum, Butomus*) latrorse. Their endothecial wall thickenings are of the Girdle Type (Limnocharitaceae, Alismataceae) or Spiral Type (the remaining families). The pollen grains are sulcate (Aponogetonaceae and Butomaceae) but are provided with two to numerous foramina (Limnocharitaceae, Alismataceae) or they may have a very thin exine lacking distinct apertures (all Hydrocharitaceae). In the marine taxa (*Thalassia* and *Halophila*), they are dispersed in the water as moniliform chains. Pollen grains (as in *Elodea*) may also float to the water surface (surface pollination).

The gynoecium consists of numerous carpels, rarely of $3+3$ (Butomaceae, Aponogetonaceae) or 2 or 3 only (some Alismataceae and Hydrocharitaceae). The carpels are free from each other in the hypogynous flowers; in the epigynous flowers of Hydrocharitaceae they are weakly united inside the receptacle to form a compound unilocular ovary, often with more or less intruded partitions (CRONQUIST 1981).

The stylodium of each carpel is inserted apically, laterally or (in many Alismataceae) basally on the carpel; in Hydrocharitaceae the stylodia are frequently two-lobed. The stigmatic surface is Dry. Each carpel contains one to many anatropous or rarely campylotropous or orthotropous ovules. The placentation is basal in Aponogetonaceae and Alismataceae and laminar-dispersed in Butomaceae, Limnocharitaceae and Hydrocharitaceae. The ovules are crassinucellate. A parietal cell is cut off from the archesporial cell in the ovules of all families except Alismataceae and Limnocharitaceae, in which, however, the nucellar epidermis divides periclinally, to form a nucellar cap (JOHRI 1935a, b, c). There is also the difference between the family groups that, whereas in Aponogetonaceae, Butomaceae and Hydrocharitaceae the embryo sac formation is of the *Polygonum* Type, it is of the *Allium* Type in Alismataceae and Limnocharitaceae. Endosperm formation is helobial or more rarely nuclear.

The fruits in the taxa with hypogynous apocarpous flowers are follicles or (Alismataceae) achenes while in Hydrocharitaceae the fruits are fleshy and berry-like but generally dehiscent, opening regularly by longitudinal slits, or irregularly. The seeds have a thin testa and consist primarily of the embryo, which is straight in Aponogetonaceae, Butomaceae and Hydrocharitaceae but curved in Limnocharitaceae and Alismataceae. The marine taxa of Hydrocharitaceae have a macropodous embryo. The basic chromosome number varies much: $x = 5, 6, 7, 8, 9, 10, 11, 12$ or $13$.

*Chemistry*. The chemical constituents are rather incompletely known. Whereas anthocyan pseudobases are known in Aponogetonaceae, Butomaceae and Hydrocharitaceae, they seem to be lacking in Alismataceae and Limnocharitaceae. Phenolic glycosides with caffeic acid, chlorogenic acid and quercetin occur in the order. Cyanogenic compounds are known at least in *Vallisneria* and *Alisma*. The rhizome is sometimes conspicuously rich in starch.

*Distribution*. The geographic distribution of the order is worldwide. Hydrocharitaceae, Limnocharitaceae and Alismataceae are concentrated in tropical and subtropical regions of both the Old and the New World, whereas Aponogetonaceae, which is also primarily tropical, is absent from the Americas. The monotypic Butomaceae is found in temperate Eurasia. All families are most frequent in marshy habitats or in shallow freshwater, in swamps, on riversides and lakesides and in rice fields; the marine genera, *Thalassia* and *Halophila*, occur along pantropical to temperate coasts in both the New and the Old World.

*Relationships*. It is obvious from the above that Alismataceae and Limnocharitaceae deviate from the other families in a number of attributes, all of which are supposedly derived: they have laticifer-like schizogenous lacunae with "latex" and biforaminate to multiforaminate pollen grains, a parietal cell is not cut off but the nucellar epidermis divides to form a nucellar cap, embryo sac formation is of the *Allium* Type, the embryo is strongly curved (horseshoe-like), and they lack anthocyan pseudobases. It is equally obvious that the two families are closely related, in the evolutionary sense comprising sister-groups. Whether they are best treated as subfamilies of one family or as separate families is a matter of opinion; in their multiovulate carpels with laminar-dispersed placentation and the follicular fruit the Limnocharitaceae superficially resemble Butomaceae much more than the Alismataceae, but to what extent the shared attributes are inherited from a common ancestor or have arisen in them through convergence cannot be settled.

The Hydrocharitaceae deviate considerably in the inferior ovary which causes the fruit characters to be strongly dissimilar. However, this all depends on a single circumstance and has not been accorded great weight by us.

There seems to be a great concentration of ancestral attributes in Aponogetonaceae, a family to some extent bridging the gap between Alismatales and Zosterales, and by other authors often treated in the latter order. Among the ancestral attributes, in our view, are the well-differentiated leaves, sometimes with reticulate venation, the undifferentiated (though generally numerically reduced) tepals, the generally $3 + 3$ stamens and the $3$–$6$ carpels (i.e. with less profound secondary multiplication of initials than in most other Alismatales), the sulcate pollen grains, the marginal (-basal) placentation (laminar-dispersed placentation is probably derived) and the follicular fruit. In the spicate inflorescences the Aponogetonaceae approach the Juncaginaceae of Zosterales (next to which they are generally placed).

We wish to refer here to the multiple similarity between properties of alismatalean and nymphaealean families, which are more fully discussed on pp. 42 and 53. There have been attempts to consider the Alismatales as the most primitive extant group of monocotyledons and even to consider them dicotyledonous (EL GAZZAR and HAMZA 1975), and alternatively to consider the Nymphaeales as monocotyledons with a cleft cotyledon (HAINES and LYE 1975). However, there is good evidence (see p. 96) that the Alismatales and other Alismatiflorae are closely allied to the Ariflorae. The cuneate protein bodies in the sieve tube plastids, the single cotyledon, the lack of perisperm, the lack of ellagic acid, and the parallel stomata are good reasons for keeping the Alismatiflorae far apart from the Nymphaeiflorae.

*Key to the Families*

1. Flowers epigynous . . . . . . **Hydrocharitaceae**
1. Flowers hypogynous . . . . . . . . . . . 2
2. Embryo straight; pollen grains sulcate . . . . . . . . . . . 3
2. Embryo horseshoe-curved; pollen grains with 2 to numerous apertures . . . . . . . . . . . . . 4
3. Erect plant with upright linear leaves . . . . . . . . . . . . . **Butomaceae**
3. Floating or submerged aquatics, at least some leaves with broadened lamina . . . . . . . . **Aponogetonaceae**
4. Each carpel with numerous ovules on a laminar-dispersed placenta; fruits follicular . . . . **Limnocharitaceae**
4. Each carpel with a single (rarely a few) ovule(s) at its base; fruits generally indehiscent: achenes . . . **Alismataceae**

Fig. 136. Aponogetonaceae. **A–B** *Aponogeton subconjugatus.* **A** Specimen, the leaves in somewhat folded, unnatural condition. **B** Flower. (HEPPER 1968). **C–I** *Aponogeton ranunculiflorus,* a species where the spike forms a pseudanthium. **C** Plant. **D** Inflorescence. **E** The inflorescence, with some tepals removed. **F** Flower, lateral view. **G** Flower, from above. **H** Gynoecium, one carpel opened to show basal-axile placentation. **I** Cross-section of leaf. (GUILLARMOD and MARAIS 1972). **J–K** *Aponogeton distachyus.* **J** Inflorescence. **K** Flower. (ECKARDT 1964). **L** *Aponogeton dinteri,* pollen grain. (ERDTMAN 1952). **M** *A. angustifolius,* floral diagram. **N** *A. fenestralis,* floral diagram. (**M–N** from ECKARDT 1964)

# Aponogetonaceae J.G. Agardh (1858)   1:47
(Fig. 136)

Perennial, freshwater aquatics with starch-rich, sympodial rhizomes or corms. The leaves are situated basally on the shoot. They are loosely sheathing at the base and are long-petiolate with a linear to oblong-elliptic submerged or floating lamina, which is sometimes fenestrate. The main veins are few, parallel or arching and converging apically; they are connected by transverse secondary veinlets. Articulated laticifers with contents of tannins (proanthocyanins) or oil are present (cf. Alismataceae and Limnocharitaceae, where the "laticifers" are of different construction). Vessels, with scalariform perforation plates, may occur in the roots. Stomata are present on the lamina of floating leaves and are paracytic. The leaf petioles and inflorescence axes are provided with a system of air lacunae (TOMLINSON 1982).

The inflorescence is a simple, bifurcate or sometimes up to ten times forked spike borne on a peduncle and enclosed by a caducous (rarely persistent) spathe-like bract in the early stage (cf. the Arales). Rarely, the inflorescence is condensed, few-flowered and organized as a flower (Fig. 136D) (GUILLARMOD and MARAIS 1972). Flowering takes place above water level and pollination is probably chiefly by insects, nectar secretion taking place from the sides of the carpels.

The flowers are generally bisexual, rarely unisexual, hypogynous and usually zygomorphic. They generally have one or two, rarely up to six, tepals, but tepals may be lacking altogether. The tepals are generally white, but may be lilac or pale purple, and are free from each other; occasionally, as in *A. distachyus* they are relatively large (Fig. 136K). The stamens are $3+3$ or rarely more, in three or four circles. The anthers are tetrasporangiate and extrorse to latrorse and longitudinally dehiscent. The pollen grains are separate, sulcate, foveolate and two- or three-celled at dispersal.

Each flower has $(2-)3-6(-9)$ carpels which are nearly free or basally coherent to a variable extent. Each ovary merges into a short style. There are one to eight ovules attached marginally on the base of the carpel. They are anatropous and crassinucellate; a parietal cell is cut off from the primary archesporial cell, but there are no divisions of the epidermis. Embryo sac formation is of the *Polygonum* Type and endosperm formation helobial.

The gynoecium develops into a multifollicle, where the individual follicles are nearly free from each other. Ripening often takes place under water. The embryo is straight. $x = 8$.

*Chemistry,* see under the order.

The single genus, *Aponogeton* (47) is distributed over tropical to temperate regions of the Old World (VAN BRUGGEN 1968, 1969, 1970, 1973). *A. distachyus,* in Southern Africa, has fleshy two-branched white spikes, the flowers each with a single large tepal. The inflorescences when young can be eaten as salad or pickled and the plant is also grown as an ornamental. The starch-rich tubers of other species are also eaten by man or his livestock (mostly pigs). Several species of *Aponogeton* are grown as aquarium plants, e.g. *A. crispus* from Ceylon and *A. madagascariensis* (= *A. fenestralis*), from Madagascar, which has net-like perforated leaf blade. – In *A. ranunculoides* (Fig. 136) from Africa, several flowers are contracted into a pseudanthium.

**Butomaceae** L.C. Richard (1815)   1:1
(Fig. 137)

An erect, rather large, perennial herb with a horizontal rhizome. The rhizome is monopodial, thick, dorsiventral, aerenchymatous, rich in starch and provided with tanniniferous cells. The roots are the only parts provided with vessels, but these have simple perforation plates. The leaves are linear, erect, nearly distichous, glabrous, triangular in transverse section, more or less spirally twisted, sheathing at the base, and basally provided with intravaginal squamules. They are inserted at the top of the rhizome and thus all radical. The stomata are paracytic. Laticifers are lacking (see STANT 1967). The aerial stem represents an erect, leafless, glabrous scape terminating in a pseudo-umbel of few to rather numerous, long-pedicellate flowers. These are bisexual, actinomorphic, and hypogynous. There are $3+3$ tepals of more or less similar size and shape, white to purplish brown, the outer often somewhat green (a tendency towards sepaloid appearance). The stamens are $6+3$ in number, with erect filaments and longitudinally dehiscent, latrorse anthers. The pollen grains are sulcate and reticulate (ARGUE 1971) and three-celled when dispersed.

The carpels are six (see SINGH and SATTLER 1974), all in one cycle and free from each other unless at the very base. The styles are short and have a ventrally slightly decurrent stigmatic region. Nectar glands are present on the basal lateral sides of the carpel, corresponding to the septal nectaries (DAUMANN 1970). The placentation is laminar-dispersed, the rather numerous ovules being scattered on the lateral sides of the carpels. The ovules are anatropous, crassinucellate, with a parietal cell cut off from the archesporial cell (ROPER 1952; cf. Alismataceae and Limnocharitaceae, where this is not the case). Embryo sac formation follows the *Polygonum* Type and the endosperm formation is helobial (Fig. 137M). The carpels develop into follicles, the seeds of which have a straight embryo. $x = 13$.

*Chemistry.* Proanthocyanins occur in the tannin cells. Cyanogenesis is not known in the family.

The Butomaceae lack the schizogenous ducts of Alismataceae and Limnocharitaceae and the laticifers of Aponogetonaceae and they have a straight embryo, crassinucellate ovules and *Polygonum* Type embryo sac; in all these features they appear to represent a rather primitive level. As in many other Alismatales there are divisions of the carpel (and stamen?) initials, a condition which is considered to be a secondary specialization. The character of the numerous laminar-dispersed ovules, which they share with Limnocharitaceae, is probably also a specialization.

The family is probably most closely allied to Hydrocharitaceae (though lacking epigyny) and Limnocharitaceae-Alismataceae (but lacking several of their specializations). The connections with Aponogetonaceae are uncertain.

## Limnocharitaceae Takhtajan ex Cronquist (1981) 3:11    (Fig. 138)

Perennial, decumbent or erect, either rooted or free-floating herbs, somewhat fleshy and rich in aerenchymatous tissue. The roots are concentrated on a long or short rhizome and provided with vessels having scalariform perforation plates (vessels otherwise lacking). The stems (apart from the peduncle of the inflorescence) are creeping or (*Butomopsis*) short; when creeping (stolons) they are sympodial, with nodes at which aggregations of leaves are borne, in the axil of one of which the sympodium continues (see further details in TOMLINSON 1982). The stems and leaves are glabrous; stomata may occur on the non-submerged parts of stems and leaves, and are paracytic; secretory ducts ("laticifers") are abundant in leaves and stems as elongated cavities surrounded by a sheath (TOMLINSON 1982). The leaves are (spiro-)distichous. They are differentiated into a lanceolate to elliptic or oval lamina, a petiole and a leaf base. The lamina has arching converging main veins separate from the base upwards, with minor transverse veinlets between them.

The inflorescence is a panicle-derived, cymose pseudo-umbel (as in Butomaceae), each with rather few (to one) bracteate, long-pedicellate flowers; in *Hydrocleys* the flowers and foliage leaves are inserted at the same nodes (see Fig. 138 G).

The flowers are bisexual and hypogynous. There are 3 + 3 perianth members, the outer green and sepaloid and valvate, the inner generally larger, petaloid and white to rose or bright yellow. There are three or six to numerous stamens, which usual-

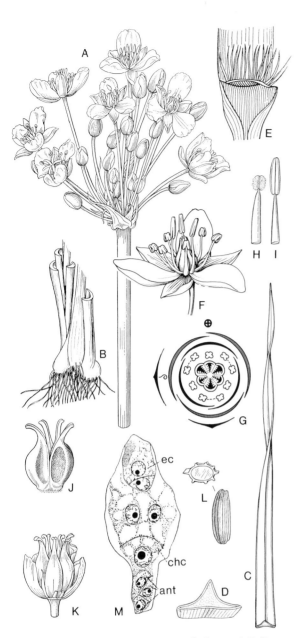

**Fig. 137.** Butomaceae. *Butomus umbellatus.* **A** Inflorescence. **B** Rhizome and base of aerial shoot. **C** Apical part of (twisted) leaf. **D** Leaf, transverse section. **E** Leaf base showing row of intravaginal squamules. **F** Flower. **G** Floral diagram. **H–I** Stamens of outer and inner whorl, respectively. **J** Carpels. **K** Carpels in fruit stage. **L** Seed, *above* in transverse section. **M** Helobial endosperm formation, early stage. (**A, D, H, I, J, L** ROSS-CRAIG 1973; **B, E, K** EMBERGER 1960; **C, F, G** LARSEN 1973a; **M** HOLMGREN 1913)

*Butomus* (1) *umbellatus,* "Flowering Rush", is widely distributed over parts of Eurasia, growing in riverside and lakeside habitats. The starch-rich rhizomes are eaten locally in Russia.

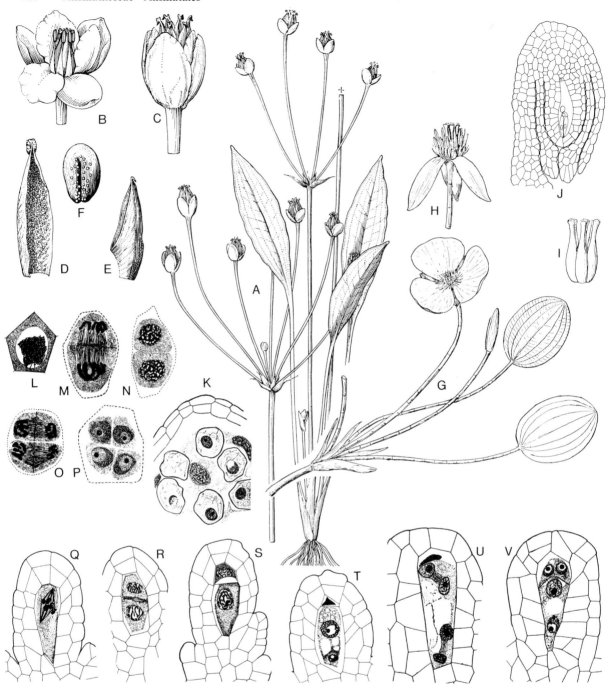

**Fig. 138.** Limnocharitaceae. **A–F** *Butomopsis lanceolata.*
**A** Plant with inflorescence. **B** Flower at anthesis.
**C** Flower in fruiting stage. **D** Carpel, opened to show
laminar-dispersed placentation, flowering stage. **E** Follicle. **F** Seed. (CARTER 1960). **G–I** *Hydrocleys nymphoides.*
**G** Node with inflorescence and foliose leaves. **H** Flower,
inner, petaloid tepals removed. **I** Gynoecium. (CABRERA
1968). **J–V** *Hydrocleys nymphoides,* embryological details. **J** Ovule with mature embryo sac. **K** Anther with
plasmodial tapetum. **L–P** Microsporogenesis, the successive type. **Q–V** Embryo sac formation; note that this
is of the *Allium* Type, two nuclei of the tetrad (**T**) taking
part in its formation. (JOHRI 1938)

ly develop centrifugally. The anthers are longitudinally dehiscent and latrorse (-extrorse). The pollen grains are tri- to polyforaminate or inaperturate, and three-celled when dispersed.

The carpels are three to nine or more (up to 20
in *Limnocharis*). They are situated in a single cycle,
and are almost free from each other. The styles
are short, the stigma is slightly decurrent, and the
sides of the ovary are basally nectariferous, as in
Butomaceae, and as in that family there are numerous ovules situated on a laminar-dispersed pla-

centa. The ovules are anatropous or campylotropous. Unlike Butomaceae, but as in Alismataceae, a parietal cell is not cut off from the primary archesporial cell, and a parietal tissue is thus not formed, but the nucellar epidermis divides periclinally to form a nucellar cap (divided epidermis seen in Fig. 138, Q–S). Embryo sac formation follows the *Allium* Type (JOHRI 1938). The carpels each develop into a follicle opening towards the centre of the flower. The seeds have a horseshoe-shaped embryo. $x = 7$, 8 or 10.

*Chemistry.* There are no tannins, anthocyans or anthocyan pseudobases in Limnocharitaceae.

*Distribution.* The family has a pantropical and pansubtropical distribution, growing in wet places, marshes, rice fields, etc.

*Butomopsis* (4) and *Limnocharis* (2) are rooted and emergent, *Hydrocleys* (5) at least frequently free-floating. *Limnocharis* and *Hydrocleys* are New World genera, whereas *Butomopsis* occurs in the Old World.

*Limnocharis flava,* indigenous in tropical America, now a widespread marsh plant, has become a common rice-field weed. It has numerous broad carpels. *Hydrocleys* consists of aquatics with elliptical-orbicular floating leaves, and is indigenous in South America. *Hydrocleys nymphoides,* the "Water Poppy", with decorative yellow flowers, is an aquarium plant.

There are numerous differences between Butomaceae and Limnocharitaceae in anatomical details, such as secretory ducts (TOMLINSON 1982, table on p. 96), pollen morphology, embryology and seed shape, whereas in all these features the Limnocharitaceae approach very closely the Alismataceae. It is with great hesitation that we distinguish the latter two as separate.

## Alismataceae Ventenat (1799)    12:90
(Figs. 135 and 139)

Rhizomatous perennial (rarely annual) herbs, marsh plants or aquatics, the latter generally rooted in the bottom. The leaves are spirally set in a basal rosette, generally differentiated into a lanceolate, elliptic, sagittate or hastate blade, a petiole of variable length, and a sheathing, open base, but subaquatic leaves generally linear and ribbon-like without this differentiation. Venation principally as in Limnocharitaceae. The stomata are paracytic or rarely tetracytic; secretory ducts are as in Limnocharitaceae. Vessels are present in

the roots only and have simple or scalariform perforation plates (see TOMLINSON 1982).

The inflorescence terminates a leafless scape. The inflorescences are paniculate, often complex and with verticillate branches, but often with one verticil only and then pseudo-umbellate; rarely with few flowers or only one. The branches are enveloped by bracteal leaves.

The flowers are bracteate, long-pedicellate, bisexual or (e.g. in *Sagittaria*) unisexual and are hypogynous. The tepals are $3 + 3$, the outer green and sepaloid, the inner petaloid and generally white or sometimes pale rose, deciduous, wider than the outer. The tepal or stamen bases may be nectariferous in *Echinodorus*. There are six (rarely three, *Wiesneria*) stamens in one whorl or many in successive whorls, originating centripetally. The anthers are extrorse and longitudinally dehiscent. The endothecial walls have the Girdle Type cell wall thickenings. The pollen grains are bi- to polyforaminate, with 9–29 apertures (ARGUE 1976) or rarely, in *Caldesia,* with two or three apertures, dispersed in the three-celled stage.

The carpels are from three in number to very numerous, either in one whorl or in successive cycles, covering an elevated floral axis; they are free or in *Damasonium* centrally united. They develop from three original initials, upon which numerous secondary initials are generally superimposed (SINGH and SATTLER 1972, 1974, 1977). The carpels have an apical, lateral or gynobasic style, and are often slightly curved (KAUL 1976). The stigma is apical or slightly decurrent, probably always with a Dry surface. Nectar secretion generally occurs from the basal parts of the sides of the carpel. There is usually only one basal ovule in each carpel locule, rarely a few. The ovules are anatropous to amphitropous, weakly crassinucellate, without a parietal cell cut off from the primary archesporial cell, but with the nucellar epidermis dividing periclinally. Sometimes there is a weakly developed integumentary tapetum. Embryo sac formation is of the *Allium* Type, and endosperm formation is helobial or, perhaps, sometimes nuclear.

The carpels develop into nutlets or rarely, as in *Damasonium,* into follicles. The seeds have a strongly curved embryo. $x = (5-)7-11(-13)$.

*Chemistry.* The chemistry is little known. Anthocyanins and leucoanthocyanins are lacking, and *Alisma* is reported to be cyanogenetic.

*Distribution.* The distribution is concentrated mostly in the Northern Hemisphere, but is subcosmopolitan, with most taxa present in the tropics

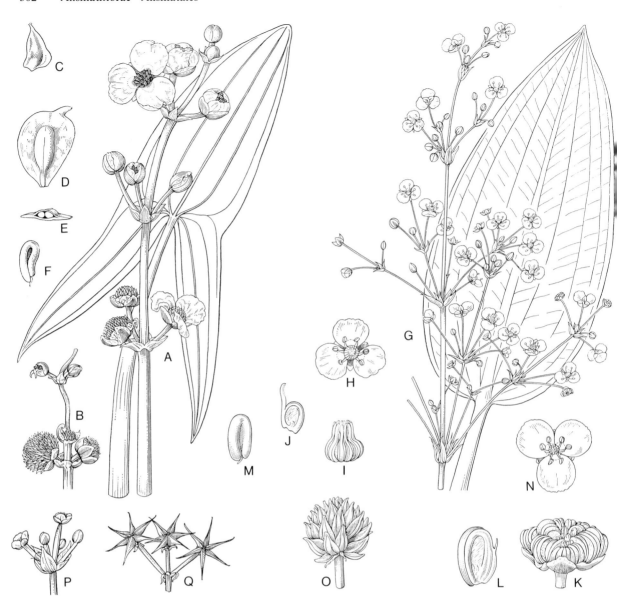

**Fig. 139.** Alismataceae. **A–F** *Sagittaria sagittifolia*. **A** Inflorescence and leaf. **B** Some flowers in the fruiting stage. **C** Carpel at anthesis. **D–E** Achene, E in transverse section, the two surfaces representing two sections of the horseshoe-shaped embryo. **F** Seed. **G–M** *Alisma plantago-aquatica*. **G** Inflorescence and leaf. **H** Flower. **I** Gynoecium of same. **J** Individual carpel. **K** Fruit collection. **L** Individual achene. **M** Seed. **N–O** *Baldellia ranunculoides*. **N** Flower. **O** Fruit collection. **P–Q** *Damasonium alisma*. **P** Inflorescence. **Q** Three flowers in fruiting stage, the carpels basally united and ventrally dehiscent. (ROSS-CRAIG 1973)

and subtropics. Most species grow in marshes or submerged in water (rooted in the bottom).

*Alisma* (9) (BJÖRKQVIST 1968) is indigenous in the Northern Hemisphere. The leaves generally have a lanceolate to cordate lamina, the inflorescences are paniculate with two or more verticils, and the carpels are situated in one whorl and have a gynobasic style. *A. plantago-aquatica* is a common lakeside and riverside species in the temperate regions of Eurasia and Africa. – *Echinodorus* (ca. 45) has most of its species in the Neotropics. In this genus the carpels are numerous and elevated on the floral receptacle (as in *Anemone*). – *Sagittaria* (20) is likewise predominantly a New World genus, although for example *S. sagittifolia* ranges to northern Europe. The plants are monoecious

and have sagittate to lanceolate leaves. – Other genera are the European *Luronium* (1), with *L. natans,* an aquatic with floating leaf blades, and *Damasonium* (6) in the Mediterranean, Western

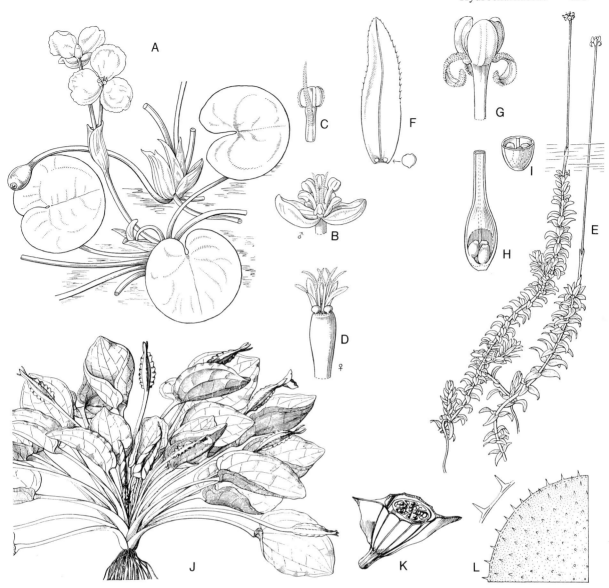

North America, and Australia. *D. stellatum* has flowers with a whorl of indehiscent or irregularly dehiscent, centrally fused carpels, arranged in a stellate configuration. Each carpel contains one to many seeds, if many, placentation is laminal.

**Hydrocharitaceae** A.L. Jussieu (1789)    16:100 (Figs. 140–142)

Perennial or annual herbs often with an elongate rhizome or stolon, frequently submerged and rarely free-floating. The leaves are sometimes rosulate; they are alternate, sometimes distichous, or opposite or verticillate. They may be differentiated into a lanceolate, elliptic or cordate-ovate lamina, a pe-

**Fig. 140.** Hydrocharitaceae. **A–D** *Hydrocharis morsus-ranae.* **A** Flowering shoot. **B** Male flower. **C** Stamen of the inner whorl with appendage. **D** Female flower with perianth removed, showing styles and staminodes (with appendages). **E–I** *Elodea canadensis.* **E** Flowering branches, male (to the *left*) and female. **F** Leaf with orbicular intravaginal squamules. **G** Female flower with recurved stylar branches. **H** Ovary, longitudinal section. **I** Same, transverse section (**A–I** ROSS-CRAIG 1973). **J–L** *Ottelia alismoides.* **J** Female plant. **K** Fruit, transverse section. **L** Part of spinulose inaperturate pollen grain. (**J** DEN HARTOG 1957; **K** HEPPER 1968; **L** ERDTMAN 1952)

tiole and a sheathing leaf base. The main veins in such leaves are distant, arching from the base and connected by fine transverse veins. Sometimes the leaves have only one distinct vein (the midvein), sometimes the veins are three or five; in *Halophila* the venation is clearly pinnate. Quite often, however, the leaves (when submerged) are uniformly linear or elongate, ribbon-like, often with obtuse apex, and have parallel veins. In *Thalassia* some of the leaves are reduced to their hyaline sheaths. Stipules are rarely present, as in *Hydrocharis,* where they are conspicuous. Unicellular prickle hairs are often present on the leaf margins, and when stomata are present, which is common on the aerial parts, these are paracytic. Vessels are restricted to the roots or may be totally lacking; if present they have scalariform perforation plates (ANCIBOR 1979). Laticifers or schizogenous secretory ducts are lacking.

**Fig. 141.** Hydrocharitaceae. **A–F** *Enhalus acoroides.* **A** Plant. **B** Male inflorescence with spathe, longitudinal section, showing numerous flower buds on a conical axis. **C** Male flower bud. **D** Male flower, opened. **E** Female flower with recurved outer tepals and long erect inner tepals; styles visible in the centre. **F** Dehisced fruit, longitudinal section showing three seeds. **G** *Vallisneria gigantea,* specimen with female flowers. (A–G DEN HARTOG 1957) **H** *Sagittaria,* embryogeny, *Caryophyllad* Type. (After SOUÉGES 1931)

The flowers are solitary, paired, or situated in panicle-derived inflorescences, which are generally umbel-like (*Hydrocharis*) (see KAUL 1970). Often the flowers are long-pedicellate and emerge from axils of leaves in a rosette or from a rhizome, like the female flower of *Vallisneria* or the flowers of *Ottelia;* in other cases the female flowers are more or less concealed in the leaf axils as in the water-pollinated *Thalassia* and *Halophila*. In the bud stage they are enclosed by one or more separate or fused spathal bracts. The flowers are generally unisexual, rarely bisexual (species of *Blyxa, Elodea,* and *Ottelia*), actinomorphic and (the female or bisexual) epigynous. They are insect-pollinated, wind-pollinated, or water-pollinated, in the latter case either water-surface-pollinated (*Vallisneria, Enhalus*) or pollinated under water (*Thalassia, Halophila*) (see further on pp. 306–307). Strong dimorphism in size and appearance of the flowers occurs in several genera with unisexual flowers (*Vallisneria, Enhalus,* etc.), and the male flowers are sometimes detached from the plant at the onset of anthesis to become free-floating during pollination, or reach the water surface by prolongation of the peduncles.

The perianth consists of 3 + 3 or 3, rarely 2, tepals; when 3 + 3 the outer may be sepaloid and the inner petaloid, and are white, rose-coloured, purple, blue, yellow or yellow-clawed, or they may all be more or less similar and semipetaloid. The stamens are generally numerous, developing centripetally, but rarely three or two in number. The anthers are generally extrorse. The innermost or outermost staminal homologues are generally staminodial and often secrete nectar at their bases. The pollen grains are inaperturate, more or less globose and dispersed in the three-celled or two-celled stage. In the marine genera *Thalassia* and *Halophila* they cohere in moniliform bands.

The female flowers are epigynous; the carpels of the ovary are (2–)3–6(–20) in number and incompletely separated, opening centrally into a common chamber (Fig. 140K). The styles are of the same number as the carpels although often bifurcate to the base, appearing to be double the number of the carpels; sometimes the stylodia are basally fused. The stigmatic surface is Dry. The placentation is laminar-dispersed, with the few to numerous ovules scattered on the walls and on the incomplete partitions when present, though sometimes only in the basal part (as in *Elodea*).

The ovules are anatropous or rarely orthotropous, crassinucellate, sometimes weakly so, with a parietal cell cut off from the primary archesporial cell. Embryo sac formation is of the *Polygonum* Type; the endosperm formation is helobial.

The fruit is fleshy and berry-like but generally dehiscent, splitting up irregularly or stellately (e.g. *Thalassia*). It generally ripens below the water surface. The seeds are few to numerous, with a straight or (in *Halophila*) slightly curved, rarely macropodous embryo. $x = 7$–$12$.

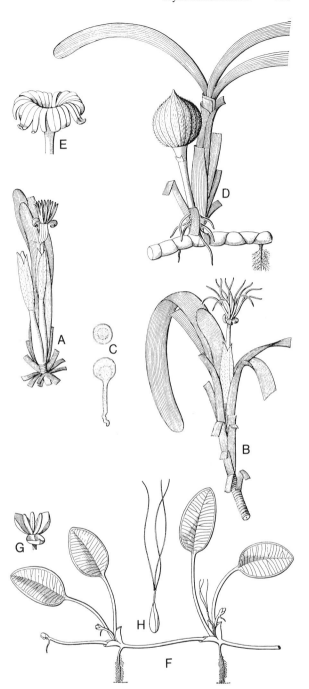

**Fig. 142.** Hydrocharitaceae. **A–E** *Thalassia hemprichii.* **A** Branch of male plant in flowering stage. **B** Branch of female plant. **C** Pollen grain, *below* in germinating stage. **D** Branch of fruiting plant. **E** Opening fruit. **F–H** *Halophila ovalis.* **F** Branch of female flowering plant. **G** Male flower. **H** Female flower. (DEN HARTOG 1957)

*Chemistry.* Cyanin glucosides, anthocyanidins, caffeic acid and chlorogenic acids are known in the family.

*Distribution.* The Hydrocharitaceae are mainly distributed in tropical and subtropical marshes and freshwater and brackish water habitats. Three genera are marine aquatics: *Enhalus, Thalassia* and *Halophila.* Genera which reach into temperate regions are *Hydrocharis, Stratiotes, Elodea, Hydrilla, Limnobium, Egeria* and *Vallisneria.*

*Taxonomy.* The family is highly variable, and genera of the family have previously been separated as Elodeaceae, Vallisneriaceae, Halophilaceae and Thalassiaceae. We assume here that Hydrocharitaceae is a natural unit, but this can be questioned, depending on how great importance one attributes to the single character of epigyny, which might not be monophyletic among hydrocharitaceous plants. *Halophila,* for example, is unusual in the pinnate leaf venation. The Hydrocharitaceae may be divided into five subfamilies as follows.

**Subfamily Hydrocharitoideae**

In this subfamily the flowers usually have a perianth differentiated into a sepaloid and a petaloid whorl and a gynoecium of 6–15 carpels. The placentae are usually long-intrusive. They are marsh plants and freshwater aquatics.

*Ottelia* (ca. 20) is a mainly palaeotropical genus with one species in Brazil. The genus consists of rosette herbs. Some species have leaves that are differentiated into a petiole and a lamina with arching veins. The flowers, which are bisexual or unisexual, are solitary or situated in umbel-like clusters or spikes subtended by an involucre of two fused or free bracts. – *Hydrocharis* (3) is an Old World genus extending to Australia. The floating leaves are differentiated into a petiole and an orbicular or reniform lamina; the female flowers are sessile and the male borne on long pedicels. – *Stratiotes* (1) *aloides,* in Europe and north-western Asia, is a large, freely floating herb with rosettes of linear, serrate leaves. The flowers are unisexual and subtended by a spathe of two free bracts.

**Subfamily Vallisnerioideae**

The flowers in this subfamily have a double perianth (the inner sometimes reduced or rudimentary). The female or bisexual ones generally have three to six carpels. Their leaves are blade-less, often ribbon-like. – *Enhalus* (1) *acoroides* is a marine aquatic rhizomatous herb, growing along coasts from the Indian Ocean to Australia and Polynesia. It is dioecious and has considerable floral dimorphism. The small male flowers are detached just before anthesis and, with the perianth reflexed, rise to the surface where they float about. Pollination occurs when the female flowers are lying horizontally on the surface, and is assisted by the tidal water movements. The rather large seeds may be eaten boiled or roasted and the fibres of the leaf sheaths are used to make fishing nets. – *Vallisneria* (3) is a pantropical and pan-subtropical freshwater genus with considerable dimorphism between the male and the female flowers. The male flowers are small, enclosed within a spathe and borne on an elongated inflorescence stalk; the female flowers are larger, solitary and subtended by a spathe of two united bracts. Before anthesis the male flowers, as in *Enhalus,* are detached and float to the water surface; there the flowers open, the perianth becoming reflexed so that only the tips of its segments rest on the water. The flowers then drift about in the wind and their stamens contact the stigmas of the female flowers and thus effect pollination. The long peduncles of the female inflorescence form a spiral after fertilization and draw the developing fruits below the water surface where they ripen.

**Subfamily Hydrilloideae** (= Elodeoideae)

*Elodea* (ca. 12, St. John 1965) is indigenous to America but is widely spread as a freshwater weed (especially *E. canadensis* and *E. nuttallii*). Like the African genus *Lagarosiphon* (9), *Elodea* comprises submerged herbs with rather short verticillate leaves and unisexual or bisexual flowers, which are water-pollinated. – *Blyxa* (9) is a palaeotropical genus, whereas *Hydrilla* (1) *verticillata* is an Eurasian species; the latter is superficially similar to *Elodea canadensis,* but with male flowers floating to the surface and the female flowers having a long, tubular ovary. *Blyxa octandra,* which is entomophilous, exhibits pollen presentation via the petals, which mimic the stigmas of the female flowers.

**Subfamily Thalassioideae**

This consists only of the marine genus *Thalassia* (2), which are rhizomatous herbs having ribbon-like leaves and solitary unisexual flowers with three uncoloured tepals. The male flowers have 3–12 stamens. The pollen grains are inaperturate, globose and stuck together into moniliform chains, but form pollen tubes before reaching the stigmas.

Pollination takes place in the water. The 6–8-carpellary fruit is a fleshy, globose, irregularly stellately dehiscing capsule. *Thalassia* is distributed along the coasts of tropical America, eastern Africa, southern Asia, Indonesia and Australia. It is ecologically very important for its great production of biomass.

### Subfamily Halophiloideae

Like the previous subfamily this is monogeneric, consisting of the marine aquatic genus *Halophila* (8). This is a creeping herb with narrow, monopodially branching rhizomes and paired, sessile or petiolate, pinnately veined leaves. The sessile inflorescences are one-flowered or have one flower of each sex. The male flowers have three tepals and three stamens, and the female have a ring representing the perianth and a three- to five-carpellary gynoecium with filiform stylodia. The pollen grains are globose and cohere in moniliform chains; they form pollen tubes before reaching the stigma. The fruit is ovoid or globose and opens by decay of the pericarp. The embryo is somewhat curved. Unlike *Thalassia* this group lacks tannin contents. – *Halophila* is widely distributed along tropical and subtropical coasts.

# Order Najadales

*Eight Families:* Scheuchzeriaceae, Juncaginaceae (incl. Lilaeaceae), Potamogetonaceae (incl. Ruppiaceae), Zosteraceae, Posidoniaceae, Zannichelliaceae, Cymodoceaceae, Najadaceae.

Perennial or annual herbs comprising freshwater, brackish or marine aquatics, or, in the first two families, bog, marsh or lakeside plants. Many taxa have well-developed and often starch-rich rhizomes, and in *Triglochin* (Juncaginaceae) and some *Potamogeton* spp. (Potamogetonaceae) there is sometimes a bulb-like underground organ. Vessels are usually totally absent, but may occur in the roots (Scheuchzeriaceae, Juncaginaceae, some Potamogetonaceae) and then have scalariform perforation plates. The leaves are opposite (rarely verticillate) or alternate, and then mostly distichous, and are either linear or subulate, rarely almost ensiform, or when floating, differentiated into petiole and lamina, the lamina with acrodromous-parallel venation. Submersed leaves with a broad, sessile lamina occur for example in Potamogetonaceae, sometimes in connection with pronounced heterophylly where there are petiolate floating leaves. Stipule-like lobes or sometimes large stipules ("stipular sheaths") occur in some families. Intravaginal squamules are constantly present (developed as hairs in *Scheuchzeria*). Stomata are usually lacking but present in some mainly terrestrial taxa, and are then of paracytic or, rarely, tetracytic types. Laticifers in the form of elongated subepidermal cells or cell files occur in *Lilaea* (Juncaginaceae) (TOMLINSON 1982).

The inflorescences are frequently spike-like or spicate, determinate, rarely racemose, or indeterminate; alternatively the flowers may be clustered in axillary cymose groups in the leaf axils or be axillary and solitary. There are generally spathes or spathal bracts subtending the flower clusters. The flowers are bisexual or unisexual. They are generally naked, but in Scheuchzeriaceae the 3 + 3 similar outer floral members probably represent a true perianth, and there is also a hyaline perianth at least in the male flowers of Najadaceae. In most taxa of Juncaginaceae, Potamogetonaceae, Zosteraceae and Posidoniaceae, however, the flat, tepal-like structures, which take the place of a perianth, are considered to be connective appendages, and are most obviously so in Posidoniaceae. Other interpretations are mentioned and commented on below. The flowers are pollinated by wind or water (perhaps, in *Triglochin*, also by insects), and nectar secretion is probably wholly lacking. The stamens, one to six in number, are very variable, from filiform and thin with terminal microsporangia to broad and laminar, with long lateral microsporangia situated below the apex. The anthers have four or two (or rarely in *Najas* one) microsporangia. Endothecial thickenings are of the Spiral or Girdle Types. Anther wall formation is of the Monocotyledonous or, in *Najas,* the Reduced Type. The pollen grains are constantly inaperturate. In some marine taxa they lack exine altogether and become filiform. They are dispersed in the three-celled or, in Juncaginaceae and Zosteraceae, two-celled stage.

The gynoecium consists of from one to six carpels, which, when two or more, are free from each other except in certain Juncaginaceae (*Triglochin*). Septal and other nectaries are unknown in the order. The stylodia are sometimes divided to the base or lobate (Najadaceae, Posidoniaceae). Each carpel contains a single ovule, rarely two, as in Scheuchzeriaceae. The ovules are anatropous and inserted basally (Scheuchzeriaceae, Juncaginaceae, Najadaceae), campylotropous and inserted laterally (some Potamogetonaceae) or orthotropous and inserted apically (the other families). A parietal cell is usually formed. Embryo sac formation generally conforms to the *Polygonum* Type and endosperm formation is either helobial or nuclear.

The carpels mature into nutlets (achenes), drupes or follicles except in *Triglochin* of the Juncaginaceae where they form the mericarps of a schizocarp. The embryo is straight or somewhat curved, sometimes macropodous, and often chlorophyllous. $x = 7$–$12$.

*Chemistry.* Sugar and starch are often accumulated in the rhizomes. Polyphenolics occur in some groups; tannins (probably pyrogallol derivatives) are common in *Potamogeton.* Anthocyanins and anthocyanin pseudobases occur occasionally in the order. The red pigment often present in Potamogetonaceae is rhodoxanthin, a carotenoid compound. Cyanogenic compounds (tyrosine pathway) occur at least in Scheuchzeriaceae and Juncaginaceae.

*Distribution, Ecology.* The order is widely distributed especially in the Northern Hemisphere. Juncaginaceae are represented by two endemic genera in Australia and one in South America, and may have a long history in the Southern Hemisphere, but the possibly rather primitive, monotypic Scheuchzeriaceae is confined to the Northern Hemisphere. Potamogetonaceae, Zosteraceae, Zannichelliaceae and Najadaceae are sub-

cosmopolitan families, Cymodoceaceae tropical and Posidoniaceae a disjunct Atlantic, Mediterranean and Australian family.

The members of Najadales are extraordinary in their adaptation to aquatic biotopes. Apart from a few members of Hydrocharitaceae this is the only order of monocotyledons containing marine taxa. In connection with the aquatic life there are great specializations and reductions in the vegetative parts and also reductions in the floral parts. Many taxa are marine sea grasses (DEN HARTOG 1970).

*Relationships, Taxonomy.* Quite often the Aponogetonaceae are placed in Najadales, which may be justified by their similarity in habit to Potamogetonaceae and their lack of differentiation of sepaloid and petaloid tepals. However, in an evolutionary sense Aponogetonaceae exhibit a number of ancestral features that make them more on a level with Butomaceae and Hydrocharitaceae, and their frequently numerous stamens may be one reason for referring them to Alismatales.

The infrafamilial relationships in Najadales are uncertain. One model expressing their possible relationships is presented in the diagram on p. 99, but there are strongly supported alternatives to this. The Najadaceae, with a basal anatropous ovule, are difficult to associate with the other advanced, aquatic taxa; they have been treated as a separate order previously by DAHLGREN (1975) and THORNE (1976, 1981, 1983). Ruppiaceae are here included in Potamogetonaceae, which we think is justified in spite of the eventually long stipes, the smaller connective appendages (? tepals) and the dimerous flowers. We have also included *Lilaea* in Juncaginaceae, but do this with some reservations after the report, by TOMLINSON (1982), of laticifers in *Lilaea* but not in (other) Juncaginaceae. There is no doubt that Scheuchzeriaceae, Juncaginaceae s.str. and *Lilaea* together represent a cluster of closely related genera.

*Morphological Problems.* One perhaps yet unsolved problem is connected with the so-called tepals in some families of Zosterales. The stamens in some taxa, such as *Posidonia* (Fig. 147 F) have a broad, flat appendage extending beyond the thecae, and ending in a tip. Whether each of the four petal-like structures in *Potamogeton* and *Groenlandia* are of the same nature, i.e. wholly stamen-derived, is uncertain, as is also whether the large appendages of *Phyllospadix* represent in fact tepals (as in some species of *Aponogeton*) or if they are also to be regarded as "connective appendages". In Juncaginaceae, the tepal-like structures outside the stamens are often considered "connective appendages", because those associated with the inner stamens are clearly situated inside the outer stamens, and such a sequence as $T_3A_3T_3A_3G_{3+3}$ would be out of place in monocotyledons.

We have not made any attempt of our own to resolve this problem, but feel that this would be highly challenging as a morphogenetic project; the fact that we have chosen to call the structures "connective appendages" in Juncaginaceae, Potamogetonaceae and Zosteraceae does not imply any firm standpoint.

*Key to the Families*

1. Flowers with three (to six) carpels; fruit follicular; ovules two per carpel; pollen grains dispersed in dyads; intravaginal squamules developed as hairs  . .  **Scheuchzeriaceae**
1. Flowers rarely with three carpels and fruits then not follicular; ovules one per carpel; pollen grains dispersed as monads; intravaginal squamules not with the appearance of hairs  . . . . . . . 2
2. Emergent marsh plants; leaves all basal  . . . . . . . . . . .  **Juncaginaceae**
2. Submerged aquatics; leaves scattered on the branches  . . . . . 3
3. Plants growing in fresh or brackish waters, not marine; pollen grains not growing out into filiform threads ("pollen tubes")  . . . . . . . . 4
3. Plants marine; pollen grains growing out into filiform threads and dispersed as such  . . . . . . . 6
4. Flowers bisexual, emerging from the water and pollinated by wind, or on the water surface  . . . .  **Potamogetonaceae**
4. Flowers unisexual, pollinated under the water surface  . . . . . . . . 5
5. Ovule basal, erect; pistil one per flower and provided with two to four elongate stylar branches  . . . .  **Najadaceae**
5. Ovule apical, pendulous; pistils generally three to four per flower, each with one style  . . .  **Zannichelliaceae**
6. Flowers bisexual, in spikes that are not flattened; style irregularly and profusely branched  . . . . .  **Posidoniaceae**
6. Flowers unisexual, either in flat spikes or in groups or solitary in leaf axils; style not as above  . . . . . . . . . 7
7. Flowers on one side of a flat spike axis  . . . . . . . . . . . . .  **Zosteraceae**
7. Flowers borne in cymose axillary groups or in pairs or solitary in the leaf axils  . . . . . . . .  **Cymodoceaceae**

## Scheuchzeriaceae Rudolphi (1830)   1:1
(Fig. 143)

A low (10–35 cm) perennial rush-like herb with a sympodial rhizome and with an erect aerial stem bearing a number of alternate, linear, terete leaves. The leaves are sheathing at the base, striated by the parallel veins and have a conspicuous apical pore (indicating the canaliculate origin of the leaf). The leaf sheath has a membranous margin; at its base there is an axillary row of densely placed hairs replacing the intravaginal squamules. Vessels (with scalariform perforations) are confined to the roots. The axis contains two cycles of vascular bundles, the outer connected by sclerenchyma to form a continuous ring (TOMLINSON 1982). Air lacunae are present in the leaves. Apart from the axillary hairs the plant is glabrous. Tetracytic stomata are present on axis and leaves. Tannin cells are also present, and crystals of oxalate occur in the mesophyll.

The inflorescence is a very sparse, raceme-like panicle, the lateral flowers of which are situated in the axils of more or less bract-like leaves (with a short lamina or none). The flowers are bisexual and wind-pollinated. They have six relatively small, inconspicuous, yellow-green tepals, six spreading to drooping stamens with extrorse, tetrasporangiate anthers (with a short connective tip), and generally a tricarpellary gynoecium. The pollen grains are dispersed as dyads; each is three-celled. The three (rarely six) carpels are distinct or centrally slightly coherent, ovoid, and have a decurrent extrorse stigma on the upper part along the midvein. The stigmatic surface is Dry. There are one or usually two ovules attached basally and axially. They are anatropous, with a parietal cell cut off from the archesporial cell. Embryo sac formation is of the *Polygonum* Type and endosperm formation is helobial (STENAR 1935). The seeds have an incurved radicle. $x = 11$.

*Chemistry. Scheuchzeria*, like both *Triglochin* and *Lilaea* of the Juncaginaceae, contains the cyanogenic glucoside triglochinin (RUIJGROK 1974).

The single genus *Scheuchzeria* consists only of *S. palustris,* growing in bogs in arctic and temperate regions of the Northern Hemisphere.

The genus is distinctive by the intravaginal hairs (instead of squamules as in Juncaginaceae), the indisputable tepals (disputable in Juncaginaceae), the pollen grains dispersed as dyads (monads in Juncaginaceae), the rhizomes (in Juncaginaceae there is usually a corm or a bulb-like stem

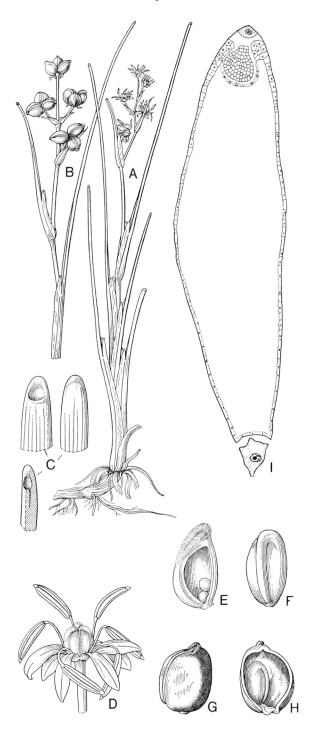

**Fig. 143.** Scheuchzeriaceae. *Scheuchzeria palustris.*
**A** Flowering plant. **B** Stem with infructescence. **C** Leaf apex with pouch, *below* in longitudinal section. **D** Flower. **E** Carpel with part of the wall removed to show the two basally inserted ovules. **F** Seed. **G** Carpel in fruit stage. **H** Same in longitudinal section. **I** Endosperm of the Helobial Type, with the chalazal chamber retained in the uninucleate stage while the micropylar has divided many times and cell walls have been formed. (**A–H** ROSS-CRAIG 1973; **I** STENAR 1935)

base), the leaves scattered on the axis (mainly radical in Juncaginaceae), the tetracytic stomata (paracytic in Juncaginaceae), the helobial endosperm formation (nuclear in Juncaginaceae), and the occurrence of two ovules per carpel (one in Juncaginaceae). Of these characters, the pollen dyads, the tetracytic stomata and the intravaginal hairs are probably to be regarded derived states. On the basis of these differences and the terete leaves (also in some Juncaginaceae) with an apical pore, we consider *Scheuchzeria* distinct enough to form a separate family.

**Juncaginaceae** L.C. Richard (1808)   5:20
(Fig. 144)

Annual or more often perennial herbs growing in marshes or on lake shores or, sometimes in strongly saline, sea-shore habitats; some species are aquatic or subaquatic, and immersed at least temporarily, but all are rooted in the substrate. They have a short vertical, frequently cormous rhizome or sometimes a bulb-like base; the leaves are chiefly or wholly radical; the scape is (largely) leafless and ends in a spicate inflorescence. The leaves are distichous or spiro-distichous, with a sheathing base, sometimes with distinct hyaline margins or "ears", and a dorsiventral, terete or sometimes (as in *Tetroncium* and some species of *Triglochin*) laterally compressed "lamina", which does not have an apical pore (as has *Scheuchzeria*). Intravaginal squamules are present in the leaf axils. Vessels are restricted to the roots and have scalariform perforation plates. Crystals are few and tannin cells lacking. Only in *Lilaea* are laticifers developed, where they appear as conspicuous hypodermal secretory canals. Stomata are present on leaves and inflorescence axis and are mostly paracytic.

The flowers are borne in a dense or sparse spike or raceme. They are usually ebracteate but bracts may be present in connection with the lowest flowers; in *Lilaea* the lowest (female) flowers are situated at the base of the axis. The flowers are inconspicuous, bisexual or unisexual and wind-pollinated. They are trimerous (*Triglochin, Cycnogeton*), dimerous (*Tetroncium, Maundia*) or monomerous (*Lilaea*). The stamens are (4+4 or) 3+3, 2+2, 3 or one or (in female flowers) lacking. Outside each stamen or staminode there is a tepal-like structure which can be interpreted either as a tepal or as an appendicular extension of the stamen, the latter interpretation being supported by the

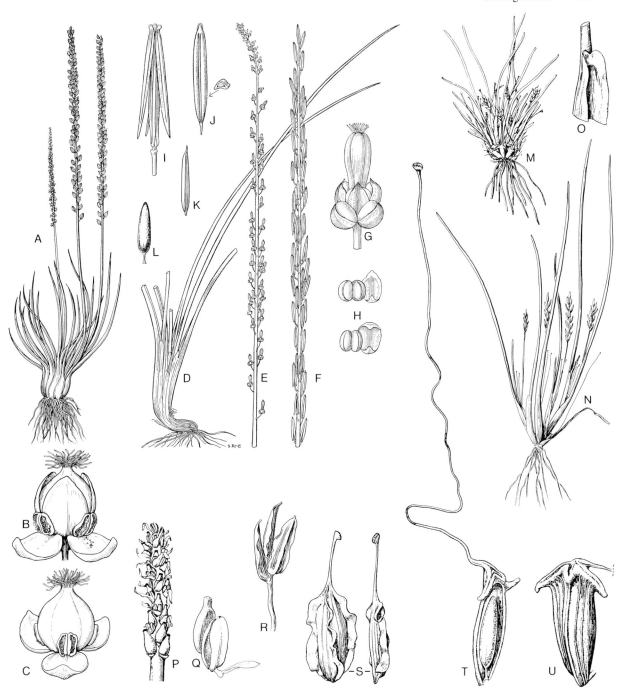

**Fig. 144.** Juncaginaceae. **A–C** *Triglochin maritimum*. **A** Plant. **B** Complete flower. **C** Same with outer whorl of stamens and tepal-like appendages removed, to show inner whorl of stamens. (LARSEN 1973a). **D–L** *Triglochin palustre*. **D** Leaf rosette and rhizome. **E** Inflorescence. **F** Infructescence. **G** Flower. **H** Inner and outer stamen, each with tepal-like appendage. **I** Schizocarp splitting up from the base. **J–K** Mericarp. **L** Seed. (ROSS-CRAIG 1973). **M–U** *Lilaea scilloides*. **M–N** Two plants from different situations. **O** Leaf base with hyaline, ligule-like margin. **P** Inflorescence with male and bisexual flowers. **Q** Bisexual flower consisting of one carpel and an anther, with the appendage on a short stalk. **R** Fruits. **S** Fruits of bisexual flowers in spike. **T–U** opened fruit of female flower situated in leaf rosette. (**M, O–Q, S–V** CORREA 1969; **N** and **R** HITCHCOCK et al. 1969)

fact that the inner whorl (at least in several species of *Triglochin*) is inserted distinctly inside the outer whorl of stamens. This structure is rather bract-like in the bisexual and male flowers of *Lilaea* and it is generally broad and convex and rather stiffer than the tepals of *Scheuchzeria;* in *Tetroncium* it is even slightly variegated. Apart from these dubiously androecial scales, each stamen consists of a (very) short filament and an extrorse, longitudinally dehiscent, tetrasporangiate anther. The endothecial thickenings are of the Spiral Type. The pollen grains are globose, inaperturate and dispersed as monads in the two-celled (GARDNER 1976) or three-celled stage. The plants are mainly wind-pollinated (occasionally water-pollinated).

The carpels number 3 + 3, 2 + 2 or 1, i.e. generally equal in number to the stamens (even when separate from them in unisexual flowers). In the largest genus, *Triglochin*, there are three fertile carpels alternating with three sterile ones, all being fused centrally to form a syncarpous pistil (which in the fruit stage breaks up into three mericarps, see below). In other genera the 3 + 3 or 2 + 2 carpels vary from nearly distinct (*Cycnogeton*) to more or less fused centrally (*Tetroncium, Maundia*), but becoming separate in the fruit stage. In *Lilaea* the pistil is monocarpellary (suspected by CRONQUIST 1981, to be tricarpellary) and different between the basal female flowers, with very long styles, and the spicate flowers, where the style is shorter. In the other genera the style is very short or lacking, often with an irregularly lobate stigmatic surface, which is Dry.

There is usually one basal-axile, anatropous ovule in each locule or monocarpellary pistil, but in *Maundia* the ovule is subapical and rather orthotropous. The ovule is crassinucellate, with a parietal cell cut off from the archesporial cell. As far as is known, embryo sac formation is of the *Polygonum* Type and endosperm formation is nuclear (AGRAWAL 1952). The fruit conditions are variable. In *Triglochin* the three sterile carpels are filled with parenchymatous tissue; the fruit develops into a schizocarp in which the fertile carpels loosen from the others acropetally, as one-seeded, dry mericarps. In *Maundia* and *Tetroncium* the carpels likewise loosen from each other and form separate achenes, in *Cycnogeton* the carpels form individual achenes. The monomerous pistil in *Lilaea* develops into a triquetrous, three-winged achene. The seeds have a straight or nearly straight embryo. $x = 6$, 8 or 9.

*Chemistry.* As with Scheuchzeriaceae, the taxa of Juncaginaceae are cyanogenic, containing the cyanogenic glucoside triglochinin. Crystals of oxalate occur in the tissues.

*Distribution.* The family Juncaginaceae is subcosmopolitan, especially in its genus *Triglochin* which ranges all over the world with several species in Australia as well as the Mediterranean. *Tetroncium* and *Maundia* are endemic in South America, and so is *Lilaea,* which has now spread over the world.

*Triglochin* (15) is distributed over all continents and grows in fresh-water or salt marshes, on sea shores, in wet sand and even submerged in rivers. The leaf bases are sometimes swollen and form a true bulb, as in *T. bulbosum.* The carpels are 3 + 3, of which 3 are fertile, and all are fused centrally. The fruit is a schizocarp.

*Lilaea* (1) *scilloides* (New World; introduced in Australia and Portugal) is a *Juncus*-like herb, which differs from *Triglochin* in having polygamous, monomerous flowers. There are two kinds of female flowers: some are situated together with bisexual and male (unistaminate, dithecous) flowers in a spike-like pedunculate inflorescence, while others are solitary in the leaf axils (enclosed in the leaf sheaths) but have long filiform styles. The species is ecologically rather specialized, growing in annual or vernal pools, and having aquatic and terrestrial phases in its life cycle.

According to one theory (BURGER 1977), the monomerous flowers of *Lilaea* represent a relatively primitive type in the family, while in the other genera, e.g. *Triglochin,* several such monomerous flowers have become aggregated to form tetra- or hexamerous synanthia (the "flowers" in the traditional sense).

## Potamogetonaceae Dumortier (1829)    3:ca. 100
(Figs. 145–146)

Perennial or rarely annual, glabrous aquatic herbs rooted in the bottom and with leaves distributed along the branches and alternate (distichous), opposite or rarely in verticils of three. The stems are weak, generally slender, sympodial (*Potamogeton, Groenlandia*) or monopodial (*Ruppia*). The leaves vary from linear to broadly ovate and are often sessile, but the floating leaves are petiolate in some species. The base is sheathing, generally tubular, sometimes with a hyaline margin and in many species projecting into stipule-like appendages ("stipular sheath"), which either cohere with the sheath or are wholly free from it. The lamina is provided with dense or sparse, parallel or arch-

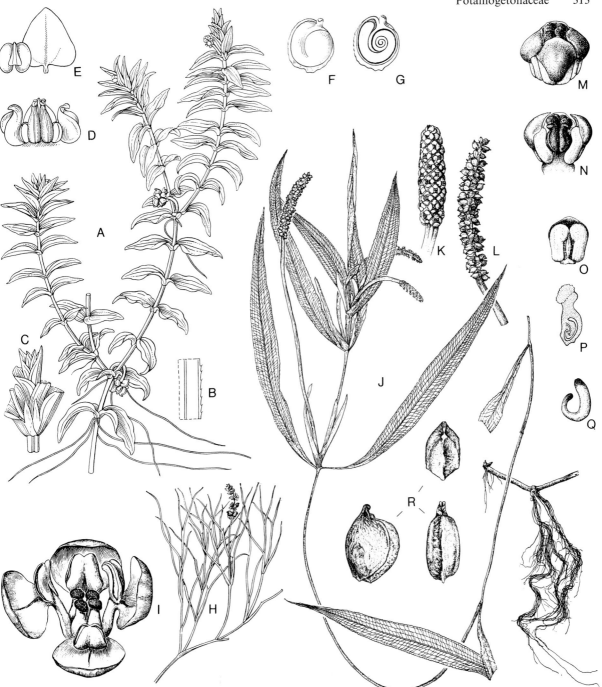

ing-converging veins, the broad leaves often with distinct transverse veinlets between the main longitudinal veins. Some species of *Potamogeton* exhibit a marked heterophylly between non-petiolate submerged and petiolate floating leaves, the latter having also a different texture of the lamina. The leaf axils are always provided with intravaginal squamules. Either there are no vessels or these are restricted to the roots and have scalariform perforation plates. Stomata are lacking; cells with dark tanniniferous contents are common.

**Fig. 145.** Potamogetonaceae. **A–G** *Groenlandia densa*. **A** Part of plant. **B** Leaf margin. **C** Young shoot with base of flowering branch, stipules visible. **D** Flower, one stamen and tepal-like appendage removed, shown in **E**. **F** Seed. **G** Same in longitudinal section. (ROSS-CRAIG 1973). **H–I** *Potamogeton pectinatus*. **H** Branch. **I** Flower. (CORREA 1969). **J–R** *Potamogeton schweinfurthii*. **J** Branch. **K** Inflorescence. **L** Infructescence. **M** Flower. **N** Same with one stamen and tepal-like appendage removed, shown in **O**. **P** Carpel, longitudinal section. **Q** Embryo of seed. **R** One of the four drupes seen in different views. (LISOWSKI et al. 1978)

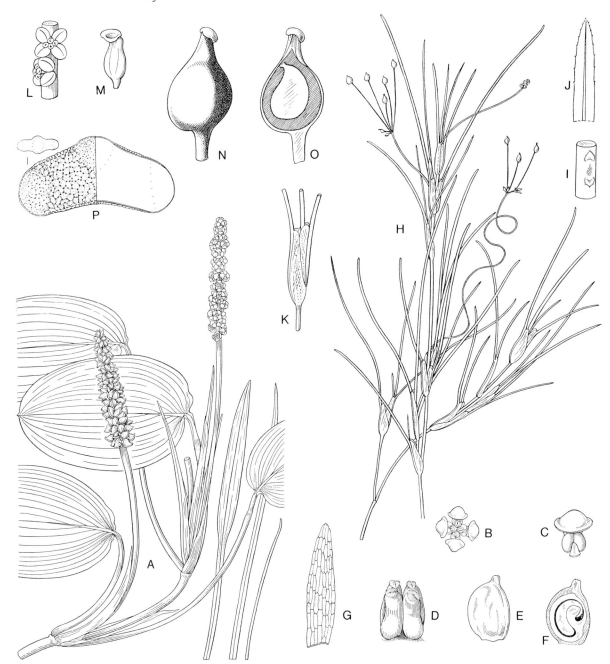

**Fig. 146.** Potamogetonaceae. **A–F** *Potamogeton natans.*
**A** Branch with flowering and fruiting spikes and with
leaves from young shoot. **B** Flower. **C** Stamen and its
tepal-like appendage. **D** Gynoecium. **E** Drupe. **F** Same,
longitudinal section. (Ross-Craig 1973). **G** *Potamogeton
perfoliatus,* intravaginal squamule. (Solereder and
Meyer 1933). **H–O** *Ruppia spiralis.* **H** Branches with
one flowering and two fruiting spikes, the latter with
long-stipitate fruitlets. **I** Inflorescence axis with anthers
and carpel removed, showing connective appendages.
**J** Leaf apex. **K** Leaf bases. **L** Two flowers showing
anthers and carpels. **M** Carpel. **N** Fruit, the long stipe
not included. **O** Same, longitudinal section. (Ross-Craig
1973). **P** *Ruppia spiralis,* pollen grain. (Erdtman 1952)

The inflorescences are pedunculate spikes, which
have only two flowers in *Groenlandia* and *Ruppia*
but several to many in *Potamogeton.* The flowers
are dimerous, trimerous or (generally) tetramerous. According to the commonest interpretation
they lack a perianth, but outside each stamen there
is generally a broad, fleshy, flat to convex, generally brownish member which is regarded as an appendage of the stamen. Alternatively, this is interpreted as a tepal. The two outer appendages are
inserted at the abaxial top of the short filament
outside the base of the outer anthers, the inner

ones at the filament base (Fig. 145I). The anthers are extrorse and longitudinally dehiscent; in *Ruppia* the thecae are rather distant on the broad connective. The endothecial cells have the Girdle Type of wall thickening. The pollen grains are globose, elliptic or (*Ruppia*) elongate, isobilateral (Fig. 146P) and three-celled when dispersed, which is by wind or water (on the water surface). The carpels are generally four, though the number may vary between three and eight. They are separate from each other and obliquely ovoid or pear-shaped, with a very short style and a slightly extended stigmatic surface of the Dry Type. The solitary ovule is ventral in the carpel (Figs. 145P, and 146F), anatropous or varying (ontogenetically) from anatropous to campylotropous, crassinucellate and with a parietal cell. Embryo sac formation is of the *Polygonum* Type and endosperm formation is helobial (MURBECK 1902).

The carpels develop into drupes, achenes or rarely (*Groenlandia*) berries, each with one seed having a slightly curved, more or less macropodous embryo. $x = 8$–$10$ (*Ruppia*) or $13$–$15$ (*Potamogeton*).

*Chemistry.* The Potamogetonaceae are rich in tannins, which probably consist of pyrogallol derivatives. They also contain C-glycosyl-flavones, but lack anthocyanins. The red pigment found in many taxa consists of rhodoxanthin.

*Distribution.* The distribution is subcosmopolitan.

*Potamogeton* (ca. 100) consists of freshwater and brackish water aquatics. The spikes are multilateral and the flowers are several to many and tetramerous, rarely penta- or hexamerous, with well-developed petaloid scales. The branching system is sympodial and the rhizome well developed. The spikes emerge from the water during flowering and are elongate-globose. The leaves are very variable, and inserted alternately, (with distichous phyllotaxy), or are opposite on part of the shoot. They are sessile and broad and ribbon-like or narrow and filiform, or are petiolate with a more or less broad lamina. The sheath may have hyaline margins only or there may be stipule-like appendages, up to several centimetres long, as in *P. natans*. The fruits have a hard endocarp and are best described as drupelets (or achenes). *Potamogeton* includes many common freshwater aquatics, such as *P. perfoliatus* (Fig. 146). *P. natans* with floating petiolate leaves and *P. gramineus,* with floating petiolate leaves and sessile submerged leaves, are both common in Europe. *P. pectinatus* is a widely distributed species with narrow leaves and small, few-flowered spikes.

*Groenlandia* (1) *densa* deviates in being annual and in that all leaves of the erect shoot are opposite, the flowers are only two per spike and the fruits have a soft endocarp (i.e. are baccate). The single species is distributed in Europe, West Asia and North Africa.

*Ruppia* (1–7) has a complicated taxonomy on the species level and is sometimes considered to represent one variable species, but more often a few species.

The branches are monopodial, and the alternate or subopposite leaves are linear and have a midvein only. The sheath has an expanded hyaline margin, but not with stipular sheaths. The spikes have long peduncles which are spirally twisted especially before and after flowering; they consist of few, often only two, flowers, in which the tepal-like scales (or tepals?) are lacking, the stamens are two in number with an expanded connective, and the carpels are generally four or more. The ovule is campylotropous and apically inserted. The stipes of the carpels elongate considerably during ripening so that the drupelets of each flower resemble an umbel. Pollination takes place on the water surface, the pollen grains being isobilateral, slightly curved with a smooth area (laesura) in each corner (Fig. 146P). *R. cirrhosa* (= *R. spiralis*) and *R. maritima*, both of which are subcosmopolitan, grow in shallow brackish to saline or alkaline waters, often in river estuaries. *R. filifolia* grows in freshwater in the Andes up to an altitude of ca. 4,000 m.

In spite of some conspicuous diagnostic character states *Ruppia* is in good agreement with other Potamogetonaceae.

**Posidoniaceae** Lotsy (1911)   1:3   (Fig. 147)

Perennial, monoecious, marine, wholly submerged herbs with a thick horizontal, flattened, monopodially branching rhizome. The leaves are distichous on the rhizome. They are linear and strap-shaped, to almost terete, apically truncate or rounded, parallel-veined or with only a midvein, with abundant fibres and basally with an amplexicaul sheath bearing a short curved ligule. Stipule-like lobes are often present at the base. The rhizomes are covered with fibres of old leaf sheaths (when detached these fibres are rolled by the waves into balls that can be found on the sea shores: "*Posidonia* Balls").

Vessels, hairs and stomata are lacking. In all vegetative parts of the plants there are scattered cells

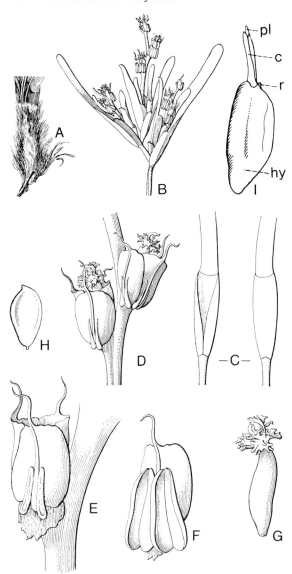

**Fig. 147.** Posidoniaceae. *Posidonia oceanica.* **A** Base of plant, showing fibrous remains of leaf sheaths. **B** Inflorescence. **C** Leaf sheath, inner and dorsal side. **D** Part of inflorescence with two flowers. **E** Flower after removal of ovary and thecae showing the wide connectives. **F** Stamen with dehisced thecae. **G** Pistil with laciniate stigma. **H** Fruit. **I** Germinating fruit (*hy* hypocotyl; *r* raphal part; *c* cotyledon; *pl* plumule). (**A** ASCHERSON and GÜRKE 1889; **B–H** DEN HARTOG 1970; **I** EMBERGER 1960)

"spike" seems to be terminated by a flower. The spikes are subtended by two to four short leaves with broadened sheaths. The flowers are bisexual, rarely male; they lack a perianth and consist of three stamens and a single, unilocular and (probably) monocarpellary pistil. The stamens lack a distinct filament; they consist of a broad, shield-like plate ("connective"), ending as a prominent slender tip, and have two basal, bisporangiate thecae on the outside near the midvein (Fig. 147F). The pollen grains are globose when released (in the water) but soon grow out to become filamentous (pollen tubes), in which state they are dispersed. They are three-celled. The pistil has an elongate ovary and an irregularly, many-lobed, spreading stigma. There is one apical, pendulous, orthotropous, ovule, the detailed embryology of which is not known. The fruit is slightly fleshy, having a spongy pericarp, but dehisces to release the elongate seed (TOMLINSON 1982). This has a straight, macropodous embryo. No chromosome number has been reported.

*Chemistry.* The plants are rich in tannin cells (with proanthocyanins); starch is present in the rhizomes. Crystals are lacking.

*Posidonia* (3) has a disjunct distribution. *P. oceanica* occurs along the Atlantic and Mediterranean coasts, whereas *P. australis* and *P. ostenfeldii* occur along the coasts of Australia and Tasmania. *P. oceanica* inhabits exposed to moderately sheltered localities from the water surface down to a depth of 30–40 m and forms extensive submarine meadows (DEN HARTOG 1970). *P. ostenfeldii* has narrow, subfiliform leaves. The fibres of *P. oceanica* and *P. australis* can be used for packing material, the latter even for coarse fabrics and sacks.

The family obviously comes closest to Potamogetonaceae and Zosteraceae.

## Zosteraceae Dumortier (1829)    3:18    (Fig. 148)

Perennial, wholly submerged, marine (to brackish water) herbs with a creeping, monopodial rhizome, from which erect often flattened shoots come out. This bears distichous, ribbon-like, green parallel-veined foliage leaves, subtended basally by bracteal leaves. The foliage leaves are apically rounded or emarginate and have a closed (*Zostera*) or open sheath. The margin of the sheath is often provided with hyaline auricles and its upper margin is ligulate.

Vessels, stomata and lignin are lacking, whereas tannins are plentiful (TOMLINSON 1982). The inflo-

which deposit brown products (tannins) which may appear "foam-like" (TOMLINSON 1982).

The flowers are situated in a cluster of spike-like inflorescences, each with three to five flowers; whether these are racemose (TOMLINSON 1982) or cymose (CRONQUIST 1981) is difficult to determine; the individual flowers are ebracteate and each

**Fig. 148.** Zosteraceae. **A–H** *Zostera marina*. **A** Base of shoot. **B** Branch with inflorescences. **C** Leaf apex. **D** Spike enclosed in leaf sheath. **E** Apex of spadix with alternating male and female flowers. **F** Female flower, considered here to consist of one carpel with a bibrachiate stylodium. **G–H** Seed. (Ross-Craig 1973). **I–K** *Phyllospadix torreyi*. **I** Part of plant. **J** Inflorescence in leaf sheath. **K** Embryo, roots upwards. (Hitchcock et al. 1969)

rescences are axillary, flattened spikes or spadices with the reduced flowers arranged on one side. Each spadix is enfolded by the lower part of the subtending leaf and is exserted only in the fruiting state. The flowers are unisexual, those of each sex alternating in two rows on the flat axis when the plants are monoecious (*Zostera, Heterozostera*); *Phyllospadix* is dioecious.

The male flowers consist of one stamen with two bisporangiate thecae, which can also be interpreted (as by Tomlinson 1982) as two bisporangiate stamens; these are connected by a narrow flap of (? connective) tissue. Marginal outgrowths on the axis, so-called retinacula, each of which when present almost encloses a stamen, are present in most taxa (lacking in *Zostera* subgen. *Zostera*) and are conspicuous in *Phyllospadix*. The nature of the retinaculum is uncertain; it has been variously interpreted as a bract, tepal (cf. *Aponogeton distachyus*)

or a staminal appendage. The pollen grains, as in Posidoniaceae, lack exine and when released in the water soon elongate to become filamentous and up to 2 mm long (Cronquist 1981). They are two-celled or three-celled when dispersed.

The female flowers have a pistil which is probably monocarpellary (interpreted as bicarpellary by Cronquist 1981), and is unilocular, curved apically and ending in two stylodial branches. The locule contains one apical, pendulous, orthotropous ovule, in which a parietal cell is not cut off, but there are periclinal divisions in the nucellar epidermis. Embryo sac formation follows the *Polygonum* Type and endosperm formation is reported to be nuclear (K.V.O. DAHLGREN 1939). The fruit is small, thin-walled and ovoid, ellipsoidal or crescent-shaped; its wall may persist or fragment. The seed has a macropodous, slightly curved embryo lacking a radicle. $x = 6$ or 10.

*Distribution.* The distribution includes temperate seas of both hemispheres but a few species range into tropical waters. *Zostera* has the widest distribution, *Heterozostera* is found on the coasts of South America and Australia and *Phyllospadix* mainly on the coasts of the Pacific.

*Zostera* (12) is widely distributed in salt and brackish water of both hemispheres. Subgen. *Zostera* consists of four Northern Hemisphere species which have closed leaf sheaths and no retinacula, e.g. *Z. marina,* which is abundant on sandy and muddy bottoms. Dried plants of this species can be used as packing material. Subgen. *Zosterella* (8) has open leaf sheaths and retinacula, as do the following genera, but the retinacula are membranous and smaller than the ovaries or stamens. This subgenus occurs in both hemispheres and also in tropical waters. Here belongs *Z. notii* (= *Z. nana*).

*Heterozostera* (1) *tasmanica* has wiry shoots up to 1 m long, often profusely branching in the apical parts. – *Phyllospadix* (5) species are coarse plants with long leaves and adhesive root hairs. The plants are dioecious and have either male or female flowers in the spadices which project out of the leaf sheaths and are provided with larger retinacula than in the other genera. The retinacula are present also on the female spikes. The vegetative parts are more sclerenchymatous than in *Zostera* and the sheaths when decaying leave woolly fibres.

The flowers in *Zostera* have been interpreted as bisexual, each with one stamen and one pistil, in which case the floral axis would have to be merged with the spadix. Even when they are considered unisexual, there are different interpretations: the male as having two stamens rather than one, the female as having two carpels rather than one and the retinaculum as being anything from a bract to a tepal or a connective appendage.

Zosteraceae seem to be most closely allied to the two previous families, but deviate in the flat inflorescence, the reduced flowers, which are extremely difficult to interpret, and also the different embryology (e.g. endosperm formation).

**Zannichelliaceae** Dumortier (1829)   4:7 or more
(Fig. 149)

Perennial or annual, submerged, brackish water and freshwater aquatics with a slender creeping, generally sympodial rhizome and with a filamentous, much branched, ephemeral stem with narrow, linear, one- to three-veined leaves which are up to 10 cm long, alternate, distichous, rarely almost opposite or, in *Zannichellia,* situated in pseudowhorls of three to four. Besides the green linear foliage leaves there are also scale leaves corresponding to a sheath. The foliage leaves generally have an open sheath with a membranous margin, but in *Zannichellia* the sheath is separate from the green blade. A ligule is frequently present. The intravaginal squamules are often only two in number at each node.

The flowers are terminal and arranged in complex, cymose aggregates; they are unisexual, the male and female at least usually on the same plants. Tepals are lacking (*Zannichellia, Vleisia*) or forming a small, trilobate basal structure (*Althenia, Lepilaena*).

The male flowers have a single filament with 4, 8 or 12 microsporangia, and probably in the last two cases represent two or three fused stamens. The connective often projects beyond the microsporangia. The pollen grains, which are globose and minutely spinulose, are dispersed in the three-celled stage. The female flower has a small, cupule-like or three-lobed perianth and one to nine mutually free carpels; these have an ovoid ovary and a short or long style dilated apically into a symmetrically or asymmetrically funnel-shaped stigma, which is fimbriate in *Lepilaena.* The locule contains one apical, pendulous, orthotropous (to anatropous) ovule. A parietal cell is cut off from the archesporial cell; embryo sac formation (in *Zannichellia*) is of the *Allium* Type, and endosperm formation is helobial (VIJAYARAGHAVAN and KUMARI 1974; LAKSHMANAN 1965).

The fruits are usually an achenes with an inner sclerotic layer, but in *Althenia* they are dehiscent. The seed consists of an embryo which is spirally coiled, with a thick hypocotyl and a very small radicle. $x = 6$–8.

*Chemistry.* Tannins and crystals are probably lacking in the family. Flavonoid sulphates are known in *Zannichellia.*

*Distribution.* The family occurs in fresh and brackish, but not in saline, waters all over the world.

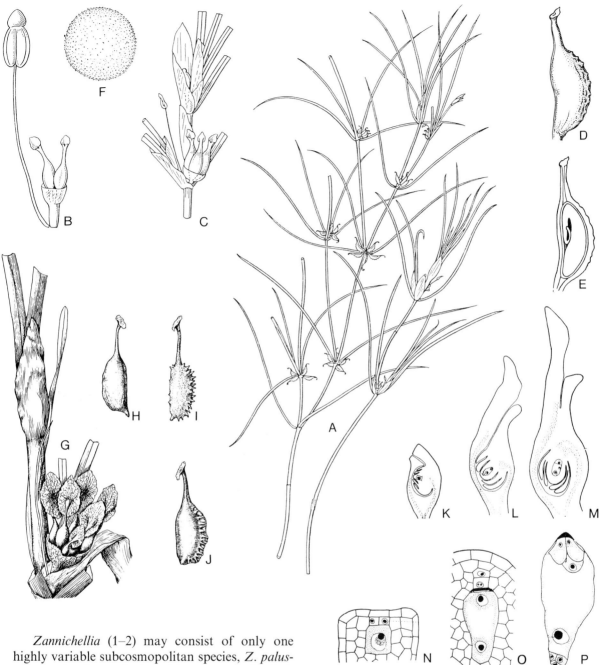

**Fig. 149.** Zannichelliaceae. *Zannichellia palustris.* **A** Branches in flowering and fruiting stages. **B** Female flower with its perianth cup and, at the base, a male flower, consisting of a single stamen. **C** Part of branch with terminal inflorescence and lateral branches. **D** Achene. **E** Same, longitudinal section. **F** Pollen grain. **G** Leaf axil with female flower. **H–J** Fruits, **I** with the outermost layer removed. **K–M** Ontogenic development of carpel, longitudinal section, to show the successively more apical-pendulous position of the ovule. **N–P** Embryo sac formation; parietal cells seen in **N–O**. (**A–E** Ross-Craig 1973; **F** Erdtman 1952; **G–J** Correa 1969; **K–P** Vijayaraghavan and Kumari 1974)

*Zannichellia* (1–2) may consist of only one highly variable subcosmopolitan species, *Z. palustris,* varying in particular with respect to the fruit (see Fig. 149D, H, I, J). The genus is characterized by having leaves without attached sheath, but with a ligular sheath. The male flowers lack a perianth and have four to eight microsporangia; the female flowers have a cup-shaped envelope (perianth?). *Zannichellia* sometimes forms conspicuous mats of vegetation. – *Vleisia* (1), a South African genus, has male flowers with eight microsporangia and vestigial lateral appendages and female flowers the carpel of which has a strongly enlarged, oblique, funnel-shaped stigma. – *Althenia* (1–2), mainly in

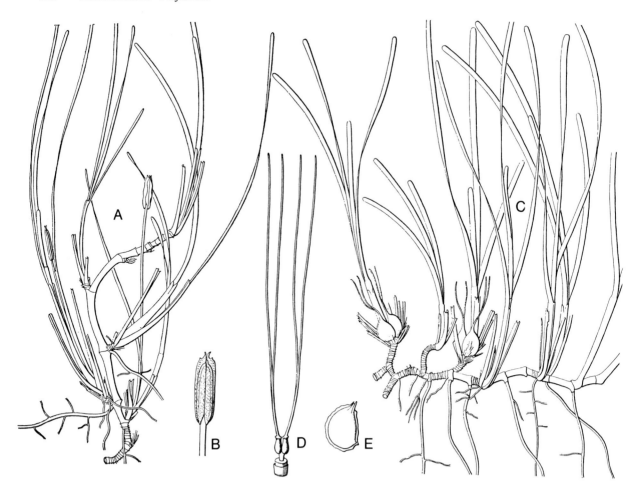

the Mediterranean, has a leaf sheath with broad hyaline margins, male flowers with a three-lobed perianth and 2 microsporangia, and female flowers with three tepals and long, trumpet-shaped styles. – *Lepilaena* (4), in Australia, Tasmania and New Zealand, has fimbriate styles and truncate leaf apices; it varies strongly in number of microsporangia and, as in *Althenia,* in perianth.

The family is very distinct, probably most closely related to Cymodoceaceae, which is sometimes included in Zannichelliaceae.

**Cymodoceaceae** N. Taylor (1909)   5:16
(Fig. 150)

Perennial, submerged, entirely marine, dioecious plants with creeping rhizomatous axes and erect branches, the rhizome being either monopodial, herbaceous and leafy (*Cymodocea, Halodule*), monopodial, herbaceous and provided with scale leaves (*Syringodium*) or more or less sympodial, woody and provided with scale leaves (*Amphibolis,*

**Fig. 150.** Cymodoceaceae. *Cymodocea nodosa.* **A** Habit of male plant. **B** Male flower, consisting of two stamens attached to each other. **C** Female plant with already enlarging fruits. **D** Female flower consisting of two carpels each with a bibrachiate stylodium. **E** Fruit. (DEN HARTOG 1970)

*Thalassodendron*) (TOMLINSON 1982). The leaves are distichous; the foliage leaves are narrowly or broadly ribbon-like or, rarely (*Syringodium*), terete, three to several-nerved, apically generally truncate or emarginate and basally with an open sheath and provided with a ligule. Aerenchyma is developed in the leaf blade. There are also scale leaves which represent the sheath only, situated on the rhizome, and bracts associated with the flowers.

Vessels, stomata and hairs are lacking. Tannin cells are present.

The flowers are usually solitary and situated at the ends of short erect shoots or on lateral branches of erect shoots; in *Syringodium* they are in cymose groups. A perianth is lacking. The male

flowers are sessile or stalked, and consist of two partly fused filaments and two extrorsely dehiscent, tetrasporangiate anthers which are dorsally more or less fused; in *Halodule* they are of different lengths. Usually there is an apical prolongation from the connective. The pollen grains are often released in the water and they elongate to become filamentous and up to 1 mm long (DUCKER et al. 1978); they are three-celled.

**Fig. 151.** Najadaceae. **A–G** *Najas marina*. **A** Flowering branch. **B** Leaf base. **C** Leaf apex. **D** Inflorescence. **E** Male flower with outer envelope and stamen enclosed in a thin, tubular perianth. **F** Female flower with tribrachiate stylodium. **G** Fruit. **H–N** *Najas flexilis*. **H** Flowering and fruiting branches. **I** Leaf base. **J** Leaf apex. **K** Male flower, early stage. **L** Male flower, later stage; the lobes at the apex belong to the inner floral envelope. **M** Female flower. **N** Fruit. (All from ROSS-CRAIG 1973)

The female flowers are sessile or stalked and consist each of two mutually free carpels, the stylodia of which are long, slender and simple (*Halodule*), bibrachiate (most genera) or tribrachiate (*Amphibolis*). The carpels each contain one apical, pendulous, orthotropous ovule. This is crassinucellate. Endosperm formation (in *Cymodocea*) is nuclear. The fruit is indehiscent, either with a stony endocarp (*Cymodocea, Halodule, Syringodium*) or viviparous (*Amphibolis, Thalassodendron*). The seed is often specialized. $x = 7$ (*Cymodocea*).

*Chemistry.* Tannin cells (containing proanthocyanins) are abundant especially in the epidermis of the leaf blades.

*Distribution.* The family for the most part occurs in tropical and subtropical waters, but *Amphibolis* is temperate and confined to Australia, and *Cymodocea nodosa* is Mediterranean. They are all sea grasses that often form extensive stands, "submarine meadows".

Three genera have monopodial rhizomes and stalked male flowers, viz. *Cymodocea* (4), *Halodule* (6) and *Syringodium* (2). – *Cymodocea* has a disjunct distribution (Indo-Pacific, Western Australian, Mediterranean-East-Atlantic) and has ribbon-like, rather narrow leaves, often with marginal teeth. – *Halodule,* in tropical waters of both the Old and the New World, has even narrower leaves (<3 mm) without marginal teeth. The carpels have a single stylodial branch. – In *Syringodium* (2) in the Caribbean and Indo-Pacific, the leaf blades are terete.

Two genera have sympodial, woody rhizomes and sessile male flowers, viz. *Thalassodendron* (2), from East Africa to Eastern Malaysia and Western Australia, with rounded leaf apices with coarse teeth and bistylodial carpels, and *Amphibolis* (2) in Australia and Tasmania, with broad, apically bidentate leaves and tristylodial carpels (DUCKER et al. 1977).

The above grouping may justify division into two subfamilies.

The family is quite distinct and its inclusion in Zannichelliaceae, with which it is dubiously most closely related, is precluded by a number of differences (see TOMLINSON 1982, p. 401).

**Najadaceae** A.L. Jussieu (1789)   1:40
(Fig. 151)

Annual or perennial, monoecious or dioecious, submerged aquatic plants rooted in the bottom. Stems slender, much-branched, with leaves subopposite or in verticils of three. The leaves are linear, narrow and one-veined, with an open, sheathing base, and are sometimes widened, with hyaline sides, at the base. The margins are entire or denticulate, each tooth ending as a brownish point. The stems lack mechanical tissue, and vessels, stomata and hairs are also lacking. For each leaf with axillary flowers there are generally two leaves without flowers.

The flowers are small, unisexual and situated in groups in the leaf axils, usually within a leaf verticil. The male flowers are generally enclosed by two thin hyaline envelopes. The outer (lacking in *N. graminea*) is tubular or cupular and sometimes has two to four short brown teeth. The inner, which is tight and flask-shaped, encloses the single stamen of the flower, and is partly fused with this; it is provided with two relatively large and apically thickened lobes (Fig. 151 L). These envelopes are often interpreted as a spathe and a perianth, respectively, but this is uncertain. The stamen has a short filament and four microsporangia or, in *N. flexilis,* only one microsporangium. When mature the inner wall in the anther breaks down. Anther wall formation follows the Reduced Type, and the endothecium is lacking. The pollen grains are released apically. They are spheroidal and three-celled.

The female flowers are only rarely enclosed by an envelope, which then corresponds to the outer envelope of the male flowers. The pistil, which is presumably monocarpellary, has a unilocular ovary and a style divided some distance from the base into two to three stigmatic branches and sometimes into some shorter non-stigmatic lobes as well. There is only one, basal, anatropous ovule. A parietal cell is cut off from the primary archesporial cell. Embryo sac formation is according to the *Polygonum* Type, and endosperm formation is nuclear (or helobial or cellular; see SWAMY and LAKSHMANAN 1962).

The fruit is a thin-walled achene with the pericarp tightly enclosing the seed, which has a straight embryo. $x = 6$ or 7.

*Chemistry.* Tannins and crystals occur in small quantities.

*Najas* (ca. 40) is subcosmopolitan in tropical to boreal, fresh or brackish waters, where the plants may form dense mats. The variable European species *N. marina* has dentate leaves.

# Superorder Bromeliiflorae

*Six Orders:* Velloziales, Bromeliales, Philydrales, Haemodorales, Pontederiales and Typhales.

From small to quite large herbs, rarely woody tree-like plants with a considerable trunk (*Vellozia, Puya*) or with a short to long underground stem: a rhizome which is rarely shortened to form a corm or tuber. The roots arise opposite the xylem strands in the stem, and the root-hair cells in the cases studied (except *Anigozanthos*) are of the same length as other root epidermal cells. Velamen has not been recorded. Vessels are present in the roots and have scalariform or simple perforation plates, or both.

The leaves are either spirally set or distichous. They are sheathing basally, linear or sometimes differentiated into a pseudopetiole and a lamina, which may be lanceolate or cordate. In some orders (Philydrales and Haemodorales) they may be equitant-ensiform (ARBER 1922a, 1925). The ptyxis is variable. The venation is parallel or arching-convergent, in some Pontederiaceae pinnate with parallel-converging laterals. The leaves in some groups (Pontederiaceae, Typhaceae) are strongly aerenchymatous, in others (Bromeliaceae and Velloziaceae) succulent or sclerenchymatous or both. Stipule-like structures are rarely present, and some Pontederiaceae have prominent ligules.

The stems frequently have vessels, then nearly always with scalariform perforation plates, and such vessels may also be present in the leaves (Typhales, Velloziales, many Bromeliales and some Haemodorales). Laticifers are perhaps lacking (see under Haemodorales, however). The sieve tube plastids lack starch and protein filaments, having cuneate protein crystalloids only. Silicic acid is found only in Bromeliales, as solitary spherical crystals in epidermal cells, whereas oxalate raphides seem to be present in all orders, although they may be present only in certain tissues (e.g. in the tapetum in Philydraceae). The starch in vegetative parts consists of simple, centric or eccentric grains. Whereas the Typhales are glabrous, uniseriate hairs occur in some Haemodorales, Philydrales and Pontederi-

ales, and very complex, peltate or stellate hairs in Bromeliales, the latter being of great importance in water uptake. Intravaginal squamules are lacking. The stomatal complexes always include subsidiary cells, either only two, parallel to the guard cells, or in addition two terminal and sometimes another two, so that the stomata are paracytic, tetracytic or hexacytic. The wall divisions of the cells adjacent to the stoma proper are often oblique (TOMLINSON 1974). Epicuticular wax, when clearly oriented, is of the *Strelitzia* Type.

The inflorescences probably include both determinate and indeterminate (racemose) types, although the latter appear to be strongly predominant (and probably universal in Bromeliales, Philydrales and Pontederiales); in Typhales the inflorescences are highly complex (U. MÜLLER-DOBLIES 1969; D. MÜLLER-DOBLIES 1970). In Bromeliales the inflorescence bracts are often large, stiff and brightly coloured.

The flowers are trimerous, bisexual or unisexual, hypogynous or epigynous (rarely semi-epigynous in Haemodoraceae), actinomorphic or zygomorphic, and vary from strongly adapted to entomogamy or ornithogamy (many Bromeliaceae) to strongly adapted to anemogamy (Typhales). The perianth consists of 3 + 3 tepals or is generally derivable from this condition, although there is great variation and specialization in the perianth: in Bromeliales there is generally differentiation into small-sized and sepaloid outer tepals and petaloid inner tepals; in Velloziales, Pontederiales and Haemodorales both whorls are petaloid, sometimes with marked zygomorphy in the last two orders; in Philydrales the tepals are also petaloid, the three upper being fused into a broad upper lip, whereas the laterals (outer whorl) are fairly small, and the inner median again fairly large, forming a lower lip; in Typhales the tepals are inconspicuous or transformed into hairs in connection with wind dispersal. The tepal colour in the animal-pollinated taxa is generally bright. Perigonal nectaries are lacking, but septal nectaries occur in all orders except Philydrales and Typhales. Various degrees of fusion of tepals occur in some orders, in particular in Philydrales (see above) and some Haemodorales.

The stamens are 3 + 3 in number, but by division of initials many in some Velloziales or by reduction 5 or fewer (down to 1) in other orders. Their filaments are free and only rarely hairy and the anthers are dithecous and tetrasporangiate, basifixed or dorsifixed, introrse, and generally longitudinally dehiscent (poricidal in *Monochoria* of Pontederiales). Anther wall formation is of the Monocotyledon Type, and the endothecial wall thickenings are generally spiral, but girdle-shaped in Pontederiales. The tapetum is glandular-secretory in Velloziales, Philydrales, at least most Bromeliales and, perhaps, some Pontederiales, but amoeboid in Haemodorales, Typhales and some Pontederiales, with the tapetal cells varying much in number of nuclei. Microsporogenesis is successive. The pollen grains may cohere in tetrads (occasionally in Velloziales, Bromeliales, Philydrales and Typhales). They are sulcate or bi- or trisulculate (Pontederiales), inaperturate (some Velloziaceae), ulcerate (Typhales) or two- to oligoforaminate (some Bromeliales, some Haemodorales), and are always dispersed in the two-celled stage.

The carpels are generally three in number, but in Typhales are generally reduced to one. In most cases the fusion of the carpels, when three are present, is almost complete so that there is a trilocular or very rarely unilocular ovary and a simple style, with a capitate, punctiform or trilobate stigmatic apex. The stigmatic surface is Dry or (in many Bromeliaceae and some Haemodoraceae) Wet. The placentation is axile, laminar-dispersed (Velloziales) or rarely (*Heteranthera* and *Hydrothrix* of Pontederiaceae, *Philydrum* of Philydraceae) parietal. In the generally monocarpellary pistil of Typhales there is one apical pendulous ovule. The ovules are otherwise generally several to many, but solitary in some genera of Haemodorales and Pontederiales.

The ovules are generally anatropous, but either orthotropous or more rarely hemianatropous in Haemodorales. They are crassinucellate, sometimes weakly so, but tenuinucellate in Velloziales, where a parietal cell is not formed either, as in the other orders. Embryo sac formation is perhaps always monosporic, of the *Polygonum* Type. Characteristic for all members of the Bromeliiflorae is that endosperm formation is of the Helobial Type, and that the chalazal chamber of the endosperm is very restricted in size and cell number; cell walls are formed sooner in the chalazal chamber than in the micropylar, and often the chalazal chamber

lacks starch grains which are copious in the micropylar endosperm.

The syncarpous fruits are capsules, berries or rarely nutlets; in Typhales the monocarpellary fruits are drupaceous (Sparganiaceae) or achene-like follicles (Typhaceae). Arils and similar structures are lacking. The seeds are generally brownish, very rarely black (*Wachendorfia* in Haemodorales), but then probably do not have phytomelan. The storage tissue consists of starchy endosperm; the starch grains are all similar and compound (HAMANN 1961). Perisperm and chalazosperm are lacking. The endosperm is non-ruminate and always contains copious starch. The embryo is generally linear, of the *Urginea* Type in shape, also in the wind-pollinated Typhales. In Haemodorales the embryo may be small and rather of the *Trillium* Type (somewhat curved in *Wachendorfia*). For chromosome numbers, see under the orders.

*Chemistry*. Steroidal saponins are only rarely recorded in the Bromeliiflorae (from Bromeliaceae). Also chelidonic acid is evidently rare, with the notable exception of Haemodoraceae, where it occurs in several (all?) genera. Cyanogenic compounds (tyrosine pathway) are known in Pontederiales and Typhales. Several flavonols and flavones are known in Bromeliaceae, including C-glycoflavones, but tricin and sulphurated flavonols are lacking. HARBORNE (1982) notes that the Bromeliaceae possess 3,5-diglucoside and 3-rutinoside-5-glucoside, the former being present in Commelinaceae and both in Iridaceae, but these details may not be of great phylogenetic significance. One test of isoenzymes of dehydraquinate hydrolyase (DHQ-ase) proved to be negative. All taxa studied (by HARRIS and HARTLEY 1980) of the Bromeliiflorae, except one of Velloziaceae, showed UV-fluorescence in the cell walls, indicating the presence of the same compounds as are found in the Commeliniflorae, Zingiberiflorae and Areciflorae.

*Parasites*. The fungal parasites do not seem to contribute any considerable information of phylogenetic value, although the fact that *Uromyces sparganii* attacks both *Sparganium* (Typhales) and *Acorus* (Arales) has sometimes been taken as an indication of relationships.

*Distribution*. The orders of Bromeliiflorae show a rather marked concentration in the Southern Hemisphere. The Bromeliales are wholly concentrated in South and Central America, the Velloziales are also most variable in South America although they also occur in Africa, and the Pontederiales have their centre of variation in the Neotropics although some (e.g. *Monochoria*) occur in the Old World. The Haemodorales are best represented in Australia and Southern Africa, but genera such as *Xiphidium* and *Lachnanthes* are American. The Philydraceae have their centre in Australia, but reach through eastern Asia to Japan. The Typhales are subcosmopolitan.

*Relationships*. Several of the families treated here in separate orders are often placed in the Liliiflorae or Liliales, viz. Haemodoraceae, Velloziaceae, Philydraceae and Pontederiaceae. In DAHLGREN and CLIFFORD (1982) we also noted that they agree with the Liliiflorae in more features than with the Commeliniflorae. The majority of these features are connected with the insect pollination syndrome (or at least the lack of wind pollination syndrome), viz. petaloid tepals and presence of nectaries; others are considered as plesiomorphies, viz. sulcate pollen grains, presence of oxalate raphides, several or numerous ovules on an axile placenta, and dehiscent fruit (see HAMANN 1961).

However, most Bromeliiflorae share with the Commeliniflorae some rather marked presumably derived attributes, viz. epicuticular wax of the *Strelitzia* Type, presence of UV-fluorescent compounds in the cell walls, endosperm with copious starch, and stomatal complexes with well differentiated subsidiary cells.

Another interesting circumstance is that the Bromeliiflorae have a kind of helobial endosperm characteristic of this group (HAMANN, personal communication), and which probably evolved in their ancestors, a kind lacking in the Zingiberiflorae (where helobial endosperm is also rare).

Therefore we presume that the ancestors of the Bromeliiflorae belonged to the same branch as did the ancestors of both Commeliniflorae and Zingiberiflorae (possibly also those of the Areciflorae). Unlike most Commeliniflorae they had retained the entomophilous syndrome and in this they are still "liliifloran". One exception is the Typhales which, in spite of sharing several conspicuous features with other Bromeliiflorae (the particular kind of helobial endosperm, the oxalate raphides, the amoeboid tapetum, etc.), are adapted to wind pollination. (See the chapter on Evolution within the Monocotyledons.)

The strong resemblance between many Bromeliales and the Heliconiaceae, Strelitziaceae and Musaceae in the firm, boat-shaped and brightly coloured inflorescence bracts we consider to be a similarity by convergence (adaptation to bird pollination). Likewise, we do not consider the rather strong resemblance between Velloziaceae and Hypoxidaceae in general habit, epigyny and paracytic stomata to reflect a close phylogenetic relationship: the Hypoxidaceae are typical members of the Asparagales and have phytomelaniferous seeds. Even stronger, perhaps, is the superficial similarity between the South African genus *Lanaria*, which we refer to Tecophilaeaceae, and for example *Dilatris* of Haemodoraceae, both being densely hairy taxa with "corymbose" inflorescences. Only details such as embryology (DE VOS 1963) have been able to give strong evidence so far that *Lanaria* is tecophilaeaceous.

A problem which is not yet settled is whether the Burmanniales might also belong with the Bromeliiflorae, a possibility pointed out by HAMANN (personal communication). They have small seeds recalling in shape those of Philydraceae; the seeds before they are quite ripe may have starchy endosperm; endosperm formation is helobial; the endosperm has a small chalazal chamber as is typical of the Bromeliiflorae; and vessels have been recorded in autotrophic *Burmannia*. However, there are no UV-fluorescent compounds in the cell walls and the stomata are anomocytic, which is not the case with any Bromeliiflorae.

The closest relatives of the Bromeliiflorae, with the possible exception of Burmanniales, are undoubtedly the Zingiberiflorae.

# Order Velloziales

*One Family:* Velloziaceae.

Perennial plants, woody and either shrubby or arborescent, or herbaceous. The stem is simple or, seemingly, dichotomously branched, and is covered with persistent leaf sheaths. The roots are situated on the subterranean continuation of the stem in a dense fascicle. In *Xerophyta humilis* the subterranean stem may be described as a rhizome. The root vessels have simple perforation plates.

The leaves are spirally set, not distichous, the green, younger ones being crowded at the ends of the branches. They are linear, dorsiventral, flat, parallel-veined, entire or often dentate-spinulose along the margins, and sheathing at the base, and they lack stipule-like structures. Adventitious roots are characteristically formed at the nodes of the aerial stem, and penetrate the fibrous old leaf bases, a condition corresponding to that found in Bromeliaceae and *Kingia* of Dasypogonaceae. The stomata are principally paracytic, with two cells outside and parallel to the stoma, but sometimes tetracytic; they are sunk in longitudinal grooves on the lower side of the leaves. Trichomes of various types, unicellular or uniseriate, occur in the family and tend to occur in tufts. The leaves have more or less prominent sclerenchyma strands accompanying the veins, while vessels, with scalariform perforation plates, are only occasionally present in the stem; such vessels are usually present in the leaves (cf. woody taxa in the Liliiflorae). Tannin cells may by present and rhomboidal crystals occur in the leaves of some species. Laticifers are lacking. Crystal raphides seem to occur at least in some genera (AYENSU 1974).

The flowers appear, solitary or few, from the apex of the stem. Each peduncle-like "scape" probably represents a pedicel. Whether the flowers are terminal or not does not seem to have been investigated. The flowers are epigynous, actinomorphic and bisexual. Their 3 + 3 petaloid tepals are usually brightly coloured: blue, violet, purple, white or yellow, and are free or fused basally. In some genera the tepals bear separate or fused "coronal" appendages opposite the stamens, which are more or less fused with the tepals. These appendages undoubtedly have their origin in filament lobes as in Amaryllidaceae. The stamens are 3 + 3 or often, in *Vellozia,* by division of the stamen initials, more (18–66); in the latter case the stamens are in groups of three or other numbers. The filaments are narrow or flat and variously expanded

(see above), and in some *Barbacenia* species end in a tip on each side of the anther. The anthers are basifixed or peltate and normally long and linear and they dehisce introrsely, latrorsely or (very rarely) extrorsely with a longitudinal slit. The tapetum is secretory and microsporogenesis is successive. The pollen grains are solitary or dispersed in tetragonal tetrads (species of *Vellozia* only); they are sulcate or rarely inaperturate (AYENSU and SKVARLA 1974) and irregularly rugose in *Vellozia* but reticulate in *Barbacenia* and some other genera. They are probably mostly dispersed in the three-celled stage.

The ovary is trilocular, often globose and variously pubescent on the outside, with glands or scale-like or other hair structures. Septal nectaries are present in the ovary walls. The straight, slender and simple style generally widens apically into a trilobate, capitate or clavate stigma. Placentation is central and somewhat variable; the placentae being sessile, laminar or stalked, often globose or bibrachiate. They always bear numerous ovules. These are anatropous and weakly crassinucellate or nearly tenuinucellate. No primary parietal cell is cut off from the archesporial cell, but periclinal divisions in the nucellar epidermis usually occur. Embryo sac formation seems to conform to the *Polygonum* Type, and endosperm formation is helobial (DE MENEZES 1976).

The fruit is capsular, frequently hard, and opens irregularly or with apical fissures. It contains numerous seeds. The ripe seeds are relatively small, larger in *Vellozia* than in *Xerophyta,* and probably disseminated by wind. Their testa consists mainly of the thick outer epidermis of the outer integument, which is red from phlobaphenes; the inner integument is thin and unpigmented. The seeds contain copious endosperm, which in its outer layers contains aleurone and lipids and in the inner layers plenty of isodiametric (not bean-shaped) starch grains. The embryo is several times shorter than the endosperm and narrowly ovoid in shape.

*Chemistry.* Little is known on the chemistry of Velloziaceae. Tanniniferous cells are sometimes present, and crystals, including raphides, most likely of calcium oxalate, are likewise present.

*Distribution.* Most species of the family occur in the New World, mainly in South America, with the centres in Brazil, but the genus *Xerophyta* occurs also in the Old World, growing in eastern Africa from South Africa to Ethiopia and on Madagascar. The members inhabit more or less arid regions, and often grow in mountains with grass or shrub vegetation. Some species form veritable

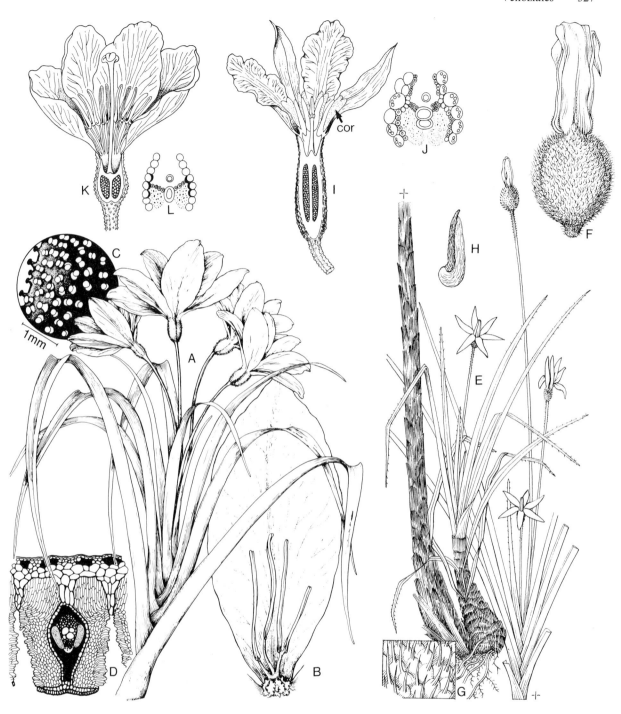

miniature "lily trees" up to 1 or 2 m, or very occasionally up to 6 m tall.

*Relationships.* Rarely has a family been so variously interpreted as regards its phylogenetic affinity as the Velloziaceae. Sometimes it has been considered most closely allied to the Bromeliaceae, and there is much to support this view: the general growth pattern, certain vegetative features, including the para-, tetra- (or hexa)cytic stomata, floral

**Fig. 152.** Velloziaceae. **A–D** *Vellozia stipitata.* **A** Flowering branch. **B** Tepal and attached staminal group. **C** Glands from the ovary surface. **D** Part of transverse section of leaf. (SMITH and AYENSU 1976). **E–H** *Vellozia schnitzleinia.* **E** Old branch (base to the *left,* top to the *right*) and young branch. **F** Young fruit. **H** Seed. (HEPPER 1968). **I–L** Comparison of subfamilial features of subfamily Vellozioideae (**K–L**) lacking a corona and with a single bundle sheath of leaf, and subfamily Barbacenioideae (**I–J**) with a corona (*cor*) and a double bundle sheath. (DE MENEZES 1980)

details, phytomelan-free seeds with endosperm having copious starch, etc. This view was advocated, for example, by HUBER (1969); he later treated the family as a separate order, Velloziales, next to Bromeliales in his Bromeliiflorae (HUBER 1977). An affinity to the Haemodorales can plausibly also be postulated.

The other views mainly stress affinity with either the Hypoxidaceae (AYENSU 1973) or the Amaryllidaceae – or both – within the Liliiflorae (two families here placed in Asparagales). DE MENEZES has recently (1980) remarked that "there is no family with a floral morphology closer to that of Velloziaceae than the Amaryllidaceae". In claiming this, DE MENEZES stresses in particular the close agreement in filamental and androecial corona structures: in fact there is an impressive series of parallels between the families in this regard. One detail that should be pointed out in particular is the occasional presence, in *Barbacenia* and *Xerophyta,* of flat filaments ending in two pointed lobes, one on each side of the anther, such as is found in some Amaryllidaceae, Alliaceae and Hyacinthaceae. Obstacles to acceptance of the latter affinity are the para- to tetracytic stomata of Velloziaceae (anomocytic in the three families mentioned), the lack of a leafless mesomorphic scape in Velloziaceae, the lack of a bulb, the quite different leaves, the different kind of epicuticular wax, the richly differentiated trichome structures, and the phytomelan-free seeds with a starchy content.

The Hypoxidaceae have a superficially more similar habit, they differ from nearly all other Asparagales in having paracytic stomata, and they likewise have epigynous flowers, whereby they may appear to be more likely candidates as relatives of the Velloziaceae. However, the totally different, larger seeds with a thick phytomelan layer and with a different testal construction, and the starch-free endosperm, argue against this.

It seems best to relate the Velloziales provisionally to the Bromeliales, but to note that the matter is still uncertain.

**Velloziaceae** S.L. Endlicher (1841)   5–6:360
(Fig. 152)

Description, see under the order.

The family has its centre in Brazil and adjacent countries in South America, where the two large genera *Vellozia* and *Barbacenia* occur, but it is also represented, by many species of *Xerophyta*, in Africa.

The family is divisible according to filament shape, stamen insertion, etc. The most recent classification, followed here, is that of DE MENEZES (1980). A different, possibly more natural classification is proposed by SMITH (1962) and SMITH and AYENSU (1974, 1976).

**Subfamily Vellozioideae**

This is characterized by flowers lacking a corona derived from the filaments and by leaves having a single bundle sheath. It includes the two genera *Vellozia* (100) and *Xerophyta* (incl. *Talbotia*; 55).

*Xerophyta,* distributed in central South America as well as over the eastern half of Africa and Madagascar, is characterized by six stamens, which may be clavate or capitate. – Related is *Vellozia,* which seems to be more derived in some respects. The stigmas are horizontal or reflexed and the stamens are nine or more in number. Typical coronal appendages are lacking although there may be filamental lobes ventral to the anthers. The pollen grains may cohere in tetrads. Some species in this entirely American genus are rather tall, with a trunk up to 6 m (DE MENEZES 1980), while others are small and some are rather delicate, e.g. the almost grass-like *V. exilis* and *V. grisea.* In some species, treated in the section *Radula,* the perigone tube is 2 cm long or more, while in other *Vellozia* species it is much shorter.

**Subfamily Barbacenioideae**

Subfamily Barbacenioideae is characterized by having coronal appendages in the flower and leaves with a double bundle sheath. It includes three or four genera.

The dominant genus in this subfamily is *Barbacenia* (100), which generally has glabrous leaf laminas (with or without ciliate margins), six stamens, and a clavate stigma. Coronal appendages (Fig. 152) are always present in this genus, but vary much in shape. Other, smaller genera, likewise in the New World, are *Pleurostima* (?), *Burlemarxia* (1) and *Aylthonia* (15).

# Order Bromeliales

*One Family:* Bromeliaceae.

Mainly rosette herbs but also herbs or semi-woody plants with leaves on a lengthened stem, and more rarely rosette trees, the maximum size being achieved by *Puya raimondii,* which reaches a height of up to 10 m, with a trunk up to 4 m. A climbing stem occurs in species of *Pitcairnia.* Most taxa are, however, trunkless, mostly medium-sized herbs, growing terrestrially or to a great extent as epiphytes. The stem continues vertically downwards in a subterranean part clothed with a sometimes thick periderm. Adventitious roots appear in the cortex of the upper parts of the stem. The roots generally have vessels with scalariform (rarely, as in species of *Pitcairnia* and *Vriesia,* simple) perforation plates. In the epiphytic plants, the roots often serve as climbing roots and their vascular system is then often reduced, water uptake being effected by the peltate hairs.

The leaves are usually spirally set (distichous only in some *Tillandsia* species), mostly dorsiventral, thick, stiff and horny, and supplied with a broad sheathing base. They frequently form a vase-shaped rosette in which water accumulates and from which it is absorbed ("tank plants"). The leaves are linear, broadly subulate, lanceolate, or rarely ovate; they often end as a thorn and may be laterally dentate or spiny. Stipules and similar structures are lacking. Internal water-storing mesenchymatous tissue is generally present.

The vascular strands in stems and leaves often but not consistently contain vessels, which then have exclusively scalariform perforation plates (CHEADLE 1955b). Characteristic of the family are the water-absorbing peltate scales, which have a uniseriate stalk, more or less sunk in a groove of the epidermis, and a multicellular stellate to plate-like shield which is often pressed against the epidermis (BENZING et al. 1978). Stomata are mainly localized in furrows. They generally have two longitudinal subsidiary cells outside the stoma cells (i.e. a paracytic basic pattern) but in addition are frequently surrounded by four other epidermal cells. Sacs with bundles of calcium oxalate raphides are abundant in all parts and, in addition, spherical silica bodies of variable size, although mostly small, occur in epidermal cells of the leaves and stem. Moreover, yellowish resin- or oil-like droplets may be secreted in the parenchyma of the leaf veins.

The inflorescences are terminal on erect leafy stems, which come out from the rosettes and are well delimited from the lower part of the plant. The leaves on these axes are more or less bracteal. Inflorescence construction seems to be mainly racemose: simple or compound spikes, racemes or thyrses, which often have conspicuous distichous and brightly coloured bracts subtending the flowers or lateral inflorescence branches. The flowers are hypogynous or epigynous and generally bisexual and actinomorphic, rarely somewhat zygomorphic. The perianth consists of 3+3 tepals, which are more or less unequal. The outer are generally much smaller than the inner, often hyaline or greenish and either free or more or less connate with each other. The inner tepals are petaloid and have various, bright colours, blue, violet, reddish, yellow, white, greenish; they are free or basally somewhat connate, and sometimes have "ligular" appendages on each side of the stamens (cf. the Velloziaceae). The aestivation is imbricate or (clockwise) contorted.

The stamens are 3+3 in number. Their filaments are narrow and free from each other and are inserted at the base of the tepals when these are free, and adnate to them when they are connate. The anthers are basifixed or peltate, introrse and often long and linear; their dehiscence is longitudinal. The tapetum is secretory and microsporogenesis is successive. Pollen tetrads occur in at least two genera (*Cryptanthus* and *Hohenbergia*).

The pollen grains are sulcate in most taxa but in several genera of subfamily Bromelioideae they are biforaminate or rarely tri- or polyforaminate. They are dispersed in the two-celled or rarely in the three-celled stage.

The ovary is trilocular and the style slender, with three often contorted, but sometimes small, commissural stigmatic branches or lobes. Septal nectaries are present. The stigmatic papillae have a Wet or less commonly Dry surface. The centrally inserted placentae bear many or rarely few ovules, which may or may not have a chalazal appendage. The ovules are anatropous, or rarely campylotropous, and crassinucellate. A primary parietal cell is cut off from the archesporial cell and forms some parietal tissue. Embryo sac formation conforms to the *Polygonum* Type, and endosperm formation is helobial (BILLINGS 1904).

The fruit is variable, either a septicidal or, more rarely, a loculicidal capsule or a berry. The seeds are rather small. The testa is smooth or fleshy, rarely appendiculate or provided with a wing or pappus-like process. The seeds contain a well-developed endosperm and a cylindrical embryo, which is basal and either peripheral or almost axile in relation to the endosperm. This contains plenty of starch grains and has lipids and aleurone at the periphery.

*Chemistry*. Beyond what is said above on raphides, silica bodies, starch, etc., the family is not very well known chemically. Leaf waxes and mucilage, contained in schizolysigenous ducts, occur in the family, and steroidal saponins occur as well (reported for *Hechtia,* at least). Chelidonic acid is not reported for the family. Soluble starch occurs in the ovaries and other parts. The great concentration of sugar in the swollen inflorescence of *Ananas* is the reason for its use as "fruit".

*Distribution*. All Bromeliales are indigenous to the Americas with the exception of one species doubtfully native in West Africa. In America the family makes up a conspicuous part of the epiphyte vegetation in monsoon forests, rain forests and mist forests, and terrestrial genera are important in various dry habitats. In the Andes bromeliaceous taxa ascend to 4,000 m.

*Relationships*. The Bromeliaceae has been referred to the Liliiflorae or Commeliniflorae, which is to be expected, as the family shares attributes with both.

Commeliniflorous features, in particular, are the usual presence of vessels in the stems and leaves, the accumulation of silica bodies in the epidermis, the mealy starch endosperm, and the small peripheral embryo. Liliiflorous features are, for example, the presence of bundles of crystal raphides, the petaloid inner whorl of the perianth, the most frequently two-celled pollen grains, the anatropous ovules, the helobial endosperm formation and the occurrence (in some case at least) of steroidal saponins.

Bromeliales seem to be most closely related to such orders as Velloziales, Pontederiales, Haemodorales and, perhaps, also Philydrales (but also perhaps to Xyridaceae, Rapateaceae and Juncaceae, which are members of the Commeliniflorae).

**Bromeliaceae** A.L. Jussieu (1789)   51:1.520
(Figs. 153–154)

Description as for the order.

The family is to be regarded as a "climax group". Within its combination of characters it has reached an extraordinarily rich variation. The specializations in water economy (the sheaths, the xeromorphic leaf features, the water-absorbing

peltate hairs, the water storing tissue, etc.) are the features that are most striking.

The floral syndrome of ornithogamy is prevalent in the family, and is obviously expressed in the frequency of bright primary colours, and the participation of bracts and leaf bases in this display. Although most such adaptive features are related to hummingbirds (Trochilidae), syndromes clearly related to other types of birds are found in the terrestrial *Pitcairnia* and *Puya*. Flowers adapted to Lepidoptera are also found, for example in *Tillandsia lindenioides*.

The Bromeliaceae are traditionally divided into three subfamilies, a division that will also be followed here. There are some differences between the subfamilies in growth and habitat; thus the Pitcairnioideae are mostly terrestrial, the Tillandsioideae mostly epiphytic and the Bromelioideae a combination of both.

The geographic concentration of Bromeliaceae is in tropical and Andean America from Patagonia and Tierra del Fuego in the south to eastern Virginia and Mexico in the north. *Pitcairnia feliciana* occurs in West Africa, where it grows on the coast. It probably represents an example of long-distance dispersal if it has not been carried there by man.

**Subfamily Pitcairnioideae**

This is characterized by hypogynous flowers and capsular fruits, the seeds of which are generally provided with wings or other appendages, but not with a hair tuft. This subfamily includes mainly terrestrial plants, many of which have spiny leaf margins. It consists of ca. 13 genera with ca. 420 species.

*Navia* (6) in Brazil, Guayana and Colombia, are rosette plants with a compact, almost sessile inflorescence. The flowers lack "coronal" structures and the capsules have reticulate seeds without appendages; in the latter character the genus is exceptional in this subfamily.

Among the more important genera are *Pitcairnia* (185), a variable, largely Andean genus with conspicuous, slightly zygomorphic flowers and septicidal capsules. Many species grow in rocky places; a few are epiphytes. Some, like *P. ferruginea,* are large and have a stout, almost woody base and a much-branched inflorescence; other species are small. In *P. heterophylla* some leaves are small and spiny, others larger and spineless.

*Puya* (90), *Dyckia* (75) and seven other genera have capsules opening septicidally as well as loculicidally, and their flowers are actinomorphic or

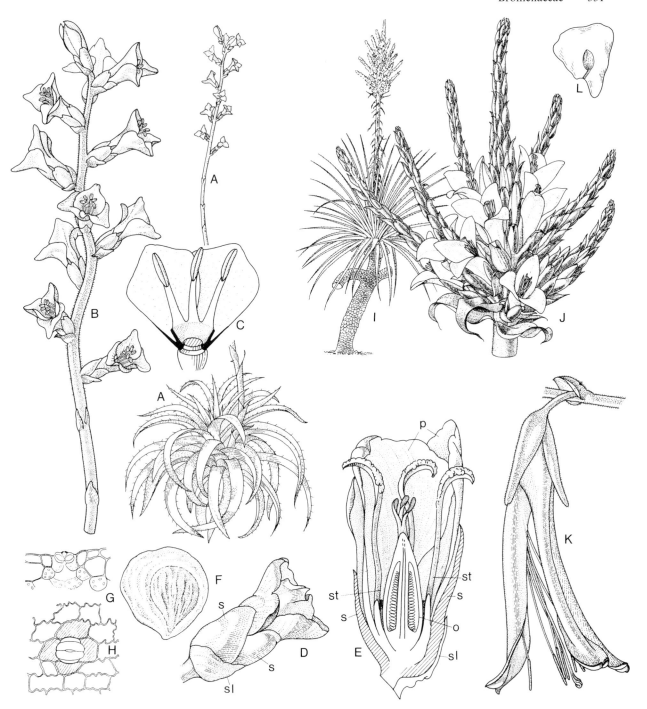

(rarely in *Puya*) slightly zygomorphic. – *Puya* includes some species which reach an altitude of up to 4,000 m. An alpine, Andean species is *P. raimondii,* which has a trunk up to 4 m high and with its terminal leaf rosettes and branched inflorescences reaches a height of about 10 m. – *Dyckia* differs from *Puya* in having basally connate inner tepals. It has spicate inflorescences with small bracts.

**Fig. 153.** Bromeliaceae. **A–C** *Dyckia remotifolia.* **A** Plant. **B** Inflorescence. **C** Two of the inner tepals and three stamens, showing the basally fused filaments. **D–F** *Dyckia* spp. **D** Flower. **E** Same, longitudinal section (*sl* subtending bracteate leaf; *s* members of the sepaloid outer tepal whorl; *p* members of the petaloid inner tepal whorl; *o* ovary). **F** Seed. **I–J** *Puya chilensis.* **I** Habit. **J** Part of inflorescence. **K–L** *Puya ferruginea.* **K** Flower. **L** Seed. (**A–L** from RAUH 1981). **G–H** *Pitcairnia xanthocalyx,* stoma, in transverse section and in surface view. (TOMLINSON 1969)

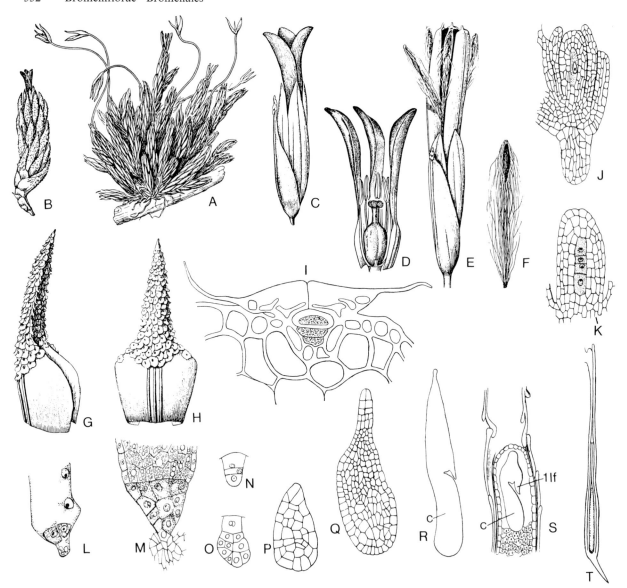

**Fig. 154.** Bromeliaceae. **A–H** *Tillandsia pedicellata.*
**A** Plant in its epiphytic position. **B** Lateral shoot, flowering stage. **C** Flower with its subtending bract. **D** Flower
with sepaloid outer tepals removed, and inner tepals and
stamens opened up to show pistil. **E** Capsule in the stage
of dehiscence. **F** Seed. **G–H** Leaf in different views.
(Correa 1969). **I–T** *Tillandsia usneoides.* **I** Peltate hair.
**J** Ovule. **K** Nucellus with linear megaspore tetrad.
**L** Basal part of embryo sac showing polar nuclei and
antipodals. **M** Endosperm, demonstrating the relatively
small, starchless chalazal chamber. **N–Q** Embryogeny
(*Asterad* Type). **R** Embryo (*c* cotyledon). **S** Part of mature seed with embryo and part of endosperm (*c* cotyledon; *1 lf* first leaf). **T** Seed. (**I** Solereder and Meyer
1929; **J–T** Billings 1904)

## Subfamily Tillandsioideae

This subfamily, like subfamily Pitcairnioideae, has
hypogynous or rarely (*Glomeropitcairnia*) half-epigynous flowers, and their seeds are provided with
a hair tuft. Tillandsioideae consist mostly of epiphytic plants. Their leaves generally have entire,
spineless margins. The subfamily includes only 8
genera, but about 675 species.

The largest genus is *Tillandsia* (390), which is
centred in tropical and subtropical America, but
ranges from southern U.S.A. in the north to southern Chile in the south. The species vary in size
and range from herbs with large rosettes or stiff
leaves to the small, copiously branched, greyish
*T. usneoides* ("Spanish Moss"), which lacks roots
and festoons the branches of the trees in great
quantities resembling the lichen *Usnea*. In recent

years it has become used as packing material! (Plants such as this, relying largely on traces of nutrients contained in rain water and the rinsings from the outer parts of the tree crowns, have been described as "air plants"). Many species, which often have bluish inner tepals and rather large inflorescence bracts, are grown as greenhouse ornamentals. – *Vriesia* (115) consists of species which often have wide, smooth leaves and spikes in which the distichous bracts are often large and brightly coloured. The flowers are often fleshy and have tepals with "coronal" appendages. Many are grown as ornamentals. – *Guzmannia* (85), unlike the two preceding genera, has connate inner tepals. A "corona" is lacking. The inflorescences are spicate, and may have coloured bracts.

## Subfamily Bromelioideae

Subfamily Bromelioideae is characterized by epigynous flowers and baccate fruits. The seeds lack hairs or appendages. Some 30 genera with ca. 425 species belong to this subfamily. Many are terrestrial plants but a great number are epiphytes. Their leaves are often serrate-spiniferous.

Here belong *Bromelia* (36) in South America and the West Indies, a genus with serrate leaves and partly fused inner tepals, and *Fascicularia* (5) with rigid leaves and an almost sessile, compact inflorescence inserted in the centre of the leaf rosette. – *Nidularium* (30) is a "tank plant" with the flowers opening just above the surface of the water contained in the rosette, the inner leaves of which may be red at the base. Some species are grown as ornamentals, as are many of *Aechmea* (135) which has fringed "coronal" structures. – *Ananas* (3), likewise with tepals having "coronal" structures, is a terrestrial genus in which fruits and inflorescence parts fuse and swell into a fleshy, sugar-rich fruit aggregate. The genus has its centre in Brazil. *A. comosus*, "Pineapple", is not known as wild in the West Indies, where it was possibly introduced by the earliest inhabitants. – *Billbergia* (50) consists mainly of epiphytes, many highly ornamental, like the commonly cultivated *B. nutans*. While most of the previous genera of the subfamily have foraminate pollen grains, they are sulcate in this genus. The flowers have short outer and long inner tepals; the latter have "coronal" structures and tend to be weakly zygomorphic. Many species of *Billbergia* and *Nidularium* hold a reservoir of water in leaf rosettes serving as breeding places for mosquitoes, some species of which breed nowhere else.

## Order Philydrales

*One Family:* Philydraceae.

Erect, perennial, from small to rather large, pubescent herbs with short, sometimes tuber-like rhizomes (or corms) on which the roots are fascicled. The roots have vessels with scalariform perforation plates, but vessels are lacking in stems and leaves (at least in *Orthothylax*). The leaves are cauline and alternate, basally distichous, linear, flat and parallel-veined. They are sheathing at the base, otherwise compressed, unifacial and ensiform (as in *Iris*). Stipule-like and "ligular" structures are lacking. The stomata are paracytic but besides the two subsidiary cells parallel to the guard cells there are two or four less distinct subsidiary cells. Silica bodies and crystal raphide bundles seem to be lacking in the vegetative parts, but oxalate styloids occur. Small glandular hairs and longer, uniseriate hairs are frequent, especially on the leaf sheaths and stem of most taxa. Laticifers are lacking.

The inflorescences are simple or compound spikes, with the flowers in axils of, and sometimes partly fused with, rather large bracts.

The flowers are basically trimerous, hypogynous and strongly zygomorphic. The tepals are petaloid, yellow or white in colour. The two lateral (upper) tepals of the inner whorl are fused with the median (upper) tepal of the outer to form a broad, large upper lip, which in *Philydrum* may be tridentate at the apex, and has three rather prominent veins. The median (lower) tepal of the inner whorl likewise forms a large, lower lip, while the lateral tepals of the outer whorl are small, petaloid and sometimes fused with the solitary functional stamen. Perigonal nectaries are lacking.

The androecium is reduced to a solitary stamen situated medially in the lower central part of the flower. It is considered to correspond to the median stamen of the inner whorl in an imaginary complete ancestor. Basally its filament, which is glabrous, may be fused with the inner tepals. The anther, which may have a broad connective, is peltate, two-celled and tetrasporangiate. It varies from introrse to extrorse and generally becomes helically coiled in anthesis. It dehisces longitudinally. The tapetum is secretory. Oxalate raphides are often present in the tapetal cells. Microsporogenesis is successive, and the pollen grains, which are dispersed singly or, in *Philydrum,* as tetrads, are sulcate or triaperturate (ZAVADA 1983) and two-celled.

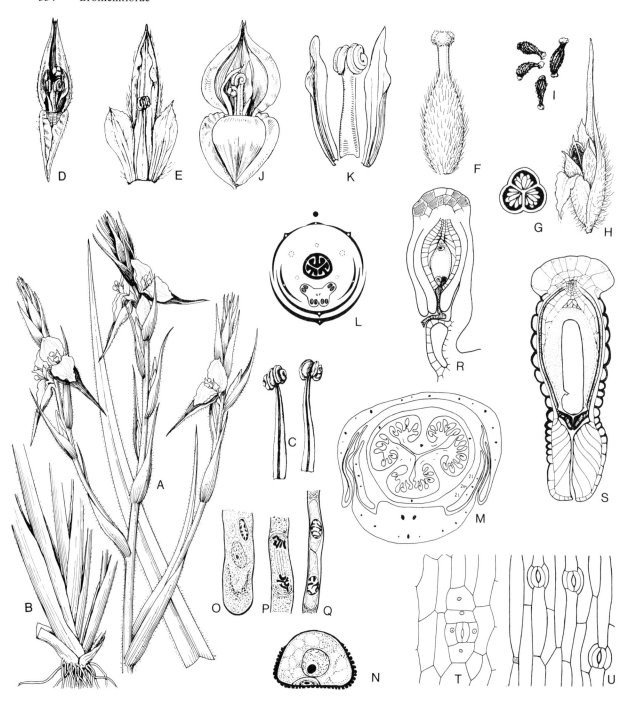

**Fig. 155.** Philydraceae. **A–L** *Philydrum lanuginosum.* **A** Inflorescence and part of leaf. **B** Base of shoot. **C** Stamen in different views. **D** Floral bud opened up. **E** Flower, lower lip and pistil removed. **F** Pistil. **G** Ovary, transverse section. **H** Capsule. **I** Seeds. **J** Complete flower. **K** Lateral tepals of the inner whorl and functional stamen. **L** Floral diagram. **M** Flower, transverse section at ovarial level. **N** Pollen grain from late tetrad stage. **O–Q** Pollen tube at different stages showing division of the generative cell into two sperm cells. **R** Ovule in the state of being fertilized, longitudinal section. **S** Ripe seed, longitudinal section, orientation as for **R**. **T** Stomatal complex. (A SIMS 1904; **B, E** and **G–I** YANG 1978, partly redrawn; **D** and **J–T** HAMANN 1966). **U** *Philydrella pygmaea,* leaf epidermis with stomata. (HAMANN 1966 a)

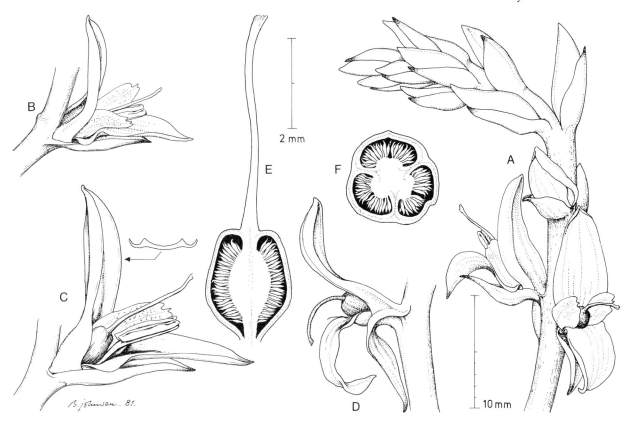

The pistil is tricarpellary and trilocular in all genera except *Philydrum,* where it is unilocular (although trilocular at the base) and the placentae are parietal, far-intrusive and bibrachiate. Septal nectaries are lacking. The style is simple and the stigma capitate to trilobate with a Dry and papillate surface. There are numerous ovules, which are anatropous and crassinucellate. A primary parietal cell is cut off from the archesporial cell and forms a parietal tissue. Embryo sac formation follows the *Polygonum* Type and endosperm formation is helobial. In the endosperm there is unusually rapid division of nuclei and wall formation in the small chalazal chamber, a type of endosperm found also in the Typhales.

The fruit is a loculicidal capsule or rarely irregularly dehiscent. The seeds are narrowly pyriform and terete and have a twisted epidermal pattern on the main body. A number of cells, with tannin-like contents, at the chalazal end enlarge to form a "cap"-like body, which must not be confused with an elaiosome; at the apical end the cells of the outer layer of the outer integument enlarge as well to form a caruncle-like prolongation. In the ripe seeds the endosperm is copious and rich in starch, fatty oils and crystalline bodies, probably protein. The starch grains are of two types, large

**Fig. 156.** Philydraceae. *Orthothylax glaberrimus.* **A** Apex of inflorescence with flowers and flower buds. **B** Flower. **C** Same, with one of the lateral tepals of the inner whorl removed to show the single stamen. **D** Flower in early fruiting stage, one lateral tepal removed. **E** Pistil, longitudinal section, showing the very numerous ovules. **F** Ovary, transverse section. (Orig. B. JOHNSEN)

and bean-shaped to elliptic and small and isodiametric. The embryo is linear and straight and nearly as long as the endosperm. $x = 8, 11$.

*Chemistry.* The order is poorly known as regards the chemical contents.

*Distribution.* It is distributed in Eastern Asia, Indomalaysia and Australia. Most taxa grow in marshy habitats.

*Relationships.* The family Philydraceae has been thoroughly analyzed by HAMANN (1966a), who considers it homogeneous and, like HUBER (1969), finds it to approach most closely the Pontederiales and, although probably less closely, the Haemodoraceae. [Very recent evidence indicates a possibly close affinity also to the Burmanniales, although the studies in these conditions are not completed (HAMANN, pers. communication).]

The superficial floral similarities to the Orchidaceae are due to convergent evolution.

**Philydraceae** Link (1821)   4 : 5   (Figs. 155–156)

Description as for the order. The family ranges from Southern Japan through South-Eastern Asia to Southern Australia, but not Tasmania.

*Helmholtzia* (3) in Indonesia and North-Eastern Australia consists of large rhizomatous herbs with branched inflorescences with white flowers and indehiscent fruits.

*Philydrum* (1) *lanuginosum* is the commonest species and ranges almost through the whole distribution of the family. It is likewise a rather tall herb but has a more tuberous "caudex" and a less branched, often simple spike of yellow flowers. The fruit is a loculicidal capsule. – There are two more monotypic genera, *Orthothylax* (Fig. 156) and *Philydrella,* the former from Eastern Australia and the latter from South-Western Australia.

# Order Haemodorales

*One Family:* Haemodoraceae.

Terrestrial, green, mostly erect herbs with subterranean short rhizomes, stem-tubers or bulbs. The roots are fasciculate and fibrous, sometimes swollen and tuberous, as in *Tribonanthes* (PATE and DIXON 1981), and their vessels have scalariform or, in particular, simple perforation plates. Stolons are rarely present.

The aerial stems are from short up to more than 1 m high and are generally more or less pubescent. The stems as a rule are leafy at least on the basal parts, the leaves being distichous, linear, *not* differentiated in petiole and lamina, sheathing at the base, and either dorsiventral and triangular or flat in transection or, in certain genera of both subfamilies, equitant-ensiform, often prominently parallel-veined. Ptyxis is sometimes plicate. The leaf tips are generally cylindrical. Stipule-like or ligule-like structures are lacking.

The hairs in Haemodoraceae subfamily Haemodoroideae are unicellular bristle hairs mingled with longer, uniseriate hairs; in subfamily Conostyloideae they are frequently multicellular and branched. The stomata are paracytic (STEBBINS and KHUSH 1961), with oblique divisions in the neighbouring cells. Silica bodies are lacking, but crystal raphides contained in mucilage cells are present in the order. Laticifers are perhaps lacking, although the rhizome of *Dilatris,* if cut, exudes a bright red juice, containing haemocorin. Vessels are often lacking in stems and leaves, in particular in subfamily Conostyloideae, but are sometimes present (at least in *Lachnanthes* and *Xiphidium*) in which case they have scalariform perforation plates (CHEADLE 1968, 1969).

The inflorescence varies from sparse and pauciflorous to much-branched and multiflorous; sometimes it is even restricted to a single flower. It represents a panicle, a thyrse with cymose lateral branches, or a raceme. The inflorescence axes are usually hairy or, more rarely, glabrous (in species of *Haemodorum, Barberetta*).

The flowers are bisexual, trimerous, hypogynous, epigynous or hemiepigynous and from nearly actinomorphic to strongly zygomorphic. The 3 + 3 perianth members are persistent and either biseriate (in subfamily Haemodoroideae) or uniseriate (in subfamily Conostyloideae), in which case they are more or less fused into a short or long, straight or curved tube. The tepals are similar in the two whorls and are generally pubescent on the outer (back) side. They vary in colour from violet to red, orange, yellow or, in certain parts, also green or black. In particular in certain bird-pollinated genera (e.g. *Anigozanthos*) the flowers are very brightly two- or three-coloured. Pollination is also effected by insects or even small mammals. Genera with both free or nearly free tepals and with tubular flowers show zygomorphy, in the former case with differences in size of perianth members and occurrence of basal spots (nectar guides), while in the latter case there may be a deep slit between the lowest two lobes, so that all six lobes seem to form an upper lip. Perianth nectaries are lacking.

The stamen number may be three (most genera of subfamily Haemodoroideae) or six (subfamily Conostyloideae and *Phlebocarya*). When the stamens are three the outer staminal whorl is missing. The filaments, which are glabrous, are free from each other, but are often attached to the inner side of the perigone tube. The anthers are elongate, either basifixed (often sagittate-basifixed) or peltate, and they dehisce longitudinally. In some cases (*Conanthera*) an apical process, entire or divided, is present on the anther tip. The tapetum in *Anigozanthos* has been reported to be amoeboid, but it is not certain that this is typical of the family. Microsporogenesis is successive. The pollen grains, which are single, are either sulcate (subfamily Haemodoroideae) or have two, three or four, rarely up to eight, circular apertures (mainly subfamily Conostyloideae) (RADULESCU 1973a). They are dispersed in the two-celled stage.

The pistil is trilocular, with septal nectaries present in the ovary. The style is simple and entire, terminating in a small, capitate or trifid stigma. The surface of this is Dry (e.g. in *Xiphidium*) or Wet (e.g. in *Anigozanthos* and *Wachendorfia*). The ovary has three locules in each of which are one to many mostly downward-directed, orthotropous (or in *Wachendorfia,* hemianatropous) and crassinucellate ovules. A primary parietal cell is always cut off from the archesporial cell and forms a parietal tissue which is several cell layers in thickness. Embryo sac formation in all cases known conforms to the *Polygonum* Type. The embryo sac has large antipodal cells. Endosperm formation is helobial.

The fruit is a loculicidal capsule with one to several seeds per locule. In Haemodoroideae, the seeds are ovoid and kidney-shaped in *Wachendorfia,* where they also have a caruncle. In the other genera they are irregularly tetrahedral without appendages (*Schiekia, Xiphidium*) or flat and orbicular (coin-like), sometimes with an annular margin (*Haemodorum, Lachnanthes, Dilatris*). In Conostyloideae they are ovoid-elliptic or slightly cristate on one side, but occasionally they are angular-polygonal or tetrahedral (and in their longitudinally ridged surface they slightly resemble the seeds in Taccaceae). The outer integument of the seed coat is two-layered, and either colourless or with brown or (in *Wachendorfia*) black pigment in the outer epidermis, but it is never encrusted with phytomelan. It consists of well-defined cells in subfamily Haemodoroideae, but of collapsed cells in the members of subfamily Conostyloideae studied. The inner integument consists of depressed cells with brownish contents. The endosperm is well developed, and clearly differentiated into a peripheral aleurone and lipid layer, in which the cell walls are also partly thick and cellulose-storing, and an inner zone of cells with thin, unpitted walls and containing starch grains. The starch grains are simple and bean-shaped or, in *Wachendorfia,* compound (often of four partial grains). The embryo is small, globose to ovoid, and mostly only ca. 1/5 of the length of the endosperm or less.

*Chemistry.* Oxalate raphides and chelidonic acid are widely distributed in the order. The occurrence of other compounds is incompletely known. The subterranean parts of genera in subfamily Haemodoroideae often contain red pigments, which consist of a glycoside, haemocorin. Alkaloids and steroidal saponins are not reported to occur in the order. Starch grains occur in the vegetative parts as well as in the endosperm.

*Distribution.* Haemodorales have a Southern Hemisphere distribution with the main centre in Australia and other centres in South Africa and South (to North) America. *Lachnanthes* occurs in Atlantic North America.

*Relationships.* The order is defined largely by the syndrome of embryological characters (tapetum, pollen, ovule and seed characters). Its members have been confused in most previous literature with members of the asparagalean families Hypoxidaceae, Tecophilaeaceae and Cyanastraceae, but the order has been defined largely thanks to works by DE VOS (1956), HUBER (1969) and HUTCHINSON (1959, 1973), who circumscribed it nearly as here. Recently SIMPSON (1983) has clarified the circumscription of Haemodoraceae in a pollen-morphological study.

The affinity of Haemodorales hitherto has been somewhat unclear. A close affinity to the families Hypoxidaceae and Tecophilaeaceae (Asparagales) is *not* likely. The features shared with these families are rather the result of convergent evolution. Nor does there seem to be such a close affinity to Taccales as was proposed by HUBER (1969).

The closest affinity is probably with the Philydrales, Pontederiales, Velloziales and Bromeliales. HUBER (1977) finds Haemodorales so isolated that he places the order in a separate superorder, Haemodoriflorae.

## Haemodoraceae R. Brown (1810)    14:100
(Fig. 157)

Description and distribution as outlined above, under the order.

The family is divisible into two subfamilies.

### Subfamily Haemodoroideae

The members of this subfamily are glabrous or hairy herbs, the hairs being simple in the sense that they are unicellular or consist of a single cell row. The starch grains in the vegetative parts are simple. The flowers have free or almost free tepals, which are biseriate, those of the outer whorl conspicuously overlapping those of the inner whorl. In androecial characters the subfamily is advanced, the outer whorl of stamens being totally absent. The pollen grains are sulcate. The flowers may be either hypogynous or epigynous. The seeds are variable (see under the order), but are not longitudinally ridged; they are hairy in *Wachendorfia*, glabrous in the other genera.

This subfamily, which is the largest in number of genera, occurs in all three of the Southern Hemisphere continents and consists of the genera *Barberetta, Dilatris, Haemodorum, Lachnanthes, Pyrrhorhiza, Schiekia, Xiphidium* and *Wachendorfia*.

*Wachendorfia* (5) in the Cape Province of South Africa, consists of herbs from ca. 20 cm to 1 m high, with subterranean tubers with red tissue. Their leaves are equitant-ensiform, plicate and more or less hairy. The inflorescence is a panicle or thyrse with rather large hypogynous flowers, which have yellowish to violet, spreading tepals and three equal stamens. Each locule of the ovary contains a single ovule, which develops into a globose-ovoid seed covered with short, black hairs. *W. paniculata* is a large, stream-side species with deep yellow flowers. – *Dilatris* (5), likewise in the Cape Province, has smaller, mostly blue-violet, epigynous flowers in a corymbose inflorescence. In this genus, and also in a third African genus,

*Barberetta* (1), the locules contain a single ovule, which matures into a flat seed.

The other genera in the subfamily have two or more ovules in the locules, and seeds with simple starch grains. They are distributed in Australia (*Haemodorum*) and America (four genera). – *Haemodorum* (20), from Australia, New Guinea and Tasmania, are herbs with glabrous, thyrsoid inflorescences. The flowers are actinomorphic and epigynous or hemiepigynous. – Of the American genera, with one or few species each, may be mentioned *Xiphidium* (2), ranging between tropical South America and Mexico, with hypogynous flowers, and the white woolly *Lachnanthes* (1) *tinctoria,* from Florida to Massachussetts, which has epigynous flowers.

### Subfamily Conostyloideae

This subfamily for the most part consists of pubescent herbs. Their hairs are branched and made up of several cell rows. The starch grains in the vegetative parts are compound, but those in the endosperm are simple. The tepals may be almost free but are more often fused into a tube which may be deeply split medially on the lower side (*Anigozanthos*). They are all inserted, as it seems, in a single whorl. In contrast to the Haemodoroideae, the stamens are six in number and are generally attached far up in the perianth tube. The pollen grains are also more complicated than in the previous subfamily, having two, three or more (up to eight) apertures, which are rounded. The flowers vary from hypogynous to epigynous, and have glabrous, ovoid to tetrahedral, longitudinally furrowed seeds, which have a seed coat with colourless outer integument. This subfamily is more uniform than Haemodoroideae. It is restricted to Western Australia. It consists of the genera *Anigozanthos, Blancoa, Conostylis, Macropidia, Phlebocarya* and *Tribonanthes*.

The only hypogynous genus is *Tribonanthes* (5), in which the anthers have a conspicuous apical connective appendage. – Of the other, more or less epigynous, genera may be mentioned *Macropidia* (1), which has almost actinomorphic, yellow, black-hairy, tubular flowers with long filaments and one-seeded locules. The fruit is a tripartite schizocarp with large seeds. – The related well-known genus *Anigozanthos* (12), the "Kangaroo Paw", is frequently cultivated for ornament. It has equitant-ensiform leaves and subepigynous, zygomorphic, tubular, six-toothed flowers with a deep lower slit. The colours of the flowers, which are

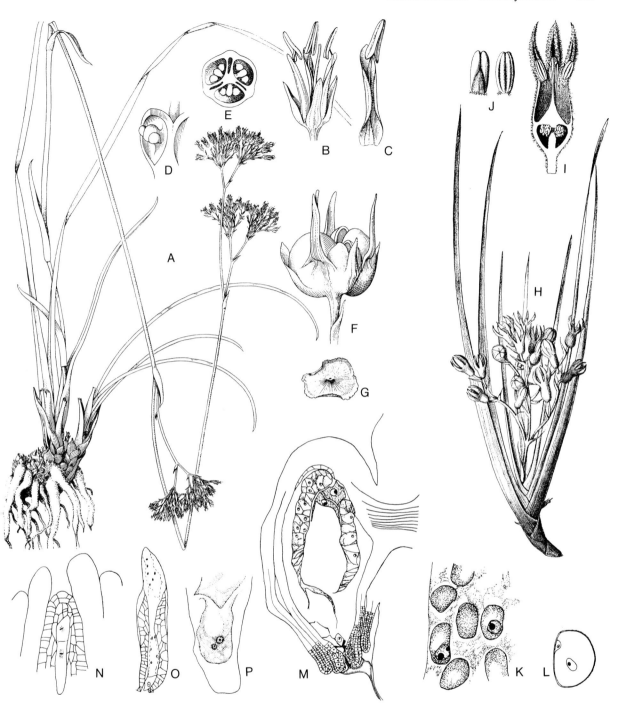

**Fig. 157.** Haemodoraceae. **A–F** *Haemodorum corymbosum.* **A** Plant. **B** Flower. **C** Tepal and opposite, attached stamen. **D** Sectioned ovary to show insertion of ovules. **E** Ovary, transverse section. **F** Fruit with concrescent perianth and persistent filaments, loculus dehiscing on top, that in front more or less abortive. **G** *Haemodorum planifolium,* seed. (**A–G** VAN STEENIS 1954). **H–J** *Conostylis phathyrantha.* **H** Plant. **I** Flower, longitudinal section. **J** Stamen in different views. (PAX and HOFFMANN 1930). **K–L** *Wachendorfia paniculata.* **K** Amoeboid tapetum. **L** Two-celled pollen grain. **M** Ovule (hemianatropous). **N–P** Helobial endosperm formation, **O–P** showing the very delayed and minimal chalazal chamber of the endosperm. (**K–M** DELLERT 1933; **N–P** DE VOS 1956)

hairy on the outer side, are unusual, often green, red, yellow or blackish in various combinations, being thus attractive to their bird pollinators. The fruits are capsular and several-seeded. *A. flavidus* and *A. pulcherrimus* are frequently grown in gardens. – *Conostylis* (40) has a more capitate inflorescence. The variation of the flowers in this Western Australian genus is great; the tepals vary for example from being almost free from each other to connate into a long tepal tube; the placentae and seed numbers are also very variable.

*Pauridia* (2) is probably a member of Hypoxidaceae (see under that family).

Another genus, sometimes referred to Haemodoraceae, is *Lanaria* (1), which is here included in Tecophilaeaceae mainly on the basis of its embryological characters and its phytomelaniferous and starchless seeds.

# Order Pontederiales

*One Family:* Pontederiaceae.

Rhizomatous or stoloniferous annual or perennial plants including small to fairly large swamp or aquatic herbs, generally with a sympodial branch system. The plants may be submerged, floating or rooted and emergent. The branches are mostly spongy and aerenchymatous, and erect or ascending. The leaves are generally distichous and differentiated into a considerable sheath which envelops the stem, a petiole of sometimes considerable length (as in *Heteranthera* and *Pontederia*) and a relatively broad, broadly linear to lanceolate, ovate or even broadly cordate lamina; in *Hydrothrix* the leaves are dissolved into undifferentiated, filiform bands situated in pseudo-whorls. The sheath may be widened and continue in the petiole base. In several genera a ligule-like membrane is present at the top of the leaf sheath. Peculiar inflated petioles filled with aerenchymatous tissue are characteristic of *Eichhornia crassipes,* an adaptation to its partly free-floating habit. The main venation of the leaf lamina is parallel or (in *Pontederia*) pinnate with densely placed laterals ending along the leaf margin.

Trichomes are lacking on the vegetative parts. The stomata are of paracytic type. Laticifers are lacking. Oxalate raphides are present, at least in some genera, but solitary crystals are more common. Tanniniferous idioblasts may be present on the leaf petioles. The starch grains in the vegetative parts are simple (centric and oval) as well as compound (HAMANN 1961). Vessels with scalariform perforation plates are present in the roots, and often (but not always) in the stems, while the leaves are vessel-less.

The inflorescences are terminal racemes, spikes or thyrses (rarely unifloral), often subtended by a spathe-like leaf. The flowers are large to fairly small, entomogamous or autogamous and blue, lilac, white or rarely yellow. They are hypogynous, trimerous, and from almost actinomorphic to distinctly zygomorphic, in the latter case with the medial inner tepal directed upwards, larger than the others and provided with basal spots or streaks

Fig. 158. Pontederiaceae. **A–E** *Eichhornia crassipes.* **A** Plant. **B** Stamen. **C** Pistil. **D** Capsule. **E** Seed. **F–I** *Pontederia lanceolata.* **F** Plant. **G** Flower. **H** Stamen. **I** Pistil. **J** Nutlet. (**A–J** from CABRERA 1968). **K** *Pontederia cordata,* pollen grain. (ERDTMAN 1952)

(nectar guide). All of the 3 + 3 tepals are petaloid, blue, violet, white or rarely yellow, and are often basally fused into a tube. Cleistogamous flowers occur in *Heteranthera*.

Triheterostyly, with three different lengths of the style, occurs in species of *Eichhornia, Pontederia* and *Reussia* (Cook et al. 1978), while homostylous species occur in the same genera and in others. *E. azurea* and *P. rotundifolia* are trimorphically heterostylous and this is accompanied by strong trimorphism of the pollen grains and self-incompatibility, while in *E. crassipes* populations often consist of only one or two morphs, pollen trimorphism is weaker and self-compatibility is high (Barrett 1978).

There are 3 + 3 stamens, rarely by reduction (in some *Heteranthera*) only three or even, in *Hydrothrix* and some *Heteranthera,* a single stamen; in the last-mentioned case the functioning stamen is the adaxial member of the inner whorl; it is supplemented with two staminodes. Dimorphic stamens occur in *Monochoria, Heteranthera* and *Scholleropsis*. The filaments are free from each other and sometimes (at least in species of *Eichhornia*) bear glandular hairs, which are then also present on the style. The anthers are introrse and basifixed or peltate, and dehisce by longitudinal slits or in *Monochoria* by apical pores. The endothecial thickenings are of the Girdle Type, which is a difference from nearly all Liliiflorae studied (except a number of orchids), where they are of the Spiral Type. The tapetum is generally of the Secretory Type, but is sometimes (as in *Monochoria hastifolia*) reported to be amoeboid. Microsporogenesis is successive, and the pollen grains are single (Banerji and Gangulee 1937). The pollen grains are bi- or trisulculate and elongate (T.S. Rao and R.R. Rao 1961), and are distributed in the three-celled stage.

The pistil is tricarpellary and has a superior, trilocular ovary, although sometimes, as in *Pontederia,* only one locule is fertile. Septal nectaries occur in some genera, such as *Eichhornia* and *Pontederia,* while there is no nectar secretion in *Heteranthera* (Daumann 1965). The style is simple and bears an apical, capitate to trilobate stigma. The stigma surface is of the Dry Type. The placentae as a rule bear several or many ovules, but at least in *Pontederia* and *Reussia* one ovule only. The ovules are anatropous and crassinucellate; a parietal cell, forming parietal tissue, is cut off from the archesporial cell. Embryo sac formation conforms to the *Polygonum* Type, and endosperm formation is helobial.

The fruit is a capsule or nutlet (*Pontederia* and *Reussia*). After flowering the floral axis curves downwards, and the fruits mature in the water (hydrocarpy). The seeds are ovoid and longitudinally furrowed. The inner integument of the testa, which is strongly compressed, contains a red-brown pigment, and the endosperm contains copious starch, the starch grains being simple, both large bean-shaped and small and isodiametric. The linear, straight embryo is of nearly the length of the endosperm.

*Chemistry*. Among chemical characters it may be noted that neither steroidal saponins nor chelidonic acid have been recorded. Cyanogenic compounds are known both in *Eichhornia* and *Monochoria*.

*Distribution*. Pontederiales is a pantropical-subtropical group with a centre in America: *Pontederia, Reussia, Zosterella, Hydrothrix* and *Eurystemon* being confined to the Americas, while *Eichhornia* and *Heteranthera* occur in both the New and the Old World; *Monochoria* is found in the Old World tropics and *Scholleropsis* is eastern Asiatic.

*Relationships*. The Pontederiales seem to have a very central position among the orders with paracytic stomata and a starch-containing endosperm. The order is undoubtedly related to the Philydrales, and less closely to the Haemodorales, Velloziales and Bromeliales.

However, there is also evidence that the Zingiberales, through the primitive Musaceae and related families, approach the Pontederiales. In some Zingiberales there are paracytic stomata; in the families with five to six stamens oxalate raphides are present; the leaves also show a differentiation similar to that in *Pontederia,* and it is also in the five to six-staminate families of Zingiberales that there is copious, starchy endosperm (in other families the endosperm is almost completely substituted by perisperm and chalazosperm).

In addition, the Pontederiaceae exhibit certain features met with in the anemogamous Typhales; they are marsh or water plants, oxalate raphides occur in both orders and spathal bracts subtend the inflorescences (which are more contracted and no doubt more compound in the Typhales, however). The tapetum tends to be amoeboid and endosperm formation is of the Helobial Type. It should be stressed that some of these features found in Typhales constitute differences between this order and most Commeliniflorae, where Typhales could, logically, be placed on grounds of their anemogamy.

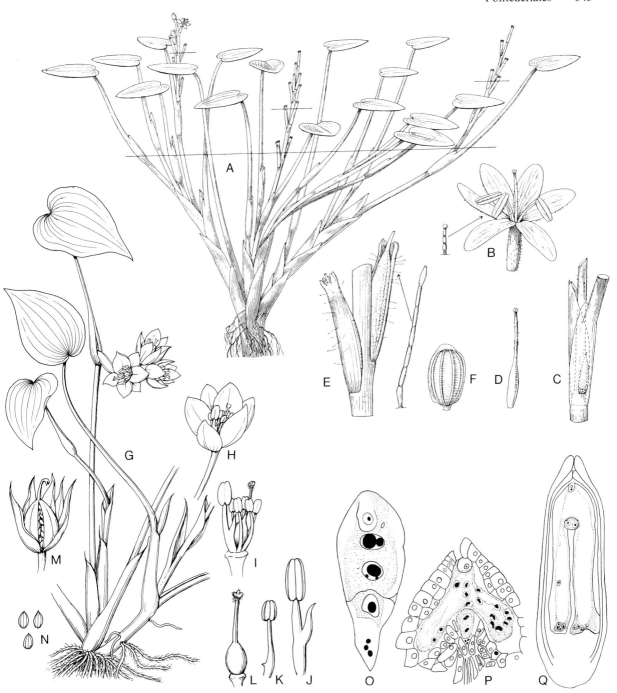

**Fig. 159.** Pontederiaceae. **A–F** *Heteranthera callifolia.* **A** Flowering plant. **B** Flower and hair from abaxial side of tepal. **C** Leaf base. **D** Pistil. **E** Capsules enclosed in perianth tubes, and hair from perianth tube. **F** Seed. (VERDCOURT 1968). **G–N** *Monochoria vaginalis.* **G** Plant. **H** Flower. **I** Androecium and gynoecium, showing stamen dimorphism. **J** Large stamen with filamental appendage. **K** One of the small stamens. **L** Pistil. **M** Capsule. **N** Seeds. (Redrawn from YANG 1978). **O** *Monochoria vaginalis,* early stage of helobial endosperm formation. **P–Q** *Monochoria korsakowii,* later stages of endosperm, note the chalazal chamber in the centre, the flanges of the micropylar chamber reaching far below on each side. (SCHNARF 1931)

**Pontederiaceae** Kunth in Humboldt (1816)   9:34
(Figs. 158–159)

Description and distribution are given under the order above.

The family is found in swamps, shallow lakes, rivers and other inland waters in the tropics.

The following genera may be mentioned.

*Eichhornia* (6), "Water Hyacinth", is indigenous in tropical and subtropical America; one species, *E. diversifolia,* also in tropical Africa. The genus consists of succulent herbs with thick stems and leaf petioles sometimes filled with porous, spongy aerenchyma. The large, decorative, blue flowers are somewhat zygomorphic and have six stamens and a trilocular ovary. – *E. crassipes,* with fusiform, swollen petioles, propagates rapidly in cultivation and has spread in many tropical waters in the world, the dense floating colonies causing problems to shipping in rivers and lakes. It may, however, become a high-yielding crop in the future in controlled cultures and especially those associated with sewage effluent. In addition it is a decorative ornamental.

*Monochoria* (5) is a paleotropical, Asiatic-Australian genus with spreading tepals and six stamens, of which the lower median one is generally larger than the others. – *Heteranthera* (15) in tropical America and Africa has three (rarely one) functional stamens. The perianth is long-tubular, and has six spreading blue tepal lobes. In this genus the inflorescence sometimes has only one to three flowers. – A single functional locule and one stamen occur in the Brazilian annual herb *Hydrothrix* (1–2), which is submerged and has leaves dissolved into filiform bands (see above).

*Pontederia* (4) and the related *Reussia* (3), both American, have spicate inflorescences with rather small, two-lipped flowers. Although these have six stamens, the pistil has only one fertile locule which contains a single ovule. The fruit is a nutlet.

# Order Typhales

*Two Families:* Sparganiaceae and Typhaceae.

Perennial, erect or floating, rhizomatous herbs, growing in moist habitats and usually inundated at least basally. The rhizomes form a sympodial, creeping, more or less horizontal and often thick branch system and are rich in starch. This is deposited as simple grains, up to 20 μ wide. Lateral branches of the rhizome, after a number of internodes, ascend as erect aerial shoots. The horizontal part of the rhizome bears distichous, scale-like leaves. The vascular strands of stems and leaves as well as of roots, contain vessels with scalariform perforation plates.

The aerial stems bear a variable number of distichous, long and linear, sessile and dorsiventral leaves with a sheathing base. The basal leaves, forming a transition between the scale-like rhizome leaves and the long, linear, aerial leaves, bear a widened sheath with stipule-like lobes ("ears"). The leaves are flat or keeled, basally often triangular in transection, with a flat or concave adaxial side. The leaf tissue is partly loose and aerenchymatous. The stomata are paracytic, with the subsidiary cells cut off with oblique divisions. Trichomes are lacking. Calcium oxalate is deposited as various types of crystals: as raphide bundles (KRONFELD 1887) in mucilage-filled cells and sacs, as styloids in cell rows above the sclerenchyma bundles, and as crystal druses or single rod-shaped crystals. So-called myriophyllin cells with darkish contents are also common.

The stem is terete or triangular, branched or unbranched, with the leaves basally concentrated or in Sparganiaceae often continuing far up, the upper ones subtending lateral inflorescences. All members of the order are monoecious, with compound, basally female and apically male inflorescences. In Sparganiaceae the globose inflorescence units are two or three times compound heads, racemose in nature; in Typhaceae, where the dense inflorescences are formed from ring-shaped meristems, the number of ramifications is uncertain, but here also the inflorescence is compound with secondary or tertiary branches, not a simple spike as sometimes thought (see U. MÜLLER-DOBLIES 1969 and D. MÜLLER-DOBLIES 1970).

The flowers are basically di- or trimerous; in Sparganiaceae they have (1–)3–4(–6), more or less equal, greenish tepals, which in Typhaceae are replaced by a variable number of long setae. These are, no doubt, derivatives of tepals, as is shown by the common presence of vascular strands. One or more hairs may emerge from a common base, but they are often inserted in from one to four irregular whorls. They are often dilated apically and may be variously lobed or dissected.

The stamens in the male flowers vary between six (or even eight) and one, partly according to the position of the flower. The filaments are narrow-filiform and the anthers are basifixed, extrorse (where this can be observed), and widest in the apical part where the microsporangia are kept apart by the broadened connective. The tapetum represents a transition between glandular and amoeboid, the tapetal cells becoming bi- to multi-nucleate before they dissolve and form a periplasmodium. Microsporogenesis is successive. While the pollen grains are usually single, they cohere in T-shaped or tetragonal tetrads in *Typha latifolia*. They are ulcerate with the rounded aperture diffusely delimited, and have a reticulate surface. The pollen grains when dispersed are two-celled or possibly (in some Sparganiaceae) three-celled.

In the female flower the pistil is superior and normally monomerous (with one locule and one stylodial branch only) or pseudomonomerous (with one fertile and one empty locule and two stylodial branches). The stylodium (style) is short; in dimerous gynoecia the stylodial branches are free almost from the base. The stigma is elongate, obliquely decurrent and has a Dry papillate (Sparganiaceae) or non-papillate (Typhaceae) surface.

Each locule normally contains a single apical, pendulous, anatropous and apotropous, crassinucellate ovule. A primary parietal cell is cut off from the archesporial cell and forms a parietal tissue. Embryo sac formation conforms to the *Polygonum* Type and endosperm formation is helobial (ASPLUND 1968, 1972; U. MÜLLER-DOBLIES 1969). In the originally small chalazal chamber the endosperm nuclei may be free up to the 16-nucleate stage before wall formation sets in, or wall formation may occur at an earlier stage.

The fruit in Sparganiaceae is drupaceous or, when dry, approaching a nutlet; it is globose or pyramidal in shape, with a dry, spongy or rarely fleshy exocarp and a hard, smooth or longitudinally furrowed endocarp. In Typhaceae the fruit is small, "stipitate", fusiform and achene-like, but finally dehiscent and therefore follicular. The seeds fill up the locule of the fruit. Both integuments in *Sparganium* develop into a thin membrane which separates the endosperm from the fruit wall, except at the micropylar end where the integumentary cells become thick-walled and form a conical double structure, a "seed lid". In Typhaceae, the outer integument of the seed coat consists of compressed, thin-walled cells, while the inner integument consists of thicker-walled cells.

The endosperm is richly developed and "mealy". That of the micropylar chamber contains lipids, aleurone and starch nutrients, which are all largely lacking in the cells of the chalazal chamber. The endosperm is enclosed by a one- or two-layered nucellar covering. The inner part of the endosperm contains the straight, cuneate or fusiform embryo which is about 3/4 as long as the endosperm. The basic chromosome number in the order is $x = 15$ throughout.

*Chemistry.* The chemistry of Typhales is little known. Steroidal saponins, chelidonic acid, cyanogenic compounds or alkaloids have not been reported in the order. As mentioned above, starch is present in the vegetative parts, in particular in the rhizomes and micropylar endosperm, and oxalate is present in various forms, including raphides. Silica bodies are sometimes present. Cells with mucilage are common; these contain polyphenolics, e.g. leucoanthocyans and catechin (and tannins).

*Distribution.* Typhales (according to ECKARDT 1964) were widely dispersed in the Northern Hemisphere during the late Cretaceous and in the Tertiary. Sparganiaceae are mostly North temperate-subarctic, reaching however into the Southern Hemisphere through Malaysia to Southern Australia and New Zealand. Typhaceae are distributed in both hemispheres and occur in tropical regions as well as temperate, including South Africa, Patagonia and Australia.

*Relationships.* The position of Typhales among the monocotyledons has been a matter for controversy, and the evidence is still subject to divergent interpretations. The presence of vessels in stems and leaves, the starch-rich endosperm, the stomata with two subsidiary cells, and the absence of steroidal saponins may be taken to indicate a position in Commeliniflorae, and a position in this complex next to the Cyperales, where silica bodies are nearly absent and endosperm formation is likewise helobial, can perhaps be justified.

However, in the light of the marshy habitats preferred, the distichous leaves, the scalariform perforation plates of the stem vessels, the presence of oxalate raphides, the paracytic stomata, the complete flowers (in Sparganiaceae), the helobial endosperm formation (even the modification of it!) and the starchy endosperm, one can see that there is equally strong, or stronger, evidence to relate the Typhales to the previous group of orders, among which the Philydrales (which likewise lack septal nectaries) and the Pontederiales should be considered in particular.

The adaptation to wind pollination which has occurred in Typhales – if regarded as having evolved independently from that in the Cyperales and the Poales lines – would explain the aberrant syndrome of attributes in which the Typhales deviate from, for example, Pontederiales or Philydrales. They include:

- the unisexual flowers (a common wind-pollination attribute),
- the reduction of the perianth, and in particular its non-petaloid character,
- the lack of nectaries (because of anemogamy),
- the great number of male flowers exposed in the upper part of the inflorescence complex,
- the consequently more complex inflorescence systems,
- the smooth, ulcerate pollen grains (note the parallel with grasses),
- the reduction in carpel number,
- the reduction of ovule number in the ovaries to one,
- the enlarged, in this case decurrent, stigmatic surface of the carpel.

All these character states have positive selective value for a wind-pollinated plant and thus should be estimated as a syndrome. Although the individual attributes mentioned must not be ignored in a phylogenetic discussion, they must at least be estimated as adaptive in character and be considered as secondary and mutually linked states. With this perspective one can more easily appreciate a probable affinity to orders like Pontederiales, Philydrales and Haemodorales, each of which has been subject to other types of selection, favouring animal pollination.

The character syndrome hypothesis is presented in DAHLGREN and CLIFFORD (1981), but otherwise seems to have been largely neglected.

Another, although less likely view, was expressed by HUTCHINSON (1959, 1973), who regarded the Typhales as closely allied to the Xanthorrhoeaceae (incl. Dasypogonaceae). A similar conclusion was reached by D.W. LEE and FAIRBROTHERS (1972) on serological grounds (see Phytochemistry).

Furthermore, the Typhales have often been included in or associated with Pandanales, which may be connected mainly with the unisexual flowers and complex inflorescences (see under Pandanales).

The Typhales also show a number of impressive similarities to the Arales. Some of these are very superficial and include features such as the inflorescence; yet as shown above there is not a true spadix in Typhales. Other similarities are factual, such as the starchy rhizomes, the floral structure of *Sparganium* which resembles various Araceae, the similar floral reductions, the abundance of crystal raphides, the starchy endosperm (only in certain Araceae), the amoeboid tapetum and – a character stressed by THORNE (1976) – the fact that *Sparganium* and *Acorus* (Araceae) share the same rust parasite, *Uromyces sparganii*. Differences are the presence of vessels in stems and leaves of Typhales; the different, less differentiated leaves; the stoma type (mostly tetracytic in Araceae); placentation and the endosperm formation. A possible connection between these two orders does not seem to be as likely as the one advocated above.

**Sparganiaceae** Rudolphi (1830)    1:20
(Fig. 160)

Erect herbs growing on marshy ground as well as decumbent and floating aquatics. The leaves, which are distributed along the stem, are linear and flat or sometimes at least basally triangular (in transection); they are weak and floating in aquatic forms. Their upper side is often minutely dotted by "myriophyllin cells" filled with mucilage and tannins.

The inflorescence system consists of one or several globose compound female inflorescences in the lower part and one or more, likewise globose and compound male inflorescences in the upper part. These may be situated on one main axis or in a branched system (as in *Sparganium erectum*), and are subtended by the upper leaves, though sometimes (by concaulescence) emerging at some distance above the leaf axils.

The flowers are anemogamous and have a uniseriate or biseriate perianth of generally three to four (actually one to six) greenish, inconspicuous, thick or usually thin tepals of more or less spatulate shape, the distal part being fleshy or cupular. The male flowers are generally ebracteate. They normally have three stamens, but the number varies from one to six according to the position of the flower in the inflorescence. Their filaments are generally free, but sometimes fused basally. The pollen grains are free from each other.

The female flowers are generally bracteate. Their pistil in most cases is monomerous or pseudomon-

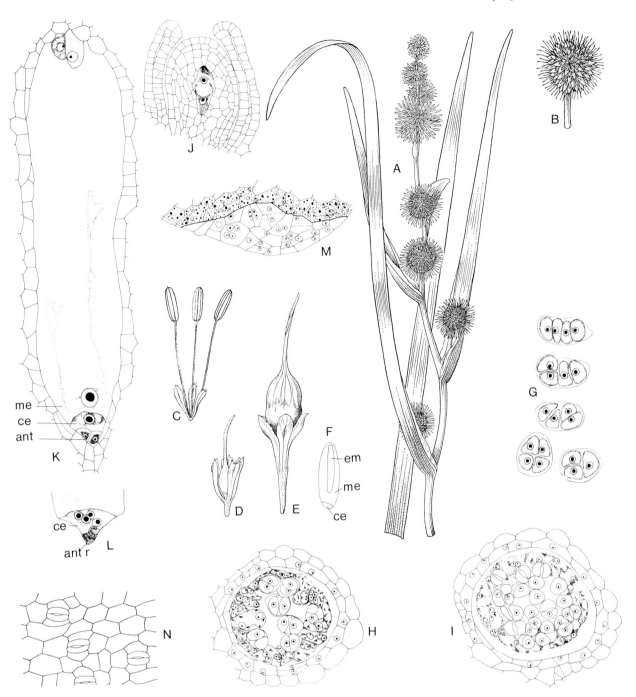

**Fig. 160.** Sparganiaceae. **A–E** *Sparganium emersum* ( = *S. simplex*). **A** Inflorescence and distal part of leaf. **B** Female head. **C** Male flower. **D** Female flower. **E** Fruit. (ROSS-CRAIG 1973). **F–N** *Sparganium erectum,* embryological details. **F** Organisation of seed (*ce* chalazal part of endosperm; *me* micropylar part of endosperm; *em* embryo). **G** Different types of pollen tetrad configurations (microsporogenesis successive). **H–I** Successive stages in the amoeboid tapetum; note the frequently several nuclei in each tapetal cell before they form a peri-

plasmodium. **J** Part of ovule (outer integument not shown). **K** Embryo sac just after the first division of the endosperm nucleus (*ant* antipodals; *ce* chalazal chamber of endosperm; *me* micropylar chamber of endosperm). **L** Chalazal chamber (*ce*) and antipodal remains (*ant r*) at later stage. **M** Chalazal endosperm and base of micropylar endosperm, showing that the starch grains are present in the latter only. **N** Stomata. (**F–L** U. MÜLLER-DOBLIES 1969; **N** SOLEREDER and MEYER 1933)

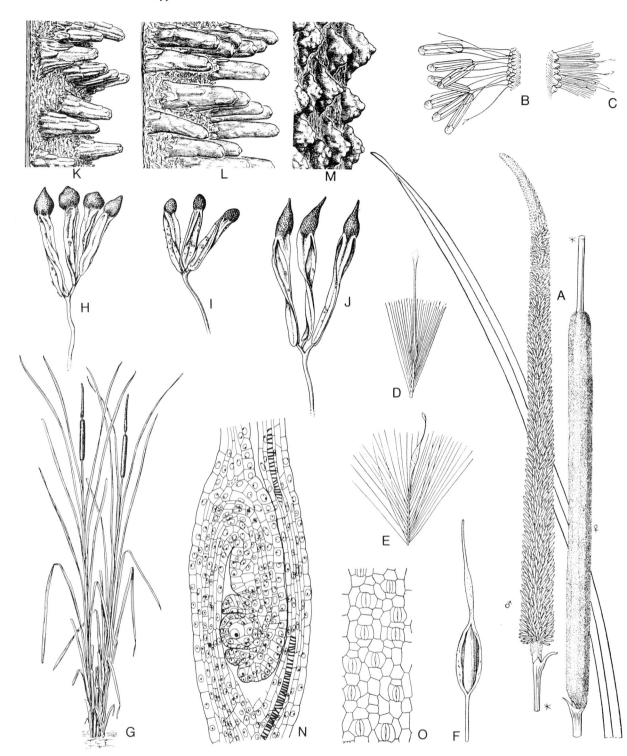

**Fig. 161.** Typhaceae. **A–G** *Typha angustifolia*. **A** Male and female parts of inflorescence and distal part of leaf. **B** Male flowers in position on axis. **C** Female flowers in position on axis. **D** Female flower. **E** Young fruit subtended by hairs on floral axis. **F** Fruit (follicular). **G** Habit of plant. (Ross-Craig 1973). **H** and **K** *Typha domingensis*, male flowers and axis of inflorescence with lateral minor axes. **I** and **L** *T. angustifolia*, same details. **J** and **M** *T. subulata*, same details. (**H–M** Correa 1969). **N–O** *Typha angustifolia*. **N** Ovule. **O** Stomata. (**N** D. Müller-Doblies 1970; **O** Solereder and Meyer 1933)

omerous, in the latter case actually bicarpellary with one empty locule. In abnormal cases the pistil has two or three fertile locules and as many stylodial branches (being bi- or tricarpellary). The stigmas have a Dry and papillate surface.

The fruit is drupaceous and one-seeded. The mainly thin seed coat is thickened at the micropylar end into a "seed lid" which fills a hollow formed by the pericarp.

The single genus *Sparganium* (20), "Bur Reed", is distributed mainly in the Northern Hemisphere. Its species grow in streams and lakesides, in stagnant or slow-running water, or in reed swamps. A few species are floating. They occur in the temperate, boreal and subarctic zones. Two species reach Australia and New Zealand in the south. – *S. erectum* is widely distributed; it is an erect, tall herb with a branched inflorescence system; its flowers have thick perianth members with a dark apex. *S. friesii, S. minimum* and other species are constantly or temporarily floating. The rhizomes of *S. eurycarpum,* in North America, can be eaten as a vegetable.

## Typhaceae A.L. Jussieu (1789)   1:15
(Fig. 161)

Erect herbs with straight, stiff and terete vertical stems. The leaves are mostly flat, but rarely may be triangular in transection. Mucilage cells are richly present on the inner side of the leaf sheaths, and myriophyllin cells are present in all vegetative parts.

The inflorescence system consists mostly of two superimposed, cylindrical parts, a female below and a male above. Their nature is described under the order above. Pollination is by wind. The male flowers are bracteate, each furnished with a variable number of simple, lobed or forked hairs, one or more of which are homologous with a perianth member. These hairs are inserted in one or more, somewhat diffuse whorls. Each flower possesses, as a rule, one to three stamens, the filaments of which continue directly into the pedicel and which are fused basally for a variable distance, sometimes almost to the base of the anthers. These are slightly broadened apically, where the widened connective reaches beyond the microsporangia. The pollen grains are free or rarely united in tetragonal or T-shaped tetrads.

The female flowers consist of a floral axis with diffuse whorls of perigone hairs and a monocarpellary, stipitate pistil with a relatively long style and a widened, decurrent stigma with non-papillate, Dry surface. The unit of dispersal is the complete flower, the perigonal hairs of which serve as a "pappus". The style is persistent and the one-seeded fruit (follicle) dehisces after dispersal. The seeds have a thicker seed coat than those of Sparganiaceae.

*Distribution.* Typhaceae are very widespread and may be regarded as subcosmopolitan, though they are absent from some tropical regions and the arctic.

*Typha* (15), "Cat Tail" or "Reed Mace", grows mostly along stream sides and on shores of freshwater lakes. *T. latifolia* and *T. angustifolia* are common Eurasian species. The leaves of the latter are sometimes used for matting. The leaves of *T. elephantina,* "Elephant Grass", mainly in India, have a similar use in making mats, baskets and small boats. The pollen is eaten locally, and the species can be grown as a soil binder on river banks. The rhizomes of certain *Typha* species can be eaten as emergency food, and the dried inflorescences are sometimes used for ornamental purposes.

# Superorder Zingiberiflorae

*One Order:* Zingiberales.

Perennial herbs or rarely unbranched trees, generally with sympodial rhizomes, which are often thick and rich in starch. The roots arise opposite the xylem strands of the rhizomes. The root hairs develop from cells much shorter than other epidermal cells. Velamen is lacking. Vessels are present in the roots and generally have scalariform perforation plates, simple in many Marantaceae, and mixed simple and scalariform in other Marantaceae and in Cannaceae, Costaceae and Musaceae.

The stems are usually herbaceous but woody in some Strelitziaceae; they always lack secondary thickening growth but rarely may have periderm. The leaves are alternate, generally distichous, sometimes spirally inserted. They are basally provided with an open (or in Costaceae closed) tubular sheath and generally have a large and often broad lamina which may be basally constricted into a variably long petiole. They are always dorsiventral, supervolute in bud, linear to oblong-lanceolate or ovate, with a prominent midvein and pinnate-parallel secondary veins; these are close or sparse and arching towards the margin. The leaves are often torn along the secondary veins, especially in Musaceae and Strelitziaceae, and may then appear "compound". Sometimes, as in Lowiaceae, there are fine transverse veinlets between the laterals. The petiole may be short or up to 1 m long. A ligule is present in Zingiberaceae.

Vessels are present in the stem only in Marantaceae, and in single genera of some other families and, except in a few Marantaceae and Strelitziaceae, they have scalariform perforation plates. Articulated laticifers occur in Musaceae only. Characteristic probably of all members of the Zingiberales are the sieve tube plastids which contain starch grains as well as cuneate protein bodies, a condition otherwise rare in the monocotyledons. Raphides occur in cells and sacs of Lowiaceae, Musaceae, Strelitziaceae and Heliconiaceae, but not in the other families. Silica bodies enclosed in cells occur in all families and are associated with the vascular bundles, but only rarely (in Zingiberaceae) occur in epidermal cells. Hairs are generally lacking but may occur and are then mostly unicellular. The stomata are paracytic or there may be additional subsidiary cells around them making them tetra- and hexacytic. Epicuticular wax, if regularly oriented, is of the *Strelitzia* Type. Systems of air canals in the leaf petiole and axis occur in several families (TOMLINSON 1969).

The inflorescences are terminal or more rarely axillary (as in *Ravenala*) on the erect stems or are terminal on short shoots from the rhizome. They are pedunculate or non-pedunculate and racemose, consisting either of racemes or spikes or of sometimes very complicated thyrses. The flowers or the lateral branches of the inflorescence are often subtended by large, conspicuous, brightly coloured bracts (often an indication of ornithogamy), but bracts may rarely be lacking.

The flowers are generally bisexual and always epigynous and either zygomorphic or asymmetric. The perianth consists of 3 + 3 tepals, the inner, at least, generally petaloid to hyaline, the outer sometimes green and sepaloid or hyaline. The outer may be free from each other or connate to form a tube which is sometimes spathe-like; generally they are different from the inner tepals, but two of the latter may be similar to and fused with the outer (in Heliconiaceae and Musaceae) to form a lip-like structure. The inner tepals are highly variable in appearance, two of them being sometimes fused, as in Strelitziaceae, into a specialized structure or one of them forming a labellum, as in Lowiaceae; in other cases the inner tepals, like the outer, are often less conspicuous than the petaloid staminodes where these occur.

Stamens or stamen homologues are six or fewer, but all are rarely functional as in *Ensete* and *Ravenala*; generally either five are functional and one is lacking, or only one is functional and the others are represented by coloured, flat, petaloid staminodes, or are lacking. Variously shaped nectaries occur at (or sometimes on) the base of the staminodes in Zingiberaceae. Here true septal nectaries are lacking but the nectaries may still be gynoecial.

The petaloid staminodes are often conspicuous; in Zingiberaceae the two lateral of the inner whorl are fused to form a labellum sometimes rather like the labellum formed by the inner median tepal in Lowiaceae. The anthers are basifixed in most Zingiberales but dorsifixed in some Zingiberaceae; they are introrse, often flat and longitudinally dehiscent. In many taxa the connective projects beyond the microsporangia. The anthers are always longitudinally dehiscent. Anther wall formation follows the Monocotyledon Type and the endothecial thickenings are of the Spiral Type (though sometimes atypical). The tapetum is generally glandular-secretory, but is reported to be amoeboid in some Zingiberaceae and in Cannaceae. The pollen grains nearly always have a very thin or virtually no coherent exine and are usually classified as "inaperturate", rarely sulcate (*Zingiber, Dimerocostus*) or spiraperturate (*Tapeinochilos*) or foraminate (*Costus, Monocostus*). They are nearly always two-celled when dispersed.

The pistil consists of three fused carpels, generally with a single, long, slender style and a capitate or trilobate stigma, which constantly has a Wet and generally a papillate surface. The ovules are solitary in Marantaceae and Heliconiaceae, but are more frequently numerous; they are situated on an axile or basal-axile placenta. Of the three locules one or two may be empty, or the locules are confluent and there are parietal placentae (Zingiberaceae: Globbeae and some Costaceae). Septal nectaries, opening on top of the ovary, are present in all families except Zingiberaceae.

The ovules are generally anatropous, rarely campylotropous (some Marantaceae) or orthotropous (a few Zingiberaceae); they are crassinucellate and a parietal cell is always cut off from the primary archesporial cell and may form a parietal tissue. Embryo sac formation is nearly always of the *Polygonum* Type, and endosperm formation is nuclear except in Zingiberaceae and Costaceae where it is helobial.

The fruit is generally a loculicidal capsule, but rarely may form a schizocarp, nutlet or berry. The seeds are nearly always arillate although the aril is rudimentary in Musaceae and Cannaceae and perhaps lacking in Heliconiaceae (by reduction). The seeds are generally provided with an operculum next to the radicle (a similarity to the Commelinaceae). The endosperm is fairly well developed and starchy in Lowiaceae, Musaceae, Strelitziaceae and Heliconiaceae where it is generally complemented with a well-developed nucellar tissue

(perisperm); it is, however, generally starchless and substituted functionally by a well-developed starchy perisperm in Zingiberaceae, Costaceae, Cannaceae and Marantaceae, in the last three of which there is also a well-developed chalazosperm. The starch is mealy, with compound and sometimes hard starch grains. The embryo is linear or sometimes capitate. For chromosome numbers, see under the families.

*Chemistry.* Steroidal saponins are rare in the Zingiberiflorae, and chelidonic acid is reported only in Cannaceae. Cyanogenesis, on the other hand, is known in several families, and follows the tyrosine pathway. Several kinds of flavonols are known in Zingiberaceae but few are particularly widespread in the order. C-glycoflavones are common in Marantaceae. Isoenzymes of dehydroquinate hydrolyase are not reported in the order. HARRIS and HARTLEY (1980) reported the occurrence of compounds giving UV-fluorescence in several families in the order, which indicates that the Zingiberiflorae are on the same evolutionary branch as Bromeliiflorae and Commeliniflorae. The Zingiberales are characterized by richness in essential oils.

*Distribution.* The Zingiberiflorae make up a tropical-subtropical group found in all warm regions of the world. Thus, Zingiberaceae and Marantaceae are large pantropical families. Heliconiaceae are primarily and Cannaceae exclusively South American. Lowiaceae occur in China, Malaysia and the Pacific Islands, and Musaceae are also paleotropical, whereas Strelitziaceae are South American-South African. This scattered distribution indicates a considerable age, but the fossil records are not older than the Maestrichtian. Most members occur in rain forest and in moist regions. The thin exine of the pollen grains indicates humid conditions. Pollination is by animals: insects, birds and bats.

*Relationships, Taxonomy.* The Zingiberiflorae, whether treated as a separate superorder, as here, or an order in a more widely circumscribed unit, is one of the most indisputably natural suprafamilial groups. The numerous autapomorphies, which are mentioned on pp. 100 and 102, include habit, presence of aril, presence of starch in the sieve element plastids, and a well-developed perisperm tissue. The circumscription of the families may be a matter of tradition. If more broadly circumscribed, the Musaceae may include Heliconiaceae and Strelitziaceae and the Zingiberaceae might include the Costaceae. The differences will be mentioned and sometimes discussed under the separate families.

For comparison with other groups the least specialized character states should be considered in combination. These are the petiolate laminar leaves with pinnate venation, the presence of septal nectaries, of oxalate raphides, of 5(–6) functional stamens, well-developed starchy endosperm, and UV-fluorescent contents of cell walls. On this basis, our conclusion is that the Zingiberales approach the Pontederiales and related orders of Bromeliiflorae, but there are still many points of uncertainty in this respect.

### Key to the Families

1. Functional stamens five to six; endosperm well-developed; raphides present . . . . . . . . . . . . . 2
1. Functional stamen one, this often with only one theca; perisperm/chalazosperm much better developed than endosperm; raphides lacking . . . . . . . . . . . . . . 5
2. Lower, median tepal of the inner whorl developed as a large labellum; style widened into three large fimbriate stigmatic lobes; leaves of moderate size with transverse veinlets between the sparse lateral veins . . . . . . . . . **Lowiaceae**
2. Lower, median tepal not larger than the others (and sometimes the smallest); style not as above; leaves large, with dense lateral veins and no transverse veinlets visible . . . . . . . . . . . . . . . . 3
3. Flowers unisexual; leaves spirally set; fruit a fleshy berry; laticifers present . . . . . . . . . . . . . . . **Musaceae**
3. Flowers bisexual; leaves distichous; fruit with dry pericarp; laticifers lacking . . . . . . . . . . . . . 4
4. Fruit a schizocarp with one seed per carpel; seeds exarillate . . . . . **Heliconiaceae**
4. Fruit a loculicidal capsule; seeds several and arillate . . . . . . . **Strelitziaceae**
5. Functional stamen symmetrical, not petaloid, with a complete, tetrasporangiate anther; flower zygomorphic; endosperm development helobial . . . . . . . . . . . . . . 6
5. Functional stamen petaloid, asymmetrical, with a bisporangiate theca only on one side; flowers asymmetrical; endosperm formation nuclear . . . . . . . . 7
6. Leaves spirally set; ethereal oils lacking; labellum consisting of five connate petaloid staminodes . . . . **Costaceae**
6. Leaves distichous; etheral oils present; labellum consisting of two connate petaloid staminodes of the inner staminal whorl . . . . **Zingiberaceae**
7. Ovules several or many per carpel; leaves spiral, not distichous; seeds without a well-developed aril; embryo straight . . . . . . . **Cannaceae**
7. Ovules one per carpel (all carpels may not be fertile); leaves distichous; seeds arillate; embryo more or less curved . . . . . . . **Marantaceae**

## Order Zingiberales

*Six Families:* Lowiaceae, Musaceae, Heliconiaceae, Strelitziaceae, Zingiberaceae, Costaceae, Cannaceae and Marantaceae.

Description as for the superorder, above.

### Lowiaceae Ridley (1924)   1:6
(Fig. 162)

Perennial herbs with a horizontal sympodial rhizome bearing scale-like leaves, foliage leaves and few-flowered cymes. The foliage leaves are distichous, sheathing at the base, with a relatively short petiole and a broadly lanceolate lamina with a prominent midvein, from which at regular intervals lateral, arching and apically convergent veins diverge. Rather distinct transverse minor veins connect these longitudinal veins.

Vessels are present only in the roots and have scalariform perforation plates with numerous bars. The rhizomes are rich in starch grains, which are compound. The leaves lack a palisade layer, the chlorenchymatous layer consisting of a mixture of small and large cells (TOMLINSON 1969). Hat-shaped silica bodies occur in connection with the vascular bundles, and raphide sacs occur in all parts of the plant. Hairs are lacking; the stomata are paracytic, with deeply sunken subsidiary cells (TOMLINSON 1969).

The inflorescence is a branched spike with one or few flowers per branch (LARSEN 1961b) and terminates each shoot generation of the rhizome. The flowers are bracteate, bisexual, epigynous, zygomorphic and superficially orchid-like (cf. the generic name *Orchidantha*!). The outer tepals are subequal; they are fused basally into a tube, and their lobes are linear-lanceolate and pointed; of the inner tepals the median is large and differentiated to form an elliptic-spathulate or boat-shaped labellum, which may have a spotted colour pattern, the lateral inner tepals being smaller and elliptic to filiform. There are only five stamens, the position opposite the labellum being empty. The sta-

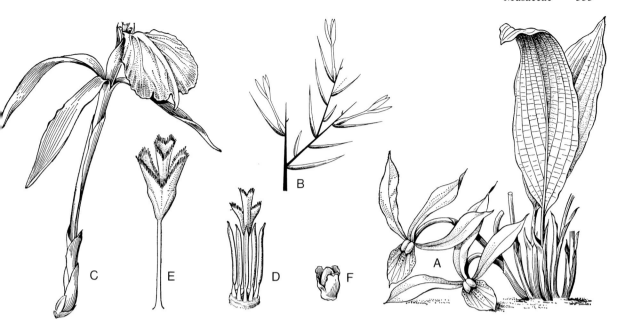

**Fig. 162.** Lowiaceae. **A** and **F** *Orchidantha maxillarioides.* **A** Plant; note the transverse veinlets of the leaves. **F** Seed with trilobate aril. (TAKHTAJAN 1982). **B** *Orchidantha laotica,* diagrammatic representation of inflorescence, showing racemose construction. (LARSEN 1961). **C–E** *Orchidantha longiflora.* **C** Flower and subtending bracts. **D** Androecium and gynoecium. **E** Style and fimbriate stigmatic lobes. (TAKHTAJAN 1982)

mens are free from each other and each has a short filament and an elongate tetrasporangiate anther ending in a shortly subulate connective tip. The pollen grains are inaperturate. The ovary is trilocular, with septal nectaries; in each locule there are numerous anatropous ovules on an axile placenta. The style is erect and it widens apically into three broad, asymmetric, flattened marginally laciniate-fimbriate stigmatic lobes (the broad lobes recalling the condition in some Iridaceae). The embryology is evidently still unknown.

The fruit is an elongate capsule with a papery pericarp and numerous seeds. The seeds, situated in two rows in the locules, are globose, endospermous and enclosed each by a trilobate aril. The starch grains of the endosperm are isodiametric, with a facetted surface (CRONQUIST 1981). $x = 9$.

*Chemistry.* The chemistry, apart from what is said above, seems to be unknown.

Lowiaceae consists of the single genus *Orchidantha* ($= Lowia$) (6) in Southern China, Thailand, Laos, Malaya and Borneo, and some of the Pacific Islands. *O. longiflora* (Fig. 162C–E) is said to be up to 160 cm tall; the other species are much smaller plants.

The position of Lowiaceae near the Musaceae, Heliconiaceae and Strelitziaceae, in the Zingiberales, is supported by the possession of raphides, hat-shaped silica bodies, the five stamens, the shape of the fruit, and the adaxial system of air cavities (though reduced) in the leaf axis.

## Musaceae A.L. Jussieu (1789)   2:42
(Fig. 163)

Large to giant herbs with a short, thick underground stem ("corm"). The large leaves are spirally set (not distichous as in the two previous families), in reality in a basal rosette but overlapping each other successively with their long sheaths so as to form a "pseudo-stem" around the terminal scape. The stem varies from ca. 60 cm (*M. lasiocarpa*) to ca. 13 m (*Ensete* spp.), and the leaves from ca. 30 cm to more than 6 m in length. The leaves have a well-developed, coarse, tubular sheath, a long petiole and an entire, often impressive, lanceolate or oblong blade. This has a prominent midrib and a densely pinnate venation with the veins curving up at the margin, joining each other to form a marginal vein. The leaf blade is easily torn along the lateral veins and ultimately may appear almost "palm-like".

Vessels are restricted to the roots and have scalariform or also simple perforation plates. The root cortex has air lacunae and the stele contains numerous scattered vessels and phloem islands, each phloem strand with several sieve tubes. Periderm is sometimes formed in the cortex of the root. Arti-

**Fig. 163.** Musaceae. **A–D** *Musa × sapientum.* **A** Plant. **B** "Pseudostem", transverse section; the central rather thin scape is enclosed by several layers of leaf sheaths. **C** Inflorescence, partly in fruit and partly in the flowering stage. **D** Fruit, "banana". (**A–C** LARSEN 1973a; **D** DEGENER 1947). **E–H** *Musa acuminata.* **E** Male, distal part of inflorescence. **F** Male flower. **G** Female flower. **H** Seed. (TAKHTAJAN 1982). **I** *Musa,* floral diagram. (LARSEN 1973a). **J–K** *Musa × sapientum.* **J** Stoma. **K** Silica cells. (TOMLINSON 1969). **L–Q** *Ensete gillettii.* **L** Plant (ca. 2 m tall). **M** Rows of young male flowers. **N** Male flower. **O** Female flower, late flowering stage. **P** Fruit, part of the wall removed. **Q** Fruit, transverse section. (HEPPER 1968). **R–S** *Ensete homblei.* **R** Seed, longitudinal section. **S** Pollen grain. (MILNE-REDHEAD 1950)

culated laticifers occur in association with the vascular bundles in all parts except the roots. Raphide sacs are widely distributed in the plant and calcium oxalate also occurs as rhombohedral crystals. Silica bodies occur in the shoot, mostly in connection with vascular bundles, and are "trough-like". Hairs are lacking. The stomata are tetracytic. The mesophyll includes a palisade of two to four cell layers. (TOMLINSON 1959, 1969).

The inflorescence axis represents the aerial stem, which is terminal on the "corm". It has little mechanical tissue and is held up by the sheathing leaf bases. It ends in an extensive thyrse, which is erect or more commonly drooping, at least in the fruit stage. The lateral cymes (cincinni) of the inflorescence are not distichous. They are dense and compact and may appear as a dense row of flowers in the axils of the cincinnal bracts, which are large, firm and boat-shaped, with the halves compressed, and often dark purplish. The flowers are subtended by hyaline, recurved bracts. They are unisexual, either female (basally in the thyrse), with non-functional stamens, or male (apically), with non-functional pistil. There are 3 + 3 tepals which are arranged so that the median inner tepal, which is distinct from the others, is directed downwards, the other five tepals being connate at least basally. The three outer tepals are often somewhat larger than the inner. Five stamens are generally developed, the median of the inner staminal whorl being missing, but all six are present in most species of Ensete. The filaments are narrow and the anthers are elongate and tetrasporangiate, as in Strelitziaceae and Heliconiaceae. The pollen grains are inaperturate and similar to those in Strelitziaceae.

The ovary is trilocular with two or more axile rows of ovules in each locule. Septal nectaries are present. The style is slender and ends in a trilobate stigma. The ovules are anatropous and crassinucellate. A very rudimentary aril has been discovered (FRIEDRICH, personal communication). The nucellar epidermis consists of somewhat elongate cells and the nucellar tissue is multicellular. A parietal cell is cut off from the archesporial cell; embryo sac formation is of the *Polygonum* Type and endosperm formation is nuclear.

The fruit is generally baccate with a tough outer skin, which is easily split open longitudinally, and with a homogeneous inner layer; less commonly the fruit is rather dry and much less fleshy. The seeds vary in number. They are operculate, as in *Heliconia,* and are superficially exarillate, but some fine hair-like structures in *Musa* have been shown by Friedrich (see above) to be a rudimentary aril. The seed is filled by starchy perisperm and endosperm and by a straight or, in *Ensete,* slightly curved, embryo. x = 9–11, 16, 17.

*Chemistry.* Compounds produced within the family are cyanogenic glucosides, indole alkaloids and 3-deoxyanthocyanins (GORNALL et al. 1979); the tannin cells may contain proanthocyanins. The laticifers carry mucilaginous contents, which on exposure to the air become dark.

*Distribution.* Musaceae is a paleotropical family ranging from Africa to Eastern Asia, Australia and the Pacific Islands. The family consists of *Musa* (ca. 35) and *Ensete* (7).

*Ensete* (Fig. 163 L–Q) can be distinguished from *Musa* in that the large leaf blade continues along the midrib down to the very sheath, by the larger bracts, the drier fruits and the larger seeds. The plants of *Ensete,* unlike those of *Musa,* are monocarpic, i.e. they flower only once, and then die. The stamens are generally six in number. The genus is more concentrated in Africa than in Asia, and generally comprises large trees up to 13 m high. *Ensete* is of limited economic value, though *E. ventricosum,* in East Africa, has edible peduncles and the leaf sheaths can be used for fibres.

*Musa,* the "Banana" genus, with a wider distribution, has a slenderer stem and a generally smaller total size. The leaf blade is not decurrent on the petiole, the bracts are relatively smaller, the stamens mostly five in number, and the fruits fleshy, with more numerous, but smaller seeds. *Musa* is extremely important for its fruits, which in the edible strains develop without fertilization of ovules (parthenocarpic). Most bananas belong to the hybrid species *M. × sapientum* (Fig. 163 A–D), the parents of which may be *M. acuminata* and *M. balbisiana,* which both occur in South-East Asia. The "Fruit Bananas" are highly variable in plant size and cultural requirements and in the size, shape (ovoid to long and crescent-shaped), colour (yellow, red, green) and sugar content of the fruit. The cultivation of banana is now extensive all through tropical and subtropical countries, including the Canary Islands. In one variety, often called "M. paradisiaca", the "Starch Banana" or "Plantain", the fruits are starchy (they are richer in sugar in the previous forms) and used in cooking, as a vegetable. *Musa textilis,* from the Philippines, yields "Manila Hemp", and is cultivated mainly on these islands.

The genus contains a number of other little-known species, some of which are quite small plants, e.g. *M. lasiocarpa,* which is less than 60 cm high and has hairy, rather dry fruits. The solitary Australian species is *M. hillii*, up to 10 m tall, with erect inflorescences but fruits less than 5 cm long, with numerous seeds.

Pollination of *Musa* is mainly by bats (see, for example, START and MARSHALL 1976), and the flowers of bat-pollinated species are functional only for one night; some species may be bird-pollinated.

## Heliconiaceae T. Nakai (1941)   1:100–120
(Fig. 164)

Erect perennial herbs, from less than one to almost 7 m tall, with creeping rhizomes. Leaves distichous, either all basal and the aerial stem consisting mainly of the inflorescence axis, which eventually protrudes through the channel formed by the closely overlapping leaf sheaths ("pseudo-stem"), or leaves inserted on an elongate aerial axis (in some members of the derived subgenus *Stenochlamys* of *Heliconia*). The leaves have a basal, tubular, sheath, a long petiole and a narrowly elliptic-oblong, entire lamina, which is apically generally caudate-acuminate and basally often truncate or cuneate. The lamina has a prominent midrib from which diverge obliquely a great number of parallel, faintly sigmoid, secondary veins arching near the margin to merge into a marginal vein. The leaves vary from only ca. 25 cm (*H. aurantia*) to 3 m long or more. Ligules are lacking.

Vessels are confined to the roots and have scalariform perforation plates. Laticifers are probably lacking, but silica bodies occur in cells associated with the vascular bundles and are rectangular with a deep central hollow (TOMLINSON 1969). The leaf blades have a palisade of two to three layers (cf. Lowiaceae). The stomata are generally tetracytic, more rarely paracytic, with two narrow deep subsidiary cells adjacent to the guard cells. The leaves are sometimes glabrous, but branched uniseriate "candelabra" hairs occur on the leaf axis and inflorescence in a number of species (TOMLINSON 1959, 1969), and unbranched thick-walled hairs occur in some species.

The inflorescence is situated terminally on an erect peduncle and consists of a large, flattened thyrse, which is erect or drooping, often with a conspicuously geniculate axis. Each lateral branch is subtended by a stiff, showy, usually boat-shaped bract, which may be broad or narrow, rather small or often quite large, and which is usually brightly coloured (dull green in Asiatic species), red and green, red, orange etc. A dense monochasial cyme, a cincinnus, of a few to many flowers, is situated in the axil of each of these bracts, and is sometimes nearly concealed in its axil. The flowers are situated in the axils of floral bracts, which are much smaller and thinner than the cincinnal bracts, being generally pale and membranaceous.

The flowers are bisexual, epigynous and strongly zygomorphic. Of the six tepals, the median in the outer whorl is nearly free from the others, which are all fused to form a five-dentate or five-lobate upper lip. The five fused and the one free tepal form a tube, which can be widened (e.g. when visited by a broad-beaked bird) without bursting. The slit is tightened by the staminode, which permits the tube to keep more nectar than would otherwise be possible. There are five functional stamens and one staminode which is subulate or to some degree petaloid. The filaments are free from each other and filiform and each bears an elongate, basifixed, tetrasporangiate anther. The pollen grains are inaperturate, with the exine restricted to small spinules; the intine is thick in the distal part and radially striate or channelled (KRESS et al. 1978; KRESS and D.E. STONE 1982; D.E. STONE et al. 1979). In shape the pollen grains are round in polar view, but asymmetric, oblate to spheroidal in lateral view, sometimes with a distal top-notch, which represents the apertural region.

The pistil has a trilocular ovary and a narrow, often slightly curved style and on its often slightly thickened apex a small, capitate to trilobate papillate stigma. Septal nectaries are present. There is only one basal-axile ovule in each locule. The ovule is anatropous and lacks an arillar structure (cf. Strelitziaceae), whereas a proliferating division takes place especially in the micropylar part of the outer integument (MAURITZON 1936). A parietal cell is cut off from the primary archesporial cell, but forms no substantial parietal layer. Embryo sac formation follows the *Polygonum* Type and endosperm formation is nuclear. The nucellus is always strongly developed.

The fruit is a (usually blue) drupe, each of the three stones of which contains a single seed. The seeds are exarillate, triangular, and provided with an operculum at the micropylar (cotyledonary) end. The seeds contain copious endosperm as well as a rich perisperm (at least the latter with starch), and a straight embryo. $x = 12$.

*Chemistry.* Apart from the raphides, calcium oxalate may occur as crystal sand. Tannin is common in unspecialized cells (TOMLINSON 1969). The rhizomes are rich in flat starch grains.

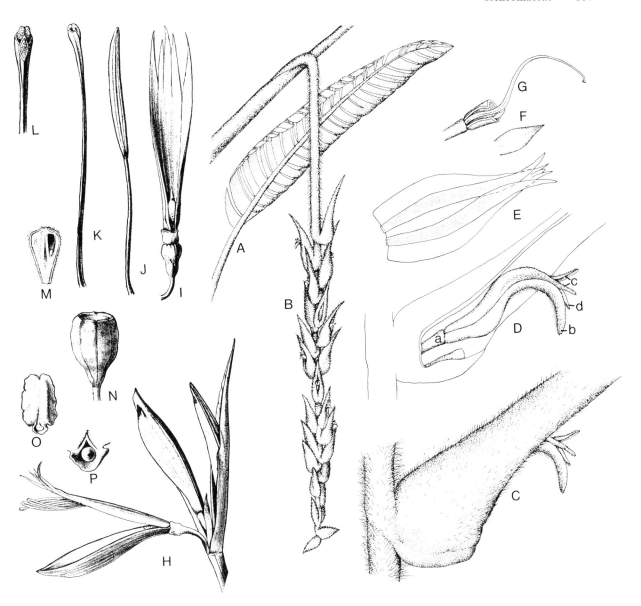

The genus *Heliconia* (100–120) is a primarily neotropical genus which forms conspicuous stands in shaded, moist tropical forests as well as in open formations, and also grows in disturbed habitats. Small but interesting groups of species (e.g. *H. indica* and *H. paka*), on Sumatra and eastwards, have dull flower colours and are probably chiropterogamous. Their present distribution was possibly achieved through an ancient, successful long-distance dispersal. Several species are grown as ornamentals.

Most species are strongly adapted to bird pollination, being visited for nectar by hummingbirds (Trochilidae) (STILES 1978). The conspicuous bracts are part of the syndrome of ornithogamy

**Fig. 164.** Heliconiaceae. **A–G** *Heliconia magnifica.* **A** Leaf. **B** Inflorescence. **C** Cincinnal bract and flower protruding at anthesis. **D** Same, the bract partly cut away, and floral bracts removed to show a flower at anthesis (*a* ovary; *b* free sepal; *c* fused sepals, partially reflexed; *d* anthers, included in apex of corolla tube). **E–F** Perianth. **E** Outer surface showing two tepals of the outer perianth whorl, fused to the partially spread open tube formed by the inner whorl, **F** being the median, free tepal of the outer tepal whorl. **G** Style and stigma. (KRESS 1981). **H–P** *Heliconia psittacorum.* **H** Flowers of cincinnal group. **I** Perianth seen from below. **J** Functional stamen. **K** Style and stigma. **L** Style apex and stigma. **M** Ovary longitudinal section. **N** Fruit. **O** Seed; note the operculum at the base. **P** Operculum. (WINKLER 1930)

(cf. Bromeliaceae) but in some *Heliconia* species they are smaller and inconspicuous.

The Heliconiaceae are obviously very closely allied to both Musaceae and Strelitziaceae, but differ in various details. (See further under these families).

## Strelitziaceae Hutchinson (1934)    3:7
(Fig. 165)

Giant (up to several metres high) or medium-sized perennial herbs or unbranched trees, with variably well-developed rhizomes or underground stems. The leaves are distichous and, as in *Heliconiaceae,* differentiated into a sheath, an often rather long petiole and a simple lamina with about the same shape and venation, although the sheath is shorter. The leaves are often torn or fragmented along the veins. They are situated on the stem, which is sometimes very short (*Strelitzia reginae*) or sometimes of considerable length. Rarely, in a population of *S. reginae,* the laminae are not developed, the leaves being then spear-like and developed in higher number.

Vessels are found in the roots and have scalariform (in *Phenakospermum* also simple) perforation plates, and in at least *Strelitzia* and *Phenakospermum* also in the stem, but not in the leaves. Laticifers are lacking. Raphide sacs are present, and spherical silica bodies are present in hypodermal cells; druse-like silica bodies are found in cells adjacent to the vascular bundles. Subterranean roots are unusual in having wide vessels scattered uniformly throughout the ground tissue together with phloem strands, each with only one, wide sieve tube (TOMLINSON 1969). Tannin is also common and stored in unmodified cells. Starch is found in the ground tissue of the woody stem and in the rhizome. The starch grains are flattened in *Ravenala* (as in *Heliconia*), otherwise spherical. The stomata are generally tetracytic to hexacytic; hairs are lacking (TOMLINSON 1959, 1960, 1969).

There is one terminal giant inflorescence in *Phenakospermum* (Fig. 165 H) and up to several, lateral ones in *Strelitzia* and *Ravenala* (Fig. 165 D). The inflorescences are thyrses with one to several lateral, distichously arranged cincinni, each one in the axil of a large, stiff, lanceolate or boat-shaped bract which generally has a bright colour. (In the commercially common *Strelitzia reginae* there is only one such bract). The flowers, as in Heliconiaceae, are bracteate, bisexual, and more or less zygomorphic. The outer tepals are separate from each other, more or less equal, linear or narrowly lanceolate. The inner tepals are variously specialized, the lateral being more or less fused with each other, the median free or connate with the others only basally. In *Strelitzia* the lateral inner tepals are fused to form a stiff, blue, arrow-shaped keel enclosing the five functional stamens and the style (Fig. 165 B).

The functional stamens are six in *Ravenala* but five in *Phenakospermum* and *Strelitzia*; in the last-named the median stamen of the inner whorl is staminodial or lacking. The filaments are filiform and the anthers linear and tetrasporangiate; in *Strelitzia* the anther has a slightly prolonged connective. The pollen grains are globose, almost exine-less but with thick intine (omniaperturate, sensu THANIKAIMONI 1978).

The ovary is trilocular, with two, or in *Phenakospermum* at least four, rows of ovules in each locule; the placentation is axile. The style is long and filiform and terminates in three linear stigmatic, papillate lobes (cf. Lowiaceae). The ovules are anatropous and crassinucellate, with an arillar initial developed from the funicular-chalazal region. A parietal cell is cut off but does not form a parietal tissue. The nucellar epidermis consists of rather elongate cells and the nucellus is very many-celled. The outer integument is not as thickened as in Heliconiaceae (MAURITZON 1936). Embryo sac formation conforms to the *Polygonum* Type and endosperm formation is nuclear.

The fruit is a woody, loculicidal capsule with numerous shiny seeds enveloped by an aril which consists of a dense hair-like covering (*Strelitzia, Phenakospermum*) or laciniate lobes (*Ravenala*) of bright colour, different in the different genera. The seeds are operculate, as in the Heliconiaceae. The seed contains copious, starchy (and mealy) perisperm and endosperm and the embryo is straight and has a large, massive cotyledon. $x = 7$, 8 and 11.

*Chemistry.* The chemical constituents are approximately as in Heliconiaceae.

*Distribution.* Strelitziaceae is distributed in tropical South America (*Phenakospermum*), temperate to subtropical Southern Africa (*Strelitzia*) and Madagascar (*Ravenala*), a highly disjunctive distribution.

*Ravenala* (1) *madagascariensis,* the "Traveller's Tree", has a woody trunk up to several metres tall. In the natural habitat, on Madagascar, it grows up to an altitude of 1,600 m, but on the high levels occurs as small, stemless plants in the indigenous forests. The leaves are strictly disti-

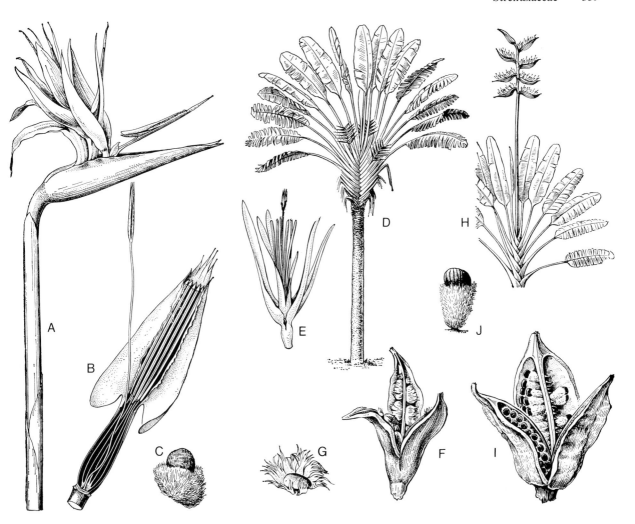

chous and arranged in a fan-shaped cluster and the inflorescences are lateral (Fig. 165 D). The tepals are only slightly zygomorphic and the aril is bright blue. The species is widely grown in the tropics as an ornamental.

*Strelitzia* (5), in Southern Africa, varies strongly in size, from sometimes less than a metre high (*S. reginae*, Cape Province), to giant tree-like herbs growing in rivulets or river valleys in Natal, which may be up to ca. 10 m tall. In this genus two of the inner tepals are fused to form a rigid arrow-like, blue-coloured keel, in which the stamens and style are enclosed. The aril is orange-coloured. A few species are grown as ornamentals, especially *S. reginae*, the "Bird-of-Paradise Flower".

*Phenakospermum* (1) *guianense* in Guiana, Brazil, Peru, etc. grows along the Amazonian watercourses and in swamp forests. It has a short stem, large leaves, and a huge terminal inflorescence. The flowers, as in *Ravenala*, are not so strongly differentiated as in *Strelitzia*. The large capsules

**Fig. 165.** Strelitziaceae. **A–C** *Strelitzia reginae*. **A** Inflorescence. **B** The fused tepals forming a keel, the enclosed stamens and the style bending upwards. **C** Seed with aril. **D–G** *Ravenala madagascariensis*. **D** Plant. **E** Flower. **F** Dehiscing capsule. **G** Seed with aril. **H–J** *Phenakospermum guianense*. **H** Plant. **I** Dehiscing capsule. **J** Seed with aril. (All from TAKHTAJAN 1982)

contain many seeds, which are arranged in several rows per locule; they have bright red arils.

*Strelitzia* is bird-pollinated, as is *Ravenala*, in which the flowers are reported to be explosive (PROCTOR and YEO 1973); *Strelitzia* is evidently adapted to larger birds than most bird-pollinated flowers. The size of the flowers and inflorescences of *Phenakospermum* is also suggestive of bird or bat pollination.

The Strelitziaceae are undoubtedly closely allied to Heliconiaceae and Musaceae (LANE 1955) in which, as in *Phenakospermum*, the inflorescence is terminal.

**Zingiberaceae** Lindley (1835)   ca. 50:1,000
(Figs. 166–167)

Perennial, rhizomatous herbs, sometimes of considerable size. The roots are generally the only
parts of the plants that contain vessels; these have
scalariform perforation plates. The sympodial rhizome is generally horizontal and creeping, often
thick and rich in starch, the starch grains being
large and simple. The shoot is erect or ascending,
generally short and unbranched. The shoots are
often differentiated into inflorescence-bearing
branches and branches with foliage leaves, both
emerging from the rhizome. The inflorescence-
bearing branches have leaves which consist of the
scale-like sheath only.

The leaves are distichous and have a long open
(or rarely closed) sheath (SPEARING 1977), often
a petiole of variable length, which may also be
lacking, and a lamina, which is generally lanceolate, oblanceolate, oblong or linear and acuminate
or obtuse. The leaf sheaths are often long, and
being superimposed, contribute to the formation
of a "pseudo-stem" like that in *Musa*. At the upper margin of the sheath there is nearly always
a membranous ligule. The pseudopetiole, if present, and the midvein contain a circle of aerenchyma strands. Venation is pinnate, with a prominent midvein and straight and parallel to arching
secondary veins.

Vessels are only rarely present in the stem (very
rarely in the whole shoot) and then have scalariform perforation plates. The stem cortex is thick
and is delimited from the pith by a circular fibre
cylinder. The vascular bundles of the cortex are
provided with fibrous sheaths, which do not make
contact with the main fibre cylinder. The main
bundle arcs of the pseudopetiole and midvein face
towards the abaxial side of the leaf. The plants
are generally glabrous but rarely have unicellular,
sunken hairs (very rarely uniseriate hairs). Bicellular hairs occur on the rhizome of *Curcuma*. The
stomata, which have asymmetrical guard cells, are
tetracytic or, more rarely, paracytic or hexacytic.
Spherical silica bodies occur in epidermal cells of
the leaves, mostly overlying vascular strands, and
silica sand may also be present in epidermal cells.
Scattered in the plants are cells containing ethereal
oils and other contents (see below), giving the
plants a characteristic smell. Raphides are generally lacking, but other kinds of oxalate crystals are
common.

The inflorescence is terminal on the shoot (whether
this is leafy or a specialized floral shoot). It is

a spike or a thyrse. Unlike the leaves of the vegetative part, the bracts of the inflorescence, when numerous, are spirally set; rarely they are few, solitary or lacking. The bracts are often densely imbricate and the inflorescence is then cone-like.

The flowers are evidently highly specialized mechanically (as is shown by the relationships of the
style to the stamen and the modifications of the
latter). They also look as if they are specialized
in respect of pollinators and adapted to a variety
of insects, including bees and Lepidoptera. In addition, obviously bird-adapted types occur in, for
example, *Hornstedtia* and *Nicolaia*; in these the
floral parts are red, narrow and convolute and
in *Nicolaia* the bracts are also red. A number of
such species carry the inflorescence at ground level. Despite the appearance of specialization, observations of pollinator visits are very sparse (B.L.
BURTT, personal communication).

The flowers are bisexual, trimerous and zygomorphic. They have 3 + 3 tepals. The outer three
are generally sepaloid, greenish or colourless, and
fused into either a trilobate cup or an asymmetrical
spathe-like structure with a slit open to the base
or almost so on one side. The inner tepals are
likewise fused basally into a tube, inside that of
the outer tepals, but are longer, equal or unequal
(the median often longer than the lateral) and generally petaloid or colourless.

The androecial whorls for the most part are transformed into staminodes, the median inner stamen
being the only one which is functional. It is dithecous and tetrasporangiate. In the outer staminal
whorl the median member is always lacking,
whereas the two lateral ones are developed as petaloid staminodes. These are often much smaller
than the lateral staminodes that belong to the inner staminal whorl, which are fused to form a
highly variable lip, "labellum", which in position
and appearance much resembles the labellum of
the orchids (although the labellum in orchids is
the median member of the inner tepal whorl). The
labellum is often bilobate and guide-marked and
it may be conspicuously stalked, as in *Hedychium*.
Sometimes, as in *Mantisia*, the lateral staminodes
of the outer whorl are short and connate to the
functional stamen and seem to be lateral lobes of
this, but more often they are elongate and situated
on each side of the functional stamen; they may
also be lacking. The single functional stamen is
generally linear, with introrse, elongate thecae situated on or below the frequently flattened end of
the stamen (Fig. 166 F). The anther dehisces by
longitudinal slits. The tapetum is glandular-secre-

tory or amoeboid (reports in *Amomum* and *Nicolaia*). Microsporogenesis is successive. The pollen grains are inaperturate or rarely (*Zingiber*) sulcate, with thick intine but thin exine or none.

Epigynous, nectariferous glands, generally two, but sometimes three or one in number, of a shape characteristic of genera or generic groups, are present at the style base. They are filiform, subulate, scale-like, wart-like, tooth-like, etc. (V.S. RAO 1963; BURTT 1972). Their morphological interpretation has been subject to various opinions; V.S. RAO (1963) considers them outgrowths of the up-

**Fig. 166.** Zingiberaceae. **A–F** *Caulokaempferia saksuwaniae*. **A** Plant. **B** Flower. **C** Flower base, showing asymmetrical outer tepal whorl. **D–E** Flower, back and front side, showing sepal-like inner tepal whorl, lateral petaloid staminodia (outer staminal whorl) and labellum and anther (the fused lateral members and the median member of the inner staminal whorl respectively). **F** Anther and stigma (LARSEN 1973b). **G–K** *Hedychium cylindricum*. **G** Flower. **H** Anther, style apex and stigma. **I** Style base and epigynous gland. **J** Ovary, transverse section. **K** Stigma. (BURTT and SMITH 1972). **L–N** *Hedychium coronarium*. **L** Inflorescence and leaves. **M** Leaf base, showing ligule. **N** Flower. (CABRERA 1968). **O** *Mantisia wardii,* flower. **P** *Globba leucantha,* flower. (O–P BURTT 1972)

per surface of the ovary (tips of vestigial septal nectaries?).

The ovary is trilocular or unilocular, in the latter case with basal-parietal placentation, very rarely with a free-central placenta (*Scaphochlamys*). The style is straight, linear-filiform and entire, resting as a rule on the functional stamen, embraced by the thecae. The apical stigma is papillate and often irregularly infundibular.

The ovules are several or many in each carpel, anatropous (orthotropous in *Hitchenia*) and crassinucellate, with a parietal cell cut off from the archesporial cell. The nucellar epidermis generally becomes multilayered and forms a nucellar cap; it also contributes to the perisperm of the seed. Embryo sac formation is of the *Polygonum* Type. Endosperm formation is helobial, but the chalazal chamber degenerates quite early, often as early as the two-celled stage. The endosperm forms a restricted tissue but may contain aleurone grains and sometimes also starch grains, although most starch is in the perisperm. The embryogeny is of the Asterad Type. The nucellar tissue develops into a starchy perisperm but the chalazal tissue is not much developed (i.e. chalazosperm none or limited).

The fruit is generally a capsule, which may be fleshy; sometimes the fruit is indehiscent and fleshy (i.e. a berry) or indehiscent and dry. The seeds have a thin and veil-like, generally lobate or laciniate aril. They also have a lid, which at least in *Amomum, Alpinia* and *Elettaria* develops from the inner epidermis of the outer integument. Most of the seed is filled up by the starchy, hard or more often mealy perisperm (with compound starch grains) surrounding the thinner endosperm (see above), which encloses the embryo, which is linear and generally tapering at the cotyledonary end. The cotyledonary strands are only two in number (more in Costaceae), and the root primordia four (8–12 in Costaceae). The embryo, unlike that of Costaceae, lacks fat bodies.

The basic chromosome number in most genera is $x = 12$, but $x$ varies between 9 and 26.

*Chemistry.* The whole plant of all members of Zingiberaceae contains cells with ethereal oils. Rhizomes, leaves or seeds may be richest in such compounds, according to the genus or species. Mono- and sesquiterpenoids and aromatic ketones are characteristic constituents of these oils, and aliphatic compounds occur in considerable amounts in some of the ethereal oils. These render a number of species of Zingiberaceae important as sources of spices, as will be mentioned further below.

Apart from this it may be mentioned that silica bodies are present in the epidermis of most or all taxa (see above). Calcium oxalate is present as various kinds of crystals but in most accounts raphides are reported to be lacking, although they do occur in the tapetal cells of *Elettaria* (PANCHAKSHARAPPA 1966, Fig. 167P).

There are single reports of steroidal saponins (*Alpinia*) and cyanogenic compounds (*Hedychium*) in the family. Tannins (from proanthocyanins) are often present. *Kaempferia* is reported to contain an alkaloid (N-methyl-anthranilic acid methyl ester); indole is known in *Hedychium*. Various flavonols are reported from Zingiberaceae (GORNALL et al. 1979), C-glycoflavones are lacking. Dihydroflavones are known in some cases, and among the anthocyanins, cyanidin and methyl derivatives are reported (HARBORNE in DAHLGREN and CLIFFORD 1982), the major glycosides being 3-glycoside and 3-rutenoside.

*Distribution.* The Zingiberaceae are pantropical. The greatest concentration of genera and species is in South-East Asia. They are chiefly forest floor plants growing in humus-rich shade or semishade habitats. Due to the great difficulty of preserving the flowers, taxonomic work is difficult and several genera are incompletely known.

BURTT and SMITH (1972) proposed a division of the family into four tribes. Their definitions will be followed here.

### Tribus Hedychieae

The distichy of the leaves is parallel to the rhizome (this is also the case in the tribe Zingibereae). The ovary is trilocular. In the androecial whorls the lateral staminodes (outer whorl) are well developed and free from the labellum (very rarely connate to it as in some species of *Kaempferia*). The style is not far exserted beyond the anther (as it is in Zingibereae), and the anther crest is not wrapped around the style. The stigma is expanded.

This tribe comprises about 16 genera, including *Boesenbergia, Brachychilum, Caulokaempferia* (Fig. 166A–F), *Curcuma, Hedychium, Kaempferia* and *Roscoea* (Fig. 167A–I).

*Hedychium* (50, Fig. 166G–N), with centres in eastern India and Malaysia, are erect plants with few to numerous stems, which may form thickets. Rarely they may become up to 4 m tall. A number of species are grown as ornamentals, and the rhizomes of *H. spicatum* is used as a source of perfume.

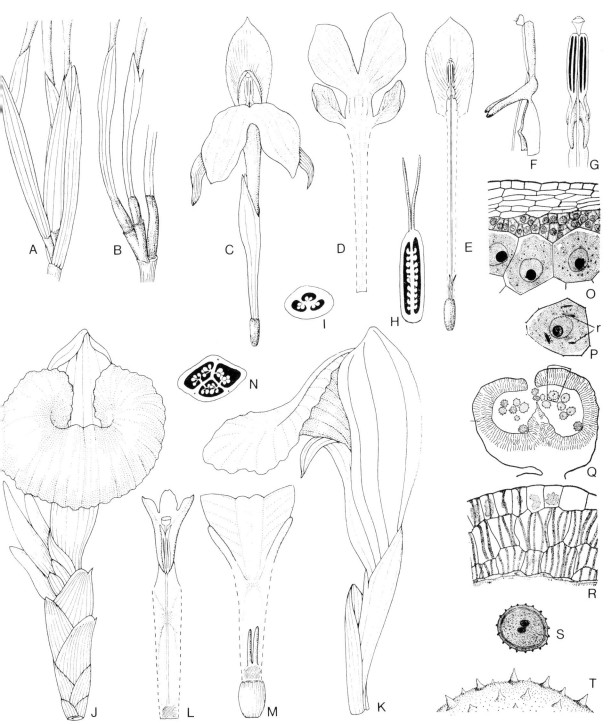

**Fig. 167.** Zingiberaceae. **A–I** *Roscoea purpurea.* **A** Basal part of inflorescence, with bracts. **B** Same, bracts removed. **C** Flower. **D** Labellum and lateral staminodes (of outer staminodial whorl). **E** Median tepal of inner whorl, tepal tube, anther and pistil. **F–G** Anther enclosing style apex. **H** Ovary, longitudinal section, with epigynous glands. **I** Ovary, transverse section. **J–N** *Aframomum luteo-album.* **J** Inflorescence. **K** Flower with bract. **L** Tube of inner perianth whorl, longitudinal section, showing anther and enclosed style apex. **M** Tube of inner perianth whorl, longitudinal section, showing lateral staminodes, base of labellum and epigynous glands. **N** Ovary, transverse section. (A–N BURTT and SMITH 1972). **O–S** *Elettaria cardamomum.* **O** Anther wall, early stage, showing archesporial tissue and tapetum. **P** Archesporial cells (*r* raphides). **Q** Anther wall, showing section where **R–S** are taken. **R** Endothecial layers. **S** Three-celled pollen grain, the sperm nuclei dark, the vegetative nucleus invisible. (PANCHAKSHARAPPA 1966). **T** *Alpinia speciosa,* part of spinulose pollen grain (ERDTMAN 1952)

*Kaempferia* (50) occurs from tropical Africa through India to Malaysia and southern China. It consists of low herbs with a short or with almost no stem, with from two to a few broad leaves, and often with large showy flowers. Sometimes the two lobes of the labellum resemble strongly the staminodes of the outer whorl (as in *K. brachystemon*). The rhizome is thick and tuber-like. The tuber-like rhizomes of *K. galanga* are used throughout Asia as a condiment, "*rhizoma galangae*", and that of *K. rotunda* is used both as a stomach medicine and a condiment. – *Boesenbergia* (50) is a large genus concentrated in the region from Indo-Malaysia to the Philippines and China. – *Curcuma* (50), in tropical parts of Asia and Australia, are rosette herbs with thick, often tuber-like rhizomes, long and often lanceolate leaves and thick, cone-like spikes with imbricate bracts, the upper of which may be enlarged and brightly coloured. The rhizome of *C. domestica* (= *C. longa*) is used as a condiment. It is an important constituent in curries and is also the source of a yellow dye (curcumin) for cotton, silk and wool. The rhizomes of other species of the genus yield starch ("East Indian Arrowroot"), or are used as pickles, and the flowers of, for example, *C. alismatifolia* can be eaten as a vegetable.

### Tribus Zingibereae

This consists only of the genus *Zingiber,* which like the Hedychieae, has the plane of the leaves in line with the rhizome. The ovary is trilocular. The style is exserted well beyond the fertile part of the anther and is enveloped by the elongate prolongation of the stamen ("anther crest"). The pollen grains are sulcate. The stigma is not expanded.

*Zingiber* (75) has thick, aromatic rhizomes and lanceolate leaves situated on erect branches longer than and different from those bearing the inflorescences. *Z. officinale,* "Ginger", indigenous in Eastern India, is cultivated all over the tropics, but mostly in Jamaica and China, for the aromatic rhizomes, used since Roman times. The peeled rhizomes can be used for marmalade, but also, when boiled, sun-dried and made into a powder, as a condiment. It is used to flavour cakes and "ginger-beer".

### Tribus Alpinieae

This is characterized by the leaves lying in a plane transverse to the rhizome. The ovary is generally trilocular (except in *Riedelia*), with axile placentae. The lateral staminodes (of the outer staminal whorl) are usually present as two small teeth at the base of the often very wide labellum, or they may even be lacking; only rarely are they petaloid or linear and may then be attached to the basal part of the labellum (Fig. 167D, M). If the style is exserted far beyond the anther the "anther crest" does not envelop it. The stigma is expanded.

This is a large tribe, which includes a number of important genera, such as *Aframomum, Nicolaia, Amomum, Elettaria, Renealmia, Alpinia* and *Riedelia.*

*Aframomum* (50; Fig. 167J–N) is restricted to tropical Africa. It has large, single flowers emerging from the rhizome; the labellum is wide and funnel-shaped, and the fruit is fleshy and flask-shaped. The stamen has a trilobate continuation beyond the anther. The seeds of *A. melegueta,* "Grains of Paradise", are used as a condiment in the manufacture of a liqueur.

*Amomum* (150), in eastern Asia and Malaysia, Papuasia and tropical Australia, are herbs of medium size or up to 5 m tall with long, leafy branches and with creeping, often slender rhizomes. The leaves are often somewhat hairy and have a long sheath and also often a long, thin ligule. The inflorescences are borne on short branches and are spicate to capitate, often few-flowered. Even though the leafy stem may be 2 m tall or more, the inflorescence remains near the ground. The seeds of *A. cardamomum* are used as a condiment, "Siam-Cardamon" (cf. *Elettaria*), and other species are also used, with various of their parts, for similar purposes.

*Elettaria* (2; Fig. 167O–S) from India and Sri Lanka to Malaysia, comprises short-stemmed herbs with one or few leaves and a prostrate elongate inflorescence which is partly subterranean. In this the flowers are situated in lateral cincinni. *E. cardamomum,* "Cardamon", is grown mainly in India and Sri Lanka, but also, for example, in Guatemala. The capsules are roundish, with small aromatic seeds, which are ground or used whole to flavour cakes, pickles, etc. – *Alpinia* (250) is a variable genus, which is probably heterogeneous in the present concept. It occurs in eastern Asia, Malaysia, and the islands of the Pacific Ocean, and comprises strong, often rather tall herbs which may have a banana-like habit. The leafy branches terminate in inflorescences (rarely the inflorescence-bearing shoots are short and emerge from the rhizome). The inflorescence is thyrsoid or spicate, of variable size. The fruits are capsular, sometimes fleshy and irregularly dehiscent. *A. galanga*, "Greater Galangal", is cultivated for the rhizomes from which an essential oil is extracted; it is used in perfumery. – *Renealmia* (70), which occurs in both Africa and America, has stellate hairs on the leaf lamina. – *Riedelia* (50) is centred in New Guinea. Small to large, sometimes epiphytic herbs with leafy shoots bearing a terminal, racemose inflorescence. The flowers have a tubular outer perianth.

*Nicolaia elatior* is the well-known "Torch-Ginger", a tropical ornamental with red flowers in a dense red-bracted capitulum-like inflorescence; it clearly displays the syndrome of ornithogamy.

### Tribus Globbeae

A small group of four genera with short rhizomes, and bizarre flowers, with relatively large lateral staminodes and with an anther that is usually long-exserted on an arched ascending filament. The ovary is unilocular with parietal placentation.

*Globba* (50–100, Fig. 166P) and *Mantisia* (2; Fig. 166O) are two related genera, the former reaching from India to China, the Philippines and New Guinea, the latter with a more restricted Indian distribution. *Globba* generally has a slender inflorescence with ovate, often violet or blue bracts and orange, yellow or whitish flowers. The long tube formed by the inner tepals, the long style and stamen, and the often arrow-shaped labellum make the genera characteristic. In *Mantisia* the lateral staminodes (outer whorl) are fused with the filament of the fertile stamen, whereas in *Globba* they arise at the level of the inner tepals. Species of *Globba* are grown as ornamentals.

The affinities of Zingiberaceae are not regarded as problematical. The family is obviously closely allied to Costaceae, and these two are often treated as subfamilies of a Zingiberaceae sensu lato. There is also evidence that these two taxa are closely allied to Cannaceae and Marantaceae and also to the other families of Zingiberales.

## Costaceae T. Nakai (1941)   4:150–200
(Fig. 168)

Herbaceous rhizomatous plants, low to medium-sized, rarely up to 5 m tall (*Costus maximus*), generally with a well-developed aerial stem, which is branched, and which bears spirally arranged leaves with phyllotaxies other than 1:2, e.g. 1:4, 1:5, 1:6 and 1:7. The rhizome is sympodial and is generally thick and fleshy, and may be tuber-like.

The leaves have a short, closed tubular sheath and an entire, lanceolate or oblanceolate to linear, generally pointed lamina, which is continuous with the sheath or is basally constricted into a short petiole. Ptyxis is supervolute. A ligule is present at the upper end of the sheath. The venation is generally pinnate-parallel, with prominent midvein and slightly arching, converging lateral veins.

Vessels occur in roots, rhizome and stem, and in the roots they may have simple perforation plates but otherwise they have exclusively scalariform perforation plates. The stem cortex is thinner than in Zingiberaceae and its fibre cylinder is undulate in transverse section (circular in Zingiberaceae). The cortical bundles also meet the fibre cylinder, which is not the case in Zingiberaceae. The aerial stem bears buds in the leaf axils, and thus a nodal plexus is developed, which is not the case in Zingiberaceae either. In the petioles and midveins the main bundle arcs face towards the adaxial side of the leaf (in Zingiberaceae towards the abaxial side). Further, in Costaceae branches and leaves are rather often hairy, the hairs being multicellular and uniseriate and not sunken at the base. (In Zingiberaceae, where hairs occur, they are nearly always unicellular and sunken). The stomata are tetracytic or paracytic. Silica occurs in both the stem and the aerial leaves, but not in the epidermal cells (which is likewise a difference from Zingiberaceae, where they occur only in the epidermis of the leaves). The silica bodies are stellate or druse-shaped, with a smooth surface. Unlike the Zingiberaceae, the Costaceae lack oil cells, and thus are not aromatic plants. Raphides are probably lacking, but calcium oxalate occurs in other shapes.

The inflorescence is terminal on the shoot and comprises an elongate or compact spike or a head. Sometimes, the inflorescences are borne on separate branches without foliage leaves. In *Monocostus* the flowers are situated in axils of well-developed leaves at the end of the shoot. The inflorescence is otherwise bracteate, the bracts being generally broad, densely imbricate and supplied with a nectariferous gland (callus) below their apex.

The flowers have 3+3 tepals, the outer greenish, fused into a tubular, trilobate collar, the inner somewhat longer, petaloid and basally fused, the median member being longer than the laterals and often upcurved. In the staminal whorls, all

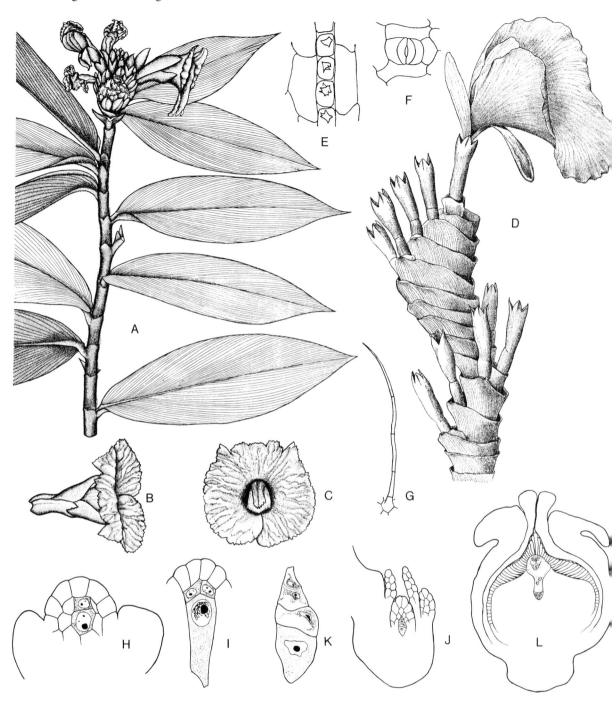

**Fig. 168.** *Costaceae.* **A–C** *Costus speciosus.* **A** Flowering branch. **B–C** Flower. (DEGENER 1940). **D** *Dimerocostus uniflorus,* inflorescence. (SCHUMANN 1904). **E** *Costus lucanusianus,* silica bodies. **F** *Costus lucanusianus,* stoma. **G** *Costus malortieanus,* trichome. (**E–G** from TOMLINSON 1969). **H–L** *Costus cylindricus,* embryological details. **H** Young ovule, a parietal cell has been cut off from the archesporial cell. **I** Later stage, with two parietal cells and a megaspore mother cell. **J–K** Ovule with megaspore tetrad (detail: **K**). **L** Ovule with complete embryo sac; note the palisade-like nucellar epidermis. (FAGERLIND 1939)

members but one, the median of the inner whorl, are developed as petaloid staminodes and all five are fused to form a conspicuous excentric labellum, which varies from five- or three-lobed to subentire, and is often undulate and white or conspicuously coloured. The functional stamen is opposite the middle of the labellum; it is flat and more or less petaloid, and bears two introrse, mutually somewhat distant thecae on the adaxial surface, generally far below the apex. A large number of prominent vascular bundles enters this stamen, the middle one being the most prominent of all. Anther dehiscence is by longitudinal slits and the tapetum is glandular-secretory (BOEHM 1931; PAN-CHAKSHARAPPA 1962). The pollen grains of Costaceae are the only ones in the order, beside those in *Zingiber,* that are acetolysis-resistant and possess distinct apertures (PUNT 1968; D.E. STONE et al. 1981), being sulcate (*Dimerocostus*), spiraperturate (*Tapeinocheilos*) or foraminate, with 5–16 foramina (*Costus, Monocostus*).

The pistil has a trilocular or, by the abortion of one locule, bilocular ovary with numerous ovules on axile placentae. The ovules are anatropous and crassinucellate; the nucellar epidermis remains one-layered and does not take part in the formation of the perisperm tissue.

Endosperm development is helobial; the chalazal chamber is coenocytic and remains at the time when the embryo is differentiated. The embryogeny is reported to follow the Caryophyllad pattern (in Zingiberaceae the Asterad, which is the common type in this and most related orders).

The fruit is a capsule or is dry and indehiscent; on its apex the outer perianth remains as a crown. The seeds are provided with a bulbous mound at the micropylar end and a lid (operculum). The protecting (hard part) of the seed coat is formed by the inner epidermis of the testal layer only. The seed is filled with a strongly developed perisperm and a well-developed chalazosperm (the latter not or poorly developed in Zingiberaceae), whereas the endosperm is little developed and contains fat only. The embryo is cylindrical, rather short, and has a cotyledonary sheath; it contains fat as storage material. $x = 8$ or 9.

*Chemistry.* The most conspicuous chemical attribute of the Costaceae is the absence of essential oils (which characterize the Zingiberaceae). Steroidal saponins are reported from *Costus,* as also are cyanogenic compounds. Flavonoids on the whole are few, although C-glycoflavones have been reported.

Much of the embryological information is taken from PANCHAKSHARAPPA (1962, 1970), by whom Costaceae was found distinct enough to deserve subfamily rank (under Zingiberaceae). We are in agreement with him and other taxonomists that the group should be treated as a separate family.

*Distribution.* Costaceae are pantropical in distribution, but very clearly centred in Central America, although the genus *Tapeinocheilos* has its centre in Eastern Malaysia. They are most frequent in rain forests and other moist shady situations and are terrestrial.

There are four genera, the genus *Costus* being the wholly dominating one.

*Costus* (125–175) is a common genus covering almost the whole distribution area. Some species are robust plants, several metres high, whereas in the other genera the flowers terminate short leafy shoots. Subgenus *Costus,* with ca. 100 species, in both the Old and the New World, includes a number of tall forms, some of which are conspicuous and may be grown as ornamentals. In the subgenus *Paracostus* the foliage leaves are solitary or few and may be broad, with arching-converging veins that are separate all from the base of the lamina (see MAAS 1972).

*Tapeinocheilos* (20) is concentrated in Malesia, ranging to Northern Australia. Its ovary is bilocular or has, at most, a fairly small third locule. The labellum is also smaller than in *Costus.*

Two further, small genera, *Dimerocostus* and *Monocostus* occur in the New World, in tropical American and Peru, respectively.

As has been emphasized in the description, Costaceae on a close examination turn out to differ strongly from Zingiberaceae s.str. Pollen grains with a distinct (but thin) exine and apertures, have been considered to represent ancestral attributes which are puzzling in the view of the other, fairly advanced attributes of the family. It could also be regarded as neotenous (D.E. STONE et al. 1981).

## Cannaceae A.L. Jussieu (1789)    1:ca. 50
(Fig. 169)

Herbaceous, rhizomatous herbs from ca. 50 cm to 5 m tall, with spirally set (not distichous) leaves on an erect aerial stem terminating in an inflorescence. The rhizome may be horizontal and branched, but is generally short, compact, starch-rich and tuber-like. The vascular strands of the roots contain vessels with simple perforation plates or with scalariform perforation plates with few bars, while vessels are lacking in stems and leaves.

The stem is terete and often robust. The leaves are often large. They have an open sheath and a linear, lanceolate or oblanceolate and pointed lamina which often tapers basally into a variably long pseudopetiole. The venation is pinnate; the lateral veins are parallel and arching-convergent.

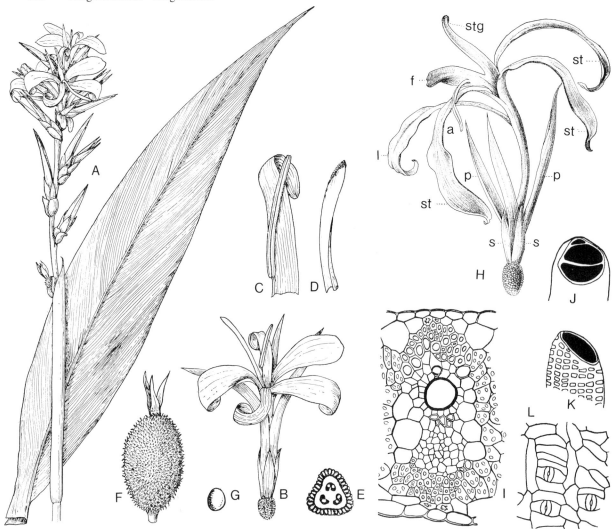

**Fig. 169.** Cannaceae. **A–D** *Canna glauca*. **A** Inflorescence and leaf. **B** Flower. **C** Functional stamen; note that this is petaloid and has only one developed theca. **D** Style and stigma (CABRERA 1968). **E–H** *Canna indica*. **E** Ovary, transverse section. **F** Fruit. **G** Seed. (YANG 1978). **H** *Canna indica,* flower (*s* tepals of the outer whorl; *p* tepals of the inner whorl; *st* staminodes of the outer staminal whorl; *l* labellum, a staminode of the inner staminal whorl; *a* anther half, and *f* the flat petaloid part of the single functional stamen; *stg* stigma on top of the likewise rather petaloid style. (WINKLER 1930). **I–L** *Canna* spp. **I** Part of leaf with vascular strand, note position of silica bodies. **J–K** Perforation plates of root vessels. **L** Stomata. (TOMLINSON 1969)

Stems and leaves are generally glabrous, but sometimes for example the leaf sheath may be pubescent. The stomata are paracytic, tetracytic or hexacytic. Oxalate crystals occur as needle-like, rod-like or plate-like single crystals, not as raphides. Silica is lacking in the epidermis, but occurs as druse-like bodies in cells adjacent to vascular bundles. Secretory canals with mucilage occur in the rhizome and the aerial stem, but do not enter the leaves. Cells with tanniniferous contents occur in the whole plant (TOMLINSON 1961 a, 1969).

The inflorescence is a spike or thyrse, in the latter case the lateral components are, as a rule, two-flowered cymules. The inflorescence bracts vary in size, but are generally green and inconspicuous, and often rather small. The flowers are asymmetrical. The perianth consists of 3 + 3 tepals. The outer are much smaller than the inner. Those of each whorl are subequal or somewhat different in length. The outer tepals are ovate to lanceolate, greenish or subpetaloid, generally purplish and free from each other; the inner are often several times as long as the outer and are lanceolate or linear, petaloid, one of them often being shorter than the other two; they are more or less connate basally and fused with the stamen and the staminodes into a tube.

In the androecium only the median member of the inner whorl is a functional stamen; this is flat and fleshy-petaloid, with only one theca.

A variable number (one to four) of petaloid staminodes are developed, but the stamen homologues are maximally five in number. Always present is one petaloid staminode of the inner staminal whorl, which is wider than the others and is sometimes called the labellum. The functional theca is lateral on the stamen, often inserted far below its apex, bisporangiate, long, linear, dehiscing with a longitudinal slit. In *C. indica,* which has secondary pollen presentation, the filament enfolds the style. The tapetum is alternatively reported to be amoeboid and secretory. The pollen grains (KRESS and D.E. STONE 1982), are globose, spinulose and inaperturate (or rather omniaperturate).

The inferior ovary is covered with radiate, densely situated emergences which make the surface warty. It is trilocular with axile multiovulate placentae. The style is straight, flat and fleshy-petaloid. In *C. indica* the stigma is on the margin at the apex; pollen from the anther (see above) is deposited proximal to this and adheres only to the area on which it is first deposited and, if transferred there, to the stigma (P.F. YEO, personal observation). The ovules are anatropous and crassinucellate; embryo sac formation is of the *Polygonum* Type, endosperm formation is nuclear and embryogeny is of the Asterad Type. The nucellar tissue is strongly developed, especially at the chalazal end.

The fruit is a hard capsule with a warty surface, occasionally crowned by the persistent tepals of the outer whorl. The seeds are ovoid, rather smooth, black and provided with a collar. They are embedded in loose hairs, but these develop from the funicle rather than chalaza and are often not regarded as a true aril (the very divergent statements as to whether an aril is present may depend on the interpretation of these hairs). When the seeds are shed these hair tufts remain on the capsule wall. The seeds have a distinct germination slit (sometimes regarded as an "operculum"), and a micropylar collar. The seed coat is formed by the testal part, especially the hard layer formed by its outer epidermis and the layers beneath this; it has stomata. The endosperm comprises a thin but starchy layer and is surrounded by the richly developed perisperm, which is hard and also rich in starch. The embryo is straight and linear; secondary roots and plumule are present on the mature embryo. $x = 9$.

*Chemistry.* Oxalate is deposited as crystals of various shapes except as raphides, and silica occurs as druses. Cyanogenic compounds are reported to occur. The family is poor in flavonoids but proanthocyanins occur in the "tannin cells".

*Distribution.* Cannaceae are indigenous in tropical and subtropical parts of the Americas, and are widely dispersed (by man only?) also in the Old World. The plants prefer moist, humus-rich ground in forests, along riversides and in other habitats; some species even grow in swampy places.

The only genus is *Canna* (ca. 50), which contains a variety of species, often with much slenderer flowers than the massive ones of *Canna × generalis,* which is one of the most commonly grown garden ornamentals in the tropics (it is a hybrid derived from several species and often known by the name of one of them, *C. indica*). The colours of the flowers in the genus vary from yellow to bright red or purple. There is also great variation in the rhizome, which may be slender or thick and branched or compact and tuber-like. *C. edulis,* "Queensland Arrowroot", is a minor crop in South-Eastern Queensland. *C. tuerckheimii,* from Guatemala, reaches a height of 4–5 m and has a 12 cm long staminodial tube.

The closest relationships of Cannaceae are considered to be with Marantaceae, although the family has a very distinct position.

**Marantaceae** Petersen in Engler & Prantl (1888) 30:170   (Fig. 170)

Terrestrial, perennial rhizomatous herbs, rosulate or with an aerial stem which varies much in length, sometimes up to several metres high and liana-like. The rhizomes are sympodial, often starchy, and may have scale-like leaves. The vascular system of the roots is provided with vessels that have simple perforation plates; scalariform perforation plates are sometimes present as well.

The aerial shoots are diversified and show regularity in the internode length, certain internodes being frequently suppressed, so that the leaves become distributed in sequences. The first leaf of each lateral branch is a two-keeled prophyll, the next one or few are restricted to a sheath and the following are fully differentiated as described below. Low-growing species are rosulate, with the leaves concentrated basally. The inflorescence is pedunculate, mostly coming out from the centre of the leaf rosette in the rosulate species and at the apex

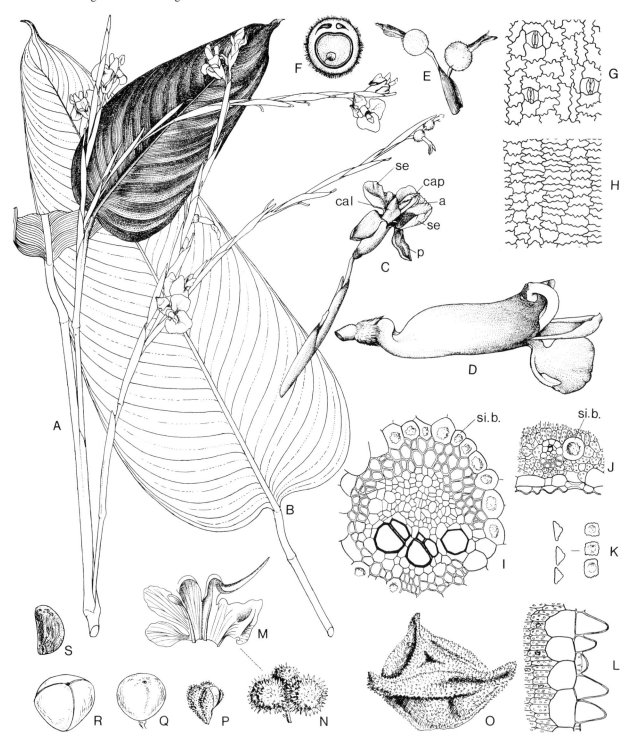

**Fig. 170.** Marantaceae. **A–F** *Marantochloa cuspidata.*
**A** Inflorescence. **B** Leaf. **C** Flower (*p* tepal of inner
whorl; *se* petaloid staminodes of outer staminal whorl;
*cap* cucullate staminode, inner staminal whorl; *cal*
callose staminode, same whorl; *a* functional anther).
**D** Flower with perianth and staminodes of outer staminal whorl removed, showing the recurved style apex
with stigma projecting out of the staminodial tube, the
functional anther turning downwards, the callose staminode forming the upper and the cucullate staminode
the straightly projecting part. **E** Fruits. **F** Fruit, transverse section, two locules empty. (MANGENOT and DE-
VILLER 1965). **G–H** *Ischnosiphon aruma,* epidermis from
abaxial and adaxial side of leaf. **I** *Sarcophrynium bra-
chystachyum,* vascular strand from rhizome, transverse
section (*si.b.* silica body). **J** *Marantochloa mannii,* silica
cell in leaf transection. **K** *Donax* sp. and *Marantochloa*
sp., hat-shaped silica bodies from epidermal cells, differ-

of the leafy branches in the large-growing and sometimes richly branched taxa. Sometimes the stems are scandent, with sequences of long internodes.

The leaves are generally distichous and are differentiated into an open sheath, a distinct, often rather long petiole (which is sometimes winged) and a lamina, which is entire and pointed, obtuse or rounded. A "pulvinus", i.e. a terete callous section, is intercalated between the lamina and the petiole or sheath. A ligule is lacking. The leaf has a prominent midvein and pinnate parallel secondary veins, which arch and fuse marginally, some of them, at intervals, being thicker than the others. The leaf surface is often variegated, especially in juvenile plants.

The stems and leaves are glabrous or frequently more or less hairy, the hairs being nearly always unicellular. Stomata are paracytic or tetracytic and have asymmetric guard cells. Schizogenous air lacunae are nearly always present in the petiole and midrib of the leaves. Calcium oxalate is present as solitary crystals, not as raphides. Silica bodies occur in all green parts and are generally found in cells associated with veins, but may also occur in other mesophyll cells. They are generally druse-like, rarely hat-shaped. Silica never occurs in cells of the epidermis. Tannins are often present in large modified cells or in "normal" cells (TOMLINSON 1961 b, 1969).

The inflorescences are thyrses. They are terminal or lateral on the shoot, and vary much in complexity. The lateral, cymose components of the thyrses may be few-flowered or many-flowered, the ultimate floral groups being as a rule two-flowered or rarely one-flowered cymules, each with a prophyll on the dorsal side and sometimes with more phyllaries on the ventral side.

The flowers often have one or two dorsal bracteoles. They are bisexual and asymmetric due to the presence and shapes of the petaloid staminodes and the single functional stamen, which is asymmetric. In the flower pairs present in most inflorescences the two flowers are mirror images.

The tepals are 3 + 3. The members of the outer whorl are distinct and non-petaloid, often rather short. The members of the inner whorl are generally connate basally into a tube; they are more or less petaloid and variable in colour, the median member often longer than the others and somewhat hood-like.

The outer staminal whorl consists of one or two staminodes (or is rarely totally missing), the median member being always absent. These staminodes when present vary from broad, flat and petaloid to subulate.

The inner staminal whorl consists of two petaloid staminodes and one stamen, which possesses only one, lateral theca. One of the staminodes is more or less hood-like (*staminodium cucullatum,* or "labellum") and the other is fleshy and firm (*staminodium callosum*); the filament of the functional stamen varies from narrow to broad and partly petaloid. The members of the outer staminal whorl are fused with each other to form a tube which is longer than the inner tepals, with which they are fused basally.

The functional theca is elongate and it dehisces longitudinally. Microsporogenesis is successive (SCHNARF 1931). The tapetum does not seem to have been described. The pollen grains are large, spherical, symmetrical and without sporopollenin (exine), the wall consisting of the intine only. They are two-celled when dispersed (checked for *Maranta* only; BREWBAKER 1967).

The ovary is tricarpellary and trilocular, one or two of the locules being often empty and compressed. Sunken septal nectaries are present. The ovule is solitary in each locule and inserted basally towards the centre. The style is terminal, narrow, basally fused with the tube that is formed by the inner tepals; it is often slightly widened and lobed in the apical part which is hooked. The stigma is located on the inner face of the somewhat infundibular style apex.

The pollen is released from the anther at an early stage and becomes deposited in a subapical cavity on the outside of the curved part of the style (secondary pollen presentation). Before the flower opens the style is held under tension in the hood-like sheath of the cucullate staminode. When the pollinator, which is generally a bee, forces its head between the callous and cucullate staminodes in its search for nectar (secreted in the septal nectaries), it pushes aside the appendage of the cucullate staminode, so that the style becomes released (with a "puff"). The pollen previously deposited on the style is then transferred to the insect. In general,

---

ent views. **L** *Calathea zebrina,* adaxial surface of leaf lamina, transverse section. (**G–L** TOMLINSON 1969). **M** *Hypselodelphys poggeana,* androecium. (KOECHLIN 1964). **N** *Hypselodelphys poggeana,* fruit. **O** *Hypselodelphys violacea,* fruit. **P** *Trachyphrynium braunianum,* fruit. **Q** *Sarcophrynium prionogonianum,* fruit. **R–S** *Megaphrynium macrostachyum,* fruit and seed. (**N–S** HEPPER 1968)

the family appears to be adapted to insect pollination.

The ovules are anatropous, becoming campylotropous to amphitropous in later stages. They are crassinucellate and the nucellus is well developed, especially in the chalazal region. In the course of development chalazal tissue, the chalazosperm, grows into the increasing nucellar tissue, the perisperm forming a frequently branched strand in it. The outer integument is thick while the inner is very thin. A "micropylar collar", in the form of a ridge around the micropyle, is formed by divisions in the central part of the outer integument; periclinal divisions outside this ridge give rise to the operculum. The nucellar epidermis becomes palisade-like and forms a cap-like structure. The primary archesporial cell cuts off a parietal cell, but its divisions do not produce any considerable tissue. Embryo sac formation is of the *Polygonum* Type, and endosperm formation is nuclear in the cases studied. Embryogeny is insufficiently known. In the course of development the embryo becomes linear, curved and of considerable size, and the endosperm is used up; the storage tissue, which is enclosed by and envelops the long, curved embryo, consists of perisperm and chalazosperm.

The fruit is generally a loculicidal capsule with one to three seeds, the pericarp varying much in thickness and texture. In some taxa it remains indehiscent, the pericarp being dry and sometimes hard, or else fleshy.

There is only one seed per locule; it is comparatively large, subglobose or conical-ellipsoid, with a smooth to rugose surface. The seed coat consists of the testal layers only. The micropylar part forms an operculum. In dehiscent fruits the seeds are provided with an aril formed by outgrowths from the outer epidermis of the micropylar part of the outer integument. The aril is lacking or transformed to a pulp in indehiscent fruits. The perisperm/chalazosperm is starchy.

The basic chromosome numbers range between $x = 4$ and $x = 14$ or more; for most genera the chromosome numbers are surprisingly poorly known.

*Chemistry.* The rhizomes contain copious amounts of starch (see under *Maranta,* below). Silicic acid occurs as hat-shaped or druse-like bodies. Calcium oxalate is frequent but evidently is never present as raphides. Saponins and chelidonic acid are probably lacking, whereas cyanogenic compounds occur, at least, in some genera (e.g. *Maranta, Phrynium* and *Thalia*). Flavonoids are richly present, C-glycoflavones being found more often than in the other families of the order, but in Mar-

antaceae luteolin/apigenin, cyanidin, delphinidin, methyl derivatives, anthocyanins and flavonol sulphates are also recorded (see GORNALL et al. 1979; HARBORNE in DAHLGREN and CLIFFORD 1982). Isoenzymes of dehydroquinate hydrolyase (DHQase) were found by BOUDET et al. (1977) to be lacking from one genus investigated. Some genera contain considerable amounts of wax.

*Distribution.* The Marantaceae have a pantropical distribution but with a clear Neotropical concentration. The find of fossil remains reminiscent of Marantaceae in Eocene deposits of England indicates that the distribution may have been much wider in earlier times. At present ca. 350 out of the 450 species occur in the New World (200 belonging to *Calathea*). There are few species (ca. 35) in Africa, mainly in the western parts, and ca. 50 species, at least, in Asia, between Sri Lanka and the Solomon Islands.

The Marantaceae are terrestrial and grow in tropical rain-forest margins and in clearings. Some require moist or swampy habitats.

*Taxonomy.* No satisfactory subdivision has been proposed for the family. A partly artificial division (PETERSEN 1889) was based on whether the ovary has three fertile locules (tribus Phryneae) or one only (tribus Maranteae), but according to ANDERSSON (1981a), this is not satisfactory. ANDERSSON (1981; unpublished) proposes a few tentative groups as follows.

**The Phrynium Group.** In this the cymules of the inflorescences are generally condensed (brachyblastic), the flowers lack bracteoles, the corolla tube is generally moderately long, and there are (one or) two staminodes present in the outer staminal whorl. Here belong, perhaps, five genera, all in the Old World (Africa, Asia).

Among these are *Phrynium* (20) which are rosulate herbs in South-Eastern Asia.

**The Calathea Group.** This also has brachyblastic cymules, but the flowers possess bracteoles, the tube formed by the inner tepal whorl is long to very long, and in the outer staminal whorl there is only one petaloid staminode. This group may also include five genera, but all are distributed in the Americas.

*Calathea* (more than 200) are low, rosulate herbs, with simple to moderately branched inflorescences, with spiciform to capitate components. The fruit is mostly three-seeded and dehiscent and the seeds arillate. Pollination details for some species have been worked out by KENNEDY (1978). Upon its release from the cucullate staminode the style increases its curvature. It is so positioned that its stigmatic part scrapes the recess under the insect's head (where the proboscis is held when folded) removing any pollen; immediately afterwards the subapical stylar cavity makes contact with the same part of the insect, depositing its own load of pollen. The pollen is quite inaccessible to the insect. The genus is widely distributed in the American tropics, and the species taxonomy is not yet clear. Many species are cultivated as indoor foliage plants. – Another genus, *Ischnosiphon* (35), consists of rosulate or caulescent plants with one-seeded fruits. Some species are quite tall. The spathal bracts of the inflorescence branches are rolled before anthesis. This genus is also widely distributed in tropical America.

**The Donax Group** has less condensed cymules, the flowers possess bracteoles, which are small and sclerotic, the tube formed by the inner tepals is short, and two petaloid staminodes are present in the outer staminal whorl. Here belong about six genera, all in the Old World; several of the small genera are restricted to Africa.

*Donax* (3–4) are tall plants with cane-like stems crowded with clusters of leaves. The fruits are indehiscent with exarillate seeds. *Donax* occurs in Malaysia, Indochina, the Philippines and Polynesia. – *Hypselodelphys* (4) and *Trachyphrynium* (1) are examples of African genera of scandent, much-branched herbs, and *Sarcophrynium* (3) is an African genus of small rosulate herbs (see Fig. 170). The fruits may be dehiscent with arillate seeds, or indehiscent, with exarillate seeds; in *Sarcophrynium* the fruit is baccate with a mucilaginous aril.

**The Maranta Group** differs from the *Donax* Group mainly in that the bracteoles are lacking or rudimentary and that the floral tube is longer. This group consists of about four genera, and oc-

curs in Africa (e.g. *Marantochloa*) and South America (e.g. *Maranta*).

*Marantochloa* (15) are shrub-like, up to ca. 3 m tall, with branched panicle-like inflorescences. The fruits are dehiscent, with arillate seeds. The genus occurs from West Africa to Tanzania and Sudan.

*Maranta* (15–20) are from small and decumbent to erect and rather tall herbs, which occur in humid parts of tropical America. Some are grown as indoor foliage plants. *Maranta arundinacea* is cultivated as a crop plant (West Indian Arrowroot) because of the high-quality starch obtained from the rhizomes; it is indigenous in the Caribbean region.

**The Myrosma Group,** with variably condensed cymules, agrees in most features with the *Maranta* Group but has a short or very short perianth tube. Here belong ca. six genera, all in the New World tropics. – In another group, *the Montagma Group*, the tube formed by the inner tepals is very long. These genera are also South American. Many of these plants are low herbs with rosulate leaves.

*Thalia* (5–7), which is difficult to refer to a group, is an American genus of marsh and swamp plants, some small and rosulate, others caulescent and rather tall. The inflorescences are often rather branched, with deciduous spatheal bracts. The flower has only one staminode in the outer staminal whorl, but this is large and showy, and the callous staminode is also large, with a narrow petaloid rim. The fruit is indehiscent and functions as a nutlet. A population in Africa is probably introduced.

The closest phylogenetical relationships of Marantaceae are generally considered to be with Cannaceae in the first place, and secondly with Costaceae and Zingiberaceae, which is logical on the basis of the collected evidence.

# Superorder Commeliniflorae

*Four Orders:* Commelinales (incl. Eriocaulales and Xyridales), Hydatellales, Cyperales (incl. Juncales) and Poales (incl. Restionales).

Autotrophic, mainly herbaceous, perennial or annual plants of variable size, but also some woody and partly arborescent plants (e.g. bamboos, *Microdracoides* in Cyperaceae). Perennials, generally rhizomatous. Lateral roots appearing opposite xylem strands or (which is unique in monocotyledons) often also opposite phloem strands (VAN TIEGHEM 1887; VAN TIEGHEM and DULIOT 1888). The root-hair cells are either of the same size and shape as ordinary epidermal cells or are much shorter than these. Velamen is lacking. The roots generally have vessels with simple, or scalariform as well as simple perforation plates.

The stems are variously constructed; in most grasses they are hollow but have compact, often swollen nodes. The leaves are alternate, distichous, tristichous or with some other phyllotaxy, but are never opposite or verticillate. They are generally sheathing at the base. The sheath is open or closed. The leaf blade is linear to lanceolate or ovate and has parallel or arching-converging venation. Transverse veinlets occur in some Commelinaceae. Stipular lobes are only rarely present, whereas a ligule is frequently present as an adaxial continuation of the sheath. The ligule is usually hyaline but sometimes transformed into a rim of hairs. Rarely, a contraligule is present on the abaxial side of the sheath. The lamina normally continues directly from the sheath, but in some genera it is basally constricted into a pseudopetiole which is sometimes considerable (e.g. in various Commelinaceae and Poaceae subfamily Bambusoideae). Unifacial-ensiform or compressed leaves occur at least in Xyridaceae and Anarthriaceae and some Juncaceae and Centrolepidaceae. The ptyxis is involute, supervolute or rarely plicate (see ARBER 1922a, 1925).

The stems generally have vessels (lacking in *Cartonema* of Commelinaceae); simple or mixed simple and scalariform perforation plates dominate, but in some groups the perforation is mainly scalariform (see CHEADLE and KOSAKAI 1980, 1982). Also the leaves generally contain vascular tissue with vessels (although absent in Restionaceae, at least) and in them scalariform perforation is more frequent than in the stem vessels. Laticifers are lacking. The sieve tube plastids lack starch grains and protein filaments. Silicic acid in the form of silica bodies is widespread in the superorder, but is lacking in several families (e.g. Juncaceae (?), Hydatellaceae, Centrolepidaceae and Mayacaceae) and is rare in others. Silica bodies occur either in epidermal cells, as in Rapateaceae, Poaceae and Cyperaceae, or in cells connected with the vascular strands. Their shape is highly variable and frequently characteristic of the family or generic group. Oxalate raphides are lacking in all families but Commelinaceae.

Uniseriate hairs with few to several cells are widespread in the order and include the "microhairs" of the grasses (see under Poaceae, and Fig. 193J–O) but many taxa are glabrous. Usually the stomatal complexes are either paracytic (often, as in grasses, with rather specialized subsidiary cells widely different from the strictly filed, much larger ordinary epidermal cells) or, as in the Commelinaceae, tetracytic or hexacytic; only in Hydatellales are they anomocytic, which contributes to the uncertainty of the position of this order in the Commeliniflorae. The divisions of the cells around the stomata are generally even (TOMLINSON 1974). In many taxa the plants are strongly sclerenchymatous, and aerenchyma is strongly developed in some groups, especially in Cyperales. The epicuticular wax, when structured and oriented, is always of the *Strelitzia* Type.

The inflorescences are often borne on a conspicuous peduncle, which is terminal or lateral; they are determinate or indeterminate and extraordinarily diverse. A paniculate type is probably basic. Very often small terminal units of the inflorescence branchlets are developed as spikelets, which are determinate or indeterminate, in the first case often terminated by a female flower, in the latter case consisting of bisexual, unisexual (and often also neuter) florets, i.e. small flowers, all lateral and frequently in a distichous arrangement. The spikelets are often arranged, as if they were flowers, in "panicles", in determinate, spike-like aggregates, or otherwise. In Commelinaceae, the inflorescence is a thyrse with cymose lateral components. Spikes or spikelets have probably originated independently in Poales, Cyperales and Commelinales (Rapateaceae and Xyridaceae). In Eriocaulaceae the inflorescences have a construction similar to the heads of Asteraceae; they have a flat, plate-like axis subtended by scale-like phyllaries, and small erect flowers often with tubular "petals" (i.e. the inner, subpetaloid tepals).

The flowers are bisexual or often unisexual, trimerous or especially in Eriocaulaceae frequently dimerous, hypogynous, actinomorphic or zygomorphic, and are either insect-pollinated or, very frequently, wind-pollinated. Some are autogamous. The perianth frequently consists of $3+3$ (or $2+2$) tepals. In Commelinales there is always very strong differentiation between the outer tepals, which are sepaloid or hyaline, and the inner, which are petaloid and often basally fused into a tube. In the other orders the tepals are non-petaloid, and may either be inconspicuous and bract-like or may be transformed into fleshy scales (lodicules in grasses) or into bristles or hairs (many Cyperaceae) or may be lacking altogether. Perigonal and septal nectaries are lacking. The entomogamous flowers offer pollen or, rarely, are provided with staminodial attractants (Xyridaceae, Eriocaulaceae).

The stamens are generally $3+3$ $(2+2)$ or 3 (2) in number; occasionally they are reduced to one stamen or (in unisexual flowers) are wholly staminodial or lacking. Very rarely, as in some grasses, there are numerous stamens. The filaments are narrow, distinct or basally fused and generally glabrous, but hairy in many Commelinaceae; in most wind-pollinated taxa they are long and pendulous. The anthers are tetrasporangiate or (in Centrolepidaceae, most Restionaceae and many Eriocaulaceae) bisporangiate, introrse or latrorse, and longitudinally dehiscent or (in Mayacaceae and Rapateaceae) poricidal. Anther wall formation is of the Monocotyledon Type. The wall thickenings of the endothecium are usually of the Girdle Type, rarely of the Spiral Type. The tapetum is glandular-secretory except in Commelinaceae and *Abolboda* of Xyridaceae, where it is amoeboid-plasmodial. Microsporogenesis is successive, except in Cyperales and Rapateaceae, where it is simultaneous. Pollen tetrads occur in Cyperales; here all four pollen grains of the tetrad

develop in Juncaceae and Thurniaceae but only one does so in Cyperaceae, the other three degenerating at an early stage and becoming incorporated in the wall of the functional pollen grain. Post-reductional meiosis occurs in Cyperales. The pollen grains have variable aperture conditions; they are very often ulcerate (several families), sulcate (Commelinaceae, most Rapateaceae, Hydatellaceae), spiraperturate (Eriocaulaceae), or rarely sulculate, foraminate or inaperturate. Often, in the wind-borne pollen grains, the surface is smooth. The pollen grains are generally three-celled when dispersed, but two-celled in at least many Commelinales, in Hydatellales and in some Cyperaceae of Cyperales.

The pistil is generally tricarpellary or (in, for example, many Eriocaulaceae) bicarpellary and syncarpous, but in Hydatellales and Centrolepidaceae it is monocarpellary. When the carpels are two or three there may be equally many stylodia or stylodial branches, but sometimes the style is simple and bears a three-lobed or capitate stigma. The stigmatic surface is Dry, rarely Wet, as in some Commelinaceae. The ovary is trilocular, bilocular or unilocular and the placentation in the bi- or trilocular ovaries varies from axile or (rarely) parietal to basal or apical; in the monocarpellary ovaries it is apical.

The ovules vary from many down to one (the latter mainly in wind-pollinated taxa). The ovules are anatropous or, especially in many Poales and various Commelinales, orthotropous, more rarely hemianatropous or campylotropous; they are crassinucellate or tenuinucellate. A parietal cell and, subsequently, a parietal tissue are formed in Cyperales and many Commelinales, but not in the other orders. Periclinal division of the nucellar epidermis gives rise to a nucellar cap in many Poaceae (Poales). Embryo sac formation is nearly always of the *Polygonum* Type, and endosperm formation is nuclear in all members except in Juncaceae (unknown in Thurniaceae) and *Abolboda* of Xyridaceae, where it is helobial, and in Hydatellales, where it is cellular. The embryo is generally formed according to the Asterad Type except in Cyperales, where it follows the Onagrad Type; in the grasses the walls are laid down in an oblique way, which may justify recognition of a grass subtype (of the Asterad Type).

The fruits of the bi- or tricarpellary pistils are often capsules, as in most Commelinales, Juncaceae and Thurniaceae of Cyperales and in most Restionaceae (and two small related families) of Poales, but are otherwise indehiscent and form nutlets or caryopses (in which the testa is fused with the pericarp), or rarely drupes or berries (Joinvilleaceae, Flagellariaceae, a few Poaceae). Where the pistils are monocarpellary [Centrolepidaceae, Hydatell- aceae (?)], the fruitlet is follicular. Arils or similar structures, e.g. elaiosomes, occur in some Comme- linaceae, in Rapateaceae, and also in some Res- tionaceae. The seeds are never phytomelan-coated. They are generally filled, for the most part, with an endosperm with copious amounts of starchy, mealy, non-ruminate endosperm; this is substi- tuted by oil in some grasses. A notable exception from this normal condition is that in Hydatellales, where the nucellar tissue develops into a starchy perisperm, and where the endosperm is restricted to relatively few cells next to the broad embryo. The embryo is generally lens-shaped and broad, i.e. of the *Xyris* Type, or mushroom-shaped to capitate, of the *Scirpus* Type, but in some Junc- aceae and Cyperaceae it may be rather elongate (approaching the *Urginea* Type). In the grasses it is extraordinary in being situated laterally to and outside the endosperm; here it varies much in out- line but has a characteristic morphology (Fig. 199). The seedling is often, as in the grasses, pro- vided with a coleoptile, which is often interpreted as representing the cotyledon.

*Chromosome Numbers.* Chromosome numbers are given under the families. Of special note is that the centromeric activity in the chromosome of Cyperales is not restricted to one definite point. Fragmentation of chromosomes therefore does not severely disturb their function, which leads to a great degree of aneuploidy.

*Chemistry.* Besides silica (see above) some Commeliniflorae, in particular the Rapateaceae and some members of Cyperaceae and Centro- lepidaceae, store considerable amounts of alumin- ium. Calcium oxalate occurs in Commelinaceae. Steroidal saponins are rare in this superorder; tri- terpenes (and triterpene saponins) are probably commoner. Chelidonic acid is reported in some grasses, but is otherwise not known in the Comme- liniflorae. Cyanogenesis is very common and known in many grasses and Cyperaceae, but also in other groups. Among the flavonoids, tricin is known in many Poaceae and Cyperaceae and some Restionaceae. C-glycoflavones, luteolin/apigenin and sulphurated flavonoids have similar distribu- tions in the superorder. 5-0-Me-flavones are char- acteristic of the Cyperales. Isoenzymes of dehydro- quinate hydrolyase (DHQ-ase) have been proved to be present in all taxa studied of the Commelini- florae (in Poaceae, Cyperaceae, Juncaceae) except

in the solitary species of Commelinaceae studied (BOUDET et al. 1977). It was found by HARRIS and HARTLEY (1980) that UV-fluorescent constituents occur in cell walls all over the Commeliniflorae, all taxa studied being positive. This fluorescence is associated with the presence of bound ferulic acid, p-coumaric acid and diferulic acid.

*Parasites.* Also various groups of fungal para- sites, smuts and rusts are concentrated on members of the Commeliniflorae, more than on other groups of monocotyledons. Many of them occur on grasses only, others on members of Cy- peraceae and Juncaceae. Certain fungal species or groups of fungal species attack closely related gen- era of grasses. Also, host specificity of various in- sects such as chinch bugs (SLATER 1976) and plant lice (EASTOP 1979) probably yields useful indica- tions of phylogenetic relationships among the Commeliniflorae. Thus the psyllid subfamily Li- viinae and the aphid tribe Saltusaphidini among the plant lice attack taxa of Juncaceae and Cypera- ceae, which may be taken as a support for relation- ship between these families (see DAHLGREN and CLIFFORD 1981).

*Distribution.* The distribution of the Comme- liniflorae is worldwide and so is the distribution of some of its families, in particular Poaceae, Cy- peraceae and Juncaceae, whereas the Commelin- aceae are pantropical-subtropical. Among these families both the Juncaceae and Commelinaceae have their centre of diversity in South America, and on this continent, especially in the Guiana Highland, we find the centres of Rapateaceae (one genus in Africa), Xyridaceae (*Xyris* with wide dis- tribution in the tropics) and Thurniaceae. Also the Eriocaulaceae and Mayacaceae have their greatest variation in South and Central America.

Whereas the Commelinales and possibly the Cy- perales thus have a South American concentration and probably a West-Gondwana origin, the Poales are concentrated in the Old World with Restion- aceae shared chiefly between South Africa and Australia (*Leptocarpus* extending to South Amer- ica), Centrolepidaceae centred in Australia, Tas- mania, New Zealand and Malaysia, and Hydatell- aceae centred in Australia; Ecdeiocoleaceae and Anarthriaceae, which can be separated from Res- tionaceae, are also Australian. This is of interest because fossil restionaceous pollen is known be- fore grass pollen, in the late Cretaceous. Flagellari- aceae and Joinvilleaceae are South-East Asian–Pa- cific, and the Bambusoideae of the grasses also have their main concentration in Asia.

We conclude that the earliest differentiation of the Commeliniflorae probably took place on the southern continents splitting off from the Gondwana continent around or before the mid-Cretaceous and that the Commelinales and Cyperales complexes were most probably differentiated primarily in South America, while Poales and Hydatellales differentiated in the Australian sector. The rather late, explosive diversification of both Poaceae and Cyperaceae, during the Tertiary, is probably connected with their ability for vegetative reproduction and their physiological-ecological plasticity.

*Relationships.* The relationships within the superorder seem to be relatively uncontroversial, except for Hydatellaceae, which is separately discussed on pp. 32–33, and by HAMANN (1975, 1976), DAHLGREN and CLIFFORD (1982) and DAHLGREN and RASMUSSEN (1983).

The naturalness of the Commelinales, with a petaloid inner tepal whorl, is disputable but probable, and is supported by the geographic pattern. As we have stated in the evolutionary chapter, there are difficulties in finding firm grounds for family groupings in this order, and we refrain from splitting Commelinales into several minor orders, which would be the alternative.

Much more evidence supports the monophyly of our Cyperales, and their close interrelationships are acknowledged in most current classifications, although CRONQUIST (1981), possibly on phenetic grounds, prefers to treat Poaceae and Cyperaceae in his Cyperales. He claims: "Although the Cyperaceae and Poaceae have traditionally been associated in older systems of classification, in recent years many authors have taken each family to form a separate, unifamilial order. This latter view has been conditioned in large part by the obvious relationship of the Poaceae to the Restionaceae, and the Cyperaceae to the Juncaceae". He also remarks that it is not difficult to point out a series of differences between Poaceae and Cyperaceae, and we find it regrettable that he refrains from breaking up this highly unnatural ordinal unit.

Poales and Restionales were regarded as distinct orders at the onset of our comparative study (DAHLGREN and CLIFFORD 1982), but proved difficult to uphold, which was emphasized in particular by CRONQUIST (personal communication), who prefers to widen the circumscription of the present Poales.

The closest relationships of the Commeliniflorae among the monocotyledons are probably with the Zingiberiflorae. There are some interesting similarities in seed morphology, e.g. the presence of an operculum, between Commelinaceae and some Zingiberales, which we have not evaluated here. There is also obviously a close relationship with the Bromeliiflorae. The basically paracytic stomata, the type of structured epicuticular wax, the presence of UV-fluorescent compounds in the cell walls and the starchy endosperm are some conspicuous similarities which we interpret as synapomorphies for Commeliniflorae, Zingiberiflorae and Bromeliiflorae.

The highly puzzling links between the Areciflorae (palms) and Commeliniflorae are also probably of significance, and future analysis may prove that there is a much closer affinity between these superorders than is admitted in any of the present systems of classification (see, for example, DAHLGREN 1983a; and THORNE 1983).

# Order Commelinales

*Five Families:* Commelinaceae, Mayacaceae, Rapateaceae, Xyridaceae and Eriocaulaceae.

Perennial or less commonly annual, mesophytic to somewhat xerophytic, usually terrestrial herbs. Some are aquatic (e.g. Mayacaceae), some are epiphytic or twining, and many live in marshy habitats.

The roots vary from thin and fibrous to uniformly thickened, tapering or fusiform. They may arise opposite the xylem or phloem poles in the underground stem. The root-hair cells are shorter than other epidermal cells of the roots.

The root vessels vary considerably in the order and may have only scalariform perforation plates, as in Mayacaceae and Rapateaceae, but generally the vessels have simple perforation plates or there are vessels with simple and vessels with scalariform perforation plates mixed together. Rhizomes dominate in some families but are rare in Commelinaceae. Tubers occur in *Cartonema* (Commelinaceae).

The branching is sympodial or monopodial. The leaves are alternate (rarely subopposite or subverticillate), spirally arranged or distichous, in some families generally situated in rosettes, either on ground level or (in many Eriocaulaceae) on the ends of sparse branches. The leaf lamina is simple and entire, often rather succulent, linear to lanceolate, rarely ovate, and generally pointed, in Mayacaceae often apically bidentate; it is generally dorsiventral but terete-subulate or even ensiform in some Xyridaceae. The leaf base is sheathing except in Mayacaceae; the sheath is open in Rapateaceae, Xyridaceae and Eriocaulaceae, but closed in Commelinaceae. In Rapateaceae the lamina is basally asymmetrical and seems to form the continuation of only one side of the sheath. In Commelinaceae the leaf lamina is frequently separated from the sheath by a distinct pseudopetiole. The ptyxis of the flat leaves is generally supervolute or involute, rarely curved or plicate. The venation consists of parallel or arching-converging veins; there is often a distinct midvein (the only vein in Mayacaceae), and a variable number of less distinct lateral veins; fine cross-veins occur in some Commelinaceae. Stipules and ligules are very rare.

The vascular system of the stem, except in *Cartonema* (Commelinaceae), contains vessels, which in Commelinaceae generally have simple perforation plates only, in Mayacaceae and some Eriocaulaceae and Rapateaceae scalariform perforation plates only, while vessels of both kinds are mixed in the other families of the orders. Also the veins of the leaves generally contain vessels, where the conditions are similar, although simple perforations are slightly less frequent. Silica bodies occur in epidermal cells of the Rapateaceae and some genera of Commelinaceae but occur also in non-epidermal cells of some members of Eriocaulaceae (TOMLINSON 1966, 1969). Raphides, contained in raphide sacs forming continuous series (raphide canals), are present in all members of Commelinaceae except *Cartonema,* but are lacking in the other families.

Hairs of several types occur in the order, including glandular microhairs. The hairs are generally uniseriate, and some cells in these may be thin-walled and even branched (*Palisota*). T-shaped hairs are common as, for example, in Eriocaulaceae. Uniseriate hairs with a terminal mucilage-filled cell occur in Xyridaceae, while the Rapateaceae have only non-glandular, uniseriate, two- to five-celled hairs. The stomata are paracytic or, in Commelinaceae, generally tetracytic or hexacytic.

The inflorescences are terminal or axillary or both and are extraordinarily variable. When they are terminal, as in Mayacaceae and many Commelinaceae, they may be leaf-opposed as a consequence of sympodial branching. In Xyridaceae the inflorescences consist of a single or complex spike, in Rapateaceae of spikes or clusters of spikes subtended as a rule by two spathal leaves, and in Eriocaulaceae of button-like capitulae, whereas in Mayaceae the flowers are solitary and morphologically terminal. In Commelinaceae the flowers are arranged in a thyrse, the lateral components of which are helicoid cymes, which are sometimes reduced to a pair of sessile cincinni.

The flowers are bisexual or frequently unisexual. In Commelinaceae, bisexual flowers may occur together with functionally male flowers, together with functionally female flowers or together with functionally unisexual flowers of both sexes (the plants then are andromonoecious, gynomonoecious or polygamomonoecious). The flowers are actinomorphic with two distinct trimerous perianth whorls; dimerous flowers, frequent in Eriocaulaceae, are, however, necessarily disymmetric. The outer tepals are always very different from the inner; in Commelinaceae and Mayacaceae they are usually sepaloid (rarely white or with other colour, and then sometimes petaloid). In Xyridaceae, Rapateaceae and Eriocaulaceae they are sepaloid to hyaline or bracteal. They are free from each other or sometimes partly fused and are mostly lanceolate. The inner tepals are petaloid, of various colours (but not bright red), imbricate in bud, deliquescent after a few hours, free or rather often connate into a tube (in which the outer tepals do not take part), sometimes clawed and equal or unequal. Nectaries and nectar are lacking except in certain Eriocaulaceae.

The stamens occur in one or more often two whorls. When there is only one whorl this is the inner. The stamens may all be equal, but in some Commelinaceae they may be dissimilar, either all fertile or only one, two or three fertile stamens, accompanied by five to three staminodes; in Xyridaceae and Eriocaulaceae the outer whorl may be present as conspicuous staminodes or lacking. The filaments are narrow and either free or basally connate, sometimes adnate to the tube formed by the inner tepals. In many Commelinaceae the filaments are provided with conspicuous hairs (supplying food for the insect vectors). These hairs are uniseriate and often moniliform. The anthers are basifixed or dorsifixed, introrse or latrorse, and tetrasporangiate or rarely (some Eriocaulaceae) bisporangiate. They dehisce longitudinally or, in Mayacaceae, Rapateaceae and a few Commelinaceae, with pores or pore-like slits, which in Rapateaceae vary between one and four in number. The endothecial thickenings are of the Girdle or perhaps more rarely of the Spiral Type; they deserve further study. The tapetum is amoeboid-plasmodial in Commelinaceae (DAVIS 1966; HEMARADDI 1981) and Abolboda of Xyridaceae (HAMANN, personal communication) but in other taxa glandular-secretory. Meiotic divisions are successive, except in Rapateaceae, where they are simultaneous (HAMANN, personal communication), and the microspore tetrads are usually isobilateral or decussate, but tetrahedral in Rapateaceae. The pollen grains are dispersed separately. They are sulcate, circumsulcate, bisulculate, or spiraperturate (see under the families), never ulcerate, and only rarely (some Xyridaceae) inaperturate. They are dispersed in the two-celled or three-celled state. The pistil is syncarpous and may be trilocular, bilocular or unilocular. Usually the style is terminal and simple, at least basally, but in Eriocaulaceae and Xyridaceae it is tri- or bibrachiate, sometimes nearly from the base. In Xyridaceae and Eriocaulaceae there are sometimes two or three appendages on the style. The stigmatic surface is Dry or Wet (Commelinaceae) or Dry (one case in Eriocaulaceae). The placentation is axile (most Commelinaceae, many Rapateaceae), parietal (Mayacaceae, some Xyridaceae), basal (most Xyridaceae, some Rapateaceae) or apical (Eriocaulaceae). The ovules are generally many per carpel, but constantly solitary in Eriocaulaceae, and are anatropous, hemianatropous or orthotropous; they are generally crassinucellate, but tenuinucellate in Eriocaulaceae and some Xyridaceae, in which families a parietal cell is *not* cut off from the archesporial cell, as is the case in the other families. Embryo sac development is generally of the *Polygonum* Type although the *Allium* Type, at least, has been reported in some Commelinaceae (reports of other types need to be verified). Endosperm formation is nuclear, rarely, in *Abolboda,* helobial (HAMANN, personal communication). Embryogeny follows the Asterad Type.

The fruits in most taxa are trilocular or unilocular, loculicidal and trivalved or dehiscing by a lid. In a few taxa they are indehiscent and baccate (some Commelinaceae). The seeds vary much in shape and sculpture and have a punctiform to linear hilum. In Commelinaceae they are provided with a generally dorsal cap-like callosity, the operculum, which covers the embryo. An elaiosome is sometimes present, and in Commelinaceae arillate seeds are reported in three genera. The testa at least in Commelinaceae is derived from both integuments. Often the outer epidermis of the inner integument (tegmen) is fairly well developed, whereas the outer integument (testa) may be thin and in Eriocaulaceae may persist mainly as a row of small hooks. The endosperm has copious mealy starch. The starch consists of compound starch grains, composed of many components, or, in Rapateaceae, possibly of simple starch grains (HAMANN 1961). The embryo is generally small and often little differentiated, lenticular or conical (*Xyris* Type) or in many Commelinaceae capitate (*Scirpus* Type). It is located at the micropylar end of the seed.

*Chromosome Numbers.* The basic chromosome numbers range from $x = 4$ to $x = 29$; in Commelinaceae $x = 6–16$ are most frequent; in the other families numbers of $x = 8–17$ have been found.

*Chemistry.* Silica is rather rarely accumulated except in Rapateaceae, where it is constantly present as simple or complex crystalloids in some epidermal cells. In this family aluminium is also accumulated in richer quantities than in other monocotyledons (CHENERY 1950). Raphides are restricted to Commelinaceae. Steroidal saponins are reported only in one genus (*Cyanotis*) of Commelinaceae; alkaloids and chelidonic acid seem to be lacking. Cyanogenesis is perhaps insufficiently known; it is reported from a couple of genera in Commelinaceae.

The principal flavonoids in Commelinaceae are C-glycoflavones (DEL PERO DE MARTINEZ 1981);

while flavones are otherwise rare or absent. Flavonols and tannins are found scattered in the order. Tricin or sulphurated flavonoids are not reported. The vegetative parts of the plants are poor or at least not particularly rich in starch.

*Distribution.* Within the order the Commelinaceae are widely distributed in the tropics and extend into the temperate regions of the Northern and Southern Hemispheres. The other families are strongly concentrated in South America, where they may have found a refugial region and/or a centre of differentiation in the Guayana Highlands.

*Relationships, Evolution.* Some speculations may be made on the differentiation within the order (see also the chapter on Evolution within the Monocotyledons).

The Commelinaceae, unlike other Commelinales, possess raphides which we interpret here as probably ancestral. The Commelinaceae also appear less specialized in their growth, with dispersed cauline leaves in which transverse veinlets may occur, in the thyrsoid inflorescences and in the axile placentation. On the other hand, the family is possibly advanced in having a closed leaf sheath, amoeboid tapetum and operculate seeds. It is less concentrated in the South American continent than are the other families of the order and it probably diverged early from the ancestral stock of this order.

Mayacaceae may or may not represent an aquatic offshoot of the Commelinaceae, where poricidal anthers have appeared in a few genera. The narrow, one-veined, apically bidentate leaves without a distinct sheath are probable specializations to the aquatic life as are the unifloral inflorescences. Further knowledge of embryological details may help to show the relationships of this unigeneric family.

The remaining three families are largely rosette plants, in which the flowers are densely crowded in spikes (clusters of spikes) or heads, the developmental sequence of which is acropetal-centripetal.

Among these families, the Xyridaceae seem somewhat heterogeneous. *Abolboda,* which was distinguished on contentious grounds by AIRY SHAW (1973) as a separate family, now turns out to be possibly distinct on the basis of quite different, embryological characters, viz. helobial endosperm formation and amoeboid tapetum (HAMANN, personal communication). However, awaiting further evidence, we have not proposed any separation of the genus. *Achlyphila* has a more complex inflorescence structure than the other genera and has clearly zygomorphic flowers.

Rapateaceae shares with Mayacaceae the poricidal dehiscence of the anthers, but considering the profound differences, this similarity may have arisen by convergence (poricidal anthers occur also in a few genera of Commelinaceae). They have more complex inflorescences than have most Xyridaceae, which may be an ancestral feature, but show several specializations (having asymmetric leaf bases, epidermal silica cells, spathal envelopes around the inflorescences and simultaneous microsporogenesis).

Even more special are the Eriocaulaceae, with a wide range of vegetative forms but with a leaf rosette and long peduncles bearing the button-shaped heads with probably largely geitonogamous (neighbour-flower pollination) condition. The pollen grains in Eriocaulaceae, where known, are always spiraperturate. Some of the relatively few genera of Eriocaulaceae have numerous species, *Eriocaulon* perhaps 400, *Paepalanthus* perhaps 480, *Syngonanthus* perhaps 200 (the figures in all genera are uncertain), and the family on these grounds can be regarded as a climax group.

*Key to the Families*

1. Leaf sheath distinct and closed (raphides present; tapetum amoeboid; seeds operculate) . . . . . **Commelinaceae**
1. Leaf sheath distinct or indistinct, when distinct always open (raphides lacking; tapetum glandular, *Abolboda* excepted; seeds not operculate) . . . . . . . . . . . . . . . 2
2. Aquatics with leaves evenly distributed on the branches; leaves one-veined, without distinct sheath, often apically bidentate; flowers solitary . . . . . . . . . **Mayacaceae**
2. Terrestrial plants or subaquatic rosette plants; leaves apically pointed or obtuse, not bidentate, with a well-differentiated sheath; veins, if distinct, more than one; flowers in spikes, clusters of spikes or heads . . . . . . . . . 3
3. Flowers in spikes or clusters of spikes: several ovules on axile or basal placentae (anatropous or hemianatropous; pollen grains not spiraperturate) . . . . . . . . . . . . . 4
3. Flowers in button-like disc-shaped heads (reminiscent of those in Asteraceae), ovules one per locule (apical-pendulous, orthotropous; pollen grains spiraperturate) . . . . **Eriocaulaceae**
4. Anthers poricidal; leaf bases oblique (epidermal cells with silica bodies present) . . . . . . . . **Rapateaceae**
4. Anthers longitudinally dehiscent; leaf bases symmetrical (no epidermal cells with silica bodies) . . . . . **Xyridaceae**

# Commelinaceae R. Brown (1810)    50: ca. 700
(Figs. 171–173)

R. B. FADEN

Perennial or less commonly annual, mainly mesophytic herbs, which are usually terrestrial. The only regularly epiphytic taxa are *Belosynapsis vivipara* of the Deccan Peninsula, India, and the two species of the neotropical, bromeliad-like genus *Cochliostema*. A few species are aquatic or twining. True twiners occur in the neotropical genus *Dichorisandra* and, perhaps, in *Cochliostema,* and in the Asiatic genera *Streptolirion, Spatholirion* and *Aëtheolirion,* and perhaps, *Porandra.* The roots are generally thin and fibrous, rarely uniformly thickened, tapering or fusiform, and then the swollen part sessile or stalked. Rhizomes are uncommon but occur in, e.g. *Palisota* and *Dichorisandra.* Stolons are frequent in various forest genera, e.g. *Pollia.* Bulbs occur in *Cyanotis.* The corms referred to in some species of *Cyanotis* (BRENAN 1968; FADEN 1974) are of uncertain structure. The tubers in species of *Cartonema* seem to be morphologically distinct from other storage organs in the family.

Sympodial growth occurs regularly in *Commelina, Hadrodemas, Streptolirion* and other genera. Lateral branches emerge either from the tops of the leaf sheaths or perforate them. Prophyllar buds and branching occur in species of *Aneilema, Dichorisandra* and *Tinantia.* The leaves are alternate (rarely subopposite or subverticillate), spirally arranged or distichous, or sometimes mainly in terminal or basal rosettes. When a mixed phyllotaxy occurs, the differentiation is usually between the vegetative and reproductive shoots; the primary shoots perhaps always have a spiral phyllotaxy. The leaves have a closed sheath. The lamina is often somewhat succulent. Frequently it is separated from the sheath by a distinct pseudopetiole. The ptyxis is involute, convolute or rarely plicate, and is usually specific to the genus, although both convolute and involute ptyxis have been observed in *Commelina* and *Aneilema.* In the leaves several veins are usually discernible, and transverse veinlets may be clear or obscure.

Vessels with simple perforation plates are usually present in all organs (TOMLINSON 1969). However, vessels are lacking from the stems and leaves of *Cartonema.* Raphide sacs in continuous series (raphide canals) occur in all genera except *Cartonema.* Silica is present in the epidermis of several genera. Starch, when present in the vegetative parts, consists of simple, excentric grains. Uniseriate hairs including glandular microhairs are common (lacking only in *Cartonema* and *Triceratella*). The stomata are tetracytic or hexacytic, but paracytic stomata are characteristic of *Triceratella* and have also been noted in *Cuthbertia* (TOMLINSON 1966, 1969).

The inflorescences may be terminal or axillary or both; when terminal they may be leaf-opposed. The basic inflorescence type is the thyrse, composed of a central axis to which are attached several to many helicoid cymes (cincinni). It is generally bracteate and bracteolate. In many genera the thyrse is variously reduced. In the tribe Tradescantieae of ROHWEDER (1956) it is composed of a pair of sessile cincinni fused back to back. Sometimes these double cincinni are grouped into larger, compound inflorescences. Compound inflorescences may also be formed by the close association of a terminal thyrse with adjacent lateral thyrses, as in *Floscopa* and *Stanfieldiella.* Inflorescences may be partly or entirely enclosed in or closely subtended by spathaceous bracts, as in *Commelina, Rhoeo* and *Streptolirion.* Solely axillary inflorescences are relatively uncommon; such inflorescences may emerge from the tops of the sheaths (*Cochliostema* and *Rhoeo*) or perforate them (*Amischotolype, Buforrestia, Campelia, Coleotrype, Geogenanthus* and *Porandra*).

The attachment of the cincinni may be alternate, subopposite or subverticillate. A fully developed cincinnus consists of a sterile basal portion (cincinnus peduncle or stipe) and a distal fertile portion, the sympodial cincinnus axis. Each cincinnus generally is subtended by a bract (lacking in the upper cincinni of *Spatholirion, Commelina* and *Commelinopsis*). The spathe of *Commelina* and other genera is a modified cincinnus bract. Persistent, pedicel-opposed bracteoles, arranged in two ranks, are usually present along the cincinnus axis (bracteoles lacking or caducous in some or all species of *Cochliostema, Commelina, Dictyospermum, Murdannia, Palisota, Spatholirion, Streptolirion* and *Tricarpelema*).

The flowers are all bisexual or there may be bisexual and male, or rarely bisexual and female, or bisexual, male and female flowers. These kinds, when different, are produced in the same inflorescences, sometimes in a specific sequence. Bisexual and functionally male flowers of the same plant generally differ solely by the abortion of the gynoecium in the male flowers, but other differences may include pedicel length, stamen or staminode number, or length and curvature of some of the stamen filaments. Cleistogamous flowers are produced by some species of *Commelina* and *Tinantia.*

The flowers are actinomorphic or zygomorphic and generally pentacyclic, with two distinct, trimerous perianth whorls, the outer usually sepaloid (rarely white or coloured and then sometimes petaloid), the inner always petaloid. The outer, sepaloid tepals are imbricate in bud, free or occasionally connate, generally similar (the median sometimes dissimilar). The inner tepals are free or occasionally basally connate into a tube, clawed or not, and equal or unequal. Their colour is usually blue, violet, purple or white but yellow in *Cartonema* and *Triceratella*. Basally connate inner tepals occur in *Coleotrype, Cyanotis, Tradescantia* (some sections), *Weldenia* and *Zebrina*. The flowers are strongly scented in *Cochliostema* and in some species of *Aploleia, Callisia, Palisota* and *Tripogandra*. Weaker scent has been noted in some *Aneilema* species (FADEN 1975).

The stamens are usually free but they may be adnate to the inner tepals and/or some filaments may be basally connate. There may be 3 + 3 equal stamens, but in some genera the stamens are dissimilar but all fertile. Commonly, however, there are only one to three fertile stamens and two to four staminodes. When the functional stamens are reduced to three or fewer they may all belong to the outer staminal whorl (*Anthericopsis, Aploleia, Murdannia* and *Tripogandra*), or may all belong to the inner staminal whorl (*Palisota*), or they may represent the upper three (posticous) (*Cochliostema*) or the lower three (anticous) (*Aneilema, Commelina, Commelinopsis, Phaeosphaerion, Pollia, Polyspatha, Rhopalephora* and *Tricarpelema*). A reduced stamen number (three or fewer) without staminodes occurs in *Aploleia* and *Dictyospermum* and in some species of *Callisia* and *Murdannia*. Staminodes, when present, may be of a characteristic form for a genus: densely bearded and lacking an anther in *Cochliostema* and *Palisota;* with bilobed anthers in *Aneilema* and *Rhopalephora,* trilobed anthers in *Murdannia* or cruciform anthers in *Commelina*. Some of the filament bases are regularly connate in some or all species of *Aneilema, Cochliostema, Coleotrype, Floscopa, Murdannia, Rhopalephora* and *Tinantia*. The filaments may be glabrous or bearded, the filament hairs being uniseriate and often moniliform. The anthers are tetrasporangiate, usually dithecous, and basifixed or dorsifixed, sometimes versatile. They generally dehisce longitudinally, the dehiscence being introrse, latrorse or rarely extrorse, but sometimes they open by means of apical pores or pore-like slits. In some or all species of *Cyanotis* the dehiscence is functionally by basal pores.

The tapetum is periplasmodial, with the contents of the tapetal cells invading the anther cavity at an early stage. Microsporogenesis is successive and the pollen grains are generally two-celled when dispersed; three-celled in *Floscopa scandens* and *Commelina attenuata* (CHIKKANNAIAH 1965; DAVIS 1966; HEMARADDI 1981; OWENS 1981). The pollen grains are sulcate, but extra apertures have been reported in species of *Tinantia* and *Zebrina* (ROWLEY 1959; ROWLEY and DAHL 1962; POOLE and HUNT 1980). Dimorphic pollen grains are recorded in species of *Tripogandra* and *Aneilema* (R.E. LEE 1961; HANDLOS 1970, 1975; FADEN 1975; MATTSSON 1976).

The pistil is syncarpous and generally trilocular or bilocular, the style simple, and the stigma terminal, small to capitate, circular to triangular or trifid or, rarely, penicillate (OWENS and KIMMINS 1981). The stigmatic surface is either Wet or Dry (HESLOP-HARRISON and SHIVANNA 1977; OWENS and KIMMINS 1981). Variation in stigma papillae is described by OWENS and KIMMINS (1981). The ovary may be sessile or stipitate. Placentation is generally axile, and the ovules are uniseriate or biseriate, numbering one to ca. 20 per locule.

The ovules vary from orthotropous to hemianatropous (reportedly campylotropous in *Tinantia pringlei*; PARKS 1935) in most genera but completely anatropous in *Cartonema* (GROOTJEN 1983). They are bitegmic and usually crassinucellate (tenuinucellate in *Cyanotis cucullata*) (HEMARADDI 1981). The micropyle is generally formed by both integuments, but it may be formed by only the inner integument, or it may be lacking (HEMARADDI 1981). Embryo sac development is generally of the *Polygonum* Type, but the *Allium* Type also occurs (and *Adoxa, Oenothera* and *Scilla* Types have been reported; DAVIS 1966; HEMARADDI 1981). Endosperm formation is nuclear (DAVIS 1966).

The fruit is usually a trivalved, trilocular or bivalved, bi- or trilocular, loculicidal capsule, and its valves are usually persistent. Sometimes it is indehiscent and then usually baccate or bacciform. Fruiting pedicels of *Aneilema umbrosum* are deciduous at their bases (FADEN 1975). The pericarp is fleshy and the fruits baccate in *Palisota* and *Phaeosphaerion*. The pericarp is parchment-like in *Commelinopsis* and crustaceous in the berry-like fruits of *Pollia*. The seeds are usually uniseriate, but biseriate seeds are characteristic of *Cochliostema, Dichorisandra, Geogenanthus, Pollia, Siderasis, Spatholirion* and *Triceratella* and also occur in some species of *Murdannia* and *Palisota*. The

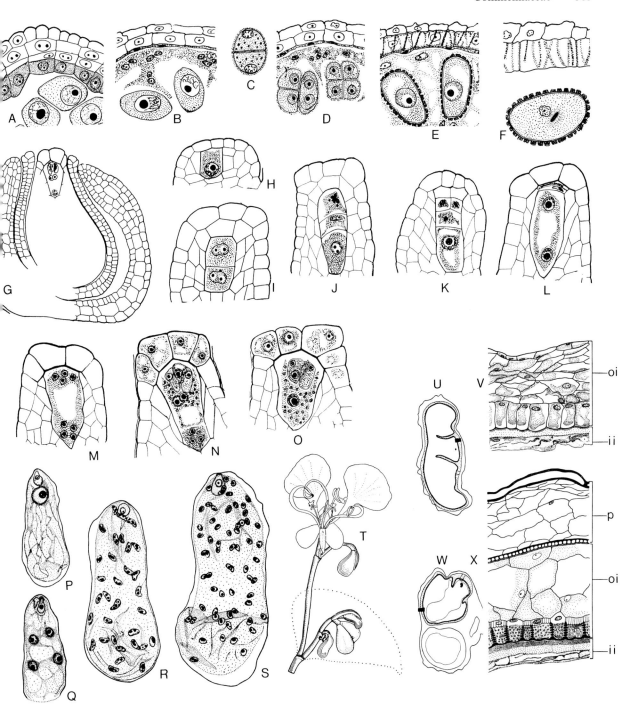

**Fig. 171.** Commelinaceae. **A–O** *Murdannia simplex.*
**A–F** Anther wall development and microsporogenesis.
**A** Anther wall (Monocotyledon Type, with early stage
of tapetal layer. **B** Tapetum already in plasmodial stage,
middle layer thin. **C** Microsporogenesis, following the
successive type, after the first division. **D** Same after
the second division, with tetrads surrounded by plasmodial tapetum. **E** Later stage with spiral thickenings in
endothecium and separate microspores. **F** Anther wall
at stage of dehiscence, with two-celled pollen grain.

**G–O** Ovule, meiosis and embryo sac formation, *Polygonum* Type; note the enlarged cells of the nucellar epidermis. (CHIKKANNAIAH 1964). **P–X** *Commelina forska-
laei,* nuclear endosperm formation (**P–S**), open flower
and cleistogamic subterranean flowers (on branch encir-
cled by *dots*) and seed in unripe (**U–V**) and ripe
(**W–X**) conditions with wall layer (*oi* outer integument
or testa; *ii* inner integument or tegmen; *p* pericarp layer).
(S.C. MAHESHWARI and BALDEV 1958)

seeds are small and hard, with a punctiform to linear hilum and dorsal to lateral (rarely terminal) cap-like callosity, the operculum (embryotega or embryostega), which covers the embryo (also in *Cartonema*; GROOTJEN 1983). The testa is derived from both integuments (GROOTJEN and BOUMAN 1981 a; HEMARADDI 1981). An aril is present on the seeds of *Amischotolype, Dichorisandra* and *Porandra*. The seeds of *Aëtheolirion* are winged.

*Chromosome Numbers.* The cytology of the Commelinaceae has been surveyed by JONES and JOPLING (1972). Further details of several genera and evolutionary trends in the karyotype within the family are discussed by FADEN and SUDA (1980). Chromosome counts have been published for all but seven of the genera and for approximately 37% of the species of Commelinaceae. Basic numbers range from $x = 4$ to $x = 29$ with $x = 6$–16 being most frequent. Chromosome size is very diverse; it is commonly characteristic for a genus. In general, the tradescantioid genera have much larger chromosomes than the commelinoid genera. *Cartonema* has very small chromosomes. Aneuploidy within genera is common at the diploid level. It is apparently infrequent in polyploids (except in *Cymbispatha*). Aneuploid series of four or more basic numbers occur in *Aneilema, Commelina, Cyanotis, Cymbispatha, Murdannia* and *Tinantia* (FADEN and SUDA 1980; FADEN, unpublished). Within these series the evolutionary direction has been towards reduction in basic number and increase in chromosome size. Symmetric karyotypes may be either primitive or advanced, according to the genus.

*Chemistry.* The chemical conditions are outlined under the order. The principal flavonoids are flavone C-glycosides (MARTINEZ 1981). Two distinctive anthocyanin patterns have been found in the Commelinaceae (STIRTON and HARBORNE 1980).

*Pollination.* The insect-pollinated Commelinaceae usually offer pollen as a food reward. Dimorphism of stamens and, in *Tripogandra,* pollen represents a division of labour between fertilizing and food stamens. The former are often concolorous and the latter yellow or have enlarged yellow connectives, while producing little pollen. Moniliform hairs on the filaments are believed to delude insects to scraping them as if gathering pollen (VOGEL 1978b). All Commelinaceae appear to be insect-pollinated or autogamous. A survey of self-compatibility and self-incompatibility in the family has been made by OWENS (1981).

*Usage.* The family is of little economic importance. Species of a number of genera are commonly cultivated. Some species are eaten (mostly by domesticated animals, sometimes by man); some are used locally as medicines or in rites. Two forms of *Tradescantia* are reputed to be useful indicators of low levels of radiation (GROSSMAN 1979).

*Distribution.* The Commelinaceae occupy the entire range of the order. The three major centres of taxonomic diversity are: tropical Africa; Mexico and northern Central America; and the Indian subcontinent. Only six genera (*Aneilema, Buforrestia, Commelina, Floscopa, Murdannia* and *Pollia*) have indigenous species in both the New World and the Old World (FADEN 1978).

*Taxonomy.* No satisfactory scheme has been proposed for the arrangement of the genera into tribes or subfamilies. Previous arrangements have been summarized by BRENAN (1966) and FADEN (1975). The two traditional tribes Commelineae and Tradescantieae have been variously circumscribed by different workers. All of these treatments are highly artificial. PICHON (1946) broke new ground by using ten tribes. BRENAN (1966) proposed 15 quasi-tribal groups of genera. The treatment can be revised on various grounds: morphology (HUNT 1971; FADEN 1975); anatomy (TOMLINSON 1966, 1969); cytology (K. JONES and JOPLING 1972; FADEN and SUDA 1980); and palynology (POOLE and HUNT 1980). However, it has not yet been superseded by a more natural system.

All workers are in general agreement that the Australian endemic genus *Cartonema* is the most eccentric in the family. Some authors have even segregated it as a unigeneric family Cartonemataceae (PICHON 1946; HUTCHINSON 1959, 1973; TOMLINSON 1966, 1969), but most workers retain it within the Commelinaceae (HAMANN 1961, 1962b; BRENAN 1966; CRONQUIST 1968, 1981; TAKHTAJAN 1969; ROHWEDER 1969; DAHLGREN 1980; GROOTJEN 1983).

**Subfamily Commelinoideae**

*Tribus Tradescantieae*

The genera of Tradescantieae sensu ROHWEDER (1956) include *Tradescantia* (ca. 60), *Callisia* (8) and *Tripogandra* (21). The inflorescence is composed of fused pairs of cincinni which are sub-

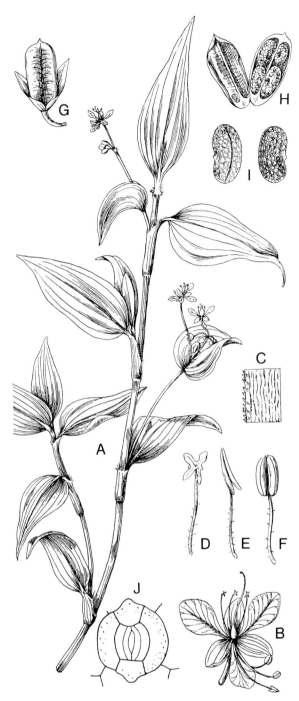

**Fig. 172.** Commelinaceae. **A–I** *Commelina nudiflora.* **A** Branch. **B** Flower. **C** Leaf margin. **D–F** Stamen variation in the same flower, **D** being non-functional. **G** Capsule. **H** Same opened to show loculicidal dehiscence. **I** Seeds. (HUTCHINSON and DALZIEL 1936). **J** *Commelina tuberosa,* stoma. (SOLEREDER and MEYER 1930)

have broad connectives and latrorse dehiscence. *Tripogandra* is unique in this group because of its zygomorphic flowers, dimorphic stamens (three of which are sometimes staminodial), some of which have extrorse anther dehiscence, and verrucate, sometimes dimorphic pollen (POOLE and HUNT 1980). *Tradescantia* has long-pedicelled flowers with six equal stamens and usually bearded filaments. *Callisia* has subsessile flowers with three to six stamens and always glabrous filaments. This group is confined to the New World and is centred in Mexico and Central America. Species of *Callisia, Rhoeo, Tradescantia* (including *Setcreasea*) and *Zebrina* are commonly cultivated.

A group of chiefly North American genera with tradescantioid affinities includes *Tinantia* (ca. 12), *Thyrsanthemum* (3) and *Gibasis* (14). These genera do not have the cincinni in fused pairs. *Tinantia* is the only one of these genera with zygomorphic flowers and dimorphic stamens. All of its species are annuals. *Thyrsanthemum* species are robust perennials with thyrsiform inflorescences. In *Gibasis* the individual cincinni have long stipes and short axes and are solitary, in pairs or whorls, but the inflorescence never has a central axis. Species of *Gibasis, Hadrodemas* and *Tinantia* are often cultivated.

The chiefly or exclusively South American genera with tradescantioid relationships are *Cochliostema* (2), *Dichorisandra* (ca. 35), *Geogenanthus* (4) and *Siderasis* (1). Except for *Siderasis,* a rosette plant, these mainly forest genera have a large number of unusual characters: epiphytic habit in *Cochliostema;* climbing habit in some species of *Cochliostema* and *Dichorisandra,* basal inflorescences in *Geogenanthus* and some species of *Dichorisandra;* fringed inner tepals in *Cochliostema* and *Geogenanthus;* poricidal anther dehiscence and arillate seeds in *Dichorisandra.* This group of genera is united by the possession of an unusual karyotype (K. JONES and JOPLING 1972; FADEN and SUDA 1980) and biseriate ovules and seeds. All of these genera have species cultivated to a limited extent, chiefly as greenhouse ornamentals.

Certain Old World genera also have tradescantioid affinities: *Belosynapsis* (4), *Cyanotis* (ca. 50), *Amischotolype* (ca. 20), *Porandra* (2), *Coleotrype* (9), *Palisota* (ca. 15) and possibly *Spatholirion* (2), *Streptolirion* (1) and *Aëtheolirion* (1). *Belosynapsis* and *Cyanotis* have succulent foliage and cincinni in the axils of leaves or foliaceous bracts (as well as terminal). The inner tepals are free in *Belosynapsis* but basally connate in *Cyanotis.* The forest genera *Amischotolype, Porandra* and *Coleotrype*

tended by small to large, sometimes spathaceous bracts. The gynoecium has three equal locules each with two (rarely one) ovules. Except for *Tripogandra* the flowers are actinomorphic and the anthers

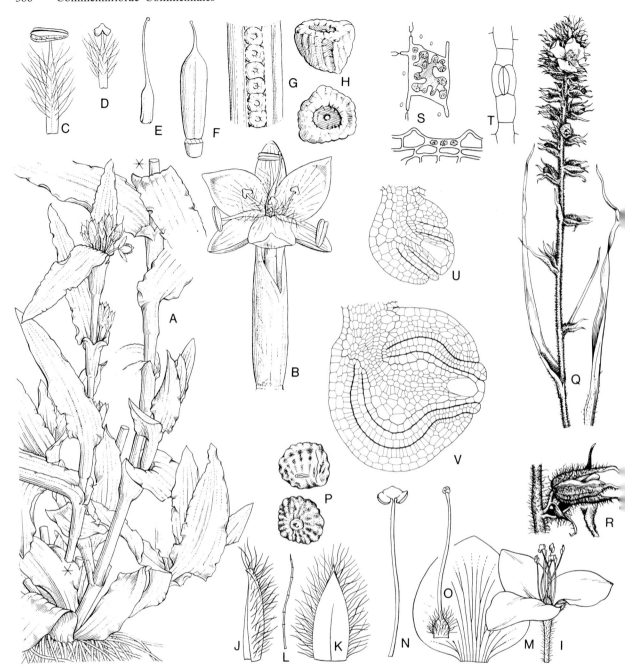

**Fig. 173.** Commelinaceae. **A–H** *Murdannia axillaris.*
**A** Plant. **B** Flower, the anthers of the inner staminal
whorl staminodial. **C** Functional stamen. **D** Staminode.
**E** Pistil. **F** Young fruit. **G** Fruit opening up. **H** Seeds.
(Brenan 1962). **I–P** *Tradescantia sillamontana.* **I** Flower.
**J–K** Tepal of outer whorl ("sepal"). **L** Hair from this.
**M** Tepal of inner whorl (petal). **N** Stamen. **O** Pistil.
**P** Seeds. (Hunt 1976). **Q–R** *Cartonema spicatum.* **Q** Inflorescence and part of leaf. **R** Capsule enclosed in
"calyx". (Tomlinson 1970). **S** *Forrestia marginata,* silica
in leaf epidermis, surface view and section. **T** *Cartonema
philydroides,* stoma. (S–T from Solereder and Meyer
1929). **U–V** *Stanfieldiella imperforata,* ovule in two
stages. (Grootjen and Bouman 1981a)

have congested axillary inflorescences which perforate the sheaths. The Asiatic *Porandra* (Hong
1974) differs from the Afro-Asiatic *Amischotolype*
by its scandent habit and poricidal anther dehiscence. The African *Coleotrype* differs by the basally fused inner tepals. The African genus *Palisota*
has large thyrsiform inflorescences, three stamens,
(the inner whorl) alternating with three or two staminodes (the outer whorl), and a baccate fruit.
The small Asiatic genera *Spatholirion, Streptolirion* and *Aëtheolirion* are all twiners (except one
species of *Spatholirion*) with thyrsiform inflores-

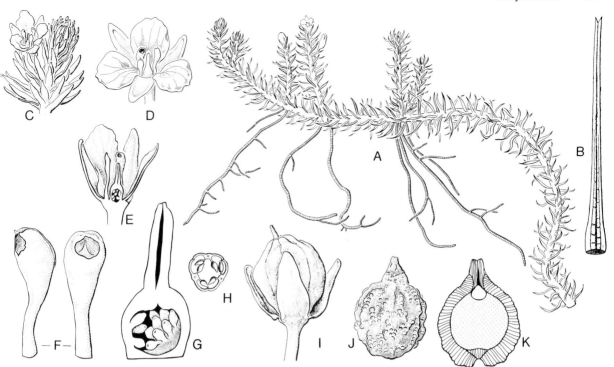

**Fig. 174.** Mayacaceae. **A, C–G, I–K** *Mayaca fluviatilis*. **A** Branch. **C** Branchlet with flower. **D** Flower. **E** Same, longitudinal section. **F** Stamen, in different views; note the poricidal dehiscence. **G** Pistil, longitudinal section. **I** Young fruit, enclosed by outer tepals. **J** Fruit. **K** Same, longitudinal section, showing embryo on top of the copious endosperm. (THIERET 1975). **B** and **H** *Mayaca baumii*. **B** Leaf. **H** Ovary, transverse section. (PILGER 1930)

cences. *Aëtheolirion* has elongate fruits, winged seeds and flowers with three to five stamens. *Spatholirion* has ebracteate upper cincinni; *Streptolirion* has all of the cincinni bracteate. Both have six stamens. Species of *Belosynapsis, Cyanotis* and *Palisota* are sometimes cultivated.

### Tribus Commelineae

The commelinoid genera are chiefly or exclusively Old World in distribution. They are not readily divisible into groups. In the nearly cosmopolitan *Commelina* (ca. 170) and its two small, neotropical derivatives *Commelinopsis* (ca. 4) and *Phaeosphaerion* (ca. 2) – separated by their bacciform and baccate fruits, respectively – the inflorescences are enclosed in spathes and the flowers are zygomorphic with three anticous, fertile stamens and three (or two) posticous staminodes. The major genera with thyrsiform inflorescences are: *Aneilema* (ca. 62), mainly African, flowers with three anticous, fertile stamens and three (or two) posticous staminodes, bracteoles usually cup-shaped; *Murdannia* (ca. 50), mainly Asiatic, three (or two) fertile stamens (outer whorl) alternating with three (rarely four or no) staminodes; *Floscopa* (ca. 20), mainly African, six (or five) fertile stamens differentiated into anticous and posticous sets, capsules stipitate, two-seeded; and *Pollia* (17), mainly paleotropical, fruits bacciform, indehiscent, lustrous, blue or lead-coloured. *Commelina* is occasionally cultivated and includes some important weeds.

### Subfamily Cartonemoideae

Two genera with no clear affinities within the family except perhaps with one another, the Australian *Cartonema* (6), and the African *Triceratella* (1), are characterized by their grass-like foliage, glandular-capitate hairs on the vegetative parts, and yellow, actinomorphic flowers. *Cartonema* species are perennials with raceme-like inflorescences; *Triceratella* is an annual with leaf-opposed, simple cincinni.

## Mayacaceae Kunth (1842)   1: ca. 10   (Fig. 174)

R.B. FADEN

Mayacaceae are freshwater, perennial, aquatic herbs with a *Lycopodium*-like habit. The multi-ranked, spirally arranged leaves lack sheaths. The lamina is linear to filiform with entire margins and commonly a bidentate apex. The stomata are paracytic. Longitudinal air channels are present in all of the vegetative organs. Hairs are lacking except in the leaf axils (TOMLINSON 1969).

The flowers are solitary and morphologically terminal, but they soon appear lateral because of the sympodial growth of the shoots. They are all perfect and actinomorphic with two distinct perianth whorls, an outer sepaloid whorl composed of three equal, free, subvalvate members, and an inner petaloid whorl composed of three equal, free, imbricate, shortly clawed members. Nectaries and nectar are lacking. The three stamens represent the outer staminal whorl. The filaments are slender and glabrous, the anthers basifixed, tetrasporangiate and usually tetratheous or sometimes ditheous (HORN AF RANTZIEN 1946). Dehiscence is by means of apical pores or pore-like slits, sometimes at the end of a tubular apical appendage. Staminodes are lacking. Microsporogenesis and tapetum are unknown. The pollen grains are sulcate, finely reticulate and two-celled. The gynoecium consists of a single pistil with a sessile, superior, tricarpellary, unilocular ovary. The carpels are positioned opposite the inner tepals, and thus all the floral whorls alternate with each other (STÜTZEL, personal communication). The style is simple and terminal, the stigma terminal and capitate to slightly trifid. The orthotropous ovules are biseriate and are attached to three parietal placentae. The fruits are capsular and trivalved, dehiscing halfway between the placentae. The seeds are ovoid to globose with a dorsal operculum (or embryostega) and mealy endosperm.

There are approximately ten species of *Mayaca* in warm temperate and tropical America and one in South-Western tropical Africa.

## Xyridaceae C.A. Agardh (1823)   4: 270
(Fig. 175)

Small or medium-sized, mostly perennial, tufted herbs with a generally corm-like, short and thick vertical rhizome, more rarely (as in *Achlyphila*) with a long horizontal and creeping rhizome. Some species of *Xyris* have a bulb-like base, the corm being enveloped by a few, rarely only two, sheathing scales with reduced laminae. Each shoot generation generally ends in a single, unbranched scapose inflorescence, the growth being sympodial by the development of lateral shoots. However, in rare cases the shoot grows continuously (being a monopodium), forming new leaves in the centre and developing axillary scapose inflorescences, as in the Rapateaceae. The vascular tissue of the roots, stems and leaves possesses vessels with simple perforation plates.

The leaves are always tufted in a basal leaf rosette. They are distichously arranged or have some other phyllotaxy, and are generally flat, either equitant-ensiform or dorsiventral, more rarely subterete or slightly compressed. The leaves have a sheathing, sometimes broad base. In *Xyris* the sheath is sometimes supplied with a membranous ligule at the upper end. The leaf apices in *Abolboda* may have a blunt apical appendage. Stomata are of the paracytic type and are not sunken. The leaf sheath in, for example, *Xyris* often remains as a protective envelope at the shoot base. Hairs are frequently present in *Xyris,* where uniseriate glandular hairs with a bulb-shaped, mucilage-filled terminal cell may occur inside the leaf sheaths and the leaf margins may bear unicellular unbranched hairs. Silica bodies are probably lacking, as are also oxalate raphides, but small oxalate crystals or crystal aggregates may be present.

The inflorescence is usually solitary (rarely is there more than one spike, as in *Abolboda paniculata*), borne on an erect scape which is usually leafless, but in *Abolboda* supplied with one or a few pairs of opposite, bracteal leaves. The scape is usually stiff, having a closed inner cylinder of fibres. The inflorescence usually represents a simple, elongate or sometimes head-like spike. Generally, the flowers are borne singly in the axils of the upper of the bracts. The bracts are variable in number, imbricate, glumaceous, tough and firm, brown to black and often shiny, and with a scarious margin and a rounded, retuse or pointed apex. *Achlyphila* deviates in having two flowers situated inside each main bract, each flower being situated in the axil of a minute secondary bract, while the terminal floral group consists of three flowers representing a cymose floral triad. It thus appears that the xyridaceous inflorescence is derived from a condensed panicle with cymose components.

The flowers are ebracteolate, sessile or indistinctly (in *Achlyphila* distinctly) pedicellate, hypogynous, actinomorphic or sometimes zygomorphic, and constantly timerous. The outer tepals ("sepals") are usually free but sometimes the lateral ones are basally fused in the median plane. In *Achlyphila, Abolboda* and *Orectanthe* all three outer tepals are subequal, but in the latter two genera the median sepal may be lacking (caducous ?). The inner tepals ("petals"), are zygomorphic in *Orectanthe* but otherwise actinomorphic; they are long-clawed, equal or subequal, petaloid and relatively

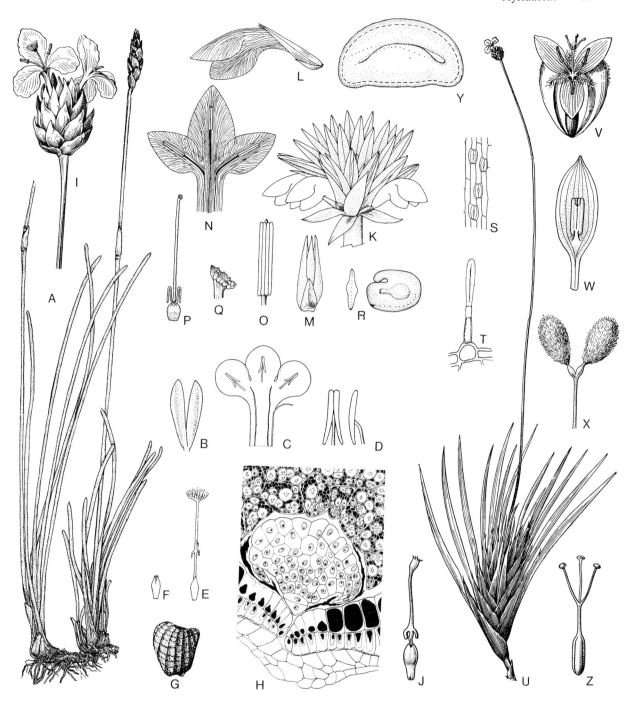

**Fig. 175.** Xyridaceae. **A–G** *Abolboda linearifolia.* **A** Plant. **B** Outer, sepaloid tepals. **C** Inner, petaloid tepals. **D** Anther in different views. **E** Pistil. **F** Young fruit. **G** Seed. (MAGUIRE and WURDACK 1958). **H** *A. americana,* embryo and part of endosperm and seed coat. (CARLQUIST 1960). **I** *A. poarchon,* inflorescence. **J** *A. poeppigii,* pistil. (**I–J** TAKHTAJAN 1981). **K–R** *Orectanthe sceptrum.* **K** Inflorescence. **L** Flower; note the zygomorphy. **M** Bract and outer, sepaloid tepals. **N** Inner, petaloid tepals (with attached stamens), spread open. **O** Anther. **P** Pistil. **Q** Stigma. **R** Seed in different views. (MAGUIRE and WURDACK 1958). **S** *Xyris involucrata,* leaf stomata. (TOMLINSON 1969). **T** *X. caroliniana,* hair from leaf sheath. (SOLEREDER and MEYER 1929). **U** *X. witsenioides,* habit. **V** *X. lacerata,* flower. **W–X** and **Z** *X. blepharophylla.* **W** Tepal and attached stamen. **X** Staminode. **Z** Pistil. (**U–X** TAKHTAJAN 1982). **Y** *X. rupicola,* pollen grain. (ERDTMAN 1952)

large, with ovate or narrowly elliptic lobes. They are usually yellow, rarely white or (in *Abolboda*) blue. The inner tepals may be free or fused into a tube. In *Xyris* they are practically free but held together by the generally bifurcate staminodes which alternate with the inner tepals. Staminodes are present in *Xyris* only, where they may be simple or bifid or, rarely, quadrifid. The branches are often tail-like and plumose and are thought to facilitate pollination by collecting pollen from the adjacent anthers and presenting it to the small visiting insects (mainly Hymenoptera).

While the outer stamens are staminodial or absent, the inner are all functional. They are equal and have narrow and often short filaments and peltate, often elongate, tetrasporangiate anthers dehiscing latrorsely by longitudinal slits. The tapetum is glandular-secretory or, in *Abolboda*, amoeboid. Microsporogenesis is successive. The pollen grains are variable in the family: smooth and provided with one or two, sometimes irregular, colpi in *Xyris;* they are smooth and inaperturate in *Achlyphila,* and spinulose and inaperturate in *Abolboda* and *Orectanthe.* They are dispersed in the three-celled or sometimes also in the two-celled stage.

The pistil is trilocular or more often (in *Xyris*) unilocular or incompletely trilocular. The style is apically tribrachiate except in *Achlyphila* where the style is simple and entire to the top. In *Abolboda* and *Orectanthe* the style bears two or three large appendages (cf. some Eriocaulaceae). Placentation is central-axile in trilocular ovaries, whereas in *Xyris* it is either parietal (in a few species the placentae are far-intrusive) or basal.

The ovules are anatropous to incompletely campylotropous or orthotropous (in *Xyris*), bitegmic and more or less tenuinucellate. The archesporial cell functions directly as the megaspore mother cell, but in species of *Xyris* the nucellar epidermis may divide periclinally. Embryo sac formation probably always conforms to the *Polygonum* Type (studied only in *Xyris*). Endosperm formation is nuclear in *Xyris* or probably helobial in *Abolboda* (TIEMANN and HAMANN, personal communication).

The fruits are small loculicidal capsules enclosed by the persistent perianth. They open by three longitudinal slits or rarely (some Australian species of *Xyris*) by a lid. The seeds are ovoid, ellipsoidal or globose, and longitudinally ridged, sometimes with a small "tip" at each end. Wings and appendages are lacking in *Xyris,* but are rarely present in *Abolboda.* The endosperm is copious and is rich in aleurone and compound starch grains, sometimes also in fatty oils. The embryo is small and lens-shaped and little differentiated, located in the micropylar part of the seed.

*Chromosome Numbers.* The chromosome number is known only in *Xyris* and seems somewhat geographically correlated, $x = 9$ being found in North American species, $x = 13$ in Australian and $x = 17$ in Asiatic-African species.

*Chemistry.* The absence of oxalate raphides and the probable absence of silica bodies should be noted. Tannins occur with the flavonols quercetin and kaempferol; also anthocyanins are known to occur in the family.

*Distribution.* Xyridaceae is a mainly tropical and subtropical family. Most species of the largest genus, *Xyris,* are concentrated in North and South America, fewer in Australia, Asia and Africa. *Abolboda* and *Orectanthe,* as well as the fairly recently discovered genus *Achlyphila,* occur in South America. The plants generally grow in marshes or at least on moist ground. The few species with bulb-like bases and those with a monopodial shoot system can endure less damp or even arid habitats. (From MAGUIRE and WURDACK 1958.)

*Xyris* (ca. 250), the distribution of which is mentioned above, is characterized by having distichous, mostly ensiform to terete leaves with stomata on both sides, and leafless scapes with simple inflorescences. These have numerous imbricate, rounded bracts. The outer tepals are unequal, the median being larger than the others, hood-like and caducous as soon as the inner tepals open, the laterals persistent, boat-shaped to spatulate and often hairy or ciliate on the margins. The basally nearly free inner tepals are yellow or white. The pollen grains are monosulcate or bisulculate and smooth. In the unilocular or incompletely trilocular ovary *Xyris* is more advanced than the other genera. – *Xyris* is divisible into three subgenera. Subgen. *Nematopus* (ca. 110) is mainly South American, and has a unilocular ovary with a basal placenta and a capsule opening with three fissures extending to the apex. – Subgen. *Xyris* is of about the same size and has a distribution covering almost the whole range of the genus. It has three parietal placentae and a capsule similar to subgen. *Nematopus.* – Subgen. *Pomatoxyris,* finally, is small, with only ca. 12–15 species, and is concentrated in Western Australia. The placentae are parietal but far-intrusive, so that the ovary becomes almost trilocular. The capsule splits only basally, along the carpel midveins, but is coherent at the apex, and is often shed as a cap. – The genus lacks economic importance, although the leaves of *X. melanocephala* are used in Indonesia to make mats.

*Achlyphila* (1) in Venezuela differs from *Xyris* in several respects, such as the shortness of the scape, the absence of hairs, the constant absence of a ligule on the leaf sheath, the very interesting inflorescence with flower pairs (see above), the inaperturate pollen grains, and the absence of staminodes. In several respects *Achlyphila* seems to be more "primitive" than *Xyris,* e. g. in the inflorescence, in the almost actinomorphic outer tepals, the lack of a perianth tube, the trilocular ovary, and the anatropous ovules.

*Abolboda* (17) and *Orectanthe* (2) are both South American, the former with its centre in the Amazonas region, the latter in Venezuela and Guayana. They have sometimes been distinguished as a separate family because of a number of common features: the spirally set, bifacial (dorsiventral) leaves (which have a colourless hypodermal layer absent in other Xyridaceae), the stomata restricted to the abaxial leaf surface, the fewer more distant bracts of the inflorescence, the two or three similar outer tepals, the inner tepals fused basally into a tube, the spinulose inaperturate pollen grains, the lack of staminodes, the presence of appendages on the style, and the trilocular ovary with anatropous ovules. In addition the inner tepals of *Abolboda* are blue (or white) instead of yellow. However, as is apparent from the above description, several features of these genera are shared with *Achlyphila,* and the differences hardly justify family distinction.

The taxonomic position of Xyridaceae has generally not caused much debate. The family is somewhat similar to and probably closely related to Eriocaulaceae, even though the reduction in the number of functional stamens and possibly also the occasional presence of stylar appendages may have developed by convergence. Xyridaceae shares several important features with Eriocaulaceae, such as the habit, the kind of stomata, the basally concentrated leaves, the "capitate" inflorescences, the heterochlamydeous perianth and some embryological characters.

Xyridaceae is advanced in the simply perforated vessels found also in the leaves, and in the glumaceous inflorescences. It seems as if the "simple spikes" in most Xyridaceae may be secondary, possibly derived from a strongly condensed paniculate inflorescence consisting of lateral cymose floral triads. A transitional stage would be the *Achlyphila* inflorescence. This could suggest an affinity with Commelinaceae and related families.

## Rapateaceae Dumortier (1829)    16:80
(Figs. 176–177)

Perennial, often rather large, herbs with a short and compact, usually simple, leafy stem continuing in a thick, short, slightly fleshy, vertical rhizome. The leaves are crowded in a basal rosette, approximately tristichous in phyllotaxy. They are flat, often large, parallel-veined and sheathing at the base. The sheaths are V-shaped and generally asymmetrical, so that the midvein becomes displaced to one side. The lamina is a continuation of one side of the sheath. It is linear to lanceolate, often rather firm and dry in texture, rarely terete. The junction between sheath and lamina is constricted, rarely petiole-like, as in *Saxofridericia.*

The stomata, as in some Xyridaceae (*Abolboda* and *Orectanthe*) are restricted to the abaxial leaf surface. They are paracytic, of poaceous type, but not sunken, nor organized in regular files. Unbranched, uniseriate hairs with one to three short basal and one to two long distal cells may occur. Tannin cells are sometimes present in the hypodermis of leaves and scape. Oxalate raphides are lacking, and calcium oxalate on the whole seems to be absent in the family. Silica bodies occur in the form of rounded druses, in epidermal cells of leaves and scape. Simple (rarely compound), oval starch grains are often abundant in the stem but rare in leaves. Vessels with scalariform or simple perforations are present in the vascular tissue in roots and stem, but are lacking in the leaves.

The inflorescence is situated on an axillary and generally solitary leafless scape, which is usually rather long, rarely (in *Maschalocephalus*) so short that the inflorescence is situated in the leaf sheaths. The inflorescence is surrounded usually by two, often large and basally broad, spathal leaves with numerous parallel veins (these leaves are lacking in *Stegolepis*). They may be fused and rarely enclose the inflorescence completely. The inflorescence consists of a cluster of many (rarely one or two) pedunculate or sessile "spikelets", situated on a short, compact inflorescence receptacle.

The "spikelets" bear a single terminal flower subtended by numerous, imbricate, glumaceous bracts. The flower is bisexual, trimerous and actinomorphic or almost so. The outer tepals ("sepals") are imbricate, usually free, sometimes basally connate, indurated, bract-like and hyaline at the base. The inner tepals are ephemeral, fragile and petaloid, yellow or rarely reddish in colour. They are usually connate at the base into a short to rather long tube. The lobes are lanceolate, ovate

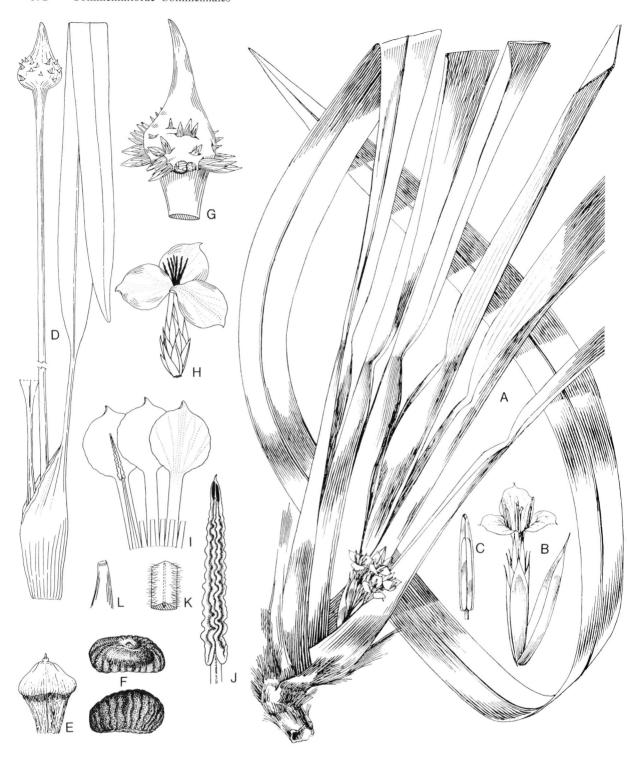

**Fig. 176.** Rapateaceae. **A–C** *Maschalocephalus dinklagei,* the only African member in the family. **A** Plant. **B** Flower. **C** Poricidal anther. (HEPPER 1968). **D–F** *Saxofridericia grandis.* **D** Part of plant. **E** Capsule. **F** Seeds. **G–L** *Saxofridericia spongiosa.* **G** Inflorescence, the flowers bursting through the spathal envelope. **H** Flower subtended by bracts. **I** Whorl of inner petaloid tepals and one stamen. **J** Anther. **K** Part of filament (hairy). **L** Pistil apex with stigma. (**D–L** MAGUIRE and WURDACK 1958)

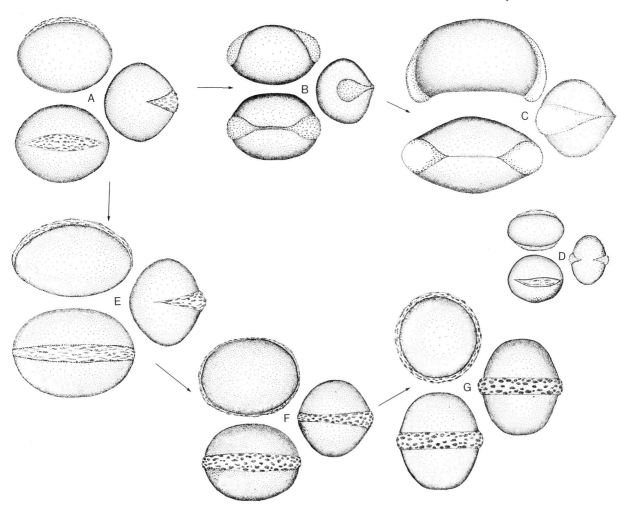

or obovate, and yellow, reddish, or sometimes with brown or violet spots.

The stamens are 3 + 3 in number. Their filaments are usually connate at the base and/or attached to the tube formed by the inner tepals. The anthers are usually very long, subbasifixed and introrse but dehiscing by one, two or four apical pores or by short apical slits. A short apical connective appendage is present in Rapateae. The endothecial thickenings are of the Spiral Type, the tapetum is secretory, with mainly binucleate tapetal cells. Microsporogenesis is probably simultaneous, with tetrahedral microspore tetrads (TIEMANN and HA-MANN, personal communication). The pollen grains (CARLQUIST 1961) are usually sulcate but in *Spathanthus* bisulculate and in *Rapatea* and two more genera zonisulculate. (Possible evolutionary relationships within the family are shown in Fig. 177.) The pollen grains are probably two-celled when dispersed.

The pistil is tricarpellary, and the ovary globose, trilocular or (in *Spathanthus*) incompletely trilocu-

**Fig. 177.** Rapateaceae. Trends of possible evolution in the pollen, the ancestral type considered to be the sulcate (**A**). The examples given here are all taken from extant taxa of the family, as follows. **A** *Guacamaya superba* (sulcus fusiform). **B** *Monotrema aemulans* (sulcus widened into a germ pore at each end, narrowed in the middle). **C** *Windsorina guianensis* (as the previous but aperture wholly constricted in the middle). **D** *Spathanthus unilateralis* (bisulculate), derivation uncertain. **E** *Schoenocephalium cucullatum* (sulcus widened at the ends). **F** *Duckea cyperaceoideae* (unevenly circum-sulculate). **G** *Rapatea* sp. (evenly circum-sulculate). The evolutionary interpretation is ours. (All from CARLQUIST 1961)

lar, with an erect simple style and a punctiform or small, capitate stigma. The ovules are solitary to several in each locule and are inserted centrally or basally in the ovary. They are anatropous and crassinucellate, and their micropyle is downwardly directed. A parietal cell is cut off from the archesporial cell. Embryo sac formation in the genera studied conforms to the *Polygonum* Type. Endosperm formation is nuclear.

The fruit is a loculicidal capsule in which all three locules or, by abortion of ovules, only one or two locules are fertile. The seeds are ovoid, crescentic or prismatic, with or without appendages, and have a smooth, striate, muriculate or faintly reticulate surface. At the chalazal end they may bear a conical or cap-like, spongy white caruncle or elaiosome. The outer layer of the testa is thin and light in colour; the inner consists of thicker-walled and darker cells. The endosperm is copious and rich in simple starch grains. The embryo is small, lens-shaped and undifferentiated and is situated at the micropylar end, with the flat side facing the endosperm.

*Chromosome Numbers.* The chromosome number, known only in the African *Maschalocephalus dinklagei,* is $n=11$.

*Chemistry.* Rapateaceae is the group of monocotyledons accumulating most aluminium. Small quantities of starch may occur in the rhizomes. Otherwise, the absence of oxalate crystals, the common presence of crystal druses of silica and the presence of tannins seem to be characteristic.

*Distribution.* Rapateaceae, with few exceptions, are restricted to Northern South America, where the family has its centre on the sandstone massifs of the Guayana Highlands. It extends into Southern Venezuela and the Central and Southern Amazonas. The plant grows on moist or marshy ground. Most of the taxa have a restricted distribution. *Maschalocephalus* makes an exception in the family, as it occurs in Liberia, West Africa (BALDWIN 1950).

Rapateaceae may be divided into four groups which may deserve subfamilial or at least tribal rank.

### Tribus Saxofridericieae

This tribe has flowers with obovate-obcordate, exserted, clearly yellow inner tepal lobes. The locules contain several ovules on axillary placentae. The seeds are prismatic, pyramidal or crescentic. The pollen grains are always monosulcate, with a fusiform aperture. Packets of parenchymatous tannin cells are present in cortical and stelar regions, while mucilage canals and cavities are lacking. The vessels, found in roots and stems, have scalariform perforation plates.

Here belong five genera. Among them is *Saxofridericia* (9) in South America, mostly in Guayana; large herbs, ca. 1–2 m high, in which the anthers bear apical appendages and dehisce by a single pore. The scapes, which are usually solitary, generally bear two large bracts and a massive head of "spikelets". – Even larger is *Stegolepis,* mostly restricted to Guayana, which has ebracteate inflorescences or inflorescences bearing membranous scales. The plants usually bear several scapes. The related *Epiphyton* (2) is epiphytic.

### Tribus Schoenocephalieae

Tribus Schoenocephalieae has flowers with lanceolate, reddish or red-spotted inner tepals. The ovules and seeds are as in Saxofridericieae. Also in this tribe the pollen grains are monosulcate, with a fusiform sulcus. Tannin cells and vessels are as in Saxofridericieae, but mucilage cells and canals are present. The tribe consists of three genera, *Schoenocephalium* (5) and two which are monotypic, all in Guayana. In *Schoenocephalium* the anthers are apically cruciform and dehisce by four pores.

### Tribus Rapateae

Tribus Rapateae has locules with only one ovule and the seeds are oval or oblong and longitudinally striate, and sometimes have a papillate-mitriform appendage. The involucre, which consists of one or two bracts, considerably surpasses the inflorescence. The pollen grains are bisulculate or zonisulculate. Tannin idioblasts may be present in stele or cortex. The vessels, found in roots and stems, have scalariform perforation plates. This tribe consists of four genera, all in South America, with the centre in Guayana.

*Spathanthus* (2) is interesting in having bisulculate pollen grains and a single spathaceous inflorescence bract, to which the inflorescence is attached. The ovary is incompletely trilocular with only a single ovule reaching maturity. The plants are somewhat araceous in habit. *Rapatea* (20), has two involucral bracts. The anthers bear appendages and the pollen grains are zonisulculate. The genus consists of large herbs with tough, ribbon-like leaves. Two smaller, related genera have grass-like leaves.

### Tribus Monotremeae

Tribus Monotremeae agrees in many respects with tribus Rapateae, e.g. in having one ovule per locule, and in having oval or oblong seeds, which are, however, not striate, but white-granulate, and supplied with an applanate appendage. The involucral leaves are also as a rule shorter than the inflorescence. The pollen grains are monosulcate, with the sulcus widened at the ends and narrow in the middle. The stem epidermis lacks silica bodies, and the hypodermal fibre strands found in other tribes are lacking. The vessels of roots and leaves have simple perforation plates. This tribe consists of four genera, of which *Maschalocephalus* is West African, the others being concentrated in the Guayana Highlands.

Whereas the American genera have scapose inflorescences, the scape in *Maschalocephalus* (1) is very short, so that the inflorescence is found in the leaf sheaths. In other respects the genus is typical of the family. The anthers, for example, dehisce by a subapical pore. – Of the other genera may be mentioned *Monotrema* (5) with sessile "spikelets" in a capitate inflorescence subtended by two bracts.

The closest affinities of Rapateaceae are, as a rule, considered to be with the Xyridaceae. The pollen morphology indicates similarity between Rapateaceae and *Xyris,* the stomata in Rapateaceae and Xyridaceae are also similar, as are the seeds with the undifferentiated embryo, the copious and starch-rich endosperm and probably also the seed coat. Moreover both families have a South American centre of variation. One problem is the homology of the rapateaceous "spikelets" which have a single terminal flower. Should these correspond to the whole scapose inflorescence of Xyridaceae, the scapes in the families are not homologous, but of a higher order in Rapateaceae.

It has been suggested that Rapateaceae may be closely allied to Cyperales (Juncaceae, and especially Thurniaceae). Undoubtedly there are great similarities, for example in the anatropous ovules, the starch-rich endosperm, the undifferentiated embryo, the habit, and the South-American distribution. Moreover, the unifloral "spikelets" of Rapateaceae might be compared to the flowers in Juncaceae which are frequently subtended by three or more bracteoles. However, the heterochlamydeous perianth (homoiochlamydous in *Thurnia*), separate pollen grains (tetrads and grains with rounded apertures in Juncaceae) and other differences suggest that Rapateaceae are probably remote from the Cyperales.

# Eriocaulaceae Desvaux (1828)    13:1,150
(Fig. 178)

Mainly small or very small annual or perennial herbs, generally 30 cm high or less, only rarely (within *Eriocaulon* and *Paepalanthus*) up to more than a metre tall. A stout trunk up to 80 cm tall is found in *Eriocaulon macrocalyx* (STÜTZEL 1984). Horizontal or vertical rhizomes or stolons occur occasionally. The roots are densely set on these or on the stem base. Vessels are present in the vascular strands of roots, stems and leaves. They have simple perforation plates in the roots, and simple and scalariform perforation plates in stems and leaves.

The leaves are spirally set on the stem, often densely so, crowded in rosettes, or dispersed along pseudodichotomously branched stems. In aquatic forms they may be submerged. The leaves are grass-like; narrow (in *Eriocaulon macrocalyx* up to 15 cm broad); pointed; flat, terete or caniculate; linear or occasionally thinly filiform; and distinctly or indistinctly sheathing at the base. They are often thin and semi-transparent, occasionally "fenestrate" by having air spaces alternating with lamellae of chlorenchyma. Rarely, as in the larger species of *Paepalanthus,* they are stiff and thick. The leaf apices are variable, and often characteristic of the species. The stomata are paracytic, of the grass type. Hairs are sometimes present; they are either three-celled and often equally or unequally T-shaped, or long and uniseriate. Raphides are lacking, but druses or needle-like or prismatic oxalate crystals sometimes occur. Silica bodies are generally absent.

The inflorescences are solitary or clustered at the ends of the branches, generally terminating erect or spreading, often thin scapes (peduncles). The heads (and peduncles) are basally enclosed by a sheathing leaf. They are capitate-globose, semiglobose or cylindrical, racemose in construction (STÜTZEL 1974), and below the flowers bear variably numerous, imbricate, spirally set, nerveless, bracteal phyllaries. These vary considerably in shape, colour and hairiness. Pollination is apparently effected by insects. Self-pollination within the heads is probably common (STÜTZEL 1982). The flat or conical inflorescence axis is glabrous or bears long hairs between the flowers between which there are also usually thin scales (often lacking in *Syngonanthus*).

The flowers are sessile or shortly pedicellate, generally small or very small (sometimes less than 0,25 mm in length), and constantly unisexual.

Male and female flowers normally occur in the same heads, the male more often peripherally, the female centrally, although the reverse condition also occurs. (For regularity here, see STÜTZEL 1982, 1984; STÜTZEL and WEBERLING 1982.) Rarely, as in some species of *Eriocaulon*, the heads are almost unisexual. Variously shaped glandular or other hairs occur on the bracts, the inflorescence axis and the flowers.

The perianth is actinomorphic or median-zygomorphic, trimerous or more rarely dimerous, heterochlamydeous, and in the female flowers hypogynous.

The outer perianth members are membranous-hyaline, with two or three members free or partly fused, in *Eriocaulon* sometimes fused into a single, more or less spathaceous structure. The inner tepals are different in shape and frequently whitish. Sometimes they are reduced to hairs in the female flowers. In the male flowers they may be fused basally with the floral axis to form a column between the outer tepals and the inner tepal lobes. The margins and adaxial apical parts of the tepals are often fringed with short papillate hairs. Inner tepal lobes which are placed peripherally in the head may be longer than the others. Sometimes the inner tepals are fused into a tubular or funnel-shaped structure with short or inconspicuous lobes. Various hairs may be found in its mouth or on the inner side, where conspicuous, nectar-secreting glands are frequently situated.

The male flowers have one or two staminal whorls, each with two or three stamens, with narrow filaments. The anthers are tetra- or bisporangiate, basifixed, introrse and longitudinally dehiscent. The anther colour, white, black or brownish, is often characteristic of the species. The tapetum is glandular-secretory. Microsporogenesis is probably always successive (although sometimes stated, as in DAVIS 1966, to be simultaneous). The pollen grains are spheroidal and peculiar in being spiraperturate, with a minutely and sparsely spinulose surface. They are shed in the two-celled or three-celled stage. A rudimentary gynoecium in the form of three small columns may be present in the centre of the male flowers.

The female flowers have a central, bi- or tricarpellary and bi- or trilocular pistil, which terminates in a more or less short style with two or three long stigmatic branches. The stigmatic surface is Dry. In all genera except *Eriocaulon* and *Mesanthemum* there are also alternating with the stylar branches as many or twice as many stylodium-like appendages which are totally or apically beset with papillae. These may resemble stylar branches, with stigmatic hairs, while the true stigmatic areas can be displaced into a commissural position.

The locules constantly contain a single basal, orthotropous and tenuinucellate ovule, in which the archesporial cell functions directly as the megaspore mother cell. Embryo sac formation conforms to the *Polygonum* Type. Endosperm formation is reported to be nuclear (which is the likely condition, although the basal position of the endosperm nucleus according to DAVIS 1966, suggests helobial endosperm formation).

The fruits are small, bi- or trilocular, thin- and dry-walled loculicidal capsules. Each locule contains, at the most, a single seed. This is ellipsoidal with circular transection. The seed coat consists of two cell layers. The outer is thin and hyaline and generally persists only as rows of small hooks on the seed. The inner consists of low, tabular, hexangular cells, yellow to red-brown in colour. The endosperm is rich in compound starch-grains and fills up most of the seed. The small, undifferentiated, lens-shaped embryo is situated at the micropylar end.

*Chromosome Numbers*. The basic chromosome number in *Eriocaulon* is $x = 8$, but $x = 10$ has also been reported.

*Chemistry*. Chemically the family is characterized by the variable types of oxalate crystals. Silica bodies have been observed only in *Paepalanthus xeranthemoides*, where they occur together with sclereidal cells in the subepidermal leaf tissue. Starch is reported only in the endosperm. It is interesting that various flavonols have been reported in the family (including myricetin and kaempferol), but no other flavonoid compounds at all.

*Distribution*. The Eriocaulaceae are pantropical, and form a frequent constituent in montane shallow pools or swamps, especially on sandy ground. The leaf rosettes are often inundated. Exceptional xerophytes, growing in only temporarily wet places occur, however, in the genus *Paepalanthus*. These are protected against desiccation by their densely hairy and/or hard leaves. The greatest concentration of species is in South America and Africa. *Eriocaulon* is quite widespread with outposts in western Europe. *Lachnocaulon* is North American.

The family may be divided according to whether there are two functional whorls of stamens or one, whether the anthers are tetra- or bisporangiate (di- or monothecous), and whether the tepals are free or fused, laminar or reduced to hairs, etc. The

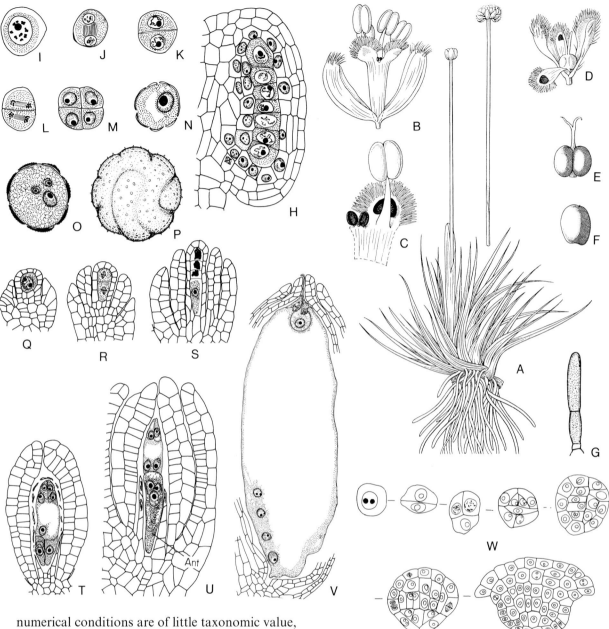

numerical conditions are of little taxonomic value, dimerous and trimerous flowers occurring in *Erio-caulon* as well as *Paepalanthus* and *Syngonanthus*. Some of the genera are surprisingly rich in species, but there is great uncertainty in the species number.

*Eriocaulon* (250–400) has a tropical and sub-tropical distribution with a great Asiatic concentration but rich representation also in Africa and South America. One species, *E. septangulare*, oc-curs in eastern U.S.A. and western Europe (Ire-land, etc.). The genus has four or six stamens and free inner tepals supplied with conspicuous nectar glands. Most taxa have a very short stem and are rosette herbs with linear, grass-like leaves. The

**Fig. 178.** Eriocaulaceae. **A–E** and **P** *Eriocaulon aquati-cum* (= *E. septangulare*). **A** Plant and inflorescence. **B** Male flower. **C** Part of inner tepal whorl and one stamen; note the dark glands. **D** Female flower. **E** Pistil. **F** Seed. **G** Hair from bract. **P** Pollen grain: (**A–F** Ross-Craig 1973; **G** Solereder and Meyer 1929; **P** Erdtman 1952). **H–O** and **Q–W** *Eriocaulon hookerianum*. **H** Mi-crosporangium, longitudinal section, early stage show-ing in centre archesporial cell row and outside this the tapetal layer; **I–M** Stages in successive micro-sporogenesis. **N** Microspore. **O** Pollen grain. **Q–S** Meio-sis in the tenuinucellate ovule. **T** Embryo sac. **U** Same, with enlarged antipodals. **V** Early stage of nuclear endosperm formation. **W** Stages in the embryogeny. (**H–O** and **Q–W** Arekal and Ramaswamy 1980)

heads are generally white-hairy. A few species are dimerous, e.g. *E. longipetalum* in Africa, which is only about 1 cm tall, and the above-mentioned *E. septangulare*. Most species are trimerous or mixed di- and trimerous.

*Mesanthemum* (6–10) is distinguished by the tubularly fused inner tepals of the male flowers, while the inner tepals in the female flowers are basally free but upwards fused into a tube. In this tropical African genus the leaves may be rather broad and the heads are densely white-hairy.

*Paepalanthus* (200–480), which is mainly South and Central American, has di- or trimerous flowers with only one staminal whorl. Its tepals lack nectar glands. In the male flowers the tepals are free. The genus is vegetatively very variable: the leaves may be in a basal rosette or distributed along elongate decumbent to ascending branches, and the scapes may be few or numerous, sometimes many in umbel-like clusters.

*Blastocaulon* (3) in Brazil are small moss-like herbs with filiform scapes, and *Tonina* (1) *fluviatilis* is an American floating species with scattered leaves and small heads with few flowers, the inner tepals of which are absent or reduced to hairs. – *Syngonanthus* (80–200), like *Paepalanthus,* is mainly South–Central American, but has a few African species. It has male flowers with three stamens and female flowers where the inner tepals are free basally and apically, but fused in the middle. This genus is also variable in habit, with a short or long leafy stem. The root is mostly thick and spongy.

The members of Eriocaulaceae are of little economic value. The rigid but slender scapes with their button-like heads are often dyed and sold as everlastings.

The position of the family has usually been regarded as close to Xyridaceae (incl. Abolbodaceae), Rapateaceae, Restionaceae and Centrolepidaceae. It is interesting to note that stylar appendages similar to those in most members of Eriocaulaceae also occur in *Abolboda*. The numerous similarities to *Abolboda* and other Xyridaceae in vegetative characters, in the compact inflorescences and in floral and embryological (mainly seed) characters indicate a not too remote relationship. On the other hand, Eriocaulaceae seem to be very distinct, and placement in a separate order may well be justified (as in DAHLGREN 1975). It is apparent that Eriocaulaceae still needs much intensive investigation. Much of the information in the above description is based on very incomplete and few studies.

# Order Hydatellales

*One Family:* Hydatellaceae.

[The following description is based mainly on HAMANN (1975, 1976) and CUTLER (1969).]

Minute, annual (or in *Trithuria* perhaps sometimes perennial) herbs which are temporarily immersed in freshwater swamps. Rhizomes occasionally seem to occur, at least in *Trithuria,* where they may bear short multicellular hairs. The leaves are basally concentrated, tufted, thin and filiform, and lack a distinct sheath. Each plant has several culms (scapes). The stomata are anomocytic and thus lack subsidiary cells. Hairs are apparently lacking on stems and leaves. Oxalate raphides and silica bodies are also absent. The vascular strands of roots and culms (and rhizomes) have vessels with scalariform perforation plates, whereas in the leaves they have tracheids only.

The inflorescence is terminal on the short culm and is subtended by two to four to six glume-like, hyaline bracts. Male and female flowers either occur together in bisexual inflorescences (in *Trithuria* and rarely in *Hydatella*) or are distributed in separate unisexual inflorescences on the same specimen (generally in *Hydatella*). The flowers (according to the current interpretation) are minute, unisexual, and naked, i.e. without perianth. The male flowers consist of a single stamen, the female of a single, stipitate pistil. Several flowers, either of both sexes or of the same sex, are crowded together in the inflorescence on a somewhat elongate axis. Apart from the bracts mentioned above, there are no leaves in the inflorescence.

The stamens have a linear filament and a basifixed tetrasporangiate anther, with sulcate (or ulcerate), two-celled pollen grains. The pistils are unilocular and are monomerous or pseudomonomerous (possibly tricarpellary in *Trithuria,* see below). They are often slightly stipitate and utricle-like, and contain a single, apical, pendulous, anatropous-apotropous ovule, the micropyle of which turns upwards. The ovary is crowned by some "stigmatic" hairs, each consisting of a single cell row. The ovules are bitegmic and tenuinucellate. The embryo sac formation does not seem to be known. The embryo sac is relatively broader than in Centrolepidaceae, and antipodal cells seem to be absent in *Hydatella* (perhaps they degenerate at an early stage?). Remarkable is the endosperm formation which is cellular, a unique condition in monocotyledons outside Arales (reported also in Thismiaceae, but highly unlikely there).

The fruit is small, and has a membranous pericarp. It dehisces by two or three slits to form two or three valves in *Trithuria,* which may indicate a bi- or tricarpellary nature, but in *Hydatella* remains indehiscent. In the seed coat the epidermal cells of the outer integument become enlarged and develop a thick outer wall. The endosperm consists of few starchless cells, and in the mature seed persists only as a scanty remainder at the seed apex and contains no nutrients whatsoever. Instead, the nucellar tissue develops into a copious, starchy *perisperm.* The embryo is peripheral in the seed, minute, lens-shaped and incompletely organized. The chemistry is unknown.

*Distribution.* Hydatellaceae is a small, bigeneric family restricted to Australia, Tasmania and New Zealand.

*Morphological Interpretations.* The above-mentioned interpretation of the flowers as composed of one stamen or one pistil only is the presently accepted and most likely one. According to another interpretation, the "inflorescences" could be unifloral (as in some juncaceous genera, like *Marsippospermum* or *Rostkovia*), and thus the flower would be bisexual in *Trithuria* and mostly unisexual in *Hydatella.*

*Relationships.* The Hydatellaceae have only recently been distinguished as a separate family (HAMANN 1976). Formerly the genera were placed in the Centrolepidaceae, which they closely resemble in growth, general appearance and the apparently multiflorous synanthia of very reduced flowers (see Reduction, pp. 32–33).

The Hydatellaceae differ conspicuously, however, from the Centrolepidaceae in having anomocytic instead of paracytic stomata, in lacking distinct leaf sheaths and in having monosulcate instead of ulcerate pollen grains, totally different pistils with "stigmatic hairs" consisting of simple cell rows, tetra- instead of bisporangiate anthers, anatropous-apotropous instead of orthotropous ovules, cellular instead of nuclear endosperm formation, a copious perisperm with starch and hardly any endosperm (at least not with nutrients) instead of a starch-rich endosperm, and a very differently constructed seed coat. This in combination indicates that the Hydatellaceae could possibly be totally unrelated to the Centrolepidaceae, which is decidedly poalean in most of its features (e.g. in the stomata, pollen grains, starchy endosperm, scutellum-like cotyledon). The Hydatellaceae would then have attained their resemblance to Centrolepidaceae through convergence; i.e. as an adaptation to inundated ground

with a short period suited for flowering and seed setting. The occurrence in Hydatellales of some features which appear to be more *primitive* than those in Centrolepidaceae also supports this independent status. These characters are the tetrasporangiate anthers and the monosulcate pollen grains. Also the anomocytic stomata may be basic in the family. This could reflect a liliifloran affinity. If one considered the cellular endosperm to be a primary state one could even consider an aralean origin.

An alternative (HAMANN, personal communication) would be to accept the similarity to the Centrolepidaceae as expressing phylogenetic affinity. An extreme adaptation to a short period for anthesis and seed-setting would account for the following features:

- small size with reduction of leaves (also leaf sheaths) and stems
- consequent absence of anatomical attributes, such as vessels in the leaves, silica bodies, and subsidiary cells
- strong reduction of floral parts leading to flowers with either a stamen or with a pistil only
- reduction of all parts of the pistil, so that even a style is lacking
- an acceleration of the formation of nutrient tissue by transfer of its function to the already formed nucellar tissue (endosperm can only be formed after fertilization).

Yet the tetrasporangiate anthers and monosulcate pollen grains strongly argue against a derivation from the Centrolepidaceae as such. A more distant origin might well be supposed. This is the reason for treating the Hydatellaceae in a separate order.

Thus, the phylogenetic position of Hydatellaceae is still a riddle. It is so different from other monocotyledonous orders that its inclusion even in any superorder will be most strained. The cellular endosperm formation agrees with the Ariflorae, the perisperm-rich seeds with the Zingiberiflorae, the graminoid habit and vessel-containing stem with the Commeliniflorae, and the anomocytic stomata and monocolpate pollen grains perhaps best with the Liliiflorae.

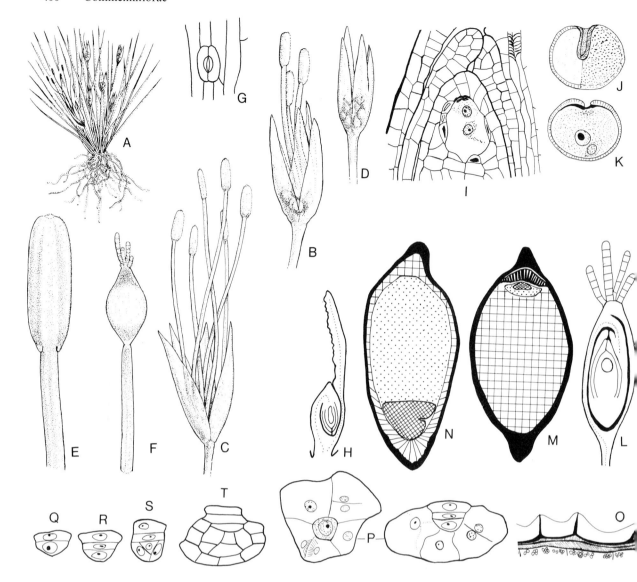

**Fig. 179.** Hydatellaceae and (**H** and **N**) Centrolepidaceae. **A–F** *Hydatella inconspicua*. **A** Plant with male and female inflorescences. **B** Bisexual inflorescence. **C** Male inflorescence. **D** Female inflorescence. **E** Stamen (=male flower). **F** Female flower; note the stigmatic uniseriate hairs. (EDGAR 1966). **G** *Trithuria submersa,* stoma. (CUTLER 1969). **I–M** and **O–T** *Hydatella inconspicua.* **I** Part of ovule showing some divisions in nucellar epidermis. **J–K** Pollen grain (probably dispersed in the two-celled state). **L** Ovary, longitudinal section. **M** Seed; note that the *checkered* part filling up most of the seed, is perisperm, the *dotted* part is endosperm and the *densely cross-hatched* part is the embryo. **O** Seed wall. **P** Embryo and endosperm (formed the cellular way). **Q–T** Embryo formation. **H** *Brizula gracilis* (Centrolepidaceae), monocarpellary pistil. **N** *Centrolepis polygyna* (Centrolepidaceae), seed, for comparison with *Hydatella,* with corresponding shading. (**I–T** HAMANN 1975)

**Hydatellaceae** U. Hamann (1976)   2:7
(Fig. 179)

Description, see under the order.

The family consists of the two genera *Hydatella* (4) and *Trithuria* (=*Juncella*) (3) only. *Hydatella* has two species in Western Australia, one in Tasmania and one in New Zealand, while *Trithuria* is restricted to Australia. They grow in permanently or temporarily inundated habitats, e.g. on lake shores.

The two genera differ in the fruit characters and in sex conditions (see above), and in the number of "stylar hairs" which are more numerous in *Hydatella* than in *Trithuria*.

# Order Cyperales

*Three Families:* Juncaceae, Thurniaceae and Cyperaceae.

Perennial or annual, often grass-like herbs; when perennial generally rhizomatous, rarely with a thick, woody trunk (in *Microdracoides* of Cyperaceae up to 1.5 m high) and in a single species (of *Gahnia,* Cyperaceae) with a slenderer, woody scandent stem up to several metres. The plants are often tufted and sometimes cushion-like. The rhizomes may be horizontal, ascending or erect. Runners occur in some species.

The roots are generally fibrous but may also be short and fleshy. They arise opposite either xylem or phloem strands. The root hairs are produced on particular, short epidermal cells. Vessels are always present in the roots and vary from elongate with oblique, scalariform perforation plates to short, with simple perforation plates or both kinds may occur together (CHEADLE 1955 a).

The rhizomes are of variable construction and at least usually have sympodial growth. They often contain copious amounts of starch. The aerial stems are terete or often trigonous, and are often aerenchymatous.

The leaves are generally tristichous (distichous only in a few genera of each of Juncaceae and Cyperaceae), and are either concentrated in a basal rosette or distributed on the aerial stems. They generally have a basal sheath which is closed or more rarely open. Stipule-like membranous lobes occur in several genera, and in some Cyperaceae there is a membranous adaxial ligule; in *Scleria* of Cyperaceae there is even a contraligule (see under the family). The leaf blades are as a rule continuous with the sheath but may be basally contracted into a pseudopetiole. In outline they generally range from linear to filiform and are sometimes large and *Pandanus*-like, in rare cases rather broad; they are generally flat and dorsiventral (in which case they may be keeled) or terete, but rarely they are laterally compressed. They are never shed from the sheaths (as happens in some Poaceae). At least some of the leaves in many taxa consist of the sheath only. In some taxa the leaves contain air canals or cavities which may be partitioned by transverse or longitudinal septa or diaphragms. The leaves in Thurniaceae are peculiar in having inverted bundles of a kind unknown in other plants. Somewhat similar ones, facing each other, are found in ensiform leaves of *Iris,* Iridaceae, and other monocotyledons.

Stomata may occur on both surfaces of the leaf, but more commonly on the adaxial than on the abaxial side. They are mainly paracytic, with dumb-bell-shaped guard cells, rarely tetracytic in Cyperaceae. The epidermal cells, except for the stomata, are equal or in Cyperaceae unequal, some of the shorter cells in Cyperaceae containing silica bodies (see below). Uniseriate hairs occur in Cyperaceae but are rather rare; the hairs in *Luzula* of Juncaceae are multicellular. Raphides and other oxalate crystals are lacking. Silica bodies are lacking in Juncaceae but are present in the two other families; in Thurniaceae they are small and occur several together in epidermal cells, in Cyperaceae generally cone-shaped, with their bases resting on the inner periclinal walls of epidermal short cells. The cones are solitary or several, and subsidiary cones may arise in a ring on the larger cones. In certain Cyperaceae there are wedge-shaped, bridge-shaped or warty silica bodies, the latter free or attached to the anticlinal walls. Vessels are present in rhizomes, aerial stems and leaves, and on the average more often have scalariform rather than entire perforation plates in Juncaceae and Thurniaceae than in Cyperaceae. Chlorenchyma organization in Cyperaceae differs as between $C_4$ and $C_3$ pathway plants (see family).

The aerial axes are procumbent to erect, and branched or unbranched. The inflorescences are often complex, frequently contracted into one or more dense, often subglobose clusters of flowers or spikelets. They are morphologically terminal, but frequently pushed aside by a subtending leaf into a lateral position. The branches of the inflorescences are determinate or indeterminate and in Cyperaceae generally aggregated into small spikes, sometimes terminated by a female flower (for spikelet structure in Cyperaceae, see under this family). In Juncaceae the inflorescences generally consist of mono- or dichasial cymes, or the flowers may rarely be solitary and terminal.

The flowers are sessile or pedicellate and sometimes supplied with one or more prophylls. They are bisexual or frequently unisexual and basically trimerous, although reductions in perianth members, stamens and carpels occur. The tepals number $3+3$; they are generally bracteal (stiff and inconspicuous) and in colour green, brown, black, purple or rarely white. Sometimes the tepals are replaced by a variable number of bristles or numerous hairs, or are completely lacking. Nectar secretion is lacking.

The stamens are $3+3$ (most Juncaceae, Thurniaceae) or 3 to one only (some Juncaceae, Cyperaceae). Rarely, as in *Evandra* of Cyperaceae, the stamen number is 12–20. The filaments are thin, of variable length, generally free from each other and glabrous. The anthers are basifixed, tetrasporangiate, dithecous, generally elongate, introrse to latrorse, and longitudinally dehiscent. The endothecial thickenings are of the Spiral Type or, in many Cyperaceae, of the Girdle Type, and the tapetum is glandular-secretory. Microsporogenesis is simultaneous, and the tetrahedrally arranged pollen grains always cohere in tetrads. Postreductional meiosis has been found to occur within Juncaceae and Cyperaceae. In Juncaceae and Thurniaceae all four pollen grains of the tetrad reach full development, whereas in Cyperaceae three of the microspores degenerate and become incorporated in the pollen grain wall of the fourth, functional pollen grain (a case otherwise known in angiosperms only in some Epacridaceae). Besides the distal sulcus there are in some Cyperaceae a number of less distinct lateral apertures or pseudoapertures (lacunae). The pollen grains are dispersed in the three-celled or in Cyperaceae sometimes in the two-celled stage.

The pistil is tricarpellary or bicarpellary and trilocular or unilocular, with the style generally divided from near the base or the middle into three or two, sometimes twisted, stylodial branches. The stigmatic papillae have a Dry surface. The ovary in Juncaceae generally contains few (3) to numerous axile, parietal or basal ovules in three distinct locules or (the septa being at least partially dissolved) in one locule; in Cyperaceae the locule is solitary and there is constantly one basal ovule. The ovules are anatropous and crassinucellate. A parietal cell is cut off from the archesporial cell and forms a parietal tissue. Embryo sac formation is of the *Polygonum* Type, frequently with ephemeral antipodal cells. Endosperm formation is helobial in Juncaceae, unknown in Thurniaceae and nuclear in Cyperaceae. Embryogeny is of the Onagrad Type.

The fruit is a capsule (most Juncaceae and Thurniaceae) or a nutlet or rarely a drupe (Cyperaceae). The seeds are ovoid, fusiform or elongate, in Thurniaceae subulate at both ends, in many Juncaceae with appendages or wings and not infrequently with an elaiosome. The endosperm is starchy and the embryo relatively small, basal and ovoid to disc-shaped or mushroom-shaped, the embryo shape being often typical of large genera or groups of genera (though variable in *Scirpus* s. lat.).

*Chromosome Numbers.* The occurrence in this order of chromosomes not with one distinct centromere but with the centromeric activity dispersed along the chromosome has led to marked aneuploidy, with many numbers, though some numbers are particularly frequent. In Juncaceae the commonest basic number for *Juncus* is $x = 20$ and for *Luzula* $x = 6$.

*Chemistry.* As mentioned above, silica is deposited in most Cyperales except Juncaceae, while calcium oxalate is not stored. Starch is often richly present in the vegetative tissues and always so in the endosperm. Cyanogenesis is known in both Juncaceae and Cyperaceae, but not as frequently as in grasses. Steroidal saponins and chelidonic acid are lacking. Among the flavonoids, tricin is frequently present at least in Cyperaceae, and C-glycoflavones, luteolin/apigenin, and sulphurated flavonols occur in both Juncaceae and Cyperaceae, in which respect these families agree with grasses (and palms!). The red pigments in Juncaceae were found by FREDGA et al. (1974) to consist of glycosides of luteolinidin, and not ordinary anthocyanins.

*Parasites.* Among the parasitic fungi, certain species of the uredinalean genus *Puccinia* (sect. *Caricinae*) attack taxa of Cyperaceae and Juncaceae, and the ustilaginalean genus *Entorrhiza* is a parasite on these two particular families, whereas several other genera, including *Anthracoidea, Cintractia, Planetella, Schizonella, Orphanomyces* and *Cintractiella*, attack Cyperaceae only. It is notable that very few genera of fungi attack members of Poales as well as members of Cyperales. SAVILE (1979) on the basis of the parasites in *Puccinia* and *Uromyces* and their uredinospore morphology came to the conclusion that those on Juncaceae were more derived than those on Cyperaceae (and thus that Juncales might be derived from Cyperales).

Host specificity that supports the proposition that Juncaceae and Cyperaceae are closely related is also found among insects, in particular some plant lice, e.g. the psyllid subfamily Liviinae and the aphid tribe Saltusaphidini (EASTOP 1979).

*Distribution.* Both Juncaceae and Cyperaceae are cosmopolitan or subcosmopolitan families inhabiting extraordinarily variable habitats, in particular mires, lake shores and sea shores, but they also occur in forests, woodland and grassland. The Thurniaceae are restricted to Guayana and Brazil. Within the Juncaceae most genera, as well as most sections within the largest genus, *Juncus,* are concentrated in South America. Thus it is likely that the order has its origin in South America, as is more obviously the case with the Commelinales.

*Relationships.* There is strong evidence that the Cyperales are a monophyletic unit and that there is close relationship between the three families: the usually tristichous leaves, mainly paracytic stomata, simultaneous microsporogenesis, pollen tetrads, post-reductional meiosis, diffuse centromeres, parietal cell and parietal tissue, Onagrad embryo development, presence of 5-O-Me-flavones and shared parasites. In floral construction the Juncaceae, with a well-developed perianth, bicyclic androecium and often trilocular ovary developing into a frequently many-seeded capsule, are less derived than are the Cyperaceae. This is also reflected in the pollen tetrads, where all four pollen grains develop, a stage more basic than one where three degenerate, as in Cyperaceae. Further, absence of silica bodies is probably here a more ancestral state than occurrence of silica, as in Cyperaceae, because the silica bodies in the latter family are (generally) unlike those in all other monocotyledons. This contradicts the conclusion of SAVILE (1979), based on *Puccinia* and *Uromyces* parasites, that the Juncaceae could be derived from cyperaceous ancestors. His fungal evidence when carefully evaluated is also rather brittle. The position of Thurniaceae is uncertain, partly because of insufficient knowledge of embryology (e.g. endosperm formation). THORNE (1983) considers this little family as scarcely distinct from Juncaceae.

The position of Cyperales within the Commeliniflorae is somewhat uncertain. It seems that the three main orders (Commelinales, Poales and Cyperales) are largely parallel and differentiated at an early stage. Yet on chemical and other grounds Cyperales seems to be closest to the Poales.

### Key to the Families

1 A. Each pistil with at least three, generally more, ovules; fruit generally dehiscent; pollen grains in tetrads, all four well-developed . . . . . . . . . . . . . . . . . 2
1 B. Each pistil with only one, basal ovule; fruit always indehiscent; pollen grains in pseudomonads, consisting of tetrads in which three microspores have degenerated and only one pollen grain comes to development . . . **Cyperaceae**
2 A. Seeds not subulate-pointed at both ends; silica bodies lacking . . . . **Juncaceae**
2 B. Seeds subulate-pointed at both ends; silica bodies present (inflorescence globose, subtended by spreading spathal bracts) . . . **Thurniaceae**

**Juncaceae** A.L. Jussieu (1789)    8:300
(Figs. 180–181)

Perennial or annual, small to fairly large, grass-like herbs, very rarely with woody trunks (as in old specimens of the South African genus *Prionium*). The perennials are generally rhizomatous, with a vertical, ascending or horizontal, short to quite long and sympodial rhizome. The rhizome and also the pith of the stem base are often rich in starch.

The leaves are either all basally concentrated and the cauline stem leafless, or the leaves may be variously distributed along the cauline stem. They are nearly always tristichous, rarely distichous (in *Distichia* and *Oxychloë*). They are linear, subulate, filiform or rarely lanceolate in outline, and are flat, canaliculate, angular, terete or laterally compressed in section. The sheath is either open or, as in *Prionium* and *Luzula,* closed. Sometimes the leaves are reduced to the sheath only, and in certain species the culms comprise the main photosynthetic tissue. Membranous, stipule-like lobes may be present on each side of the top of the sheath and sometimes these are confluent and form a ligule. Stems and leaves are frequently strongly aerenchymatous, and air canals are common in the leaves, sometimes being partitioned by distinct septa (as in *Juncus* sect. *Septatae*).

The stomata are paracytic, with the subsidiary cells much smaller than the other, longitudinal and filed epidermal cells. Hairs are largely lacking, but ciliate hairs are present in *Luzula;* they are unusual in being multiseriate, arising from multicellular mounds on the leaf margin.

Silica bodies are lacking. Rounded or elongate cells, probably with tannin contents (formed by leucoanthocyanin and catechin) occur in the vegetative parts. Vessels in roots, stems and leaves have scalariform or simple perforation plates or both kinds of vessels are mixed.

The culm (stem) is generally erect but sometimes procumbent (or procumbent-ascending in some aquatic forms). It is leafy or leafless (inflorescence bracts ignored) and unbranched or generally sparingly branched. The inflorescence, which is subtended by one or more spathal bracts, is terminal, but may be pushed aside by an erect foliaceous bract and appear lateral. The inflorescence bracts in such cases may be terete, like the culm, and may seem to be its direct continuation. The inflorescence is paniculate-racemose, sometimes with the appearance of a corymb or anthela, and may be variously contracted. Sometimes the flowers are

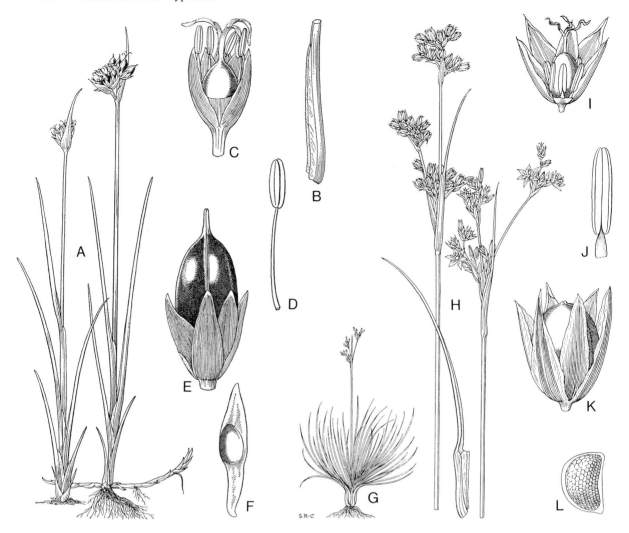

densely aggregated into heads and ebracteolate, and in some taxa the flowers are few, or even solitary and terminal, as in *Marsippospermum, Rostkovia* and other genera.

The flowers are sessile or pedicellate and generally supplied with two or more prophylls situated dorsally and ventrally in the median plane. They are trimerous, actinomorphic and generally bisexual (unisexual in *Distichia,* for example). The tepals are 3 + 3, green to white, brown, purplish brown or black, or often (partly) hyaline, obtuse or pointed, and from equal to conspicuously unequal between the two whorls (the inner whorl rarely lacking).

The stamens are generally 3 + 3 in number, but the inner whorl may be absent. They are free from each other, and generally have narrow, sometimes rather short, filaments, and introrse or latrorse anthers. An apical connective tip is sometimes present (in *Rostkovia, Marsippospermum* and *Oxy-*

**Fig. 180.** Juncaceae. **A–F** *Juncus castaneus.* **A** Rhizome with two aerial shoots. **B** Leaf sheath. **C** Flower, one tepal and two stamens removed. **D** Stamen. **E** Capsule. **F** Seed; note the terminal appendages. **G–L** *Juncus squarrosus.* **G** Habit. **H** Inflorescences and leaf. **I** Flower, one tepal removed. **J** Stamen. **K** Fruit. **L** Seed. (All from ROSS-CRAIG 1973)

*chloë*). The tapetal cells usually remain uninucleate. The pollen grains are always united in tetrahedral tetrads, within which all grains (in contrast to the Cyperaceae) are functional. They are ulcerate, with the exine faintly or not at all sculptured.

The pistils are unilocular or, quite often (many *Juncus* species, *Prionium, Andesia, Oxychloë*), trilocular, and apically have a single but often short style and three generally long and mostly twisted (rarely short) stylodial branches. The ovules are numerous on parietal or central placentae or, in

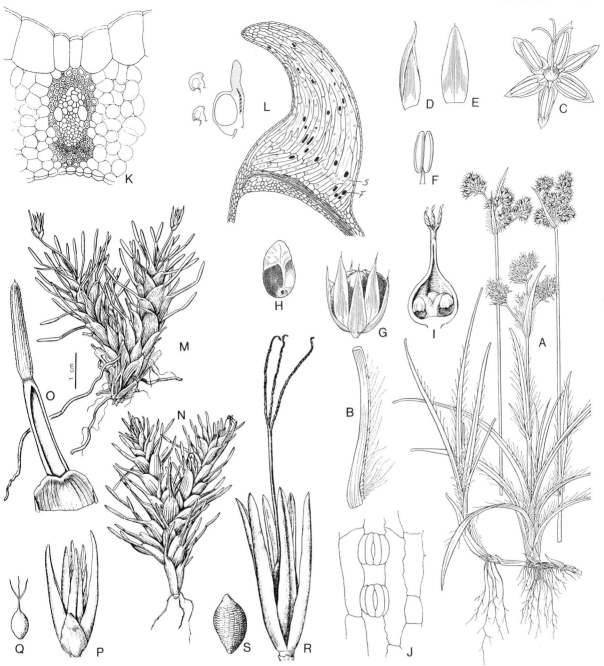

**Fig. 181.** Juncaceae. **A–H** and **J–K** *Luzula campestris.*
**A** Plant and two inflorescences in fruit stage. **B** Leaf
base. **C** Flower front view. **D** Tepal of outer whorl.
**E** Tepal of inner whorl. **F** Stamen. **G** Capsule. **H** Seed
with white elaiosome. **J** Stomata. **K** Part of leaf in transverse section; note absence of silica bodies. (**A–H** from
ROSS-CRAIG 1973; **J–K** CUTLER 1969). **I** *Luzula luzu-
loides,* pistil, longitudinal section. (HAMANN in MEL-
CHIOR 1964). **L** *Luzula pilosa,* elaiosome, development
from ovule to seed (*s* starch grain; *f* oil drop). (BRESINSKI
1963). **M–S** *Patosia clandestina.* **M** Male plant. **N** Female plant. **O** Leaf; note the pseudo-petiole between
sheath and lamina. **P** Male flower. **Q** Rudimentary pistil
from male flower. **R** Female flower. **S** Seed. (CORREA
1969)

*Luzula,* are three only, and basally inserted. Unlike
most other Commeliniflorae, the Juncaceae are reported to have helobial endosperm formation (observed in *Juncus, Luzula* and *Distichia*), with free
nuclear divisions in the large micropylar chamber,
while the chalazal chamber is one- or few-nucleate
only.

The fruit is generally a loculicidal capsule, rarely
indehiscent, and somewhat cartilaginous (*Oxychloë andina*). The pericarp varies from rather hard
to thin and parchment-like, as in species of *Luzula.*
The seeds are ovoid, or rarely fusiform, globose,

or angular and are variable in size; their chalazal end is often prolonged into a narrow, conical appendage, white in colour, while the outer integument may form a tube around the rest of the seed. In some species of *Luzula,* the seeds have an elaiosome formed from an obturator, or a hyaline appendicular body formed from the chalaza. The seed coat in *Juncus* is formed by both integuments and is longitudinally furrowed; in *Luzula* the shiny seeds expose the inner integument, while the outer integument forms an easily swelling mucilaginous layer which dries into a thin membrane. The endosperm is rich in starch and the embryo is small, basal and ovoid, with a large terminal cotyledon. The small seeds of many *Juncus* species are probably dispersed by wind, but the larger ones in this genus and in certain *Luzula* species are epizoically dispersed by the sticky jelly-like substance produced in the opened fruits; other species of *Luzula* are dispersed by ants attracted by the elaiosomes.

*Chromosome Numbers.* The commonest basic chromosome numbers are $x = 6$ for *Luzula* and $x = 20$ for *Juncus,* with descending aneuploidy in some groups. The low number in *Luzula* is doubtless a derived condition.

*Chemistry.* The subterranean parts of the perennial taxa accumulate much starch. Among the flavonoids, 5-O-methyl-flavones are a major component, and flavone bisulphates are also occasionally produced. In these characters Juncaceae agree with Cyperaceae and Poaceae. Tannin idioblasts, rounded to elongate in shape, occur in the family and seem to consist of quercetin and kaempferol derivatives. (See also the ordinal description.)

*Distribution.* Juncaceae consists of eight genera. Although *Juncus* and *Luzula,* the largest genera, are widely distributed in the Northern Hemisphere as well as in other parts of the world, the smaller genera are most strongly represented in the Southern Hemisphere, *Prionium* being South African, *Andesia, Marsippospermum, Oxychloë* and *Distichia* South American with Andean concentration, and *Rostkovia* subantarctic. Also the greater concentration of sections within the genus *Juncus* falls in South America, and thus there are good reasons to believe that the group has a South American origin.

*Prionium* (1) *serratum,* growing in streams mainly in the Cape Province of South Africa, is a large tufted plant which when undisturbed may form a trunk more than 1 m high. The large, linear and serrate leaves are tristichous, and slightly reminiscent of those of *Cladium* (Cyperaceae); they have closed sheaths. The metre-high inflorescence is a much-branched panicle. The bisexual flowers have trilocular pistils with almost free stylodia and many ovules per locule.

*Juncus* (ca. 225) is a very variable genus with a wide distribution. It has glabrous and flat, terete or compressed leaves, nearly always with open sheaths. The flowers, which are sparsely distributed or variously aggregated in clusters, are generally bisexual and have a tri- or unilocular ovary with several to many seeds. These are variable but never supplied with an elaisosome (caruncle). The outer integument of the testa is persistent or sometimes shed. *Juncus* species, together with taxa of Cyperaceae, are important consituents of the moist grassy habitats of the world, especially in the Northern Hemisphere. The tough stems of *J. acutus, J. effusus* and *J. glaucus* are sometimes used to make mats. Several sections or subgenera can be distinguished in the genus. One of these, sect. *Genuini,* consists of frequently large herbs with leafless culms. The largest subtending leaf of the inflorescence is erect and forms a continuation of the culm, pushing aside the inflorescence. Here belong common species like *J. effusus* and *J. filiformis.* In sect. *Septati,* the terete or canaliculate leaves are filled with aerenchyma separated at intervals by other tissue. An example of this is *J. articulatus.*

Among the smaller genera should be mentioned *Rostkovia* (1) *magellanica,* which has an Australian-subantarctic distribution including Tierra del Fuego, Falkland Islands, S. Georgia, Tristan da Cunha, Campbell Island and New Zealand. Its culms have a single, terminal, bisexual flower, which has equal, lanceolate tepals. The capsule is woody, unilocular, and contains numerous seeds. – *Marsippospermum* (3) with a similar, Chilean, Patagonian and New Zealand distribution has smaller inner than outer tepals, and thin-walled, almost completely trilocular capsules. – Three other genera are Andean cushion plants, *Andesia* (1) with bisexual, and *Oxychloë* (2) and *Distichia* (3) with unisexual flowers (dioecious genera) and with tri- or unilocular ovaries. In the latter two genera the leaves are often distichous.

*Luzula* (65) has a worldwide distribution. It consists of perennial herbs with basally concentrated leaves, which are, at least basally, ciliated with long hairs. The leaf sheaths are closed. The flowers are bisexual and have more or less scarious, white or more often brownish tepals, and $3 + 3$ or, more rarely, three stamens. The ovary is unilocular, with three basal ovules. The outer integument of the seed coat is thin and forms a mucilage layer, while the seed coat is formed by the inner integument which is smooth and shiny. The seeds are elliptic, with or without an elaiosome, which aids in ant dispersal.

## Thurniaceae Engler (1907)   1:2   (Fig. 182)

Rather large, tough, rhizomatous herbs with basally concentrated, long, linear coriaceous leaves, which are flat or canaliculate and smooth or serrulate-hispid along the margins. The culm is leafless except for a variable number of long leaves subtending the head-like, racemose, multiflorous inflorescence. Spheroidal silica bodies occur scattered in cells of the parenchyma and epidermis. The vessels in all parts are rather primitive, elongate, with oblique end walls and scalariform perforation plates. The leaf bundles are unique in monocotyledons. They occur in pairs above each other in a leaf transection, the lower with phloem upwards, and the upper (normal) with the phloem downwards, so that the phloem sides in each pair face each other. The stomata are paracytic or tetracytic.

The flowers are small, densely concentrated, and have a short, swollen and spreadingly puberulous pedicel. The tepals are more or less equal, and are thin, narrow, hyaline, persistent and obtuse. The stamens have thin, relatively long filaments and elongate anthers. The pollen grains are dispersed in tetrahedral tetrads and are ulcerate, as in Juncaceae.

The pistil is trilocular and has an elongate, trilocular ovary, a short style and three long stylodial branches. Each locule contains one to several central, ascending, anatropous-apotropous ovules. The capsule is triangular, loculicidal and three-seeded. The seeds are hispid and are supplied at each end with a subulate hispid process. The nucellus with the enclosed endosperm is free from the seed coat except at the micropylar end.

Thurniaceae contains the single genus *Thurnia* (2), which occurs in tropical South America: Guayana and Brazil.

Reasons for keeping *Thurnia* as a separate family are the peculiar vascular strands of the leaves, the presence of silica bodies and the peculiar terminally subulate and hispid seeds with the endosperm free from most of the tegminal layer of the seed coat. We consider it distinct enough to deserve family status. It may turn out to approach most closely the Rapateaceae or Xyridaceae.

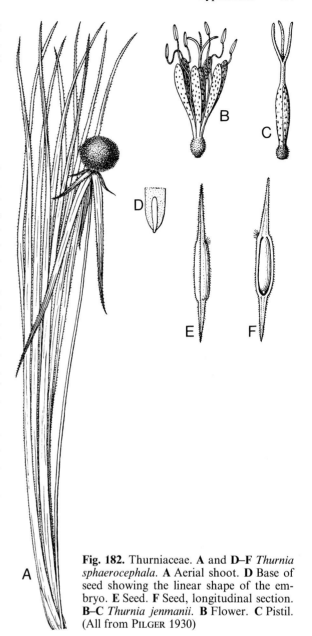

Fig. 182. Thurniaceae. **A** and **D–F** *Thurnia sphaerocephala*. **A** Aerial shoot. **D** Base of seed showing the linear shape of the embryo. **E** Seed. **F** Seed, longitudinal section. **B–C** *Thurnia jenmanii*. **B** Flower. **C** Pistil. (All from PILGER 1930)

## Cyperaceae A.L. Jussieu (1789)   96:9.300 (Figs. 183–188)

Annual or perennial, often grass-like herbs, sometimes with woody scandent stems (up to 10 m in a species of *Gahnia*) or dwarf and tree-like with a trunk up to 1.5 m (*Microdracoides*), or rarely lianes.

The roots of most species are fibrous but they may also be short and fleshy (dauciform). Perennial plants are commonly rhizomatous, and in some species the rhizome may grow out of the soil and

RHYNCHOSPOROID                                                    CYPEROID

SCIRPOID

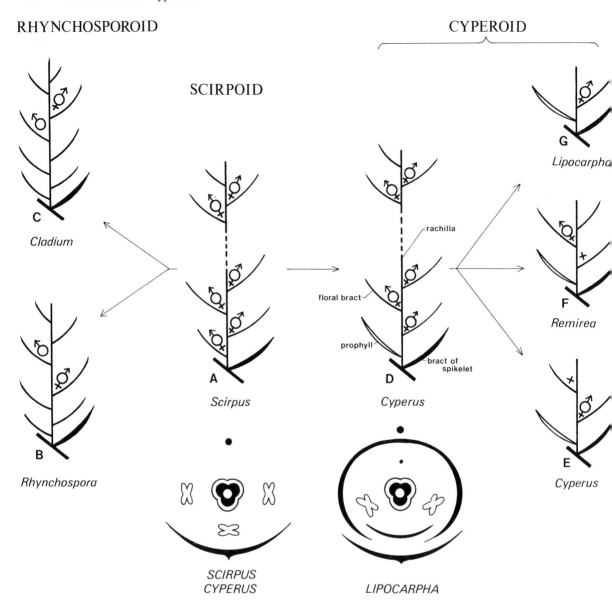

C
*Cladium*

rachilla

floral bract

prophyll

A
*Scirpus*

B
*Rhynchospora*

D
*Cyperus*

bract of
spikelet

G
*Lipocarpha*

F
*Remirea*

E
*Cyperus*

SCIRPUS
CYPERUS

LIPOCARPHA

produce an aerial stem. Adventitious roots are developed on such stems and ramify through the leaf bases to the soil.

Vessels with simple or scalariform perforation plates occur in all organs. The stems are mostly solid, but sometimes hollow or transversely septate; they are often trigonous but may be two- or more-sided or terete. In some species of *Scleria* the flowering culms attain a height of 4 m and those of *Cyperus papyrus* 5 m. The leaves are generally three-ranked, rarely distichous or polystichous. They are either borne basally or along the stem. Except in *Cymophyllus* they all have a sheathing base; the margins of the sheath are usually fused, rarely open (*Coleochloa*). The lamina

**Fig. 183.** Cyperaceae. Possible evolutionary relationships between the Scirpoid, Rhynchosporoid and Cyperoid Types of spikelets, together with floral diagrams of *Scirpus, Cyperus* and *Lipocarpha*. Spikelet diagrams adapted from KOYAMA (1961) and floral diagrams from KERN (1974). × indicates vestigial flower

**Fig. 184.** Cyperaceae, Caricoid spikelets. Possible evolutionary relationships between some spikelet diagrams for *Kobresia* and *Uncinia* (adapted from KERN 1958) and *Carex* (from KOYAMA 1961). The prophyll varies from spathal to utricular, as shown in the sketches adjacent to the diagrams. In *Uncinia* and *Carex* the reduced spikelets are arranged in spikes. Floral diagrams of *Carex* from KERN (1974)

is as a rule sessile but may be contracted into a pseudopetiole (as in species of *Mapania*). The lamina is mostly linear or setaceous but may be large and *Pandanus*-like, or very reduced or absent. It is never shed from the sheath. The leaves are generally flat or round in transverse section, rarely unifacial. Auricles are never developed, but an adaxial ligule is present in several genera. A tongue-like contraligule, opposite the leaf blade, is developed in *Scleria* (cf. similar structure in *Cocos,* Arecaceae). The first leaf on the axillary shoots is a two-keeled prophyll; it is inserted adaxially and lacks a blade.

Stomata may occur on both surfaces of the leaf, but more commonly on the adaxial than on the abaxial side. They are mainly paracytic but are tetracytic in a few Mapanieae and Rhynchosporeae. Almost always they are aligned in parallel files, but are irregularly distributed in some Mapanieae. Silica bodies are commonly present in the epidermal cells overlying the vascular bundles. They are nearly always cone-shaped, with their bases resting on the inner periclinal wall. There may be several cones in each cell and subsidiary cones often arise around the larger cones. Other forms of silica body occur especially in the Mapan-

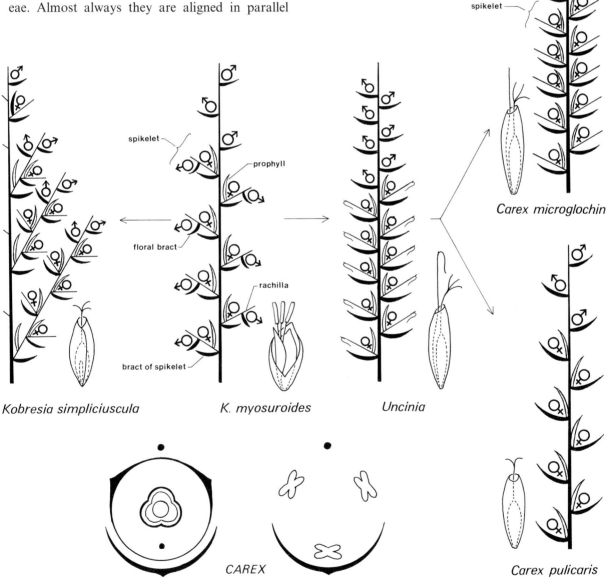

*Kobresia simpliciuscula*     *K. myosuroides*     *Uncinia*

*Carex microglochin*

*Carex pulicaris*

CAREX

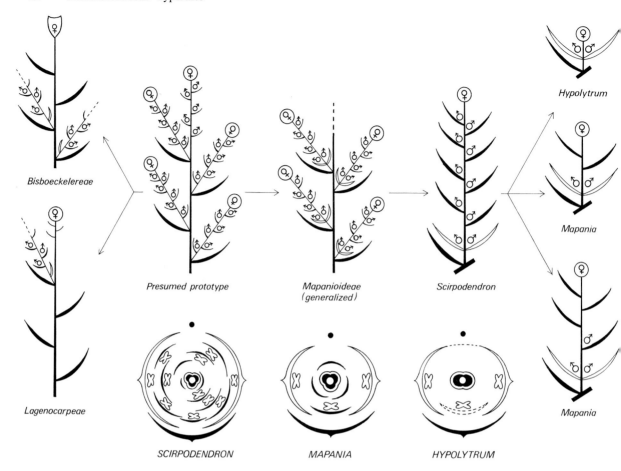

**Fig. 185.** Cyperaceae. Possible evolutionary relationships between some pseudanthium types in the Mapanioideae and Sclerioideae, together with floral diagrams of some genera of Mapanioideae. Pseudanthial diagrams adapted from KOYAMA (1971), except for Bisboeckelereae (EITEN 1976a), and floral diagrams from KERN (1974)

ioideae; these include: (a) wedge-shaped, so called because in transverse section they appear to be triangular with their bases embedded in the anticlinal or outer periclinal walls; (b) bridge-shaped, which in surface view may resemble the wedge-shaped but differ in that the wedges are connected by an arched span; (c) warty and either free and nodular or attached to anticlinal walls.

As seen in transverse section the mesophyll chlorenchyma is arranged in one of several ways: as in Poaceae it may be either radiate (i.e. the cells are radially arranged around the vascular bundles, Fig. 200 B) or non-radiate (Fig. 200 A, C), the disposition reflecting the $C_4$ and $C_3$ pathways of photosynthesis, respectively. If non-radiate the chlorenchyma may be irregularly arranged or organized into a palisade. In many species the mesophyll is interrupted by air cavities.

Within the Cyperaceae the inflorescence units are extremely contracted, thereby making difficult the analysis of the branching pattern. As shown by EITEN (1976a) the ultimate branches of the inflorescence for most taxa are monopodial and she believes that this will eventually prove to be the case for the inflorescences of all taxa (EITEN 1976b).

As with the Poaceae the ultimate axes of the inflorescences are called spikelets because they bear sessile flowers in the axils of bracts. However, it is useful to distinguish between terminal and lateral spikelets, because these often differ in structure (EITEN 1976a). Thus a terminal spikelet does not arise in the axil of a bract nor does it have a prophyll (Fig. 185). Unless otherwise qualified the term "spikelet" refers to a lateral spikelet. The term "prophyll" is used in Cyperaceae for the lowest bract on the axis of a spikelet (rachilla) when it is sterile (not subtending a flower) and is morphologically distinguishable. The flower lacks a prophyll but if the first floral bract in a spikelet is suppressed the prophyll of the spikelet comes to subtend the first flower. The term "glume" may refer to a floral bract or a spikelet bract, and its use is here avoided.

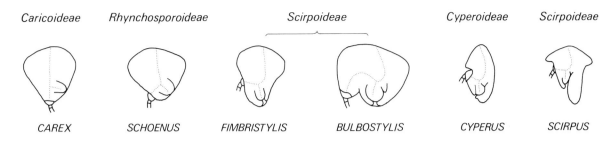

| Caricoideae | Rhynchosporoideae | Scirpoideae | | Cyperoideae | Scirpoideae |
| --- | --- | --- | --- | --- | --- |
| CAREX | SCHOENUS | FIMBRISTYLIS | BULBOSTYLIS | CYPERUS | SCIRPUS |

The simplest of the spikelets is the Scirpoid (Fig. 183 A) Type, in which all but the most distal bracts bear flowers and there is no prophyll. From this basic pattern several others may be derived by progressive reduction and sterilization. By sterilization of the proximal and distal bracts and the loss of stamens or pistils within individual flowers the Rhynchosporoid spikelet arises. Within this type two subtypes may be recognized according to whether the male flower occurs below (Cladiinae, Gahniinae) or above (Rhynchosporineae) the hermaphrodite flower (Fig. 183 B, C).

An alternative series of reductions gives rise to the Cyperoid Type of spikelet (Fig. 183 D). Here the first bract of the spikelet not only lacks a flower but is distinguishable as a prophyll. Loss of bracts and loss of flowers give rise to various kinds of few-flowered spikelets (Fig. 183 E–G).

A further form of highly reduced spikelet is the Caricoid Type. The flowers here are always unisexual with the solitary female flower in the axil of the prophyll of the spikelet, as described above (Fig. 184). Male flowers may be borne distally on the rachilla (Fig. 184, *Kobresia*), or the rachilla may be bare and extend beyond the female flower as with *Uncinia* and *Carex microglochin* (Fig. 184) or may even be missing (*Carex pulicaris*) (Fig. 184). Male flowers occur in bisexual spikelets, in simple terminal spikes and in distal spicate portions of axes which proximally bear reduced lateral spikelets (Fig. 184, all examples). Male flowers in this last situation are interpretable as spikelets consisting of nothing but the stamens of a male flower or, alternatively, these distal parts of the inflorescence (spikes) lack lateral branches (spikelets) and are truly spicate with naked male flowers in the axils of bracts. The Caricoid spikelet, like the Cyperoid, can be derived readily from the Scirpoid.

Extreme reduction within the inflorescence is presumed to have given rise to the pseudanthium of the Mapanioideae. On this theory the pseudanthium comprises a single terminal female flower and a series of lateral male flowers each subtended by a scale-like bract. Both male and female flowers lack a perianth and the males are reduced to a single stamen. The presumed prototype of the mapanioid pseudanthium is illustrated in Fig. 185. It is assumed that the generalized Mapanioid Type of pseudanthium arose from the prototype as a result of the shortening of the internode between the two lowermost florets so as to bring them opposite to one another. At the same time their bracts became differentiated from the remainder. Such changes produced the type of pseudanthium present in *Scirpodendron* and from which other types may be derived by a reduction in the number of male flowers (Fig. 185, *right*). Loss of the

**Fig. 186.** Cyperaceae. The morphology of the embryo in different genera seen in lateral view. (Redrawn from van der Veken 1965)

distal male flowers without an attendant loss of their bracts gives rise to a *Mapania*-like pseudanthium and the loss of the bracts as well as of the flowers produces a *Hypolytrum* Type pseudanthium.

The opposite insertion of the bracts bearing the two proximal male flowers is characteristic of the mapanioid spikelet, as is their strongly keeled cross-section. Unlike the Cyperoid and Rhynchosporoid prophylls they are almost always fertile.

In another line of evolution from the presumed prototype of the Mapanioideae, the terminal pseudanthia are reduced to solitary female flowers subtended by a few sterile bracts while the lateral pseudanthia are reduced to a few proximal male flowers and one or more sterile distal bracts (Fig. 185, left). In the Bisboeckelereae the bracts nearest the female flower form a utricle, whereas they remain sterile but independent in the Lagenocarpeae.

In the inflorescences in Cyperaceae, as in the Poaceae, the spikelet and not the flower represents the basic unit for classification of inflorescence. Most inflorescences comprise more than one spikelet. They vary from large leafy panicles (*Gahnia*) to densely contracted heads (*Gymnoschoenus*). The spikelets may be fascicled or arranged in spikes (*Carex*, Fig. 184). Especially in *Cyperus* and allied genera the spikelets form a more or less contracted head subtended by a series of leafy bracts borne atop a leafless culm. The spikelets and pseudanthia are themselves often arranged into secondary spikelets which are then known as pseudo-spikelets.

As already indicated, the flowers may be unisexual or bisexual. Dioecism is rare but occurs in all subfamilies except Mapanioideae. The perianth when present consists of bristles, hairs or scales. In *Eriophorum* the perianth hairs elongate after anthesis. In *Oreobolus* the scales (Fig. 188 J), which are in two alternating trimerous whorls, resemble the perianth in the Juncaceae. The stamens are mostly one to three, but may be numerous (12–20 in *Evandra*). The anthers dehisce introrsely or latrorsely. The filaments are mostly free except in a few species of *Carex* where they are connate. The connective often projects beyond the anthers as an apiculus. Microsporogenesis is simultaneous, the four microspores being retained within a common wall. Three of them degenerate so that at maturity the pollen grain appears to be single; it is known as a "pseudomonad" (Fig. 187 E–J). The grains are shed in the two- or three-celled stage.

The "pollen grains" (more correctly pseudo-monads) are mostly pear-shaped but may be spheroidal or flattened-triangular (*Mapania humilis*). There is generally a single distal ulceroid aperture and often a ring of transversely located lateral lacunae, rarely four ill-defined foramina. The tectum is perforate, the lacunae arising where it is broken into small frustillae, usually forming an areolate sculpturing (FAEGRI and IVERSEN 1964). Spheroidal pollen pseudomonads occur in some genera in three of the subfamilies, Mapanioideae (*Hypolytrum* and *Mapania*), Sclerioideae (*Calyptrocarya*) and Rhynchosporoideae (*Rhynchospora alba*). Lacunae are lacking from the pollen grains of most Mapanioideae but occur in *Lepironia mucronata*. Pear-shaped pollen grains with a distal aperture and lateral lacunae (*Carex* Type) are known only from Cyperaceae (ERDTMAN 1952).

The ovary is bi- or tricarpellary with a solitary basal, anatropous, crassinucellate ovule. The stylodia are fused basally into a single style which in some genera is swollen at the base into a stylar head. This thickened style base may be retained on the apex of the fruit. There are generally two to three stigmas, rarely as many as eight (*Evandra*) or one only (*Syntrinema*).

The fruit is most commonly a nut but drupes are formed in *Scirpodendron*, *Mapania* and *Cladium*. In some species of *Carex* (*Indocarex*) an apparent berry is formed by the fleshy base of the utricle. In the tribe Lagenocarpeae and possibly the Mapanieae the apparent fruit may be a compound structure. The fructification in these tribes, it is argued by KOYAMA (1971), is a fruit invested in a tightly adnate or fused perigynium homologous with the hypogynous disc of the Sclerieae (see p. 417). The correctness of this interpretation awaits a detailed study of the development of such fructifications, for the anatomical structure of the mature organ is not clear. The seed remains free of the fruit wall and has a mostly capitate embryo embedded in the starchy endosperm (VAN DER VEKEN 1965). The embryo, unlike that in the majority of grasses, is not visible through the pericarp.

There is considerable variation in embryo structure as may be seen from Fig. 186. Similar embryo types are found throughout in homogeneous genera such as *Cyperus, Schoenus* and *Eleocharis,* whereas in the heterogeneous *Scirpus* there are several embryo types, each characteristic of its section within the genus. Within the subfamilies the *Carex* Type embryo is uniformly present in the Caricoideae and the *Schoenus* Type is dominant in the Rhynchosporoideae. In comparison with the embryos of the Scirpoideae, those of the Rhynchosporoideae are poorly differentiated (VAN LECKE 1974).

*Pollination.* Pollination is mostly anemogamous but insect pollination may occur in some species with white or coloured bracts or where the upper leaves are conspicuously yellow as in the South African genus *Ficinia* or white as in the American genus *Dichromena.* The blue anther filaments of the South African *Chrysithrix* may also serve to attract pollinators. According to SCHULTZE-MOTEL (1966) *Rhynchospora* may also be insect-pollinated.

*Dispersal.* Dispersal is achieved by a variety of means. Several species have bouyant diaspores. These include *Carex pumila* (corky utricle), *Remirea maritima* (prolonged corky rachilla), *Capitularia* (hollow axis), *Cyperus cephalotes* (corky pericarp), *Cladium mariscus* (air cells in pericarp) and *Carex* spp. (inflated utricles). The hairs and barbed perianth members of several species are probably adaptations for epizoic dispersal. In *Uncinia* the hooked rachilla acts as a hair trap and has been observed attached to feathers. The nutlets of many species are eaten by birds although only a few, such as those of *Gahnia,* appear to be adapted for this purpose. It is likely that the drupes of *Scirpodendron, Mapania* and *Cladium* as well as the fleshy diaspores of *Carex baccans* are also eaten. All species of *Lepidosperma* and some of *Carex* have elaiosomes, the former on the nutlets, the latter on the utricle, and are ant-dispersed.

*Chromosome Numbers.* The chromosome numbers in many genera are variable and often large. In *Carex* they range from $2n=16$–$112$ and are variable in some species, e.g. in *C. lanceolata* $2n=26$–$80$ (FEDOROV 1969). No numbers are available for Mapanioideae and only one, $2n=26$, for Sclerioideae (*Scleria* sp.). The centromere, at least in several Cyperaceae, is not localized, being instead "diffuse" as is characteristic of *Luzula* (Juncaceae).

*Distribution.* The family is cosmopolitan, growing in most habitats but less often associated with well-drained than with poorly drained sites, such as swamps, bogs, riversides, tidal flats and pools. Some species of *Scirpus* are submerged aquatics – *S. confervoides* always and *S. fluitans* often.

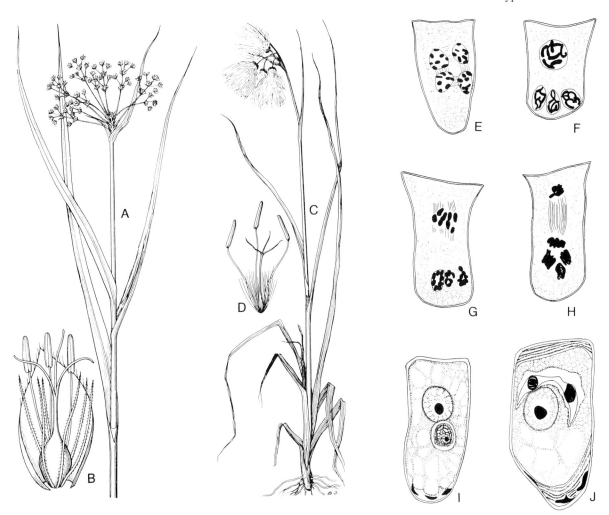

**Fig. 187.** Cyperaceae. **A–B** *Scirpus sylvaticus*. **A** Habit. **B** Flower and its subtending bract. **C–D** *Eriophorum angustifolium*. **C** Habit. **D** Flower. (**A–D** Larsen 1973a) **E–J** Development of pollen tetrad. Three of the tetrad nuclei become peripherally situated (**F**) and gradually degenerate and are finally incorporated in the wall of the pollen grain (**J**), while the fourth microspore forms a viable pollen grain. The first division (**G–H**) results in a vegetative and a generative cell (**I**), and the generative cell divides into two sperm cells before the dispersal of the pollen grain (or tetrad). (Maheshwari 1950, after Piech 1928)

*Morphological Discussion.* It has been assumed above that the Mapanieae possess pseudanthia. The flowers are here very reduced, the males comprising a single anther and the females a solitary pistil. Possibly because of the simplicity of the flowers, earlier workers interpreted the pseudanthium as a flower and expressed surprise that it should, as with *Scirpodendron* (Fig. 13), have scales subtending the anthers. The relationship between the pseudanthia of *Scirpodendron* and those of *Hypolytrum* through a reduction series has long been accepted. The pseudanthium of the Mapanieae having been accepted as a "flower", it was to be expected that attempts would be made to equate such "flowers" with those occurring elsewhere within the family. In this regard *Lipocarpha* was to play an important role.

As can be seen from Fig. 185 the reduced pseudanthia of *Hypolytrum* bear a superficial resemblance to the flowers of *Lipocarpha* (Fig. 183G). The resemblance is strengthened if it is assumed that the two keeled proximal bracts of *Hypolytrum* correspond to the prophyll in *Lipocarpha*. Such a two-bract origin for prophylls was proposed by Goebel (1898–1901) and has been widely accepted, but, as pointed out by Tomlinson (1970), it is not necessary that all prophylls arise in this way. Rejection of the two-bract origin of the prophyll removes much of the strength of the argument for deriving the flowers of *Lipocarpha* from *Hypolytrum* Type ancestors.

However, similar viewpoints have been advocated in recent times by Kern (1974) who, accepting that the Mapanieae have a pseudanthium, equated it with a *Lipocarpha*- and hence a *Cyperus*-type flower (Fig. 183). According to such an interpretation the *Cyperus* flower, together with its bract, becomes a pseudanthium. In reaching this conclusion Kern was perhaps overinfluenced by his belief that the tropical Mapanieae are primitive within the family and that the other forms need to be derived from that tribe. He also felt that the lack of vestiges of the opposite sex in the unisexual flower of the Caricoideae indicated their evolution from unisexual *Mapania*-like pseudanthia rather than from bisexual flowers. In this connection it should be noted that the three small swellings around the base of the ovary in *Uncinia* (KUKKONEN 1967) may represent aborted stamens.

KERN's hypothesis adds unnecessary complications to the situation and should be reconsidered, in particular the very dubious assumption that the prophyll of *Lipocarpha* and the proximal spikelet bracts of *Hypolytrum* are homologous.

That reduction within the *Mapania*-type spikelet has given rise to perianth-like structures is, however, evident. Thus the sterile hypogynous scales in *Lepironia* are sometimes whorled rather than spirally inserted and in some *Mapania* species the members of the whorls may be fused.

*Classification.* The reduced flowers and the highly condensed inflorescences of the Cyperaceae render their classification difficult on macromorphology alone. Considerable attention has been accorded the spikelet, but the difficulties of dissection and the lack of suitable magnifications, especially in the past, have often led to conflicting interpretations. Further anatomical, cytological and embryological data are likely to result in an emending of the classification, which is at present dominated by the consideration of spikelet structure. Five subfamilies are recognized here.

**Subfamily Scirpoideae**

The Scirpoideae (24: ca. 2,200) are herbaceous plants lacking a terminal flower. Most bracts of the spikelets subtend flowers, and these are mostly bisexual, with or without perianth, with one to three anthers, and a pistil with two to three stigmas. The silica bodies of the epidermal cells are almost exclusively conical, and the chlorenchyma is predominantly radiate. The subfamily is widespread from the tropics to polar regions.

In a group of genera centred around *Scirpus* (Scirpeae) the spikelets lack prophylls, and the bracts are usually spirally inserted.

*Scirpus* (incl. *Isolepis, Bulboschoenus, Schoenoplectus, Trichophorum* and *Holoschoenus;* ca. 200) is a probably heterogeneous, subcosmopolitan assemblage of species (cf. embryology above). It comprises a main group of Scirpoideae remaining after the more or less homogeneous genera have been delineated (KERN 1974). *S. juncoides* is a weed of rice paddies. – *Eleocharis* (ca. 150), also subcos-

mopolitan, has culms with a single terminal spikelet; its perianth bristles bear retrorse barbs. *E. palustris* is common in European fresh waters and *E. tuberosa* ("Chinese Water Chestnut") is widely cultivated as a vegetable in South-Eastern Asia and Malaysia. – *Fimbristylis* (300), mainly in tropical and subtropical parts of Malaysia and Australia, lacks a perianth and the style base is deciduous. *F. littoralis* is a weed of rice paddies. – The related *Bulbostylis* (100), common in tropical regions, is similar but retains the style base. Though it is sometimes included in *Scirpus* the embryo structure is different and argues for generic separation. – In *Ascolepis* (15) of America and Africa, each flower is enclosed in a utricle formed by its subtending bract. The inflorescence of *A. protea* resembles that of the Asteraceae. – *Eriophorum* (21) (Fig. 187 C–D) is widespread mainly in the colder areas of the Northern Hemisphere. All are "Cotton Grasses" with the perianth bristles elongating in fruit. – *Crosslandia* (1) from South-West Australia has female flowers at the base of the plant and male spikelets in heads (cf. *Alexgeorgia* of the Restionaceae). – *Fuirena* (40), in the tropics and subtropics, has glumes hairy on the abaxial surface.

In another important group with about eight genera, centred around *Cyperus,* the prophyll is distinctly two-keeled.

*Cyperus* is sometimes segregated into several genera based on whether the rachilla is persistent or not, on the direction of compression of the nutlet, on the number of style branches (two or three) and on whether or not the floral bracts are winged. Opinion is much divided as to the validity of the segregate genera though several are gaining recognition in modern literature.

*Cyperus* (incl. *Torulinum;* 380) is a widespread genus of the tropics and warm temperate regions. The culms of *C. papyrus* ("Papyrus") have long been used for making paper and its rhizomes are furthermore edible; *C. esculentus* ("Chufas") is cultivated in Africa for its edible tubers. The tubers of *C. bulbosus* and *C. usitatus* are also edible. *C. rotundus* ("Nut Grass") is a pernicious weed throughout the tropics, the tubers being the main source of infestation. Several other species, including *C. difformis, C. haspan* and *C. iria,* are weeds of rice paddies. *C. involucratus* is grown as a house plant. – *Kyllingia* (60), in the tropics and subtropics especially of Africa, includes several species which are weeds of lawns. – *Pycreus* (100) occurs in temperate and warm temperate regions, whereas *Mariscus* (200) is tropical and subtropical.

**Fig. 188.** Cyperaceae. **A–G** *Scleria barteri*. **A** Inflorescence-bearing branch. **B** Leaf margin. **C** Spikelets. **D** Fruit, with subtending glumes. **E** Cupule. **F** Nutlet. **G** Stamen. (HUTCHINSON 1933). **H–L** *Oreobolus kuekenthalii*. **H** Old, large specimen. **I** Small specimen. **J** Flower in late flowering stage, note the scale-like perianth segments and the three stamens. **K–L** Young fruit removed from perianth. (KERN 1974) For *Oreobolus* see also Fig. 14 C

## Subfamily Rhynchosporoideae

The Rhynchosporoideae (33: ca. 550) are herbs with sometimes coarse or rarely even woody stems (*Gahnia*). As in Scirpoideae the spikelets lack a terminal flower. Several of the upper and lower bracts are usually without axillary flowers. The spikelets are one- or two-flowered, one flower male or female and the other bisexual, or less commonly one male and the other female (*Dulichium*), or the single flower bisexual (*Oreobolus*). Rarely the plants are dioecious (*Caustis*). The stamens are usually one to three but sometimes numerous

(*Evandra*). The stigmas are two to eight and a perianth is present or absent. The silica bodies are mostly conical but wedge-shaped types occur in *Gymnoschoenus*. The chlorenchyma is usually irregular in its disposition, only rarely forming a palisade (*Dichromena, Gymnoschoenus* and *Rhynchospora* spp.).

Members of this subfamily are widespread from the tropics to the polar regions.

*Oreobolus* (10) (Fig. 188 H–L) grows in swampy or rocky places in the cooler areas of South America, Southern Australia, New Zealand and the mountain tops of Malaysia, Northern South America, Hawaii and Tahiti. It forms a separate group in the subfamily having solitary bisexual flowers and a bract-like perianth in two trimerous whorls.

*Dulichium* (1) *arundinaceum,* though now restricted to boreal North America, is known in fossil form in Europe (SCHULTZE-MOTEL 1966). In this and one more genus the lower flowers of the spikelet are male and the upper female. The perianth consists of bristles and the rachilla extension, when present, is small and easily overlooked (SCHULTZE-MOTEL 1959).

One group of about 30 genera, including *Rhynchospora,* is characterized by having hermaphrodite flowers in the basal part of the spikelet and male in the upper part. Here belong, for example, *Rhynchospora* (200), principally from the New World but sparingly represented in the Old World (absent from the Arctic) and *Schoenus* (100), mainly in Australia and South-East Asia but with a few species in Europe (*S. nigricans, S. ferrugineus*) and South America.

In another group of genera which includes *Cladium* and *Gahnia* the lower flowers of the spikelets are male and the upper hermaphrodite. – *Lepidosperma* (50) is a mainly Australian genus readily recognized by its isobilateral distichous leaves and short fleshy perianth members. – *Gahnia* (30) occurs mainly in Australia, but is widespread in Malaysia, Southern China, New Zealand and throughout Oceania to Hawaii. It has brightly coloured nutlets which are suspended by the persistent anther filaments. – *Tetraria* (35), mostly in extratropical South Africa, has a few species in Australia and one in Borneo. – *Maecherina* (45) is widespread throughout the tropics and subtropics. It is distinguished from the subcosmopolitan *Cladium* (2) by its two-ranked leaves.

## Subfamily Mapanioideae

Subfamily Mapanioideae (14:200) consists mostly of perennial herbs, some coarse and *Pandanus*- or *Dracaena*-like. The flowers are unisexual and lack a perianth; the female consists of a solitary pistil terminating an axis, the male of a solitary stamen in the axil of a bract. The male flowers are borne proximally on the same axis as the female, thereby constituting pseudanthia. The pseudanthia are arranged in short spikes (pseudospikelets). The two lowermost bracts of the pseudanthium are opposite, strongly keeled and except in *Displasia* bear stamens. A few of the distal bracts are sometimes sterile. In older literature the bracts are called hypogynous squamulae. The pollen grains usually lack lateral lacunae. The fruit sometimes forms a drupe with a fleshy or corky pericarp.

With respect to their leaf anatomy some members of the subfamily exhibit characters rare or absent elswhere in the family. Thus in a few genera the stomata are tetracytic and are randomly orientated. Wedge- and bridge-shaped silica bodies occur in some genera. The chlorophyll parenchyma is generally arranged as a palisade and is only rarely weakly radiate.

The Mapanioideae occur mainly in the wet tropics but extend to Tasmania in the south (*Chorizandra*).

In some genera the male flowers are subtended by scales and the leaf blades are dorsiventral. They include *Mapania* (50), a pantropical genus with three stigmas (Fig. 185), *Hypolytrum* (80), also pantropical, with two stigmas; and *Scirpodendron* (1) *ghaeri* (Figs. 13 and 185), a coarse *Pandanus*-like plant growing in coastal swamps from Ceylon to Polynesia; the fruit is eaten in Samoa. Three genera in the Southern Hemisphere of the Old World are characterized by male flowers subtended by scales and unifacial leaves or no leaves at all. – *Lepironia* (1) *articulata,* with a rush-like habit, grows in freshwater swamps throughout the Old World Tropics, including Madagascar, but is absent from Africa. It is used as a source of matting and is sometimes cultivated for this purpose in Malaysia (KERN 1974). – *Chrysithrix* (65) occurs mainly in South Africa with a pair of species in South-Western Australia. Its inflorescence is a single pseudanthium borne on a leafless scape.

Two more genera with male flowers subtended by antrorsely barbed bristles may belong to the Mapanioideae, viz. *Syntrinema* (1) *braziliense,* in Brazil, with solitary stigma, and *Micropapyrus* (1) *viviparoides,* also in Brazil, with two stigmas. From her analyses of the inflorescences of these species EITEN (1976 b) concluded that each is made up of pseudanthia and hence they belong to Mapanioideae. Bristles subtend all flowers of the former genus but only the female flowers of the latter, where male flowers lack bristles and scales. The plants of both genera are small grass-like herbs up to 25 cm tall, unlike the remaining members of the subfamily. Until embryological, anatomical and pollen characteristics are available the affinities of these genera must be somewhat conjectural.

## Subfamily Sclerioideae

Subfamily Sclerioideae (70:364) also consists mostly of perennial herbs, sometimes coarse, or rarely lianas (*Scleria boivinii*). The leaves are sometimes provided with a contraligule (Fig. 188 A). The spikelets are unisexual, the terminal comprising a solitary female flower, with or without a minute prolongation of the rachilla, the laterals male with a prolonged rachilla (Fig. 185). Tetracytic stomata are rare (in, for example *Lagenocarpus*).

In the *Scleria* Group (Sclerieae) the silica bodies are either conical or assume other forms; they are often deposited against the anticlinal walls of the epidermis in such a way as to form pairs with those in adjacent cells. The chlorenchyma is usually arranged as a palisade, only rarely being weakly radiate. The members of this subfamily are mostly savannah and forest species widespread in tropical and warm temperate regions, extending northwards to Japan. – *Scleria* (200) is a widespread tropical and subtropical genus (Fig. 188 A–G). Its female spikelets have a slightly prolonged rachilla and their flowers are subtended by a trilobed hypogynous disc, swelling or cupule (perigynium) that remains free of the pericarp and is shed with the fruit. The fruit is globular and sculptured and the embryo is large and well developed (MEERT and GOETGHEBEUR 1979). In *S. sumatrensis* the perigynium almost envelops the fruit.

In eight genera, including *Bisboeckelera* and *Calyptrocarya*, the rachilla of the terminal spikelet is not prolonged and the female flowers are subtended by a disc or enclosed in a utricle formed of two fused bracts (*Bisboeckelera*). The embryo is small and poorly developed. These genera, *Calyptrocarya* (5) and *Bisboeckelera* (8), are mainly South American.

In another group of ten genera the female flowers are subtended by neither hypogynous discs nor a utricle, but instead hairs, scales or bristles are present. The fruit is unusual in that the mesocarp apparently degenerates so that at maturity it appears to be invested in a utricle formed by the exocarp. These genera occur mainly in grassland communities in the tropics of Africa and South America.

*Lagenocarpus* (75) is widespread in both Africa and South America. – *Microdracoides* (1) *squamosus,* on isolated mountains of Western Africa, has the columnar habit of *Xanthorrhoea* (see p. 157), the trunk being up to about 1,5 m tall.

## Subfamily Caricoideae

Subfamily Caricoideae (5:ca. 2,000) consists of perennial or annual herbs with unisexual flowers borne in the same or different spikelets. If female and male flowers occur in the same spikelet the female flower is solitary and proximal in the spikelet and the male flower(s) distal. The stamens are three and the stigmas two or three in number. In some genera female spikelets have an extended sterile rachilla. The female spikelets have a well-developed prophyll, which is open and spathe-like (*Kobresia, Elyna,* Fig. 189 I) or closed to form a utricle which encloses and is shed with the fruit (*Carex, Uncinia*) (Fig. 184). In *Uncinia* the rachilla projects beyond the utricle and a hook is formed at the tip by a bract bending over the apex (KUKKONEN 1967; Fig. 184, 189). The male spikelets lack a prophyll. The leaves are usually ligulate, and their chlorenchyma never forms a palisade.

*Carex* (ca. 2,000) is a subcosmopolitan, very variable genus. Its inflorescence varies from fairly simple to prolifically branched and compound. The female spikelets have a closed utricle and the rachilla is generally short and wholly enclosed in the utricle (rarely protruding at its apex as in *C. microglochin,* Fig. 184). The female spikelets (utricles) are often situated in female spikes of higher order(s), whereas the male flowers are situated in male spikes in the distal part of the compound inflorescence (*Heterostachyae*). In large groups of the genus the male and female spikelets are found together, with the male flowers in the proximal or the distal parts of the spikes (2nd order) or otherwise (*Homostachyae,* Fig. 189 A). The carpels and consequently the stigmatic branches may be three (*Tristigmaticae*) or two (*Distigmaticae*), the nutlets, and often the utricles as well, being triangular or flattened, respectively. In certain tropical species (subgen. *Indocarex*) the branching is more complicated. The utricle in exceptional cases (e. g.

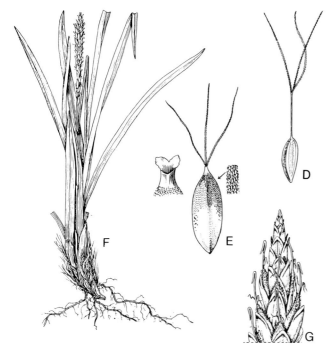

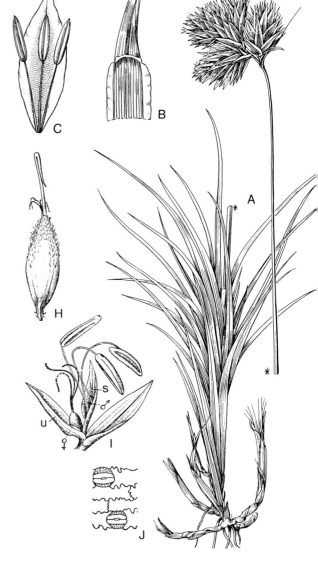

**Fig. 189.** Cyperaceae. **A–F** *Carex praeclara.* **A** Shoot. **B** Leaf base. **C** Male flower. **D** Pistil in utricle. **E** Fruit in utricle, to the *left* opening of utricle, to the *right* surface structure of utricle. (NELMES 1940). **F–H** *Uncinia brevicaulis.* **F** Shoot. **G** Part of female spike with hooked rachillae. **H** Utricle with fruit and rachilla. (CORREA 1969). **I** *Elyna bernardii,* spikelet, showing utricle homologue (*u*) and its axillary female flower and a male flower in the axil of a bract (*s*). (WETTSTEIN 1924). **J** *Carex* sp., stomata. (METCALFE 1971)

*C. baccans*) is fleshy, but normally the nutlet is dispersed within a membranous utricle, which often makes the diaspore bouyant and permits effective water dispersal. The genus is an extraordinarily important component in the vegetation of marshes and fens but is also frequent in woodland, grassland, heath and sea shore vegetation in the Northern and part of the Southern Hemisphere (e.g. South America, Australia and New Zealand). Its importance is partly taken over by Restionaceae in South Africa and Southern Australia. – *Elyna* (6) and *Kobresia* (50) are widespread in the Northern Hemisphere, *Kobresia* being most abundant and diverse in Central Asia and the Himalayas. – *Schoenoxiphium* (15) may be congeneric with *Kobresia* but is essentially African with one species in Sumatra.

*Uncinia* (30) is widespread in subantarctic regions and the mountains of the West Pacific Islands, throughout Eastern Australia northwards to the Philippines and in the New World extending to Mexico and the West Indies; it is absent from Africa.

*Cymophyllus* (1) *fraseri* of the Eastern United States has broad leaves and a single spike. Often included in *Carex* it differs in having broad leaves not differentiated into a sheath and lamina. Its white utricles and anthers contrast markedly with the green of its large tongue-shaped leaves.

# Order Poales

*Seven Families:* Flagellariaceae, Joinvilleaceae, Poaceae, Ecdeiocoleaceae, Anarthriaceae, Restionaceae and Centrolepidaceae.

Annual or perennial, small to large herbs or arborescent plants up to more than 30 m tall. Most members comprise grasses and grass-like plants. They may be minute to small or moderate-sized annuals without rhizomes, or rhizomatous, often very densely tufted plants, mat-like, spreading plants, slender vines (e.g. *Flagellaria*), floating semi-aquatics, or large, woody, mostly slender bamboo trees, and thus are greatly variable but with a rather distinctive combination of characters.

The roots are formed opposite both xylem and phloem poles, and are often fibrous, rarely thick. Sometimes they have endomycorrhiza. The root hairs generally but not constantly arise from epidermal cells much shorter than other root epidermal cells. Vessels are always present in the roots and generally, except in Centrolepidaceae, have simple perforation plates.

Rhizomes are found in the majority of the perennial species and may be vertical or ascending, sometimes quite short, or they may be horizontal, short or long, and they often provide effective vegetative reproduction. The aerial stems are procumbent, ascending or erect, with solid or especially in Poaceae frequently hollow internodes and often with prominent nodes. The stems are terete or flattened, generally branched primarily at the base, usually leafy but sometimes with few or no cauline leaves.

The leaves with few exceptions are distichously (sometimes spirodistichously) inserted. Frequently they are concentrated at the base and then may be densely tufted, with sometimes fibrous or swollen leaf bases. The leaves are differentiated into a leaf sheath and a lamina which is generally continuous with the sheath, but sometimes narrows basally into a pseudopetiole, which in some bamboos may reach a considerable length. The sheath is open or closed for some or all of its length. At its upper extremity there may be lateral, membranous, stipule-like lobes, but more often, as in nearly all grasses, there is an adaxial, membranous ligule of variable shape and length, rarely substituted by hairs. In a few grasses there is also, on the side of the sheath opposite the base of the leaf blade, a contraligule. The leaf blade may be flat, canaliculate, inrolled and falsely terete, truly terete, or longitudinally ribbed to plicate; its shape varies from filiform or narrowly linear to lanceolate. The margins are generally even. In some leaves the blade may be strongly reduced, or undeveloped and, as in Ecdeiocoleaceae and many Restionaceae, the blade may be short or lacking altogether on the cauline leaves, most or all of the photosynthesis being carried out by the cauline stem.

The vascular strands of the stems are scattered, but more densely situated in the peripheral part, and the vessels may have simple or scalariform perforation plates, or occur as a mixture of both types. Vessels are generally also present in the leaves. Raphides are lacking. Silica bodies are present in epidermal short cells in most members of the order, and are frequently characteristic of families or tribes. They may also, e.g. in Restionaceae, occur in mesenchymatous cells in connection with vascular strands. They are, however, lacking from all Centrolepidaceae, many Restionaceae and Anarthriaceae, while in Ecdeiocoleaceae there is only "silica sand" (i.e. numerous minute bodies). The stomata are paracytic and of the so-called Grass Type, i.e. with dumb-bell-shaped guard cells and subsidiary cells, much smaller than the ordinary epidermal cells which are arranged in files (Fig. 193P). Uniseriate hairs of various kinds are frequent in the order, though rare in Restionaceae. Most grasses possess two- to few-celled small hairs, so-called "microhairs" (Fig. 193Q), which may correspond to two to five-celled hairs sometimes present in the Restionaceae and Commelinaceae. Branched, uniseriate hairs occur in Joinvilleaceae. The epicuticular wax when structured and having a clear orientation is of the *Strelitzia* Type.

The inflorescences in Poales are complicated and often difficult to interpret, and generally represent or are derivable from panicles (Flagellariaceae, Joinvilleaceae) or spikelets in panicles. The spikelets represent small spikes, in this order normally with distichous bracts, the lower of which are generally empty (subtending glumes), and a number of glumes (lemmas), each subtending a flower (in grasses called a "floret"). In Poaceae, according to current interpretation, the floret has a two-keeled prophyll, the palea, opposite the lemma. The lemma and palea enclose most of the remaining parts of the floret. The highly variable spikelet structure in Poaceae is described below under this family. The flowers are also situated in spikelets in Restionaceae, Anarthriaceae, Ecdeiocoleaceae and Centrolepidaceae and may or may not be supplied with one (or two) prophylls (bracteoles). In Centrolepidaceae the individual flowers of the spikelets are much reduced and are unisexual, the monocarpellary pistils of the female ones being often fused with each other, whether on different levels or on the same level. Restionaceae are generally dioecious, the female inflorescence and often the whole female plant being widely different from the male.

The flowers are bisexual or unisexual; in the latter case the flowers are monoecious or dioecious. Quite often a spikelet contains combinations of bisexual with female or male flowers, and rarely there are female and male spikelets in different parts of the inflorescence (as in *Zea* and related genera).

The perianth may consist of 3 + 3 tepals, which are then normally bracteal, often chaffy, whitish, yellowish, green, brown, purple or black, similar or sometimes conspicuously unequal in size and distinct or slightly connate at the base. The tepals are glabrous or adaxially pubescent. Reductions of the tepals occur in some Restionaceae, where one or both of the whorls are reduced to one member only or wholly suppressed, and in Centrolepidaceae, where tepals are always lacking. In the Poaceae, according to most interpretations, the outer tepal whorl is lacking and the inner is generally represented by one, two or three (generally two) rather small so-called lodicles; these may be thick and fleshy, membranous or chaffy, of any shape from lanceolate-ovate to obcuneate, entire or toothed, and glabrous or ciliate; rarely there are no lodicles at all (see further under Poaceae).

The stamens are generally 3 + 3, three, two or one in each flower. They have a glabrous and often thin filament of variable length, and a frequently pendulous, tetrasporangiate (and dithecous) or bisporangiate (and monothecous) anther with introrse or latrorse longitudinal dehiscence. The anthers are dorsifixed or basifixed and when basifixed sometimes sagittate. The endothecial thickenings are generally of the Girdle Type, but there is a record of the Spiral Type in Restionaceae (*Hypolaena*). The tapetum is glandular-secretory with, as a rule, binucleate tapetal cells. Microsporogenesis is probably always successive, and the pollen grains are dispersed separately (not in tetrads). The pollen grains are ulcerate in all members, with a small and even ulcus or sometimes (as in Centrolepidaceae and some Restionaceae) a relatively large one with an uneven margin (CHANDA 1966). The pollen grains are generally dispersed in the three-celled stage but sometimes in the two-celled state.

The pistil consists of three carpels or in most grasses and Ecdeiocoleaceae probably of two carpels and in Centrolepidaceae of one carpel only, judging by the number of stylodial branches. The ovary is trilocular in Flagellariaceae, Joinvilleaceae, Anarthriaceae, and part of Restionaceae, bilocular in Ecdeiocoleaceae, and unilocular in Poaceae and Centrolepidaceae. There is generally a terminal style, which, at a variable distance from the base, is divided into as many branches as there are carpels; in at least one species of *Joinvillea* the style is lacking and the three lobes are situated on top of the ovary. The style is also usually lacking in many Poaceae, the two stylodial branches being then separate to the base. The stigma surface is papillate and Dry. In Centrolepidaceae the monocarpellary pistils have their ovaries fused to form complex ovaries and, subsequently, follicles (see below). The locules constantly contain only one ovule, which is apically inserted and orthotropous in most families, but in the grasses varies from lateral (or nearly basal) to apical and from (anatropous), hemianatropous or campylotropous to orthotropous. The ovules are tenuinucellate or weakly crassinucellate. A parietal cell is probably never cut off from the archesporial cell. However, in most grasses the nucellar epidermis divides periclinally to form a nucellar cap. Embryo sac development is generally of the *Polygonum* Type, although the *Allium* Type has been reported in Flagellariaceae. Endosperm formation is always nuclear, and embryogeny of the Asterad Type or, in Centrolepidaceae, a modification of the Onagrad Type (HAMANN 1962a). A special condition is found in the grasses where new cell walls are laid down obliquely in a peculiar fashion in the young embryo.

The fruit is a capsule, drupe, nutlet or, in most grasses, a caryopsis. Where it is a capsule the seeds are sometimes, in some Restionaceae, provided with an elaiosome and obviously ant-dispersed. The endosperm is copious and mealy, with starch, and the embryo is generally lenticular and central near the micropylar end of the endosperm or, in the grasses, laterally inserted, well outside the endosperm; it may then be short or extend for a considerable distance alongside the endosperm. The construction of the grass embryo is described under Poaceae, below.

*Chromosome Numbers.* The chromosome numbers will be given under the families. According to RAVEN (1975) a basic chromosome number of

$x = 7$ can be considered likely to have been the original.

*Chemistry.* Silica is often deposited in epidermal cells, rarely in internal cells. Calcium oxalate crystals occur rarely, as in Flagellariaceae and some Poaceae, but never as raphides. Starch, and to a great extent sugars (fructosans), are often found in rich quantities in rhizomes and culms, *Saccharum* (Poaceae), for example, being exploited for sugar. Saponins are known in some grasses and two of these records are steroidal saponins (*Avena, Sorghum*), but they are decidedly rare. Chelidonic acid, though rare in the order, is also known in a few grass genera. Cyanogenetic compounds, however, are very common in grasses and are also known at least in *Flagellaria,* but have been absent from the members of Restionaceae studied. Various alkaloids are scattered, but occasional, in Poaceae: isoquinoline, pyrrolizidine and indole alkaloids. Among the flavonoids, tricin is common in Poaceae and occurs also in some Restionaceae; C-glycoflavones are likewise common in grasses, and sulphurated flavonoids are common in this group. Luteolin/apigenin are also reported in a number of grasses. Proanthocyanidins and flavonols are poorly represented. For Restionaceae the flavonoid spectrum is slightly different, with 8-oxy-flavonols and 8-oxy-flavones. As mentioned under the superorder, the cell wall fluoresces in UV light, a response associated with the presence of bound ferulic acid, p-coumaric acid and diferulic acid (HARRIS and HARTLEY 1980).

*Parasites.* The grasses are attacked by a variety of fungal and insect parasites, and the host-specificity conditions of various rust and smut genera, and genera of insects, are potentially useful indicators of relationships among grass genera (see, for example, NANNFELDT 1968; HOLM 1969; SAVILE 1954, 1971). The remaining families are either less investigated or less susceptible, for there are relatively few records for them.

*Fossils.* Pollen grains referable to Restionaceae are reported from the late Cretaceous, in the Maestrichtian, whereas typical grass pollen is younger. As grass pollen, if present at all, would be likely to occur in considerable quantities, this evidence is important. The grasses obviously began their expansion about the beginning of Tertiary, and their cosmopolitan distribution is thus relatively late in time.

*Distribution.* The distributional patterns of Poales are therefore quite interesting. The Restionaceae today are strongly centred in South Africa and Australia, with outliers in Indochina, New Zealand and Chile. Anarthriaceae and Ecdeiocoleaceae, which are obviously closely related to Restionaceae, are restricted to South-Western Australia. The Centrolepidaceae are distributed in parts of Australia, Tasmania, New Zealand and South-Eastern Asia, including Malaysia. These groups accordingly indicate an Eastern South-Hemispheric origin, but some of them, e.g. Restionaceae, may have been more widely distributed earlier. The unigeneric Flagellariaceae and Joinvilleaceae, which show even more original features in flower and inflorescence construction, are both Old World groups, *Flagellaria* occurring from Africa through southern Asia to Australia, whereas *Joinvillea* occurs in South-East Asia, New Caledonia and Fiji. The Poaceae are almost cosmopolitan, and occur in a variety of habitats, though they are not marine aquatics and only rarely epiphytes. The present distribution of most of their subfamilies is tropical-centred.

*Relationships, Taxonomy.* Within the Poales the families are divisible into groups. Two families, Flagellariaceae and Joinvilleaceae, have a paniculate inflorescence of a less complicated structure, with a complete and in our view "original" floral organization, with $3 + 3$ tepals, $3 + 3$ stamens and a tricarpellary, trilocular pistil. They are both tropical Old World families. Joinvilleaceae shows trends in stem structure, sheath and silica bodies which approach grasses, although the flowers are not in spikelets. A probable synapomorphy for the two families is the drupaceous fruit. We consider these as an early branch from the ancestors of the Poales.

Many supposedly original features are met with in the Anarthriaceae, Ecdeiocoleaceae and Restionaceae, which might be collectively treated in one family, Restionaceae s.lat. In some members of this group the ovary is trilocular and the fruit is capsular, which we assume is the original condition, and the perianth is frequently present as $3 + 3$ bract-like tepals. The anthers in members of this group have gone through a partial reduction, and lost one theca, and the monothecous condition is also met with in Centrolepidaceae, with even more reduced, naked, unisexual flowers in advanced pseudanthial spikelets. The Centrolepidaceae are probably derivatives of primitive Restionaceae.

Poaceae approach the Joinvilleaceae and primitive Restionaceae. They generally have bisexual flowers and the anthers are tetrasporangiate and dithecous. Like the Restionaceae they have flowers arranged in spikelets. The fruit is always indehiscent, and only rarely is the seed separate from the pericarp. The grasses are unique in several features, such as their embryogeny and embryo construction.

The similar facies and floral structure of the Anarthriaceae, Ecdeiocoleaceae and Restionaceae suggest that they are a group of related families, a viewpoint supported also by their joint possession of peg-cell-type chlorenchyma cells.

Nonetheless they differ markedly in a number of anatomical features, such as culm anatomy as seen in transverse section (CUTLER and AIRY SHAW 1965; CUTLER 1969). The vascular bundles of the members of the Restionaceae are enclosed within a cylindrical band of sclerenchyma. In contrast, the vascular bundles of the Ecdeiocoleaceae are isolated from one another, each being surrounded by a separate sheath of sclerenchyma. The culm anatomy in the Anarthriaceae is different again from that of both of the other two families; as in the Ecdeiocoleaceae, each bundle is here surrounded by its own sclerenchyma sheath, but this forms a girdle extending to the culm epidermis. – The culms of *A. prolifera* differ from those of the other species in the family in having some of the bundles incorporated in a single sclerenchyma cylinder, while others occur in the cortex, each with its own sclerenchyma sheath.

The pollen grains in Restionaceae, Flagellariaceae, Joinvilleaceae and Anarthriaceae are scrobiculate; in this they differ from the pollen grains of the Poaceae and Ecdeiocoleaceae (LINDER and FERGUSON, personal communication).

Until more comparative embryological and cytological data are available opinions on the interrelationship of these families must remain speculative.

The position of Poales within the superorder is obviously separate from the Cyperales, which in our view are monophyletic, and presumably Poales are not direct derivatives of the Commelinales, the petaloid inner tepals of which may be secondary. We assume that the connection between the groups is as sketched in the evolutionary chapter (p. 103).

*Key to the Families*

1 A. Flowers not in spikelets but separate, in more or less richly branched panicles; fruits drupaceous (tepals six, stamens six and pistils trilocular) . . . . . . . . . . . 2
1 B. Flowers in spikes or spikelets, which are generally assembled in often complicated inflorescences; fruit a capsule, nutlet, achene, follicle, or, rarely, berry (the combination of six tepals, six stamens and a trilocular ovary never found) . . . . . . . . . . . . 3
2 A. Stem solid; leaf-blade cirrhose, ending in a "flagellum" functioning as tendril; leaf sheath closed . . . . . . . . . . . **Flagellariaceae**
2 B. Stem hollow; leaf-blades not cirrhose (more or less plicate); leaf-sheath open . . . . . . . **Joinvilleaceae**
3 A. Anthers tetrasporangiate, dithecous . . . . . . . . . . . . . . 4
3 B. Anthers bisporangiate, monothecous . . . . . . . . . . . . . 7
4 A. Fruit indehiscent; tepals three or fewer, generally small ("lodicles"); embryo asymmetrical at one side of the micropylar end, outside the endosperm . . . . **Poaceae**
4 B. Fruit dehiscent; tepals generally 3 + 3, bract-like, chaffy; embryo lenticular, centrally placed at the micropylar end of the endosperm . . . . . . . . . . 5
5 A. Stem lacking a common sclerenchymatous sheath . . . . . . . . . . . 6
5 B. Stem with a continuous cylinder of sclerenchyma (*Lyginia* and *Hopkinsia*) . . . . . . . **Restionaceae**
6 A. Pistil bilocular, with two stylodial branches . . . . . . . **Ecdeiocoleaceae**
6 B. Pistil trilocular, with three stylodial branches . . . . . . . **Anarthriaceae**
7 A. Wiry, jointed herbs; flowers generally with some tepals; stamens in male flowers usually three in number; pistil with three stylodial branches . . . . . **Restionaceae**
7 B. Small, tufted plants; flowers naked; male flowers consisting of but one stamen; pistil monocarpellary with one stylodial branch, but two or more pistils frequently fused . . . . . . . **Centrolepidaceae**

**Flagellariaceae** Dumortier (1829)   1:4
(Fig. 190)

Lianas with narrow, solid, cane-like stems arising from a diffuse sympodial rhizome, and climbing by means of involute tendrils which terminate the circinnate leaves. Branching of the apical part of

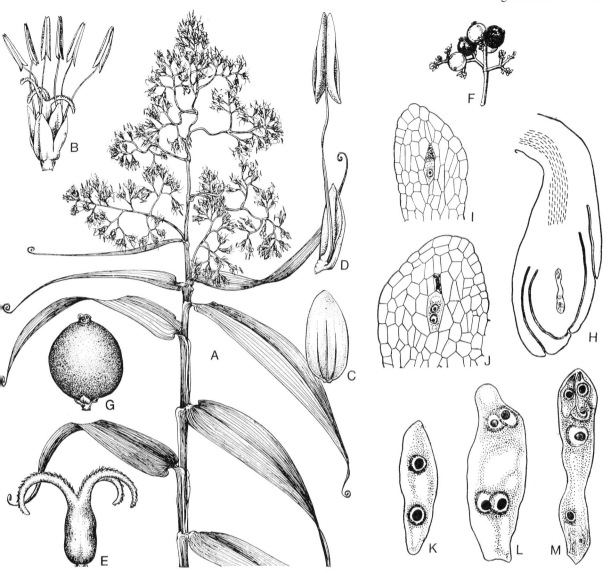

the stem is frequently by equal dichotomy (as with *Nipa,* Arecaceae), and axillary buds are lacking. The leaves are distichous. The leaf sheaths are closed and the lamina and sheath are connected by a short pseudopetiole. The lamina may be up to about 30 cm long (including tendril) and about 2 cm broad.

The epidermal cells have irregular walls and the stomata are of the grass type. Globose silica bodies occur in fibrous cells above and below the vascular bundles but never in the leaf epidermis. Secretory canals occur in the mesophyll. Vessels with simple and scalariform perforation plates occur in the stems and leaves (the roots are unknown anatomically).

The flowers are bracteate, almost sessile and borne separately (not in spikelets) in a terminal panicu-

**Fig. 190.** Flagellariaceae. **A–E** and **G** *Flagellaria guineensis*. **A** Branch with inflorescence. **B** Flower. **C** Tepal. **D** Tepal and opposite stamen. **E** Pistil. **F** Panicle. **G** Fruit. (NAPPER 1971). **F** and **H–M** *Flagellaria indica*. **F** Part of fruiting inflorescence. **H** Ovule. **I–J** Formation of tetrad; two tetrad nuclei are here considered to take part in the embryo sac, which thus would be of the *Allium* Type. **K–L** Further development into embryo sac. (**F** LI 1978; **J–M** SUBRAMANYAM and NARAYANA 1972)

late inflorescence. They are bisexual (rarely unisexual) and trimerous with two whorls of equal or subequal, half-petaloid, whitish tepals. The stamens are in two whorls and some may be reduced to staminodes. The anthers are tetrasporangiate and sagittate-basifixed with latrorse dehiscence. The pollen grains are two-celled at maturity. They

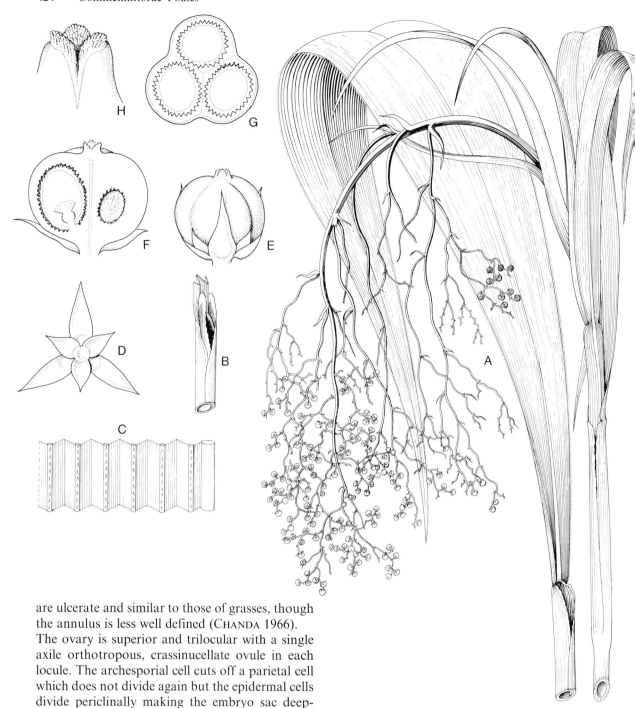

are ulcerate and similar to those of grasses, though the annulus is less well defined (CHANDA 1966).

The ovary is superior and trilocular with a single axile orthotropous, crassinucellate ovule in each locule. The archesporial cell cuts off a parietal cell which does not divide again but the epidermal cells divide periclinally making the embryo sac deepseated. Embryo sac development is of the *Allium* Type (SUBRAMANYAM and NARAYANA 1972). The style is short and divided into three stigmatic branches with plumose upper surfaces.

The fruit is a red or black drupe up to 10 mm across with one, rarely two seeds, each containing a minute embryo embedded in a copious starchy endosperm. $x = 19$.

*Chemistry.* Cyanogenesis is known in *Flagellaria.*

**Fig. 191.** Joinvilleaceae. **A, C–H** *Joinvillea plicata.* **A** leaf and branch with infructescence. **C** Section of leaf to show its plicate nature. **D** Perianth from above with larger outer and smaller inner tepals. **E** Fruit. **F** Same in longitudinal section, the section cutting through one seed with its basal embryo and, peripherally, through another. **G** Fruit, transverse section. **H** Stigmatic part of fruit. **B** *Joinvillea ascendens,* upper part of leaf sheath showing hyaline stipules. (Redrawn from GAUDICHAUD-BEAUPRÉ 1841)

*Distribution.* The genus is widely distributed in tropical Africa, Asia, Australia and the Pacific Islands.

*Flagellaria* (4) is the only genus. It occurs in humid but not constantly swampy forests and in rain-forest margins. Used for basket work and roof frames to which thatch is attached.

## Joinvilleaceae D.F. Cutler & Airy Shaw (1965)
1:2   (Fig. 191)

Reed-like, erect plants up to 5 m tall with un-branched, slender, erect, hollow stems arising from a congested sympodial rhizome. The leaves are distichous with an open sheathing base broadening abruptly into a plicate, grass-like lamina up to 1 m long and 15 cm broad with a rough surface bearing short prickle hairs and branched filamentous hairs. The lamina is not cirrhose. Membranous stipule-like lobes, and sometimes a ligule, are present at the top of the leaf sheath. The stomata are paracytic and grass-like. Silica is abundant throughout the plant including the epidermis, where it is deposited in the cell walls. Silica bodies are common in cells outside the sheaths of the vascular strands. Vessels with simple or scalariform perforation plates occur in roots, stems and leaves.

The inflorescence is a much-branched terminal panicle. The flowers are hermaphrodite and bracteate and have caducous bracteoles. The perianth is actinomorphic and consists of two trimerous whorls of bract-like tepals, of which the outer are much larger than the inner and 3–4 mm long. The stamens are 3 + 3 and have tetrasporangiate, basifixed, sagittate, latrorse anthers. The pollen grains are ulcerate with an annulus of the same kind as in grasses (CHANDA 1966).

The ovary is trilocular, each locule with a single, apical-axile, orthotropous ovule. There are three short stylodial branches, sometimes slightly connate into a style at the base, or three more or less sessile papillate stigmatic areas (Fig. 191 H). The embryology is unknown.

The fruit is a red, yellow or black drupe about 5 mm across. It contains one to three seeds, each with a minute discoid embryo in a copious starchy endosperm. $x = 18$.

The only genus, *Joinvillea* (2), is widely distributed from sea level to about 1,500 m on mountains in Malaysia and throughout the Pacific, in secondary forests, ridge forests and roadside thickets (NEWELL 1969). In Fiji the roots are said to be used as a poultice on wounds.

## Poaceae Barnhart (1895) (Gramineae)
750:10,000   (Figs. 192–210)

Annual or perennial, often tufted, erect to procumbent, often rhizomatous or stoloniferous, sometimes scandent or even tree-like plants exhibiting a great array of life forms. The annuals vary from minute, only a few centimetres tall (*Mibora minima*), to tall and leafy, up to 5 m (*Sorghum membranaceum*), and the tufted perennials show a similar variation in size. The rhizomatous perennials may produce unbranched herbaceous shoots (the inflorescence is branched), as in *Cynodon dactylon*, or all intergrades between these and giant woody, branching shoots that may attain heights up to more than 30 m, as in the bamboo *Dendrocalamus brandisii*.

Vessels occur in roots, stems and leaves and have simple or, some of them more rarely, scalariform perforation plates.

The aerial stems are known as culms, and generally terminate in an inflorescence. The culms usually die back after flowering and in perennials buds usually develop from the basal nodes or from the rhizomes. Some of the woody perennial bamboos are monocarpic and flower profusely in certain years, after which the plants die.

The root systems are fibrous and rarely form tubers (*Puelia*); roots may develop into thorns, such as those arising from the aerial nodes of several bamboos. The root hairs may arise from cells much shorter than the other epidermal cells, with which they alternate, as in the Pooideae, or from cells of the same size and shape as other epidermal cells. The shape of the endodermal cells may also be taxonomically significant; the walls in the *Arundo* Group are so thickened that the lumen becomes U-shaped in transverse section, whereas in the *Bambusa* Group it is O-shaped.

The culm nodes, at least when young, are brittle and supported by the successive leaf sheaths. The outline of the culm in transverse section is usually circular or elliptic, rarely square (as in *Chimonobambusa quadrangularis*). The internodes are either solid, as in *Zea,* or, more often, fistulose, as in *Triticum*. Secondary growth is lacking even in the largest bamboos. Intercalary meristems occur at the bases of each internode, but are concerned only with shoot elongation and may become active in, for example, raising fallen culms. The basal internodes of the culms in some species become swollen and filled with food reserves, forming corms or tubers (species of *Arrhenatherum, Phleum* and *Poa*) and acting as perennating organs.

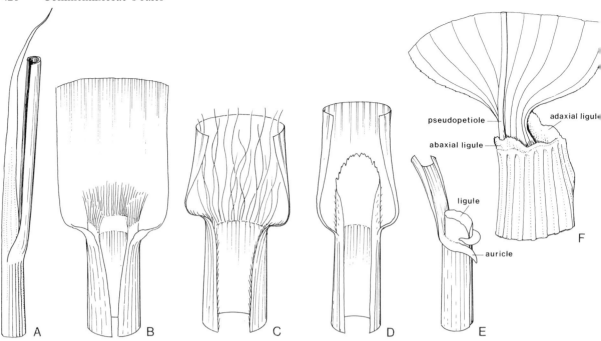

**Fig. 192.** Poaceae. Ligule forms. **A** Ligule membranous, long (*Muehlenbergia montana*). **B** Ligule membranous, with a ciliate margin (*Panicum dichotomiflorum*). **C** Ligule a rim of cilia (*Panicum virgatum*). **D** Ligule membranous, long (*Oryzopsis miliacea*). **E** Ligule membranous, short, auricles well developed (*Lolium multiflorum*). **F** Pair of ligules at the base of a pseudopetiole (*Swallenochloa subtesselata*). (A–E redrawn from Cronquist et al. 1977; F from McClure 1973)

The manner of branching is described as intravaginal if the young shoot grows up within the sheath, and as extravaginal if it bursts through the base of the sheath. In many bamboos the flowering culms differ from the vegetative in that their leaves consist of the sheath only.

The leaves are always two-ranked, except in *Micraira,* where they are three-ranked. They generally have a long, sheathing base and a more or less well-developed lamina. In many perennial species the lamina at maturity disarticulates from its sheath which remains attached to the stem. In the woody species the sheaths, also, may be ultimately shed but in the herbaceous species the leaf (if not disarticulated), or the whole shoot, decays as a single unit.

The margins of the sheaths may be quite free from one another or fused from the base upwards for varying distances. In many woody species the basal sheaths are armed with short irritating hairs. At its junction with the sheath the blade usually develops a ligule which arises from its adaxial surface. The ligule may be a membrane, a membrane tipped with hairs, or a rim of hairs (Fig. 192A–E). Its form is often useful for identifying species, particularly as it is left attached to the sheath if the lamina disarticulates. Many Bambusoideae, some Arundinoideae and a few Stipoideae have in addition an abaxial ligule (Fig. 192F). The pair of ligules is then reminiscent of the double hastula developed on the pseudopetiole in some fan palms, though it should be noted that the ligules are close to the sheath in Poaceae and close to the lamina in Arecaceae.

The lamina is smooth or corrugated on one or both surfaces and is longer than broad, and in most non-tropical grasses parallel-sided, tapering distally but not basally, at its junction with the sheath. A few grasses have broad, ovate or lanceolate leaf laminae resembling the leaves of Commelinaceae or the laminae in Zingiberaceae. In many tropical species (in particular of the Bambusoideae) the laminae taper gradually towards the junction with the sheath and are often stalked (pseudopetiolate). The pseudopetioles are usually short, as in most Bambusoideae (Fig. 192F), but may be up to 10 cm, as in *Panicum sagittifolium.* Pseudopetioles may be associated with pulvini which permit sleep movements (*Strephium*) or generate resupinate laminae (*Phaenosperma*).

Rarely with pseudopetiolate leaves the base of the lamina is strongly developed, producing a sagittate leaf. In the absence of a pseudo-petiole the base of the lamina may be produced into two claw-like

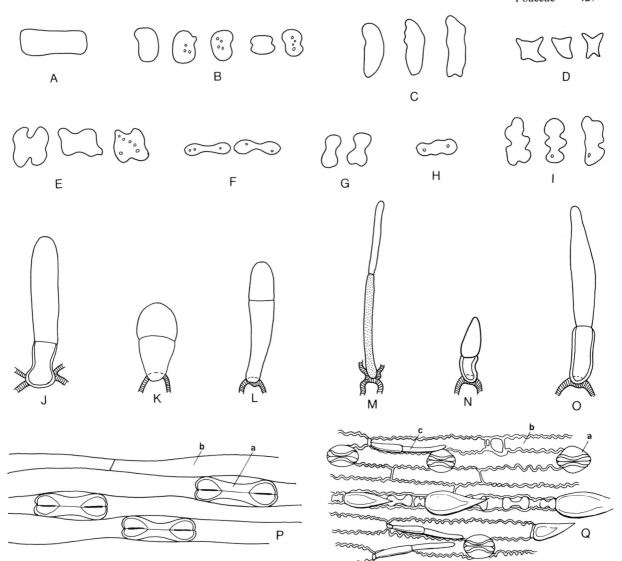

lobes, referred to as auricles (Fig. 192E). Occasionally the whole lamina is rigid and pungent, being terete with an adaxial groove. The ptyxis is convolute or plicate.

Prophylls, which are strongly two-keeled and have free margins, are always present as the first leaf of a lateral branch and arise so that they are adjacent to the parent axis. They lack a lamina and are often hidden in the sheath of the leaf subtending the branch on which they are borne.

Most grasses have short epidermal hairs which may be filled with opaline silica making the leaf harsh to the touch. Long and soft hairs also occur. Stickiness due to glandular hairs occurs in a number of species. In addition extra-floral nectaries occur rarely, principally along the leaf margin, as in some species of *Eragrostis*.

In the subfamilies other than Pooideae there occur minute thin-walled hairs, "microhairs", which are

**Fig. 193.** Poaceae. **A–I** Silica bodies in grass epidermis, the longitudinal axis of the leaf arranged to lie across the page. **A** Rectangular (*Ammophila arenaria*). **B** Saddle-shaped (*Chloris robusta*). **C** Large-narrow (*Bromus fibrosus*). **D** Acutely angled (*Heteranthoecia guiniensis*). **E** Cross-shaped (*Euchlaena perennis*). **F** Dumb-bell-shaped (*Aristida longiflora*). **G** Transversely dumb-bell-shaped (*Leersia oryzoides*). **H** Crenate (*Brachiaria distichophylla*). **I** Transversely crenate (*Olyra longifolia*). (METCALFE 1960). **J–O** Microhairs, **J–L** with inflated, **M–O** with non-inflated distal cell. **J** *Euchlaena mertonensis*. **K** *Sporobolus wrightianus*. **L** *Eragrostis chloromelos*. **M** *Loudetia* sp. **N** *Zizania latifolia*. **O** *Panicum miliaceum*. (METCALFE 1960). **P–Q** Subsidiary cells and long cells. **P** *Festuca distachya* (*a* subsidiary cell, rectangular or dome-shaped; *b* long-cell, straight-sided). *Aristida hordeacea* (*a* subsidiary cell triangular; *b* long-cells with sinuous walls; *c* microhair, distal cell not inflated). (JACQUES-FÉLIX 1962)

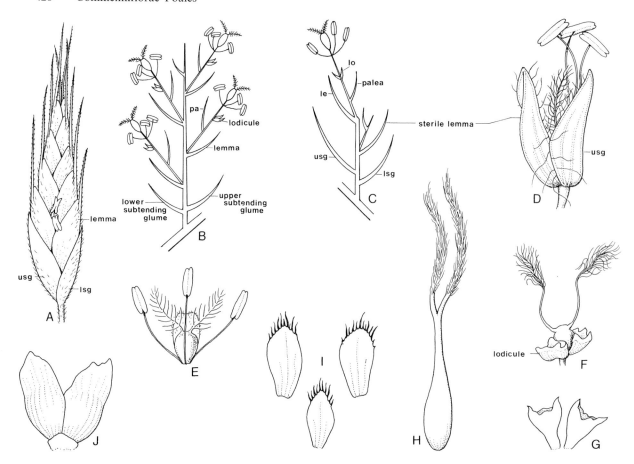

**Fig. 194.** Poaceae. Spikelet and flower structures. **A** *Bromus mollis,* complete spikelet (*lsg* lower subtending glume; *usg* upper subtending glume). **B** Indeterminate spikelet, generalized (*pa* palea). **C** Determinate spikelet, generalized (*lsg* lower subtending glume; *usg* upper subtending glume; *le* lemma; *lo* lodicule). **D** *Paspalum dilatatum,* upper subtending glume (*usg*) and sterile lemma enclosing floret; lower subtending glume not developed. **E** *Bromus mollis,* flower. **F** *Paspalum dilatatum,* flower in late anthesis, stamens shed. **G** *Panicum hallii,* lodicules. **H** *Hierochloë redolens,* pistil. **I** *Swallenochloa subtesselata,* (a bamboo) lodicles. **J** *Bromus mollis,* lodicules. (**A, D, F, H** and **J** redrawn from CLIFFORD and WATSON 1977; **B** POTZTAL 1964; **C** Orig.; **E** HUBBARD 1954; **G** CRONQUIST et al. 1977; **I** McCLURE 1973)

generally two-celled but in some Bambusoideae may have three or more cells. In *Enneapogon* and *Spartina* they appear to function as glands. The apical cell may or may not be inflated and may collapse into a thread-like structure (Fig. 193 J–O). Similar hairs are met with in Commelinaceae.

Features of the leaf transection are described separately below.

The leaf epidermis consists of files of long and short cells with their long axis parallel to the length of the leaf. The cells may have straight or sinuous walls (Fig. 193 P and Q). The stomata are paracytic, of the "Grass Type", with dumb-bell-shaped guard cells (Fig. 193 P) and subsidiary cells which are triangular or dome-shaped or parallel-sided (Fig. 193 P–Q). The short cells, in particular those above the vascular bundles, may be filled with a silica body whose shape is reasonably constant for each tribe (Fig. 193 A–J); thus the *Chloris* Group have mainly saddle-shaped and the *Panicum* Group largely cross-shaped silica bodies.

Basically the grass inflorescence is a panicle of spikelets or pseudo-spikelets (see below). Its form is determined by the number and relative length of the inflorescence branches and their degree of branching. In general, grass inflorescences are described in terms applicable to other flowering plants by regarding the spikelet as if it were a flower. The inflorescences are mostly panicles but are sometimes reduced to spikes or racemes. Rarely the spikelets are embedded in the inflorescence axis. With certain genera (e.g. *Setaria*), some of the inflorescence branches fail to produce spikelets, forming instead bristles or some form of involucre to protect or assist in dispersing the fruit.

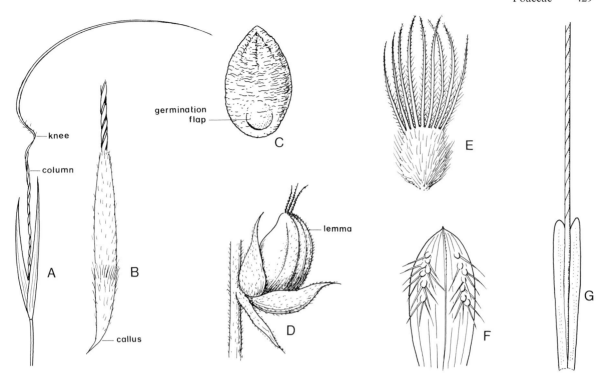

The basic unit of the inflorescence is the *spikelet,* which comprises one or more flowers and their attendant bracts (Fig. 194 A–E). Each floral axis arises in the axil of a bract (*lemma*) and usually bears a two-keeled prophyll (the *palea*), two or three lodicles, stamens and a pistil. The lemma and flower with its palea are known collectively as a *floret.* The spikelet axis is known as the *rachilla,* and generally bears two rows of florets; those towards its apex may fail to develop fully, often being male or sterile, so that then the floret consists only of lemma and palea.

The first bracts on the spikelet axis are often lemma-like. They lack axillary flowers and often have a protective function. These are the *subtending glumes.*

There is a great deal of diversity in spikelet structure within the family, mainly depending on number of florets, variation in size and shape of the subtending glumes, lemmas and paleas, and the sexuality of the floret(s). The subtending glumes and lemmas in particular may bear accessory structures, such as hairs, bristles or awns (Fig. 195 E and F), which are of importance in dispersal. Awns when present can usually be equated with modified leaf blades. Some awns are coiled and sensitive to changes in humidity (Fig. 195 A), whereas others are straight and either sensitive or indifferent to humidity changes. Extreme reduction of the lemma is met with in the *Andropogon* Group, where the entire lemma can be converted

**Fig. 195.** Poaceae. Variation in lemma structure. **A** Spikelet of *Stipa trichophylla,* long geniculate awn of lemma projecting from between subtending glumes. **B** Mature lemma of *Stipa trichophylla* showing callus. **C** Lemma of *Setaria glauca,* showing "germination flap". **D** Spikelet of *Leptaspis cochleata,* lemma inflated and invested with short prickle hairs. **E** Lemma of *Enneapogon polyphyllus,* awns not geniculate. **F** Lemma of *Centosteca lappacea* with retrorse hairs on abaxial surface. **G** Base of lemma of *Pogonachne racemosa* with column of terminal geniculate awn. (**A, B, E** Gardner 1952; **D, F, G** Bor 1960; **C** Hubbard 1954)

into an awn (Fig. 195 G). Rarely, paleas may also be awned.

Reduction leads to unifloral spikelets with a short or prolonged rachilla or to spikelets with two florets and no prolongation of the rachilla. In the latter case the upper floret is usually bisexual whereas the lower is male or wholly reduced, with or sometimes even without the palea. Spikelets with two florets, the lower of which is male or reduced, are described as "determinate" (Fig. 194 C) whereas other spikelets, usually with several florets, are described as "indeterminate" (Fig. 194 B).

In certain of the Bambusoideae the apparently sterile proximal lemmas bear vegetative buds and not flowers in their axils. These buds are called *pseudo-spikelets* because their axillary buds often develop into further pseudo-spikelets (Fig. 196 A, E).

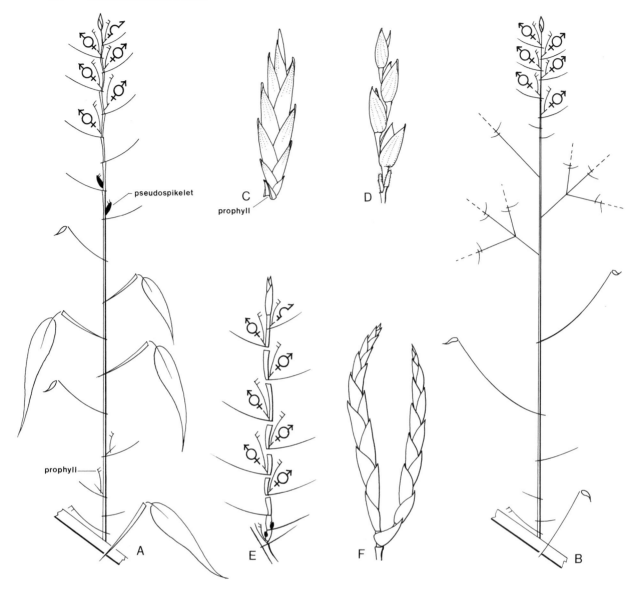

**Fig. 196.** Poaceae. **A** Indeterminate inflorescence branch of *Bambusa multiflora* with lower bracts of spikelet bearing buds (diagrammatic). **B** Determinate inflorescence branch of *Sasa veitchii* with lower bracts of spikelets lacking buds (diagrammatic). **C** Pseudospikelet of *Bambusa multiflora* with a two-keeled prophyll. **D** Spikelet of *Sasa veitchii* with two small empty glumes at its base. **E** Primary pseudospikelet of *Guadua angustifolia* (diagrammatic). **F** Two pseudospikelets of *Guadua angustifolia,* that to the *right* developed from a bud at the base of the other. (Redrawn from McClure 1966)

Spikelets and pseudo-spikelets terminate the axes of the inflorescence. When spikelets terminate the inflorescence branches all further growth of the inflorescence is inhibited, and so it is functionally determinate (Fig. 196 B). In contrast, if pseudo-spikelets terminate the inflorescence branches their vegetative buds may produce a whole succession of secondary and higher order pseudo-spikelets, and so the inflorescence is indeterminate (Fig. 196 B).

Upon the ripening of the fruit the spikelet is detached either below the subtending glumes, so that it falls as a unit, or above the glumes, in which case the rachilla may also divide into segments at the base of each lemma, so that each floret with one rachilla internode represents the diaspore. The florets may also disarticulate singly, leaving the rachilla projecting out from between the subtending glumes. The inflorescence in rare cases may break into segments comprising several spikelets or the whole inflorescence may become detached to act as a diaspore.

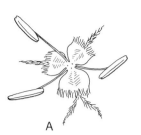
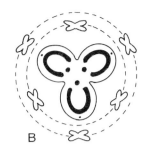
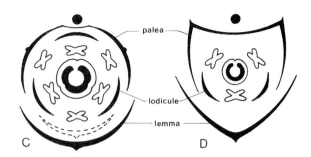

Fig. 197. Poaceae. **A** Tricarpellary gynoecium (as an aberration) of a hybrid *Triticum*. **B** Partial floral diagram based on **A**. **C** Floral diagram of *Arundinaria*. **D** Floral diagram of the majority of Poaceae. (**A–B** redrawn from MIÈGE 1936; **C** Orig., **D** redrawn from POTZTAL 1964)

The floral perianth is considered to be represented by members of the inner whorl, the *lodicles*. It is most completely developed in genera such as *Arundinaria* (Figs. 197 C, 203 M), where the three members are relatively large. The lodicles are never more than a few mm long but are quite diverse in shape, texture and indumentum. They are divisible into three types: (1) membranous and entire with margins toothed, (2) entire and glabrous or ciliate, or (3) obcuneate and fleshy throughout, but sometimes with a few hairs on the distal margins (Fig. 194 G, I, J). Where there are two anterior lodicles, which is the commonest condition, it is assumed to have arisen by suppression of the posterior lodicle. In a few species there is a single lodicle due to fusion of the anterior two. Lack of lodicles occurs occasionally in all major groups of grasses. The opinion that *Streptochaeta* has very large lodicles is no longer tenable, as the structures so regarded are better interpreted as bracts (SÖDERSTROM 1981). Where flowers have more than three lodicles, as in *Ochlandra,* the extra members appear to have been derived from aborted stamens.

The anthers are tetrasporangiate and borne on long, slender filaments, with a basal to medial attachment to the connective, allowing considerable flexibility at anthesis. Rarely the connective projects beyond the microsporangia as a glabrous or hirsute appendage. The stamen number is 3 + 3 or, by the suppression of the inner whorl, more often 3; less often it is 1, 2, 4 or 5. Rarely, as with *Ochlandra,* there are up to 170 stamens per flower. Among the Bambusoideae the stamens are sometimes monadelphous.

Microsporogenesis is simultaneous and the mature pollen grains are three-celled. The pollen grains are more or less spherical with a single operculate aperture surrounded by a raised rim. The surface is spinulose, with uniform or variably sized spinules, which may be evenly distributed or grouped into definite areas via an incised reticulum, patterns that show little taxonomic correlation. In

*Pariana* the pollen grains are tubercular (TUTIN 1936).

The pistil is unilocular and uniovulate. The style bears two (very rarely one, three or four), stigmatic lobes (Fig. 194 H) or there are two wholly separate feathery or papillose stylar branches (Fig. 194 E and F). Where there are two styles a stub of tissue is often present which may represent the base of the third style. In *Zea* the two stylar branches are fused for most of their length.

In species of *Bromus* the ovary top bears a hairy fleshy appendage (Fig. 194 E and Fig. 199 H), whose function is unknown but which should not be confused with the hairy apex on the ovary of *Triticum* (Fig. 198 A and B) and other taxa.

The ovule is anatropous, hemianatropous, campylotropous or orthotropous with two integuments (Fig. 199 F), the inner of which usually forms the micropyle (Fig. 199 F). A parietal cell is never cut off, but in most grasses (except Pooideae) the nucellar epidermis divides periclinally to form a nucellar cap. The embryo sac is of the *Polygonum* Type and embryogeny conforms to the Asterad Type. Secondary multiplication of the antipodal cells is common. Starch is the principal storage product of the endosperm, the starch grains being either simple or compound (in different grasses). The majority of grasses have a solid endosperm but in a few the central region remains liquid indefinitely. In *Phaenosperma* the endosperm is ruminate and in *Melocanna* and related genera it is missing due to its absorption by the viviparous embryo.

The testa and pericarp are usually fused, the fruit forming a caryopsis (plural: -es) or "grain". In relatively few members the fruit is a nutlet with

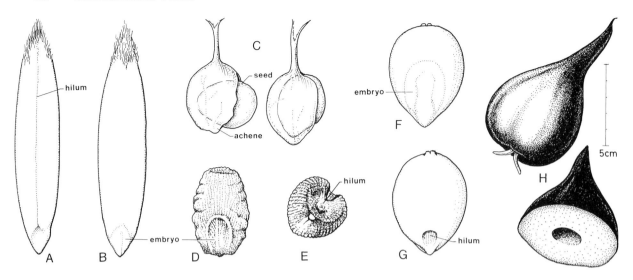

a thin pericarp readily separable from the seed (Fig. 198 C). In some Bambusoideae the fruit is a nut with the upper part of the fruit hard and woody. Very rarely the pericarp is fleshy, as with the bamboo *Melocanna baccifera,* where the berry has the size and texture of a pear (Fig. 198 H). In *Eleusine,* and *Acrachne,* which have a membranous nutlet, the seed is sculptured (Fig. 198 D and E), whereas in the related *Eragrostis* the fruit is a caryopsis with a pericarp that is sculptured reflecting the texture of the enclosed seed. Hairs may be present on the fruit apex (Fig. 198 A and B).

The embryo is lateral to the endosperm, and varies much in length relative to the endosperm (Fig. 198 B, D and F). In absolute size the grains vary from large as in maize to very diminutive, weighing as little as $12 \times 10^{-6}$ g in *Ectrosia* species. The hilum, which is usually visible through the pericarp, is either linear and as long as the grain, or round and rather short (Fig. 198 A and G).

The embryo is unique among monocotyledons in that it is not a single axis terminating at opposite poles in a cotyledon and radicle. Instead there is a well-developed *scutellum* appressed to the endosperm and bearing laterally an axis terminated at one end by a plumule protected by a coleoptile and at the other by a *coleorrhiza* protecting an included root. Growing out from the axis opposite the scutellar node there may be a further structure, the *epiblast* (Fig. 199 A). As shown by Reeder (1957) the variation in embryo form is of considerable taxonomic importance.

Here it is assumed that the scutellum is the cotyledon and the coleorrhiza the radicle. The epiblast is interpreted as a reduced form of the collar that grows up around the germ pore of the cotyledon in many other

**Fig. 198.** Poaceae. Seeds and fruits of some grasses. **A–B** *Bromus unioloides,* different views, hilum linear. **C** *Urochondra setulosa,* achene, shedding seed. **D–E** *Acrachne verticillata,* seed, embryo large, **E** showing punctiform hilum. **F–G** *Alloteropsis semialata,* caryopsis, **F** showing the large embryo, **G** the punctiform hilum. **H** *Melocanna baccifera,* berry exhibiting precocious germination, with roots coming out; *below* berry in section, embryo removed. (**A–B** and **D–G** Clifford and Watson 1978; **C** Bor 1960; **H** McClure 1966)

Commeliniiflorae. The coleoptile is accepted as the first leaf of the shoot. In *Jouvea, Streptochaeta* and *Zizania* the coleoptile is an open sheath with a single midrib and in *Zizania,* furthermore, the coleoptile has a well-developed lamina.

Other interpretations of the embryo structure abound (Negbi and Koller 1963) but most may be neglected because they are not based on developmental studies. An alternative interpretation of the grass embryo however, that cannot be lightly dismissed, is that of the "German School". Here it is argued the scutellum and coleoptile are both parts of the cotyledon, the coleoptile representing the tubular sheath and the scutellum the "blade", which is transformed into an haustorial structure (Goebel 1923; Troll 1954; Pankow and von Guttenberg 1957).

In certain grasses the scutellum and coleoptile become separated from each other by a prolongation of the axis, the "mesocotyl"; thus in these grasses the scutellum and coleoptile in spite of being (according to the "German School") parts of the same structure, the cotyledon, emerge at different levels. Their connection can sometimes be seen from the course of the vascular strands where the mesocotyl is moderately long as in *Avena* or *Triticum* (Pankow and von Guttenberg 1957).

The epiblast is now said to be an outgrowth of the root neck of the first lateral root, which represents the downward-directed root previously thought to be the radicle, while the radicle in grasses is taken to be utterly reduced (Tillich 1977).

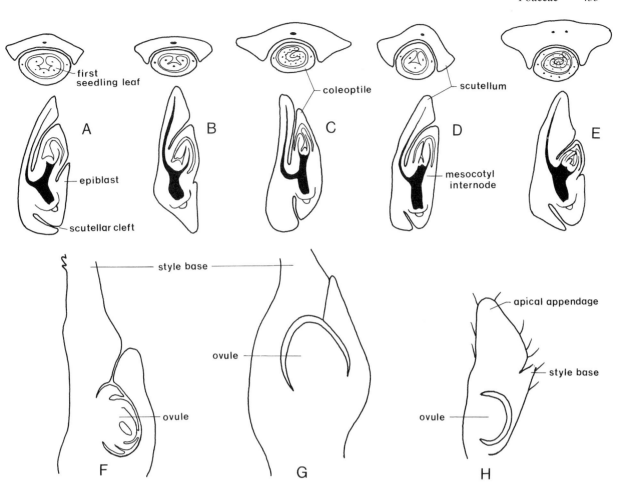

**Fig. 199.** Poaceae. Embryo structure and placentation of some grasses. **A–E** *Upper row* transverse section through coleoptile region of embryo; *lower row* embryo in median sagittal section. **A** *Eragrostis pectinata* (chloridoid). **B** *Deschampsia caespitosa* (pooid). **C** *Hackelochloa granularis* (panicoid). **D** *Phragmites australis* (arundinoid). **E** *Arundinaria tecta* (bambusoid). (REEDER 1957). **F–G** Vertical section of ovary passing through placenta. **F** *Zizania aquatica.* **G** *Zea mays.* **H** *Bromus unioloides.* (**F** redrawn from WEIR and DALE 1960; **G–H** redrawn from WALKER 1906)

On germination the scutellum remains within the fruit wall and it is the coleoptile that emerges. The next leaf to develop usually has a lamina except in most of the Bambusoideae. When present, the lamina is either long and narrow and held erect or is broader and disposed at an angle to the soil. In the species with indurated paleas and lemmas the root often emerges through a "germination flap" in the lemma (Fig. 195C).

*Features of Leaf Transection.* The anatomy of the leaves offers much of taxonomic and diagnostic value and it is useful to consider the anatomy from two viewpoints, that of the leaf as seen in transverse section and that of the epidermis as seen in surface view.

As seen in transverse section, the grass leaf exhibits considerable diversity in the arrangement of its mesophyll tissue and the structure of its vascular bundles. The differing organizations are strongly correlated with the $C_3$ and $C_4$ pathways of photosynthesis.

In plants following the $C_3$ pathway the bundle sheath as seen in transverse section is always double. The inner sheath cells are smaller than those of the outer sheath, which differ little in size from those of the mesophyll, but have somewhat thinner walls and contain fewer plastids. Starch formation takes place in the mesophyll cells which show no particular alignment with respect to the sheath. There are always more than four chlorophyll-bearing cells between adjacent sheaths (Fig. 200C).

Plants following the $C_4$ pathway have bundle sheaths that are either single or double. When double, the cells of the inner sheath are uniformly thick-walled. The cells of the outer or, when single, only sheath are usually larger than those of the adjacent mesophyll, have somewhat thicker walls and abundant plastids. Starch formation occurs in the sheath cells and the mesophyll cells tend to be radially disposed ("radiate") with respect to the sheath. There are four or fewer chlorophyll-bearing cells between adjacent bundles (Fig. 200B).

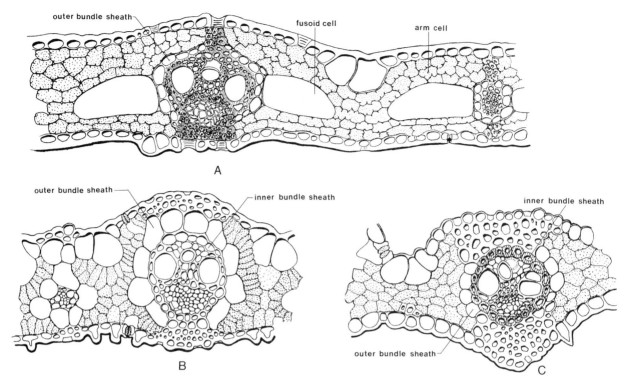

**Fig. 200.** Poaceae. Transverse sections of leaves. **A** *Streptogyne crinita*. **B** *Chloris gayana* (mesophyll radiate). **C** *Agrostis mannii*, (mesophyll irregular). (All redrawn from JACQUES-FÉLIX 1962)

The conspicuous ring of green sheath cells of the $C_4$ plants has led to their anatomy being described as Kranz Type (from the German: garland or wreath). The *Aristida* Group differs from the remaining grasses in that the majority of its members has two Kranz sheaths. Among the Kranz species it is possible to recognize two major groups in terms of the shape of the sheath cells as seen in paradermal section. The cells are either prolonged (PS) in direction of the bundle or are square or elongated at right angles (MS) to the bundle.

In the $C_4$ species in particular the mesophyll may be interrupted by plates of colourless cells traversing the leaf at right angles to its surface, thereby dividing it into separate photosynthesizing compartments. Deeply penetrating fans of bulliform cells and associated colourless cells achieve the same purpose in many other grasses with the Kranz syndrome. Other mesophyll cell types of considerable taxonomic significance are arm cells (ratchets) and fusoid cells, both of which are highly diagnostic of the Bambusoideae.

Arm cells are chlorophyllous and have strongly invaginated walls; fusoid cells lack chlorophyll and are embedded in the mesophyll adjacent to vascular bundles (Fig. 200 A).

*Interpretation of the Flower.* The structure of the gynoecium has been variously interpreted, most assuming it to have arisen from a tricarpellary ovary with parietal placentation. The basis of this belief is the occurrence of grass flowers with three stigmas and further support for the argument is found in the number of vascular bundles in the ovary wall.

Alternative explanations are that the grass pistil is the sole surviving carpel of an apocarpous gynoecium or the sole surviving carpel of a reduced tricarpellary ovary with axile placentation (Fig. 197 B). Support for each of these interpretations comes from the rare reports of tricarpellary ovaries within the family and from comparisons with the ovary structure in supposedly related families such as Flagellariaceae. Well-documented accounts of tricarpellary ovaries are available for *Festuca elatior* (NEES 1830) and hybrids of *Triticum* (MIÈGE 1936). In the former the carpels are fused except for the stylodia, whereas in the latter they are fused if at all only at the base (Fig. 197 A).

The account of the flower presented above assumes it to be haplochlamydeous and accepts the palea to be a prophyll and the lodicles to be perianth members. An alternative view of the palea is that it represents two fused sepals thereby with the lodicles making the flower dichlamydeous (SCHUSTER 1910). The floral diagram of an *Arundinaria* flower interpreted in this way is given in Fig. 197 C. A quite different interpretation regards both the palea and lodicles as bracts and hence the flower as achlamydeous. Support for this view comes from developmental studies which indicate that the third posterior lodicle when present arises higher on the axis than the anterior lodicles and so represents yet another bract. That the posterior lodicle when present is usually smaller than the anterior lodicles is a reflection of its later development.

Suppression of either the stamens or pistil results in unisexual florets. These may occur along with the bisexual florets in the same spikelets (most members of the *Panicum* Group), or may be arranged in unisexual spikelets within the same or different inflorescences on the same plant (the *Zea* Group), or may occur in different plants

(*Distichlis*). Another form of partial suppression leads to the formation of cleistogamous florets. In these the lodicles, anthers and stigmas are usually smaller than those in chasmogamous florets (*Microlaena*). Such florets may replace the chasmogamous florets or may occur together with them but on separate specialized inflorescences, such as those concealed in the leaf sheaths of *Cleistochloa*. Such hidden inflorescences are known as cleistogenes and are widespread in the family. Rarely, cleistogamous flowers are produced on subterranean inflorescences (*Amphicarpum*).

Complete suppression of the florets and the conversion of the spikelet into a bulbil occurs in several alpine species (e.g. *Festuca, Deschampsia*). The basal bulbils of *Poa bulbosa* behave in many ways like seed, for they can be dried out and stored for later planting.

*Pollination.* Pollination is essentially anemophilous, but the role of insects cannot be ignored, as they have been widely reported as gathering grass pollen. In the subtropics *Zea* pollen is often employed as a protein source by commercial apiarists. Nectar glands occur on the inflorescence branches and bracts of a few grass genera. The majority of grasses have protandrous florets, but a few, and these mostly lack lodicles, are protogynous.

*Dispersal.* The grain in most grasses does not fall separately from the plant but is shed variously enclosed in its palea, lemma, glumes or associated with other parts of the inflorescences.

Wind dispersal is assisted by the development of hairs on the glumes (*Poa*), rachilla (*Phragmites*), pedicels just below the glumes (*Imperata*) or on the awns (*Stipa*). Whole inflorescences (*Spinifex*) or parts of inflorescences (*Iseilema*) may behave as tumble weeds.

Although hairs will also serve as floats, most aquatic grasses are otherwise modified for water dispersal. With many the spikelets are embedded in a fleshy axis with disarticulation into sections, each of which acts as a float bearing its attendant spikelet (*Parapholis*). Exceptionally as with (*Thuarea*) the whole inflorescence becomes invested in a bract which serves as a float.

Animal dispersal is achieved in a variety of ways, of which that associated with a callus is one of the most important. The term callus may be applied to any thickened structure situated at the base of the lemma or below the subtending glumes. The callus may be rounded or cushion-shaped but is usually prolonged downwards as a sharp appendage. Often the callus is furnished with stiff hairs arranged retrorsely with respect to the point of the appendage (Fig. 195 B). The association of the callus and an awn results in a highly efficient diaspore which drills into the flesh and fur of animals, thereby ensuring dispersal especially by man and domestic animals. The close association of awn and callus is also very important for the establishment of seedlings. Adhesion to animals is also achieved by hooked hairs developed on the glumes or lemmas, by the development of burrs through the sterilization of inflorescence branches to form barbed involucres (*Cenchrus*) and by the development of sticky awns (*Oplismenus*) or inflorescence branches (*Eragrostis*). Rarely, the fruit bursts open to shed a sticky seed (*Sporobolus*).

Fleshiness is uncommon, occurring only in a few bamboos such as *Melocanna* where the ovary develops a fleshy pericarp (Fig. 198 H) or where the subtending glumes and sterile lemma become swollen up with oil (*Lasiacis*). Fleshy appendages in the form of elaiosomes are found in a few species widely separated in the taxonomic system, for example *Melica* and *Rottboellia*. The ingestion of grain by birds or other animals is common and a proportion of the grain often survives to reappear in the faeces.

The accidental dispersal of grasses in recent times has been much assisted by man as a consequence of trade or travel. For example about 50 species of the cosmopolitan largely subtropical *Eragrostis* (300) (Fig. 210 L–T) have been collected growing wild in the British Isles, though the genus is not endemic there (RYVES 1980). The geographical distributions given below do not allow for such introductions, which have served to extend widely the ranges of hundreds of species.

*Chemistry.* The occurrence of flavonoids is widespread with luteolin, glycoflavones, flavone 5-glucosides and tricin being well represented. In addition flavone sulphates occur in about 16% of the species investigated (DAHLGREN and CLIFFORD 1982). Alkaloids are also widely distributed in the family (HEGNAUER 1963). Some are tyramine derivatives (tyramin, hordenin), others are derived from tryptophane (gramin, 3-aminomethylindole) and a third group is based on a pyrrolizidin skeleton (lolin, norlin, lolinin). None of the alkaloids is unique to the family. Soluble nitrate and oxalate sometimes accumulate in toxic amounts. Silica is abundantly deposited throughout the plant. It is mostly deposited as a hydrate and assumes characteristic shapes in leaf epidermal cells. Several *Stipa* species contain an unknown compound which induces sleep when the plant is eaten.

*Cytology.* Cytologically the grasses are very diverse, variation occurring in both the basic numbers encountered and the size of the chromosomes. Polyploidy occurs extensively within the family and most haploid numbers within the range from 4 to 90 have been reported. Two groups of chromosome sizes may be recognized, large and small chromosomes.

Each of these is usually associated with a characteristic basic number – large, 7; small, 6, 9, 10 or 12. Exceptions to the general rule are numerous in basic numbers and their polyploids.

As might be anticipated in a family rich in high chromosome numbers, aneuploidy and apomixis are of common occurrence and introduce an element of complexity for the plant breeder. Many of the supposed sugarcane hybrids reported, and especially those involving bamboos, may have been derived from apomictic embryos whose development was stimulated by pollination.

*Distribution.* The grasses are cosmopolitan. The distributions of the subfamilies and generic groups are given below.

*Classification.* Historically the classification of the grasses has been based on exomorphic characters derived from the spikelet, seed and flowers. The adequacy of this approach was first questioned by AVDULOV (1931), who on the basis of his cytological studies showed that genera with several-flowered, laterally compressed spikelets comprised two groups rather than one as had been previously assumed. At about the same time support for this view came from the studies of PRAT (1931) on leaf anatomy and seedling morphology. Since that time many other anatomical studies have reinforced the belief that endomorphic characters are the most useful for defining subfamilies. Of these probably the most important is that of REEDER (1957) on embryo structure in relation to taxonomy. For a detailed review both of past classifications and the characters useful for that purpose reference may be made to PRAT (1960).

Here we recognize seven subfamilies but do not venture to give a tribal classification; instead we take groups of apparently related genera or occasionally single genera for consideration. We deal first with the subfamily Bambusoideae, in which (it should be noted) we include some groups which are never thought of as bamboos. The remaining subfamilies are dealt with in alphabetical order.

In order to facilitate orientation the seven subfamilies and the groups within them that are mentioned are listed below.

**Table 4.** Subdivision of Poaceae

| | |
|---|---|
| Subfamily Bambusoideae | |
| The *Oryza* Group | Subfamily Panicoideae |
| The *Anomochloa* Group | The *Arundinella* Group |
| The *Bambusa* Group | The *Panicum* Group |
| The *Dendrocalamus* Group | The *Saccharum* Group |
| The *Phyllostachys* Group | The *Sorghum* Group |
| The *Arundinaria* Group | The *Andropogon* Group |
| | The *Zea* Group |
| Subfamily Arundinoideae | The *Neurachne* Group |
| The *Aristida* Group | etc. |
| The *Arundo* Group | |
| *Micraira* | |
| *Lygeum* | Subfamily Pooideae |
| The *Ehrharta* Group | The *Triticum* Group |
| | The *Agrostis* Group |
| Subfamily Centostecoideae | The *Poa* Group |
| *Zeugites* and *Centosteca* | The *Sesleria* Group |
| | The *Avena* Group |
| | The *Melica* Group |
| Subfamily Chloridoideae | |
| The *Pappophorum* Group | Subfamily Stipoideae |
| The *Chloris* Group | The *Stipa* Group |
| The *Muhlenbergia* Group | |
| The *Eragrostis* Group | |

## Subfamily Bambusoideae

Annual or mostly perennial grasses which are tufted or rhizomatous. Sometimes scramblers or lianes. Many are woody and some attain great heights. The leaves are usually broad and pseudo-petiolate and frequently possess an abaxial as well as an adaxial ligule. The lower culm leaves usually have very much reduced blades. Arm cells and fusoid cells are present in the mesophyll of most species and microhairs (sometimes three- or more-celled) occur widely. The inflorescences are various and in some groups pseudo-spikelets replace the spikelets. There are one to four (often three) lodicles. These are strongly vascularized, membranous and often ciliate. The stamens are usually one to six (many in *Ochlandra*), mostly free but in some groups fused. The pistil has three (1–6) stigmas and the fruit is an achene, berry or nut. The first leaf of the seedling usually lacks a blade but if one is present it is generally narrow and erect. The chromosomes are mostly small and in multiples of 11 or 12 (7, 9, 10, 15).

The tall woody species are largely restricted to the monsoon tropics and the broad-leaved herbaceous species to the rain forest. Herbaceous swamp species extend from the tropics to the temperate regions. Many are mountain species, growing at or above the snowline. Widely distributed but absent from high latitudes.

On the basis of habit the subfamily may be divided into herbaceous and woody genera. Amongst *the former* may be recognized two groups, one largely confined to open swampy places, the other to shady well-drained sites.

***The Oryza Group.*** These genera are mostly allies of *Oryza* and include about 20 genera, a few of which lack fusoid cells. – *Oryza* (25) is widely cultivated principally as *O. sativa* ("Rice") with thousands of cultivars which are adapted to cultivation from the moist tropics to the warm-temperate regions. The place of origin of the crop is obscure and several parent species may be involved. With respect to their use three different groups of cultivars may be recognized, long-, medium- and short-grained. The former takes longer time to mature and is less glutinous when cooked than are the others. The long-grained types grow in warmer and the short-grained in cooler climates. From the viewpoint of cultivation two types are recognized: paddy and upland. The former is raised in seed beds and is transplanted into flooded fields which are drained about 4 months later, when the crop is harvested. The upland varieties

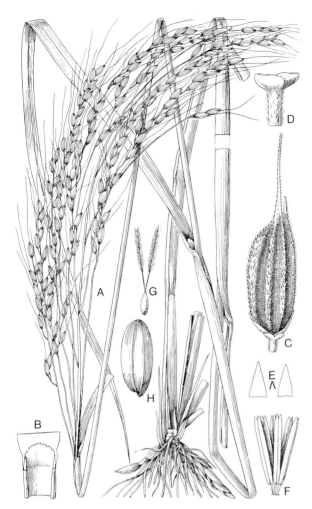

Fig. 201. Poaceae. *Oryza australiensis.* **A** Plant. **B** Ligule. **C** Caryopsis in spikelet. **D** Lower and upper subtending glumes of spikelet. **E** Lodicules. **F** Androecium (6 stamens). **G** Pistil. **H** Caryopsis. (HUBBARD 1934)

are grown like other cereals. Rice is the prime cereal crop, providing the basic diet for more than half the world's population. Outside the tropical countries the crop is widely grown under irrigation in the subtropics and warm-temperate regions including the United States, Spain, Italy and Australia.

*Zizania* (3) from Asia and America has two economic species; grain of *Z. aquatica* ("Annual Wild Rice") is gathered as a minor cereal in the U.S.A.; *Z. latifolia* is cultivated in Eastern Asia where the young shoots are harvested as a vegetable ("Kau Sun") and the older leaves used for matting. *Leersia* (15) cosmopolitan, is not widely cultivated except for the American *L. hexandra* ("Swamp Rice Grass") which has been introduced as a forage

in some countries, although it tends to become weedy.

***The Anomochloa Group.*** Among the shade-loving herbaceous genera is the peculiar grass *Anomochloa* (1) *marantoides* which is endemic to Brazil.

There are difficulties in interpreting the flowers of *Anomochloa* in terms of conventional floret structure. The flowers are bisexual with four stamens and an ovary surmounted by a single undivided style. These reproductive structures are surrounded by a broad coriaceous flask-shaped organ terminating in a large beak. In turn this is subtended by a broad laterally compressed and keeled membranous bract. It has been suggested that the flask-shaped organ may be a palea and the membranous bract a lemma. If this is so, glumes are lacking. Each "floret" is shortly pedicellate and is borne in the axil of a small bract. Groups of one to three "florets" are enclosed in large sheath-like spathes on opposite sides of the axis to form a spike-like inflorescence. In contrast to the floral structure, which is difficult to interpret, the leaf anatomy is typically bambusoid. The caryopsis has a small embryo and linear hilum, further indications of its bambusoid affinities.

An equally peculiar rainforest herbaceous genus is *Streptochaeta* (3), also from Brazil. Here the flowers are borne singly and subtended by 12 spirally arranged bracts. The lower of these bracts are relatively short and have deeply erose tips and sometimes bear buds in their axils. Of the upper six bracts three have been interpreted as paleas and three as lodicles. However, even though they vary somewhat in size, they otherwise resemble one another closely. Accordingly, there is little merit in regarding them as other than bracts. On this interpretation the flowers lack lodicles, palea and lemma. As some of the lower bracts subtend buds the whole structure may be regarded as a pseudo-spikelet bearing a solitary flower (SODERSTRÖM 1981).

Other herbaceous rain-forest genera are *Pariana* (35) of South America, which has unisexual flowers organized into spikelets arranged in verticils along a spike, and *Streptogyne* (2) which occurs in the tropics of both the Old and New Worlds, and whose long styles are persistent and retrorsely barbed.

Another group of genera occurring in both the Old and New Worlds includes *Olyra* (20), which grows in open lowland rain forest throughout the tropics. It has unisexual flowers which are arranged into strongly dimorphic inflorescences.

**Fig. 202.** Poaceae. **A–L** *Puelia coriacea*. **A** Plant. **B** Upper part of leaf sheath showing ligule and pseudopetiole. **C** Spikelet. **D** Palea of female floret. **E** Lodicule of female floret. **F** Lower glume. **G** Palea of male floret. **H** Lemma of male floret. **I** Lodicule of male floret. **J** Androecium. **K** Pistil. **L** Female floret and basal glume. (CLAYTON 1967). **M–T** *Swallenochloa subtesselata.* **M** Branch. **N** Upper part of leaf sheath showing ligule and contraligule. **O** Ultimate branchlets of inflorescence with transitional glumes attached. **P** Spikelet. **Q** Palea. **R** Lodicules. **S** Anther. **T** Pistil. (MCCLURE 1973)

*Puelia* (5) in the rain forests of tropical Africa is typical of a small group of genera with monadelphous stamens (Fig. 202 A–L).

The *woody genera* may be grouped according to whether they have ordinary spikelets only or produce pseudo-spikelets. Those with pseudo-spikelets usually have six or more stamens and may be conveniently further grouped in terms of their fruit types.

**Fig. 203.** Poaceae. Some morphological features of subfam. Bambusoideae. **A–D** *Bambusa forbesii.* **A** Upper part of leaf sheath showing auricles and ligule. **B** Pseudospikelet with two spikelets in axils of lower glumes. **C** Diagram of pseudospikelet showing position of joints in the rachilla, spikelets in axils of lower glumes not dissected. **D** Pistil, with short stylodial branchlets. (HOLTTUM 1967). **E–Q** *Arundinaria gigantea* ssp. *tecta.* **E** Leafy culms and base of leafy culm. **F** Developing branch complement. **G** Midculm branch complement. **H** Inflorescence showing paniculate branching. **I** Part of inflorescence in flowering stage. **J** Abnormally long spikelet. **K** Immature spikelet. **L** Single floret with attached axis segment. **M** Lodicles. **N** Stamen. **O** Pistil. **P** Caryosis, embryo and hilum views. **Q** Same, longitudinal section. (MCCLURE 1973)

***The Bambusa Group.*** The fruits of *Bambusa* (Fig. 203 A–D) and related genera are caryopses, those of some species closely resembling grains of wheat (*Triticum*). The hollow woody culms of the pantropical *Bambusa* (70) may attain heights of 25 m and possess great strength. They are widely used as a source of timber and may also be cut into sections and serve as water pails, milk churns etc. In the internodes amorphous silica may be deposited. Known as tabashir it is employed medicinally. – *Guadua* (30) of tropical South America is a source of prime timber fulfilling the same role as does *Bambusa* in Asia. The two genera may be separated on the basis of the palea, winged in *Guadua* and non-winged in *Bambusa,* but are often regarded as congeneric.

**The Dendrocalamus Group.** The fruits of *Dendrocalamus* and its allies are nuts. As with the *Bambusa* Group some species grow quite tall, up to 30 m having been reported. – *Dendrocalamus* (20) from South-East Asia, *Gigantochloa* (20) of Indomalaysia and *Oxytenanthera* (1) from Africa all provide important timber species. In addition the young shoots of, for example, *Gigantochloa verticillata* are edible; *Dendrocalamus* is also a source of tabashir and is furthermore the basis of the Indian paper industry. A few genera have fleshy or crustaceous fruits. Included here is *Melocanna* (2) from Indomalaysia which is a source of timber and is widely planted for that purpose. *M. baccifera* has fleshy fruits the size of a pear (Fig. 198 H) and these are edible if baked. – *Schizostachyum* (35) from South-East Asia is also harvested for timber.

**The Phyllostachys Group.** One group of Old World genera includes some species with pseudospikelets, others with spikelets. They usually have three rather than six stamens and are further characterized by their glumes possessing small but well-formed deltoid laminae. They are mostly cane-like plants up to about 12 m and several species are cultivated as ornamentals. – Included here are *Phyllostachys* (40), *Shibataea* (6), *Sinobambusa* (8) and *Chimonobambusa* (14).

**The Arundinaria Group.** The remaining woody genera have the flowers in spikelets and the glumes lack blades. They also usually have flowers with three instead of six stamens. *Arundinaria* and its allies have pistils with three stigmas. They are widely distributed in the tropics and subtropics of both the Old and New Worlds. Several species of *Arundinaria* (100) are cultivated, some as a source of poles and fishing rods, some for paper manufacture and others as garden ornamentals. The young shoots may serve as a vegetable. – A number of other genera including *Pseudosasa* (3), *Sasa* (ca. 20) and *Pleioblastus* (ca. 30), all from Asia, have species that are cultivated as ornamentals.

Closely related to *Arundinaria* is *Arthrostylidium* (25) which differs in having a thick rhizome and a pistil with two, rarely three stigmas. The genus is mainly restricted to the New World, growing especially in mountain forests. Some species have pendulous vine-like culms climbing on trees or shrubs. Unlike most other bambusoid grasses many species of *Arthrostylidium* have leaves with very narrow laminae.

The genus *Chusquea* (70) has a pistil with two stigmas and, like *Arthrostylidium,* is restricted to the New World, growing at high altitudes from Cuba to Chile. Unlike most other Bambusoideae the rachilla of the spikelets is not prolonged. A few species are cultivated as ornamentals.

## Subfamily Arundinoideae

These are annual or perennial species with a tufted, rhizomatous or stoloniferous habit. Some species are coarse and bamboo-like. The leaves are narrow and the ligule is membranous or cartilaginous or may form a rim of cilia. There is considerable variation in the shapes of silica bodies encountered but dumb-bell-shaped types predominate. Microhairs are usually present. Whereas arm cells sometimes occur, fusoid cells are always absent. The leaf anatomy may be either of the Kranz or non-Kranz Type. The flowers usually have 0–2, rarely 3 lodicles and these are fleshy in texture. There are 1–6 stamens and the pistil has 2 stigmas. The chromosomes are small and mostly in multiples of 12 (7, 10, 11, 13).

The members of the subfamily grow in a variety of habitats from the cool-temperate regions to the tropics.

The majority of species in the subfamily are included in one of two major groups. They share little in common with respect to spikelet structure or leaf anatomy but have similar embryology.

**The Aristida Group** is a group of three genera with single-flowered spikelets and lemmas generally bearing a trifid awn. Included here is *Aristida* (350), widespread in the tropics and subtropics. Its species are mainly tufted grasses, several of which are important components of savannah and open forest communities.

**The Arundo Group.** This consists of about 10 genera of mainly robust grasses with large paniculate inflorescences and several-flowered spikelets. Three of the genera, *Arundo, Phragmites* and *Gynerium,* have arm cells.

*Cortaderia* (15), from South America and New Zealand, is a genus of large tussock grasses, one of which, *C. selloana* ("Pampas Grass"), is widely planted as an ornamental. – The Eurasian *Arundo* (12) is also planted as an ornamental; one species, *A. donax* ("Giant Reed"), is used to make reeds for musical instruments and for fencing etc. – *Phragmites* (3) is cosmopolitan; *P. australis* (= *P. communis*) ("Common Reed") may be used to make brooms (panicles), thatching and matting (culms), food and medicine (young shoots); in addition it is suitable for paper pulp and bread may be made from the young crushed rhizomes. The species is becoming economically important as "energy crop", for fuel, etc.

Ecologically *Phragmites* is an important component of swamp vegetation and forms extensive floating fens at the mouth of Danube. The vegetative reproduction of *P. australis* has led to its becoming a weed from as far North as Finland to as far South as New Zealand.

**Micraira, Lygeum and the Ehrharta Group.** Three further genera or groups of genera are worthy of mention. – *Micraira* (14) from tropical and subtropical Australia has spirally inserted leaves, a condition unique in the family. – *Lygeum* (1) *spartioides* from the Mediterranean has an inflorescence that is a single spikelet. The spikelet lacks subtending glumes and the lemmas are fused into a tube for half their length. The species is harvested as a source of esparto. – The genera centred on *Ehrharta* are largely restricted to the Southern Hemisphere, excluding South America. They have three-flowered spikelets of which only the terminal is fertile, the two lower being reduced to sterile lemmas. Some *Ehrharta* (27) species are useful components of native pastures and several are widespread temperate weeds.

## Subfamily Centostecoideae

These are annual or perennial, rhizomatous or stoloniferous grasses whose leaves are pseudo-petiolate with laminae exhibiting a conspicuous tesselate venation as in many Bambusoideae. Unlike members of that subfamily, the lodicles here are fleshy instead of membranous and the hilum punctiform instead of linear. They are grasses of woodlands or forests in tropical and subtropical regions. – *Zeugites* (10) occurs in the New World and *Centosteca* (4) (Fig. 195F) in the Old World.

## Subfamily Chloridoideae

These are annual or perennial herbs that are tufted or rhizomatous (stoloniferous) and sometimes robust. The leaves are mostly narrow and the ligule is hair-fringed or reduced to a rim of hairs. The silica bodies are mostly saddle-shaped and microhairs are always present. Their distal cell is usually somewhat inflated and sometimes glandular. The mesophyll lacks both arm cells and fusoid cells and has a Kranz type PS anatomy. The spikelets have one to several florets and the rachilla is usually prolonged. There are 0–2 lodicles and these are fleshy and obcuneate. The stamens are one to three in number and the pistil has two stigmas. The fruit is a caryopsis or utricle. The chromosomes are small and occur mostly in multiples of 9 or 10.

The subfamily occurs mainly in tropical and subtropical areas with a markedly seasonal rainfall. Many species are adapted to saline or alkaline soils.

Two major categories of genera may be recognized within the subfamily: those with several (nine or more)-nerved lemmas and those with few (one to three)-nerved lemmas.

**The Pappophorum Group.** This represents the genera with nine- or more-veined lemmas, and has rather few species. Included there are the genera *Pappophorum* (7) and *Enneapogon* (30), in which the nerves of the lemma protrude as awns (Fig. 195 E) and the spikelets fall as a unit with the awns forming a pappus-like crown; *Uniola* (2) of coastal dunes in Eastern North America and the West Indies has large spikelets, resembling those of an unawned *Bromus*; *Orcuttia* which is endemic to California is unusual in that its leaves lack a ligule.

The genera with one- to three-nerved lemmas form two relatively natural groups on the basis of their inflorescence structure. One group has the spikelets arranged in spikes, the other in various forms of panicle.

**The Chloris Group,** which has spikelets in spikes, consists of several genera centred on the genus *Chloris* (90), which is widespread. One of its species, the African *C. gayana* ("Rhodes Grass") is widely planted as a pasture species. The closely related *Cynodon* (10) is also widespread and provides several pasture and lawn species; *C. dactylon* ("Couch", "Dhoub", "Bermuda Grass") is cultivated but has a tendency to weediness and may be regarded as the most serious weed of the grass family; *C. plectostachyus* ("Giant Star Grass") is widely planted in pastures. Species of *Dactyloctenium* (10) are also cultivated as lawn grasses; *D. australe* ("Durban Grass") in South Africa and *D. radulans* ("Sweet Smother Grass") as a shade lawn grass in North-Eastern Australia; *D. aegypticum* ("Crow's Foot") is a widespread weed of the tropics and subtropics. *Buchloë dactyloides* ("Buffalo Grass") of North America is planted for winter grazing and erosion control.

*Eleusine* (10) is cosmopolitan in warm regions. One species in Africa and India, *E. coracana* ("Finger Millet", "African Millet", "Korogan", "Ragi") is cultivated as a minor cereal. The related *E. indica* ("Crow's Foot") is a serious weed in the tropics and subtropics. *Leptochloa* (25), widespread in the tropics and subtropics, has at least two species, *L. panicea* and *L. chinensis*, that are weeds.

Of particular importance in the stabilization of tidal mud flats is the genus *Spartina* (10), whose species grow along the coasts of temperate eastern America, Tristan da Cunha, Europe and Mediterranean Africa. The hybrid *S.* × *townsendii* ("Townsend's Cord Grass") is presumed to have arisen spontaneously at Southampton from a cross between the European *S. maritima* ("Cord Grass") and accidentally introduced plants of the North American *S. alterniflora* ("Smooth Cord Grass"). Both *S.* × *townsendii* and its amphidiploid, *S. anglica* ("Common Cord Grass"), are widely planted as mud binders; *S. arundinacea* is used for bedding and thatching in Tristan da Cunha.

The genera of Chloridoideae with paniculate inflorescences have been traditionally regarded as comprising two groups, one with single-flowered, the other with two- or more-flowered spikelets. The view-point that they are distinct is increasingly being challenged but for convenience will be accepted here.

***The Muhlenbergia Group,*** with single-flowered spikelets is dominated by two large genera. One of these is the genus *Muhlenbergia* (125), which is restricted to the Americas save for a few species in Western Asia, and the other is *Sporobolus* (100), which has a universal distribution. In the former the lemma is awned and the fruit is a caryopsis; in the latter, the lemma is unawned and the fruit is a utricle.

Most species of *Sporobolus* ("Drop Seed") are palatable forage grasses but seldom occur in sufficient abundance to be significant native pastures; the seeds of many species were harvested in bygone times for food. One species, *S. virginicus* ("Salt-Water Couch"), is planted as a sand binder. – The New World *Epicampes* (10) has one species, *E. macroura* ("Rice Plant"), which is harvested in the mountains of Mexico and whose roots are exported for broom making.

***The Eragrostis Group.*** The group of genera with two- or more-flowered spikelets is centred on the large genus *Eragrostis* (300) (Fig. 210L–T) which, though universal in distribution, has few economic species. One species, *E. tif* ("Tiff") from

Abyssinia, is grown as a cereal on poor soils in East Africa and as a hay crop in South Africa. – A genus with a widely disjunct distribution is *Distichlis* (14) which has one species in Southern Australia, the others being American.

### Subfamily Panicoideae

These are annual or perennial herbs, either tufted or rhizomatous and sometimes with robust woody culms. Rarely they are shrubs. The leaves are broad or narrow and the ligule is a membrane, a ciliate membrane or a rim of hairs. Sometimes it is absent. The silica bodies are cross- or dumbbell-shaped and microhairs are always present. Neither arm cells nor fusoid cells occur in the mesophyll and the leaf anatomy is usually of the Kranz Type, both the PS and MS subtypes occurring. The flowers usually have two fleshy cuneate lodicules and one to three (or six) stamens. The pistil usually has two stigmas.

The subfamily is widely distributed from the tropics to the warm temperate regions.

Within the subfamily it is possible to recognize three major subgroups of genera in terms of the mode of disarticulation of the spikelets and the relative degrees of induration of the subtending glumes and the palea and lemma.

***The Arundinella Group.*** The genus *Arundinella* (55) can be taken as representative of a group of about ten genera with spikelets that disarticulate above the subtending glumes and have awned lemmas. The group is widely distributed in savannahs and woodlands throughout the tropics and subtropics.

Those genera with spikelets that disarticulate below the subtending glumes, and so fall as a single unit with the glumes, comprise two large well-defined categories, here with three and two groups, respectively. In one the palea and lemma are indurated and the subtending glumes are membranous whereas in the other the conditions are reverse.

***The Panicum Group.*** About 100 genera belong to the former category, including the *Panicum* Group, which centres about *Panicum* (500–600), a widespread tropical to warm-temperate genus. Many of its species are important pasture grasses or cereals and some are aggressive weeds. The African savannah species *P. maximum* ("Green Panic", "Guinea Grass") and the Southern African *P. coloratum* ("Makarikari Panicum") are widely planted for pastures in the tropics and subtropics; the former is sometimes a pest in sugar cane fields; *P. miliaceum* ("Proso Millet") and *P.*

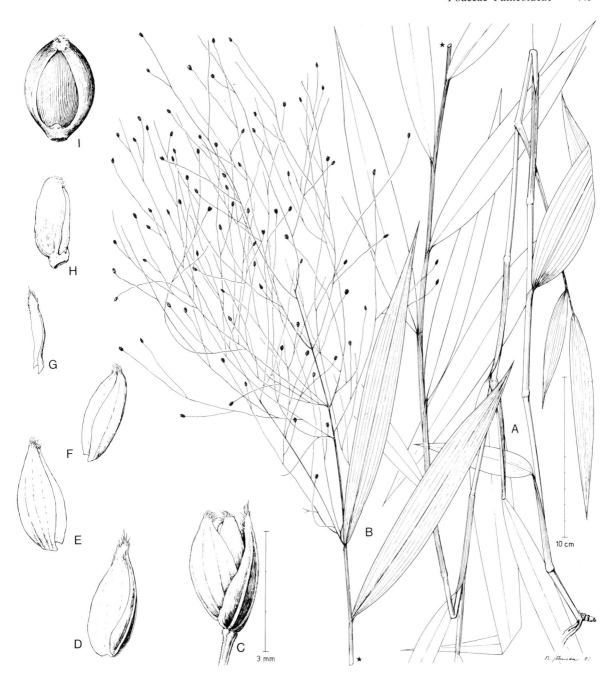

Fig. 204. Poaceae. *Panicum bambusiculme*. A Culm.
B Inflorescence. C Spikelet. D Lower glume. E Upper
glume. F Lemma. G Palea. H Ovary of immature flower.
I Lemma and palea of mature spikelet. (FRIIS and VOL-
LESEN 1982)

*sumatrense* ("Sawan") are cultivated as cereals in
India while *P. sonorum* ("Sauwi") is grown as a
cereal in Mexico (DE WET 1981). – *Brachiaria* (50),
largely confined to the tropics, provides a similar
range of economic species; the pantropical *B. mu-
tica* ("Para Grass") and the African *B. brizantha*
("St. Lucia Grass") are widely planted in tropical
pastures though each has a tendency to weediness;
*B. ramosa* ("Anda", "Horra") is cultivated as a
cereal in Southern India. – *Paspalum* (250), largely
from tropical South America, provides several spe-

cies for tropical and subtropical pastures. These
include *P. dilatatum* ("Dallis Grass"), *P. urvillei*
("Vasey Grass"), *P. vaginatum* ("Salt-Water
Couch") which is also an efficient sand binder,
*P. distichum* ("Water Couch") and *P. commersonii*

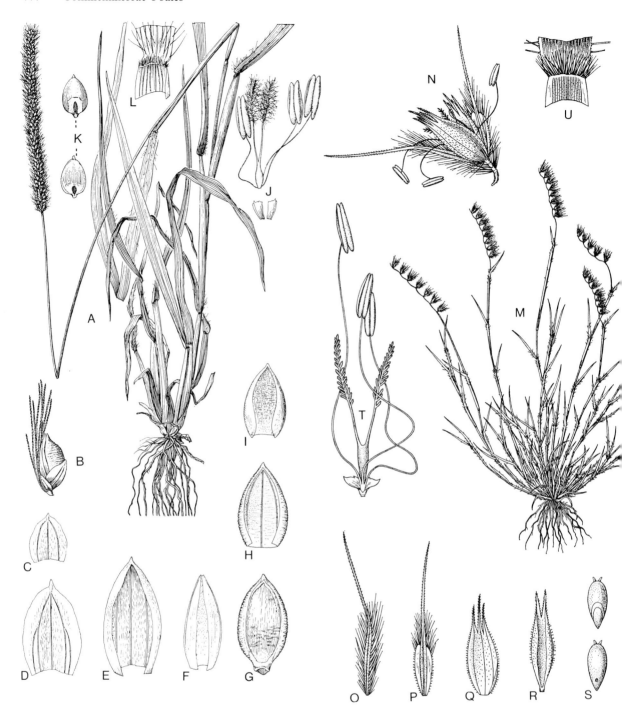

**Fig. 205.** Poaceae. **A–L** *Setaria glauca.* **A** Plant. **B** Spikelet; the bristles here represent the transformed axis of the inflorescence. **C** Lower glume. **D** Upper glume. **E** Lower lemma. **F** Palea. **G** Caryopsis, ventral aspect. **H** Upper lemma. **I** Palea. **J** Flower. **K** Caryopsis in different aspects. **L** Junction of sheath and lamina with hair-type ligule. **M–U** *Melanocenchris jacquemontii.* **M** Plant. **N** Spikelet. **O** Lower glume. **P** Upper glume. **Q** Lemma. **R** Palea. **S** Grain in different aspects. **T** Flower, much enlarged. **U** Junction of sheath and lamina with hair-type ligule. (All from Bor 1968)

("Scrobic"). Of these *P. dilatatum* is reported as a weed of plantation crops in many places, while *P. conjugatum* ("Sour Grass") is a weed of wet places throughout the tropics. *P. scrobiculatum* ("Khodo Millet") is cultivated as a cereal in India.

*Axonopus* (2), also from the New World, is widely cultivated in the wetter tropics. Both *A. affinis* ("Narrow-Leaved Carpet Grass") and *A. com-*

*pressus* ("Broad-Leaved Carpet Grass") are sown for pastures and lawns, though both tend to weediness.

The pantropical to warm-temperate *Digitaria* (400) provides many economically important species. The African *D. decumbens* ("Pangola Grass") is widely sown in tropical pastures; *D. didactyla* ("Blue Couch") from the Mascarene Islands is planted both for pastures and lawns in the subtropics; *D. cruciata* var. *esculenta* is cultivated as a cereal in India and Assam; *D. sanguinalis* ("Crab Grass", "Manna") is a minor cereal in Kashmir but elsewhere is a weed, being reported as noxious from 50° N to 40° S; *D. exilis* ("Fonio") and *D. iburna* ("Black Fonio") are cultivated as cereals in Africa; *D. adscendens* is a widespread tropical weed, and *D. scalorum* is a weed of Eastern Africa and India.

*Stenotaphrum* (3), a coastal genus of the Old World, is a useful sand binder and *S. secundatum* ("Buffalo Grass") is employed as a lawn grass in coastal areas in subtropical and warm-temperate regions – *Echinochloa* (30) is a pantropical genus several of whose species are harvested for food, but only *E. frumentacea* and *E. colona* are cultivated for that purpose. Through the range 50° N to 40° S both *E. crusgalli* ("Barnyard Millet") and *E. colona* are troublesome weeds. – The dioecious genus *Spinifex* (2) is a common sand binder on beaches of South-Eastern Asia, Indonesia, Australia and the Pacific. So also is *Thuarea sarmentosa*. The seed of the former is distributed by detached inflorescences blowing along the beaches and the latter has an inflorescence that at maturity folds over on itself and forms a float for seed dispersal.

A few genera in this group have spike-like inflorescences with the spikelets enclosed by bristles. Whereas with *Setaria* the bristles remain attached to the inflorescence when the spikelets are shed at maturity, with *Cenchrus* and *Pennisetum* the spikelets fall surrounded by an involucre of bristles. – *Pennisetum* (130) is widespread in the tropics and subtropics. The prostrate *P. clandestinum* ("Kikuyu Grass") from Kenya has inflorescences hidden within the upper leaves. It is widely planted in the subtropics and temperate regions. *P. purpureum* ("Elephant Grass", "Napier Grass"), native to the high savannahs of central Africa, is commonly planted as a pasture grass in the subtropics. *P. typhoides* ("Pearl Millet") is cultivated for its grain from India to Southern Europe and as a pasture elsewhere. The species *P. purpureum, P. polystachya* and *P. pedicellatum* have all been reported as weeds of the tropics and subtropics. Several species are grown as ornamentals in temperate gardens. – The closely related *Cenchrus* (25) is widespread on sandy soils and beaches of the Old World tropics; *C. ciliaris* ("Buffel Grass") is an important pasture species for the monsoon tropics; *C. echinatus* ("Burr Grass") is a weed of the tropics and subtropics. – The tropical to temperate genus *Setaria* (140) has few economically important species. *S. italica* ("Fox Tail Millet") is widely cultivated in Southern Europe, Japan and North China as a cereal as is *S. glauca* ("Korali") in India. *S. anceps* (syn. *S. sphacelata*) ("Setaria Grass") from Africa is widely planted as a tropical pasture species and *S. palmifolia* (tropical) has broad fan-plaited leaves and is grown as an ornamental. Two species, *S. verticillata* ("Foxtail", "Pidgeon Grass") and *S. viridis* ("Pidgeon Grass", "Bottle Grass") are troublesome weeds.

A small group of genera otherwise related to *Panicum* but with laterally compressed spikelets includes *Melinis* (20). One species, *M. minutiflora* ("Molasses Grass"), has hairs which secrete a viscid oil. It is planted for erosion control and is a useful pasture species in tropical and warm-temperate regions. – *Rhynchelytrum* (40), a predominantly African genus, has been introduced into most warm countries. *Rhynchelytrum repens,* though ornamental, has weedy tendencies and is of little economic importance.

Most of the about 100 genera possessing spikelets with indurate subtending glumes and membranous paleas and lemmas have the spikelets arranged in pairs on the inflorescence branches, one being sessile, the other stalked (Fig. 207 N).

**The Saccharum Group.** In *Saccharum* and its allies the stalked and sessile spikelets are both hermaphrodite. – *Saccharum* (12) is indigenous to the Old World. *Saccharum officinale* or its interspecific hybrids with *S. barkeri* ("North Indian Canes"), *S. sinense* ("Chinese or Japanese Canes") or *S. spontaneum* ("Asian Canes") are the source of much of the world's sugar. The culms of *S. arundinaceum* and *S. ravennae* (Fig. 206 J–U) are used as timber and the upper leaf sheaths of *S. bengalense* provide a valuable fibre. – The related *Imperata* (2) is widespread, one species, *I. cylindrica* ("Blady Grass") (Fig. 206 A–I), being weedy especially in areas subject to annual firing; its leaves are used for thatching and may also be used for paper making. – *Eulaliopsis* (2), from South-Eastern Asia, is also fire-resistant and is used for paper making, *E. binata* ("Sabai, Baib") being gathered for this purpose.

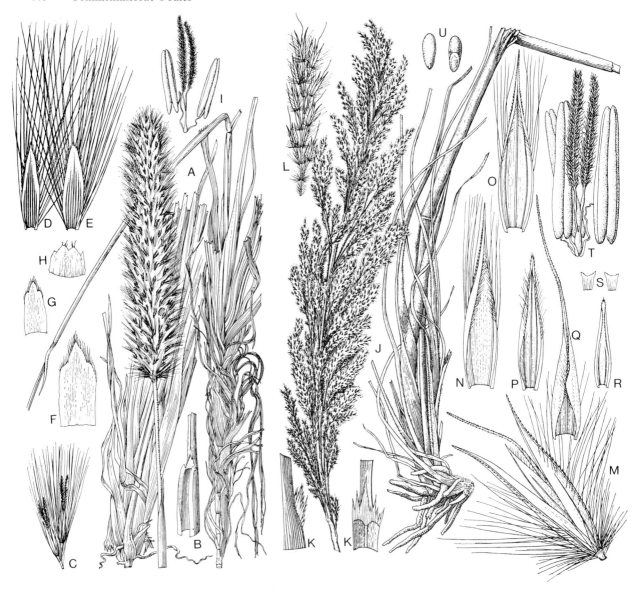

**The Sorghum Group.** In *Sorghum* and about 16 related genera the stalked spikelet is male or further reduced instead of being bisexual. – *Sorghum* (60) (Fig. 207 O–V) is a pantropical genus with several economically important species of which the following are cultivated as cereals: *S. caffrorum* ("Kaffir Corn"), *S. dochna* ("Sorgho"), *S. durra* ("Durra"), *S. nervosum* ("Kaoliang"), *S. roxburghii* ("Shallu"), *S. saccharatum* ("Sugar Sorghum") and *S. vulgare* ("Grain Sorghum", "Indian Millet"). The grain of several of these species is also brewed into beer and wine and the stems of *S. saccharatum* are, in addition, crushed for sugar and syrup. Both *S. sudanense* ("Sudan Grass") and *S. almum* ("Columbus Grass") are widely planted for pasture; *S. halepense* ("Johnson Grass") is also planted for pastures but may be-

**Fig. 206.** Poaceae. **A–I** *Imperata cylindrica.* **A** Plant. **B** Junction of sheath and lamina with ligule. **C** Pair of spikelets. **D** Lower glume. **E** Upper glume. **F** Lower lemma. **G** Upper lemma. **H** Palea. **I** Flower. **J–U** *Saccharum ravennae.* **J** Base of plant and inflorescence. **K** Junction of sheath and lamina with ligule in different aspects. **L** Portion of inflorescence. **M** The sessile spikelet. **N** Lower glume. **O** Upper glume. **P** Lower lemma. **Q** Upper lemma. **R** Palea. **S** Lodicles. **T** Floret with lodicles removed. **U** Caryopsis in different view. (All from BOR 1968)

come a serious weed on account of its extensive rhizome system. – The Old World *Vetiveria* (10) has one widely cultivated species. *V. zizanioides* ("Khus-Khus") from the rhizomes and roots of which a perfumery oil is extracted.

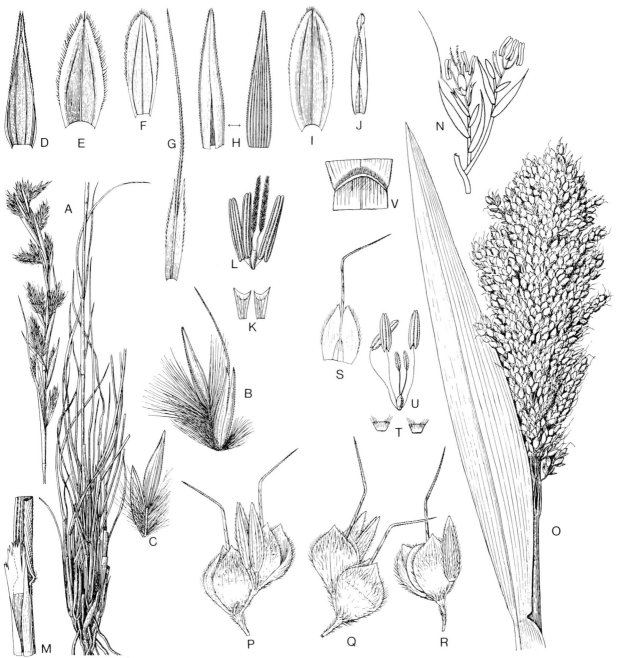

**Fig. 207.** Poaceae. **A–M** *Cymbopogon olivieri.* **A** Plant. **B** Sessile and stalked spikelets. **C** Sessile and stalked homomorphous spikelets. **D** Lower glume of sessile spikelet. **E** Upper glume of sessile spikelet. **F** Lower lemma of sessile spikelet. **G** Upper lemma of sessile spikelet. **H** Lower glume of stalked spikelet. **I** Upper glume of stalked spikelet. **J** Lemma of stalked spikelet. **K** Lodicles. **L** Flower, lodicles removed. **M** Junction of sheath and lamina with laciniate ligule. (BOR 1968). **N** *Andropogon,* spikelet pair, schematically. (POTZTAL 1964) **O–V** *Sorghum bicolor.* **O** Inflorescence and leaf. **P–R** Spikelet pair in different views. **S** Upper lemma of functional spikelet. **T** Lodicles. **U** Flower, lodicles removed. **V** Junction of sheath and lamina, with hair-fringed ligule. (BOR 1968)

Related to the *Sorghum* Group of genera are those in which the inflorescence branches are subtended by leafy spathes rather than small bracts or no bracts.

**The Andropogon Group.** Here the largest and most representative genus is *Andropogon* (120) which is widespread in both the New and Old World and although several species contribute to unimproved pastures none is cultivated. – Many species of the Old World *Cymbopogon* (60) (Fig. 207 A–M) have aromatic foliage and several

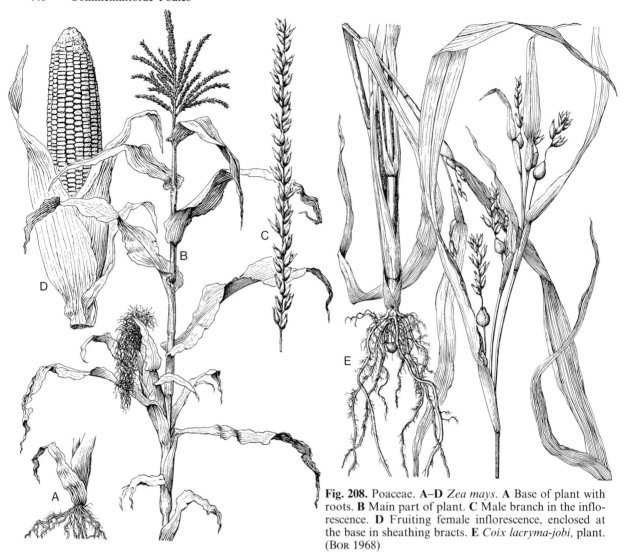

**Fig. 208.** Poaceae. **A–D** *Zea mays.* **A** Base of plant with roots. **B** Main part of plant. **C** Male branch in the inflorescence. **D** Fruiting female inflorescence, enclosed at the base in sheathing bracts. **E** *Coix lacryma-jobi,* plant. (Bor 1968)

are exploited commercially for their oil; *C. martinii* ("Motia", "Sofia") is a source of palmerosa and ginger grass oil, *C. flexuosus* ("East India Grass") produces perfumery oils and *C. citratus* ("West Indian Lemon Grass") is used to flavour curries and is a minor source of perfumery oil.

Among the species with widespread distributions are *Themeda triandra,* whose range extends from South Africa through Arabia and India to Australia (where it is known as *T. australis*) and *Heteropogon contortus* ("Black Spear Grass"), which is widespread in the tropics and subtropics. In many areas the latter species restricts sheep raising because the awned diaspores drill through the wool into the skins of the animals.

*The Zea Group.* This consists of about eight genera also related to *Sorghum* but with unisexual spikelets grouped either into different inflorescences or into different parts of the inflorescence, as is the case with the genus *Zea.* Among this group the female spikelets are often protected by a modified bract, which separates at maturity together with the included grain, thereby producing a false fruit. Where, as in *Zea mays,* the female spikelets are arranged in a spike, the even number of rows of grain produced reflects its origin from several racemes bearing paired spikelets having become fused back-to-back.

The Mexican genus *Zea* (1) (Fig. 208 A–D) is now widely planted and has several hundred cultivars within its single species, *Zea mays* ("Maize", "Indian Corn", "Mealies"). Used chiefly as a cereal the grain is locally fermented in the Andes to produce chicha or grain whisky. – Its close relative *Euchlaena* (2), also from Mexico, includes *E. mexicana* ("Teosinte") which is widely cultivated both as a grain and fodder crop. It is probable that *Euchlaena* has arisen as a hybrid between *Zea* and *Tripsacum.* – *Coix (1) lacryma-jobi* ("Job's Tears") from South-Eastern Asia (Fig. 208E) is cultivated as a grain, for fodder and as an ornamental; furthermore its bony involucres are used as beads.

**The Neurachne Group and Other Groups.** In addition to the above major groups of panicoid grasses there are a few genera or small groups of genera worthy of comment. – *Neurachne* and two more Australian genera are similar to the genera in the *Panicum* Group but differ in that the upper subtending glume exceeds the upper floret in length. – *Isachne* (60) is a pantropical genus with spikelets that disarticulate above the subtending glumes. – *Cyphochlaena* (2) and *Lecomtella* (1) are endemic to mountainous areas of Madagascar and the Comores. The former is reported to have flowers with six stamens and the latter to have a pair of ear-like growths on the somewhat elongated internode of the rachilla.

## Subfamily Pooideae

These are mostly herbaceous annuals or tufted or rhizomatous perennials. The leaves are narrow with a membranous ligule. The silica bodies are mostly crenate or roundly elliptical and microhairs are entirely lacking. Neither arm cells nor fusoid cells are present and the leaf anatomy is non-Kranz. There are usually two lodicles and they are generally fleshy with membranous tips. When fleshy they are usually fused. The stamens vary in number from one to three and the pistil has two stigmas. The chromosomes are usually large and in multiples of 7 (or 5, 8 or 9).

The distribution is largely temperate with a few species occurring on mountain tops in the tropics.

In a recent computer-based study of the subfamily, MACFARLANE and WATSON (1981) have redefined its constituent tribes and confirmed that it contains two major groups of genera distinguished clearly on the basis of the starch grains from the endosperm. In one group the starch grains are simple and the ovary usually hairy, whereas in the other the starch grains are compound and the ovary is usually glabrous.

**The Triticum Group.** To the first of these groups belongs *Triticum* and its allies. With these genera the inflorescence is usually a spike.

Several species included here are cultivated as cereals. – The temperate genus *Hordeum* (20) which includes *H. vulgare* ("Barley") is widely cultivated to 70° N. and tolerates saline conditions better than other cereals. – *Secale* (5), predominantly from the Mediterranean and Central Asia, includes *S. cereale* ("Rye") also cultivated to 70° N; and the North temperate *Triticum* (20) has several cultivated species: *T. vulgare* ("Bread Wheat"), *T. turgidum* ("Rivet", "Poullard", "Cane Wheat"), *T. durum* ("Durum Wheat"), *T. dicoccum* ("Emmer") *T. monococcum* ("Einkorn Wheat"). "Mummy Wheat", allegedly from the Egyptian tombs, is generally considered a variety of *T. turgidum.* Grain from such tombs has always proved inviable when tested.

Two species of the North-temperate and South American *Elymus* (70) are cultivated in pastures, *E. canadensis* ("Canada Wild Rye") in North America and *E. junceus* ("Russian Wild Rye") in Asia. – *Agropyron* (100–150), a widespread temperate genus, provides a number of forage species of which one, *A. repens* ("Couch", "Twitch Grass", "Quackgrass") can be a troublesome weed on account of its creeping of fibres.

Whisky is obtained by fermentation from the grains of rye and beer from those of barley.

In contrast to the above genera *Bromus* (50) and *Brachypodium* (10) have paniculate or spicate inflorescences. – The former has partly fused leaf sheaths. It includes a number of economically important species including *B. inermis* and *B. unioloides,* both widely cultivated as fodder plants. In pre-European times *B. mango* ("Mango") of South America was cultivated as a cereal – the only cultivated cereal not an annual. – *Brachypodium* has leaf sheaths with free margins. Its species are members of grassland and woodland communities and none are commercially exploited.

The genera with compound starch grains are morphologically very diverse and many tribes have been defined to cope with the diversity. Most of these tribes have been segregated on the basis of single attributes, whereas they seem best defined in terms of suites of characters. Modern studies indicate that floret number per spikelet and the relative lengths of the glumes and rachillas are not useful diagnostic characters.

**The Agrostis Group.** The genus *Agrostis* is representative of a group of about 60 genera with small spikelets, a grain with an endosperm possessing lipids and paleas that are often rounded instead of two-keeled. – *Agrostis* (200) is a cosmopolitan but chiefly North-temperate genus. Two of its species, *A. canina* ("Velvet Bent") and *A. stolonifera* var. *palustris* ("Marsh Bent"), are widely planted as lawn grasses. The Eurasian *A.*

*tenuis* ("New Zealand or Colonial Bent") is planted as a pasture; *A. gigantea* ("Red Top") is cultivated as a soil binder or haycrop in the United States. – *Trisetum* (75), largely Asian, contributes to unimproved pastures and *T. flavescens* ("Golden Oatgrass") is a useful fodder grass. – *Hierochloë* (30), a temperate and tropical mountain-top genus, has no significant economic species but the scented *H. odorata* ("Holy Grass") has been used for the weaving of mats and baskets in North America. – *Anthoxanthum* (20), also mainly temperate, like *Hierochloë* is strongly scented with coumarin; *A. odoratum* ("Sweet Vernal Grass") from Europe has a limited role as a pasture species in cool-temperate regions. – *Holcus* (8) is a North-temperate genus with one South African species; *H. lanatus* ("Yorkshire Fog", "Velvet Grass") from Eurasia is often regarded as a weed but is widely used as pasture in Chile and elsewhere. – *Ammophila* (2) is also North-temperate and principally grows on seashores; *A. arenaria* ("Marram") is widely planted as a sand binder in temperate climates. – *Phalaris* (20) with species in both the North- and South-temperate regions provides several pasture species; *P. arundinacea* ("Reed Canary Grass") is also used to control stream bank erosion; *P. minor* ("Small Canary Grass") from the Mediterranean is grown principally in South America; *P. tuberosa* ("Toowoomba Canary Grass"), also from the Mediterranean, is widely planted in Southern Australia. – *Lagurus* (1) *ovatus* ("Hare's Tail") from the Mediterranean has inflorescences that are often used for dried floral arrangements.

**The Poa Group.** Closely related to the *Agrostis* group are about 50 genera centred on *Poa*. These too have small spikelets but the endosperm in some lacks lipids and the paleas are usually strongly two-keeled. – The largest genus in this groups is *Poa* (300) which is cosmopolitan. *P. annua* ("Annual Meadow Grass") is a weed of temperate lawns and is possibly the world's most widely distributed angiosperm, being introduced even into Graham Land of Antarctica. *P. pratensis* ("Kentucky Blue Grass") of Europe is widely planted as a pasture and lawn grass in temperate regions. – *Festuca* (80) is also cosmopolitan; the temperate Eurasian *F. arundinacea* ("New Zealand Fescue", "Tall Fescue") is widely cultivated as a pasture species, the temperate Eurasian *F. pratensis* ("Meadow Fescue") is a pasture species in the Northern Hemisphere, and so is the European *F. rubra* var. *commutata* ("Chewing's Fescue") which is also grown in New Zealand; the Eurasian

*F. rubra* var. *rubra* ("Red Fescue") is also widely cultivated; *F. glauca* ("Blue Fescue") is cultivated as an ornamental. – The Eurasian-African *Cynosurus* (3–4) has one species, *C. cristatus* ("Crested Dog's Tail") which is planted in New Zealand and Europe for pasture. – The species of the north-temperate and South American genus *Briza* (20) are known as "Quaking Grasses", and the inflorescences of several species are used in dried floral arrangements. – The closely related temperate European and New World genus *Vulpia* (25–30) forms wild hybrids with *Festuca,* several of which have well-established binomial names. Several species of *Vulpia* are temperate weeds. – *Alopecurus* (50) is a genus of Northern and Southern temperate regions; *A. pratensis* ("Meadow Foxtail") is a pasture in Europe and North America. – The temperate Eurasian and North African *Lolium* (12), "Rye Grasses", has three economic species, the Mediterranean *L. multiflorum* (Fig. 209 A–B), widely planted as a temperate pasture, *L. perenne* also widely planted as a temperate pasture and the Mediterranean *L. rigidum,* widely planted in Southern Australia. In addition, the hybrid *L. perenne* and *L. multiflorum* is favoured for its winter growth. Another Mediterranean species, *L. temulentum* ("Darnel"), is a weed of winter cereal in the temperate zone and can be particularly troublesome if its grain is harvested with the crop, for the grain is frequently infected with a fungus that poisons the resulting flour. – The temperate Eurasian *Dactylis* (5) provides the pasture species *D. glomerata* ("Cocksfoot", "Orchard Grass"). – The genus *Puccinellia* (110) is largely restricted to salt marshes and alkaline soils throughout its range which is North-temperate except for a few South African species; *P. airoides* ("Alkali Grass") is occasionally cultivated in the U.S.A.

**The Sesleria Group.** A further group of genera with small spikelets is centred on the genus *Sesleria.* In these the endosperm always lacks lipids. In addition the leaf margins and the style bases are often connate. *Sesleria* grows mainly in Western Asia, extending to the Mediterranean and Europe. The genus shows a preference for calcareous soils.

There are two groups of genera with large spikelets. In one group of about eight genera the endosperm always has lipids and both the lodicles and leaf-sheath margins are free. The genus *Avena* is typical of this group (Fig. 209 A–L).

**The Avena Group.** – *Avena* (70) is widespread in temperate regions and on tropical mountain tops; *A. sativa* ("Oats") is cultivated in the tem-

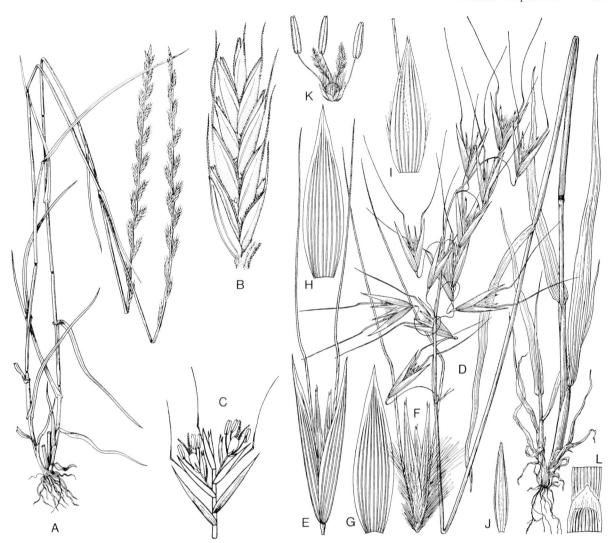

**Fig. 209.** Poaceae. **A–B** *Lolium multiflorum*. **A** Plant. **B** Spikelet. (CABRERA 1968). **C** *Avena*, schematical presentation of spikelet. (POTZTAL 1964). **D–L** *Avena ludoviciana*. **D** Plant. **E** Spikelet. **F** Spikelet with glumes removed. **G** Lower subtending glume. **H** Upper subtending glume. **I** Lemma, the bristle cut. **J** Palea. **K** Flower. **L** Ligule. (BOR 1968)

perate zone to about 70° N; *A. fatua* ("Wild Oats") is a serious weed; *Helictotrichon* (90) is restricted to the temperate region of the Old World save for a few American species; *Arrhenatherum* (6) of the Mediterranean and Europe has one species, *A. avenaceum* ("Oat Grass") that is a valuable pasture and hay species. Its stems are often swollen and tuberous at the base. *Danthoniastrum* (1) from South-East Europe is unusual in that it has flowers with three lodicles.

*The Melica Group.* The second group of about nine genera with large spikelets may be represented by the genus *Melica*. Here the endosperm, unlike that in the first group, lacks lipids and both the sheath margins and lodicles are usually fused. – The cosmopolitan *Glyceria* (40) is largely restricted to wet habitats and has several species which provide useful fodder; *Melica* (70) is also largely cosmopolitan (absent from Australia) but grows mainly in woodlands.

## Subfamily Stipoideae

*The Stipa Group.* The only group of the subfamily. Mainly perennial tussock, shrub or reed grasses, some of which grow to 3 m tall. They have one-flowered spikelets and lemmas that usually terminate in a hygroscopic awn. Disarticulation takes place above the subtending glumes and many of the species have three lodicles. The chromosomes are small and occur in multiples of 18, 12, 13 and 15. The species are widespread in temperate and cool temperate areas and especially dry steppes.

**Fig. 210.** Poaceae. **A–J** *Stipa capensis*. **A** Plant. **B** Junction of sheath and lamina, with very short ligule. **C** Spikelet. **D** Lower glume. **E** Upper glume. **F** Lemma. **G** Palea. **H** Fruiting spikelet. **I** Flower. **J** Caryopsis. (Bor 1968). **K** *Stipa*, spikelet, schematically. (Potztal 1964). **L–T** *Eragrostis cilianensis*. **L** Plant. **M** Spikelet; note the characteristic, distinct lateral veins of the lemmas, cf. **P**. **N** Lower glume. **O** Upper glume. **P** Lemma. **Q** Palea. **R** Flower. **S** Grain. **T** Junction between sheath and lamina with hair-type ligule. (Bor 1968)

Though the Stipoideae and the Pooideae share similar embryo characteristics, they appear not to be closely related. All genera of the Stipoideae have small chromosomes and many have flowers with lodicle characters which suggest affinities with the Bambusoideae, a viewpoint supported by the microhairs on the leaves of *Nardus* and some species of *Stipa,* and the presence of an abaxial ligule in some *Stipa* species. However, the lack of either arm cells or fusoid cells, and the well-developed lamina on the first seedling leaf militate against including the Stipoideae among the Bambusoideae, though their affinities lie within that part of the family.

The genus *Stipa* (300) (Fig. 210A–J) is often an important component of grassland vegetation and several species contribute significantly to the grazing potential in some areas. A few species have leaves which contain a narcotic that induces sleep in grazing animals. The Mediterranean *S. tena-*

*cissima* ("Esparto Grass") provides fibre for rope, sails, mats and paper making and the central European *S. pennata* ("Feather Grass") is cultivated for its decorative inflorescences, the lemmas of which have feathery awns up to 35 cm long.

*Nardus (1) stricta* is a wiry tufted perennial whose inflorescence is a one-sided spike. The species is common in overgrazed pastures throughout North-temperate Eurasia, and is widely distributed now in similar habitats elsewhere.

*Ampelodesmos (1) mauritanica* is a tall reed grass occurring mainly in North Africa. Its young leaves make a useful fodder and the older leaves are woven into matting or used as a source of fibre. The species is sometimes cultivated as an ornamental and has become a weed in California.

## Ecdeiocoleaceae D.F. Cutler & Airy Shaw (1965)
1:1  (Fig. 211)

Rhizomatous herbs with simple, erect, cylindrical solid culms up to 1 m tall. The leaves are restricted to the base of the culm save for a single, bladeless leaf midway along the flowering culm. The leaves are distichous, consisting of open bladeless sheaths up to 10 cm long. The rhizomes are glabrous, with overlapping scales.

The stomata are paracytic. Silica is present as sand in the chlorenchyma of the culm but is otherwise absent. Vessels are lacking from the leaves but are present in the culms. Their perforation plates are oblique and simple or scalariform. The chlorenchyma is made up of peg cells. The sclerenchyma is poorly developed and does not form a continuous cylinder in the culm as in Restionaceae.

The plants are dioecious or monoecious with the flowers arranged in single- or several-flowered spikelets aggregated into a terminal ovoid spike with numerous imbricate bracts, each larger than the perianth. The flowers have no bracteoles.

The tepals are 3 + 3 and glumaceous; the members of the same whorl are unequal, two members of the anterior whorl being strongly compressed laterally and ciliate at the apex. The male flowers have three to four stamens on separate filaments. The anthers are basifixed and tetrasporangiate, and dehisce latrorsely. The pollen grains are ulcerate with an annulus and an operculum as in Poaceae. The ovary in the male flowers is rudimentary. The pistil of the female flowers has a bilocular ovary, each locule with a single apical, pendulous orthotropous ovule. The two stylodia are long, each with stigmatic papillae extending nearly to its base; there are three minute staminodes in the female flowers. The fruit is a one- to two-seeded nutlet. The embryology is unknown.

*Ecdeiocolea* (1) is endemic to South-Western Australia. The species has no commercial value. – L.A.S. Johnson and Briggs (1981) claim the existence of a second, undescribed genus.

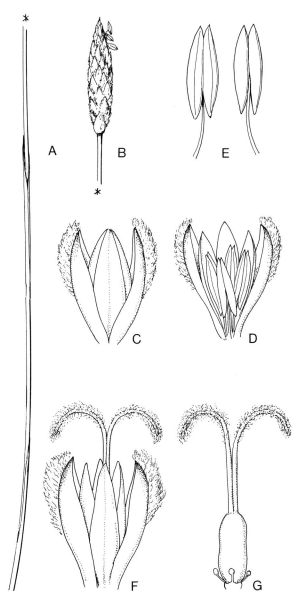

**Fig. 211.** Ecdeiocoleaceae. *Ecdeiocolea monostachya.* **A–B** Culm with spike. **C** Male flower. **D** Same, with one of the inner tepals removed. **E** Stamen, front and back view (the anthers are introrse). **F** Female flower. **G** Same, with tepals removed to show staminodes. (Redrawn from Diels and Pritzel 1905)

## Anarthriaceae D.F. Cutler & Airy Shaw (1965)
1:7  (Fig. 212)

Rhizomatous herbs to about 1 m tall; the rhizomes are covered with overlapping, glabrous scales. The aerial stems are solid or fistulose and either much branched and leafy, bearing many inflorescences (e.g. *A. prolifera*), or are unbranched and leafless, terminating in a single inflorescence (e.g. *A. squar-*

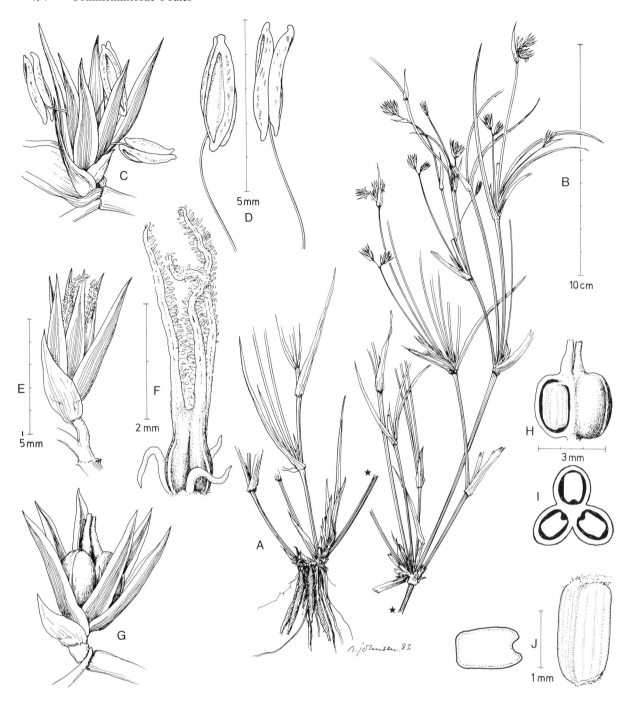

**Fig. 212.** Anarthriaceae. *Anarthria prolifera.* **A–B** Basal part and branch of male plant. **C** Male flower. **D** Stamen in different views. **E** Female flower. **F** Pistil and staminodes. **G** Fruiting flower. **H** Capsule, one locule in longitudinal section. **I** Capsule transverse section. **J** Seed, to the *left* in transverse section. (Orig. B. JOHNSEN)

*rosa*). In the latter case the leaves arise directly from the rhizome and are markedly equitant. The leaves are linear or filiform, laterally compressed or equitant, rarely ligulate, with a sheathing base whose margins are unfused. The two keeled prophylls are often conspicuous. The stomata are paracytic and grass-like. Silica is absent. Vessels are present in leaf and culm, those in the culm having scalariform perforation plates. The chlorenchyma is composed of peg cells. The sclerenchyma does not form a vascular cylinder in the culms but occurs as girders opposite the vascular bundles (CUTLER 1969).

The plants are dioecious or rarely monoecious. The inflorescence is a lax panicle of single-flowered spikelets each branch of which is subtended by an elongated leaf-like bract. Also each flower is subtended by a single bract and sometimes has one or two bracteoles.

The tepals are 3 + 3, more or less similar and glumaceous. The male flowers have three free or connate stamens with tetrasporangiate, basifixed anthers dehiscing latrorsely. The pollen grains are ulcerate with an annulus about the pore and are scrobiculate. The ovary in the male flowers is rudimentary or absent.

The ovary of the female flowers is trilocular, each locule bearing a single apical, pendulous, orthotropous ovule. The three stylodia each have a decurrent stigma. The female flowers may have three staminodes inserted opposite the inner tepals, or staminodes may be lacking. The embryology is unknown. The fruit is a one-seeded capsule or a nutlet.

*Anarthria* (7) is endemic to South Western Australia, growing in sandy or swampy soils. No species has been exploited commercially.


**Restionaceae** R. Brown (1810)    40:400
(Figs. 213–214)


Rhizomatous or tufted, rush-, sedge- or bamboo-like plants up to about 4 m tall but sometimes scramblers. Rhizome with overlapping scale leaves, glabrous or densely woolly with multicellular hairs. The aerial shoots are simple or branched and sometimes flexuose; they may be terete, quadrangular, polyhedral or compressed and are either solid or fistulose. In some genera (*Restio, Loxocarya, Elegia*) fascicles of short branches may occur in the leaf axils.

The leaves are sheathing and mostly distichous. Distally the sheath may taper or be rounded without any evident lamina. When present the lamina is only weakly developed, being short and tongue-like or a short, often deciduous mucro. The boundary between the sheath and lamina is marked by a short membranous adaxial ligule in a few species (*Restio complanatus*) or by the extension of the sheath margins as two membranous lobes on each side of the lamina. The sheath margins are never fused. The sheaths may be closely appressed to the stem or widely inflated; they are persistent or deciduous according to the species concerned.

The stomata are paracytic, often with grass-type guard cells. Silica is widespread in the tissues as sphaeroidal nodules or as amorphous granular material. It is rarely present, as in leaf epidermis of *Lepyrodia* and *Thamnochortus*. Vessels with scalariform, simple or reticulate perforation plates are present in the culms; vessels with only simple perforations occur in the rhizomes and roots, whereas vessels are absent from the leaves. The chlorenchyma is often composed of peg cells; these have peg-like outgrowths or projections from the anticlinal walls, the pegs from adjacent cells being opposed. (See CUTLER 1966, 1969.) The sclerenchymatous tissue in all Restionaceae forms a closed cylinder in the culm (an important autapomorphy of the family as conceived in its strict sense).

The plants are dioecious, less commonly monoecious; or the flowers are bisexual (*Lepyprodia hermaphrodita*). The flowers are small and arise in the axil of more or less scarious bracts; one or two bracteoles are present in a few species. In most genera the flowers are arranged in spikelets with rigid imbricate subtending bracts, the lower of which are often sterile. In other genera the flowers are arranged in spike-like panicles with the glumes not or scarcely imbricate. Spikelets when present have one to many flowers. There may be considerable sexual dimorphism in the inflorescences of the male and female plants, while the male inflorescences of allied species may be confusingly similar. In *Alexgeorgia*, female plants produce flowers at or below ground level with only the stylodia emergent, whereas the male plants produce aerial panicles (Fig. 213).

The tepals are mostly 3 + 3, rarely 2 + 2, 3 + 2, 3 + 0 or even absent (as in male flowers of *Restio* and *Loxocarya*). They are usually free or the inner members are connate at the base, sometimes enlarging after anthesis. In the male flowers the stamens are mostly three or two, or rarely one, inserted opposite the inner whorl of tepals. The filaments are sometimes united into a column. The anthers are dorsified or basifixed (*Lyginia*), bisporangiate or in *Lyginia* and *Hopkinsia* tetrasporangiate. In some genera the apex of the connective is exserted into an apiculus (*Hypodiscus*). Dehiscence of the tetrasporangiate anthers is latrorse, that of the bisporangiate introrse. A rudimentary ovary may be present in the male flowers. The pollen grains are scrobiculate and ulcerate, with or without an annulus (CHANDA 1966; LADD 1977), but with the margins of the ulcus always

**Fig. 213.** Restionaceae. *Alexgeorgia subterranea,* a genus of two species described as late as 1976 (CARLQUIST 1976) from Western Australia; note the strong dimorphism between the female plant (to the *left*), where the floriferous shoots only just reach above the ground, and the male, where the inflorescence is branched and compound. (Illustration placed at our disposal by J.P. JESSOP, Adelaide)

differentiated from the interapertural walls (LINDER, personal communication). They are two- or three-celled when shed.

In the female flowers the ovary is sessile or shortly stalked and has one to three locules each bearing a single, pendulous, orthotropous, tenuinucellate ovule. There are one style or three free or variously connate stylodia with papillose or shortly feathery stigmas on their inner surfaces. Staminodes may be present or absent.

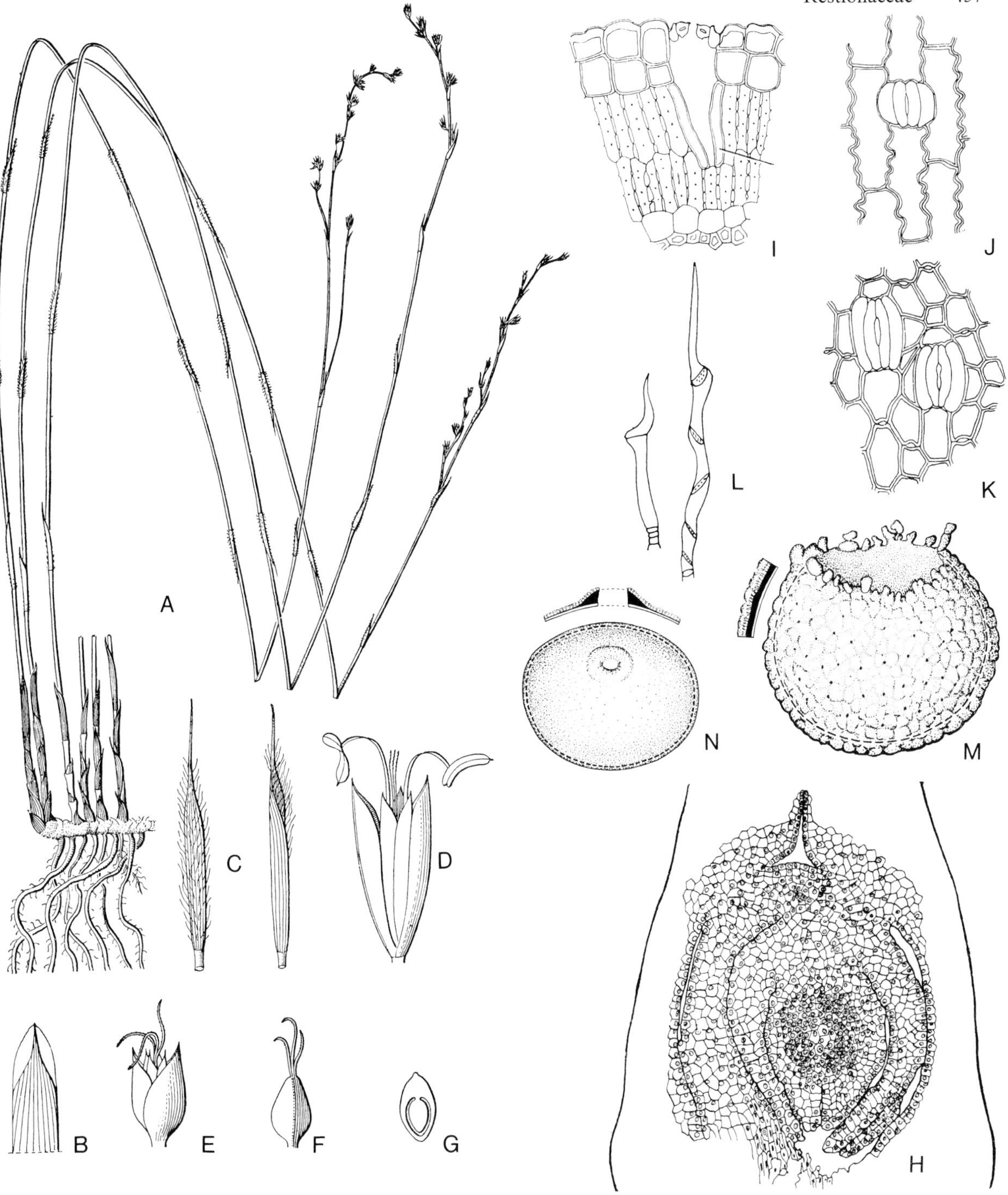

The fruit is a loculicidal capsule or is indehiscent (a nutlet). The seed has a small lenticular or obovate embryo in a copious starchy endosperm. The embryo sac is of the *Polygonum* Type and the endosperm of the Nuclear Type. In several species there is secondary multiplication of the antipodals. Adaptations to wind or animal dispersal occur in some genera; thus the fruits of *Restio, Thamnochortus, Staberoha* and some species of *Calopsis* are shed along with their perianths which aid in

**Fig. 214.** Restionaceae. **A** and **C–D** *Leptocarpus barbatus.* **A** Plant. **C** Cauline sheath leaf. **D** Bisexual flower. **B** and **E–G** *Leptocarpus disjunctus.* **B** Apex of leaf sheath. **E** Female flower. **F** Pistil. **G** Young fruit, longitudinal section. (**A–G** from Bakker 1958) **H** *Hypodiscus aristatus,* ovule. (Krupko 1962) **I–J** *Elegia obtusiflora,* stoma in transverse section and surface view. **K** *Chondropetalum marlothii,* stomata. (**I–K** from Cutler 1969) **L** *Loxocarya pubescens,* trichomes. (Solereder and Meyer 1929) **M** *Hypolaena lateriflora,* pollen grain. **N** *Staberhoa cernua,* pollen grain. (**M–N** from Erdtman 1952)

wind dispersal; a few South African genera (*Hypodiscus, Willdenowia, Cannomois*) have nuts with an elaiosome attractive to ants; in *Restio egregius* the fruit is shed enclosed in the perianth which develops a basal elaiosome. Germination of the seed is epigeal.

*Chromosome Numbers.* The chromosome complement is variable with reported basic numbers of $x = 7, 8, 9, 11, 12$ and $13$ (FEDOROV 1969).

*Chemistry.* Biochemically the family is not well known. Flavonols are widespread but flavones and flavone c-glycosides and flavonol sulphates occur only infrequently (HARBORNE 1979). Tricin has not been recorded (DAHLGREN and CLIFFORD 1982), and neither have any cyanogenic compounds.

*Distribution.* The Restionaceae are widely distributed in the Southern Hemisphere, excluding South America, where a single species occurs. They also extend into Southeastern Asia. They grow in sandy or peaty soils subject to seasonal rains or flooding or in open areas and savannahs in arid regions, rarely in salt marshes. There are two major centres of diversity, one in Southern Africa and the other in South-Western Australia. There are no genera in common between these two areas if *Restio, Hypolaena* and *Leptocarpus* are regarded as unnatural assemblages, which is obviously the case. An indication of the heterogeneity of *Restio* s. lat. as presently defined (in Africa as well as Australia) is afforded by the pollen, which displays remarkable variation (LINDER and FERGUSON, personal communication), and the culms which possess protective cells in the African taxa, while many of the Australian taxa have pillar cells (CUTLER 1969).

*Relationships, Taxonomy.* Two schemes have been proposed for the recognition of subfamilies. In one those species where the fruits are bi- or trilocular capsules, are separated from those with unilocular nuts (MASTERS 1967). The distinction is not absolute and species of a few genera transgress the boundary. In the other the species with tetrasporangiate anthers are separated from those with bisporangiate anthers (BENTHAM and HOOKER 1883). Of the two schemes the latter is supported by L.A.S. JOHNSON and BRIGGS (1981). Not only are the subfamilies defined unambiguously, but they exhibit a greater geographical coherence. The genera with tetrasporangiate anthers (*Lyginia, Hopkinsia*) are lacking from Southern Africa and occur in South-Western Australia. The South African genera seem to form a coherent group, whereas the others are paraphyletic. The family is prob-

ably most closely related to the two endemic families Anarthriaceae and Ecdeiocoleaceae.

The most widespread genus is *Leptocarpus* (25), which occurs in Australia, New Zealand, Chile, New Guinea, and Timor, with a disjunction to Malaysia and South-Eastern China. Other large genera are *Restio* s.str.(?), *Elegia* (ca. 35) and *Thamnochortus* (ca. 37) all in Southern Africa.

The family is only of minor economic importance. Species of *Chondropetalum* and *Thamnochortus* are employed for thatching; *Restio tetraphyllus* is cultivated as an ornamental and several others are used for matting or making brooms (*Restio triticeus, Elegia capensis* and *Cannomois virgata*). Several species furnish dry floral ornaments (*Thamnochortus, Elegia* and *Cannomois*) traded as "Cape Grasses". The Restionaceae form an important component of the vegetation in the South-Western Cape Province, South Africa, and so are of great importance in ensuring a good natural pasturage.

## Centrolepidaceae S.L. Endlicher (1836)  4:30–35 (Fig. 215)

Small, tufted, annual or perennial plants which resemble small grasses or even mosses. The perennial plants are rhizomatous. Roots, rhizomes and stems have vessels with scalariform perforation plates in their vascular strands, and such vessels or tracheids are found in the vascular strands of the leaves. The leaves are clustered at the base of the culms or are imbricately inserted on the stems. The leaves are linear, lanceolate or setaceous and basally have a distinct, membranous, broadened, open sheath. The apex is often acute and hyaline. Stomata are paracytic and of the Grass Type. Multicellular hairs are present in certain species of all the genera and are either unbranched or branched. Inclusions in the form of rectangular (silica?) bodies are present in some epidermal cells of *Gaimardia australis* and granular material in *G. setacea*. The sclerenchyma is poorly developed and never forms a closed cylinder.

Each plant has several or many culms which end in a complex spike-like or head-like inflorescence bearing two to several distichous, glume-like bracts. These may be hairy or glabrous and are sometimes trilobate or end in a narrow bristle. Each bract encloses a group of male flowers or one to several female flowers, or, most frequently, one or more bisexual pseudanthia. Each of these is a spikelet or rhipidium of one to two male flowers and one to several superposed or collateral fe-

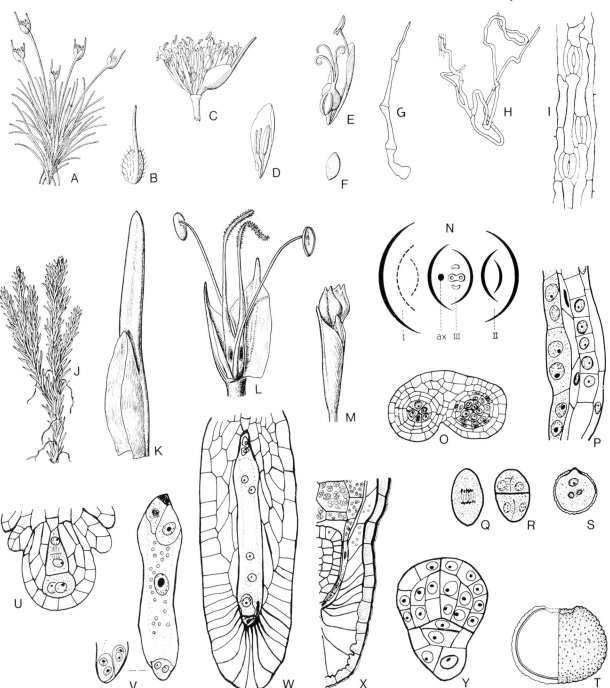

**Fig. 215.** Centrolepidaceae. **A–G** *Centrolepis fascicularis.* **A** Plant. **B** Spathal bract. **C** Inflorescence, one spathal bract removed. **D** Floral group consisting of two laterally fused female flowers and one male flower. **E** Similar group with three fused monocarpellary female flowers and one male flower. **F** Seed. (DING HOU, 1957). **H** *Gaimardia fitzgeraldii*, hair. **I** *Pseudalepyrum ciliatum*, stomata. (**H–I** from CUTLER 1969) **J–N** *Gaimardia australis.* **J** Branch. **K** Leaf. **L** Spikelet with two fused monocarpellary female flowers and two male flowers (one stamen each). **M** Opening double fruit consisting of carpels of two adjacent female flowers. **N** Diagram of spikelet.

(**J–M** from CORREA 1969; **N** from DING HOU 1957) **O–Y** Embryological details. **O** Bisporangiate anther, transverse section (*Brizula pumilio*). **P** Anther wall showing glandular, binucleate tapetal cells (*Brizula pumilio*). **Q–S** Stages of successive microsporogenesis and ripe pollen grain (*Centrolepis aristata*). **T** Mature pollen grain (same species). **U** Ovule (same species). **V** Embryo sac (*Brizula pumilio*). **W** Nuclear endosperm formation (*Centrolepis glabra*). **X** Part of seed showing half of embryo and some of the starchy endosperm (*Centrolepis polygyna*). **Y** Embryo (*Centrolepis minima*) (**O–S** and **U–Y** from HAMANN 1975, **T** from ERDTMAN 1952)

male flowers; especially when collateral, the female flowers are more or less connate. The pseudanthia frequently have one to three small, hyaline bracts or bracteoles.

The stamens (= male flowers) have a narrow filament and a bisporangiate (monothecous), dorsifixed, versatile anther which opens by a longitudinal slit. The pollen grains, unlike those of Hydatellaceae, are ulcerate and three-celled. They are superficially similar to those in many Australian Restionaceae (CHANDA 1966), but detailed investigation shows that the fine structure is very different (LINDER and FERGUSON, personal communication). Thus there is not a proper annulus, the scrobiculi are different, and there are no granules on the apertures.

The ovaries (= female flowers) are utricle-like, monocarpellary and unilocular, with one apical, pendulous, orthotropous ovule. The single, apical style is filamentous, consisting in its upper part of one-sided stigmatic papillae. The ovule is tenuinucellate, and no parietal cell is cut off from the archesporial cell. The nucellar epidermis does not divide periclinally. The embryo sac is formed according to the *Polygonum* Type; it is much longer than broad and has three well-developed, long-persistent, antipodal cells, generally with doubled nuclei. Endosperm formation is nuclear; the endosperm finally becoming multicellular. The fruit is a little follicle with a membranous pericarp which dehisces longitudinally, emitting the seed. Collateral follicles are generally fused into a capsule-like collective fruit. The seed has a rather thin seed coat, which consists mainly of the inner epidermis of the inner integument, while the outer integument is very thin. There is hardly any nucellar tissue; this remains only as a one-layered cap of palisade-like cells at the micropylar end (cf. Hydatellaceae), while the endosperm is copious and starch-rich. The embryo is minute and obconical.

These details are widely different from those in the two genera of Hydatellaceae which until recently were included in Centrolepidaceae. Both families are chemically almost uninvestigated.

*Distribution.* Centrolepidaceae are distributed in part of Australia, in Tasmania, New Zealand, Malaysia and South-Eastern Asia. A single species, *Gaimardia australis,* also enters subantarctic South America. The Centrolepidaceae is thus largely an Australasian family. It is doubtful whether the South American occurrence is ancient.

The small annual genera *Brizula* (5) and *Aphelia* (1), both with Australian-Tasmanian distributions, have unisexual and bisexual inflorescences respectively. The female flowers in these genera are separate, and not fused in groups, and the leaves are setaceous and basally concentrated.

*Centrolepis* (25) has bisexual inflorescences generally subtended by two bracts which may be apically extended. The inflorescences seem to consist of rhipidial units and contain female as well as male flowers. The female flowers are often inserted at different levels and are partly united along the stipe and inner part of the ovary. The species grow in sandy, moist habitats, most of them in Tasmania and Southern and Western Australia; a few species reach, for example, to the Philippines and Borneo.

*Gaimardia* (2) has regularly constructed synanthia with two male and two female flowers, the latter partly fused and on the same level. One species is found in New Zealand, the other in Tierra del Fuego, Cape Horn, the Falkland Islands, etc.

The Centrolepidaceae agree well in most features, such as endosperm formation, starchy endosperm, occasional presence of silica bodies, stomatal construction, etc., with the other families of Poales (in particular the Restionaceae), where they obviously have their proper position. The doubts as to their relationships to Hydatellales are mentioned under that order.

It should be stressed that the synanthia (i.e. what we interpret as groups of flowers above) are sometimes regarded as flowers, so that *Gaimardia,* for example, is considered to have bisexual and dimerous flowers. This interpretation is at present usually ignored.

# Superorder Cyclanthiflorae

*One Order:* Cyclanthales

Perennial, rhizomatous, acaulescent or short-stemmed herbs or shrubs; they are terrestrial or rarely epiphytic, or root-climbing lianas with a long slender stem often up to 8 m long, up to 30 m in *Thoracocarpus.* Acaulescent and short-stemmed genera have adventitious roots from the rhizome and aerial stem; the lianas have short climbing roots in great number and also, in some genera, long, unbranched rope roots. Air cavities occur in the root cortex. The vascular strands of the roots contain vessel tracheids or vessels with scalariform perforation plates. The stems generally develop cork (except in *Cyclanthus*); their vascular strands are without vessels.

The leaves are spirally set or rarely distichous and are petiolate with a sheathing base. The lamina is bifid or more rarely flabelliform-parted or entire and has one to three costae (a distinct central costa is found in all genera except *Cyclanthus*) and plicate ptyxis. In some genera the lamina is thick, coriaceous and lustrous.

The stomata, present on both sides of the lamina, are arranged in rows and are tetracytic, with two lateral and two terminal subsidiary cells. Mucilage canals occur in the vegetative parts of all genera except *Cyclanthus* where they are restricted to the inflorescence, and tannin cells occur in all genera except *Cyclanthus,* whereas in this genus there are laticiferous vessels (WILDER and HARRIS 1982). Crystal raphides are widely distributed, styloids occur in particular in *Evodianthus,* and crystal druses in the leaves of *Cyclanthus.* Silica bodies are lacking. The vascular strands of the leaves contain vessel tracheids or vessels with scalariform perforation plates.

The pedunculate inflorescences are axillary or terminal spadices (WILDER 1977a, b). Each spadix is subtended by two to four (to eight) lanceolate-ovate spathes which vary in size and colour (green, white, red or yellow) and enclose the spadix when young. The spadix appears to be unbranched and varies from long-cylindrical to ovate or even spherical; it contains from rather few to very numerous flowers, which in most genera are arranged in a chess-board mosaic, each female flower being surrounded by four male ones. These groups form a shallow spiral from the base to the apex of the spadix. *Cyclanthus* deviates in having the female and male flowers laterally fused into alternating cycles (rarely partly in spirals), the individual flowers being indistinguishable.

The male flowers in most genera consist of a central or eccentric pedicel, a floral receptacle with a symmetrical or asymmetrical, simple (in *Evodianthus* seemingly double) perianth, often reduced on one side or rarely wanting, and generally 10–20 (–150) stamens. The tepals vary in shape and may have an adaxial secretory glandule. The stamens are basifixed, generally elongate, tetrasporangiate, and usually have a thin, rarely basally widened connective; sometimes the anthers have an apical secretory glandule. Basally, the filaments are often widened and bulb-like. In *Cyclanthus* the male flowers are reduced and represented by a cycle with numerous stamens in four rows. The tapetum is secretory and the tapetal cells become two-, three-, four- or more-nucleate. Microsporogenesis is successive. The pollen grains are two-celled and elongate, rarely crescentic. They are sulcate or sulcoidate in most genera, ulcerate in many species of *Asplundia* and ulceroidate in taxa of *Sphaeradenia* and in *Cyclanthus. Thoracocarpus* has biforaminate pollen grains, and in *Carludovica* the pollen grains have a single pore at one end, while the distal face is provided with a longitudinal groove.

The female flowers in most genera are tricyclic, generally tetramerous, with one perianth whorl, one whorl of staminodes and one of carpels. They are epigynous or half-epigynous, nearly hypogynous only in some species of *Sphaeradenia* and *Stelestylis.* The tepals are usually distinct or rarely indistinct, often fleshy and free or basally connate. Opposite them are the four staminodia which are obtusely subulate, vermiform or filiform, (1–)3–5(–10) cm long and white to red or yellow, and sometimes supplied with an apical rudimentary anther. The pistil is four-carpellary and unilocular and contains four parietal or apical placentae (rarely one apical placenta) with numerous ovules. The four stylodia may be separate or basally fused into a short style; often the stylodia are very short or lacking and the four stigmas, which are laterally compressed or flat, broad and fleshy, then become subsessile. In *Cyclanthus* the flowers are reduced and laterally fused into pistillate cycles, the ovaries forming a continuous cavity with numerous parietal ovules. These cycles are separated from the male ones by thin wings which may represent tepals. Grooved lamellae on the inner side of these are staminodial.

The ovules are anatropous and weakly crassinucellate to tenuinucellate. A parietal cell is never cut off from the archesporial cell, which functions directly as the megaspore mother cell. In all genera except *Cyclanthus* the nucellar epidermis divides periclinally to form a nucellar cap. Embryo sac formation is of the *Polygonum* Type, and endosperm formation is helobial in the cases studied.

The fruits are berries, which in the early stages usually cohere more or less within the spadix but are free in a few genera. In *Carludovica* they cohere as large flaps and when ripe are rolled off the nodding inflorescence, being bright orange-coloured on the inside; in most of the other genera the berries loosen successively and become free. In *Cyclanthus* each of the fruiting cycles forms a syncarp, the hollow interior of which is filled with seeds; when ripe the syncarps break into halves and the seeds emerge from their openings.

The seeds are elliptic, ovoid, rectangular, or crescentic in some species of *Sphaeradenia,* and are provided with long terminal appendages in *Stelestylis*. The seed coat is variable in thickness. The endosperm is copious and its cells contain fatty oils and protein in all genera. In *Dicranopygium,* both endosperm and embryo contain starch. The embryo is small to medium-sized, linear-cylindric and straight or rarely curved.

*Chemistry.* The chemistry of Cyclanthales is incompletely known. Cyanogenic compounds are known in *Carludovica*. HARRIS and HARTLEY (1981) reported absence of UV-fluorescent compounds in the cell walls of *Carludovica,* but found p-coumaric acid in small quantities.

*Distribution.* The order contains a single family, Cyclanthaceae, which is restricted to the Neotropics, ranging from Southern Mexico in the north through Central America, the West Indies (mainly the lesser Antilles) and Northern South America as far south as between central Bolivia and South-Eastern Brazil. The centre of the order

is in the region of Colombia, where two endemic genera occur.

*Relationships.* Cyclanthales is generally acknowledged as a taxonomically fairly isolated order, and is here treated in its own superorder, Cyclanthiflorae (sometimes called Synanthae). According to HARLING (1958), who has made an extensive study of the family, it is probably most closely allied to *Freycinetia* of the Pandanaceae (Pandaniflorae), to which it shows great similarities in flower and seed characters and also in vegetative anatomy, embryology and pollen morphology.

The leaf differentiation in *Carludovica* and some other genera is reminiscent of that in certain fan palms (Areciflorae), although it is uncertain whether the similarities are explicable by convergent evolution. Unlike palms, the Cyclanthaceae lack silica bodies but stomatal structure agrees with that in Areciflorae. The inflorescences, subtended by two to several spathes, conform to the pattern of Pandaniflorae and Areciflorae; the superficially simple spadix may be derivable from a branched type (thyrsus or panicle), the lateral branchlets of which are perhaps represented by an apical (central) female flower surrounded by four male flowers, a condition easier to associate with that in certain palms than with that in extant Pandanales. The tetramery of the flower is conspicuous in Cyclanthales, but penta-, hexa- and octomery may occur. As in most Areciflorae and all Pandaniflorae, the flowers are unisexual, but in the female flowers the staminodes sometimes have functional microsporangia which in some cases have been shown to produce normal pollen (HARLING 1946); also apical secretory glandules, as on the stamens of the male flower, occur on the staminodes of female flowers in some species of *Sphaeradenia*. This all suggests that the complete, bisexual flowers are not far away in phylogeny. The sulcate pollen grains found in several genera represent a more primitive type than the ulcerate in Pandaniflorae, and also the helobial endosperm in Cyclanthales, by which this order differs from both Pandaniflorae and Areciflorae, may represent a primitive state.

Similarities to Areciflorae are the petiolate well-differentiated (but very different) leaves, the abundance of crystal raphides, the inflorescence type and the tendency for short styles. There are, however, numerous differences which make a close affinity highly improbable.

*Cyclanthus* deviates strongly from other Cyclanthaceae and shows greater similarity to Arales than do the other members (e.g. in the presence of laticifers and the tenuinucellate ovules without a nucellar cap, but this is likely to be by convergence). We assume that *Cyclanthus* in most of its peculiarities is more derived than are other Cyclanthaceae.

Affinities with Typhales have sometimes been claimed but can no doubt be dismissed (see, for example, HARLING 1958).

# Order Cyclanthales

*One Family:* Cyclanthaceae.

See the superorder, above.

## Cyclanthaceae Dumortier (1829)   11:180
(Figs. 216–217)

Description as given for the order.

The family is divisible into two subfamilies, which differ so significantly in numerous characters that they may deserve family status.

### Subfamily Carludovicoideae

Subfamily Carludovicoideae (Fig. 216) contains all species but one of the family, and covers most of the variation described above. The leaves vary from relatively small and entire (*Ludovia, Pseudoludovia*) to bifid or flabelliform-parted (*Carludovica*); sometimes there is heterophylly with small entire and larger, bifid leaves. The sclerenchymatous ring of the roots is more deeply situated than in subfamily Cyclanthoideae, and the stem has a cork layer lacking in the latter subfamily. The leaf mesophyll is differentiated into palisade and spongy parenchyma (which it is not in Cyclanthoideae), mucilage canals occur in stem, leaf petioles, peduncles and inflorescence and also tannin cells are present in the vegetative parts. In the inflorescence the flowers are arranged in groups with one female flower surrounded by four male, such groups being arranged in a spiral. The flowers are distinguishable from each other, not fused into ring-like synanthia. Embryologically the subfamily differs from the other subfamily, Cyclanthoideae, in having periclinal divisions in the nucellar epidermis, so that the embryo sac becomes situated beneath a nucellar cap.

The subfamily covers the whole distributional range of the family. Its taxa are mainly found in shady habitats, rarely in more open, often disturbed vegetation types, as is for example *Carludovica palmata*. Some are climbing lianas, and a few are epiphytic. Many species grow in moist glades or on river banks, and a few in coastal mangroves. Pollination seems to be mainly by weevils (Coleoptera) attracted by the strong scent produced by the flowers. The developing spadix produces considerable heat and the spathes may be strongly coloured. Dispersal occurs largely by ants which

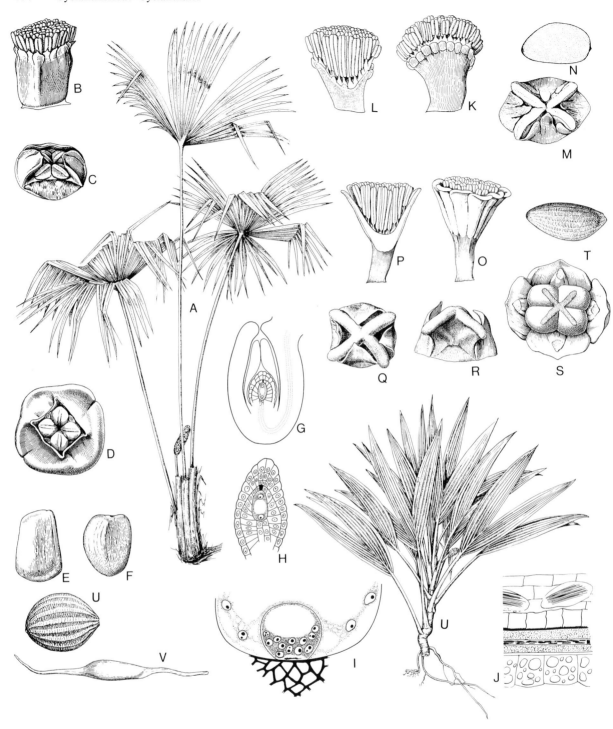

**Fig. 216.** Cyclanthaceae subfamily Carludovicoideae. **A–J** *Carludovica palmata.* **A** Plant. **B** Male flower. **C** Female flower, one tepal removed. **D** Fruiting female flower. **E–F** Seeds, different aspects. **G** Ovule. **H** Nucellus showing the nucellar cap formed by periclinal divisions of the epidermis. **I** Basal part of embryo sac, showing hypostase with thick cell walls, and the chalazal chamber of the helobial endosperm. **J** Section of seed coat, showing raphide bundles. (**A** from WILDER 1977b; **B–J** from HARLING 1958). **K–M** *Asplundia multistaminata.* **K** Male flower. **L** Same, longitudinal section. **M** Female flower. **N** *Asplundia rivularis,* seed. **O–S** *Thoracocarpus bissectus.* **O** Male flower. **P** Same, longitudinal section. **Q** Female flower. **R** Same, side view, one tepal removed. **S** Female flower, fruiting stage. **T** Seed. (**K–T** from HARLING 1958) **U** *Dicranopygium* sp., plant. (WILDER 1977a) **V** *Stelestylis surinamensis,* seed. (HARLING 1958)

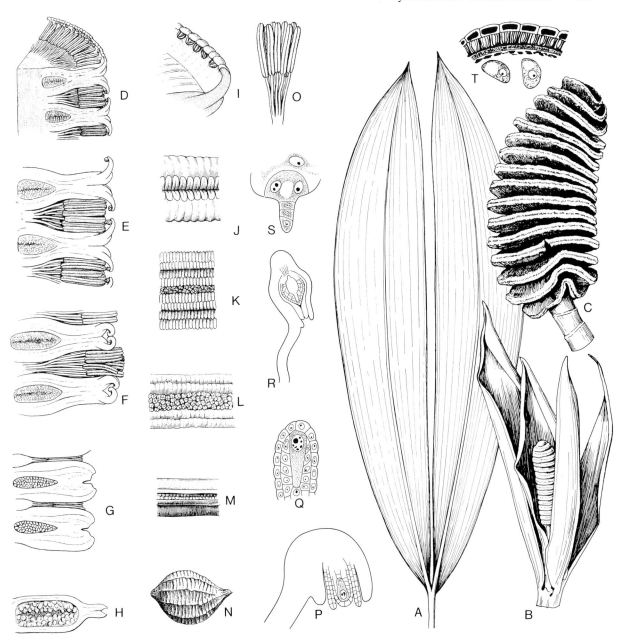

often visit the fruiting spadices, and by water for the riverside species, but birds, mammals or other animals may be attracted by the flesh of the fruit and strong fruit colour of certain taxa.

*Carludovica* (3), ranging from Guatemala to Bolivia, is a genus of tall, terrestrial short-stemmed plants with flabelliform-parted, tricostate leaves which are apically dentate or lobate (WILDER 1976). *C. palmata* is cultivated, especially in Ecuador, for the fibres, "Toquilla Straw". These are used in particular for "Panama hats" but also for making mats or baskets. – *Asplundia* (82) species, ranging from Mexico to Bolivia, are root-climbing

**Fig. 217.** Cyclanthaceae subfamily Cyclanthoideae: *Cyclanthus bipartitus*. **A** Leaf. **B** Flowering spadix. **C** Ripe fruiting spadix. **D–G** Longitudinal section of spadix in successively later stages of flowering. **H** Section through pistillate cycle in unripe fruit stage. **I** Detail of pistillate cycle showing part of staminodial lamella. **J–M** Surface view of pistillate and staminate cycles in successively older stages. **N** Seed. **O** Detail of staminate cycle (individual male flowers not recognizable). **P** and **R** Ovule in different stages. **Q** Nucellus, showing the tenuinucellate condition without periclinal divisions of nucellar epidermis. **S** Lower part of embryo sac showing antipodals and binucleate chalazal chamber of the helobial endosperm. **T** Pollen sac wall and pollen grains. (All from HARLING 1958; **A–C** drawn from photograph; **P–Q** redrawn)

lianas or terrestrial herbs or shrubs. The genus is very variable with symmetrical or asymmetrical male flowers. – *Thoracocarpus* (1) *bissectus* is a South American liana with stems up to 30 m long. The pollen grains are biforaminate. – *Dicranopygium* (44) species, ranging from Mexico to Peru, are acaulescent or short-stemmed plants with bifid leaves. – *Sphaeradenia* (38), *Ludovia* (2) and two more genera have distichous leaves with coriaceous, lustrous laminae (with crystal styloids) and apical placentation. The anthers often have apical glandulae. Both genera occur south of Panama, in Northern South America.

## Subfamily Cyclanthoideae

Subfamily Cyclanthoideae (Fig. 217) consists of the genus *Cyclanthus* (1), a tall acaulescent herb, with bipartite leaf blades, the central costa of which is forked at the base into two strong branches, each running to the apex of its segment (WILDER 1981). The stem lacks a cork layer and the roots have a sclerenchymatous ring, three to six cell layers thick, lying directly under a one-layered exodermis. The mesophyll of the leaf lamina is undifferentiated. Tannin cells are lacking and mucilage canals are restricted to the inflorescence, but non-articulate laticifers occur in all parts of the plant except the root and the internal parts of the stem (WILDER and HARRIS 1982). In the inflorescence the flowers are arranged in cycles (only rarely in a spiral), and are laterally fused. The male ones are reduced and those in each cycle form four rows of stamens which are basally fused. The pollen grains are ulceroidate. The female flowers are fused into pistillate cycles, separated from the male cycles by thin wings (=perianth?), inside which there are two larger, grooved staminodial lamellae. The carpels, in two rows per cycle, are fused laterally into a coherent ring-like cavity. The nucellar epidermis does not divide periclinally, and thus a nucellar cap is lacking.

*Cyclanthus* (1) *bipartitus,* ranges from Northern Guatemala through Central America (and the lesser Antilles) and Northern South America as far as Peru and Brazil. Its habitat is swampy rain forests, but it has spread to fairly dry disturbed habitats.

# Superorder Areciflorae

In Cooperation with K. JAKOBSEN

*One Order:* Arecales.

Trees, shrubs or climbers, rarely diminutive undershrubs. The stem is generally solitary, erect, unbranched and terminated by a crown of leaves. It varies from very short and more or less subterranean to 60 m high, the woody trunk ranging from slender and reed-like to nearly 1 m in diameter. The stem is smooth or covered to a variable extent with old leaf sheaths. More rarely it is spiny, the spines sometimes representing modified roots. Basal branching may occur, giving rise to tufted stems in caespitose palms (e.g. *Chamaerops humilis*). Dichotomous branching of aerial stems is rare (e.g. species of *Hyphaene*), but of regular occurrence in the rhizomes of *Nipa*. Branching may also occur irregularly high up on the stem (*Borassus, Cocos, Rhopalostylis, Phoenix*). Creeping, prostrate or underground rhizomes are also found in colonizing palms.

Even the largest palms with a columnar stem have no secondary growth formed from a cambium. The first growth of the young stem is only a thickening growth without internodal elongation until the stem has reached the mature diameter. This primary thickening growth occurs from meristems at the leaf bases. Later follows the elongation growth. The basal underground part of the stem when vertical usually has an inverted conical shape. Slight secondary thickening of the stem, however, may occur and is then brought about by cell divisions and cell expansions in the parenchymatous ground tissues. In the Iriarteoid palms the elongation growth is not delayed; accordingly the stems are slender at the base but supported by prop roots.

The stems in climbing palms may be up to 150 m long, always with long internodes. The stems may climb with the help of the modified prolonged leaf rachis (cirrus) armed with recurved spines, sometimes also bearing pairs of reflexed modified spine-like leaflets (acanthophylls). Some species of *Calamus* climb by means of modified, unbranched sterile inflorescence axes (flagella) likewise armed with hooks or claws.

Vessels are present in roots, stems and leaves. Usually the most specialized vessels with transverse end walls and simple perforations are found in the roots, while the vessels in stem and leaves are less specialized, frequently with oblique end walls and scalariform perforation plates.

The leaves are very variable in size, ranging from a few cm to 25 m long (in *Raphia*), and have a broad, clasping sheath and usually a distinct pseudopetiole. There are several distinctly different forms of sheath and these are constant in shape amongst the species of a genus and sometimes higher taxa (TOMLINSON 1962a). The lamina (ARBER 1922b; EAMES 1953; PERIASAMY 1962; CORNER 1966; TOMLINSON 1961c, 1962a) is usually compound, being pinnately (Feather Palms) or palmately segmented (Fan Palms) or, rarely, bipinnate (*Caryota*). A midrib is absent in truly palmately dissected leaves, but a more or less reduced rachis (costa) is present in the costa-palmate leaves, which are intermediate between the pinnate and the probably more advanced truly palmate leaves. In palmate and costa-palmate leaves there is often a ligule (hastula) which develops on either or both sides of the pseudopetiole immediately below the lamina. Such ligules are widely developed on young leaves but often shrivel away before the leaf expands (TOMLINSON 1962).

The primordial palm leaf is undivided but densely plicate and the dissection of the lamina along the folds does not occur until the leaf unfolds. After the unfolding of the leaf the margin of the primordial leaf may remain as fibrous or filiform strips (reins) connecting the apices of the segments.

Juvenile leaves of seedlings are often simple or only slightly divided. Adult palms rarely have unsegmented leaves; when present these perhaps represent persistent juvenile stages. In the pinnate or palmate leaves the pinnae are folded, at least at the base, and are either V-shaped (induplicate) or Λ-shaped, in cross-section (reduplicate). Most Feather Palms with induplicate pinnae are imparipinnate, i.e. with a terminal leaflet, while those with reduplicate pinnae are usually paripinnate.

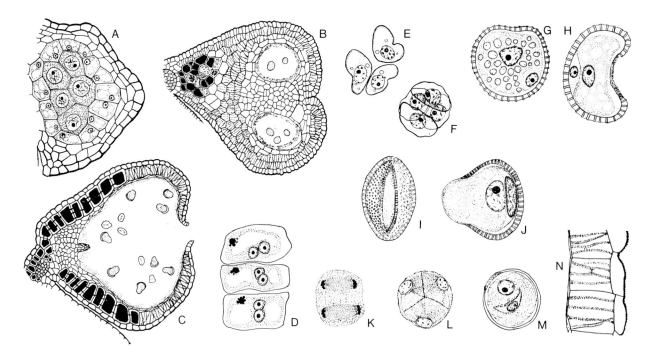

The stomata have four subsidiary cells: two terminal (one at each end of the stoma) and two lateral. Hairs are multicellular and either filamentous or peltate. The leaves are rich in sclerenchyma. Silica cells (stegmata) are found in all palms. The silica bodies are either hat-shaped to conical or spherical. Calcium oxalate usually occurs as raphides, rarely as isolated crystals or as crystal sand (TOMLINSON 1961).

The inflorescences are usually lateral, situated in leaf axils. In palms with crowded leaves they are either interfoliar or infrafoliar, i.e. situated in or at the base of the leaf crown, respectively. Rarely is the inflorescence "terminal" (in 16 genera, e.g. *Metroxylon* and *Corypha*) and then it reaches considerable dimensions, such palms being hapaxanthic. (The so-called terminal inflorescences are now better understood as aggregations of lateral inflorescences produced simultaneously in the axils of the uppermost, often reduced leaves. This is clearly shown in pairs of species such as *Metroxylon amicarum* and *M. sagu* where one species is pleonanthic, the other hapaxanthic; DRANSFIELD, personal communication). Palms with lateral inflorescences are pleonanthic. The inflorescence is usually much branched, rarely a single spike, and subtended by one or more bracts ("spathes"), which often enclose the young inflorescence system. In the much-branched inflorescences there are also several basal bracts, usually one subtending each main branch in the panicle.

**Fig. 218.** Arecaceae. Anther and pollen characters. **A** Young anther, showing archesporium, tapetal layer and two more layers inside the epidermis (*Actinophloeus macarthurii*). **B** Theca, transverse section just before dehiscence (*Arecastrum romanzoffianum*). **C** Same, just after dehiscence (*Caryota urens*). **D** Tapetal cells, glandular type (*Chrysalidocarpus lutescens*). **E–H** Microspore tetrad (incomplete in **E**) and mature pollen grain in different views (*Areca catechu*). **I–J** Pollen grain, in **J** in the state of germination (*Caryota urens*). **K–M** Microsporogenesis, simultaneous type (*Chrysalidocarpus lutescens*). **N** Endothecial layer with spiral wall thickenings (same species). (All from VENKATA RAO 1958)

The flowers are usually unisexual, and then the plants are monoecious or dioecious, but they are sometimes bisexual. They are sessile or even embedded in the axes, rarely shortly pedicellate, and either solitary or aggregated in few- to many-flowered clusters along the rachillae; sometimes they are grouped in a linear succession (acervulus) with one basal female flower and several males above, or in triads with one female and two lateral male flowers; paired arrangements also occur. The flowers are small, hypogynous, radially symmetrical and with a perianth of two more or less dissimilar, usually trimerous whorls of free or connate, often scale-like tepals. There are 3–6–9–∞ stamens (UHL and MOORE 1977, 1980) with the filaments free, connate or adnate to the perianth. The anthers dehisce longitudinally, rarely (some species of *Areca*) through apical pores. Microsporogenesis is

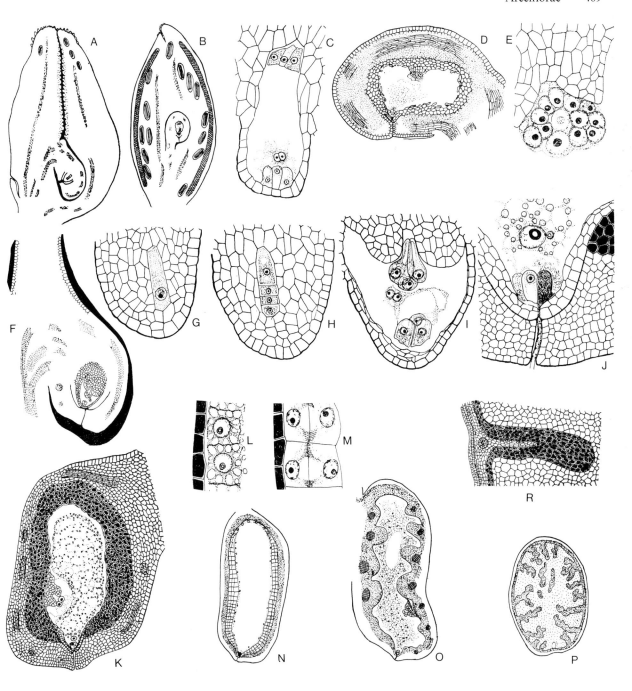

**Fig. 219.** Arecaceae. Pistil and embryological details. **A** Pistil, longitudinal section (*Chrysalidocarpus lutescens*). **B** Pistil, longitudinal section, note the abundance of raphide sacs (*Caryota mitis*). **C–E** Embryo sac in different stages showing increase in number of antipodals (*Chrysalidocarpus lutescens*). **F–I** Ovule and development of embryo sac (*Howea belmoreana*). **F** Ovule. **G** Megaspore mother cell. **H** Megaspore tetrad. **I** Embryo sac. **J** Micropylar part of embryo sac, showing entry of pollen tube; note the number of starch grains (*Actinophloeus macarthurii*). **K** Same ovule with mature embryo sac. **L–N** Wall formation in the endosperm, which is ab initio nuclear (same species). **O** Ovule, showing early stage of rumination (*Howea belmoreana*). **P** Mature seed, longitudinal section showing the ruminate endosperm and the position of the small embryo (same species). **R** Rumination in mature seed (*Caryota mitis*). (All from VENKATA RAO 1958)

either of the Simultaneous or the Successive Type. The pollen grains are usually sulcate or trichotomosulcate, rarely ulcerate, bisulculate, annulosulcate (with a ring-like, meridional aperture) or triaperturate (THANIKAIMONI 1970a, b; SOWUNMI 1972). They are glabrous or sometimes provided with warts or spinules, and they are two-celled when shed. The flowers are pollinated by wind or by insects. Different kinds of septal nectaries have been observed.

The gynoecium is apocarpous with 1–3(–4) separate carpels, syncarpous with three or rarely more locules, or pseudo-monomerous with only one fertile locule (UHL and MOORE 1971). The placenta is lateral, basal or subapical with only one ovule per carpel. The ovules are anatropous or hemianatropous, campylotropous or orthotropous and bitegmic. The archesporial cell cuts off a primary parietal cell before functioning as the megaspore mother cell. In addition, the nucellar epidermis often divides periclinally to form additional layer(s), i.e. a nucellar cap. Embryo sac formation conforms to the *Polygonum* Type, rarely to the *Allium* Type, and endosperm formation is nuclear. The fruits are usually berries or drupes with one or sometimes 2 or 3(–10) seeds. The mesocarp may be fibrous (as in *Cocos*). An endocarp is frequently well developed. The fruits are glabrous, spiny or hairy, and in some groups are covered by imbricate, reflexed scales (Fig. 222K). The seeds, which are free or united to the endocarp, contain a small embryo embedded apically, laterally or basally in the copious endosperm; a well-developed sarcotesta is sometimes present, as in most members of the subfamily Lepidocaryoideae. The endosperm is either homogeneous or ruminate and contains fats and oils, proteins in the form of aleurone grains and hemicellulose deposited as cell-wall material. Starch is never present.

*Chromosome Numbers.* The basic chromosome numbers in palms according to MOORE and UHL (1982) range from 18 to 13 along a reductional series, the primitive taxa having the higher numbers. Polyploidy is known only in one species, *Arenga caudata.*

*Chemistry.* Considering the economic importance of the palms, the chemical contents are yet poorly known. Accumulation of silica as well as calcium oxalate (for example as raphides) is a characteristic feature and an unusual combination (known in orchids, mainly five to six-staminate Zingiberiflorae, Bromeliales and a few Commelinaceae). Steroidal saponins are known in a few species, and cyanogenic compounds are common in

the palms, whereas chelidonic acid seems unknown in the group. Tannins are common in palms and occur in all parts of the plant; they apparently consist mainly of catechin and leucoanthocyanins. Among the flavonoid compounds tricin, C-glycoflavones, luteolin and flavone sulphates and 5-glucosides are all conspicuously common in the palms in which they agree with families of Poales and Cyperales in the Commeliniflorae. On the other hand, BOUDET et al. (1977) did *not* find isoenzymes of dehydroquinate hydrolyase (DHQ-ase) in the palms, as he did in Poales and Cyperales. HARRIS and HARTLEY (1981) found that the unlignified cell walls contain compounds giving UV-fluorescence, as is typical of the starch-endosperm superorders (Commeliniflorae, Zingiberiflorae and Bromeliiflorae).

True alkaloids are rare but known at least in seeds of *Areca catechu.* The palm waxes, which are often formed in rich quantities, consist of fatty acids and long-chain alcohols, usually esterified with the fatty acids. Great quantities of waxes, useful for commercial purposes, occur especially in species of *Copernicia, Ceroxylon* and *Raphia.*

Many hapaxanthic palms accumulate large amounts of starch in the pith of the trunk. Such palms (e.g. *Metroxylon*) are important sources of commercial sago. Pearl sago is prepared from sago flour by gelatinization. By wounding, many palms produce a sap rich in sugar, which may have economic use, especially that of the Sugar Palm (*Arenga pinnata*).

*Distribution.* The palms are largely pantropical with centres in the New as well as in the Old World. Of the largest subfamilies, the Lepidocaryoideae and Arecoideae have their greatest concentration in Eastern Asia and the Pacific and the Cocosoideae in Southern and Central America. Africa and Western Asia are comparatively poor in palms and in Europe there are only two indigenous species.

*Relationships.* The combination of many primitive and also very specialized trends in palms indicates a long evolutionary history, also substantiated by fossil records dating from the Cretaceous. It has been suggested that palms were among the first angiosperms and probably originated from Western Gondwanaland. The Coryphoid palms have a surprisingly large number of unspecialized characters, some even have stipitate and conduplicate carpels with an open ventral suture and some lack a distinct style, being thus similar to "primitive" carpels met with in Magnoliales.

Conventionally, the palms are regarded as most closely allied to the Cyclanthales (Cyclanthiflorae) and Pandanales (Pandaniflorae). The closest relationships of the palms in relation to other groups of monocotyledons are, however, uncertain, and are worthy of reconsideration.

The Cyclanthiflorae show features, especially in the plicate leaves, that are similar to the palms, but they have helobial endosperm formation, lack silica, and have a quite different floral construction that has much less of the "monocotyledon base pattern" in it than has the floral construction in most palms. The possibilities of a close vegetative similarity by convergence versus a similarity in detail await a thorough analysis.

The Pandaniflorae are even more doubtful as closely related to the palms. Their leaf morphology is different, they lack silica bodies, the endosperm of the seeds is starchy (in *Freycinetia*) and the floral construction dubiously similar.

Yet it seems that all three superorders, which have hitherto, perhaps too uncritically, been treated together, are more difficult to distribute elsewhere in the monocotyledons.

However, the palms make one great exception. Recent finds by BARTHLOTT and FRÖHLICH (1983) on epicuticular wax, by HARRIS and HARTLEY (1980) on UV-fluorescent organic acids of the cell walls, by HARBORNE (1973) on the spectrum of flavonoid compounds (in particular tricin and sulphurated flavonoids), and on certain parasites to an increasing degree tend to knit the palms to the Commeliniflorae. A probable link between palms and grasses was also suggested by one of us long ago, though on less decisive evidence (CLIFFORD 1970). If we assume that the palms have a common ancestry with the Pandaniflorae, one of the chief differences between Areciflorae and Commeliniflorae, viz. that the former have a starchless endosperm, is neutralized by the presence of starch in the endosperm of *Freycinetia* in Pandanaceae. The presence of raphides, lacking in Commeliniflorae, but probably present in their ancestors, and present in Bromeliiflorae and some Zingiberiflorae, suggests the possibility that the Areciflorae, and perhaps also Pandaniflorae and Cyclanthiflorae, had a common ancestry with the Commeliniflorae-Bromeliiflorae stock.

# Order Arecales

*One Family:* Arecaceae.

See the superorder, above.

**Arecaceae** C.H. Schultz-Schultzenstein (1832) (= Palmae)   212:2,780   (Figs. 218–223)

Morphology as given under the superordinal description above.

The palms comprise a huge group of tropical plants with centres in Central and Northern South America and in Eastern Asia, Malaysia, and the Pacific Islands to as far north as Japan and as far south as New Zealand. Being frequently large trees, they characterize habitats strongly and they often make up part of the understorey vegetation as subshrubs and form thickets and lianas. As will be mentioned below, their importance to man in warm regions is great and manifold.

*Evolutionary Trends in the Family.* In a recent paper, MOORE and UHL (1982) have given a thorough treatment of the variation patterns in the various structures of the palms. They also outlined the evolution of palms (and monocotyledons) and presented a number of conclusions, among which the following may be mentioned. Lateral flowers are primitive in palms; trimerous flowers are basic; flowers with whorls of fewer or more numerous parts than three are derived; the number of flowers in the inflorescence and of ovules in the ovary may increase or decrease in the course of evolution; many flowers with few ovules may be quite as primitive as fewer flowers with many ovules; fleshy, one-seeded fruits may be more primitive than dry many-seeded fruits.

Some evolutionary trends were sketched within the palms (taken from MOORE and UHL 1982)

1. branching, from sympodial to monopodial;
2. size, from moderate towards gigantism and nanism;
3. stem, from unbranched to dichotomous; from little to much strengthening; from short to long internodes;
4. leaf, from undivided to having a palmate, costapalmate, pinnately ribbed or pinnate blade; from undivided with induplicate ptyxis to divided along the adaxial rib or along the abaxial rib; from pinnate to bipinnate leaf or with pinnae once or twice divided longitudinally, from sheath open opposite petiole to closed and tubular; from marcescent to deciduous;
5. inflorescence units, from moderately branched to spicate or less frequently to more diffusely branched; from one unit per leaf axil to more than one per axil; from situated among the leaves to situated below them or above them in a compound ter-

minal inflorescence; from pleonanthic to hapaxanthic;
6. bracts, from conspicuous to small or absent;
7. flowers, from solitary, pedicellate and bracteolate
to clustered in a triad or dyad or more, or to a
monopodium of two to four;
8. flowers, from bisexual to unisexual, then associated
with polygamy or monoecism to dioecism;
9. perianth, from trimerous or tetramerous to deca-
merous or reduced and monochlamydeous; outer
tepals from distinct and imbricate to connate, inner
tepals from distinct and imbricate to either valvate
or strongly imbricate, or connate;
10. stamens, from 3 + 3 to 3 or to more than 6 (up to
ca. 950);
11. stamens, from rather slender to broad and thick;
12. pollen grains, from sulcate to trichotomosulcate, to
bisulcate or monosulculate, or bi- to triforaminate;
13. gynoecium, apocarpous or syncarpous;
14. locules, from 3 to 2–1 or 4–10;
15. ovules, from anatropous to hemianatropous, campylotropous or orthotropous;
16. fruit, from fleshy to dry and fibrous;
17. seed, from moderate-sized to small or very large;

18. endosperm, from homogeneous to invaginated or
ruminate;
19. chromosome complement, from n = 18 to successively lower numbers.

It must be emphasized that some of these courses,
against which we do not object, may be opposed
to the course in primitive monocotyledons. Thus,
for example, although fleshy fruits are primitive
in palms, they are derived in monocotyledons, the
dry fruits in palms being further derived. Although
rumination in the endosperm may be derived in
palms it could be ancestral in monocotyledons (cf.
the condition in some Dioscoreales and in some
Magnoliiflorae).
The complex variation within the family makes
a subdivision difficult. Conventionally, nine subfamilies are recognized. An attempt at a better subdivision of the infrafamilial relationships and taxonomy was made by MOORE (1973) and is followed
here in the main features.

*Key to the Subfamilies and Groups of Palms.* (Based on MOORE 1973)

1 A. Leaves with induplicate pinnae or segments (exceptions ca. 35–40% of the Coryphoideae)  . . . . . . . 2
1 B. Leaves with reduplicate pinnae or segments, usually paripinnate or sometimes with a
pseudoterminal pinna, rarely palmate or costapalmate  . . . . . . . . . . . . . . . . . . 5
2 A. Flowers borne singly or in clusters, but never in triads of two male and one female flower . . . . . . . 3
2 B. Flowers usually borne in triads of one female and two male flowers;
leaves pinnate or bipinnate . . . . . . . . . . . . . . . . . . . . . . 6. Subfamily **Caryotoideae**
3 A. Flowers pedicellate to sessile, never embedded in the inflorescence axes  . . . . . . . . . . . . . 4
3 B. At least the staminate flowers embedded in the thickened inflorescence axes;
leaves palmate or costapalmate  . . . . . . . . . . . . . . . . . . . . 3. Subfamily **Borassoideae**
4 A. Leaves palmate or costapalmate . . . . . . . . . . . . . . . . . 1. Subfamily **Coryphoideae**
4 B. Leaves imparipinnate . . . . . . . . . . . . . . . . . . . . . 2. Subfamily **Phoenicoideae**
5 A. Fruit with imbricate scales; leaves pinnate, rarely palmate or costapalmate   4. Subfamily **Lepidocaryoideae**
5 B. Fruit without imbricate scales; leaves always pinnate or pinnately ribbed  . . . . . . . . . . . . 6
6 A. Monoecious palms with the female flowers in a terminal head and with the
male flowers on lateral branches at the base of the terminal head;
female flowers with 3(–4) separate carpels . . . . . . . . . . . . . . . . 5. Subfamily **Nipoideae**
6 B. Flowers bisexual or unisexual; monoecious plants never with a female
terminal head; carpels united or gynoecium pseudomonomerous . . . . . . . . . . . . . . 7
7 A. Female flower clusters not capitate; perianth in the female flowers in two whorls . . . . . . . . . . 8
7 B. Female flower clusters capitate and with numerous spirally set dimorphic
tepals (i.e. differentiated into "sepals" and "petals");
fruit tuberculate and with five to ten seeds; dioecious . . . . . . . . . 8. Subfamily **Phytelephantoideae**
8 A. Fruit usually with a thin endocarp which never has pores  . . . . . . . . 7. Subfamily **Arecoideae**
8 B. Fruit with a thick endocarp with three or more pores  . . . . . . . . . . . 7h. The **Cocos** Group
9 A. Flowers borne singly or in acervuli, rarely in triads (*Opsiandra*)  . . . . . . . . . . . . . . 10
9 B. Flowers usually borne in triads or derivatives of triads; flowers unisexual . . . . . . . . . . . . 12
10 A. Basal flowers on the rachillae hermaphrodite, distal flowers staminate;
flowers pseudopedicellate . . . . . . . . . . . . . . . . . . . . . 7a. The **Pseudophoenix** Group
10 B. All flowers unisexual   . . . . . . . . . . . . . . . . . . . . . . . . . . . 11
11 A. Flowers with bracteate pedicel  . . . . . . . . . . . . . . . . 7b. The **Ceroxylon** Group
11 B. Flowers sessile   . . . . . . . . . . . . . . . . . . . . . . 7c. The **Chamaedorea** Group
12 A. Peduncle of inflorescence with a prophyll and more than two (in *Podococcus*
sometimes two) sterile bracts; pinnae without a distinct midrib, apex erose  . . . . . . . . . . . 13
12 B. Peduncle of inflorescence with a prophyll and 0–2 sterile bracts; pinnae
usually acute and usually with a distinct midrib  . . . . . . . . . . . . . . . . . . . 14
13 A. Flowers not embedded in the inflorescence axes   . . . . . . . . . . . 7d. The **Iriartea** Group
13 B. Flowers embedded in the inflorescence axes . . . . . . . . . . . . 7e. The **Podococcus** Group
14 A. Styles prominent   . . . . . . . . . . . . . . . . . . . . . . . 7f. The **Geonoma** Group
14 B. Styles very short or absent . . . . . . . . . . . . . . . . . . . . 7g. The **Areca** Group

**Subfamily Coryphoideae**   32:322

Small to large, bisexual, polygamous or rarely dioecious fan palms. The leaves are usually induplicately palmate or costapalmate, rarely subpinnately ribbed. The silica bodies are spherical. The inflorescences are interfoliar or terminal and usually very much branched and with one to many bracts. The flowers are pedicellate or sessile and either hermaphroditic or unisexual. Male and female flowers are not markedly different. The solitary or clustered flowers have one or two usually trimerous perianth whorls and 6–24 stamens, which are free, connate and/or adnate to the perianth. The pollen grains are usually sulcate. The gynoecium consists of (four,) three or two free or united carpels or only one. Usually only a single carpel develops into fruit, which is a one-seeded berry or drupe. Septal glands occur in syncarpous genera.

To this subfamily belong many of the supposedly primitive palms. Fourteen genera are apocarpous. The group also exhibits a wide variety of more or less specialized flowers and both wind and insect pollination occur.

The subfamily is pantropical to subtropical.

*Trithrinax* (5) from Southern America has hermaphrodite flowers with three free carpels. – The West Mediterranean *Chamaerops* (1) is also apocarpous, but the plants are dioecious or polygamous. *C. humilis* is the only reasonably common native European palm. It has spiny petioles and short, suckering trunks frequently forming dense bushy patches on dry ground in the Mediterranean. – Apocarpous also are the Asiatic *Rhapis* (12) and *Coccothrinax* (20), from Florida and the West Indies, and several other genera. – *Livistona* (28) and *Licuala* (108), from South-East Asia to Australia, and *Pritchardia* (36), from Fiji and Hawaii and other Pacific Islands, have hermaphroditic flowers with basally fused stamens and three carpels free at the bases fused in the stylar region. – The same condition of the carpels occurs in *Copernicia* (25) from tropical America and the West Indies. *C. prunifera* and *C. alba* are important sources of vegetable wax ("Carnauba"). – The costapalmate *Corypha* (8) from India to Malaysia has a terminal inflorescence 6–7 m long. The trunk is sometimes used as a source of starch. – Two costapalmate palm genera that reach far north in the United States are *Washingtonia* (2), to Southern California, and *Sabal* (14), from the Gulf states to North Carolina (*S. palmetto*). Both genera are widely cultivated as ornamentals.

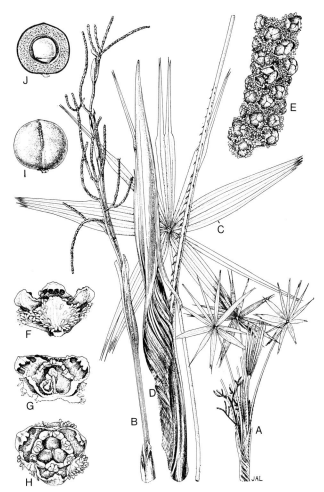

**Fig. 220.** Arecaceae. *Livistona exigua,* a dwarf member of the genus. **A** Habit. **B** Single inflorescence. **C** Leaf, showing leaflets of varying widths. **D** Leaf base showing well-developed tongue-like extension. **E** Portion of inflorescence branch, showing flower buds. **F** Flower, lateral view. **G** Flower, longitudinal section. **H** Flower, from above. **I** Fruit. **J** Same in longitudinal section. (DRANSFIELD 1977)

**Subfamily Phoenicoideae**   1:17

The only genus, *Phoenix,* is distributed in Africa and Arabia, and from South-Eastern Asia to Sumatra, and comprises solitary or tufted, dioecious, feather palms with imparipinnate leaves and induplicate leaflets. The lower leaflets are modified into spines. Silica-bodies are spherical. The inflorescences are interfoliar and covered, when young, by a single bract (prophyll). The inflorescence axes are flattened. The flowers are solitary, unisexual and wind-pollinated. Female flowers have staminodes and an apocarpous gynoecium with three carpels, two of which are usually abortive. The fruit

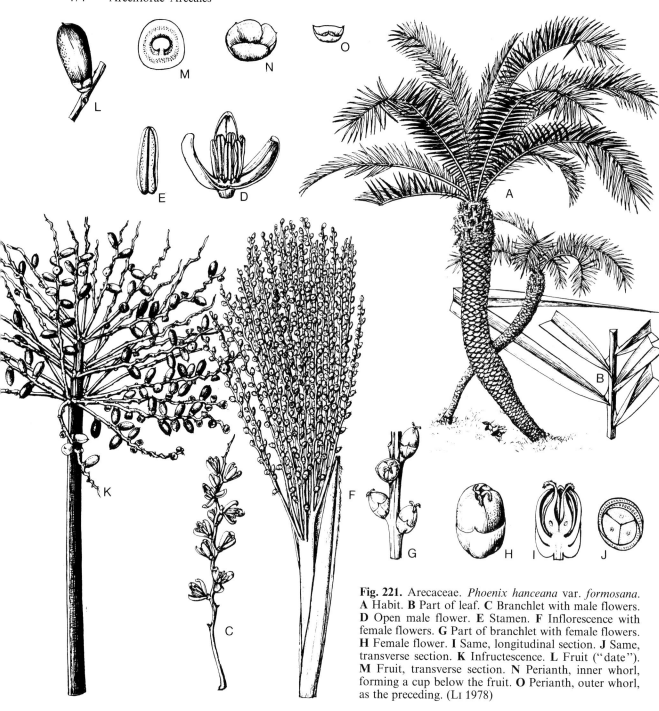

**Fig. 221.** Arecaceae. *Phoenix hanceana* var. *formosana.*
**A** Habit. **B** Part of leaf. **C** Branchlet with male flowers.
**D** Open male flower. **E** Stamen. **F** Inflorescence with
female flowers. **G** Part of branchlet with female flowers.
**H** Female flower. **I** Same, longitudinal section. **J** Same,
transverse section. **K** Infructescence. **L** Fruit ("date").
**M** Fruit, transverse section. **N** Perianth, inner whorl,
forming a cup below the fruit. **O** Perianth, outer whorl,
as the preceding. (Li 1978)

is a berry or a berry-like drupe with membranous
endocarp and a mesocarp very rich in sugar. The
storage material in the endosperm consists of
hemicellulose. Male flowers usually have six stamens. The pollen grains are sulcate. In many features *Phoenix* seems to be related to the coryphoid
palms.

*P. dactylifera,* the "Date Palm", is an important
and ancient crop plant in the desert or semi-desert
regions of northern Africa and Arabia eastwards
to the Indus. It is only known in cultivation and
its origin is uncertain. In the Middle East the Date
Palm has been cultivated for at least 5,000 years;
nowadays it is also cultivated in the southern parts
of the United States and Australia. *P. canariensis*
is an endemic of the Canary Islands and widely
cultivated in warm temperate regions. *P. theo-
phrasti* is an endemic of Crete and Turkey.

## Subfamily Borassoideae    7:56

Solitary or tufted, dioecious fan palms with induplicate palmate or costapalmate leaves. The trunk is rarely branched (many species of *Hyphaene*), and is sometimes ventricose, with one or more swellings. The silica bodies are usually spherical, but hat-shaped in *Lodoicea*. The inflorescences are interfoliar with many bracts. The flowers are unisexual. The male flowers, which are embedded in the axes, are either solitary or in cincinni and have a tubular calyx and six to many stamens which, together with the three inner, petaloid tepals, are borne on an elongated receptacle. The pollen grains are sulcate. Female flowers are either embedded or sessile and have staminodes and a trilocular gynoecium with septal nectaries. The fruit is usually a drupe with one to three seeds, each enclosed in its own bony endocarp.

The borassoid palms are, like *Phoenix*, probably related to the coryphoid palms. The seven genera have a palaeotropical distribution from Africa to Indomalaysia.

To *Borassus* (7) belongs the "Palmyra Palm" (*B. flabellifer*), which is an important crop plant in India and parts of Indonesia, used as source of sugar, wine and vinegar. The fruits are edible. – The monotypic *Lodoicea,* endemic on the islands Praslin and Curieuse in the Seychelles, has very large, bilobed, usually one-seeded fruits, "Double Coconut", up to ca. 55 cm in diameter. The seed is the largest in the Plant Kingdom. – *Hyphaene* (41) inhabits Africa, Madagascar and Arabia. *H. thebaica* is the Egyptian "Doum Palm" with a branched trunk.

## Subfamily Lepidocaryoideae    22:664

This subfamily is very variable in vegetative habit, comprising both tall, erect, solitary or tufted palms and palms with low, short and thick or subterranean stems; several genera are scandent. Aerially branching stems are frequent in *Korthalsia.* Branched underground rhizomes and also stilt roots may occur. The silica bodies are spherical. The leaves are reduplicate and entire (rarely) or pinnately dissected or rarely costapalmate or palmate (*Mauritia* and *Lepidocaryum*). Inflorescences are either interfoliar or terminal. The flowers are borne singly or in pairs and are bisexual or unisexual, the plant being polygamous, monoecious or dioecious. The number of stamens is six (–70). The pollen grains are very variable, being sulcate, bisulcate, ulcerate or with two circular apertures, and

sometimes spiny; in *Salacca* they have a nearly ring-like, meridional aperture as in *Nipa*. The most characteristic feature is the syncarpous, trilocular gynoecium, which like the fruit, is covered with reflexed, imbricate scales. The fruit is usually one-seeded, the seed usually with a sarcotesta. A thick, bony endocarp occurs in the genus *Eugeissona*.

The largest concentration of genera is in the region of South-Eastern Asia  and Western Malaysia. Only five genera are found in Africa and three in America. One of these, *Raphia,* occurs both in Africa and South America.

Only the South American *Mauritia* (21) and *Lepidocaryum* (9) are fan palms. All other genera in the subfamily are feather palms. *Mauritia* and *Lepidocaryum* are dioecious with lateral inflorescences and unisexual flowers. Wine and sago are prepared from *M. flexuosa.*

*Eremospatha* (12) and *Laccosperma* (syn. *Ancistrophyllum;* 1) from Africa and *Korthalsia* (26) from Indomalaysia always have hermaphrodite flowers. The first two have scandent stems and leaves prolonged into long cirri. In *Korthalsia* the leaf base is developed as an ochrea, which may be inflated and inhabited by ants.

*Metroxylon* (8), the "Sago Palm", from Indomalaysia and Melanesia has an erect, columnar stem and a branched rhizome. The inflorescence is terminal and the flowers are male and bisexual and borne in a paired arrangement. Just before flowering a large amount of starch is stored in the pith. *M. sagu* ( = *M. rumphii*) is especially rich in starch; a 15 year old palm yields up to 400 kg. This species is the most important source of sago starch and pearl sago.

The monoecious *Raphia* (28), from Africa and Madagascar and with one species (*R. taedigera*) in South and Central America, has an erect, columnar to usually very short stem and very large leaves, up to 25 m long. The inflorescence is terminal. Some species yield starch, wine and oil. Leaves of most species are  a source of fibre ("Raffia") used as binding material.

*Salacca* (15) from Indomalaysia is dioecious. The stem is short to subterranean and the leaves are armed with long spines. *S. zalacca* and other species have edible fruits. – The likewise dioecious *Calamus* (370), the genus of "Rattan Palms", is palaeotropical, but in Africa represented by only nine species. The usually scandent stems may be up to 150–180 m long and are supported by long, hooked flagella, which are modified lateral inflorescences, or by cirri (modified leaf tips). The stems of *C. caesius, C. manan* and other species

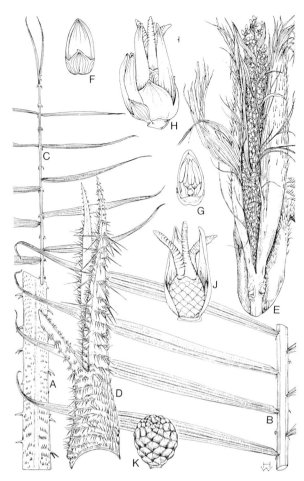

**Fig. 222.** Arecaceae. *Pogonotium divaricatum,* a member of the subfamily Lepidocaryoideae. **A** Adaxial surface of leaf petiole. **B** Mid-portion of leaf with "leaflets". **C** Apex of leaf with divaricate "leaflets". **D** Leaf sheath with two auricles. **E** Portion of leaf sheath with male inflorescence. **F** Male flower (outer tepals basally connate). **G** Same in longitudinal section. **H** Female flower with accompanying sterile male flower. **J** Female flower with outer tepals and one of the inner tepals removed. **K** Mature fruit. (DRANSFIELD 1980)

are used in furniture and basket industries. – *Daemonorops* (115) from Indomalaysia are usually also scandent, like *Calamus,* but flagella are lacking and the stems climb by cirri alone.

**Subfamily Nipoideae    1:1**

Only one genus with a single species, *Nipa fruticans,* the "Nipa Palm", a mangrove palm growing in brackish or salt water and distributed from India, Malaysia, Solomon and Ryukyu Islands to tropical Australia.

The trunk is subterranean or prostrately creeping and dichotomously branched. The leaves are erect and paripinnate with reduplicate pinnae and have shiny scales on the lower surface. Silica bodies are hat-shaped. The flowers are dimorphic and unisexual. The erect, interfoliar inflorescence is monoecious and protogynous with a terminal head of female flowers and with male flowers on lateral branches at the base of the female head. The male flowers are spirally set and have only three stamens, which are united by their filaments and connectives into a column. The pollen grains are provided with stout spines and have a meridional, ring-like aperture. The female flowers have three (to four) large, separate carpels, which are much larger than the surrounding tepals. One or more of the carpels develop into fruits, which have a fibrous mesocarp and a thick endocarp adapted for dispersal by water. *Nipa* seems to be pollinated by drosophilid flies.

The "Nipa Palm" is placed apart from the other palms and has been treated as a separate family (Nipaceae) by several authors. Fossil records from the Eocene and Miocene are known from a wider area than the present distribution of the genus, including Europe, and *Nipa* pollen has been found in Senonian strata on Borneo.

**Subfamily Caryotoideae    3:35**

Small to large, solitary or caespitose monoecious or (very rarely) dioecious palms with induplicate, imparipinnate or, in *Caryota,* bipinnate leaves. The pinnae usually have divergent nerves, a toothed apex and a triangular outline ("Fish-Tail Palms"). The silica bodies are hat-shaped. The inflorescences are interfoliar or infrafoliar and usually develop in a basipetal sequence on the vegetative stem. In basipetal species the whole shoot dies after the end of flowering.

The caryotoid palms are monoecious with unisexual, dimorphic flowers usually borne in triads of one female and two male flowers, but unisexual inflorescences occur in *Arenga* and *Wallichia.* The male flowers have 3–250 stamens and the pollen grains are sulcate and sometimes spiny. In the female flowers the inner tepals are united and the gynoecium is syncarpous and tri- to unilocular. Septal nectaries are present. The fruit is a three- to one-seeded berry or drupe with thin to cartilaginous endocarp; raphides are abundant in the mesocarp.

The caryotoid palms occur from India, Malaysia and Melanesia to tropical Australia. They are supposed to have close relationship both to the Coryphoid and the Arecoid palms.

**Fig. 223.** Arecaceae. *Arenga engleri.*
**A** Habit. **B** Leaflet. **C** Male inflorescence. **D** Male flower. **E** Same, longitudinal section. **F–G** Stamen in different views. **H** Fruiting branchlet. **I** Fruit, seen from the side. **J** Same, seen from the base. **K** Same, longitudinal section. (Li 1978)

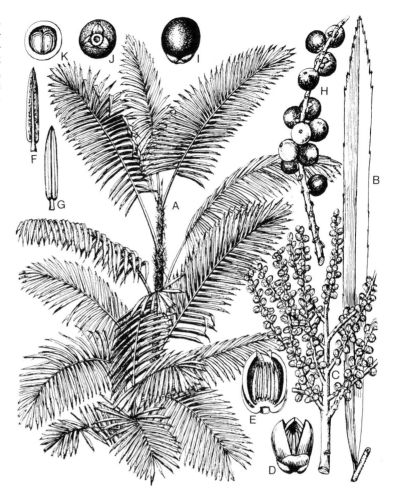

*Caryota* (12) is the only palm with bipinnate leaves. The endosperm is ruminate. *C. urens,* the "Fish-Tail Palm", is used as source of sago and fibre. Juice from the trunk is used to make sugar or toddy. – *Arenga* (17) has once pinnate leaves and usually unisexual inflorescences. *A. pinnata* (= *A. saccharifera*), the "Sugar Palm", is widely used in India and Malaysia as source of palm sugar.

**Subfamily Arecoideae**    114:1085

This is the largest subfamily of the palms, comprising 7 of the 15 major groups mentioned by Moore (1973). The subfamily is very variable in its vegetative characters. It includes small and reed-like to large, solitary, caespitose or stoloniferous, monoecious or dioecious palms, which are from erect to repent. Most of them are pleonanthic feather palms with reduplicately paripinnate leaves usually having an abscission zone at the base, but quite a number of species have entire leaves. The silica

bodies are spherical to ellipsoidal or hat-shaped. Inflorescences are interfoliar or infrafoliar, branched or unbranched and have one to several bracts; they are sometimes sexually dimorphic. The flowers are unisexual (except in *Pseudophoenix*) and more or less dimorphic. They are usually sessile and frequently embedded in the inflorescence axes; rarely they are pedicellate. The male flowers have six to many (rarely three) stamens. The pollen grains are usually sulcate, but in the *Areca* Group very variable. The gynoecium is syncarpous and trilocular to pseudomonomerous. The fruit, which contains three to one seeds, never has pores in the endocarp.

Most genera and species belong to ***the Areca Group*** (88:759), which are monoecious palms with branched or spicate inflorescences with the flowers usually borne in triads or derivatives of triads. The pollen grains are very variable, being sulcate, trichotomosulcate, ulcerate (*Dypsis*), nearly annulosulcate (*Pinanga*) or triaperturate (*Sclerosperma*), the latter condition being unique among the palms.

The gynoecium may be trilocular and triovulate (mainly New World genera), but is more often pseudomonomerous.

The *Areca* Group is predominantly palaeotropical, but in Africa is represented only by the genus *Sclerosperma* (3) in tropical Western Africa. – *Areca* (48) ranges from Indomalaysia to Northern Australia. The seeds of *A. catechu,* the "Betelnut Palm", contain alkaloids and are used as a masticatory in betel-chewing. – *Roystonea* (6) are tall palms from Florida to tropical South America. *R. regia,* the "Royal Palm", is a commonly cultivated ornamental in tropical countries. *R. oleracea* from the West Indies is one of the largest palms: the trunk may be up to 50 m high. – *Rhopalostylis* (3) from the New Zealand region and other Indo-Malaysian and Polynesian palms are frequently grown as ornamentals. – *Phloga* (2) from Madagascar may have leaves which are fanned out (plumose) and twisted to give a whorled appearance.

The *Ceroxylon* Group (4:30) and *Chamaedorea* Group (6:146) are both confined to the Neotropics and the Madagascan region. The former consists of dioecious palms with pedicellate flowers borne singly on the inflorescence axes. The gynoecium is trilocular and triovulate and the fruit has one to three seeds. The *Chamaedorea* Group consists of small to moderately large, monoecious or dioecious palms with sessile flowers borne in acervuli or singly. The gynoecium is trilocular and triovulate, but the fruit is one-seeded. The silica bodies are hat-shaped. – *Ceroxylon* (17) in Northern South America, includes *C. alpinum,* the "Wax Palm" of the Andes, which secretes wax on the trunk. – *Chamaedorea* (133) are small, mostly reedy palms in tropical America, frequently cultivated as ornamentals.

The *Pseudophoenix* Group (1:4), *Iriartea* Group (8:52) and *Geonoma* Group (6:92) are all exclusively American palms. *Pseudophoenix* is unique in the subfamily in having hermaphrodite, pseudopedicellate, spirally set flowers (distal flowers male), which perhaps indicates primitiveness. The Iriarteoid and Geonomoid palms are all monoecious with sessile or embedded, unisexual flowers principally in triads (although sterilization of elements of the triads may give unisexual inflorescences) with a trilocular to pseudomonomerous gynoecium and usually one-seeded fruits. The silica bodies are hat-shaped in the *Iriartea* Group, but spherical to ellipsoidal in the two other groups. – *Geonoma* (75) from Central and South America are partly reedy palms with short stems.

The *Podococcus* Group (1:2) is the only exclusively African group in the subfamily, and consists of the reed-like, monoecious *Podococcus,* occurring in rain forests in western Africa. It differs from the other groups in having imparipinnate leaves.

## Subfamily Cocosoideae    28:583

Small to very large monoecious pleonanthic palms with reduplicately paripinnate or pinnately ribbed leaves. Their silica bodies are either spherical to ellipsoidal or hat-shaped. The inflorescences are interfoliar, sometimes sexually dimorphic and once-branched or unbranched with the unisexual flowers borne in triads. The distal parts of the axes may have only male flowers, in pairs or singly. The flowers are sessile, pedicellate or immersed. The male flowers have six to many stamens, with sulcate or trichotomosulcate pollen grains. The gynoecium is syncarpous with 3–ca. 7 locules and 3–ca. 7 ovules. It develops into a one- or several-seeded fruit with a fleshy to fibrous mesocarp and a bony endocarp with 3 to 7 pores.

The cocosoid palms are closely related to the *Areca* Group in Arecoideae. The distinctive character is the bony endocarp with pores. Cocosoideae is predominantly neotropical, only the monotypic *Jubaeopsis* and one species of *Elaeis* being African; *Cocos* is supposed to be of Melanesian origin.

*Cocos* (1) *nucifera,* the "Coconut Palm", is an important commercial crop widely cultivated in the tropics. "Copra" is the dried endosperm used as source of coconut oil, and the fibrous mesocarp is a source of coir fibre. – *Elaeis* (2) has one species in tropical America; the other is the West African *E. guineensis,* the "Oil Palm", which has separate male and female inflorescences. The oil palm is widely cultivated, being a very important source of edible palm oil, which is also used for soap-making. The oil is obtained from the seeds and the fibrous mesocarp. – *Attalea* (30) is from Central and South America; the leafstalks of *A. funifera* are source of piassava fibre. – *Desmoncus* (59), from tropical America, are climbing palms with leaves having a prolonged rachis with recurved hooks. – *Bactris* (239) from Central and South America form dense thickets and are armed with long spines on leaves and stems. – *Jubaea* (1) *spectabilis,* the "Chilean Wine Palm", has been rendered almost extinct in the wild by being cut down to extract its sap for wine-making.

**Subfamily Phytelephantoideae**   3:15

Pleonanthic, dioecious palms usually with a short trunk. Leaves reduplicately paripinnate with persistent petioles. The inflorescences are interfoliar and sexually dimorphic, the male ones forming, as a rule, dense spikes, the females globose heads. The male flowers have numerous stamens and a perianth which in *Phytelephas* may consist of a single whorl of reduced tepals. The female flowers have three to seven tepals in the outer and seven or more tepals in the inner whorl and numerous staminodes, all set in a spiral. The syncarpous gynoecium has five to ten one-ovuled locules. The style is divided into long stigmatic branches. The fruits are clustered in a globose head and have a tuberculate exocarp and five to ten seeds.

The Central and Northern South American Phytelephantoideae is a highly specialized group of palms perhaps related to the subfamily Arecoideae. – *Phytelephas* (12) has seeds with a very hard endosperm (hemicellulose) used as vegetable ivory *P. macrocarpa* is the "Ivory Palm".

# Superorder Pandaniflorae

In Cooperation with K. JAKOBSEN

*One Order:* Pandanales.

The group consists of one family with three genera, *Sararanga, Pandanus* and *Freycinetia.*

Woody plants, branching trees or shrubs (*Sararanga, Pandanus*) or root-climbing lianas with clasping, aerial roots or scrambling, rarely reed-like, low-growing shrubs (*Freycinetia*). Many species (of *Freycinetia;* few of *Pandanus*) are epiphytic. Prop roots or stilt roots are developed from the base of the trunk or from axils of leaves (except in *Sararanga*). Pneumatophores are rare (*Pandanus palustris*).

The stems have no secondary growth or secondary vascular tissue. Increase of trunk diameter is the result of primary thickening growth alone. The stems are obconical at the base (as in most palms) and the increase of vascular tissue occurs at a higher level. In *Pandanus* the obconical zone is usually short and subterranean but in *Freycinetia* elongated owing to internodal elongation.

The genera are similar in anatomical characters. They share with only *Gahnia* (Cyperaceae) and *Dasypogon* (Dasypogonaceae) the frequent presence of compound vascular bundles (few in *Freycinetia*), which have two or three distinct (bi- or tripolar) vascular strands enclosed by a common bundle sheath, only one strand having the phloem oriented in the normal way. The metaxylem vessels are often very long and have oblique, scalariform perforation plates except in *Sararanga* where the adult stem contains tracheids only. The sieve tubes in the metaphloem have long, oblique, compound sieve plates. The sieve tube plastids have annularly organized protein filaments (which is otherwise known in *Aloë;* BEHNKE 1981a) besides the cuneate protein crystalloids.

In contrast to most palms, adult plants usually have branched stems. Branching is sympodial, the inflorescences being developed from the apical meristems. The presence of lateral buds in the axils of all foliage leaves represents another difference from the palms. These buds are normally inhibited, except after flowering. The resultant branching is pseudodichotomous. Exceptionally (e.g. in *Pandanus gemmiferus*) many lateral buds may develop into short-lived shoots. In certain species (of *Pandanus*) the main trunk branches monopodially, the inflorescences being then restricted to lateral shoots.

The leaves are alternate and tristichous except in *Sararanga,* where they are tetrastichous; but the ranks run in spirals owing to spiral growth of the stem, and so in practice are spirotristichous and spirotetrastichous, respectively. The leaves are usually aggregated at the ends of the branches. They are always undivided, long and narrow, usually rigid, coriaceous with sclerotic spines on the margins and on the midrib, which is very prominent in *Pandanus,* less prominent in *Freycinetia* and *Sararanga.* In *Freycinetia* the leaves show stronger differentiation into lamina and sheath than in the other genera. The stomata are more or less distinctly tetracytic (TOMLINSON 1965a; HUYNH 1974). In *Pandanus* the subsidiary cells of the stomatal apparatus and other epidermal cells may have papillae, which vary from simple to forked or dendritic (Fig. 225K, L), but in *Freycinetia* there are only a few simple outgrowths or none at all, and in *Sararanga* there are none. (TOMLINSON 1965a; HUYNH 1974).

Silica accumulation has not been observed in the family, but calcium oxalate is present, either as raphides in mucilaginous cells, or as single cubic crystals in isodiametric idioblasts (Fig. 225N).

All three genera are dioecious. The inflorescence is a terminal panicle in *Sararanga,* whereas in *Pandanus* and *Freycinetia* the flowers are borne in one to several spadices, which are racemosely arranged and are subtended by differently coloured spathes.

The structure of the unisexual flowers is very obscure.

In *Sararanga* the flowers are distinct, both the male and female being pedicellate, subtended by a bract, and with the stamens and carpels basally surrounded by an entire or irregularly three- to four-lobed, somewhat fleshy cup, which may represent a reduced perianth. The male flowers have many stamens with fleshy filaments. In the female flowers the fleshy pistil consists of 10–80 carpels united into a sinuous and forked double row in which the one-seeded carpels are subopposite in pairs. They have a sessile stigma and a ventral sutural pore. In this genus neither staminodes nor pistillodes have been observed.

In *Freycinetia* and *Pandanus* the perianth and usually also subtending bracts are absent, and the stamens and carpels are so closely crowded that a distinction of the flowers from one another is obscure or totally absent.

In male inflorescences of *Freycinetia* the stamens are frequently aggregated into groups, each surrounding an open, rudimentary pistillode. Each group possibly represents a single multistaminate flower with papillose filaments. In the female inflorescences the flowers are more distinct. The unilocular pistil consists of 1–12 or more united carpels with sessile stigmas and parietal placentation, each carpel containing several ovules. The pistil is frequently surrounded basally by staminodes.

Much more obscure is the floral structure in *Pandanus*. In the male inflorescences the stamens are either connate or separate. When united the stamens are in clusters of few (triads) or many, racemosely to umbellately arranged in phalanges. Frequently they have branched filaments and often an apical prolongation of the connective. In the most specialized phalanges the filaments are short or absent and the anthers crowded on the discoid, enlarged apex of the staminal column (stemonophore). A subtending bract at the base of the phalanges has been observed in *Pandanus barklyi,* and sometimes a pistillode occurs above the stamens at the top of the phalanges (e.g. in subgenus *Martellidendron*), suggesting the phalange to be equivalent to a flower. In species where the stamens are separate any trace of a flower structure is absent. In the female inflorescences the carpels are either free or connate 2–30 together into phalanges. Each carpel bears a sessile stigma or has a short style. The carpels are frequently incompletely closed, in the stigmatic region. In the phalanges the ventral suture of each carpel may be turned inwards (centripetal) and very rarely staminodes are then present at the base of the phalange, so that a flower-like structure is discerned. But the ventral suture of the united carpels may also be turned outwards (centrifugal; as in subgenus *Coronata*) or towards the apex of the inflorescence, or it may have a more complex orientation. Such phalanges may be the result of lateral fusion of more floral units. In *Pandanus* each carpel has only one ovule or, rarely (*P. freycinetioides*), a few.

The anthers have an amoeboid tapetum and the pollen grains are two-celled and ulcerate to sulcoidate. The ovules are anatropous, crassinucellate and bitegmic, with the inner integument forming the micropyle. The embryo sac formation is only known in *Pandanus* and *Freycinetia,* where it is of the *Polygonum* Type, and endosperm formation is nuclear. In some species of *Pandanus* diploid nucellar nuclei migrate into the embryo sac from the chalazal region and may perhaps contribute to agamospermy and parthenocarpy (CHEAH and STONE 1975). This migration of nucellar nuclei has not been observed in *Freycinetia*. A hypostase may be formed below the embryo sac.

The fruit is a berry in *Freycinetia,* where the heads of fruits may consist of 10–1,000 aggregated berries. In *Sararanga* and *Pandanus* the fruits are drupes, in *Sararanga* with 12–80 pyrenes per fruit. In *Pandanus* the drupes are either monocarpellary (monodrupes) and one-seeded or the drupes consist of the connate carpels of the phalanges (polydrupes), either with united or separate endocarps. The seeds are rather small with copious oily (or in *Freycinetia* starchy) endosperm and a small embryo. A strophiole may occur, apparently developed from the raphe.

*Chromosome Number*. The basic chromosome number is $x = 30$ (RAVEN 1975) with some aneuploidy.

*Chemistry*. Silica is lacking in the superorder whereas calcium oxalate occurs both as raphides and other kinds of crystals. Saponins are not known, nor are chelidonic acid or alkaloids. Cyanogenic compounds occur at least in species of *Pandanus*. The flavonoid pattern of the superorder seems to be virtually unknown. HARRIS and HARTLEY (1981) did not find any UV-fluorescent organic acids in the cell walls.

*Distribution*. The Pandaniflorae are restricted to the Old World, with centres in Madagascar (where *Pandanus* is highly variable), South-East Asia (incl. Malaysia) and the Pacific Islands, in particular Melanesia.

*Fossils*. The Pandaniflorae are known as fossil (pollen) from Upper Cretaceous, viz. Maestrichtian (MULLER 1981).

*Relationships*. The relationships of this group are subject to much uncertainty. Traditionally, it is associated with Arecales and Cyclanthales. HARLING (1958) considers that *Freycinetia* is probably the closest relative of the Cyclanthales, and it is not unlikely that it evolved from palm-like ancestors. It lacks silica and unlike palms may have a starchy endosperm. The leaves are less differentiated, than in both Arecales and Cyclan-thales.

An alternative affinity would be with members of the Commeliniflorae, but not, as sometimes suggested, with taxa such as *Scirpodendron* of Cyperaceae, which are highly derived representatives.

If they evolved from these, it would have been from primitive forms, where raphides were still retained, where pollen grains had not become fixed as tetrads or pseudomonads (as they have in Cyperales), but were separate. This latter hypothesis is compatible with a rather close relationship to Arecales and Cyclanthales which may likewise share their ancestry with the Commeliniflorae.

# Order Pandanales

*One Family:* Pandanaceae.

Description, see under the superorder.

## Pandanaceae R. Brown (1810)    3:800–900 (Figs. 224–225)

The family was divided by B.C. STONE (1972) into two subfamilies, *Pandanoideae* with the arborescent *Sararanga* and *Pandanus,* which have uniovulate carpels, and *Freycinetioideae* with the lianoid *Freycinetia,* having multiovulate carpels.

### Subfamily Pandanoideae

*Sararanga* (2). The two species have restricted distributions, *S. sinuosa* being known from New Guinea, the Solomon Islands and the Admiralty Islands, and *S. philippinensis* from the Philippines only. They grow in woody slopes near the sea but also inland on forested mountain hills and in clearings. The leaves are set in four twisted rows, and the pendent, paniculate inflorescences have flowers with a perianth-like cup, in which *Sararanga* differs significantly from the other two genera. Both species are trees, *S. sinuosa* being up to 20 m tall and having leaves up to 4.5 m long, *S. philippinensis* being somewhat smaller, 6–9 m high and with leaves up to 2.5 m long. The latter species is unique in the family in having stellate-pubescent inflorescences. The greenish to white inflorescences are assumed to be wind-pollinated. The drupes are orange-coloured or red.

*Pandanus* (500–600), the "Screw-Pines", have previously been referred to segregate genera but are now considered to constitute a single genus. This was divided by B.C. STONE (1974) into eight subgenera and 61 sections. The genus is distributed from West Africa through Central Africa, Madagascar and India to Malaysia, Melanesia, Northern Australia and Polynesia, reaching Hawaii to the north and Pitcairn to the south. The maximum concentration of species is in Malaysia, Melanesia and Madagascar. The genus is probably ancient. Many of its species are local endemics or have a very restricted distribution. The screw pines occur in a multitude of habitats: along sea shores and lagoons, on mangrove fringes, in lowland swamps, by water courses, in mountain forests or even in shady forests, in the last of which they may be understorey shrubs. Few species are epiphytic (only in subgen. *Acrostigma,* sect. *Epiphy-*

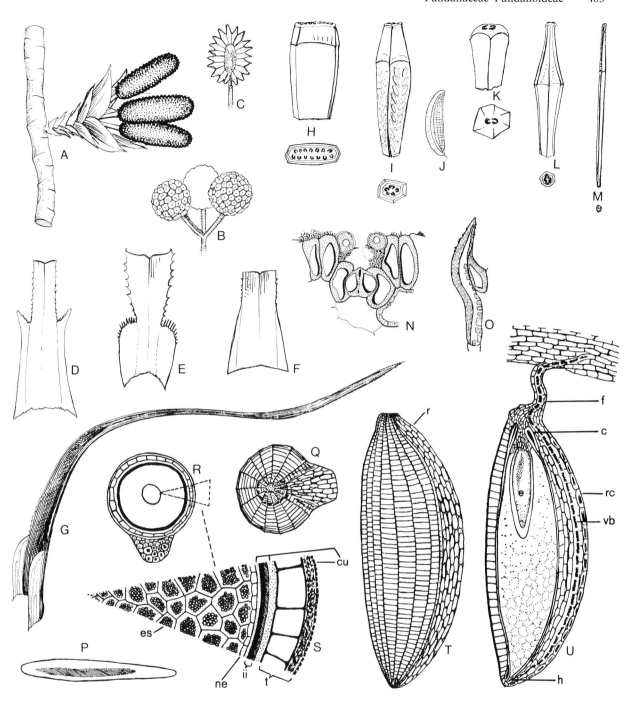

**Fig. 224.** Pandanaceae. Details of *Freycinetia*. **A** *F.* sect. *Lateriflorae*, lateral inflorescence. **B** *F.* sp., globose heads borne on a hispidulous peduncle. **C** *F.* sp., elongate head, longitudinal section. **D–F** *F.* spp., variation in leaf base. **G** *F.* sp., leaf, showing hyaline flanks of leaf base. **H–I** and **K–M** Berries of different species, consisting of two (**K, M**) to several (**H, I**) carpels, lateral and top views; **J** showing a seed. (**A–M** from B.C. STONE 1968) **N–O** *F. rigidifolia.* **N** Stoma from lower side of leaf. **O** Prickly marginal hair from leaf sheath. **P** *F. suma-* *trana,* raphide sac from leaf sheath. (**N–P** from GOVINDARAJALU and THANYAKUMAR 1977) **Q–U** *F. baueriana.* **Q–R** Seed, **Q** in top view, **R** in transverse section. **S** Part of transverse section (of **R**), enlarged (*es* endosperm, here with starch; *ne* nucellar epidermis; *ii* inner integument = tegmen; *t* outer integument = testa; *cu* cuticle). **T** Seed in profile view, raphe (*r*) to the *right*. **U** Seed, longitudinal section (*e* embryo; *f* funicle; *c* collar; *rc* raphide-cell; *vb* vascular bundle; *h* hypostase). (**Q–U** from B.C. STONE 1973)

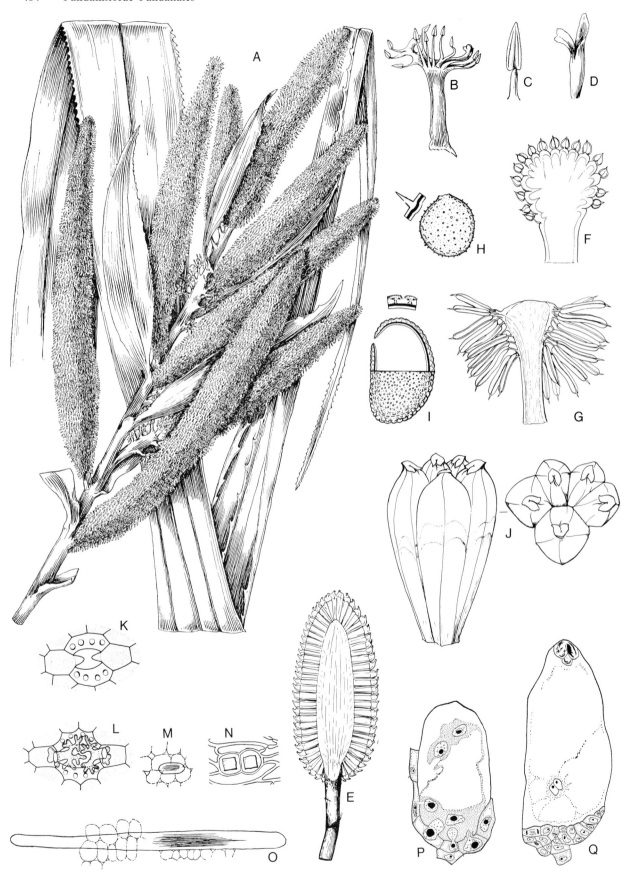

tica and part of sect. *Acrostigma* in Malaya and on Borneo). A diversity of growth forms occurs. Many are tall trees, up to 20 m high and with leaves up to 5 m long or more. The trunks are generally supported basally by large aerial roots (stilt roots), but these may be lacking in some species. Usually the stem is pseudodichotomously divided into equivalent branches, each of which ends in a dense crown of leaves. Many of the coastal *Pandanus* species have spreading branches supported by aerial roots. These branches may become detached from the main trunk and give rise to separate individuals, a type of vegetative reproduction. A striking growth form is met with in subgenus *Vinsonia* sect. *Acanthostyla* with 13 species endemic to Madagascar, growing in deep swamps and in gallery forests. The main axis in this section is a monopodial columnar trunk up to 15 m high, ending in a large rosette of leaves. It bears lateral, horizontal and forked branches with leaves smaller than those of the terminal rosette. The inflorescences are confined to these lateral branches. The habit of these screw pines is somewhat coniferoid: a unique growth form in the family as well as in the monocotyledons. Young plants of these screw pines may have a terminal rosette with leaves up to 9–10 m long and 30–36 cm broad.

It is generally assumed that *Pandanus* is wind-pollinated. But as many species have the inflorescences enclosed by coloured, white to pink spathes and odour is known from the male inflorescences at least of *P. odoratissimus*, pollination by insects and other vectors is likely, although observations are lacking. The yellow, pink or red drupes are dispersed by animals but dispersal by sea currents also occurs. In this connection it may be noted that none of the many species on Madagascar are dispersed by water.

---

Fig. 225. Pandanaceae. **A–D** *Pandanus candelabrum*. **A** Inflorescence and part of leaf. **B** Male flower. **C** Stamen. **D** Stigmas from female flower. (HEPPER 1968) **E** *P. graminifolius*, infructescence. **F** *Pandanus* sect. *Cauliflora*, male phalange. **G** *Pandanus* sect. *Bernardia*, male phalange. (**E–G** from B.C. STONE 1974). **H** *P. odoratissimus*, pollen grain. **I** *P. eydouxia*, pollen grain. (**H–I** from ERDTMAN 1952). **J** *Pandanus* sect. *Megakeura*, phalanges, side and top views. (B.C. STONE 1974). **K–L** *Pandanus* spp., stomata showing papillar protrusion (simple and complex, respectively). (HUYNH 1974). **M–O** *P. utilis*. **M** Raphide cell. **N** Paired cells containing square oxalate crystals. **O** Raphide vessel. (SOLEREDER and MEYER 1933). **P** *P. dubius*, embryo sac in 4-nucleate stage, with nucellar nuclei entering the embryo sac. **Q** *P. ornatus*, mature embryo sac with a large number of cells in the chalazal part. (**P** and **Q** from FAGERLIND 1940)

The screw pines have only local use. The leaves or the fibres from the roots are used for manufacturing mats, baskets, fishing nets or ropes. A few species have edible fruits or seeds. Fruits of *P. edulis* are used on Madagascar and the pineapple-flavoured fruits of the Malayan *P. houlletii* are edible. *P. odoratissimus*, which is cultivated, has edible seeds and an essential oil is obtained from the fragrant male inflorescences. Some species are cultivated as ornamentals.

## Subfamily Freycinetioideae

*Freycinetia* (ca. 180) occurs from Sri Lanka throughout South-Eastern Asia to Northern Australia, Polynesia and New Zealand. The genus consists of woody climbers with clasping, often flattened and sometimes long and wiry roots and leaves that are linear to ovate or obovate, usually not more than 2 m long. The leaves are basally provided with membranous auricles, the margins are usually prickly, at least near the apex, and the leaf apex is usually acuminate to caudate. The spadices are terminal either on normal shoots or on specialized leafless shoots, often emerging three or more together in a subumbellate fashion, rarely in a racemose fashion or solitarily. The lower bracts are leaf-like, the upper usually white or variously coloured (pink, orange, salmon, red, yellow, lavender), the uppermost often being fleshy (and edible!). The female spikes consist of carpels or syncarps, and the male of densely crowded solitary stamens, each with a minutely scabrid filament and a basifixed, globose to oblong or linear, blunt-tipped anther. The fruiting heads are usually red, yellow or white, consisting of berries, each with a firm, rigid apex (pileus), the basal parts fleshy. (Extracted from B.C. STONE 1982.) According to VAN DER PIJL (1956), pollination is by bats and birds. In each case the food reward is the edible sugary bracts.

The three genera of the Pandanaceae are very distinct and it may be argued whether or not they are closely related. In this work, having treated all the diverse groups of palms in one family, and *Cyclanthus* together with the genera of Carludovicoideae in one family, it would be inconsistent to divide the Pandanaceae.

It can be assumed that *Sararanga* is more ancestral in its floral construction than either of the other genera. All three genera are, however, specialized in their ways.

# References

ACKERMANN JD, WILLIAMS NH (1980) Pollen morphology of the tribe Neottieae. Grana 19:7–18

AGRAWAL JS (1952) The embryology of *Lilaea subulata* H.B.K. with a discussion on its systematic position. Phytomorphology 2:15–29

AIRY SHAW HK (1981) A new species of *Hanguana* from Borneo. Kew Bull 35:819–821

AMBROSE JD (1975) Comparative anatomy and morphology of the Melanthioideae (Liliaceae). Thesis, Cornell Univ, Ithaca, New York

AMBROSE JD (1980) A re-evaluation of the Melanthioideae (Liliaceae) using numerical analyses. In: BRICKELL CD, CUTLER DF, GREGORY M (eds) Petaloid monocotyledons. Linn Soc Symp Ser 8, Academic Press, London New York, pp 65–81

ANCIBOR E (1979) Systematic anatomy of vegetative organs of the Hydrocharitaceae. Bot J Linn Soc 78:237–266

ANDERSSON L (1976) The synflorescence of Marantaceae. Organisation and descriptive terminology. Bot Not 129:39–48

ANDERSSON L (1981a) The neotropical genera of Marantaceae. Circumscription and relationships. Nord J Bot 1:218–245

ANDERSSON L (1981b) Revision of *Heliconia* sect. *Heliconia* (Musaceae). Nord J Bot 1:759–784

ARBER A (1922a) Leaves of Farinosae. Bot Gaz 74:80–94

ARBER A (1922b) On the development and morphology of the leaves of palms. Proc R Soc Lond [Biol] 93:249–261

ARBER A (1923) On the "squamulae intravaginales" of the Helobiae. Ann Bot (London) 37:31–41

ARBER A (1925) Monocotyledons. A morphological study. Cambridge Univ Press, Cambridge

ARBER EAN, PARKIN J (1907) On the origin of angiosperms. Bot J Linn Soc 38:29–80

AREKAL GD, RAMASWAMY SN (1980) Embryology of *Eriocaulon hookerianum* Stapf and the systematic position of Eriocaulaceae. Bot Not 133:295–309

ARGUE CL (1971) Pollen of the Butomaceae and Alismataceae. Development of the pollen wall in *Butomus umbellatus* L. Grana 11:131–144

ARGUE CL (1976) Pollen studies in the Alismataceae with special reference to taxonomy. Pollen et Spores 18:161–201

ARROYO SC (1982) Anatomía vegetativa de *Ixiolirion* Fisch. ex Herb. (Liliales) y su significado taxonómico. Parodiana 1:271–286

ASCHERSON P, GÜRKE M (1889) Hydrocharitaceae. In: Engler A, Prantl K (eds) Die natürlichen Pflanzenfamilien, vol II. Englemann, Leipzig, pp 238–258

ASPLUND I (1968) Embryological studies in the genera *Sparganium* and *Typha*. A preliminary report. Sven Bot Tidskr 62:410–412

ASPLUND I (1972) Embryological studies in the genus *Typha*. Sven Bot Tidskr 66:1–17

AVDULOV NP (1931) Karyo-systematische Untersuchungen der Familie Gramineen. Bull Appl Bot [Suppl] 44:1–428 (Russian with German summary)

AYENSU ES (1966) Taxonomic status of *Trichopus*; anatomical evidence. Bot J Linn Soc 59:425–430

AYENSU ES (1973) Biological and morphological aspects of Velloziaceae. Biotropica 5:135–149

AYENSU ES (1974) Leaf anatomy and systematics of New World Velloziaceae. Smithsonian Contrib Bot 15:1–125

AYENSU ES, SKVARLA JJ (1974) Fine structure of Velloziaceae pollen. Bull Torrey Bot Club 101:250–266

BAIJNATH H (1978) *Jodriella*, a new genus of Liliaceae from tropical Africa. Kew Bull 32:571–578

BAIJNATH H (1980) A contribution to the study of leaf anatomy of the genus *Kniphofia* Moench (Liliaceae). In: BRICKELL CD, CUTLER DF, GREGORY M (eds) Petaloid monocotyledons. Linn Soc Symp Ser 8. Academic Press, London New York, pp 89–103

BAILLON BM (1867–1875) Histoire des plantes. Libraire Hachette, Paris

BAILLON H (1894) Histoire des plantes. Monographie des Amaryllidacées, Bromeliacées et Iridacées. Libraire Hachette, Paris

BAKKER CA (1958) Restionaceae. Flora Malesiana. Ser I Spermatophyta 5:416–420

BALDWIN JT Jr (1950) Geography of *Maschalocephalus Dinklagei*. Am J Bot 37:402–405

BAMBACIONI V (1928) Come avviene in *Fritillaria persica*, 10 sviluppo del gametofito femminile e l'aumento dei cromosomi nella regione chalazale. C R Accad Lincei Roma 66:544–546

BANERJI I, GANGULEE HC (1937) Spermatogenesis in *Eichhornia crassipes* Solms. J Indian Bot Soc 16:289–296

BARTHLOTT W (1976a) Struktur und Funktion des Velamen radicum der Orchideen. In: SENGHAS K (ed) Proc 8th World Orchid Conf, pp 438–443

BARTHLOTT W (1976b) Morphologie der Samen von Orchideen im Hinblick auf taxonomische und funktionelle Aspekte. Proc 8th World Orchid Conf, pp 444–453

BARTHLOTT W, FRÖHLICH D (1983) Micromorphologie und Orientierungsmuster epicuticularer Wachs-Kristalloide: ein neues systematisches Merkmal bei Monokotylen. Plant Syst Evol 142:171–185

BARTHLOTT W, ZIEGLER B (1981) Micromorphologie der Samenschalen als systematisches Merkmal bei Orchideen. Ber Dtsch Bot Ges 94:267–273

BATE-SMITH EC (1968) The phenolic constituents of plants and their taxonomic significance. 3. Monocotyledons. J Linn Soc (London) 60:325–356

BATYGINA TB, KRAVTSOVA TI, SHAMROV II (1980) The

comparative embryology of some representatives of the orders Nymphaeales and Nelumbonales. Bot Zh 65:1071–1087 (in Russian)

BATYGINA TB, SHAMROV II (1981) The embryology of Nymphaeales and Nelumbonales. 1. The development of the anther. Bot Zh 66:1696–1709 (in Russian)

BATYGINA TB, SHAMROV II, KOLESOVA GE (1982) Embryology of the Nymphaeales and Nelumbonales. 2. The development of the female embryonic structures. Bot Zh 67:1179–1195 (in Russian)

BEHNKE H-D (1969) Die Siebröhren-Plastiden bei Monocotylen. Naturwissenschaften 55:140–141

BEHNKE H-D (1971) Zum Feinbau der Siebröhren-Plastiden von *Aristolochia* und *Asarum* (Aristolochiaceae). Planta 97:62–69

BEHNKE H-D (1981a) Siebelement-Plastiden, Phloem-Protein und Evolution der Blütenpflanzen: 2. Monokotyledonen. Ber Dtsch Bot Ges 94:647–662

BEHNKE H-D (1981b) Sieve element characters. Nord J Bot 1:381–400

BEHNKE H-D, BARTHLOTT W (1983) New evidence from the ultrastructural and micromorphological fields in angiosperm classification. Nord J Bot 3:43–66

BENTHAM G, HOOKER JD (1883) Genera Plantarum, vol III, part 2. Reeve, London

BENZING DH, SEEMANN J, RENFROW A (1978) The foliar epidermis in Tillandsioideae (Bromeliaceae) and its role in habitat selection. Am J Bot 65:359–365

BERG RY (1958) Studies in Liliaceae, tribe Parideae. I. Seed dispersal, morphology, and phylogeny of *Trillium*. Skr Nor Vidensk Akad Oslo 11:1–36

BERG RY (1959) Seed dispersal, morphology and taxonomic position of *Scoliopus*, Liliaceae. Skr Nor Vidensk Akad Oslo 4:1–56

BERG RY (1960) Ovary, ovule, and endosperm of *Calochortus amabilis*. Nytt Mag Bot 8:189–206

BERG RY (1962a) Morphology and taxonomic position of *Medeola*, Liliaceae. Skr Nor Vidensk Akad Oslo N Ser 3:1–55

BERG RY (1962b) Contribution to the comparative embryology of the Liliaceae: *Scoliopus, Trillium, Paris* and *Medeola*. Skr Nor Vidensk Akad Oslo N Ser 4:1–64

BHANDARI NN (1971) Embryology of the Magnoliales and comments on their relationships. J Arnold Arbor Harv Univ 52:1–39

BILLINGS FH (1904) A study of *Tillandsia usneoides*. Bot Gaz 38:99–121

BJÖRKQVIST I (1967) Studies in *Alisma* L. I. Distribution, variation and germination. Opera Bot 17:1–128

BJÖRKQVIST I (1968) Studies in *Alisma* L. II. Chromosome studies, crossing experiments and taxonomy. Opera Bot 19:1–138

BJÖRNSTAD I (1970) Comparative embryology of Asparagoideae-Polygonatae, Liliaceae. Nytt Mag Bot 17:169–207

BJÖRNSTAD I, FRIIS I (1972) Studies on the genus *Haemanthus* L. (Amaryllidaceae). I. The infrageneric taxonomy. Norw J Bot 19:187–206

BLUNDEN G, JEWERS K (1973) The comparative leaf anatomy of *Agave, Beschorneria, Doryanthes* and *Furcraea* species (Agavaceae: Agaveae). J Linn Soc Bot 66:157–179

BOEHM K (1931) Embyologische Untersuchungen an Zingiberaceen. Planta 14:411–440

BOGNER J (1978) A critical list of the aroid genera. Aroideana 1 3:63–73

BOR NL (1960) The grasses of Burma, Ceylon, India and Pakistan. Pergamon, London New York

BOR NL (1968) Gramineae. In: TOWNSEND CC, GUEST E, AL-RAWI A (eds) Flora of Iraq, vol 9. Minist Agric Baghdad

BOUDET AM, LECUSSON R, BOUDET A (1975) Mise en évidence et propriétés de deux formes de la 5-déshydroquinate hydrolyase chez les végétaux supérieures. Planta 124:67–75

BOUDET AM, BOUDET A, BOUYSSOU H (1977) Taxonomic distribution of isoenzymes of dehydroquinate hydrolyase in the angiosperms. Phytochemistry 16:912–922

BOULTER D (1973a) The molecular evolution of higher plant cytochrome c. In: SWAIN T (ed) Chemistry in evolution and systematics. Butterworths, London, pp 539–552

BOULTER D (1973b) The use of amino acid sequence data in the classification of higher plants. In: BENDZ G, SANTESSON J (eds) Chemistry in botanical classification. Nobel Symp, vol 25. Academic Press, London New York, pp 211–216

BOULTER D, PEACOCK D, GUISE A, GLEAVES JT, ESTABROOK G (1979) Relationships between the partial amino acid sequences of plastocyanin from members of ten families of flowering plants. Phytochemistry 18:603–608

BRENAN JPM (1962) *Murdannia axillaris* Brenan. In: HOOKER's icones plantarum, Ser. 6, vol 4, Tab 3578. Kew, Richmond

BRENAN JPM (1966) The classification of Commelinaceae. J Linn Soc Bot 59:349–370

BRENAN JPM (1968) Commelinaceae. In: HEPPER FN (ed) Flora of West Tropical Africa, vol I, 2nd edn. R Bot Gard Kew, London, pp 22–50

BRESINSKI A (1963) Bau, Entwicklungsgeschichte und Inhaltsstoffe der Elaiosomen. Bibl Bot 126:1–54

BREWBAKER JL (1967) The distribution and phylogenetic significance of binucleate and trinucleate pollen grains in the angiosperms. Am J Bot 54:1069–1083

BROWN R (1833) Observations on the organs and mode of fecundation in Orchideae and Asclepiadeae. Trans Linn Soc 16:685–745

BRUGGEN HWE VAN (1968) Revision of the genus *Aponogeton*. I. The species of Madagascar. Blumea 16:243–263

BRUGGEN HWE VAN (1969) Revision of the genus *Aponogeton*. III. The species of Australia. Blumea 17:121–137

BRUGGEN HWE VAN (1970) Revision of the genus *Aponogeton* (Aponogetonaceae). IV. The species of Asia and Malesia. Blumea 18:457–486

BRUGGEN HWE VAN (1973) Revision of the genus *Aponogeton* (Aponogetonaceae). VI. The species of Africa. Bull Jard Bot Natl Belg 43:193–233

BURBIDGE NT (1963) Dictionary of Australian plant genera: gymnosperms and angiosperms. Angus and Robertson, Sydney

BURGER WC (1977) The Piperales and the monocots. Alternative hypotheses for the origin of the monocotyledon flower. Bot Rev 43:345–393

BURGER WC (1981) Heresy revived: the Monocot Theory of angiosperm origin. Evol Theory 5:189–225

BURTT BL (1972) General introduction to papers on Zingiberaceae. Notes R Bot Gard (Edinburgh) 31:155–165

BURTT BL, SMITH RM (1972) Key species in the taxonomic history of Zingiberaceae. Notes R Bot Gard (Edinburgh) 31:177–228

CABRERA AL (1968) Flora de la Provincia de Buenos Aires, vol 1. Coleccion Cient, INTA, Buenos Aires

CAMPBELL DH (1930) The phylogeny of monocotyledons. Ann Bot (London) 44:311–331

CAMPBELL EO (1968) An investigation of *Thismia rodwayi* F. Muell. Trans R Soc N Z Bot 3:209–219

CAPOOR SP (1937) Contribution to the morphology of some Indian Liliaceae, II. The gametophytes of *Urginea indica* Kunth. Beih Bot Centralbl 56:156–170

CARLQUIST S (1960) Anatomy of Guayana Xyridaceae: *Abolboda, Orectanthe,* and *Achlyphila.* Mem N Y Bot Gard 10:65–117

CARLQUIST S (1961) Pollen morphology of Rapateaceae. Aliso 5:39–66

CARLQUIST S (1964) Morphology and relationships of Lactoridaceae. Aliso 5:421–435

CARLQUIST S (1976) *Alexgeorgia,* a bizarre new genus of Restionaceae from Western Australia. Aust J Bot 24:281–295

CARTER S (1960) Alismataceae. In: MILNE-REDHEAD E, POLHILL RM (eds) Flora of Tropical East Africa. R Bot Gard Kew, London

CARTER S (1969) *Cyanastrum johnstonii* Baker. In: Hooker's icones plantarum, Ser 5, vol 7, Tab 3641. Dulau, London

CAVE MS (1955) Sporogenesis and the female gametophyte of *Phormium tenax.* Phytomorphology 5:247–253

CAVE MS (1966) The female gametophytes of *Lapageria rosea* and *Philesia magellanica.* Gayana Bot 15:25–31

CAVE MS (1975) Embryological studies in *Stypandra* (Liliaceae). Phytomorphology 25:95–99

CHAKROUN S, HÉBANT CH (1983) Developmental anatomy anatomy of *Aphyllanthes monspeliensis,* a herbaceous monocotyledon with secondary growth. Plant Syst Evol 141:231–241

CHANDA S (1966) On the pollen morphology of Centrolepidaceae, Restionaceae and Flagellariaceae, with special reference to taxonomy. Grana 6:355–415

CHANDA S, GHASH K (1976) Pollen morphology and its evolutionary significance in Xanthorrhoeaceae. In: FERGUSON IK, MULLER J (eds) The evolutionary significance of the exine. Linn Soc Symp Ser 1. Academic Press, London New York, pp 527–559

CHANT SR (1978) Triuridaceae. In: HEYWOOD VH (ed) Flowering plants of the world. Oxford Univ Press, Oxford, p 277

CHEADLE VI (1942) The occurrence and types of vessels in the various organs of the plant in the Monocotyledoneae. Am J Bot 29:441–450

CHEADLE VI (1943a) The origin and certain trends of specialization of the vessel in the Monocotyledoneae. Am J Bot 30:11–17

CHEADLE VI (1943b) Vessel specialization in the late metaxylem of the various organs in the Monocotyledoneae. Am J Bot 30:484–490

CHEADLE VI (1944) Specialization of vessels within the xylem of each organ in Monocotyledoneae. Am J Bot 31:81–92

CHEADLE VI (1953) Independent origin of vessels in the monocotyledons and dicotyledons. Phytomorphology 3:23–44

CHEADLE VI (1955a) The taxonomic use of vessels in the metaxylem of Gramineae, Cyperaceae, Juncaceae and Restionaceae. J Arnold Arbor Harv Univ 36:141–157

CHEADLE VI (1955b) Conducting elements in the xylem of the Bromeliaceae. Bull Bromeliad Soc 5:3–7

CHEADLE VI (1968) Vessels in Haemodorales. Phytomorphology 18:412–420

CHEADLE VI (1969) Vessels in Amaryllidaceae and Tecophilaeaceae. Phytomorphology 19:8–16

CHEADLE VI, KOSAKAI H (1971) Vessels in Liliaceae. Phytomorphology 21:320–333

CHEADLE VI, KOSAKAI H (1980) Occurrence and specialization of vessels in Commelinales. Phytomorphology 30:98–117

CHEADLE VI, KOSAKAI H (1982) Occurrence and specialization of vessels in Xyridales. Nord J Bot 2:97–109

CHEAH CH, STONE BC (1975) Embryo sac and microsporangium development in *Pandanus* (Pandanaceae). Phytomorphology Silver Jubilee, pp 228–238

CHEN SC (1979) On *Diplandrorchis,* a very primitive and phylogenetically significant new genus of Orchidaceae. Acta Phytotaxon Sinica 17:1–6

CHENERY EM (1950) Aluminium in the plant world, part II. Monocotyledons and gymnosperms. Kew Bull 1949:463–473

CHEW WEE-LEK (1972) The genus *Piper* (Piperaceae) in New Guinea, Solomon Islands, and Australia, 1. J Arnold Arbor Harv Univ 53:1–25

CHIKKANNAIAH PS (1964) An embryological study of *Murdannia simplex* (Vahl) Brenan. J Indian Bot Soc 43:238–248

CHIKKANNAIAH PS (1965) A contribution to the life history of *Floscopa scandens* Lour. J Karnatak Univ 9:110–122

CHUPOV VS, KUTIAVINA NG (1978) The comparative immuno-electrophoretic investigations of seed proteins of Liliaceae. Bot Zh 63:473–493 (in Russian)

CHUPOV VS, KUTIAVINA NG (1981) Serological studies in the order Liliales. Bot Zh 66:75–81 (in Russian)

CLAYTON WD (1967) *Puelia coriaceae* In: CLAYTON WD (ed) Hooker's icones plantarum, Ser 5, vol I, Tab 3642. Kew, Richmond

CLIFFORD HT (1970) Monocotyledon classification with special reference to the origin of the grasses (Poaceae). In: ROBSON NKB, CUTLER DF, GREGORY M (eds) New research in plant anatomy. Academic Press, London New York, pp 25–34

CLIFFORD HT, WATSON L (1978) Identifying grasses. Data, methods and illustrations. Univ Queensland Press, St Lucia

CLIFFORD HT, WILLIAMS T (1980) Interrelationships amongst the Liliatae: a graph approach. Aust J Bot 28:261–268

CODD LE (1968) The South African species of *Kniphofia.* Bothalia 9:363–513

CONOVER M (1983) The vegetative morphology of the reticulate-veined Liliiflorae. Telopea 2:401–412

COOK CDK, GUT BJ, RIX EM, SCHNELLER J, SEITZ M (1978) Water plants of the world. Junk, The Hague

COOPER DC (1936) Development of the male gametes in Lilium. Bot Gaz 98:169–177

CORNER EJH (1954) Durian Theory extended. II. The arillate fruit and the compound leaf. Phytomorphology 4:152–165

CORNER EJH (1966) The natural history of palms. Univ Calif Press, Berkeley

CORNER EJH (1976) The seeds of Dicotyledons. Cambridge Univ Press, Cambridge, pp 1–2

CORREA MN (1969) Flora Patagonica, vol II. Colleccion Cienta del Inta, Buenos Aires

CROAT TB (1981) Distribution of Araceae. In: LARSEN K, HOLM-NIELSEN L (eds) Tropical botany. Academic Press, London New York, pp 291–308

CRONQUIST A (1968) The evolution and classification of flowering plants. Nelson, London Edinburgh

CRONQUIST A (1981) An integrated system of classification of flowering plants. Columbia Univ Press, New York

CRONQUIST A, HOLMGREN AH, HOLMGREN NH, REVEAL JL, HOLMGREN PR (1977) Intermountain flora. Vascular plants of the inter-mountain West. Columbia Univ Press, New York

CULLEN J (1978) A preliminary survey of ptyxis (vernation) in the angiosperms. Notes R Bot Gard Edinburgh 37:161–214

CUTLER DF (1966) Anatomy and taxonomy of the Restionaceae. Jodrell Lab Notes 4:1–25

CUTLER DF (1969) In: METCALFE CR (ed) Anatomy of the monocotyledons, vol IV. Juncales. Clarendon Press, Oxford

CUTLER DF, AIRY SHAW HK (1965) Anarthriaceae and Ecdeiocoleaceae: two new monocotyledonous families, separated from the Restionaceae. Kew Bull 19:489–499

DAFNI A, IVRI Y (1981) The flower biology of *Cephalanthera longifolia* (Orchidaceae). Plant Syst Evol 137:229–240

DAFNI A, IVRY Y, BRANTJES NBM (1981) Pollination of *Serapias vomeracea* Briq. (Orchidaceae). Acta Bot Neerl 30:69–73

DAGHLIAN CP (1981) A revision of the fossil record of monocotyledons. Bot Rev 47:517–555

DAHLGREN R (1974) Angiospermernes taxonomi, vol I. Akademisk Forlag, København

DAHLGREN R (1975) A system of classification of the angiosperms to be used to demonstrate the distribution of characters. Bot Not 128:119–147

DAHLGREN R (1979) Angiospermernes taxonomi, vol I, 2nd edn. Akademisk Forlag, København

DAHLGREN R (1980) A revised system of classification of the angiosperms. Bot J Linn Soc 80:91–124

DAHLGREN R (1983a) General aspects of angiosperm evolution and macrosystematics. Nord J Bot 3:119–149

DAHLGREN R (1983b) The importance of modern serological research for angiosperm classification. In: JENSEN U, FAIRBROTHERS DE (eds) Proteins and nucleic acids in plant systematics. Springer, Berlin Heidelberg New York, pp 371–394

DAHLGREN R, CLIFFORD HT (1981) Some conclusions from a comparative study of the monocotyledons and related dicotyledons. Ber Dtsch Bot Ges 94:203–227

DAHLGREN R, CLIFFORD HT (1982) The monocotyledons. A comparative study. Academic Press, London New York

DAHLGREN R, RASMUSSEN FN (1983) Monocotyledon evolution: characters and phylogenetic estimation. Evol Biol 16:255–395

DAHLGREN KVO (1939) Endosperm- and Embryobildung bei Zostera marina. Bot Not 1939:607–615

DANDY JE, FOSBERG FB (1954) The type of *Amaryllis belladonna* L. Taxon 3:231–232

DAUMANN E (1965) Das Blütennektarium bei Pontederiaceen und die systematische Stellung dieser Familie. Preslia 37:407–412

DAUMANN E (1970) Das Blütennektarium der Monocotyledonen unter besonderer Berücksichtigung seiner systematischen und phylogenetischen Bedeutung. Feddes Repert Z Bot Taxon Geobot 80:463–590

DAVIS GL (1966) Systematic embryology of the angiosperms. Wiley & Sons, New York London Sydney

DEGENER O (1940) Zingiberaceae: *Costus speciosus*. In: DEGENER O, DEGENER I (eds) New Illustrated Flora of the Hawaiian Islands. N Y Bot Gard, New York

DEGENER O (1947) Musaceae: *Musa paradisiaca* var. *sapientum*. In: DEGENER O, DEGENER I (eds) New illustrated flora of the Hawaiian Islands. N Y Bot Gard, New York

DEGENER O, GRENWELL (1956) Liliaceae: *Pleome fernandii*. In: DEGENER O, DEGENER I (eds) New illustrated flora of the Hawaiian Islands. N Y Bot Gard, New York

DELLERT R (1933) Zur systematischen Stellung von *Wachendorfia*. Oesterr Bot Z 82:335–345

DIELS L, PRITZEL E (1905) Fragmenta phytogeographiae Australiae occidentalis. Beiträge zur Kenntnis der Pflanzen Westaustraliens, ihrer Verbreitung und ihrer Lebens-Verhältnisse. Bot Jahrb Pflanzengesch Pflanzengeogr 35:55–662

DILCHER JA (1974) Approaches to the identification of angiosperm leaf remains. Bot Rev 40:1–157

DILCHER DL (1979) Early angiosperm reproduction: an introductory report. Rev Paleobot Palyn 27:291–328

DING HOU (1957) Centrolepidaceae. In: Flora Malesiana, Ser I Spermatophyta 5:421–448

DOMIN K (1911) Morphologische und phylogenetische Studien über die Stipularbildungen. Ann Jard Bot Buitenzorg Ser 2. 9:117–326

DOYLE JA (1973) The monocotyledons: their evolution and comprative biology. V. Fossil evidence on early evolution of the monocotyledons. Q Rev Biol 48:399–413

DOYLE JA, HICKEY LJ (1976) Pollen and leaves from the mid-Cretaceous Potomac Group and their bearing on early angiosperm evolution. In: BECK CB (ed) Origin and early evolution of angiosperms. Columbia Univ Press, New York, pp 139–206

DRANSFIELD J (1977) A dwarf *Livistona* (Palmae) from Borneo. Kew Bull 31:759–762

DRANSFIELD J (1980) *Pogonotium* (Palmae: Lepidocaryoideae), a new genus related to *Daemonorops*. Kew Bull 34:761–768

DRANSFIELD J (1982) A reassessment of the genera *Plectocomiopsis, Myrialpis* and *Bejaudia* (Palmae: Lepidocaryoideae). Kew Bull 37:237–254

DRENTH E (1972) A revision of the family Taccaceae. Blumea 20:367–406

DRESSLER RL (1974) Classification of the orchid family. Proc 7th Orchid Conf, pp 269–279

DRESSLER RL (1981) The orchids. Natural history and classification. Harvard Univ Press, Cambridge Mass

DRESSLER RL, DODSON CH (1960) Classification and phylogeny in the Orchidaceae. Ann Mo Bot Gard 47:25–67

DUCKER SC, FOORD NJ, KNOX RB (1977) Biology of Australian seagrasses: the genus *Amphibolis* C. Agardh (Cymodoceaceae). Aust J Bot 25:67–95

DUCKER SC, PETTITT JM, KNOX RB (1978) Biology of Australian seagrasses: pollen development and submarine pollination in *Amphibolis antarctica* and *Thalassodendron ciliatum* (Cymodoceaceae). Aust J Bot 26:265–275

DUTT BSM (1970) Hypoxidaceae. In: Symposium on comparative embryology of angiosperms. Bull Indian Natl Sci Acad 41:368–372

DYER RA (1941) *Bowiea volubilis*. Flowering Plants of South Africa, vol 21. Reeve, Ashford, p 815

DYKES WR (1913) The genus *Iris*. Cambridge

EAMES AJ (1953) Neglected morphology of the palm leaf. Phytomorphology 3:172–189

EASTOP V (1979) Sternorrhyncha as angiosperm taxonomists. Symb Bot Ups 22 4:120–134

ECKARDT TH (1964) Reihe Helobiae. In: MELCHIOR H (ed) A Engler's Syllabus der Pflanzenfamilien, 12th edn. Bornträger, Berlin, pp 499–512

EDGAR E (1966) The male flowers of *Hydatella inconspicua* (Cheesem.) Cheesem. (Centrolepidaceae). N Z J Bot 4:153–158

EITEN LT (1976a) Inflorescence units in the Cyperaceae. Ann Mo Bot Gard 63:81–112

EITEN LT (1976b) The morphology of some critical Brazilian species of Cyperaceae. Ann Mo Bot Gard 63:113–199

EL GAZZAR A, HAMZA MK (1975) On the monocots-dicots distinction. Publ Cairo Univ Herb 6:15–28

EL-HAMIDI A (1952) Vergleichend-morphologische Untersuchungen am Gynoecium der Unterfamilien Melanthoideae und Asphodeloideae der Liliaceae. Ark Inst Allg Bot Univ Zuerich, Ser A 4:1–49

ELLIS CJ, FOO LY, PORTER LJ (1983) Enantiomerism: a characteristic of the proanthocyanidin chemistry of the Monocotyledoneae. Phytochemistry 22:483–487

EMBERGER L (1960) Les végétaux vasculaires. In: CHAUDEFAUT M, EMBERGER L (eds) Traité de botanique, vol II. Masson, Paris, pp 1–1539

ENGLER A (1905) Araceae: Pothoideae. In: Das Pflanzenreich, vol IV 23 B, H 37. Engelmann, Weinheim/Bergstrasse, S 4–138

ENGLER A (1908) Araceae: Monsteroideae. In: Das Pflanzenreich, vol IV 23 B, H 37. Engelmann, Weinheim/Bergstrasse, S 4–138

ENGLER A (1911) Araceae: Lasioideae. In: Das Pflanzenreich, vol IV 23 C, H 48, Engelmann, Weinheim/Bergstrasse, S 1–130

ENGLER A (1912) Araceae. Allgemeiner Teil, Homalomeninae and Schismatoglottidinae. In: Das Pflanzenreich, vol IV 23 Da, H 55. Engelmann, Weinheim/Bergstrasse, S 1–134

ENGLER A (1913) Araceae: Philodendroideae-Philodendreae. In: Das Pflanzenreich, vol IV 23 Db, H 60. Engelmann, Weinheim/Bergstrasse, S 1–143

ENGLER A (1915) Araceae: Anubiadeae, Aglaonemateae, Dieffenbachieae, Zantedeschieae, Typhonodoreae, Peltandreae. In: Das Pflanzenreich, vol IV 23 Dc, H 64. Engelmann, Weinheim/Bergstrasse, S 1–78

ENGLER A (1920a) Araceae: Colocasioideae. In: Das Pflanzenreich, vol 23 E, H 71. Engelmann, Weinheim/Bergstrasse, S 1–132

ENGLER A (1920b) Araceae: Aroideae und Araceae: Pistoideae. In: Das Pflanzenreich, vol IV 23 F, H 73. Engelmann, Weinheim/Bergstrasse, S 1–274

ENGLER A (1920c) Araceae. Pars generalis et index familiae generalis. In: Das Pflanzenreich, vol IV 23 A, H 74. Engelmann, Weinheim/Bergstrasse, S 1–71

ERDTMAN G (1952) Pollen morphology and plant taxonomy. Almqvist & Wiksell, Stockholm

ERNST A, BERNARD C (1912) Beiträge zur Kenntnis der Saprophyten Javas. 9. Entwicklungsgeschichte des Embryosackes und des Embryos von *Burmannia candida* und *B. championii*. Ann Jard Bot Buitenzorg Ser 2 10:161–188

EUNUS AM (1951) Contribution to the embryology of the Liliaceae. V. Life history of *Amianthium muscaetoxicum* Walt. Phytomorphology 1:73–79

EUNUS AM (1952) Contributions to the embryology of the Liliaceae. 3. Embryogeny and development of the seed of *Asphodelus tenuifolius*. Lloydia 15:149–155

EVERARD B, MORLEY BD (1970) Wild flowers of the world. Octopus Books, London

EYDE RH (1976) The foliar theory of the flower. Am Sci 63:430–437

FADEN RB (1974) Commelinaceae. In: AGNEW ADQ (ed) Upland Kenya wild flowers. Oxford Univ Press, London, pp 653–668

FADEN RB (1975) A biosystematic study of the genus *Aneilema* R Br (Commelinaceae). Thesis, Washington Univ St Louis

FADEN RB (1978) *Pollia* Thunb. (Commelinaceae): the first generic record from the New World. Ann Mo Bot Gard 65:676–680

FADEN RB, SUDA Y (1980) Cytotaxonomy of Commelinaceae: chromosome numbers of some African and Asiatic species. Bot J Linn Soc 81:301–325

FAEGRI K, IVERSEN J (1964) Textbook of pollen analysis. Munksgaard, Kopenhagen

FAGERLIND F (1939) Kritische und revidierende Untersuchungen über das Vorkommen des Adoxa ("Lilium")-Typs. Acta Horti Bergiani 13 1:1–49

FAGERLIND F (1940) Stempelbau und Embryosackentwicklung bei einigen Pandanaceen. Ann Jard Bot Buitenzorg 49:55–78

FAHN A (1954) Anatomical structure of Xanthorrhoeaceae Dumort. J Linn Soc Bot 55:158–184

FEDOROV A (ed) (1969) Chromosome numbers of flowering plants. Komarov Bot Inst, Leningrad

FREDGA A, BENDZ G, APELL A (1974) Red pigments in the Juncaceae family. In: BENZ G, SANTESSON J (eds) Chemistry in botanical classification. Nobel Academic Press, New York London, pp 121–122

FRIES TH CE (1919) Der Samenbau bei *Cyanastrum* Oliv. Sven Bot Tidskr 13:295–304

FRIIS I, BJÖRNSTAD IN (1971) A new species of *Haemanthus* (Amaryllidaceae) from Southwest Ethiopia. Norw J Bot 18:227–230

FRIIS I, NORDAL IN (1976) Studies in the genus *Haemanthus* (Amaryllidaceae). IV. Division of the genus into *Haemanthus* s str and *Scadoxus* with notes on *Haemanthus* s str. Norw J Bot 23:63–77

FRIIS I, VOLLESEN K (1982) New taxa from the Imatong Mountains, South Sudan. Kew Bull 37:465–479

FULVIO TE, DI, CAVE MS (1964) Embryology of *Bland-*

*fordia nobilis* Smith (Liliaceae) with special reference to its taxonomic position. Phytomorphology 14:487–499

GARAY LA (1960) On the origin of Orchidaceae. Leafl Harv Univ Bot Mus 19:57–96

GARAY LA (1964) Evolutionary significance of geographical distribution of orchids. Proc 4th Int Orchid Conf, Singapore, pp 170–187

GARAY LA (1972) On the origin of Orchidaceae, 2. J Arnold Arbor Harv Univ 53:202–215

GARDNER CA (1952) Flora of Western Australia, I. Government Printer, Perth

GARDNER RO (1976) Binucleate pollen in *Triglochin* L. N Z J Bot 14:115–116

GAUDICHAUD-BEAUPRÉ C (1841) Atlas. Voyage autour du monde exécuté pendent les années 1836 et 1837 sur la corvette Bonite, commandée par M Vaillant. Botanique Bertrand, Paris

GEERINCK D (1969) Genera des Haemodoraceae et des Hypoxidaceae. Bull Jard Bot Natl Belg 39:47–82

GEERINCK D (1970) Burmanniaceae. In: Flore du Congo du Rwanda et du Burundi. Jard Bot Nat Belg, Bruxelles

GEORGE AS (1980) *Rhizanthella gardneri* RS Rogers, the underground orchid of Western Australia. Am Orchid Soc Bull 49:631–646

GIBBS RD (1974) Chemotaxonomy of flowering plants. McGill-Queen's Univ Press, Montreal London, pp 1–4

GLÜCK H (1901) Die Stipulargebilde der Monocotyledonen. Verh Naturhist Ded Ver Heidelberg N F 7:1–96

GOEBEL K (1898–1901) Organographie der Pflanzen. Fischer, Jena

GOEBEL K (1923) Organographie der Pflanzen, vol III, 2nd edn. Spezielle Organographie der Samenpflanzen. Fischer, Jena

GOLDBLATT P (1971) Cytological and morphological studies in the South African Iridaceae. S Afr J Bot 37:317–460

GOLDBLATT P (1977a) Chromosome cytology of *Hessea, Strumaria,* and *Carpolyza* (Amaryllidaceae). Ann Mo Bot Gard 63:314–320

GOLDBLATT P (1977b) The genus *Moraea* in the winter rainfall region of Southern Africa. Ann Mo Bot Gard 63:657–786

GOLDBLATT P (1980) Polyploidy in angiosperms: monocotyledons. In: LEWIS WH (ed) Polyploidy: biological relevance. Plenum Publ Corp, New York, pp 219–239

GORNALL RJ, BOHM BA (1978) Angiosperm flavonoid evolution: a reappraisal. Syst Bot 3:353–368

GORNALL RJ, BOHM BA, DAHLGREN R (1979) The distribution of flavonoids in the angiosperms. Bot Not 132:1–30

GOVINDAPPA DA (1955) Development of the gametophytes in *Aloë ciliaris* Haw. J Indian Bot Soc 34:146–150

GOVINDARAJALU E, THANYAKUMAR S (1977) The vegetative anatomy of *Freycinetia sumatrana* Hemsl and *F rigidifolia* Hemsl along with the comparative study of shoot apex organization in *F rigidifolia* and *Pandanus tectorius* Soland. Adansonia Ser 2 17:59–76

GREILHUBER J, SPETA F (1976) C-banded karyotypes in the *Scilla hohenackeri* Group, *S persica* and *Puschkinia* (Liliaceae). Plant Syst Evol 126:149–188

GROOTJEN CJ (1983) Development of ovule and seed in *Cartonema spicatum* R Br (Cartonemataceae). Aust J Bot 31:297–305

GROOTJEN CJ, BOUMAN F (1981a) Development of ovule and seed in *Stanfieldiella imperforata* (Commelinaceae). Acta Bot Neerl 30:265–275

GROOTJEN CJ, BOUMAN F (1981b) Development of the ovule and seed in *Costus cuspidatus* (N E Br) Maas (Zingiberaceae) with special reference to the formation of the operculum. Bot J Linn Soc 83:27–39

GROSSMAN K (1979) The spiderwort: radiation monitor. Garden 1979 May/June 22–25

GUILLARMOD AJ, MARAIS W (1972) A new species of *Aponogeton* (Aponogetonaceae). Kew Bull 27:563–565

GUTTENBERG H VON (1960) Grundzüge der Histogenese höherer Pflanzen. 1. Die Angiospermen. Handb Pflanzenanat 8 3:1–315

HADLEY G (1982) Orchid mycorrhiza. In: ARDITTI J (ed) Orchid biology, vol II. Cornell Univ Press, New York, pp 83–118

HÄNSEL R, LEUSCHKE A, GOMEZ-POMPA A (1975) Aporphine type alkaloids from *Piper auritum*. Lloydia 38:529–530

HAINES RW, LYE KA (1972) Studies in African Cyperaceae. VII. Panicle morphology and possible relationships in Sclerieae and Cariceae. Bot Not 125:331–343

HAINES RW, LYE KA (1975) Seedlings of Nymphaeaceae. Bot J Linn Soc 70:255–265

HAMANN U (1961) Merkmalsbestand und Verwandtschaftsbeziehungen der "Farinosae". Willdenowia 2:639–768

HAMANN U (1962a) Beitrag zur Embryologie der Centrolepidaceae mit Bemerkungen über den Bau der Blüten und Blütenstände und die systematische Stellung der Familie. Ber Dtsch Bot Ges 75:153–171

HAMANN U (1962b) Weiteres über Merkmalsbestand und Verwandtschaftsbeziehungen der "Farinosae". Willdenowia 3:169–207

HAMANN U (1962c) Über Bau und Entwicklung des Endosperms der Philydraceae und über die Begriffe "mehliges Nährgewebe" und "Farinosae". Bot Jahrb 81:397–407

HAMANN U (1963) Neue Methoden der Dokumentation in der systematischen Botanik. Ber Dtsch Bot Ges 76/1:80–91

HAMANN U (1966a) Embryologische, morphologisch-anatomische und systematische Untersuchungen an Philydraceen. Willdenowia Beih 4:1–178

HAMANN U (1966b) Nochmals zur Embryologie von *Philydrum lanuginosum*. Beitr Biol Pflanz 42:151–159

HAMANN U (1974) Embryologie und Systematik am Beispiel der Farinosae. Ber Dtsch Bot Ges 77:45–54

HAMANN U (1975) Neue Untersuchungen zur Embryologie und Systematik der Centrolepidaceae. Bot Jahrb 96:154–191

HAMANN U (1976) Hydatellaceae – a new family of Monocotyledoneae. N Z J Bot 14:193–196

HANDEL SN (1978) New and dispersed species in the genera *Carex, Luzula* and *Claytonia*. Can J Bot 56:2925–2927

HANDLOS WL (1970) A biosystematic study of *Tripogandra* (Commelinaceae). Thesis, Cornell Univ

HANDLOS WL (1975) The taxonomy of *Tripogandra* (Commelinaceae). Rhodora 77:213–333

HARBORNE JB (1973) Flavonoids as systematic markers in the angiosperms. In: BENDZ G, SANTESSON J (eds) Chemistry in botanical classification. Nobel Symp, vol 25. Academic Press, London New York, pp 103–115

HARBORNE J (1979) Correlations between flavonoid chemistry, anatomy and geography in the Restionaceae. Phytochemistry 18:1323–1327

HARBORNE JB (1982) Flavonoid compounds. In: DAHLGREN R, CLIFFORD HT: The monocotyledons. A comparative study. Academic Press, London New York, pp 264–274

HARBORNE JB, MABRY TJ, MABRY H (eds) (1975) The flavonoids. Chapman and Hall, London

HARLING G (1946) Studien über den Blütenbau und die Embryologie der Familie Cyclanthaceae. Svensk Bot Tidskr 40:257–272

HARLING G (1958) Monograph of the Cyclanthaceae. Acta Horti Bergiani 18:1–428

HARRIS PJ, HARTLEY RD (1980) Phenolic constituents of the cell walls of monocotyledons. Biochem Syst Ecol 8:153–160

HARTOG C DEN (1957) Hydrocharitaceae. In: Flora Malesiana, Ser I. Spermatophyta 5:381–413

HARTOG C DEN (1970) The sea grasses of the world. Verh K Ned Akad Wet Afd Natuurkd 2 Reeks, deel 59/1:1–275

HARTOG C DEN, PLA F VON DER (1970) A synopsis of the Lemnaceae. Blumea 18:355–368

HEDGE IC, WENDELBO P (1972) Various new taxa and records. Notes R Bot Gard Edinburgh 31:331–350

HEGNAUER R (1963) Chemotaxonomie der Pflanzen, vol II. Birkhäuser, Basel

HEGNAUER R (1973) Die cyanogene Verbindungen der Liliatae und Magnoliatae-Magnoliidae: zur systematischen Bedeutung der Cyanogenese. Biochem Syst Ecol 1:191–197

HEGNAUER R (1977) Cyanogenic compounds as systematic markers in Tracheophyta. Plant Syst Evol [Suppl] 1:191–207

HEMARADDI B (1981) Morphological and embryological studies in some Commelinaceae. Thesis, Karnatak Univ

HENDERSON RJF (1982) Romnalda grallata, a new species of Xanthorrhoeaceae from Queensland. Kew Bull 37:229–235

HENNIG W (1950) Grundzüge einer Theorie der phylogenetischen Systematik. Deutscher Zentralverlag, Berlin

HENNIG W (1966) Phylogenetic Systematics. Univ Illinois Press, Urbana

HEPPER FN (1967a) The identity of Grains of Paradise and Melegueta Pepper (Afromomum, Zingiberaceae) in West Africa. Kew Bull 21:129–137

HEPPER FN (1967b) [1968] New and noteworthy Xyridaceae. Kew Bull 21:419–425

HEPPER FN (ed) (1968) (Various families.) In: Flora of West Tropical Africa, vol III, part 1, 2nd edn. R Bot Gard Kew, London

HERKLOTS G (1976) Flowering tropical climbers. Dawson, Folkestone

HESLOP-HARRISON Y, SHIVANNA KR (1977) The receptive surface of the angiosperm stigma. Ann Bot (London) 41:1233–1258

HICKEY LJ, DOYLE JA (1977) Early Cretaceous fossil evidence for angiosperm evolution. Bot Rev 43:3–104

HICKEY LJ, WOLFE JA (1975) The bases of angiosperm phylogeny: vegetative morphology. Ann Mo Bot Gard 62:538–589

HIJNER JA, ARDITTI J (1973) Orchid mycorrhiza: Vitamin production and requirements by the symbionts. Am J Bot 60:829–835

HITCHCOCK CL, CRONQUIST A, OWNBEY M, THOMPSON JW (1969) Vascular plants of the Pacific Northwest, vol I. Univ Washington Press, Washington

HOFMEISTER W (1861) Neue Beiträge zur Erkenntnis der Embryobildung der Phanerogamen. II. Monokotyledonen. Abh Math-Phys Kl Saechs Ges Wiss 5:629–760

HOLM L (1966) Études urédiniques, 4. Sur les Puccinia caricicoles et leurs alliés. Sven Bot Tidskr 60:23–32

HOLM L (1969) An uredological approach to some problems in angiosperm taxonomy. Nytt Mag Bot 16:147–150

HOLMGREN I (1913) Zur Entwicklungsgeschichte von Butomus umbellatus. Sven Bot Tidskr 7:58–77

HOLTTUM RE (1967) The bamboos of New Guinea. Kew Bull 21:263–292

HONG D-Y (1974) Revisio commelinacearum sinicarum. Acta Phytotax Sinica 12:459–483

HORN AF RANTZIEN H (1946) Notes on Mayacaceae of the Regnellian herbarium in the Riksmuseum, Stockholm. Sven Bot Tidskr 40:405–424

HUBBARD CE (1934) Oryza australiensis Domin. In: Hooker's icones plantarum, Ser 5, vol III, Tab 3232. Dulau, London

HUBBARD CE (1954) Grasses. A guide to their structure, identification, uses and distribution in the British Isles. Penguin Books, Harmondsworth

HUBER H (1969) Die Samenmerkmale und Verwandtschaftsverhältnisse der Liliiflorae. Mitt Bot Staatssamml Muenchen 8:219–538

HUBER H (1977) The treatment of monocotyledons in an evolutionary system of classification. Plant Syst Evol Suppl 1:285–298

HUNT DR (1971) Dichorisandra thyrsiflora. Curtis's Bot Mag 178:Tab 590

HUNT DR (1976) Tradescantia sillamontana. Curtis's Bot Mag 181:Tab 706

HUNT DR (1980) Sections and series in Tradescantia. American Commelinaceae IX. Kew Bull 35:437–442

HUTCHINSON J (1926) The families of flowering plants, vol I. Dicotyledons. MacMillan, London

HUTCHINSON J (1933) Scleria barteri Boeck. In: Hooker's icones plantarum, Ser 5, vol II, Tab 3191. Dulau, London

HUTCHINSON J (1934) The families of flowering plants, vol II. Monocotyledons. MacMillan, London

HUTCHINSON J (1959) The families of flowering plants, vol II, 2nd edn. Monocotyledons. Clarendon Press, Oxford

HUTCHINSON J (1973) The families of flowering plants, 3rd edn. Clarendon Press, Oxford

HUYNH K-L (1974) La morphologie microscopique et la taxonomie du genre Pandanus. Bot Jahrb Syst 94:190–256

HUYNH K-L (1980) La morphologie du pollen de Pandanus subgen. Vinsonia (Pandanaceae) et la signification taxonomique. Pollen Spores 22:173–189

IVRI Y, DAFNI A (1977) The pollination biology and

ecology of *Epipactis consimilis* Don (Orchidaceae) in Israel. New Phytol 79:173–177

JACQUES-FÉLIX H (1962) Graminées d'Afrique Tropicale, I. Géneralités, classification, description des genres. Inst Rech Agron Trop Cult Vivrieres, Paris

JACQUES-FÉLIX H (1982) Les monocotylédones n'ont pas de pas de cotyledon. Bull Mus Nat Hist Nat Sect B Adansonia 4 Ser 4:3–40

JAFRI SMH, EL-GADI A (1978) 57. Liliaceae. In: Flora of Libya. Al Faateh Univ, Tripoli

JOHANSEN DA (1950) Plant embryology. Embryogeny of the spermatophyta. Chron Bot Co, Waltham/Mass

JOHNSON AM (1931) Taxonomy of the flowering plants. Century Co, New York London

JOHNSON LAS, BRIGGS B (1981) Three old southern families – Myrtaceae, Proteaceae and Restionaceae. In: KEAST A (ed) Ecological biogeography of Australia. Junk, Utrecht, pp 427–469

JOHRI BM (1935a) Studies in the family Alismaceae. 1. *Limnophyton obtusifolium* Miq. J Indian Bot Soc 14:49–66

JOHRI BM (1935b) Studies in the family Alismaceae. II. *Sagittaria sagittifolia* L. Proc Indian Acad Sci 1/7:340–348

JOHRI BM (1935c) Studies in the family Alismaceae. III. *Sagittaria guayanensis* H B K and *S latifolia* Willd. Proc Indian Acad Sci Sect B 2/1:33–48

JOHRI BM (1938) The embryo sac of *Hydrocleys nymphoides* Buchen. Beih Bot Centralbl 58:165–172

JONES HA, EMSWELLER SL (1936) The development of the flower and megagametophyte of *Allium cepa*. Hilgardia 10:415–428

JONES K, JOPLING C (1972) Chromosomes and the classification of the Commelinaceae. Bot J Linn Soc 65:129–162

JONKER FP (1938) A monograph of the Burmanniaceae. Meded Bot Mus Herb Rijks Utrecht 51:1–279

JONKER FP (1939) Les Géosiridacées, une nouvelle famille de Madagascar. Recuil Trav Bot Néerl 36:473–479

JONKER FP (1948) Burmanniaceae. Flora Malesiana, Ser I Spermatophyta 4:13–26

KAPLAN DR (1973) The monocotyledons: their evolution and comparative biology. VII. The problem of leaf morphology and evolution in the monocotyledons. Q Rev Biol 48:437–457

KAUL RB (1970) Evolution and adaptation of inflorescences in Hydrocharitaceae. Am J Bot 57:708–715

KAUL RB (1976) Conduplicate and specialized carpels in the Alismatales. Am J Bot 63:175–182

KEIGHERY GJ (1982) *Lomandra* and related genera in the plant family Xanthorrhoeaceae. Aust Plants 11:275–278

KENNEDY H (1978) Systematics and pollination of the "closed-flowered" species of *Calathea* (Marantaceae). Univ Calif Publ Bot 71:1–90

KERN JH (1958) Florae Malesianae precursores XXI. Notes on Malaysian and some S.E. Asian Cyperaceae. VII. Acta Bot Neerl 7:786–800

KERN JH (1974) Cyperaceae. Flora Malesiana, Ser i Spermatophyta I/3:435–753

KLERCKER JE-F DE (1883) Recherches sur la structure anatomique de l'*Aphyllanthes monspeliensis* Lin. Bihang K Vetenskapsakad Handl 8 no 6, Stockholm

KNUTH R (1924) Dioscoreaceae. In: ENGLER A (ed) Das

Pflanzenreich, vol IV, 43. Engelmann, Leipzig, pp 1–387

KOECHLIN J (1964) Scitaminales: Musacées, Strelitziacees, Zingiberacées, Cannacées, Marantacées. In: AUBREVILLE A (ed) Flore du Gabon. Mus Natl Hist Nat, Paris

KOYAMA T (1961) Classification of the family Cyperaceae (1). J Fac Sci Univ Tokyo Sect 38/1:37–148

KOYAMA T (1971) Systematic interrelationships among Sclerieae, Lagenocarpeae and Mapanieae (Cyperaceae). Mitt Bot Staatssamml Muenchen 10:604–617

KOYAMA T (1978) Smilacaceae. In: LI, H-L (ed) Flora of Taiwan, vol V. Epoch Publ, Taipei, pp 110–137

KRAUSE K (1930) Liliaceae. In: ENGLER A (ed) Die natürlichen Pflanzenfamilien, vol 15a. Engelmann, Leipzig, pp 227–386

KRESS WJ (1981) New Central American taxa of *Heliconia* (Heliconiaceae). J Arnold Arbor Harv Univ 62:243–260

KRESS WJ, STONE DE (1982) Nature of the sporoderm in monocotyledons, with special reference to the pollen grains of *Canna* and *Heliconia*. Grana 21:129–148

KRESS WJ, STONE DE, SELLERS SC (1978) Ultrastructure of exine-less pollen: *Heliconia* (Heliconiaceae). Am J Bot 65:1064–1076

KRONFELD M (1887) Über Raphiden bei *Typha*. Bot Centralbl 29:154–156

KRUPKO S (1962) Embryological and cytological investigations in *Hypodiscus aristatus* Nees (Restionaceae). J S Afr Bot 28:21–44

KUKKONEN I (1967) Spikelet morphology and anatomy of *Uncinia* Pers (Cyperaceae). Kew Bull 21:93–97

KUPRIANOVA LA (1948) Pollen morphology of the monocotyledons. Tr Komarov Bot Inst USSR Acad Sci Ser 1/7:163–262

LADD PG (1977) Pollen morphology of some members of the Restionaceae and related families, with notes on the fossil record. Grana 16:1–14

LAKSHMANAN KK (1965) Note on the endosperm formation in *Zannichellia palustris* L. Phyton (Buenos Aires) 22:13–14

LANE IE (1955) Genera and generic relationships in Musaceae. Mitt Bot Staatssamml Muenchen 2:114–131

LARSEN K (1961a) Studies in the flora of Thailand. Liliaceae, Triuridaceae, Trilliaceae, Iridaceae, Polygonaceae. Dan Bot Ark 20/1:37–54

LARSEN K (1961b) New species of *Veratrum* and *Orchidantha* from Thailand and Laos. Bot Tidsskr 56:345–350

LARSEN K (1966) Two new Liliaceae from the Khao Yai National Park. Bot Not 119:196–200

LARSEN K (1973a) Kormofyternes taxonomi. Akademisk Forlag, København

LARSEN K (1973b) Studies in Zingiberaceae, VI. Bot Tidsskr 68:157–159

LAWALRÉE A (1945) La position systématique des Lemnaceae et leur classification. Bull Soc R Bot Belg 77:27–38

LAWRENCE GHM (1953) A reclassification of the genus *Iris*. Gentes Herbarum 8:346–371

LEAVITT RG (1904) Trichomes of the root in vascular cryptogams and angiosperms. Proc Boston Nat Hist Soc 31:273–313

LEE DW, FAIRBROTHERS DE (1972) Taxonomic placement of the Typhales within the monocotyledons:

preliminary serological investigations. Taxon 21:39–44

LEE DW, YAP KIM PIN, LIEW FOO YEO (1975) Serological evidence on the distinctness of the monocotyledonous families Flagellariaceae, Hanguanaceae, and Joinvilleaceae. Bot J Linn Soc 70:77–81

LEE RE (1961) Pollen dimorphism in *Tripogandra grandiflora*. Baileya 9:53–56

LEINS P, STADLER P (1973) Entwicklungsgeschichtliche Untersuchungen am Androecium der Alismatales. Oesterr Bot Z 121:51–63

LENZ LW (1975) A biosystematic study of *Triteleia* (Liliaceae) 1. Revision of the species of section *Calliprora*. Aliso 8:221–258

LEWIS GJ (1962) South African Iridaceae. The genus *Ixia*. J S Afr Bot 28:45–195

LI HUI LIN (1978) Flora of Taiwan, vol V. Epoch Publ Co, Taipei

LIMPRICHT W (1928) Taccaceae. In: ENGLER A (ed) Das Pflanzenreich, vol IV 42. Engelmann, Leipzig, pp 1–32

LINDLEY J (1853) The vegetable kingdom, 3rd edn. Bradbury and Evans, London

LING PING-PING (1981) Stomatal studies in Chinese Taccaceae with a discussion on its taxonomical significance. Bull Nanjing Bot Gard, Mem Sun Yat Sen 1981:20–24

LISOWSKI S, MALAISSE F, SYMOENS JJ, VELDEN J VAN DE (1978) Potamogetonaceae. In: Flore d'Afrique Centrale (Zaire-Rwanda-Burundi). Jard Bot Natl Belg, Bruxelles, pp 1–12

LOTSY JP (1911) Vorträge über botanische Stammesgeschichte. Cormophyta siphonogamia, vol III. Fischer, Jena

LOWE J (1961) The phylogeny of monocotyledons. New Phytol 60:355–387

LÜNING B (1974) Alkaloids of the Orchidaceae. In: WITMORE C (ed) The orchids – scientific studies. Ronald Press, New York, pp 349–383

MAAS PJM (1972) Costoideae. In: Flora Neotropica, Monogr No 8. Hafner, New York

MACFARLANE TD (1980) A revision of *Wurmbea* (Liliaceae) in Australia. Brunonia 3:145–208

MACFARLANE TD, WATSON L (1980) The circumscription of Poaceae subfamily Pooideae, with notes on some controversial genera. Taxon 29:645–666

MACMILLAN BH (1972) Biological flora of New Zealand. 7. *Ripogonum scandens* JR et G Forst. (Smilacaceae). N Z J Bot 10:641–672

MADISON M (1979) Protection of developing seeds in neotropical Araceae. Aroideana 2/2:52–61

MAGUIRE B, WURDACK JJ (1958) The botany of the Guayana Highland, III. Mem N Y Bot Gard 10:1–156

MAHESHWARI P (1946) The Fritillaria type of embryo sac: a critical review. J Indian Bot Soc 25:101–119

MAHESHWARI P (1950) An introduction to the embryology of angiosperms. McGraw-Hill, New York

MAHESHWARI SC (1956) The endosperm and embryo of *Lemna* and the systematic position of Lemnaceae. Phytomorphology 6:51–55

MAHESHWARI SC (1958) *Spirodela polyrrhiza*, the link between the aroids and the duckweeds. Nature (London) 181:1745–1746

MAHESHWARI SC, BALDEV B (1958) A contribution to

the morphology and embryology of *Commelina forskalei* Vahl. Phytomorphology 8:277–298

MAHESHWARI SC, KAPIL RN (1963) Morphological and embryological studies on the Lemnaceae. I. The floral structure and gametophytes of *Lemna paucicostata*. Am J Bot 50:677–686

MAHESHWARI SC, KHANNA PP (1956) The embryology of *Arisaema wallichianum,* and the systematic position of Araceae. Phytomorphology 6:379–388

MAHESHWARI SC, MAHESHWARI N (1963) The female gametophyte, endosperm and embryo of *Spirodela polyrrhiza*. Beitr Biol Pflanz 39:179–188

MANGENOT G, DEVILLER M-A (1965) In: Icones plantarum africanarum, vol VII. Inst Fr Afr Noire, Ifan-Dakar

MANN J (1980) Secondary metabolism. Clarendon Press, Oxford

MARCHANT CJ (1973) Chromosome variation in Araceae, V. Kew Bull 28:199–210

MARLOTH R (1915) The flora of South Africa, vol IV. Darter Bros & Co, Cape Town

MARTIN AC (1946) The comparative internal morphology of seeds. Am Midl Nat 36:513–660

MARTINEZ MA DEL PERO DE (1981) Los flavonoides de las Commelinaceae y su interpretación con otras familias de Monocotiledóneas. Resumenes de los trabajos presentados a las XVIII Jornadas Argentinas de Botanica, San Miguel de Tucuman, Argentina, pp 72–73 (Abstr)

MASTERS MT (1967) Synopsis of the S. African Restionaceae. J Lin Soc Bot 10:210–279

MATTSSON O (1976) The development of dimorphic pollen in *Tripogandra* (Commelinaceae). In: FERGUSON IK, MULLER J (eds) The evolutionary significance of the exine. Linn Symp Ser I, Academic Press, London New York, pp 163–183

MATTSSON O (1982) The morphogenesis of dimorphic pollen and anthers in *Tripogandra amplexicaulis*. Light microscopy and growth analysis. Opera Bot 66:1–46

MAURITZON J (1936) Samenbau und Embryologie einiger Scitamineén. Acta Univ Lund 31/19:1–31

McCLURE FA (1966) The bamboos – A fresh perspective. Harvard Univ Press, Cambridge/Mass

McCLURE FA (1973) Genera of bamboos native to the New World (Gramineae: Bambusoideae). Smithsonian Contrib Bot 9:1–148

McKELVEY SD, SAX K (1933) Taxonomic and cytological relationships of *Yucca* and *Agave*. J Arnold Arbor Harv Univ 14:76–80

MEERT M, GEOTGHEBEUR P (1979) Comparative floral morphology of Bisboeckelerieae and Cariceae (Cyperaceae) and the basis of the anthoid concept. Bull Soc R Bot Belg 112:129–143

MEIJER W, BOGNER J (1983) *Pentastemona* (Stemonaceae): the elusive plant. Nature Malaysiana 8/1:26–27

MELCHIOR H (ed) (1964) A Engler's Syllabus der Pflanzenfamilien, Bd II. Bornträger, Berlin-Nikolassee

MENEZEZ NL DE (1980) Evolution in Velloziaceae, with special reference to androecial characters. In: BRICKELL CD, CUTLER DF, GREGORY M (eds) Petaloid monocotyledons. Linn Soc Symp Ser 8. Academic Press, London New York, pp 117–138

METCALFE CR (1960) Anatomy of the monocotyledons, vol I. Gramineae. Oxford Univ Press, London

METCALFE CR (1971) Anatomy of the monocotyledons, vol V. Cyperaceae. Oxford Univ Press, London

MIÈGE E (1936) Sur la descendence et la constitution des lignées à fleurs polycarpiques dans un hybrid de *Triticum vulgare*. Ann Sci Nat Bot Sér 10/18:95–103

MILNE-REDHEAD E (1950) *Pentamenes vaginifer* Milne-Redhead (Tab 3478). *Ensete homblei* (Bequaert ex De Wild) EE Cheesm (Tab 3479). In: Hooker's icones plantarum, Ser 4, vol IV. Dulau, London

MØLLER JD, RASMUSSEN H (1984) Stegmata in Orchidales – character state distribution and polarity. Bot J Linn Soc (in press)

MOORE HE JR (1953) The genus *Milla* (Amaryllidaceae-Allieae) and its allies. Gentes Herbarum 11:262–294

MOORE HE Jr (1973) The major groups of palms and their distribution. Gentes Herbarum 11:27–141

MOORE HE Jr, UHL NW (1982) Major trends of evolution in palms. Bot Rev 48:1–69

MÜLLER-DOBLIES D (1970) Über die Verwandtschaft von *Typha* und *Sparganium* in Inflorescenz- und Blütenbau. Bot Jahrb Syst 89:451–562

MÜLLER-DOBLIES D, MÜLLER-DOBLIES U (1978) Studies on tribal systematics of Amaryllidoideae. 1. The systematic position of *Lapiedra* Lag. Lagascalia 8:13–23

MÜLLER-DOBLIES U (1969) Über die Blütenstände und Blüten sowie zur Embryologie von *Sparganium*. Bot Jahrb Syst 89:359–450

MÜLLER-DOBLIES U, MÜLLER-DOBLIES D (1975) De Liliifloris notulae, 1. Zum Merkmalsbestand von *Haemanthus* (Amaryllidaceae). Bot Jahrb Syst 96:324–327

MULLER J (1981) Fossil pollen records of extant angiosperms. Bot Rev 47:1–146

MUNOZ PIZARRO C (1959) Sinopsis de la flora Chilena. Ediciones Univ Chile, Santiago

MURBECK S (1902) Über die Embryologie von *Ruppia rostellata*. Kgl Svenska Vet-Akad Handl 36:1–21

NAGARAJA RAO A (1955) Embryology of *Trichopus zeylanicus* Gaertn. J Indian Bot Soc 34:213–221

NAHRSTEDT A (1975) Triglochinin in Aracéen. Phytochemistry 14:2627–2628

NANNFELDT JA (1968) Fungi as plant taxonomists. Acta Univ Upsal 17 (Festskrift till Torgny Segerstedt): 85–95

NAPPER DM (1971) Flagellariaceae. In: MILNE-REDHEAD E, POLHILL RM (eds) Flora of tropical East Africa. R Bot Gard Kew, London

NEES CG (1830) Etwas über die Anlage zu einer dreizähligen Frucht bei den Gräsern. Linnaea 5:679–681

NEGBI M, KOLLER D (1963) Homologies of the grass embryo – a re-evaluation. Phytomorphology 12:289–296

NELMES E (1940) *Carex praeclara* Nelmes. In: Hooker's icones plantarum, Ser 5, vol V, Tab 3403. Dulau, London

NETOLITZKY F (1926) Anatomie der Angiospermen-Samen. In: Handbuch der Pflanzenanatomie, vol X. Bornträger, Berlin

NEUENDORF M (1977) *Pardinae*, a new section of *Bomarea* (Alstroemeriaceae). Bot Not 130:55–60

NEWELL TK (1969) A study of the genus *Joinvillea* (Flagellariaceae). J Arnold Arbor Harv Univ 50:527–555

NEWMAN IV (1928–29) The life history of *Doryanthes excelsa*. Proc Linn Soc NSW LII 5:499–538

NEWTON GD, WILLIAMS NH (1978) Pollen morphology of the Cypripedioideae and the Apostasioideae (Orchidaceae). Selbyana 2:169–182

NIKITICHEVA ZI, YAKOVLEV MS, PLYUSHCH TA (1981) The development of the ovule, embryo sac and endosperm in the species of *Peperomia* (Piperaceae). Bot Zh 6:1388–1397 (in Russian)

NILSSON A (1981) Pollination ecology and evolutionary pocesses in six species of orchids. Acta Univ Ups 593:1–40

NORDENSTAM B (1982) A monograph of the genus *Ornithoglossum* (Liliaceae). Opera Bot 64:1–51

OBERMEYER AA (1983) *Protasparagus* Oberm, nom nov. new combinations. S Afr J Bot 2:243–244

OGANEZOVA GG (1981) Anatomical and morphological study in *Ixiolirion tataricum* ssp *montanum*. Bot Zh 66:702–713 (in Russian)

OGURA H (1964) On the embryo sac of two species of *Tricyrtis*. Sci Rep Tôhoku Univ Sendai 4 Biol 30:219–222

OWENS SJ (1981) Self-incompatibility in the Commelinaceae. Ann Bot (London) 47:567–581

OWENS SJ, KIMMINS FM (1981) Stigma morphology in Commelinaceae. Ann Bot (London) 47:771–783

PANCHAKSHARAPPA MG (1962) Embryological studies in the family Zingiberaceae. I. *Costus speciosus* Smith. Phytomorphology 12:418–430

PANCHAKSHARAPPA MG (1966) Embryological studies in some members of Zingiberaceae. II. *Elettaria cardamomum, Hitchenia caulina* and *Zingiber macrostachyum*. Phytomorphology 16:412–417

PANCHAKSHARAPPA MG (1970) Zingiberaceae. In: Symposium on comparative embryology of angiosperms. Bull Indian Natl Sci Acad 41:380–385

PANKOW H, GUTTENBERG H VON (1957) Vergleichende Studien über die Entwicklung monocotyler Embryonen und Keimpflanzen. Bot Stud 7:1–39

PAPNICOLAOU K, ZACHAROF E (1980) *Crocus* in Greece, new taxa and chromosome numbers. Bot Not 133:155–163

PARKS M (1935) Embryo sac development and cleistogamy in *Commelinantia Pringlei*. Bull Torrey Bot Club 62:91–104

PARMELEE JA, SAVILE BDO (1954) Life history and relationships of rusts of *Sparganium* and *Acorus*. Mycologia 46:823–836

PATE JS, DIXON KW (1981) Plants with fleshy underground store organs – a Western Australian survey. In: PATE JS, McCOMB AJ (eds) The biology of Australian plants. Univ W Aust Press, Nedlands, pp 181–215

PAX F, HOFFMANN K (1930) Amaryllidaceae. In: ENGLER A (ed) Die natürlichen Pflanzenfamilien, 2nd edn, vol 15a. Engelmann, Leipzig, pp 341–430

PEISL P (1957) Die Binsenform. Untersuchungen zur Morphologie, Ökologie, Merkmalsphylogenie und Stammesgeschichte einer bei Juncacéen, Cyperacéen und anderen Pflanzenfamilien anzutreffen den pflanzlichen Erscheinungsform. Ber Schweiz Bot Ges 67:101–213

PERIASAMY K (1962) Morphological and ontogenetic studies in palms. I. Development of the plicate condition in the palm leaf. Phytomorphology 12:54–64

PETERSEN OG (1889) Marantaceae. In: ENGLER A (ed) Die natürlichen Pflanzenfamilien, 1st edn, vol II (6). Engelmann, Leipzig, pp 33–43

PICHON M (1946) Sur les Commelinacées. Not Syst (Paris) 12:217–242

PIECH K (1928) Zytologische Studien an der Gattung *Scirpus.* Bull Acad Pol Sci [Bid] 1928:1–43

PIJL L VAN DER (1936) Fledermäuse und Blumen. Flora (Jena) 131:1–40

PIJL L VAN DER (1956) Remarks on pollination by bats in the genera *Freycinetia, Duabanga* and *Haplophragma,* and chiropterophily in general. Acta Bot Neerl 5:135–144

PIJL L VAN DER, DODSON CH (1966) Orchid flowers, their pollination and evolution. Coral Gables, Florida

PILGER R (1930) Mayacaceae (pp 33–35) and Thurniaceae (pp 58–59). In: ENGLER A (ed) Die natürlichen Pflanzenfamilien, vol 15a. Engelmann, Leipzig

POOLE MM, HUNT DR (1980) Pollen morphology and the taxonomy of the Commelinaceae: an exploratory survey. American Commelinaceae, VIII. Kew Bull 34:639–660

POSLUSZNY U, SATTLER R (1974) Floral development of *Ruppia maritima.* Can J Bot 52:1607–1612

POTZTAL E (1964) Reihe Graminales. In: MELCHIOR H (ed) A Engler's Syllabus der Pflanzenfamilien, 12th edn, vol 2. Bornträger, Berlin, pp 561–579

PRAT H (1931) L'épiderme des Graminées; étude anatomique et systématique. Ann Sci Nat Bot 14:117–324

PRAT H (1960) Vers une classification naturelle des Graminées. Bull Soc Bot Fr 107:32–79

PROCTOR M, YEO P (1973) The pollination of flowers. Collins, London

PUNT W (1968) Pollen morphology of the American species of the subfamily Costoideae (Zingiberaceae). Rev Paleobot Palynol 7:31–43

RADULESCU D (1973a) La morphologie du pollen chez quelques Haemodoraceae. Lucr Gradinii Bot Bucuresti 1972–73:123–132

RADULESCU D (1973b) Recherches morpho-palynologiques sur la famille Liliaceae. Lucr Gradinii Bot Bucuresti 1972–73:133–248

RAMASWAMY SN (1967) A contribution to the embryology of *Burmannia coelestis* Don. Proc 54th Indian Sci Congr (Hyderabad) III. (Abstr 331)

RAMASWAMY SN (1970) Burmanniaceae. In: Symposium on comparative embryology of angiosperms. Bull Indian Natl Sci Acad 41:375–379

RAO AN (1955) Embryology of *Trichopus zeylanicus* Gaertn. J Indian Bot Soc 34:213–221

RAO AN (1967) Flower and seed development in *Arundina graminifolia.* Phytomorphology 17:291–300

RAO CV (1959) Contributions to the embryology of Palmae, II. Ceroxylineae. J Indian Bot Soc 31:46–75

RAO TS, RAO RR (1961) Pollen morphology of Pontederiaceae. Pollen Spores 3:45–46

RAO VS (1963) The epigynous glands of Zingiberaceae. New Phytol 62:342–349

RAO VS (1974) The relationships of the Apostasiaceae on the basis of floral anatomy. Bot J Linn Soc 68:319–327

RASMUSSEN FN (1982) The gynostemium of the neottioid orchids. Opera Bot 65:1–96

RASMUSSEN H (1981) The diversity of stomatal development in Orchidaceae subfamily Orchidoideae. Bot J Linn Soc 82:381–392

RASMUSSEN H (1982) Developmental studies of stomata in Liliiflorae. Ph D diss, Univ Copenhagen

RAUH W (1981) Bromelien, 2nd edn. Ulmer, Stuttgart

RAUNKIAER C (1895–99) De danske blomsterplanternes naturhistorie, vol I. Enkimbladede. Gyldendal, Kjøbenhavn

RAVEN PH (1975) The bases of angiosperm phylogeny: cytology. Ann Mo Bot Gard 62:724–764

RAVEN PH, AXELROD DI (1974) Angiosperm biogeography and past continental movements. Ann Mo Bot Gard 61:539–673

REEDER JR (1957) The embryo in grass systematics. Am J Bot 44:756–768

RIGGINS R, FARRIS JS (1983) Cladistics and the roots of angiosperms. Syst Bot 8:96–101

ROHWEDER O (1956) Die Farinosae in der Vegetation von El Salvador. Abh. Auslandskd Reihe C Naturwiss 18:98–179

ROHWEDER O (1969) Beiträge zur Blütenmorphologie und -anatomie der Commelinaceen mit Anmerkungen zur Begrenzung und Gliederung der Familien. Ber Schweiz Bot Ges 79:199–220

ROPER RB (1952) The embryo sac of *Butomus umbellatus* L. Phytomorphology 2:61–74

ROSS-CRAIG S (1948) Drawings of British plants, vol 2. Bell & Sons, London

ROSS-CRAIG S (1971) Drawings of British plants, vol 28. Bell & Sons, London

ROSS-CRAIG S (1972) Drawing of British plants, vol 29. Bell & Sons, London

ROSS-CRAIG S (1973) Drawings of British plants. Vols 30–31. Bell & Sons, London

ROSSO SW (1966) The vegetative anatomy of the Cypripedioideae (Orchidaceae). Bot J Linn Soc 59:309–341

ROWLEY JR (1959) The fine structure of the pollen wall in the Commelinaceae. Grana Palynol 2:3–31

ROWLEY JR, DAHL AO (1962) The aperture of the pollen grain in *Commelinantia.* Pollen Spores 4:221–232

ROYEN P VAN (1962) Sertulum papuanum, 5. Nymphaeaceae. Nova Guinea Bot 8:103–126

RÜBSAMEN T (1983) Nectaries of the Burmanniaceae (Burmannieae). Acta Bot Neerl 32:351

RUIJGROK HWL (1974) Cyanogenese bei *Scheuchzeria palustris.* Phytochemistry 13:161–162

RYVES TB (1980) Alien species of *Eragrostis* P Beauv in the British Isles. Watsonia 13:111–117

SANDWITH NY (1938) *Brodiaea circinata* Sandwith. In: Hooker's icones plantarum, Ser 4, vol II, Tab 3350. Dulau, London

SATTLER R, SINGH V (1978) Floral organogenesis of *Echinodorus amazonicus* Rataj and floral construction of the Alismatales. Bot J Linn Soc 77:141–156

SAUPE SG (1981) Cyanogenic compounds and angiosperm phylogeny. In: YOUNG DA, SEIGLER DS (eds) Phytochemistry and angiosperm phylogeny. Praeger, New York, pp 80–116

SAVILE DBO (1954) The fungi as aids in the taxonomy of the flowering plants. Science 120:583–585

SAVILE DBO (1979) Fungi as aids in higher plant classification. Bot Rev 45:377–503

SCHAEPPI H (1939) Vergleichend-morphologische Untersuchungen an den Staubblättern der Monocotyledonen. Nova Acta Leopold N F 6:389–447

SCHILL R (1978) Palynologische Untersuchungen zur systematischen Stellung der Apostasiaceae. Bot Jahrb Syst 99:353–362

Schill R, Pfeiffer W (1977) Untersuchungen an Orchideenpollinien unter besonderer Berücksichtigung ihrer Feinskulpturen. Pollen Spores 19:5–118

Schlechter R (1926) Das System der Orchidaceen. Notizbl Bot Gart Berlin-Dahlem 9:563–591

Schlittler J (1943) Die Blütenabgliederung und die Perikladien bei den Vertretern des Anthericum-typus sowie ihre Bedeutung für die Systematik. Ber Zuerich Bot Ges 1943:491–507

Schlittler J (1949) Die systematische Stellung der Gattung *Petermannia* F v Muell. und ihre phylogenetische Beziehung zu den Dioscoreaceae Lindl. Vierteljahrsschr Naturforsch Ges Zuerich Beih 1:1–28

Schlittler J (1951) Die Gattung *Eustrephus* R Br ex Sims und *Geitonoplesium* (R Br) A Cunn Morphologisch-anatomische Studie mit Berücksichtigung der systematischen, nomenklatorischen und arealgeographischen Verhältnisse. Mitt Ber Schweiz Bot Ges 151:175–239

Schlittler J (1955) Vorläufige Mitteilung über die organphylogenetischen Zusammenhänge der wichtigsten Grundgestalten bei den Monocotyledonen mit spezieller Berücksichtigung der Liliacéen und einigen Bemerkungen zu deren Systematik. Vierteljahrsschr Naturforsch Ges Zürich 100:182–193

Schmidt OC (1935) Rafflesiaceae. In: Engler A, Harms H (eds) Die natürlichen Pflanzenfamilien, vol 16b, 2nd edn. Engelmann, Leipzig, S 204–242

Schnarf K (1929) Die Embryologie der Liliaceae und ihre systematische Bedeutung. S B Akad Wiss Wien Math Nat Kl 138:69–92

Schnarf K (1931) Vergleichende Embryologie der Angiospermen. Bornträger, Berlin

Schnarf K, Wunderlich R (1939) Zur vergleichenden Embryologie der Liliaceae-Asphodeloideae. Flora (Jena) N F 33:297–327

Schneider EL, Ford EG (1978) Morphological studies of the Nymphaeaceae. X. The seed of *Ondinea purpurea* Den Hartog. Bull Torrey Bot Club 105:192–200

Schultze-Motel W (1959) Entwicklungsgeschichtliche und vergleichend-morphologische Untersuchungen im Blütenbereich der Cyperaceae. Bot Jahrb 78:129–170

Schultze-Motel W (1966) Cyperaceae. In: Hegi's illustrierte Flora von Mitteleuropa, vol 11/1. Hanser, München

Schulze W (1971) Beiträge zur Pollenmorphologie der Iridaceae und ihre Bedeutung für die Taxonomie. Feddes Repert Z Bot Taxon Geobot 82:101–124

Schulze W (1975a) Beiträge zur Taxonomie der Liliifloren. I. Asphodelaceae. Wiss Z Friedrich-Schiller-Univ Jena Math-Nat R 24:403–415

Schulze W (1975b) Beiträge zur Taxonomie der Liliifloren. II. Colchicaceae. Wiss Z Friedrich-Schiller-Univ Jena Math-Nat R 24:417–428

Schulze W (1978) Beiträge zur Taxonomie der Liliifloren. IV. Melanthiaceae. Wiss Z Friedrich-Schiller-Univ Jena Math-Nat R 27:87–95

Schulze W (1980) Beiträge zur Taxonomie der Liliifloren. VI. Der Umfang der Liliaceae. Wiss Z Friedrich Schiller-Univ Jena Math-Nat R 29:607–636

Schumann K (1904) Zingiberaceae. In: Engler A (ed) Das Pflanzenreich, vol VI 46. Engelmann, Leipzig

Schuster J (1910) Über die Morphologie der Grasblüte. Flora (Jena) 100:213–266

Sealy JR (1938) *Tulipa borzczowii* Regel (Liliaceae). In: Hooker's icones plantarum, Ser 4, vol II Tab 3356. Dulau, London

Sealy JR (1964) *Trillium rivale* (Liliaceae). Curtis's Bot Mag 175:Tab 444

Seidenfaden G (1973) Notes on *Cirrhopetalum* Lindl. Dan Bot Ark 29/1:1–260

Seidenfaden G (1978) Orchid genera in Thailand. VI. Neottioideae Lindl. Dan Bot Ark 32/2:1–195

Seidenfaden G (1979) Orchid genera in Thailand. III. *Bulbophyllum* Thou. Dan Bot Ark 33/3:1–228

Seidenfaden G (1981) Contributions to the orchid flora of Thailand, IX. Nord J Bot 1:192–217

Seigler DS (1977) Plant systematics and alkaloids. Alkaloids 16:1–82

Semple KS (1974) Pollination in Piperaceae. Ann Mo Bot Gard 61:868–871

Shaw HKA (1981) A new species of *Hanguana* from Borneo. Kew Bull 35:819–821

Siebe M (1903) Über den anatomischen Bau der Apostasiinae. Diss Naturwiss-Math Fak Ruprecht-Karls-Univ Heidelberg

Simpson MG (1983) Pollen ultrastructure of the Haemodoraceae and its taxonomic significance. Grana 22:79–103

Sims J (ed) (1804) Curtis' Botanical Magazine, 19. London.

Singh V, Sattler R (1972) Floral development of *Alisma triviale*. Can J Bot 50:619–627

Singh V, Sattler R (1974) Floral development of *Butomus umbellatus*. Can J Bot 52:223–230

Singh V, Sattler R (1977) Development of the inflorescence and flower of *Sagittaria cuneata*. Can J Bot 55:1087–1105

Slater JA (1976) Monocots and chinch bugs: a study of host plant relationships in the Lygaeid subfamily Blissinae (Hemiptera: Lygidae). Biotropica 8:143–165

Slingsby P, Bond W (1981) Ants-friends of the fynbos. Veld Flora 67:39–44

Slob A, Jekel B, Schlatmann E (1975) On the occurrence of tuliposides in the Liliiflorae. Phytochemistry 14:1997–2005

Smith LB (1962) A synopsis of the American Velloziaceae. Contrib U S Natl Herb 35:251–292

Smith LB, Ayensu ES (1974) Classification of Old World Velloziaceae. Kew Bull 29:181–205

Smith LB, Ayensu ES (1976) A revision of American Velloziaceae. Smithsonian Contrib Bot 30:1–172

Söderstrom TR (1981) Some evolutionary trends in the Bambusoideae (Poaceae). Ann Mo Bot Gard 68:15–47

Solereder H, Meyer FJ (1928) Systematische Anatomie der Monocotyledonen, vol III. Principes, Synanthae, Spathiflorae. Bornträger, Berlin

Solereder H, Meyer FJ (1929) Systematische Anatomie der Monocotyledonen, vol IV. Farinosae. Bornträger, Berlin

Solereder H, Meyer FJ (1930) Systematische Anatomie der Monocotyledonen, vol VI. Scitamineae, Microspermae. Bornträger, Berlin

Solereder H, Meyer FJ (1933) Systematische Anatomie der Monocotyledonen, vol I. 1. Pandanales, Helobiae, Triuridales: Typhaceae-Scheuchzeriaceae. Bornträger, Berlin

Speta F (1976) Über *Chionodoxa* Boiss, ihre Gliederung

und Zugehörigkeit zu *Scilla* L. Naturkd Jahrb Stadt Linz 21:9–79

SOUÈGES R (1931) L'embryon chez le *Sagittaria sagittifolia*. Ann Sci Nat Bot 10/13:353–402

SOWUNMI MA (1972) Pollen morphology of the Palmae and its bearing on taxonomy. Rev Palaeobot Palynol 13:1–80

SPEARING JK (1977) A note on the closed leaf-sheaths in Zingiberaceae-Zingiberoideae. Notes R Bot Gard Edinburgh 35:217–220

STAFF IA, WATERHOUSE JT (1981) The biology of arborescent monocotyledons, with special reference to Australian species. In: PATE JS, MCCOMB AJ (eds) The biology of Australian plants. Univ W Australia Press, Nedlands, pp 216–257

STANT MY (1964) Anatomy of the Alismataceae. J Linn Soc Bot 59:1–42

STANT MY (1967) Anatomy of the Butomaceae. J Linn Soc Bot 60:31–60

STANT MY (1970) Anatomy of *Petrosavia stellaris* Becc, a saprophytic monocotyledon. In: ROBSON NKB, CUTLER DF, GREGORY M (eds) New research in plant anatomy. Suppl 1 to Bot J Linn Soc [Suppl] 1:147–161

START AN, MARSHALL AG (1976) Nectarivorous bats as pollinators of trees in west Malaysia. In: BURLEY J, STYLES ST (eds) Tropical trees, variation, breeding and conservation. Academic Press, London New York, pp 141–150

STAUDERMANN W VON (1924) Die Haare von Monocotyledonen. Bot Arch 8:105–184

STEBBINS GL (1956) Cytogenetics and evolution in the grass family. Am J Bot 43:890–905

STEBBINS GL (1974) Flowering plants. Evolution above the species level. Arnold, London

STEBBINS GL, KHUSH GS (1961) Variation in the organisation of the stomatal complex in the leaf epidermis of monocotyledons and its bearing on their phylogeny. Am J Bot 48:51–59

STEENIS CGGJ VAN (1954) Haemodoraceae. Flora Malesiana Ser I. Spermatophyta 5:111–113

STEENIS CGGJ VAN (1982) *Pentastemona*, a new 5-merous genus of monocotyledons from North Sumatra (Stemonaceae). Blumea 28:151–163

STENAR H (1925) Embryologische Studien. Akad Diss, Uppsala

STENAR H (1928a) Zur Embryologie der *Asphodeline*-Gruppe. Sven Bot Tidskr 22:145–159

STENAR H (1928b) Zur Embryologie der *Veratrum*- und *Anthericum*-Gruppen. Bot Not 81:357–378

STENAR H (1935) Embryologische Beobachtungen über *Scheuchzeria palustris*. L. Bot Not 88:78–86

STENAR H (1942) Zur Embryologie der *Dracaena*-Gruppe. Heimbygdas Tidskr Fornvårdaren 8/2:183–195

STERLING C (1975) Comparative morphology of the carpel in the Liliaceae: Glorieseae. Bot J Linn Soc 70:341–349

STERLING C (1977) Comparative morphology of the carpel in the Liliaceae: Uvularieae. Bot J Linn Soc 74:345–354

STEYN E (1973) 'N embryologiese ondersoek van *Romulea rosea* Eckl var *reflexa* Beg. J S Afr Bot 39:235–243

STILES GF (1978) Temporal organization of flowering among hummingbird food-plants of a tropical forest. Biotropica 10:194–210

STIRTON JC, HARBORNE JB (1980) Two distinctive anthocyanin patterns in the Commelinaceae. Biochem Syst Ecol 8:285–287

ST JOHN H (1965) Monograph of the genus *Elodea* (Hydrocharitaceae). Summary. Rhodora 67:155–180

STONE BC (1961) The genus *Sararanga* (Pandanaceae). Brittonia 13:212–224

STONE BC (1968) Materials for a monograph of *Freycinetia* Gaud. IV. Subdivision of the genus, with fifteen new sections. Blumea 16:361–372

STONE BC (1972) A reconsideration of the evolutionary status of the family Pandanaceae and its significance in monocotyledon phylogeny. Q Rev Biol 47:34–45

STONE BC (1973) Materials for a monograph of *Freycinetia* Gaudich. XIV. On the relation between *F banksii* A Cunn of New Zealand and *F baueriana* Endl of Norfolk Island, with notes on the structure of the seed. N Z J Bot 11:241–246

STONE BC (1974) Towards an improved infrageneric classification in *Pandanus* (Pandanaceae). Bot Jahrb Syst 94:459–540

STONE BC (1982) The Australian species of *Freycinetia* (Pandanaceae). Brunonia 5:79–94

STONE DE, SELLERS SS, KRESS WJ (1979) Ontogeny of exineless pollen in *Heliconia,* a banana relative. Ann Mo Bot Gard 66:701–730

STONE DE, SELLERS SC, KRESS WJ (1981) Ontogenetic and evolutionary implications of a neotenous exine in *Tapeinocheilos* (Zingiberales: Costaceae) pollen. Am J Bot 68:49–63

STOUTAMIRE WP (1974) Australian terrestrial orchids, thynnid wasps, and pseudocopulation. Am Orch Soc Bull 43:13–18

STOUTAMIRE WP (1975) Pseudocopulation in Australian terrestrial orchids. Am Orch Soc Bull 44:226–233

STRÖMBERG B (1956) The embryo-sac development of the genus *Freycinetia*. Sven Bot Tidskr 50:129–134

STÜTZEL T (1981) Zur Funktion und Evolution köpfenförmiger Blütenstände, insbesondere der Eriocaulaceae. Beitr Biol Pflanz 56:439–468

STÜTZEL T (1984) Blüten- und infloreszenzmorphologische Untersuchungen zur Systematik der Eriocaulaceae. Dissert Bot 71:1–108. Gantner, Vaduz

STÜTZEL T, WEBERLING F (1982) Untersuchungen über Verzweigung und Infloreszenzaufbau von Eriocaulaceen. Flora 172:105–112

SUBRAMANYAM K, NARAYANA HS (1972) Some aspects of the floral morphology and embryology of *Flagellaria indica* Linn. In: MURTY YS, JOHRI BM, MOHAN RAM HY, VARHESE TM (eds) Advances in plant morphology. Sarita Prakashan, Meerut, pp 211–217

SUESSENGUTH K (1921) Beiträge zur Frage des systematischen Anschlusses der Monocotylen. Beih Bot Centralbl 38:1–79

SWAMY BGL (1948) Vascular anatomy of orchid flowers. Bot Mus Leafl Harv Univ 13:61–95

SWAMY BGL (1949a) Embryological studies in the Orchidaceae. I. Gametophytes. Am Midl Nat 41:184–201

SWAMY BGL (1949b) Embryological studies in the Orchidaceae. II. Embryogeny. Am Midl Nat 49:202–232

SWAMY BGL (1964) Observations on the floral morphology and embryology of *Stemona tuberosa* Lour. Phytomorphology 14:458–468

SWAMY BGL, LAKSHMANAN KK (1962) Contributions

to the embryology of the Najadaceae. J Indian Bot Soc 41:247–267

TAKHTAJAN A (1969) Flowering plants. Origin and dispersal. Oliver and Boyd, Edinburgh

TAKHTAJAN A (1980) Outline of the classification of flowering plants (Magnoliophyta). Bot Rev 46:225–359

TAKHTAJAN A (1982) Plant Life, vol 6. Magnoliophyta or Angiospermae: Liliopsida or Monocotyledones. Moskva (in Russian)

TAKHTAJAN A (1983) A revision of *Daiswa* (Trilliaceae). Brittonia 35:255–270

THANIKAIMONI G (1969) Esquisse palynologique des Aracées. Inst Fr Pondichery Trav Sci Tech V/5:1–31

THANIKAIMONI G (1970a) Pollen morphology, classification and phylogeny of Palmae. Adansonia Ser 2 10:347–365

THANIKAIMONI G (1970b) Les palmiers: palynologie et systématique. Inst Fr Pondichery Trav Sci Tech 11:1–286

THANIKAIMONI G (1978) Pollen morphological terms: proposed definitions – 1. IVth Int Palynol Conf Lucknow (1976–77), pp 228–239

THIERET JW (1975) The Mayacaceae in the southeastern United States. J Arnold Arbor Harv Univ 56:248–255

THORNE RF (1976) A phylogenetic classification of the Angiospermae. Evol Biol 9:35–106

THORNE RF (1981) Phytochemistry and angiosperm phylogeny, a summary statement. In: YOUNG DA, SEIGLER DS (eds) Phytochemistry and angiosperm phylogeny. Praeger, New York, pp 233–295

THORNE RF (1983) Proposed new realignments in the angiosperms. Nord J Bot 3:85–117

TIEGHEM PH VAN (1887) Structure de la racine et disposition des radicelles dans Centrolepidées, Eriocaulées, Juncées, Mayacées et Xyridées. J Bot (Paris) 1:305–315

TIEGHEM PH VAN, DULIOT H (1888) Recherches comparatives sur l'origine des membres endogènes dans les plantes vasculaires. Ann Sci Nat Ser 7 8:1–666

TILLICH HJ (1977) Vergleichend-morphologische Untersuchungen zur Identität der Gramineen-Primärwurzel. Flora 166:415–421

TJADEN WL (1981) *Amaryllis belladona* Linn. An up-to-date summary. Plant Life 37:21–26

TOMLINSON PB (1959) An anatomical approach to the classification of the Musaceae. J Linn Soc Bot 55:779–809

TOMLINSON PB (1960) The anatomy of *Phenakospermum* (Musaceae). J Arnold Arbor Harv Univ 41:287–297

TOMLINSON PB (1961a) The anatomy of *Canna*. J Linn Soc Bot 56:467–473

TOMLINSON PB (1961b) Morphological and anatomical characteristics of the Marantaceae. J Linn Soc Bot 58:55–78

TOMLINSON PB (1961c) Vol 2. Palmae. In: METCALFE CR (ed) Anatomy of the monocotyledons. Clarendon, Oxford

TOMLINSON PB (1962a) The leaf base in palms – its morphology and mechanical biology. J Arnold Arbor Harv Univ 43:23–50

TOMLINSON PB (1962b) Phylogeny of the Scitamineae – morphological and anatomical considerations. Evolution 16:192–213

TOMLINSON PB (1965a) A study of the stomatal structure in Pandanaceae. Pac Sci 19:38–54

TOMLINSON PB (1965b) Notes on the anatomy of *Aphyllanthes* (Liliaceae) and comparison of Eriocaulaceae. J Linn Soc Bot 59:163–173

TOMLINSON PB (1966) Anatomical data in the classification of Commelinaceae. J Linn Soc Bot 59:371–395

TOMLINSON PB (1969) Anatomy of the monocotyledons, vol III. In: METCALFE CR (ed) Commelinales-Zingiberales. Clarendon Press, Oxford

TOMLINSON PB (1970) Monocotyledons – towards an understanding of their morphology and anatomy. In: PRESTON RD (ed) Advances in botanical research, vol III. pp 208–290

TOMLINSON PB (1974) Development of the stomatal complex as a taxonomic character in monocotyledons. Taxon 23:109–128

TOMLINSON PB (1982) Anatomy of the monocotyledons, vol VII. In: METCALFE CR (ed) Helobiae (Alismatidae). Clarendon Press, Oxford

TOMLINSON PB, AYENSU ES (1968) Morphology and anatomy of *Croomia pauciflora* (Stemonaceae). J Arnold Arbor Harv Univ 49:260–277

TOMLINSON PB, AYENSU ES (1969) Notes on the vegetative morphology and anatomy of Petermanniaceae (Monocotyledones). J Linn Soc Bot 62:17–26

TOMLINSON PB, FISHER JB (1971) Morphological studies in *Cordyline* (Agavaceae). I. Introduction and general morphology. J Arnold Arbor Harv Univ 52:459–478

TOMLINSON PB, POSLUSZNY U (1978) Aspects of floral morphology and development in the seagrass *Syringodium filiforme* (Cymodoceaceae). Bot Gaz 139:333–345

TOMLINSON PB, ZIMMERMANN MH (1969) Vascular anatomy of monocotyledons with secondary growth – an introduction. J Arnold Arbor Harv Univ 50:159–179

TRAUB HP (1957) Classification of Amaryllidaceae – subfamilies, tribes and genera. Plant Life 13:76–83

TRAUB HP (1970) An introduction to Herbert's "Amaryllidaceae, etc." and related works. In: CRAMER J, SWANN HK, HERBERT W (eds) Amaryllidaceae. Reprint edn Lehre

TRAUB HP (1972a) The order Alliales. Plant Life 28:129–132

TRAUB HP (1972b) Genus *Allium* L. – subgenera, sections and subsections. Plant Life 28:132–137

TRAUB HP (1972c) Tribe Hosteae, family Agavaceae. Plant Life 28:137–138

TRAUB HP (1983) The lectotypification of *Amaryllis belladonna* L. (1753). Taxon 32:253–267

TRAUB HP, MOLDENKE HH (1949) Amaryllidaceae: tribe Amarylleae. Am Plant Life Soc, Stanford/Calif

TROLL W (1954) Praktische Einführung in die Pflanzenmorphologie, vol I. Der Vegetative Aufbau. Fischer, Jena

TURRILL WB (1943) *Fritillaria davisii* Turill, *Fritillaria fusca* Turill. In: Hooker's icones plantarum, Ser 2, vol V, Tab 3427. Dulau, London

TUTIN TG (1936) A revision of the genus *Pariana* (Gramineae). J Linn Soc Bot 50:337–362

UHL NW, MOORE HE (1971) The palm gynoecium. Am J Bot 58:945–992

UHL NW, MOORE HE (1977) Centrifugal stamen initiation in phytelephantoid palms. Am J Bot 64:1152–1161

UHL NW, MOORE HE (1980) Androecial development in six polyandrous genera representing five major groups of palms. Ann Bot (London) 45:57–75

UNTAWALE AG, BHASIN RK (1973) On the endothecial thickenings in some monocotyledonous families. Curr Sci 42:398–400

UTECH FH (1978a) Floral vascular anatomy of *Medeola virginiana* L. (Liliaceae-Parideae = Trilliaceae) and tribal note. Ann Carnegie Mus 47:13–28

UTECH FH (1978b) Floral vascular anatomy of *Pleea tenuifolia* Michx. (Liliaceae-Tofieldieae) and its reassignment to *Tofieldia*. Ann Carnegie Mus 47:423–454

UTECH FH (1978c) Floral vascular anatomy of the monotypic Japanese *Metanarthecium luteoviride* Maxim. (Liliaceae-Melanthioideae). Ann. Carnegie Mus 47:455–477

UTECH FH, KAWANO S (1976) Biosystematic studies on *Disporum* (Liliaceae-Polygonatae). III. Floral biology of *D sessile* D Don and *D smilacinum* A Gray from Japan. Bot Mag 89:159–171

VANHECKE L (1974) Embryography of some genera of Cladiineae and the Gahniineae (Cyperaceae) with additional notes on their fruit anatomy. Bull Jard Bot Natl Belg 44:367–400

VEKEN P VAN DER (1965) Contribution à l'embryolographie systématique des Cyperaceae-Cyperoideae. Bull Jard Bot État Brux 35:285–354

VELENOVSKÝ J (1907) Vergleichende Morphologie der Pflanzen, vol II. Řivnač, Prag

VENKATA RAO C (1958) Contribution to the embryology of the Palmae. J Indian Bot Soc 38:46–75

VERMEULEN P (1966) The system of the Orchidales. Acta Bot Neerl 15:224–253

VEYRET Y (1974) Development of the embryo and the young seedling stages of orchids. In: WITHNER CL (ed) The orchids – scientific studies. Wiley & Sons, New York, pp 223–265

VIJAYARAGHAVAN MR, KUMARI AV (1974) Embryology and systematic position of *Zannichellia palustris* L. J Indian Bot Soc 53:292–302

VOGEL EF DE (1969) Monograph of the tribe Apostasieae (Orchidaceae). Blumea 17:313–350

VOGEL S (1963a) Duftdrüsen im Dienste der Bestäubung. Abh Akad Wiss Litt Mainz Math-Nat Kl H10 Jg 1962:600–763

VOGEL S (1936b) Das sexuelle Anlockungsprinzip. Oesterr Bot Z 110:308–337

VOGEL S (1978a) Pilzmückenblumen als Pilzmimeten. Flora (Jena) 167:329–366, 369–398

VOGEL S (1978b) Evolutionary shifts from reward to deception in pollen flowers. In: RICHARDS AJ (ed) The pollination of flowers by insects. Academic Press, London, pp 89–96

VOS MP DE (1948) The development of the ovule and the seed in the Hypoxideae. Part I. *Ianthe* Salisb. J S Afr Bot 14:159–169

VOS MP DE (1956) Studies on the embryology and relationships of South African genera of the Haemodoraceae: *Dilatris* Berg and *Wachendorfia* Burm. J S Afr Bot 22:41–63

VOS MP DE (1963) Studies on the embryology and relationships of South African genera of the Haemodoraceae: *Lanaria* Ait. J S Afr Bot 29:79–90

VOSA CG (1975) The cytotaxonomy of the genus *Tulbaghia*. Ann Bot 34:47–121

WAGNER P (1977) Vessel types of monocotyledons.: a survey. Bot Not 130:383–402

WALKER ER (1906) On the structure of the pistils of some grasses. Univ Stud Univ Nebr 6/3:203–217

WEBBER EE (1960) Observations on the epidermal structure and stomatal apparatus of some members of the Araceae. Rhodora 62:251–258

WEBERLING F (1981) Morphologie der Blüten und der Blütenstände. Ulmer, Stuttgart

WEIMARCK H (1939) Types of inflorescences in *Aristea* and some allied genera. Bot Not 33:616–626

WEIR CE, DALE HM (1960) A developmental study of wild rice *Zizania aquatica* L. Am J Bot 38:719–739

WENDELBO P (1970) Amaryllidaceae. In: RECHINGER KH (ed) Flora Iranica, vol 62. Akademische Druck, Graz, pp 1–8

WET JMJ DE (1981) Grasses and the culture history of man. Ann Mo Bot Gard 68:87–104

WETTSTEIN R (1924) Handbuch der systematischen Botanik. Deuticke, Leipzig Wien

WILDER GJ (1976) Structure and development of leaves in *Carludovica palmata* (Cyclanthaceae) with reference to other Cyclanthaceae and Palmae. Am J Bot 63:1237–1256

WILDER GJ (1977a) Structure and symmetry of species of the *Asplundia* group (Cyclanthaceae) having monopodial vegetative axes: *Schultesiophytum chorianthum*, *Dicranopygium* sp nov, *Asplundia rigida*, and *Thoracocarpus bissectus*. Bot Gaz 138:80–101

WILDER GJ (1977b) Structure and symmetry of species of the *Asplundia* group (Cyclanthaceae) having sympodial axes: *Evodianthus funifer* and *Carludovica palmata*. Bot Gaz 138:219–235

WILDER GJ (1981) Structure and development of *Cyclanthus bipartitus* Poit. (Cyclanthaceae) with reference to other Cyclanthaceae. II. Adult leaf. Bot Gaz 142:222–236

WILDER GJ, HARRIS DH (1982) Laticifers in *Cyclanthus bipartitus* Poit. (Cyclanthaceae). Bot Gaz 143:84–93

WILDMAN WC (1968) The Amaryllidaceae alkaloids. Alkaloids 11:307–405

WILDMAN WC, PURSEY BA (1968) Colchicine and related compounds. Alkaloids 11:407–457

WILEY EO (1981) Phylogenetics. Wiley & Sons, New York

WILLEMSTEIN SC (1983) Some transformation series in angiosperms: a flower-ecological approach. Acta Bot Neerl 32:345

WILLIAMS CA (1975) Biosystematics of the Monocotyledoneae – flavonoid patterns in the leaves of the Liliaceae. Biochem Syst Ecol 3:229–244

WILLIAMS CA, HARBORNE JB, MAYO SJ (1981) Anthocyanin pigments and leaf flavonoids in the family Araceae. Phytochemistry 20:217–234

WILLIAMS NH (1979) Subsidiary cells in the Orchidaceae: their general distribution with special reference to development in the Oncidieae. Bot J Linn Soc 78:41–66

WILLIAMSON G (1983) *Hexacyrtis dickiana* Dinter – a new record for the Richtersveld. Veld Flora 69:16–19

WILSON KA (1960) The genera of the Arales in the southeastern United States. J Arnold Arbor Harv Univ 41:47–72

WINKLER H (1930) Musaceae and Cannaceae. In: ENGLER A (ed) Die natürlichen Pflanzenfamilien, vol 15a. Engelmann, Leipzig, pp 505–541, 640–654

WIRTH M, WITHNER CL (1959) Embryology and development in the Orchidaceae. In: WITHNER CL (ed) The orchids, a scientific survey. Ronald Press, New York, pp 155–188

WIRZ H (1910) Beiträge zur Entwicklungsgeschichte von *Sciaphila* spec und von *Epirrhizanthes elongata* Bl. Flora 101:395–446

WIT HCD DE (1971) Aquarienpflanzen. Ulmer, Stuttgart

WOOD CE (1971) The Saururaceae in the southeastern United States. J Arnold Arbor Harv Univ 52:479–485

WUNDERLICH R (1936) Vergleichende Untersuchungen von Pollenkörner einiger Liliaceen und Amaryllidaceen. Oesterr Bot Z 85:30–55

WUNDERLICH R (1937) Zur vergleichenden Embryologie der Liliaceae-Scilloideae. Flora (Jena) 32:48–90

WUNDERLICH R (1959) Zur Frage der Phylogenie der Endospermtypen bei den Angiospermen. Oesterr Bot Z 106:203–293

YAKOVLEV MS, ZHUKOVA GY (1980) Chlorophyll in the embryos of angiosperm seeds, a review. Bot Not 133:323–341

YAMASHITA T (1972) Eigenartige Wurzelanlage des Embryos bei *Ruppia maritima* L. Beitr Biol Pflanz 48:157–170

YANG, YUAN-PO (1978) In: Flora of Taiwan, vol V. Epoch Publ Co, Taipei, pp 138–140

YEO PF (1968) A contribution to the taxonomy of the genus *Ruscus*. Notes R Bot Gard Edinburgh 28:237–264

YOUNG DA (1981) Are the angiosperms primitively vesselless? Syst Bot 6:313–330

ZAVADA MS (1983) Comparative morphology of monocot pollen and evolutionary trends of apertures and wall structures. Bot Rev 49:331–379

ZAVADA MS, ZU X-L, EDWARDS JM (1983) On the taxonomic status of *Lophiola aurea* Ker-Gawler. Rhodora 85:73–81

ZIMMERMANN W (1965) Ordnungen (=Reihen) Nymphaeales, Magnoliales, Ranunculales. In: Hegi's illustrierte Flora von Mitteleuropa, vol III/3, 2nd edn. Hanser, München

# Index of Names

References to illustrations printed in bold face type